WITHDRAWN BY THE
UNIVERSITY OF MICHIGAN

John Emsley & Dennis Hall
University of London King's College

A Halsted Press Book

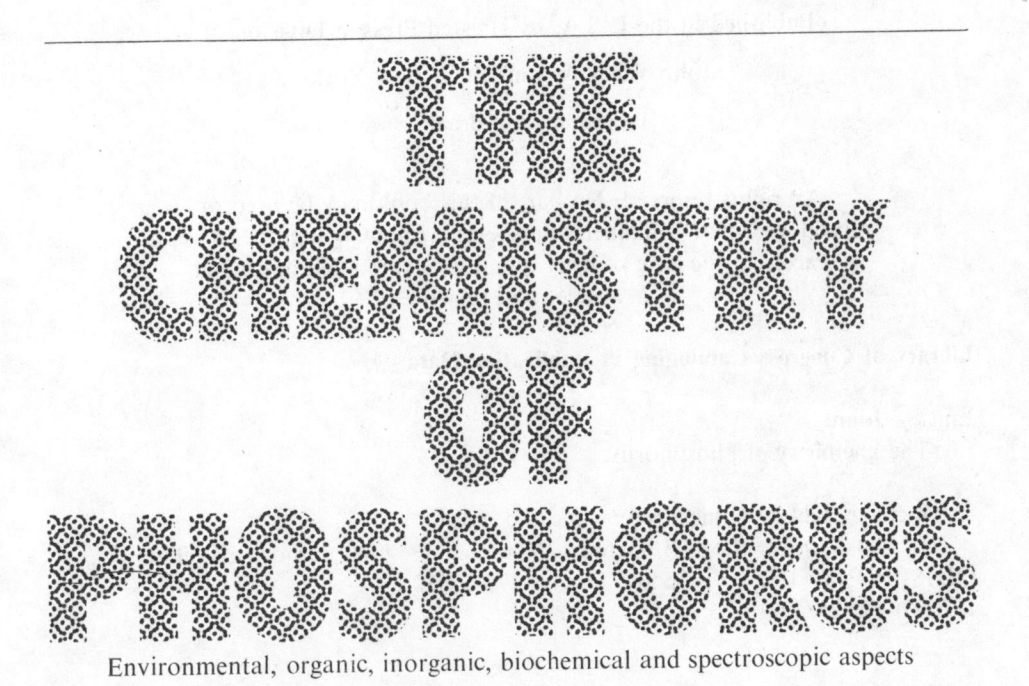

THE CHEMISTRY OF PHOSPHORUS

Environmental, organic, inorganic, biochemical and spectroscopic aspects

John Wiley & Sons
New York

Chemistry Library
QD
181
.P1
E55

Copyright © 1976 John Emsley and Dennis Hall
Published in the U.S.A. by Halsted Press, a Division of
John Wiley & Sons, Inc., New York

Printed in Great Britain

All rights reserved. No part of this book may be used or
reproduced in any manner whatsoever without written permission
except in the case of brief quotations embodied in critical
articles and reviews.

Library of Congress Cataloging in Publication Data

Emsley, John.
    The chemistry of phosphorus.

    Includes bibliographical references and index.
    1. Phosphorus. I. Hall, Dennis, 1934–   joint author. II. Title.
QD181.P1E45 1976   546'.712   75-17432
ISBN 0-470-23869-0

Acknowledgements

The authors wish to thank their wives, Jean Hall and Joan Emsley, and the following friends and colleagues for their help, advice, encouragement and patience: Catherine Cahn, John Coe, Donald Denney, Dorothy Denney, Mavis Devlin, Richard Dewing, Michael Forster, David Hogg, Marshall Smalley, Colin Taylor and Peter Watts.

Contents

	Preface	xiii

Chapter 1
Phosphorus in the environment — 1

1.1	**The phosphate cycles**	2
1.2	**The primary inorganic phosphate cycle**	5
1.2.1	Phosphate in bone	7
1.2.2	Mining of phosphates	7
1.2.3	Processing of phosphate ore	10
1.2.4	Industrial uses	11
1.3	**The land-based phosphorus cycle**	14
1.4	**The water-based phosphate cycle**	18
1.4.1	Photosynthesis and respiration	18
1.4.2	Sediment	22
1.4.3	Natural waters	23
1.4.4	The sea	24
	Problems	26
	References	27

Chapter 2
Bonding at phosphorus — 29

2.1	**Introduction**	30
2.2	**Valence-shell electron-pair repulsion theory**	31
2.3	**Single-bond lengths of phosphorus bonds**	34
2.4	**Tricoordinate phosphorus compounds**	35
2.4.1	Dipole moments (p) of tricoordinate phosphorus compounds	38
2.4.2	Isomers of tricoordinate phosphorus compounds	40
2.5	**Tetracoordinate phosphorus compounds**	42
2.5.1	Pi-bonding of tetracoordinate phosphorus compounds	43
2.5.2	Isomers of tetracoordinate phosphorus compounds	47
2.6	**Pentacoordinate phosphorus compounds**	48
2.6.1	The bonding in pentacoordinate phosphorus compounds	51
2.6.2	The topology of pentacoordinate phosphorus	58
2.6.3	Berry mechanism for polytopal rearrangement-pseudoration (ψ)	60
2.6.4	Ligands attached to pentacoordinate phosphorus and their preferred positions	65
2.6.5	Pentaccordinate phosphorus reaction intermediates	68

2.7	Hexacoordinate phosphorus anions	72
2.8	The electronegativity (χ) of phosphorus and electronic effects of attached groups	73
	Problems	75
	References	76

Chapter 3
^{31}P n.m.r. and vibrational spectra of phosphorus compounds — 77

3.1	**^{31}P nuclear magnetic resonance**	78
3.1.1	Phosphorus chemical shifts	79
3.1.2	Coupling constants	85
3.2	**Vibrational spectra of phosphorus compounds**	91
3.2.1	Group frequency correlations of phosphorus compounds	92
3.2.2	Vibrational assignments of phosphorus molecules	97
3.2.3	Hydrogen-bonding	102
	Problems	106
	References	108

Chapter 4
Tricoordinate organophosphorus chemistry — 111

4.1	**Nucleophilic reactivity**	113
4.1.1	Displacements at saturated carbon	113
4.1.2	Displacements at halogen	119
4.1.3	Attack at unsaturated centres	126
4.1.4	Miscellaneous nucleophilic displacement reactions	142
4.2	**Electrophilic reactivity**	145
4.3	**Biphilic reactivity**	152
4.4	**Dienophilic reactivity**	157
4.4.1	Dienes	157
4.4.2	α-Diketones, ortho-quinones and monofunctional aldehydes and ketones	163
4.4.3	α,β-Unsaturated carbonyl compounds	169
4.5	**Conclusion**	171
	Problems	171
	References	174

Chapter 5
Phosphorus(III) ligands in transition metal complexes — 177

5.1	Introduction	178
5.2	Phosphine, PH$_3$, complexes	179

5.3	Phosphorus trifluoride, PF$_3$, complexes	180
5.4	Phosphite, P(OR)$_3$, complexes	183
5.4.1	Orthophenylation of phenyl phosphites	185
5.4.2	Homogeneous catalysis	186
5.5	Organophosphine, PR$_3$, complexes	187
5.6	Bonding between phosphorus and metals	191
5.7	Trans effect and trans influence	198
5.8	Steric factors and P(III) ligands	202
5.9	Summary	203
	Problems	204
	References	204

Chapter 6
Pentacoordinate and hexacoordinate organophosphorus chemistry

		209
6.1	Introduction	210
6.2	Methods of preparing phosphoranes	215
6.2.1	From trivalent phosphorus compounds	215
6.2.2	By exchange reactions	222
6.2.3	From tetracoordinate compounds	225
6.3	The chemistry of phosphoranes	228
6.3.1	Acyclic phosphoranes	228
6.3.2	Cyclic phosphoranes	230
6.4	Hexacoordinate phosphorus compounds	242
6.4.1	The preparation of hexacoordinate phosphorus compounds	243
6.4.2	The physical and chemical properties of hexacoordinate phosphorus compounds	245
	Problems	248
	References	250

Chapter 7
Tetracoordinate organosphosphorus chemistry: Part 1 phosphonium salts and phosphorus ylids

		253
7.1	Phosphonium salts	254
7.1.1	Introduction	254
7.1.2	Preparation of phosphonium salts	255
7.1.3	Reactions of phosponium salts	259
7.2	Phosphorus ylids	274
7.2.1	Introduction	274
7.2.2	The structure and bonding in phosphorus ylids	274
7.2.3	The preparation of phosphorus ylids	275

| 7.2.4 | The Wittig reaction. | 279 |
| 7.2.5 | Miscellaneous reactions of phosphorus ylids | 290 |

Problems 299

References 300

Chapter 8
Tetracoordinate organophosphorus chemistry: Part 2. phosphoryl esters and related compounds 305

8.1	Introduction	306
8.2	Nucleophilic displacement at phosphorus: acyclic compounds	308
8.2.1	$S_N1(P)$ mechanism (elimination–addition)	311
8.2.2	Addition–elimination	318
8.2.3	$S_N2(P)$ mechanism	319
8.2.4	Catalysis	325
8.3	Nucleophilic displacement at phosphorus: cyclic phosphorus esters	329
8.4	Nucleophilic displacement at phosphorus: hexacoordinate intermediates	336
8.5	Phosphorylation and the design of phosphorylating agents	338
8.6	Reactions of phosphorus acids and esters not involving the phosphorus atom	343

Problems 345

References 346

Chapter 9
Phosphorus radicals 351

9.1	Introduction	352
9.2	Seven-electron phosphorus radicals	355
9.2.1	The phosphino radical	355
9.2.2	The phosphinyl radical	357
9.2.3	The phosphinium radical cation	359
9.3	Nine-electron phosphorus radicals	363
9.3.1	The phosphoranyl radical	363
9.3.2	The phosphonium radical anion ($R_3P-\bullet$)	373
9.4	Conclusion	374

Problems 374

References 376

Chapter 10
Phosphorus-nitrogen compounds 379

| 10.1 | Phosphorus–nitrogen single bonds–phosphorus amides and amines | 380 |
| 10.1.1 | Phosphorus amides | 380 |

10.1.2	Phosphorus amino derivatives	383
10.1.3	The bonding in phosphorus amides and amines; evidence of π contributions	387
10.2	**Monophosphazenes: the phosphorus-nitrogen double bond**	388
10.2.1	Synthesis of monophosphazenes	389
10.2.2	Chemical attack at phosphorus	390
10.2.3	Chemical attack at nitrogen	391
10.2.4	Physical properties of monophosphazenes and the nature of the bond	393
10.3	**The cyclodiphosphazanes**	395
10.3.1	Synthesis of cyclodiphosphazanes	396
10.3.2	Physical properties of the cyclodiphosphazanes and the nature of the bonding	399
10.4	**The linear polyphosphazenes**	401
10.4.1	Synthesis of the linear polyphosphazenes	401
10.4.2	Reactions and properties of the linear polyphosphazenes	403
10.5	**The cyclopolyphosphazenes**	405
10.5.1	The synthesis of the cyclopolyphosphazenes	405
10.5.2	The bonding in the cyclopolyphosphazenes	410
10.5.3	Physical properties of the cyclopolyphosphazenes	412
10.5.4	Nucleophilic substitution at phosphorus in the cyclopolyphosphazenes	420
10.5.5	Reactions involving nitrogen in the cyclopolyphosphazenes	431
10.5.6	Polyphosphazene high polymers	434
	Problems	437
	References	438

Chapter 11
Less-common phosphorus bonds: boron–phosphorus and phosphorus–phosphorus 445

11.1	**The boron–phosphorus bond. The boraphosphoranes**	446
11.1.1	Phosphorus-boron Lewis adducts	448
11.1.2	The cyclopolyboraphosphanes	449
11.2	**The phosphorus–phosphorus bond. The polyphosphines**	454
11.2.1	Diphosphine and linear polyphosphine derivatives	456
11.2.2	Cyclopolyphosphines	461
	Problems	468
	References	468

Chapter 12
Biophosphorus chemistry 471

12.1	**Phosphorus in life**	472
12.1.1	Adenosine triphosphate	474

12.1.2	The formation of adenosine triphosphate	479
12.1.3	Biosynthesis	488
12.1.4	Metabolic activation by adenosine triphosphate	491
12.2	**Phosphorus in death: pesticides and chemical warfare agents**	494
12.2.1	Organophosphorus pesticides	495
12.2.2	Chemical warfare agents	502

References 509

Appendix 1: The nomenclature of phosphorus compounds 511

Appendix 2: Antidotes for toxic organophosphorous chemicals - pesticides and nerve gases 521

Answers to problems 525

Index 543

Preface

The interest and research in phosphorus chemistry has expanded so rapidly in the past twenty-five years that it now occupies an important position in chemistry as a whole. This unique role of phosphorus chemistry arises because of the many industrial applications of phosphorus compounds, because of the importance of phosphorus chemistry in biological systems and because of the theoretical problems which arise in attempting to understand the multifaceted reactions and structures of phosphorus compounds.

The industrial uses of phosphorus compounds are many; some of the more important ones are their use as fertilizers, pesticides, flame retardents, and antioxidants. An understanding of phosphorus chemistry is paramount to an understanding of the role of the many important phosphorus compounds in life. Significant contributions have been made both by direct investigations of these materials and by studies of model systems. From these researches there has arisen a firm base from which new investigations can proceed with the hopes of unraveling the complex chemistry of this most important element.

This book is the first attempt to bring together under one cover the inorganic, organic and bioorganic chemistry of phosphorus. The coverage is excellent and the discussions are modern and theoretically sound. The two authors, one an expert in inorganic phosphorus chemistry and the other in organic, have collaborated in such a way that the whole topic of phosphorus chemistry is covered in a uniform manner. This book represents therefore the only single source for a thorough, complete and competent overview of phosphorus chemistry as a whole.

The presentation of the material is at such a level that individuals with a reasonable background in organic and physical chemistry will have no trouble following the chemistry and the reasoning. The text will certainly serve very satisfactorily at the senior or graduate level for a course in phosphorus chemistry. The style and exposition are such that an individual who wishes to familiarize himself with phosphorus chemistry will have no difficulty in using the book as a self-study text. Workers in the field will also find this book of value because of its coverage and timeliness. The authors are to be congratulated on their significant contribution.

April 1975

Professor Donald B. Denney
Rutgers University, New Brunswick, New Jersey

Chapter 1
Phosphorus in the environment

THE CHEMISTRY OF PHOSPHORUS

Textbooks dealing with phosphorus chemistry have, until recently, ignored its technological and environmental aspects. It was of little chemical consequence where phosphorus was mined, how it was processed, to what ends its compounds were put and what happened to its waste products. Its use in fertilizers, detergents, foods, pesticides, oils etc. went unquestioned – indeed the benefits were so great that ten years ago people were worried whether the known deposits of phosphates would last long enough. We now know that these will last at least 1500 years, even at the present rate of mining, yet the cry to reduce phosphate mining is still heard, no longer on the grounds of husbanding limited resources but for entirely different reasons. Phosphates find themselves cast in the role of major water pollutants. What has gone wrong? In fact very little, and as this chapter will show, the remedy is simple. Nature has a mechanism for dealing with excess phosphate but requires time for its effect to be felt, time which Man can give if he uses a little chemical common sense.

The aim of this chapter is to take a global view of phosphorus. By this means we can place in proper perspective the pollution charge levelled against it: which is the spoiling of certain natural waters by phosphates and in particular the destructive effect these have on the ecosystems of some lakes. This, however, is only one sector of the natural phosphorus cycle which consists of a primary inorganic phosphate cycle and two secondary *organic** cycles, one on land and the other in the sea. These phosphorus cycles have been in operation for upwards of 500 million years. Primitive man has a place in the natural cycles, but urban man tends to upset the natural rhythm by discharging his phosphorus wastes into water instead of returning them to the land. Yet provided the urban:rural population ratio is small the effect is barely noticeable. Industrial man on the other hand is predominantly urban and by virtue of this fact alone he is capable of upsetting the water cycle in his vicinity. By his very industry he is capable of affecting all the phosphorus cycles, not only the secondary ecological ones but the primary phosphate cycle as well.

The subject of phosphorus in the environment will be dealt with in terms of the natural cycles, discussing them as a whole and then in turn. In each case the undisturbed cycle will be outlined and then the effects of industrial man will be shown.

1.1 The phosphate cycles

Phosphorus rotates very slowly through the primary inorganic phosphate cycle which consists of three movements. Firstly, phosphate is leached from the land by weathering and carried by rivers to the sea; the leaching is a slow process but is helped by life on the land. The second movement is the precipitation of the phosphate in the sea as calcium phosphates (apatites) which are deposited mostly on continental shelves. And the third movement which brings the wheel full circle is the geological uplifting of these marine deposits so that once again they are back on land and exposed to weathering. This primary cycle is shown in Figure 1.1.

---

* Throughout this chapter and chapter 12 the word organic is used with two meanings – organic as of organic (i.e. carbon) chemistry, and *organic* as of organisms (i.e. compounds derived from living things). The antonym inorganic means more or less the same in both cases and should not lead to confusion.

1.1 The phosphate cycles

Figure 1.1 The primary inorganic phosphate cycle.

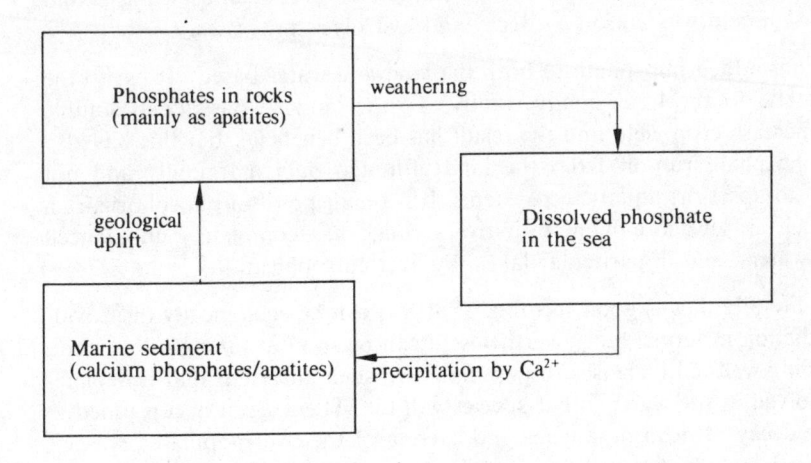

Phosphates in rocks (mainly as apatites) →(weathering)→ Dissolved phosphate in the sea →(precipitation by Ca^{2+})→ Marine sediment (calcium phosphates/apatites) →(geological uplift)→ Phosphates in rocks

Secondary cycles are the movement of phosphorus through the life cycles of the land and the sea and these consist of the uptake of phosphate by plants which may be eaten or simply die. The phosphate in them is then released again to the environment by excretion or decay. These cycles can be related to the primary cycles as shown in Fig. 1.2. The secondary cycles are linked by rivers which wash away soluble phosphates from the land and thereby replenish the phosphate of lakes and seas, from which phosphate is continually lost by sedimentation, either as precipitated calcium phos-

Figure 1.2 Schematic representation of primary and secondary phosphorus cycles.

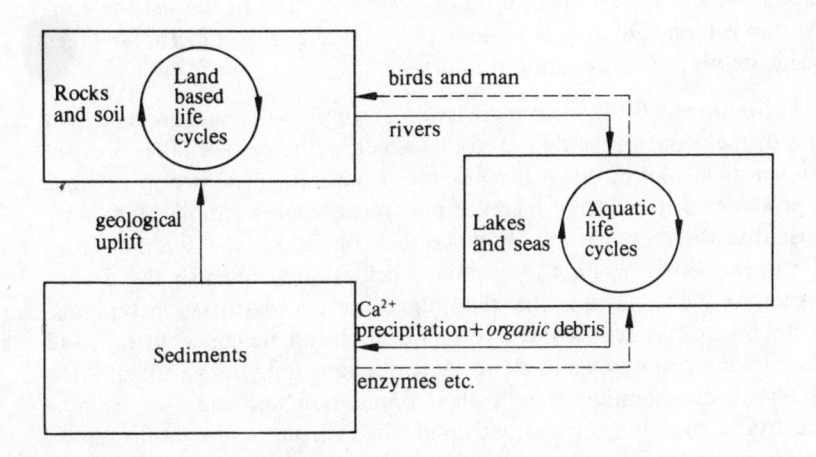

Rocks and soil / Land based life cycles — birds and man, rivers — Lakes and seas / Aquatic life cycles; Sediments ← Ca^{2+} precipitation + *organic* debris, enzymes etc.; geological uplift from Sediments to Rocks and soil.

phate or *organic* debris (excreta and dead organisms). Although the net transfer of phosphorus is in the direction of the primary cycles there are movements in the opposite direction: from water to land by fish-eating creatures such as seabirds and man, and from sediment to solution by decay and hydrolysis processes.

Industrial man has added phosphate to both the land and water based cycles; to the former deliberately, to the latter inadvertently. To the land he has added soluble phosphates to increase crop yield and the result has been beneficial, but this has also increased the phosphate run-off from the land although only marginally and not enough by itself to upset the aquatic ecosystems. It is the disposal of phosphorus-rich waste-waters such as sewage effluent into rivers which has completely unbalanced certain natural waters, and in particular lakes, by over eutrophication.

Eutrophication literally means good feeding. That this can be ecologically disastrous seems a contradiction in terms, but nevertheless it can be so in a water environment. If the species being well-fed increases in population to such an extent that it depletes the oxygen dissolved in the water, other species will die. The oxygen is consumed in coping with the decay of dead organisms and excreta of the over-populated species. The whole system is unbalanced and the result is a minor or major ecological disaster. Phosphates can have this effect because phosphates are the limiting growth factor in many systems. Carbon and nitrogen which are the other essential nutritional elements are not thought to be limiting and so the supply of these is not relevant.

When three ingredients are necessary to maintain a species the population will expand until it reaches a limit determined by the supply of one of the ingredients. The supply of this component then acts as the *limiting factor* of the population. It does not matter how much excess there is of the two other ingredients, the population responds only to the limiting one. If the amount of this falls so will the population, if it increases so will the population increase, that is until there comes a point at which its supply is so abundant that it ceases to be limiting and the role of limiting factor then passes to one of the other ingredients. For aquatic and marine life the limiting factor appears to be phosphate – carbon and nitrogen can be supplemented from the atmosphere, phosphorus cannot. Phosphate in natural waters is kept low by two processes which remove it, precipitation and the sedimentation of organic debris. In the sea the concentration of calcium is large and precipitation is probably the critical factor; in lakes the loss via organic debris is possibly more important.

If phosphate is the limiting factor then an increase in its supply will cause an explosion in the population of the species best able to use it, which in the case of lakes is algae and the result is an algal bloom. Such blooms are a natural phenomenon because certain natural processes can suddenly increase phosphate concentrations. It is only when they persist that the damage is done since they block out the sunlight from deeper plant life thus preventing photosynthesis which would replenish the deeper layers with oxygen. At the same time the dead algae use up what oxygen remains. Normally algal blooms quickly reduce the lake's natural phosphate concentration and growth ceases before any real damage is done. A continual supply of phosphate-rich river water will, however, maintain a high algal population and the lake become eutrophied. Industrial man pollutes rivers with phosphates from industrial processes, sewage and detergents, and these rivers often go on to cause eutrophication of lakes. There are no insurmountable problems to his remedying this situation.

Most rivers empty into the sea and eutrophication can disfigure estuaries, but in general the sea can deal with the concentration of phosphate simply by precipitation. Other things threaten the sea much more than phosphates and since it can regulate phosphate concentration automatically there is no long term threat to it from this sector. This is an optimistic long term outlook to set against the alarming details we shall encounter in a closer look at the three natural phosphorus cycles and the effect industrial man has on them.

1.2 The primary inorganic phosphate cycle

Most of the phosphate in this cycle is tied up as insoluble calcium phosphates (apatites) in sedimentary deposits. However before dealing with the primary cycle it is pertinent to consider the origin of the phosphate in the cycle. There are two sources – igneous rocks and meteorites. The weathering of the original rocks, the igneous rocks of the planet's surface, provided the majority of the phosphate, but meteorites, which contain 0.02–0.94% phosphorus by weight, also account for a sizeable amount* although only a small proportion of the whole[1].

Igneous rocks, such as granites, contain on average 0.1% phosphorus by weight[2]. The phosphorus exists as phosphate, either as minute crystals of apatite or as individual phosphate units in which a silicon atom of the rock silicate has been replaced by a phosphorus atom. A few igneous rocks are very rich in apatite, sufficiently so for them to be worthwhile mining for their phosphate. Although apatites are very water insoluble continual exposure to weathering gradually leaches out the phosphate. The emergence of life on the planet greatly aided the process. Plants in particular are able to extract the phosphate they require from soils and thereby start its movement through the cycle.

The supply of fresh phosphate from igneous rocks is now relatively small. Most of the surface of the Earth is covered with sedimentary rocks such as sandstones and shales which also contain 0.05–0.1% phosphorous, again as phosphate, this being deposited along with the bulk of the rock during its formation. The weathering of sedimentary rocks and soils, assisted by the life they support, also leaches out phosphate[3]. Just as with igneous rocks, there are sedimentary rocks rich in calcium phosphates and these are termed *phosphorites*. Some of these are 80% apatite and the bulk of mined phosphate comes from them.

Whether it be phosphate fresh to the primary cycle, from igneous rocks, or phosphate having already been through the cycle, from sedimentary rocks, the solid phase is predominently calcium phosphate known as apatite. The importance of this to the stability of the environment cannot be over-emphasised. Other metal phosphates are found naturally, in fact over 200 phosphate minerals are known[4,5], but only the calcium phosphates, not all of which are apatites, are important. The most commonly mined mineral is fluorapatite, of empirical formula $Ca_5(PO_4)_3F$, which may be of igneous or sedimentary origin, although the latter also includes hydroxyapatite,

---

* The daily influx of meteorites is about 100 metric tons per day which with an average phosphorus content of 0.1 % means an annual input of 35 metric tons per year. Over the lifetime of the earth, 4.6×10^9 years, the amount from this source would be *ca* 10^{11} tons.

$Ca_5(PO_4)_3OH$, and carbonate apatites, $Ca_5(PO_4; CO_3, OH)_3F$. The Florida phosphorites also contain aluminium phosphates such as crandallite, $CaAl_3H(PO_4)_2(OH)_6$, millisite, $Na_2CaAl_{12}(PO_4)_8 \cdot 8H_2O$, and wavellite, $Al_3(PO_4)_2(OH)_3 \cdot 5H_2O$.

The solubility product of $Ca_3(PO_4)_2$, $K_{sp}(=[Ca^{2+}]^3[PO_4^{3-}]^2)$ is 2.0×10^{-29}, in other words it is very insoluble. This simple approach is not very helpful in the study of phosphate in natural waters however, the problem here being the uncertainty over the amount of soluble phosphate which is free orthophosphate, i.e. PO_4^{3-}, HPO_4^{2-} or $H_2PO_4^-$, and that part which is bound phosphate, i.e. complexed with metal ions, adsorbed on colloidal particles or attached to organic moieties. Methods of distinguishing the two kinds of soluble phosphate will be discussed later in this chapter. Whatever the true concentration of orthophosphate in natural waters the concentration of Ca^{2+} is usually much higher (ten to a hundredfold that of orthophosphate) and this serves to depress proportionally the concentration of phosphate necessary to achieve saturation. The situation is complicated because calcium may also be precipitated as the insoluble carbonate or sulphate depending on conditions.

The ternary system $Ca(OH)_2-H_3PO_4-H_2O$ is more complex than is generally realized. Orthophosphoric acid is tribasic so that mono-, di- and tricalcium phosphates are possible. Of these only the monocalcium phosphate $Ca(H_2PO_4)_2 \cdot H_2O$ is soluble; in water it disproportionates to the insoluble dicalcium phosphate, $CaHPO_4 \cdot 2H_2O$. The natural form of this insoluble phase is called brushite. This is only one of several insoluble calcium phosphates. Which solid phase precipitates out depends upon the conditions prevailing[6]. In neutral solution the order of solubility is:

$Ca_5(PO_4)_3OH$ < β-$Ca_3(PO_4)_2$* < $Ca_8H_2(PO_4)_6 \cdot 5H_2O$
hydroxyapatite   whitlockite   octacalcium phosphate
                   < $CaHPO_4$ < $CaHPO_4 \cdot 2H_2O$ ≪ $Ca(H_2PO_4)_2 \cdot H_2O$
                      monetite   brushite   soluble

In acid solutions the order is different; brushite and monetite are less soluble even than hydroxyapatite.

Though in neutral solution hydroxyapatite is the least soluble, $K_{sp} = [Ca^{2+}]^5[PO_4^{3-}]^3[OH^-] = 10^{-51}$†, it does not follow that it will precipitate out. Its crystal growth appears to be a relatively slow process compared to that of brushite or octacalcium phosphate. In natural systems brushite probably precipitates first and is then slowly converted to hydroxyapatite via octacalcium phosphate – which would explain why some hydroxyapatite has a crystal form like that of octacalcium phosphate. This process of interconversion is accelerated by fluoride ions which can also slowly replace the hydroxyl groups of the hydroxyapatite which is formed, to give fluorapatite, $Ca_5(PO_4)_3F$, and this is the least soluble calcium phosphate of all. Monetite is naturally rare, not only because it is more soluble (except in acid waters), but because it too has a slow rate of crystal growth. β-Whitlockite is also rare; its crystal growth has been found to be promoted by Fe^{2+} and Mg^{2+} cations.

---

* The α-form of $Ca_3(PO_4)_2$ is unstable in contact with water.
† Other estimates put it as low as 10^{-58}.

1.2 The primary inorganic phosphate cycle

1.2.1 Phosphate in bone

Hydroxyapatite has another important role being the principal crystalline material in bone, of which it constitutes 23%, and in teeth especially the enamel which is 90% apatite. The enamel of teeth is the most crystalline of the apatites produced by living things but even this is far from perfect by crystal standards, containing as it does carbonate impurities, which make it more soluble than one would wish. Again the incorporation of fluoride strengthens teeth and for this reason it is added to toothpastes, foods, and in some localities to public water supplies – a move that was first suggested a hundred years ago[8]. Ironically the first of these, toothpaste, is also made from calcium phosphates[9] usually $CaHPO_4 \cdot 2H_2O$ which is stabilized to prevent its dehydration to $CaHPO_4$ by the addition of $Na_4P_2O_7$ (i.e. tetrasodium pyrophosphate) or $Mg_3(PO_4)_2$*. $CaHPO_4 \cdot 2H_2O$ is used in the mistaken belief[10] that it is acting as an effective abrasive and for this reason the harder $CaHPO_4$ or $Ca_3(PO_4)_2$ are sometimes used in special formulations designed for the badly stained teeth of smokers.

Bones were shown to contain calcium phosphate by Scheele as long ago as 1771 and for many years were used as the chief source of phosphorus. Their part in the natural phosphate cycle is small but they can lock up phosphate for long periods and by being so resistant to decomposition they serve to preserve traces of previous ages.

Other insoluble phosphates are $AlPO_4$, $K_{sp} = [Al^{3+}][PO_4^{3-}] = 5.8 \times 10^{-19}$, and $FePO_4$, $K_{sp} = [Fe^{3+}][PO_4^{3-}] = 1.3 \times 10^{-22}$, which play a very small part in the primary cycle even though they are less soluble than calcium phosphates in terms of the phosphate concentration of a saturated solution. The reason why they do not precipitate is that the concentration of metal cation is generally too low. In the sea for example Al and Fe concentrations are ca 4×10^{-4} times less than that of Ca. In certain lakes however thay may be responsible for precipitating phosphate.

Although the slow dissolution of calcium phosphates from the land and their reprecipitation in the sea constitute the primary cycle this is a slowly turning one. Industrial man however speeds up the process by mining the richest land deposits and converting the phosphate into soluble inorganic forms destined for a variety of uses, and then after use dumping the wastes into the sea or into lakes. This aspect of phosphorus chemistry can be seen essentially as part of the primary cycle.

1.2.2 Mining of phosphates

World production of phosphate rock in 1971 was 84 million metric/British tons (ca 93 million US tons). Over three quarters of this came from three countries – USA, Morocco and USSR – and in particular from three deposits within those countries: the Bone Valley formation in Florida, the Oulad–Abdoun basin in Morocco and the Kola peninsula in the far northern USSR. The first two are phosphorite deposits, the last is of igneous origin. Florida supplies about 40% of the world's needs from a 10 m (30 ft) thick deposit covering 5000 km² (2000 mi²). It is sufficiently near the surface to be strip mined and is interlayed with other deposits such as limestone and

---

* A typical toothpaste consists of $CaHPO_4 \cdot 2H_2O$ (45 %) as polishing agent, glycerine or sorbitol (30%) as moisturizer, water (20%), and small amounts of surfactant, binder, preservative (e.g. $PhCO_2H$) and flavouring.

sand. Table 1.1 lists the main producing countries with their output and estimated reserves in 1971.

The reserves total in Table 1.1 is low because it does not include the enormous Phosphoria formation of the Western USA. This deposit is over 700 km (450 mi) long and probably exceeds the rest of the world's deposits put together. At present only a small amount is economic to mine however. Also the total does not include large deposits recently reported in Peru[4] and Australia. However, the table shows that even at today's high rate of mining, at least 1000 years supply exists which is longer than some present day Cassandras will allow for the lifetime of industrial man. Table 1.1 also lists a third source of rock phosphate, that of guano origin. This accounts for about 3% of the world's supply at present but reserves of this are small compared to that of the other types. This phosphate rock comes from the very limited amount of phosphorus which has found its way from the sea to the land via fish-eating birds. Trivial though the amount each bird transfers, over the centuries millions of birds have built up sizeable deposits of guano, on certain islands and seaboards near

Table 1.1 World output of phosphate rock (1971)[11]*

Country of origin	Annual production /(million metric tons)	Estimated reserves /(million metric tons)	Type of ore	Percentage of world's output
USA total	37	13 000		43
Florida	32	2000	phosphorite	
Western States	4	see text	phosphorite	
North Carolina	1	10 000	phosphorite	
USSR total	20	4300		24
Kola peninsula	11	1300	igneous	
Other deposits	9	3000	phosphorite	
Morocco	15	40 000	phosphorite	18
Spanish Sahara	4	2000	phosphorite	5
Nauru	2	55	guano origin	
Togo	2	—	phosphorite	
Senegal	1.5	180	phosphorite	10
Christmas Island	1	200	guano origin	
South Africa	1	500?	igneous	
Israel	1	200	phosphorite	
Others	ca 2	5000?		
Grand total	84	85 000+		100%

* In 1974 Tunisia and China produced 4 and 3 million tons respectively, while the Spanish Sahara output fell to below 1 million tons due to 'production difficulties'.

1.2 The primary inorganic phosphate cycle

the upwelling regions of the sea where fish are especially plentiful. Guano itself has been mined as a fertilizer but undisturbed guano is slowly leached by rainwater which hydrolyzes *organic* phosphates and carries the soluble products, together with inorganic phosphates, down to underlying rocks which, if they are limestone or similar, react to form insoluble calcium phosphate. These then are the phosphate rocks of guano origin[3].

Table 1.2 Composition of high grade ores taken from ref. 4

Country of origin	% P as P_2O_5[a]	% Ca as CaO	% F	
USA (Florida)	35.2	49.2	3.8	phosphorite
Morocco	35.0	53.0	4.1	phosphorite
USSR	40.3	52.3	3.7	igneous
Pacific	39.2	53.8	3.8	guano
[$Ca_5(PO_4)_3F$	42.3	55.8	3.8]	

[a] See problem 1 at end of chapter.

All important phosphate rocks contain fluoride as the major impurity. The composition of high grade ores is shown in Table 1.2 and approximates to that of fluorapatite. Other metals such as Na, K, Mg, Sr, Ba, Ti, V, Mn, Zn and Cd are present as cations and other anions such as chromate, borate, silicate, arsenate and sulphate are also found. One metal in particular, uranium, is associated with both igneous and sedimentary rocks but not guano-derived ones[4]. Like fluoride, uranium has a special affinity for the apatite structure and is absorbed into the lattice during prolonged weathering. The uranium content is usually about 0.01–0.03% but its recovery is not yet a practical proposition. Likewise fluorine, although in increasing demand, is not recovered but rather treated as a nuisance to be removed from the phosphate and dumped. The acid process for treating phosphate rock removes about a third of the fluorine as volatile SiF_4 (the silicon coming from impurity silica in the rock) and the rest can be precipitated as sodium hexafluorosilicate, Na_2SiF_6, thus

$$F^- \xrightarrow{H_3O^+} HF \xrightarrow{SiO_2} SiF_4\uparrow \xrightarrow{F^-} SiF_6^{2-} \xrightarrow{Na^+} Na_2SiF_6\downarrow \qquad (1)$$

In this way the fluorine content is reduced to one thousandth of that of the original rock[12]. Another way to defluorinate rock is to heat it to high temperatures (*ca* 1500 °C) with silica and steam. Some rock processed in this manner is suitable for use as animal food supplement without further treatment.

Most phosphate ore is used as fertilizer after being treated with concentrated sulphuric acid, which forms "superphosphate", or concentrated phosphoric acid, which forms "triple superphosphate". In effect the rock is being converted from insoluble calcium phosphate to soluble monocalcium phosphate and the equations which summarize the chemical changes are

$$2Ca_5(PO_4)_3F + 7H_2SO_4 + H_2O \rightarrow \underbrace{3Ca(H_2PO_4)_2 \cdot H_2O + 7CaSO_4\downarrow}_{\text{"superphosphate"}} + 2HF\uparrow \quad (2)$$

$$Ca_5(PO_4)_3F + 7H_3PO_4 + 5H_2O \longrightarrow \underbrace{5Ca(H_2PO_4)_2 \cdot H_2O}_{\text{"triple superphosphate"}} + HF\uparrow \quad (3)$$

A discussion of phosphate fertilizers as such is deferred to later when the land cycle is dealt with. What we are interested in here is the conversion of phosphate rock into industrially useful starting materials and there are two processes for this – the electric furnace process which gives elemental phosphorus and the acid process which gives phosphoric acid.

1.2.3 Processing of phosphate ore

1.2.3.1 Electric furnace process

Elemental phosphorus was first prepared by Hennig Brandt in 1669 during the course of alchemical experiments with urine. Although his exact method is not known the method which developed from it, and was used throughout the following century, was to heat a mixture of boiled-down urine, sand and charcoal. Phosphorus distils from such a mixture and can be condensed under water as white phosphorus. What amazed the early chemists was the ability of this phosphorus to luminesce in the dark, a property which led to the noun phosphorus and the verb to phosphoresce. The equation which sums up the chemical reaction is

$$4PO_4^{3-} + 6SiO_2 + 10C \longrightarrow P_4\uparrow + 10CO\uparrow + 6SiO_3^{2-} \quad (4)$$

and it is essentially the same process which is taking place in today's electric furnace method.

As the demand for phosphorus grew, first bones then phosphate rock was used instead of urine but manufacture on a commercial scale only began in the 1830's. The early processes used a retort method in which phosphoric acid and charcoal were heated in clay pots and this method persisted until about 1890–95 when the introduction of the electric furnace revolutionized the industry and the retort manufacturing process disappeared. Although many technological improvements have been made to the process since then the method is basically the same and represented by (4). The ingredients are a mixture of phosphate rock (ground and fused into golf-ball size agglomerates), sand and coke, which are heated in a *ca* 70 MW electric furnace at 1500 °C. The phosphorus vapour and CO gas which are evolved are passed through an electrostatic precipitator to remove particulate matter and then to water condensers where the phosphorus collects. Yields of 90% are obtained. The fluorine compounds in the gas stream are HF and SiF_4 and these are removed in fluorine scrubbers and discarded. The condensed phosphorus is handled as a warm liquid under a layer of water, being sufficiently mobile and, at 55 °C, denser than the protective water layer[13]. The mechanism of the reaction is still not established; the silica, SiO_2, may be acting

1.2 The primary inorganic phosphate cycle

as a Lux-flood acid* or simply as a fused salt solvent or both[14]. If the former applies then eqns. (5) and (6) summarize the process; if the latter, then eqn. (7) is most appropriate.

$$4PO_4^{3-} + 6SiO_2 \longrightarrow P_4O_{10} + 6SiO_3^{2-} \tag{5}$$

$$P_4O_{10} + 10C \longrightarrow P_4 + 10CO \tag{6}$$

$$Ca_3(PO_4)_2 + 10CO \xrightarrow[C]{SiO_2} P_4 + 6CaO + 10CO_2 \tag{7}$$

Very little elemental phosphorus is required as such by industry. Rather it is converted to other more useful starting materials such as PCl_3, $POCl_3$, P_4S_{10}, P_4O_{10} and especially H_3PO_4 – half the phosphorus is converted via P_4O_{10} to phosphoric acid. This is the most important industrial phosphorus chemical, most of which however is prepared by direct acid treatment of phosphate rock.

1.2.3.2 The wet-acid process[12]

This can be summed up by eqn. (8). Finely ground rock is mixed with acid to form

$$Ca_5(PO_4)_3F + 5H_2SO_4 + 10H_2O \longrightarrow 3H_3PO_4 + 5CaSO_4 \cdot 2H_2O\downarrow + HF\uparrow \tag{8}$$

a slurry which is left to digest for several hours before the insoluble calcium sulphate is filtered off and the acid concentrated. Most HF is lost at the digestion stage when the slurry is cooled by vacuum pumping; SiF_4 which also forms from silica impurities is partly removed at this stage, the remaining SiF_6^{2-} is precipitated from the acid by addition of Na_2SO_4. Even though most of the chief impurity is thus removed, the acid from this process is far from pure, containing as it must most of the impurities present in the rock. The phosphoric acid from the wet-acid process is however eminently suitable for the manufacture of "triple superphosphate" fertilizer which is in fact how it is used. The USA phosphoric acid production in 1969 was 5 million tons, 79% of which was wet-acid and 21% acid produced from elemental phosphorus[15]. Somewhat surprisingly the former is an increasingly used process – the economics of the two processes being governed by the cost of sulphuric acid as against the cost of electricity. However the acid produced by the wet-acid method is only suitable for manufacturing fertilizer and some cattle feed supplements; the rest must come from the purer product of the phosphorus process.

1.2.4 Industrial uses

The modern pattern of commerce in phosphorus compounds began in early industrialized England. Fertilizers, baking powders and silk weighting (with stannic phosphate) were among the first uses. Today hundreds of uses are found for scores of phosphorus derivatives, although even in the highest industrialized societies fertilizers account for about three quarters of the output. In the third world this predominance rises to over 90%. The second largest use of phosphorus is in detergents, then animal feeds, metal surface treatments, deflocculants, insecticides, oil additives and so on. Phosphorus crops up in such opposites as matches (P_4S_3, strike anywhere

---

* In the Lux-flood acid–base system an acid, instead of releasing H^+, accepts O^{2-}. Silica, SiO_2, is a stronger acid than P_4O_{10} so it takes O^{2-} forming SiO_3^{2-}, and PO_4^{3-} loses O^{2-}.

matches; red phosphorus, safety matches) and flameproofing treatments,* from the harmless pleasures of fizzy colas (dilute H_3PO_4) to the horrors of nerve gases (see Chapter 12).

Phosphoric acid is the starting material for most commercial phosphorus compounds the majority output of which are ortho or polyphosphates of sodium or calcium. The acid itself is used in metal surface treatments, mainly for iron and steel[17]. For example, iron dipped into a bath of $Fe(H_2PO_4)_2$ solution, formed from iron filings and H_3PO_4, acquires a coating of hard, crystalline, insoluble $FePO_4$ which not only serves to protect the bulk of the metal but acts as an excellent base for paint. These phosphate coatings absorb oil and for this reason such treated surfaces have good anti-friction properties. Phosphorus-containing oil additives are thought to work by forming such layers on engine surfaces. Zinc dialkylphosphorodithioates have been mostly used as engine oil additives[18]. These dialkylphosphorodithioates can be prepared as shown in equation (9). The coating formed by these compounds is thought to be less brittle than pure phosphate ones.

$$P_4S_{10} + 8ROH \longrightarrow 4(RO)_2P(S)SH \xrightarrow{ZnO} 2Zn(S_2P(OR)_2)_2 \qquad (9)$$

Another use of phosphoric acid in metal treatments is in the electropolishing of aluminium products which can be achieved by dipping them into a hot bath of 85% H_3PO_4 containing a little concentrated HNO_3.[17]

Although almost all the wet-acid is used to manufacture calcium phosphate fertilizer some of the furnace process acid is also put to this use. Ammonium phosphates, $(NH_4)H_2PO_4$ and especially $(NH_4)_2HPO_4$ are prepared from ammonia and H_3PO_4; both phosphates are very soluble and make good fertilizers.

The industrial process which consumes most of the high grade phosphoric acid is the manufacture of sodium tripolyphosphate, $Na_5P_3O_{10}$, which is extensively used as a deflocculant and as a detergent "builder". Production of this exceeds one million tons a year in the USA alone[15]. Deflocculation is the opposite of aggregation. Many inorganic solids in aqueous suspensions will tend to form aggregates if left to themselves and thereby become unworkable. The addition of as little as 0.01–0.1% of sodium tripolyphosphate (which in most industries means *ca* 2–4 lbs of $Na_5P_3O_{10}$ per ton of processed solids) prevents this and keeps suspensions fine and mobile, a

aggregated   deflocculated   tripolyphosphate anion

---

* Ammonium phosphates and esters of phosphoric acid have been used for this purpose on fabrics and paper[16]. On heating they release H_3PO_4 which catalyzes the decomposition of cellulose to slow-burning carbon to such an extent that this reaction outpaces the flame-supporting decomposition, which goes via liquid and gaseous compounds. As a result the flame dies.

1.2 The primary inorganic phosphate cycle

necessary condition for the handling of slurries as in kaolin coating of paper, cement production, water-based paints, and oil drilling muds[19]. Deflocculation of soil can be beneficial in making it less permeable to water so that irrigation ditches sealed with polyphosphate transport more and absorb less water.

Polyphosphates act as deflocculants by being adsorbed onto particles which thereby become negatively charged. This makes them more attractive to the solvent, water, and repulsive to other negatively charged particles. Once adsorbed, the polyphosphate is very difficult to remove but since there is so little of it anyway, and since in most uses it does not interfere, it usually ends up by being incorporated in the product. With careful analysis the amount of tripolyphosphate can be adjusted so that virtually all is adsorbed and none passes into the waste waters of the manufacturing plant and out into the environment. The presence of Ca^{2+} and Al^{3+} in the aqueous layer will tend to keep the tripolyphosphate in solution since it acts as a good ligand towards these cations, but with care the phosphate pollution from these industries can be kept at a minimum[19].

Sodium tripolyphosphate finds extensive use in solid detergents, and potassium pyrophosphate, $K_4P_2O_7$, in liquid detergents, where they partly act as a deflocculant but more importantly as water softener, buffer and saponifier of fats and fatty acids[20]. Most detergents contain 30–50% of $Na_5P_3O_{10}$, without which the amount of surfactant would need to be increased tenfold to achieve the same cleaning efficiency. By complexing with Ca^{2+} or Mg^{2+} present in hard water the $P_3O_{10}^{5-}$ anion prevents these cations interfering with the surfactant, e.g. $CH_3(CH_2)_9C(Me)H\text{–}C_6H_4\text{–}SO_3Na$, and this is the main part they play. They also act to neutralize acids by the formation of $HP_3O_{10}^{4-}$ etc. and thereby keep the solution alkaline and this in turn facilitates removal of dirt.

Unlike the industrial use of deflocculants, the phosphates of detergents are released to the environment where they are eventually hydrolyzed to orthophosphate. Domestic dishwashing and clothes-washing machines dispose of massive amounts of phosphate to the drains most of which passes through sewage plants and into the rivers. Replacements for polyphosphates have been tried but are either biologically suspect (such as nitrogen triacetate), or very caustic (such as metasilicates)[21]. The polyphosphates perform their detergent function so well that the research effort to find replacements is somewhat misdirected. Their contribution to the phosphates in rivers is significant but their removal from detergents would not by itself solve the overall problem.

Dicalcium phosphate, $CaHPO_4$, is another major industrial product – over half a million tons is produced in the USA every year[15]. When destined for animal feed supplement it is prepared from defluorinated wet-acid and hydrated lime ($Ca(OH)_2$) but for other uses such as in toothpastes it is made from furnace-acid. For many years bone meal was used to provide phosphorus for animals in areas where the forage or pasture was phosphorus deficient. Areas such as Queensland, Uruguay, South Africa, India, and Argentina are such regions. Nowadays, $CaHPO_4$ is used[22].

Phosphates added to foods for human consumption are there not directly for nutritional purposes but for their acidity. Calcium monophosphate $Ca(H_2PO_4)_2 \cdot 2H_2O$

with NaHCO$_3$ releases CO$_2$ which aerates baked foods giving a lighter texture[9]. This particular combination works rather rapidly and dehydrated Ca(H$_2$PO$_4$)$_2$ crystals, coated with a protective layer of the sodium or aluminium salt and then fused at 220 °C, are used to give a slower release of acid. Disodium pyrophosphate, Na$_2$H$_2$P$_2$O$_7$, also gives a controlled release of acid for this purpose. Self-raising flour may contain such components as (1.5% Ca(H$_2$PO$_4$)$_2$ + 1.2% NaHCO$_3$) or (1.75% Na$_2$H$_2$P$_2$O$_7$ + 0.3% CaCO$_3$ + 1.25% NaHCO$_3$) for this purpose. Baking powder has a similar ratio (40% Ca(H$_2$PO$_4$)$_2$ + 30% NaHCO$_3$) or (35% Na$_2$H$_2$P$_2$O$_7$ + 5% Ca(H$_2$PO$_4$)$_2$ + 30% NaHCO$_3$), the remaining 30% being flour. Other foods with added phosphates are macaroni, instant puddings, processed cheeses and meats[9]. All this goes to increase the phosphorus content of urine and faeces but only marginally.

In all these ways and many others industrial man reintroduces inorganic phosphate to the primary cycle. We must now turn to the secondary *organic* land-based cycle, on which his efforts have the biggest effect.

1.3 The land-based phosphorus cycle

This is shown in Fig. 1.3 which also includes the contributions industrial man makes to the cycle. In fact it is no longer meaningful to discuss the "natural" part of the cycle in isolation since very little of the surface area of the globe now remains free from man's interference.

The natural occurrence of phosphorus in soil is quite low, but unlike the other elements essential to life, the amount remains fairly stable because the majority of the phosphorus is there as insoluble inorganic phosphates, called fixed phosphates. The soluble inorganic phosphate, called available phosphate, required for plant growth is H$_2$PO$_4^-$ and this is produced by acid conditions rising from dissolved CO$_2$, or decaying vegetation, or even by a special acid mechanism of roots. In an undisturbed ecosystem the soluble phosphate is taken up by plants and returned to the soil by the decay of those plants or from the excreta and debris of creatures living off those plants. A small amount of the soluble soil phosphate is leached out by rain and lost to the system. An even smaller amount of the fixed phosphate is eroded away by the same process. In sandy, acid, or water-logged soils the loss of phosphate by drainage from the land is more rapid, but since most phosphate is fixed, even from these soils the loss is relatively small.

In modern times the greatest loss of phosphorus from the land is as plant crops and the products of farm animals. In a simple rural economy the cropping still occurs but the waste and excreta of the users is eventually returned to the land. In an industrial/urban society the waste is tipped, the excreta ends up in the rivers. The continual transfer of crops and animal products to cities would eventually reduce the soil phosphate until the result was poorer crop yields (assuming phosphate to be the limiting factor). To counteract this, soluble phosphate is included in fertilizers[23] and the land replenished.

Superphosphate, Ca(H$_2$PO$_4$)$_2 \cdot$H$_2$O + CaSO$_4$, triple superphosphate, Ca(H$_2$PO$_4$)$_2 \cdot$H$_2$O, and ammonium phosphates, (NH$_4$)H$_2$PO$_4$ but mainly (NH$_4$)$_2$HPO$_4$, have

1.3 The land-based phosphorus cycle

Figure 1.3 The land-based phosphorus cycle.

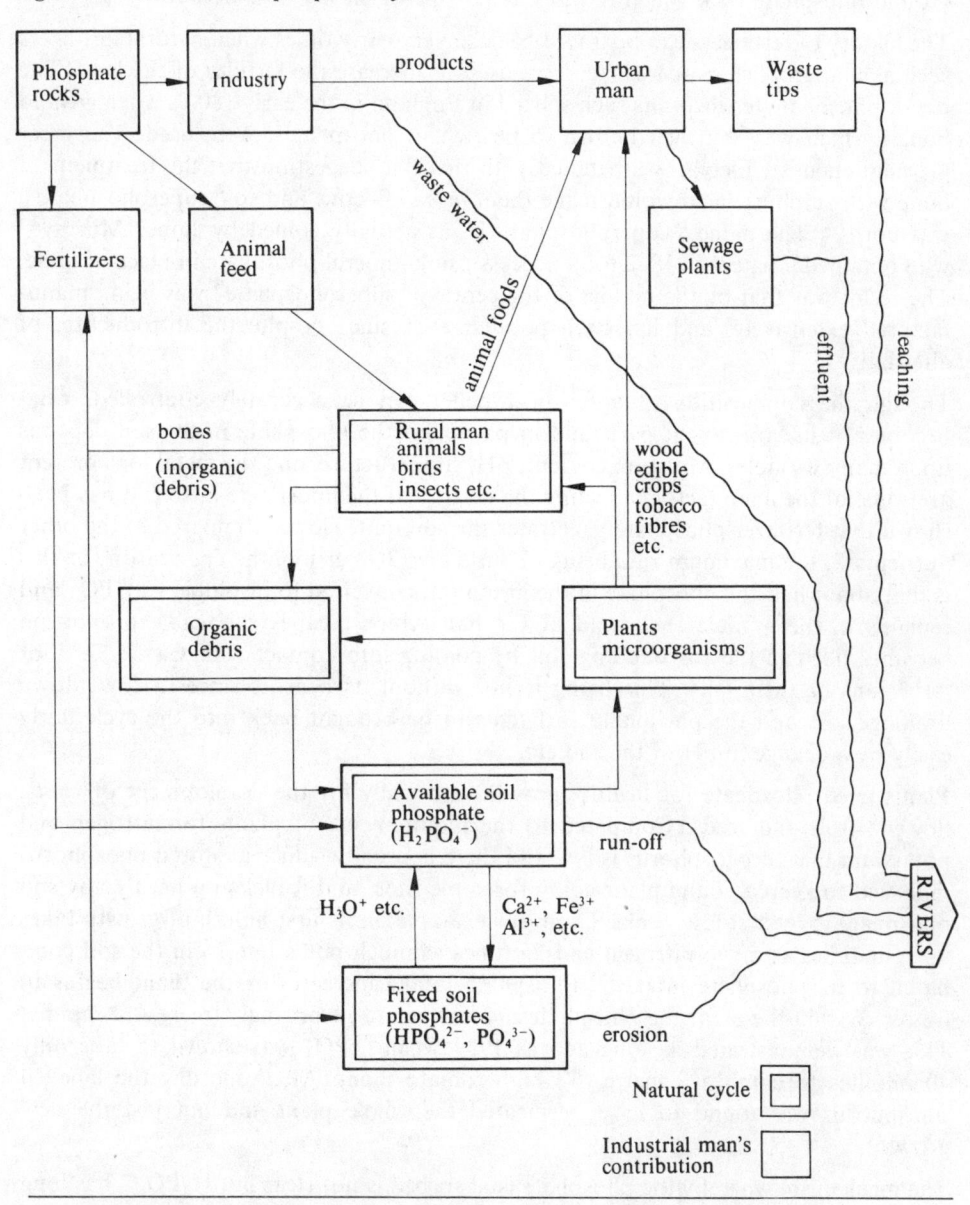

already been mentioned and these are the principle phosphate ingredients of fertilizers[23]. Ammoniated superphosphate, eqn. (10), is also used and so are several others.

$$Ca(H_2PO_4)_2 + NH_3 \longrightarrow CaHPO_4 + (NH_4)H_2PO_4 \qquad (10)$$

Most fertilizers are blended in bulk to give the proper proportions of the nutrients nitrogen, phosphorus and potassium e.g. $NH_4NO_3/Ca(H_2PO_4)_2 \cdot H_2O/KCl$ or $(NH_2)_2CO/(NH_4)_2HPO_4/KCl$, and then granulated for easier application and more controlled release. An alternative to annual treatment with phosphate is the use of ground phosphate rock which releases its phosphate slowly over decades.

The history of fertilizers can be traced back to very early times when natural fertilizers such as manure, fish, carcases etc. were used to increase the fertility of the land. The use of phosphate fertilizers as such started in England in the early 1800's when ground bones, which were by then known to be rich in phosphates, were used. The great German chemist, Liebig, is attributed with the first suggestion that the treatment of bone with sulphuric acid would make them more effective and so "superphosphate" was born[24]. The name "superphosphate" was actually coined by James Murray[25] who obtained a patent in 1842 for a process using mineral phosphate in place of bone. The result was that by the middle of the century 'superphosphate' was being manufactured extensively, and has been popular ever since despite the introduction of alternatives.

Though the composition of a fertilizer pellet can be accurately controlled, what becomes of its constituent parts and in particular the phosphate in the soil depends upon many variables. Moisture content, pH, soil structure, and the metal ions present are some of the main factors affecting the destiny of the phosphate in soil. It has been shown that fertilizer phosphate penetrates the soil quite slowly compared to the other nutrients[23], the maximum rate being about 5 cm (2") per month. The reason for this is that about half the phosphate in the granule is converted to insoluble $CaHPO_4$ and remains in the granule shell, and of the half which escapes a sizeable proportion becomes fixed in the surrounding soil by coming into contact with Ca^{2+}, Fe^{3+} or Al^{3+} ions or their salts. This fixing is not without its benefits since it slows down drainage loss and the phosphate so fixed can be brought back into the cycle fairly easily by a change in pH of the soil etc.

Plants need phosphate for healthy growth, especially for the development of roots, flowers, fruit and seeds. Compared to the requirements of plants for nitrogen and potassium that of phosphorus is low and there is usually sufficient stored phosphorus in a seed to keep a young plant going for some time, and it picks up hardly any soil phosphate in its first few weeks. Corn, for example, in its first month of growth takes up four times as much nitrogen and 25 times as much potassium from the soil compared to its phosphate intake[26]. Phosphate intake increases as the plant begins to flower. Within the plant the phosphate moves around surprisingly freely and rapidly. This was demonstrated as long ago as 1939 when $^{32}PO_4$ was shown to take only 40 minutes to climb a 2 metre (6') high tomato plant. After one day the labelled phosphorus was found to have permeated the whole plant and not just the new growth[27].

The mechanism whereby the phosphate is absorbed is not clear but $H_2PO_4^-$ has long been regarded as the form utilized by the plant. Even in alkaline soil when the concentration of $H_2PO_4^-$ is small it is thought that plant roots may create a low pH around themselves in order to effect $H_2PO_4^-$ transfer.

The amount of phosphorus removed from the land by a particular crop varies considerably. A ton of peanuts removes 4 lb of phosphorus, a ton of tomatoes 1 lb, a ton of grapes $\frac{3}{4}$ lb and a ton of sugar cane tops only $\frac{1}{2}$ lb. The percentage of phosphorus in certain plant foods is given in Table 1.3.

In addition to crops for human food, crops are grown for animal fodder. The meat from these animals is also destined for man but most of the phosphorus they consume

1.3 The land-based phosphorus cycle

Table 1.3 Phosphorus content of plant foods destined for human consumption[28]

Crop	% P content	Crop	% P content
Soy beans	0.63	Potatoes	0.05
Peanuts	0.39	Green beans	0.05
Wheat	0.34	Onions	0.04
Rice, whole	0.31	Cabbage	0.04
Peas	0.12	Grapes	0.02
Corn	0.11	Apples	0.01

is excreted and in many cases their diet is supplemented by direct addition of mineral phosphate. Even so a proportion of the land-cycle animal phosphate ends up in the urban community from which it exits via the sewers. Phosphate is transfered from the land-based cycle to the water-based cycle by rivers.

It has been estimated that 2 million tons of phosphate are washed to sea annually by natural processes, and about the same amount from man's activities. That of natural origin is leached from the soil partly from the fixed phosphate but mainly from the hydrolysis of the *organic* phosphates in the soil, arising originally from plant and animal debris and excreta worked over by insects, worms, bacteria, fungi and so forth. For instance a group of organisms called *Nucleobacter* can free phosphate from nucleoproteins in the soil. Other *organic* phosphates in the soil are sugar phosphates, phospholipids and pigments[23].

To the slow erosion of the 'fixed' phosphates and the more rapid leaching of the soluble or 'available' phosphates man adds his own contributions to the river's supply. The waste water from some industries has a high phosphate content. Sewage effluent also has a high phosphate content. Estimates of the yearly excretion of phosphorus per person vary from 2 to 4 lbs (1–2 kg)[29]. Individually this is small but for a river passing a city of a million people it represents a sizeable increase and on top of this is the detergent phosphate which may be an even larger proportion. The total may seem small even so. The amount of phosphate in the rivers passing through industrialized regions of the USA contained on average *ca* 24 mg phosphate per litre in 1966 and this level has not significantly increased since[30]. This low concentration nevertheless spells disaster for the lake or estuary into which the river drains, as we shall see. If the river drains into the sea, and the tidal flow is large, very little harm is done.

Where it is necessary to reduce the phosphate level in a river this can be done immediately following the sewage treatment* or on industrial waste water before it is

---

* Phosphate is necessary for the biological degradation of sewage whether this be aerobic (producing CO_2) or anaerobic (producing CH_4), but once this stage is complete it can be removed.

put into the river[30]. Chemically the answer is simple: the phosphate can be precipitated as one of its insoluble salts. Precipitation as $FePO_4$, by adding ferric sulphate, or $AlPO_4$, by adding alum $(KAl(SO_4)_2 \cdot 12H_2O)$ is preferred since these have minimum solubility at pH 5.5 to 7. Precipitation by adding lime (CaO) to the effluent has the disadvantage of making the pH too high and this would necessitate a further step to bring the pH down to about 7.

Other points have to be taken into account such as the disposal of the insoluble phosphate (only calcium phosphate could be recycled as fertilizer) and the necessity to filter the effluent since settling tanks are not sufficient to remove all the precipitate. Also if the water is hard water, i.e. already has a high sulphate content, other iron salts such as the chloride or aluminium salts such as sodium aluminate $(NaAlO_2 \cdot 3H_2O)$ may have to be used.

Other methods of reducing phosphate concentrations such as ion exchange and oxidation ponds have been suggested but seem to be unsatisfactory alternatives. The former is uneconomic but the latter method has its attractions. The ponds in effect become miniature eutrophic lakes from which the algae could be harvested and put to another use. The effluent water from a sewage plant can have 99% of its phosphate removed by precipitation and this would reduce the level in rivers to reasonable amounts and restore the unbalanced lake and estuary ecosystems into which the rivers empty. To prevent biological nuisances, the ideal phosphorus concentration of water entering a lake should be less than 0.05 mg phosphorus/litre[31].

1.4 The water-based phosphate cycle

With a concentration of 0.5 mg phosphate/litre a lake may suffer an algal bloom and this will reduce the inorganic phosphate concentration to 0.005 mg phosphate/litre. Blooms have been observed at concentrations as low as 0.02 mg phosphate/litre[32]. Of itself a sudden increase in the algae population is not deleterious to a body of water even when this increase is sufficient to colour the water and thereby produce the bloom. Blooming is natural when the nutrient which is limiting is suddenly in greater supply. Normally a bloom cannot be sustained since the surplus nutrient is quickly used up and the ecosystem returns to its former balance. However if a continuous supply of the limiting nutrient enters the water the bloom will be perpetuated and this spells disaster for the other species sharing the aquatic environment. Phosphate in most cases is the limiting factor and a high level of this in a river often results in this kind of situation, called *eutrophication*, when the river enters a lake or other shallow body of water. Before looking at the results of eutrophication it is necessary to review the undisturbed water-based phosphorus cycle, and before we do that we shall review the ecosystem and the basic equilibrium which governs the nutrients.

1.4.1 Photosynthesis and respiration

Photosynthesis (P) and respiration (R) are the terms used to describe the processes of production and decay of living matter, and it is the balance between these which governs the oxygen in the atmosphere or dissolved in water. The basic equilibrium

1.4 The water-based phosphate cycle

describing this in chemical terms is given by (11) for algal protoplasm, formula $C_{106}H_{263}O_{110}N_{16}P$.

$$106CO_2 + 16NO_3^- + H_2PO_4^- + 122H_2O + 17H^+ \underset{R}{\overset{\text{sunlight P}}{\rightleftarrows}}$$
$$C_{106}H_{263}O_{110}N_{16}P + 138O_2 \qquad (11)$$

The nitrogen:phosphorus ratio for the process is 16:1 which is about that of the sea, suggesting either that phytoplankton determine this ratio or they evolved to fit this ratio. In any event they serve to perpetuate the ratio. They also are responsible for renewing much of the oxygen in the atmosphere via (11). At equilibrium P=R and the steady state maintains a constant chemical composition of the water (chemostasis) and a relatively constant population of organisms (homeostasis).

Under equilibrium conditions the water environment is aesthetically pleasing. The rate of P may be small, as in fast flowing mountain streams, or very large, as in Pacific coral reefs which are teeming with life, but as long as P=R the result is harmonious.

If P and R are not equal the result is pollution. If P > R, indicating an influx of chemical nutrients, algae pollution results. If R > P the water suffers secondary pollution caused by the using up of the available oxygen so that decay produces not NO_3^- but NH_4^+ or N_2, not CO_2 but CH_4 and not SO_4^{2-} but HS^- or foul-smelling H_2S (the sulphur is not included in (11) but is a necessary element for some life processes). Oxygen in water can be used up because the P and R processes occur at different levels. If the body of water is not thoroughly mixed the oxygen produced by photosynthesis, which occurs nearer the surface, may be mostly lost to the atmosphere instead of reaching the region of decay at the bottom of the water where it is most required. Incomplete mixing results from stratification of the water due to density differences. These are caused by seasonal changes in temperature; mixing occurs in winter, stratification in summer. This applies mainly to lakes as we shall see.

The phosphate cycle in water, as on land, is controlled by the biocycle. This consists of *producer* organisms* which utilize the inorganic nutrients, the *consumer* organisms which feed on them and the *decomposer* organisms which break down the dead forms of the higher organisms. In aquatic environments the consumers are collectively termed zooplankton and consist of insects and their larvae, crustacea, fish etc. They generally feed on the phytoplankton but some are carnivorous. Bacteria and fungi are the chief decomposer organisms, aided by some of the bottom forms. Phosphate is essential to the biocycle and the water-based phosphate cycle is shown in Fig. 1.4.

The inorganic parts of the cycle consist of soluble phosphates and insoluble metal phosphates. Under most conditions the majority of the phosphates in solution will be HPO_4^{2-} although this will exist in equilibrium with $H_2PO_4^-$, PO_4^{3-} and H_3PO_4. Precipitation should occur immediately the solubility product of either calcium, ferric

---

* These are the phytoplankton, which range from single cell plants to giant seaweeds. The algae are the predominant producers being classed according to their colour, the common surface algae being blue-green the deeper algae being the red algae which are famous for their dramatic blooming (Red Sea, Nile turning to blood etc.)

Figure 1.4 The water-based phosphorus cycle.

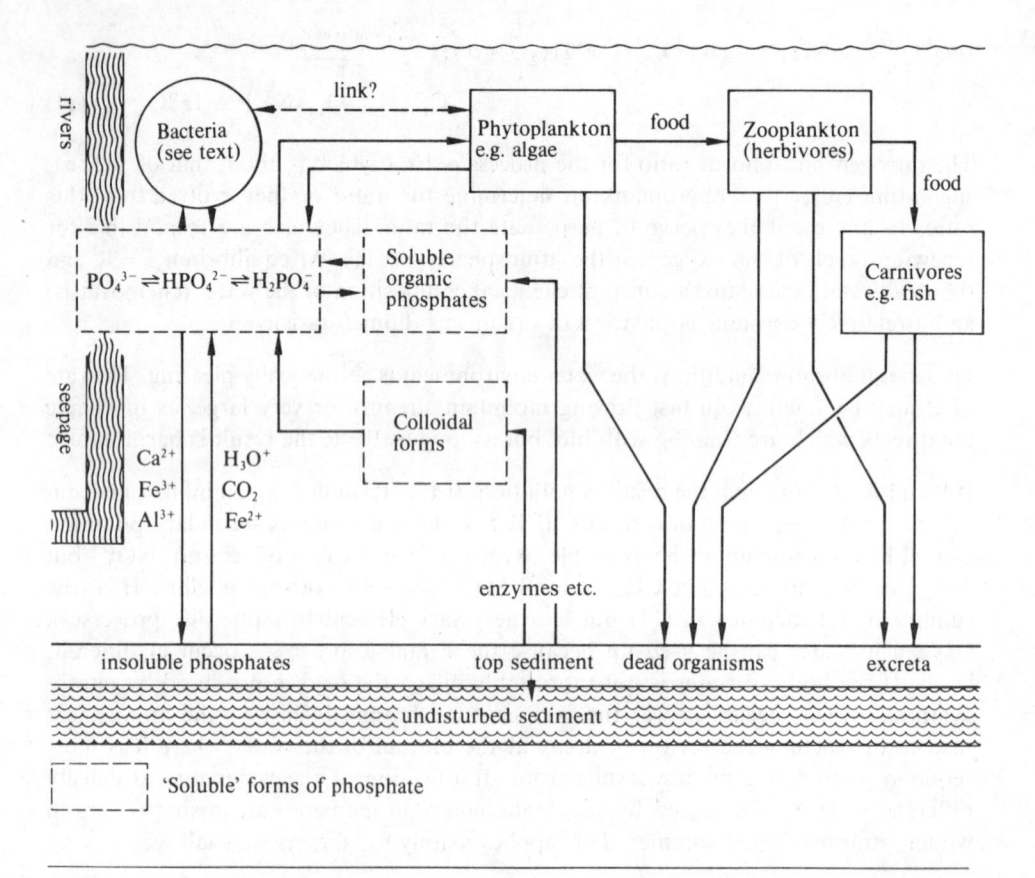

or aluminium phosphate has been exceeded but what this means in terms of phosphate concentration is not clear.

For example the solubility product of hydroxyapatite $K_{sp} = [Ca^{2+}]^5[PO_4^{3-}]^3[OH^-]$ has been calculated as *ca* 10^{-51} yet in Lake Tjeukemeer[34] the measured concentrations of these ions give a product of *ca* 10^{-46}. Either this lake is supersaturated with respect to $Ca_5(PO_4)_3OH$ or the measured concentration of one of the ions is too high and this must mean phosphate. Which brings us to one of the problems which bedevils the study of the aquatic phosphorus cycle – the difficulty of analyzing the concentrations of phosphorus in its various forms.

Some *organic* species can take up and release inorganic phosphate very rapidly. The method usually used to separate *organic* from inorganic is filtration through a membrane of known pore size usually around 0.5 µm or less. This process may in itself be destructive of some of the phytoplankton which will cause them to release phosphate. Other soluble *organic* phosphate forms may pass through such a membrane and be subsequently analyzed as inorganic phosphate[35]. This was shown by passing membrane filtered solutions down columns packed with hydrated zirconium oxide (HZO). This

1.4 The water-based phosphate cycle

compound has a high affinity for phosphate ions and should retain almost all the phosphate in such solutions. In most instances a sizeable proportion of the supposed inorganic phosphates pass through the HZO column, and in one case only 10% was retained[36]. Clearly other forms of phosphate exist in these solutions and the true concentration of orthophosphate operative in K_{sp} determinations may be an order of magnitude smaller than supposed from analyses.

Bacteria may be more important than has previously been assumed in removing phosphate from lake water. Phosphate studies with $^{32}PO_4^{3-}$ have shown the need to add antibiotics to lake water samples to destroy the bacteria before consistent experimental results can be obtained[37]. The role of bacteria is as yet unclear but it has been demonstrated that phosphate stimulates bacteria division[38]. Moreover a link is thought to exist between bacteria and phytoplankton, the former providing some growth factor for algae.

Despite these difficulties the estimates of inorganic phosphate concentrations do provide some information and show how the orthophosphate concentration changes. In Lake Geneva in May 1970 a phenomenon was observed which has been explained in terms of a solubility product being exceeded[39]. The orthophosphate concentration had built up to 0.18 mg/litre and then suddenly dropped, and this was the point at which it was thought hydroxyapatite precipitated since its solubility product had been exceeded.* Bearing in mind these difficulties with regard to inorganic phosphate, which is the simplest part of the aquatic system, we now consider the water-based phosphorus cycle in the four main types of aquatic environment – lakes, estuaries, shallow seas and oceans. Each is characterized by different kinds of water movement which affect the system quite markedly. Since the eutrophication of lakes has been the most dramatic phosphate disaster we shall spend most of the rest of this chapter dealing with this.

The first member of the food chain and the one which causes most trouble in lakes is the algae. Planktonic algae have a special relationship with inorganic phosphate. They can absorb phosphate in excess of their nutritional requirements (this is called luxury phosphate) and store it in a loosely bound, readily available state. Moreover, algae also have a method for breaking down polyphosphates and even of utilizing *organic* phosphates such as phospholipids by means of enzymes associated with their cell walls[40]. Even insoluble forms of excreted *organic* phosphorus can serve, and these sources are very important in a natural system. Zooplankton and other organisms higher up the food chain excrete a large proportion of the phosphorus they consume in their food. Some species are capable of excreting an amount of phosphorus equivalent to 100% of their body phosphorus per day, most passing through without being absorbed into the organism's metabolism. Some species however show a turnover of about 10% of their actual body phosphorus per day[40]. About half the excreted phosphorus is inorganic phosphate half is *organic* phosphate. Most can be reused by algae although some ends up as sediment.

---

* Obviously some other factor intervened to cause the phosphate concentration to fall since K_{sp} would operate so as to apply a ceiling to the phosphate concentration, excess phosphate then being precipitated. A sudden increase in $[Ca^{2+}]$ and/or $[OH^-]$ would depress $[PO_4^{3-}]$.

1.4.2 Sediment

The sediment of a lake is made up of precipitated solids, settled mud and other inorganic and *organic* debris[41]. Over a period of time the layer of sediment becomes sufficiently thick that its lower part is no longer disturbed by plants, worms and other bottom forms. To all intents and purposes any phosphorus in this part of the sediment is permanently fixed and removed from the cycle. The phosphate which ends up in this way is phosphate which is precipitated or which is adsorbed onto clay particles (cf deflocculant use of phosphates) and certain of the more stable *organic* phosphates such as humic iron phosphates which resist breakdown by bacteria etc.

While it is part of the top sediment the phosphorus can be returned to the water of the lake. Inorganic phosphates become more soluble as HPO_4^{2-} or $H_2PO_4^-$ if the pH of the lake water becomes more acid which can occur by the action of organic decomposition or dissolved CO_2. If $FePO_4$ has been precipitated and the lower waters become reducing, as in summer when the oxygen has been consumed by decay processes and the waters are not mixed, then the insoluble $FePO_4$ may be reduced to soluble $Fe_3(PO_4)_2$ and the phosphate returned to solution. *Organic* phosphates are also returned to solution and generally quite rapidly. Following the death of plankton it has been shown that 40% of the *organic* and 25% of the inorganic phosphate is released within hours and up to 90% of all the organism's phosphorus has been released within 1 day. The release is thought to be mainly automatic rather than bacterial or enzymatic, these acting instead on the liberated *organic* phosphate[42]. The small amount of body phosphorus which is not released is probably held as phosphate complexes or insoluble salts such as the phytin salts of Al^{3+}, Fe^{3+} and Ca^{2+} which are resistant to phytase.

Because some phosphorus is permanently retained by the sediment, and since phosphorus is the limiting factor for the aquatic environment, then the population of the lake will gradually decrease unless the rivers feeding it carry down fresh supplies. The loss of phosphorus was conclusively demonstrated in 1968 when Chamberlain put enough labelled $^{32}PO_4$ into a natural lake (Upper Bass Lake) to give measurable amounts of the radioisotope even when it was diluted throughout the lake[35]. Monitoring over a period of two months showed a steady decline in ^{32}P from the lake water system and it was calculated that a daily loss of 1–2% was occurring. This was conclusive proof of continuous phosphorus removal from the natural aquatic environment although laboratory experiments with ^{32}P on aquarium systems had previously shown that this should occur. The other findings concerning the movement of radiophosphorus in an aquarium microcosm are also very revealing[43].

Within 20 minutes of the addition of $H_3{}^{32}PO_4$ to such a microsystem over 90% of the added phosphorus is incorporated into plankton and particles with bacteria. The phosphate is adsorbed onto and absorbed through the surfaces of cells. This phosphate exchanges rapidly with the water and an equilibrium is set up. Phosphate moves into attached algae more slowly, reaching a maximum after about six days. The uptake by animals is slower still, depending upon their food intake. And all the time there is a downward loss of phosphate to the sediment until after 45 days this has reached about 75%. Such studies as those of Whittaker[43] reveal a complexity of transfer pattern and a rapidity of movement of phosphorus within the water-based cycle. Figure 1.4 does little more than outline the major pathways. The quantity of phosphorus being

1.4 The water-based phosphate cycle

transferred along these pathways varies in a lake according to the season because of stratification. This cycle was first investigated and described by Mortimer[44].

1.4.3 Natural waters

In summer the waters of a lake are stratified. The sun heats the surface layers thereby reducing their density and keeping them at the top, and this part of the lake is called the epilimnion. In it P, photosynthesis, is much greater than R, respiration. The lower, denser layers, called the hypolimnion, remain trapped at the bottom of the lake. Here R ≫ P, and the dissolved oxygen consumed in the degradation of organic debris is not replenished, so that the environment becomes reducing. In terms of phosphate concentration the epilimnion is very much depleted by the end of summer while the hypolimnion is enriched. And then comes the autumnal turnover.

The surface waters of the lake cool as the days shorten until there comes a day when the boundary between the epilimnion and hypolimnion disappears and the waters of the lake mix. The result is a sudden enrichment of phosphate at the surface, sufficient often to cause a natural algal bloom. The bloom cannot be sustained for long and fades quickly. With winter, algal activity falls below that of nutrient supply from rivers and sediment, and during this time the phosphate concentration of the lake builds up. The result is a spring bloom when the increasing hours and intensity of sunshine stimulate algal growth. Thus under normal conditions a lake may bloom twice a year but under normal conditions these are of relatively short duration and do not pose a threat to other life forms in the lake. Continued blooming is quite different and is an ecological disaster. The surface algae monopolize the sunlight and the dead algae monopolize the oxygen at the bottom. The lake in effect assumes the characteristics of the long-notorious duck-pond – green, slimy, smelly and undrinkable.

The constant supply of too much phosphate from polluted rivers has turned enormous lakes into seas of algae. Lake Zurich in 1910 was suffering from such eutrophication due to human sewage. Today it has regained its natural state of clean clear water supporting a balanced ecology, despite the fact that the water is still very rich in nitrate all the year round from agricultural run-off and sewage[45]. The reason is that steps have been taken to reduce phosphate concentration by not only discharging sewage effluent at the exit to the lake (this step was taken in 1912) but by treating it prior to discharge with ferric chloride to precipitate $FePO_4$.

The Lake Zurich restoration demonstrates that phosphate is probably the natural limiting factor or can be made so. Since it is the only nutrient over which Man can have effective control it *must* be made the limiting factor to control unwanted algae. Once the supply of phosphate is turned off or reduced to its natural level the process of sedimentation will quickly lower the existing phosphate concentration and within a few years the lake will regain a balanced ecosystem. All lakes eventually fill up with sediment and disappear, the rate at which this occurs depends upon the rate of eutrophication, i.e. the rate of supply of nutrients. By increasing this supply the ageing is speeded up, e.g. Lake Erie has been prematurely aged by about 15 000 years since about 1900.[46] This particular lake has been turned from a natural asset to a national embarrassment so much so that some have suggested that it has reached a stage beyond recall[47].

1.4.4 The sea

Although the sea can cope with excess phosphate there can be localized effects which work to produce algal blooms. These can occur at the mouths of rivers and fjords where the nutrient supply carried down by the river is enriched rather than diluted by the sea. The enrichment is the result of counter currents carrying the soluble products from off-shore sediment back towards the shore – see Fig. 1.5. The extent

Figure 1.5 Estuary phosphate enrichment.

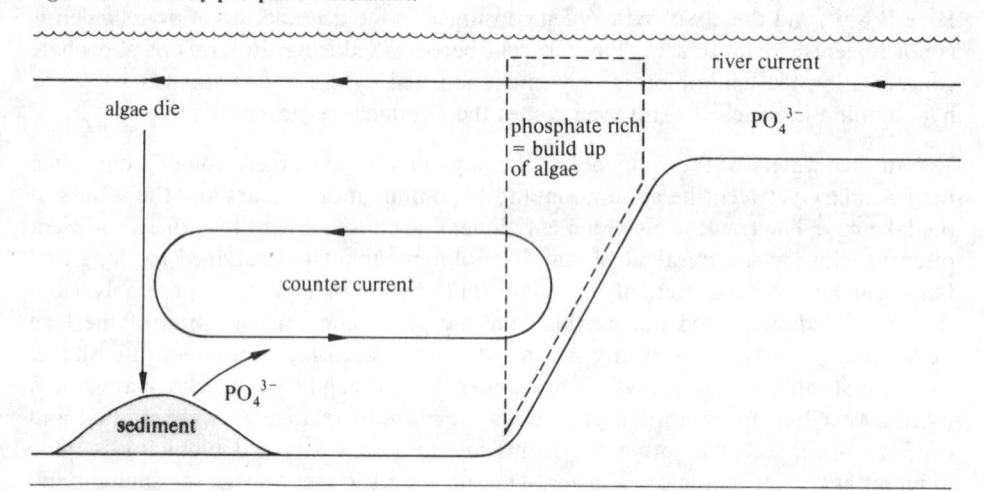

of the counter current depends on many factors such as the length of the estuary basin, its depth, turbulence etc.

Counter currents are responsible for depletion as well as enrichment and this applies particularly to the Mediterranean Sea[48]. This is the most naturally impoverished large body of water. The nutrients from its deeper waters escape into the Black Sea through the Bosporus and into the Atlantic Ocean through the Straits of Gibraltar. The cause is the evaporation of the Mediterranean which exceeds the supply of fresh water from its surrounding rivers. The deficit is made good from the Black Sea and Atlantic and the inflow of surface water from these generates a counter current outflow of deeper water carrying with it the regenerated nutrients from the sediment. The effect on the Atlantic is not noticeable but for the Black Sea the result is dramatic as Table 1.4 shows. As the countries surrounding the Mediterranean become more industrialized the Black Sea will bear the brunt.

The sea at large cannot seriously be threatened by phosphate. The concentration of phosphate in sea water is very low whilst that of Ca^{2+} is very high – a combination which serves to precipitate hydroxy or fluorapatite. It is thought that below depths of 1000 metres the oceans are in effect saturated with one of these since the phosphate concentration is stationary as Fig. 1.6 shows, although the concentration varies from ocean to ocean[48]. The total dissolved phosphorus in the sea is 9.8×10^{10} metric tons[49] (about 0.01% of the total in the primary cycle) and natural sources supply 2×10^6 tons per year from rivers. The residence time of phosphorus in the sea is thus

1.4 The water-based phosphate cycle

Table 1.4 Concentration of phosphorus (μmol/litre) in the Mediterranean and adjacent seas–Black and Atlantic[48]

Depth (m)	North Atlantic	surface water → [Gibraltar] ← deeper nutrients	Mediterranean Sea	← surface water [Bosporus] deeper nutrients →	Black Sea
1	a		0.04		—
10	0.26		—		0.12
100	0.74		0.11		1.03
300	1.05		0.32		5.55
500	1.23		0.41		6.07
1000	1.14		0.34		7.00
2000	1.09		0.32		7.46

a Surface concentration in May *ca* 0.61, August *ca* 0.12 μmol P/litre.

$9.8 \times 10^{10}/(2 \times 10^6)$ i.e. 50 000 years, which is short in geological terms especially for a non-volatile element[48]. (Na$^+$ has a residence time of 260 000 000 years or 5000 times as long.) Clearly some mechanism is removing phosphate from sea water and this is probably precipitation plus some organic debris such as fish teeth which can withstand the journey to the bottom. Most organic debris gets no further than the 1000–2000 metre zone before the phosphorus it contains is released to solution.

The concentration of phosphorus in the surface waters of the oceans is low which is why so few fish are found in most of them[50]. Contrary to popular belief, only a small percentage of the oceans is teeming with life. The regions where fish are plentiful are areas of upwelling sea water. This upwelling is part of a global movement of the oceans and brings to the surface the lower waters which are richer in phosphate. Such upwelling regions account for only 0.1% of the sea's surface but they produce 50% of the world's fish. Upwelling regions are the mid-Pacific and the Pacific coasts of America (California and Peru especially), Arabia and Antarctica. In addition to upwelling regions there are sinking regions such as the North Atlantic[48].

Coastal and continental shelf regions are also fishing areas, being relatively shallow and to some extent replenished by rivers. Most of the ocean is unmixed however. Even gales do not serve to stir up the deeper phosphate-rich layers. For the sea the paradox is that all the light is at the top and all the nutrients, and especially the limiting factor, phosphate, is at the bottom. For this reason large tracts of sea are relatively barren and likely to remain so.

The secondary phosphorus cycles are extremely complex, involving as they do many components and variables. Faced with such a collection of phosphorus compounds and their reactions the phosphorus chemist finds it difficult to be certain about the

Figure 1.6 Phosphorus in the sea taken from data in ref. 48.

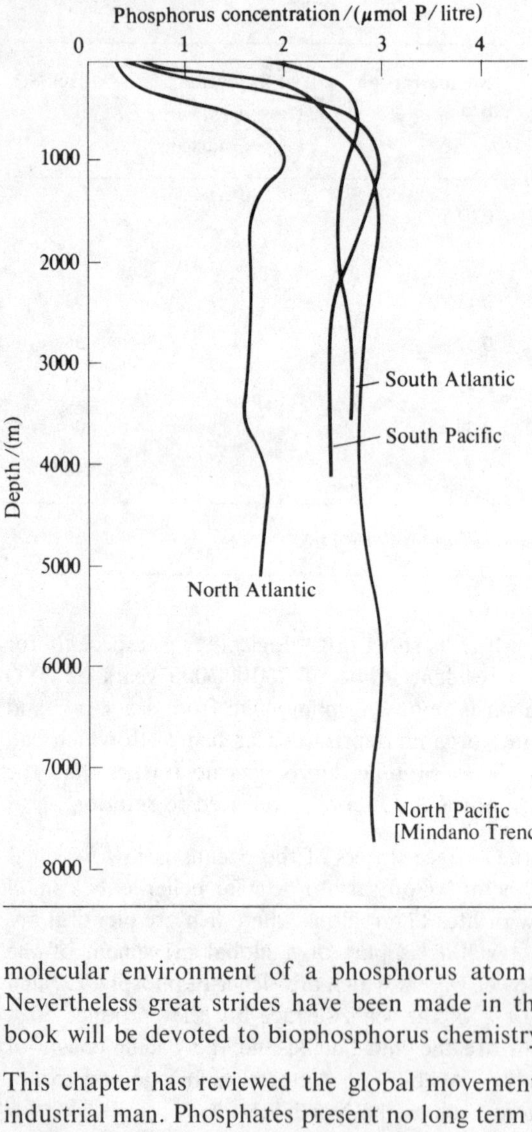

Phosphorus concentration/(μmol P/litre)
0
1
2
3
4
1000
2000
3000
4000
5000
6000
7000
8000
Depth/(m)
South Atlantic
South Pacific
North Atlantic
North Pacific
[Mindano Trench]

molecular environment of a phosphorus atom in any part of a secondary cycle. Nevertheless great strides have been made in this area and the last chapter of this book will be devoted to biophosphorus chemistry.

This chapter has reviewed the global movement of phosphorus and the impact of industrial man. Phosphates present no long term threat as a pollutant; common sense based on chemistry is the answer to phosphate pollution. And even if this is not forthcoming Nature has its own mechanism for dealing with the problem when Industrial Man has run his course.

Problems

1 It is common practice to express the composition of rocks in terms of the oxide of each element, e.g. % P as P_2O_5. In Table 1.2 on page 9 this is done for a selection of phosphate rocks and for fluorapatite. It can be seen that for $Ca_5(PO_4)_3F$ the total of

its components comes to 101.9%. Why? What else is inconsistent about this method of recording the composition of a material? Does the method have any advantages? Rewrite Table 1.2 in a more meaningful chemical way in terms of % P, % Ca and % F as the elements.

2   At depths below 1000 m the concentration of soluble phosphorus compounds in sea water is fairly constant at about 0.084 mg P/litre, and has this concentration down to the ocean floor. It has been suggested that the sea at these depths is saturated with respect to hydroxyapatite, $K_{sp} = [Ca^{2+}]^5[PO_4^{3-}]^3[OH^-] = 10^{-51}$. Is this so? The concentration of calcium in sea water is 0.400 g/litre and the pH of the sea is *ca* 8.0. Explain your results.

References

[1] C. B. Moore, "Phosphorus in Meteorites and Lunar Samples", *Env. P Handbook*, chap. 1. This abbreviation refers to the compilation of 37 review articles under the title *Environmental Phosphorus Handbook* (Editors E. J. Griffith, A. Beeton, J. M. Spencer and D. T. Mitchell), John Wiley & Sons, New York, 1973.

[2] V. E. McKelvey, "Abundance and Distribution of Phosphorus in the Lithosphere", *Env. P Handbook*, chap. 2.

[3] Z. S. Altschuler, "The Weathering of Phosphate Deposits", *Env. P Handbook*, chap. 3.

[4] D. R. Peck, "The History and Occurrence of Phosphorus", *Mellor's Comprehensive Treatise on Inorganic and Theoretical Chemistry. Vol. VIII. Supplement III. Phosphorus*, Section I, Longman, London, 1971.

[5] D. J. Fisher, "Geochemistry of Minerals Containing Phosphorus", *Env. P Handbook*, chap. 6.

[6] W. E. Brown, "Solubilities of Phosphates", *Env. P Handbook*, chap. 10.

[7] A. S. Posner, "Mineralized Tissues", *P and its Compounds*, chap. 22. This abbreviation refers to *Phosphorus and its Compounds. Vol. II. Technology, Biological Functions and Applications* (Editor J. R. Van Wazer), Interscience, New York, 1961. This is a pre-pollution era compilation of review articles written mostly by the editor and covering many topics and aspects of phosphorus chemistry which are still relevant today.

[8] F. Erhardt, *Monatsschr. Rationelle Aerzte*, **19**, 359 (1874).

[9] J. R. Van Wazer, "Food and Dentifrice Applications", *P and its Compounds*, chap. 25.

[10] Report on toothpastes, *Which? Consumer Magazine*, February, 1974.

[11] G. D. Emigh, "Economic Phosphate Deposits", *Env. P Handbook*, chap. 4.

[12] H. M. Sevens, "Wet-Process Phosphoric Acid", *P and its Compounds*, chap. 16.

[13] T. L. Hurst, "Manufacture of Elemental Phosphorus", *P and its Compounds*, chap. 18.

[14] A. D. F. Toy, *Comp. Inorg. Chem.*, (Editor A. F. Trotman-Dickenson) vol. 2, chap. 20, p. 391; Pergamon, Oxford, 1973.

[15] E. D. Jones III, "Phosphorus in the US Economy", *Env. P Handbook*, chap. 36.

[16] J. R. Van Wazer, "Miscellaneous Phosphorus Applications", *P and its Compounds*, chap. 32.

[17] J. R. Van Wazer, "Surface Treatment of Metals", *P and its Compounds*, chap. 30.

[18] G. J. J. Jayne and J. S. Elliott, *J. Inst. Petr. London*, **56**, 42 (1970).

[19] J. W. Lyons, "Distribution of Polyphosphate Deflocculants in Aqueous Suspensions of Inorganic Solids", *Env. P Handbook*, chap. 13.

[20] J. C. Harris and J. R. Van Wazer, "Detergent Building", *P and its Compounds*, chap. 27.

[21] A. L. Hammond, *Science*, **172**, 361 (1971).

[22] J. Kastelic and R. M. Forbes, "Animal Nutrition and Phosphates in Feeds", *P and its Compounds*, chap. 24.

[23] G. E. C. Mattingly and O. Talibudeen, "Progress in the Chemistry of Fertilizer and Soil Phosphorus", *Top. Phosphorus Chem.*, **4**, 157 (1967).

[24] J. Liebig, *Traite de Chimie Organique*, Fortiss, Masson et Cie, Paris, 1840.

[25] J. Murray, *Advice to Farmers*, Longman & Co., London, 1842.

[26] J. Davidson, *J. Amer. Soc. Agron.*, **18**, 962 (1926).

[27] D. I. Arnon, P. R. Stout and F. Sipas, *Amer. J. Botany*, **27**, 791 (1940).

[28] W. H. Petersen, J. T. Skinner and F. M. Strong, *Elements of Food Biochemistry*, Prentice-Hall, New Jersey, 1949.

[29] K. M. Mackenthum, "Eutrophication and Biological Associations", *Env. P Handbook*, chap. 33, table 1.

[30] J. B. Nesbitt, "Phosphorus in Wastewater Treatment", *Env. P Handbook*, chap. 35.

[31] K. M. Mackenthum, *J. Amer. Water Works Assoc.*, **60**, 1047 (1968).

[32] W. Rodhe, *Symb. Bot. Ups.*, **10**, 1 (1948).

[33] W. Stumm and J. J. Morgan, *Aquatic Chemistry. An Introduction Emphasizing Chemical Equilibria in Natural Waters*, pp 429–438, Wiley, New York, 1970.

[34] H. L. Golterman, "Vertical Movement of Phosphate in Freshwater", *Env. P Handbook*, chap. 29, table 2.

[35] W. M. Chamberlain, Ph.D. Thesis, University of Toronto, Canada, 1968. See F. H. Rigler, "Phosphorus Cycle in Lakes", *Env. P Handbook*, chap. 30, p 544.

[36] F. H. Rigler, *Limnol. Oceanog.*, **13**, 7 (1968); *Env. P Handbook*, chap. 30, pp. 548–550.

[37] Ref. 41, p 583.

[38] Ref. 45, p 585.

[39] Ref. 34, p 534.

[40] F. F. Hooper, "Origin and Fate of Organic Phosphorus Compounds in Aquatic Systems", *Env. P Handbook*, chap. 9.

[41] P. R. Hesse, "Phosphorus in Lake Sediments", *Env. P Handbook*, chap. 31.

[42] S. M. Marshall, and A. P. Orr, *J. Mar. Biol. Assoc. UK*, **34**, 495 (1955).

[43] R. H. Whittaker, *Ecological Monographs*, **31**, 157 (1961).

[44] C. H. Mortimer, *J. Soil Sci.*, **1**, 63 (1949).

[45] E. A. Thomas, "Phosphorus and Eutrophication", *Env. P Handbook*, chap. 32.

[46] A. M. Beeton, *Limnol. Oceanog.*, **10**, 240 (1965).

[47] R. H. Whittaker, *Communities and Ecosystems*, p 138–139, Macmillan, London, 1970.

[48] R. A. Gulbrandsen and C. E. Robertson, "Inorganic Phosphorus in Seawater", *Env. P Handbook*, chap. 5.

[49] M. K. Horn and J. A. S. Adams, *Geochim. Cosmochim. Acta*, **30**, 279 (1966).

[50] I. Morris, *Chem. Brit.*, 198 (1974).

Chapter 2
Bonding at phosphorus

2.1 Introduction

Phosphorus is a non-metal and its bonding is covalent. Any theory which successfully explains the covalent bonding of the first row non-metal elements should also be able to explain the bonding at phosphorus. Indeed much of the chemistry of phosphorus can be explained in such terms but there are aspects of its behaviour which cannot be dealt with by analogy with first row chemistry. In particular phosphorus can form stable pentacoordinate compounds such as PCl_5. Among the non-metals this co-ordination number is extremely rare and attention is thus focussed on any element displaying it, and in particular on any element displaying it in such variety as does phosphorus.

There are features of pentacoordination which give it a special fascination. Some of these features have meant the development of new theories to explain them (and new words such as pseudorotation and apicophilicity to discuss them). Other aspects have been explained by extending existing theories but this has led to argument. All the bonding of the first row elements can be explained without recourse to d orbitals. The most noticeable difference between first and second row elements is the larger number of bonds per atom which the latter are capable of forming. Associated with this difference, at least in the minds of most chemists, is the fact that second row elements have empty 3d orbitals and that they are in some way using these to expand their coordination sphere. But is it right to assume the two are linked? In recent years doubt has been cast on whether 3d orbitals have any significant part to play in the chemistry of phosphorus, or any second row element. The underlying theme of this chapter will be the discussion of the degree of involvement of 3d orbitals in bonding at phosphorus. For the simplest, tervalent phosphorus compounds, 3d orbitals play their smallest part; for the more complex, higher valent derivatives, 3d orbitals should be important if they are to play any part at all.

Apart from investigating the nature of a particular covalent bond it is also necessary to look at the spatial arrangement of such bonds, in other words to explain the shapes of molecules. One theory, the VSEPR theory does this extremely well and since it is built on the concept of the electron as a charged *particle* it is very simple. It allows us to introduce the geometry of phosphorus molecules without recourse to a detailed account of bonding theory.

In 1916 the great American chemist G. N. Lewis introduced his theory of covalent bonding based on the sharing of electron pairs between the bonded atoms. At that time electrons were regarded purely as particles and therefore his theory needed only to take this side of their nature into account. Even so it was a great success. Ten years after, it was found that electrons also had wave properties and it became necessary to explain the covalent bond from this point of view as well. The treatment of electrons as waves requires a much more mathematical approach to bonding, and it becomes difficult to visualize the bonding situation. However, by defining electrons around a nucleus in terms of *orbitals* of various kinds and by formulating rules for their manipulation and combination, it became possible for chemists to discuss covalent bonding in wave mechanistic terms. Indeed, so popular did this treatment become that chemists were in danger of forgetting that whatever else electrons may be, they are first and foremost particles. In its way, then, the Lewis approach is just as valid as the orbital approach – the two theories are not mutually exclusive but complementary.

2.2 Valence-shell electron-pair repulsion theory

In recent years R. J. Gillespie[1] has become the champion of the particle approach to covalent bonding, which under the title of Valence-Shell Electron-Pair Repulsion theory (VSEPR) is enjoying a growing popularity. The theory is summarized in six axioms (Table 2.1) which serve to explain the shapes of molecules, and this they do

---

Table 2.1 The axioms of VSEPR theory

Axiom 1: electrons in closed inner shells take no part in bonding; these, together with the nucleus, form the atomic cores from which molecules are built, with these being held together by the interactions of valence electrons.

Axiom 2: electrons in valence shells are physically paired by which it is meant that they are in close proximity occupying a common volume of space.

Axiom 3: electron-pairs arrange themselves about a particular atom so as to minimize Coulombic repulsions between them.

Axiom 4: bonding electron-pairs and non-bonding electron-pairs inter-repel differently; the order of repulsions is bonding↔bonding < bonding↔non-bonding < non-bonding↔non-bonding.

Axiom 5: as the size of the atom core increases (as one descends a group of the periodic table) the nucleus retains more control over its non-bonding than its bonding pairs; the effect is to exaggerate the repulsion differences of axiom 4.

Axiom 6: two and three electron-pairs can be shared between two atoms in the formation of a double or a triple bond; as such they arrange themselves so as to be equidistant from the atoms.

---

very well. The chief justification for accepting VSEPR theory is that it works; the shape of XeF_6 (distorted octahedron) was successfully predicted by it while orbital based theories went astray in their predictions*. How does VSEPR theory help in explaining the structure of phosphorus compounds?

Tervalent phosphorus has four electron-pairs in its valence shell and these adopt a tetrahedral configuration to minimize Coulombic repulsion (axiom 3). However one pair is non-bonding so the angle between it and the bonding pairs will be wider than the angle between the bonding pairs themselves (axiom 4). In addition, and by virtue of the size of phosphorus, the non-bonding pair has a proportionately larger effect than the non-bonding pair in an analogous nitrogen compound (axiom 5). The shape of a PX_3 molecule should therefore be pyramidal with a sharper angle at the phosphorus apex than the corresponding NX_3 case (Fig. 2.1). The drop in angle from NX_3 to PX_3 is not constant but finer points of detail like this are not yet consistently explained by VSEPR theory.

Tetracoordinate phosphorus cations such as PH_4^+, PCl_4^+ and PPh_4^+ have the four bonding electron-pairs tetrahedrally arranged about the phosphorus atom as predicted, but the other kind of tetracoordinate phosphorus compound, exemplified by

---

* Further theoretical support for the VSEPR method has recently been provided by the Hellmann–Feynman theorem; see B. M. Deb, *Rev. Mod. Phys.*, **45**, 22 (1973), for details.

Figure 2.1 Shapes of phosphorus and nitrogen molecules taken from ref. 1.

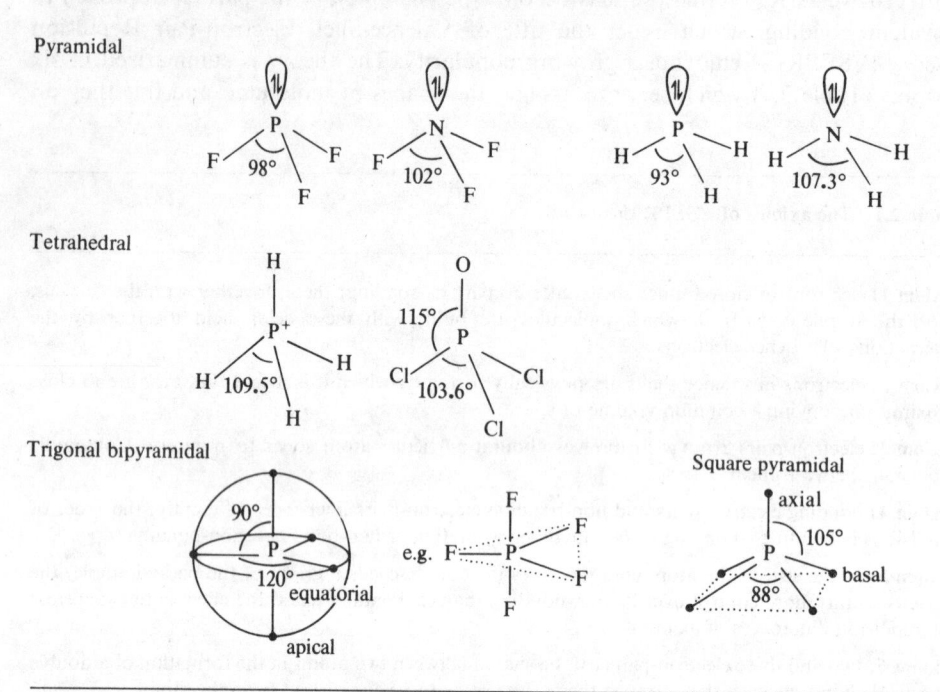

Pyramidal

Tetrahedral

Trigonal bipyramidal

Square pyramidal

POCl$_3$, presents a problem. However before discussing this class of derivatives it is necessary to consider the general situation at phosphorus when it is surrounded by five electron-pairs in its valence shell.

There is no possible way in which five such pairs can be arranged so that they are equivalent to one another. This is really a geometric fact of life and not something special about phosphorus. If the five electron-pairs were five like charges on the surface of a sphere, free to arrange themselves to minimize Coulombic repulsion the result would be the same. It would be found that the configuration they adopted would correspond in solid geometry to a structure called a trigonal bipyramid, which will be referred to from now on as *tbp* and which is illustrated in Fig. 2.1. In this arrangement three points are lying on an equator with one point at each pole or apex of the sphere. Thus in the case of five electron-pairs we would predict a *tbp* configuration (axiom 3) but with the three equatorial atoms of such a molecule nearer to the central phosphorus atom than the apical atoms. The reason for this being that an equatorial bonding-pair experiences bonding-pair↔bonding-pair repulsion from only two other electron-pairs at 90° (i.e. the apical pairs) while the apical bonding-pairs experience repulsion from three electron-pairs at 90° (i.e. the equatorial trio). Pairs at angles of 120° to each other may be discounted in this analysis since their interaction is relatively unimportant. The net result is a predicted lengthening of the apical bonds, and this is what is found. Moreover, the more electronegative a ligand attached to phosphorus, the more it will assist in this process of withdrawing its bonding pair. Consequently given the choice, an electronegative group prefers to occupy an apical position.

2.2 Valence-shell electron-pair repulsion theory

Under certain circumstances an alternative square pyramidal (*spy*) configuration – see Fig. 2.1 – is possible for five-coordinate phosphorus. The stability of *spy* relative to *tbp* is improved by having more electronegative ligands attached to phosphorus. These prefer basal sites. This alone is not enough to tip the scales in favour of *spy* and since this structure is only found in a few spirocyclic phosphoranes it would appear that certain physical restraints are necessary for its observation; see Fig. 2.6.

Returning now to the structure of the tetracoordinate phosphorus compounds of the type POX_3 or PSX_3 it can be seen that these may be explained in two ways. Either they have four electron-pairs with a donor σ-bond to the oxygen or sulphur atom, e.g. $Cl_3P \rightarrow O$, or they have five electron-pairs with two of the pairs forming a VSEPR style double bond, e.g. $Cl_3P \diamondsuit O$. In theory it should be possible to decide between the two on bond angle data since the first alternative should have angles approaching the tetrahedral angle of 109°28′ whereas the second alternative should have a much smaller ClPCl angle.

Unfortunately bond angle arguments are inconclusive as Table 2.2 shows; steric

Table 2.2 Bond angles in phosphoryl and thiophosphoryl halides from ref. 1 (table 3.5, p 52)

	∠OPX	∠XPX		∠SPX	∠XPX
POF_3	116°	102.5°	PSF_3	117.5°	100.3°
$POCl_3$	115°	103.6°	$PSCl_3$	117.3°	100.5°
$POBr_3$	111°	108°	$PSBr_3$	113°	106°

factors dominate the geometry. Even in a compound where there is definitely the formation of a donor σ bond as in the adduct $F_3P \rightarrow BH_3$ the angle ∠FPF increases by only 2° from 97.8° in PF_3 to 99.8° in the adduct; and the ∠BPF at 118° is still far from the tetrahedral angle, even more so than in POF_3.

For tetracoordinate phosphorus compounds of a formal pentavalency VSEPR theory flounders between two possible bonding arrangements, distinguishing neither the donor σ nor the double-bond situation. Perhaps it is asking too much too soon from this revised 'particular' electron theory. It looks at present as if too much has been sacrificed to achieve simplicity, but it does seem that although VSEPR theory can provide a framework for the shape of a molecule it can not provide an outline for discussing the bonding therein.

The VSEPR approach to bonding has been dealt with separately as an opener to this chapter because it still has only B-movie status in the chemical world. The main purpose of this chapter is to consider in detail the various situations which phosphorus finds itself in. These will be distinguished by the coordination numbers displayed by phosphorus. Throughout these sections it will be necessary to consider the multiplicity of the bonding and one of the ways of judging whether a particular bond has a π component is to compare its bond length with that of a recognized single bond.

2.3 Single-bond lengths of phosphorus bonds

Most second row elements display a wider range of bond lengths than first row elements in formal single-bond situations so that it is not as easy to pick out a particular value as typical. Table 2.3 lists average bond lengths and shows the range of values upon which they are based; nitrogen single-bond lengths are included for comparison.

Certain bonds seem to vary very little despite changes in valence and coordination number of the phosphorus e.g. the P—Cl bond. Others seem particularly sensitive to change such as the P—O bond, but here the situation is complicated by possible delocalization contributions from neighbouring π-bonds. Some bonds, and in particular the P—N bond are almost always disposed to forming donor $p\pi(N) \rightarrow d\pi(P)$ bonding unless the nitrogen lone pair is otherwise occupied as it is in the phosphoramidate ion, $[H_3N^+—PO_3^{2-}]$.

Some molecules in Table 2.3 are pentacoordinate and both the shorter, equatorial, and the longer, apical, bond lengths are quoted.

Table 2.3 Phosphorus single-bond lengths in pm from ref. 2 (100 pm = 1.00Å)

Bond P—X	Average[a] r(P—X)/pm	Range of r(P—X) values	Average r(N—X)/pm
P—H	142	PH_3, 142; PH_4I, 142; Me_2PH, 145.	101
P—F	153	POF_3, 152; $(NPF_2)_3$, 152; PSF_3, 153; PF_3, 154; PF_5, 157, 159; $NaPF_6 \cdot H_2O$, 158, 173.	136
P—Cl	200	PCl_4^+, 197; $(NPCl_2)_3$, 197; $POCl_3$, 199; PCl_5, 201, 207; $PSCl_3$, 202; PCl_3, 204; PCl_6^-, 204.	195
P—Br	214	$POBr_3$, 206; $PSBr_3$, 213; PBr_3, 218; PBr_4^+, 220.	—
P—I	247	PI_3, 246; P_2I_4, 248.	—
P—B	197	$P(NH_2)_3 \cdot BH_3$, 189; $(Me_2P \cdot BH_2)_3$, 194; BP, 196; $(Me_2P \cdot BH_2)_4$, 208.	145
P—C	184	$P(CN)_3$, 179; PPh_4^+, 180; Ph_2PO_2H, 181; $(NPMe_2)_4$, 181; $POMe_3$, 182; PPh_3, 183; PMe_3, 184; $(PPh)_5$, 184; PPh_5, 185, 199; $PSEt_3$, 186; $P(CF_3)_3$, 194.	147
P—N	177[b]	$(NP(—NMe_2)_2)_4$, 168[c]; $Na_3(H_3N_3P_3O_6) \cdot 4H_2O$, 168; $Me_3N_3P_3O_3(OMe)_3$, 169; $Na(H_3N—PO_3)$, 177.	146
P—P	222	$(NH_4)_2H_2P_2O_6$, 217; P_4, 221; P_2I_4, 221; $(PPh)_5$, 221; black P_n, 223; P_4S_3, 224.	177
P—O	162[d]	PO_4^{3-}, 154[e]; H_3PO_4, 157; $(PhO)_3PO$, 157; $P(OEt)_3$, 158; P_4O_{10}, 160; P_4O_6, 165.	136
P—S	209	P_4S_3, 209; P_4S_{10}, 209.	175

[a] Average of representative values; [b] longest value found – see chapter 10 for fuller discussion of P—N bond lengths; [c] exocyclic P—N bonds; [d] average values in detail are P—O—P, 161; P—O—R, 159; P—O—H, 156; P—O$^-$, 151; note P=O average 146; [e] some π character.

2.4 Tricoordinate phosphorus compounds

Phosphorus has the same kind of ground state electron configuration, [Ne]$3s^2 3p^3$, as nitrogen, [He]$2s^2 2p^3$, and consequently a certain degree of similarity in their chemistry is anticipated. But in fact there is virtually no likeness between them, and what little can be found is for the tervalent compounds NX_3 and PX_3 where at least there is similarity of structure. One of the basic underlying causes of the difference between the behaviour of phosphorus and nitrogen compounds stems from the differences, both relative and absolute, in the strengths of the bonds they form to other elements. Table 2.4 lists the mean bond enthalpies of common bonds.

Table 2.4 Bond enthalpies of tervalent phosphorus and nitrogen bonds (from ref. 3)

X	E(P—X) /(kJ mol^{-1})	/(kcal mole^{-1})	E(N—X) /(kJ mol^{-1})	/(kcal mole^{-1})
H	328 (PH$_3$)	78	391	93
F	490 (PF$_3$)	117	272	65
Cl	319 (PCl$_3$)	76	193	46
Br	264 (PBr$_3$)	63	116	28
I	184 (PI$_3$)	44	—	
B	?		375 (B(NMe$_2$)$_3$)	89.5
C	264 (PMe$_3$)	63	285	68
N	ca 290[a]	ca 70	160	38
P	209 (P$_4$)	50	ca 290[a]	ca 70
O	407 (P$_4$O$_6$)	97	201	48

[a] See chapt. 10

Although much thermochemical information is available for nitrogen compounds, so that average bond enthalpies are quoted in Table 2.4, the same is not true of phosphorus compounds, and mean bond enthalpies of individual derivatives are given. (For some bonds such as P—N, no simple compound has as yet had its ΔH_f°(g, 273.15 K) determined.) Some of the bonds in these phosphorus compounds may not be truly representative and this objection might apply particularly to the P—P bond in P$_4$ which is a tetrahedron of phosphorus atoms with bond angles of 60°. However when E(P—P) is calculated from P$_2$H$_4$, using the E(P—H) value from PH$_3$, it turns out to be 207 kJ mol^{-1} which is almost the same as that of Table 2.4.

Bond enthalpies show that for nitrogen the N—H bond is dominant. The ability of compounds with this bond to participate in hydrogen-bonding, denied to the P—H bond, means in effect that nitrogen and phosphorus have virtually nothing in common in a large area of their chemistries*.

The phosphorus bonds show the inverse relationship between bond length and bond enthalpy – Fig. 2.2. Although there is a fairly wide scatter of points, those of the halo-

Figure 2.2 Bond enthalpies vs bond lengths for phosphorus bonds.

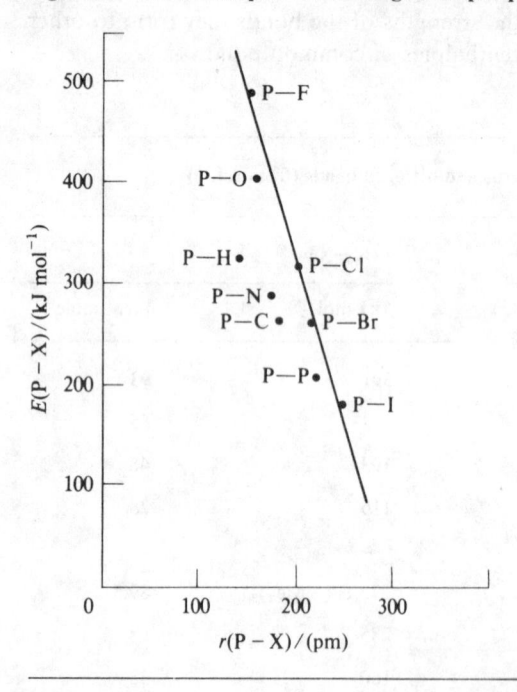

gen bonds fall on a straight line the slope of which represents an increase in enthalpy of 3.3 kJ mol^{-1} per pm decrease in bond length (0.8 kcal mole^{-1} per 0.01 Å). This ratio will be referred to later in this chapter when the different bonds of the *tbp* of such compounds as PF$_5$ are discussed.

Bond angles are always narrower in phosphine derivatives than in their nitrogen counterpart. The VSEPR theory had perhaps a too facile explanation of this difference; the usual explanation is in terms of the relative contributions of the s and p orbitals to

---

* Another major difference between them is in their relative π-bonding capacities. Nitrogen can form pπ–pπ bonds to first row elements whereas phosphorus cannot, although there are some compounds in which phosphorus is bonded to only two carbon atoms where it would seem likely that pπ–pπ bonding is present, e.g.

Phosphorus generally uses its 3d orbitals in π-bond formation but the resulting pπ–dπ link is much weaker than pπ–pπ links.

2.4 Tricoordinate phosphorus compounds

Figure 2.3 Valence atomic orbitals of phosphorus.

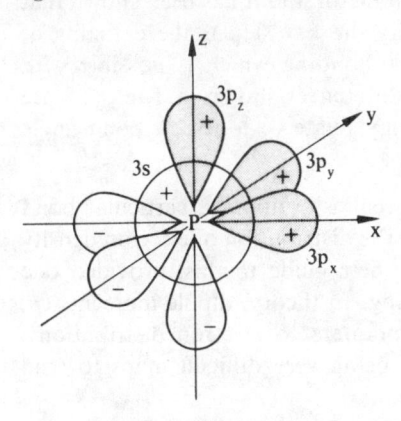

the bonding. The superimposed 3s and 3p orbitals are shown in Fig. 2.3 in which it can be seen how the three p orbitals are at right angles to one another. A molecule formed from such a phosphorus atom and using only the 3p orbitals would have an apical angle of 90°. Any wider angle can then be referred to the pure p situation by assuming the s orbital is contributing to the bonding. Thus PH_3 with an angle of 93° is said to have only a small percentage of s character in the bonding orbital, while PI_3 with an angle of 102° has much more. In the latter compound however it is probably inter-atomic repulsions which govern the angle. [Bond angles less than 90° are found in P_4 (60°) and the cyclotetraphosphines, $(RP)_4$ (ca 85°).]

If phosphorus uses its 3p orbitals for bonding then its non-bonding pair must reside in the 3s, which is spherically symmetrical and therefore relatively diffuse. Little wonder then that the lone pair is less available in PH_3 than in NH_3 as their basicities show – Table 2.5. As the s character of the bonds increases so the p character of the

Table 2.5 Basicity of phosphines and amines (pK_a values) from ref. 4 (table 8, p 25)

PH_3	– 14	NH_3	9.21
$MePH_2$	– 3.2	$MeNH_2$	10.62
Me_2PH	3.9	Me_2NH	10.64
Me_3P	8.65	Me_3N	9.76

lone-pair orbital will increase, making it less diffuse and thereby more available. Trimethylphosphine with a bond angle of 99° is a strong base. The nature of the ligands attached to phosphorus probably play as important a part in the performance of the lone pair as does the bond angle. The introduction of methyl groups for instance has a profound effect on basicity, more so than in nitrogen chemistry as Table 2.5 shows.

Basicity, quoted as pK_a, depends upon several factors one of which is the availability of the lone pair. Even this is not the most important factor and it has been shown that of the thermochemical changes involved in basicity the key step is the enthalpy of solvation. This is strongly influenced by hydrogen-bonding which is possible with N—H compounds but not P—H. Hence the big differences shown in Table 2.5 are due to solvation effects; the smaller differences may relate to lone-pair/bond angle effects.

Bond lengths, bond enthalpies and bond angles reveal a lot about a particular bond or molecule but they provide no information about the distribution of electron density within the bond or molecule. Electronegativity can be a guide to this, provided one is dealing only with σ-bonds, but is often misleading. In theory, dipole moments (p) and bond moments (μ) should be much better pointers to electron distribution[5]. However dipole moments have the reputation of being very difficult maps to read often leading the unwary traveller into a dead-end.

2.4.1 Dipole moments (p) of tricoordinate phosphorus compounds

The permanent dipole moment of a molecule arises when the centre of net positive charge in the molecule, due to the protons of the nuclei, does not coincide with the centre of net positive charge, due to the electrons. The two centres are rarely more than a few pm apart but their relative position in the molecule cannot be measured – in other words though the magnitude of the dipole can be accurately known, the direction in which it points* cannot (although it must lie along the principal axis of the molecule).

Another word of warning about dipole moments is that the value for a particular compound can vary according to the medium in which it is measured. This is demonstrated in phosphorus chemistry by the values obtained for PCl_3: 2.33×10^{-30} Cm [0.70 D] in the solid, 2.60 [0.78] in CCl_4, 2.64 [0.79] in the gaseous phase, 3.00 [0.90] in C_6H_6, and 6.35 [1.90] in dioxan. Where possible the gaseous phase value is used.

If it is assumed that each bond in a molecule has associated with it a bond moment, brought about by an unequal sharing of the bonding electrons, then the overall molecular dipole moment can be seen as the resultant of the bond moments. For pyramidal PX_3 molecules of C_{3v} symmetry the dipole will lie along the C_3 axis in (2.1),

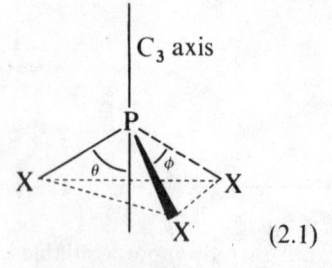

C_3 axis

X—P—X
  X   (2.1)

The dipole moment, $p(PX_3)$, can then be related to the bond moments, $\mu(P-X)$, by (1).

---

* The negative end of the dipole is the 'point' and the common symbol for a dipole moment is $\leftrightarrow$ meaning $\oplus$—$\ominus$. For bond moments the representation is $\overrightarrow{A-B}$.

2.4 Tricoordinate phosphorus compounds

The angle each bond makes with the principal axis, θ, is related to the bond angle, ϕ, by (2).

$$p(PX_3) = 3\mu(P-X) \cos \theta \tag{1}$$

$$\cos \theta = \sqrt{1 - \tfrac{4}{3} \sin^2(\tfrac{1}{2}\phi)} \tag{2}$$

The lone pair on phosphorus also has an effect on the dipole moment otherwise it is difficult to explain why $p(PF_3)$ is less than $p(PMe_3)$, as Table 2.6 shows, when it is known that the polarity of the P—F bond exceeds that of the P—CH$_3$ bond. Mauret and Fayet have shown that the problem of the lone pair can be dealt with by assigning it a moment lying along the principal axis. The net molecular dipole moment will then consist of either the sum (case A) or difference (cases B$_1$ and B$_2$) of the lone-pair

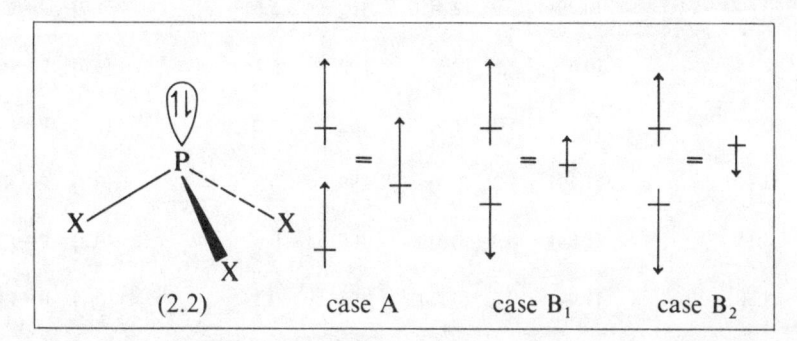

moment and the resultant of the three bond moments (2.2). When the bond moments oppose the lone-pair moment the overall dipole may point in either direction depending upon which component is the greater, hence B$_1$ and B$_2$.

Mauret and Fayet[6] estimated the lone-pair moment to be 5.3×10^{-30} Cm [1.6 D]. Values of p in Table 2.6 lower than this must be either B$_1$ or B$_2$ types. Values higher than this are almost certainly type A with the possible exception of P(OPh)$_3$ which, with its strongly electron-withdrawing groups, is almost certainly type B$_2$. For PI$_3$ the resultant of the bond moments exactly cancels out the lone-pair moment and so for this molecule $3\mu(P-I) \cos \theta = 5.3(\times 10^{-30})$ Cm. It is fortuitous that the dipole moment of PI$_3$ is exactly zero otherwise it would have been impossible to deduce with certainty the size and direction of the $\mu(P-I)$ vector. Knowing this it is possible to say that the other halogen bonds will be more polar. For PH$_3$ the resultant of the bond moments will be either 3.4×10^{-30} Cm, if we assume case B$_1$ is relevant, or 7.1×10^{-30} Cm if B$_2$ is relevant; and the respective values of $\mu(P-H)$ are 2.1 and 4.3 $\times 10^{-30}$ Cm; in Table 2.6 the former value is quoted. Similar arguments apply to PMe$_3$ and PPh$_3$ but in view of the dipole moment of PEt$_3$ which represents a case A situation these compounds are treated as if the net dipole moment of the molecule is of type B$_1$.

In terms of bond polarity and electron withdrawal from tricoordinate phosphorus the order revealed by Table 2.6 μ values is P—OPh > P—F > P—Cl > P—Br > P—I > P—H > P—Me > P—Ph > P—Et $\simeq$ P—OEt. This order is much as expected but one anomaly is the relatively low bond moment of the P—F bond. Back extrapolation from the other halides would have suggested a value for $\mu(P-F)$ of the order of 7×10^{-30} Cm [ca 2 D]. A contribution from donor $2p\pi(F) \to 3d\pi(P)$ bonding would explain the

Table 2.6 Dipole moments, p, bond angles, ϕ, and bonds moments, μ, of pyramidal phosphorus compounds[5,7]

	p ($\times 10^{-30}$ Cm)	[D]	Resultant of bond moments ($= 3\mu \cos\theta$)		ϕ	μ ($\times 10^{-30}$ Cm)	[D]	Direction of bond moment
PH$_3$	1.93	[0.58]	3.4	[1.0]	93°	2.1	[0.6]	$\overset{\leftrightarrow}{\text{P—H}}$
PF$_3$	3.44	[1.03]	8.7	[2.6]	98°	5.9	[1.8]	$\overset{\leftrightarrow}{\text{P—F}}$
PCl$_3$	2.64	[0.79]	7.9	[2.4]	100°	5.6	[1.7]	$\overset{\leftrightarrow}{\text{P—Cl}}$
PBr$_3$	2.15	[0.61]	7.5	[2.3]	100.5°	5.4	[1.6]	$\overset{\leftrightarrow}{\text{P—Br}}$
PI$_3$	0.0	[0.0]	5.3	[1.6]	102°	4.0	[1.2]	$\overset{\leftrightarrow}{\text{P—I}}$
PMe$_3$	3.97	[1.19]	1.3	[0.4]	99°	0.9	[0.3]	$\overset{\leftrightarrow}{\text{P—Me}}$
PEt$_3$	6.15	[1.84]	0.9	[0.3]	100°[a]	0.6	[0.2]	$\overset{\leftarrow\leftarrow}{\text{P—Et}}$
PPh$_3$	4.80	[1.44]	0.5	[0.15]	103°	0.4	[0.1]	$\overset{\leftrightarrow}{\text{P—Ph}}$
P(OEt)$_3$	6.07	[1.82]	0.8	[0.25]	104°	0.6	[0.2]	$\overset{\leftarrow\leftarrow}{\text{P—OEt}}$
P(OPh)$_3$	6.76	[2.03]	12.1	[3.63]	105°[b]	10.0	[3.0]	$\overset{\leftrightarrow}{\text{P—OPh}}$

[a] Estimated from PMe$_3$ value; [b] estimated from P(OEt)$_3$ value.

low value as it doubtless explains the electron-releasing nature of alkoxy ligands. However the phenoxy ligand is highly electron withdrawing due mainly to the oxygen's lack of ability to π back-bond to phosphorus due to the demands of the phenyl group.

Since it is a general observation that the fewer the number of lone pairs on a ligand the stronger the ability to donate them, this means that amino groups would have even higher $\mu(\overset{\leftarrow\leftarrow}{\text{P—NR}_2})$ values and come at the electron-releasing end of the polarity spectrum.

The big drawback to the above treatment of dipole moments is the underlying assumption that the lone pair remains unaffected by the ligands attached to phosphorus. Clearly this cannot be so as basicity measurements indicate, especially Lewis basicity data which will be dealt with in chapter 11. However μ(P—X) values turn out as anticipated, bearing in mind the possibility of π bonding, and give some idea of the electron distribution in common bonds.

2.4.2 Isomers of tricoordinate phosphorus compounds

What prevents separation of isomers of tricoordinate nitrogen compounds, such as amines NR^{1}R^{2}R^3, is their rapid interconversion. The energy barrier to this process

2.4 Tricoordinate phosphorus compounds

which is called *inversion* depends chiefly upon the size of the groups attached to nitrogen, being lowest for NH_3 (25 kJ mol^{-1}; 6 kcal mole^{-1}) and rising with the substitution of heavier methyl groups e.g. Me_3N (34 kJ mol^{-1}; 8 kcal mole^{-1}). It has been estimated that a barrier of at least 100 kJ mol^{-1}, 24 kcal mole^{-1}, would be necessary for physical separation to be feasible but no amine or nitrogen derivative comes within striking distance of this value.

Phosphine derivatives on the other hand have inversion barriers in excess of the minimum 100 kJ mol^{-1}, e.g. PH_3 has 115 and Me_3P has 133 kJ mol^{-1} energy barrier. A tricoordinate phosphorus atom with three different groups should therefore be separable into enantiomers (2.3a) and (2.3b):

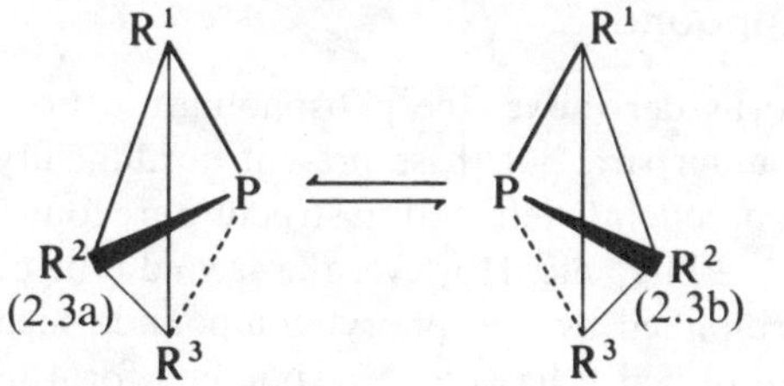

$$\text{(2.3a)} \quad \rightleftharpoons \quad \text{(2.3b)}$$

In recent years asymmetric phosphines have been prepared and isolated via tetracoordinate phosphonium salts or phosphine oxides which can be resolved and the tricoordinate isomers regenerated by methods which are stereoretentive, i.e. which preserve the configuration of the phosphine. One method of separation uses benzyl phosphonium salts $R^1R^2R^3P^+CH_2Ph$ which, after resolution into diastereomeric salts with an optically active acid anion, can be converted back to the phosphine $R^1R^2R^3P$ by electrolytic reduction. Another method uses separated $R^1R^2R^3P(O)$ enantiomers and their reaction with Si_2Cl_6 to generate the phosphine. Transition metal complexes of phosphines have been used in separation[8], for example (2.4) which is converted to a *trans* dichlorophosphineamine platinum(II) complex (2.5)

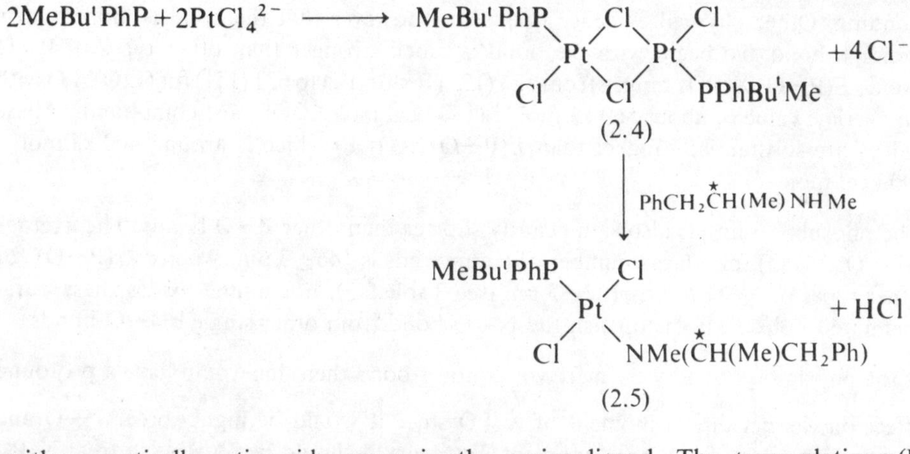

$$2MeBu^tPhP + 2PtCl_4^{2-} \longrightarrow \underset{(2.4)}{MeBu^tPhP\text{-}Pt(Cl)_2\text{-}Pt(Cl)_2\text{-}PPhBu^tMe} + 4Cl^-$$

$$\downarrow PhCH_2\overset{*}{C}H(Me)NHMe$$

$$\underset{(2.5)}{MeBu^tPhP\text{-}PtCl_2\text{-}NMe(\overset{*}{C}H(Me)CH_2Ph)} + HCl$$

with an optically active side group in the amine ligand. The *trans* platinum(II) complex will separate into two diastereomers on fractional crystallization and the $MeBu^tPhP$ enantiomers can be regenerated from the separated complexes by treatment with KCN which releases the phosphine ligands.

The mass of the phosphorus atom is probably the critical factor in explaining why its tricoordinate derivatives can be isolated as enantiomers whereas those of nitrogen cannot.

So far in this chapter there has been no need to bring any new factors into the discussion of the bonding of phosphorus compounds than were necessary for dealing with equivalent nitrogen compounds, although there were suggestions of a π contribution to the bonding in some tricoordinate phosphorus derivatives having ligands with non-bonding pairs. In discussing tetracoordinate phosphorus compounds we shall explore this aspect of the bonding more fully.

2.5 Tetracoordinate phosphorus compounds

There is a type of tetracoordinate phosphorus derivative, the phosphonium cation, which has a directly analogous nitrogen counterpart, but these present no difficulty in bonding terms. Four bonding pairs and consequently tetrahedral structure are found for cations of this class such as PH_4^+, PCl_4^+, PPh_4^+ etc. However the second type of tetracoordinate phosphorus compound, exemplified by phosphoryl compounds such as Cl_3PO, presents us with a bonding situation which has no corresponding form in nitrogen chemistry even though nitrogen compounds of the same formulae, namely the amine oxides, R_3NO, exist. Tetracoordinate phosphorus compounds of this class which are formally pentavalent, $\diagdown\!\!\!\!-\!\!P\!=\!\!,\diagup$ make up the great bulk of known phosphorus derivatives so it will be necessary to find a working model of the bonding.

The formula $X_3P=O$, whilst satisfying the valency requirements of phosphorus, prejudges the issue of the nature of the bonding of the phosphoryl link. The alternative bond, a purely donor bond, $X_3P \rightarrow O$ is a possibility on the grounds that amine oxides have this kind of bond. This $P \rightarrow O$ link could not be disproved on the evidence of bond angles when it was discussed earlier in the chapter under the VSEPR approach to bonding. Other physical evidence does show however that the bond is more than a donor σ-bond. To begin with the bond is much stronger than other types of P—O bond; $E(P-O, X_3PO)$ ranges from 523 (125) for Br_3PO to 631 (151) for $(EtO)_3PO$ with an average value of about 560 kJ mol^{-1} (134 kcal mole^{-1}) in most compounds. These values are substantially higher than $E(P-O, (RO)_3P)$ which is around 400 kJ mol^{-1} (90 kcal mole^{-1}).

The phosphoryl link is also significantly shorter than other P—O bonds. The average $r(P-O, X_3PO)$ for a large number of compounds is 146 ± 3 pm, whereas $r(P-O)$ for other types of P—O link is 162 ± 5 pm (see Table 2.3). For amine oxides these parameters do not clearly distinguish the $N \rightarrow O$ bond from other single $\sigma N-O$ bonds.

If the phosphoryl bond were merely a donor σ-bond then this would have a profound effect on the net dipole moment of X_3PO since it would be highly polar $\overset{\leftrightarrow}{P-O}$ and comparable to the P—F bond moment. In theory it should be possible to treat X_3PO molecules in the same way as X_3P molecules except that $\mu(P-O)$ would replace the lone pair moment in the calculations. Unfortunately this approach does not work and gives bond moments ranging from almost zero in Cl_3PO to almost double that of the

2.5 Tetracoordinate phosphorus compounds

phosphorus–fluorine bond in some of the molecules such as R_3PO, R = alkyl. Attempts have been made to relate the bond moment of the phosphoryl link to its multiplicity, the smaller the moment the greater its double-bond character. In other words the phosphoryl bond can and does participate in back-bonding, in effect building a donor π system onto the donor σ-bond: $\ \diagdown\!\!\!\!\!/P \underset{\sigma}{\overset{\pi}{\rightleftarrows}} O$. With more than one non-bonding pair available on oxygen it is possible to have two mutually perpendicular donor π systems and transfer sufficient electron density back to phosphorus to counterbalance the σ-bond and explain the anomalous dipole results.

2.5.1 Pi-bonding of tetracoordinate phosphorus compounds

If a π-bonding system were present in the phosphoryl bond then one would have expected supporting evidence from n.m.r. and uv spectroscopy of the kind one finds to support multiple bonds between first row elements. These π bonds are built on 2p orbitals which are compatible whereas those between second and first row elements, such as P and O, are built on 3d and 2p orbitals which are less compatible and lead to a much more diffuse π system. Such a $d\pi-p\pi$ system makes a much smaller contribution to the overall bond than does a $p\pi-p\pi$ system, and is in consequence harder to detect by spectroscopic methods. Nevertheless it can play a very important role in the chemistry of second row elements like phosphorus.

Accepting that π bonding is present in X_3PO compounds and the like does not necessarily mean accepting the donor σ/donor π explanation put forward in the above paragraphs. Other models for the bonding can be constructed. However the VSEPR and simple atomic orbital models which sufficed for tricoordinate phosphorus cannot be extended to meet the new requirements. The more sophisticated valence bond or molecular orbital approaches are needed.

A simplified valence bond model has a tetrahedral framework of σ bonds about phosphorus formed by overlap of ligand orbitals and phosphorus hybrid orbitals (sp^3 or te). This accounts for four of the valence electrons leaving the fifth which is forced to occupy one of the empty 3d orbitals. In the P—O bond the oxygen is trigonally hybridized (sp^2 or tr) with two lobes accommodating non-bonding pairs and the other electron in the remaining p orbital. Overlap between phosphorus and oxygen then produces a σ and π bond as shown in (2.6). Since phosphorus has other empty 3d orbitals apart from 3d$_{xz}$ it is conceivable that a second donor π system perpendicular to the first can be present adjusting the hybridization at oxygen to sp or di, with one lobe taking the non-involved lone pair while the other lone pair occupies the 2p$_y$ orbital in which it can now overlap with another 3d orbital (2.7).

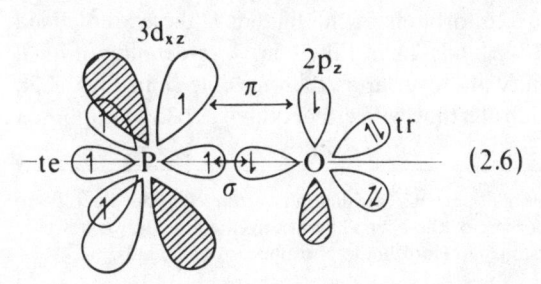

(2.7)

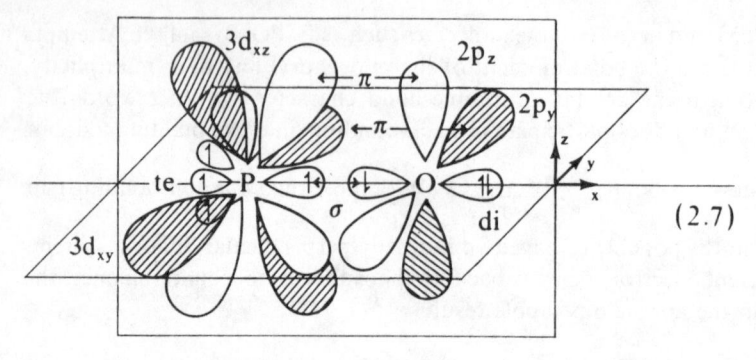

Attractive though this approach is in visual terms it is basically unsound. The energy change involved in 'promoting' an electron of phosphorus to an empty 3d orbital is prohibitively high – there is in fact an empty 4s orbital available of lower energy but this is an unattractive alternative for reasons which will be given later. If we abandon the valence bond model we leave ourselves with the molecular orbital approach which ignores the 3d orbitals and in so doing finds an answer which may be theoretically satisfying but still fails to provide an adequate explanation even so. The most acceptable compromise is in effect to return to the donor π bond model, in other words to make a clear distinction between the tetrahedral framework, based on molecular orbitals formed from only the 3s and 3p atomic orbitals of phosphorus, and any supplementary π-bonding based entirely on ligand lone pairs donating into the 3d orbitals. This approach will be developed more fully in the section on pentacoordinate phosphorus when the debate as to whether 3d orbitals are necessary for the σ framework becomes even keener.

If 3d orbitals can provide a π-bonding system, the question is which of the five 3d orbitals are suitably placed for this purpose. Some of the 3d orbitals are so positioned that they interact with the (te)^4 framework about the phosphorus atom; Fig. 2.4 illustrates this. To maximize orbital interaction away from the regions of the σ-bond framework – and this is what π-bonding is all about – it is not possible to include more than one orbital of the $3d_{xy}$, $3d_{xz}$, $3d_{yz}$ subset. In Fig. 2.4 it can be seen that for the σ framework drawn as shown, only the $3d_{xz}$ orbital meets the non-interacting requirement Fig. 2.4c. In Fig. 2.5a 2p orbitals of the ligands have been included to show how these may overlap with $3d_{xz}$ in effect producing a sort of tetrahedrally delocalized π system. Something of this sort would explain the symmetry and bonding features of the phosphate anion, PO_4^{3-}.

The other 3d subset, $3d_{x^2-y^2}$ and $3d_{z^2}$, have the lobes of their orbitals directed away from the framework although for $3d_{z^2}$ there is some interference in the xy plane. Of these two orbitals the $3d_{x^2-y^2}$ is thus slightly preferable, although neither is orientated to interact effectively with $2p_x$, $2p_y$, or $2p_z$ orbitals of the ligands if these are defined with respect to the same axes as the 3d orbitals, as in Fig. 2.5a. A re-defining of local ligand axes can circumvent this difficulty and overlap is then feasible as in Figs. 2.5b and 2.5c but only two ligands can overlap at a time with either $3d_{x^2-y^2}$ or $3d_{z^2}$ as opposed

---

* As drawn, the tetracoordinate arrangement is not exactly tetrahedrally arranged about P; to be so the bonds should be directed towards the corners of the cube. The arrangement as drawn however does more nearly reflect the bond angles in some tetracoordinate phosphorus compounds.

2.5 Tetracoordinate phosphorus compounds

Figure 2.4 The relationship of the 3d orbitals to the tetracoordinate phosphorus framework.*

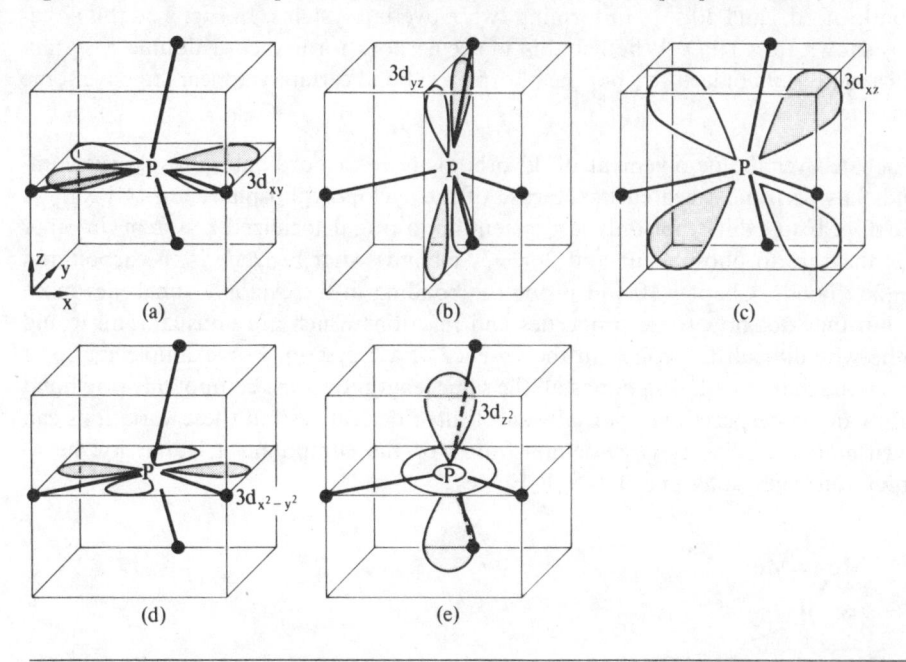

Figure 2.5 Overlap of ligand 2p and phosphorus 3d orbitals (not to scale).

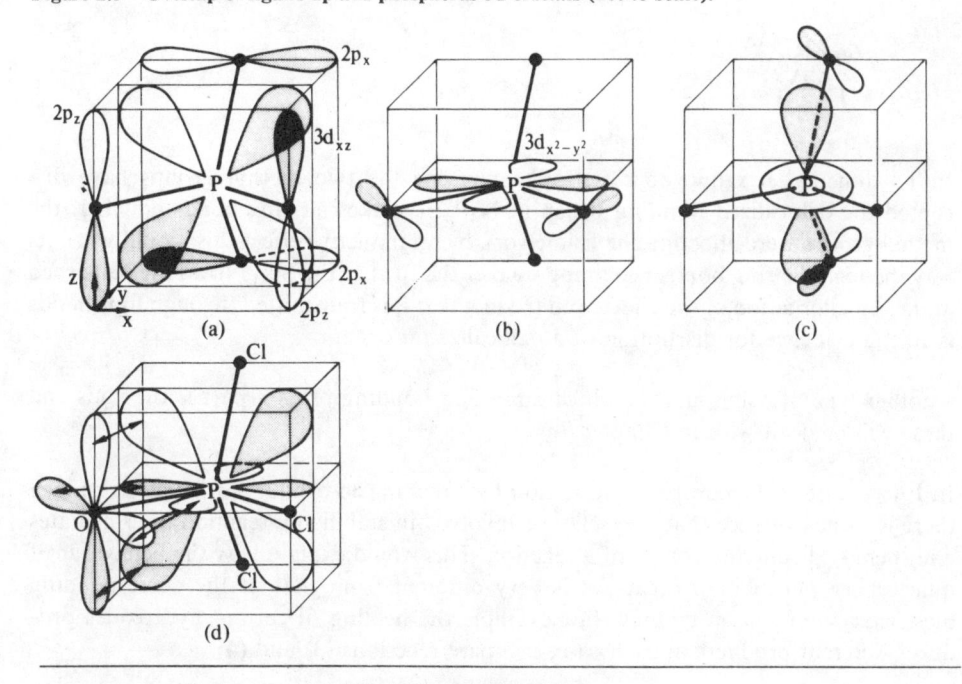

to all four for $3d_{xz}$. However there are many molecules in which only one ligand has 2p orbitals, e.g. Cl$_3$PO, or two 2p orbitals, e.g. (NPCl$_2$)$_3$, and these might well make use both of $3d_{xz}$ and $3d_{x^2-y^2}$ in forming two π overlap systems. In Fig. 2.5d this situation is shown for Cl$_3$PO. Whether this molecule does form such a double π system is still a matter of conjecture – but the phosphoryl bond certainly appears to have some π character.

The debate over the involvement of 3d orbitals in tetracoordinate phosphorus compounds has been most heated in the case of the cyclopolyphosphazenes, (NPX$_2$)$_{3,4}$. These appear to exhibit not only a π system but a ring delocalized π system, in other words they are to phosphorus and $2p\pi$–$3d\pi$ bonds what benzene is to carbon and $2p\pi$–$2p\pi$ bonds. Chapter 10 will probe the bonding in the cyclophosphazenes more fully but they do show some properties and reactions which are unusual, and would be otherwise difficult to explain in the absence of a π system. For example many of the rings are planar with ring bonds all the same length. In some compounds ring bond lengths can vary especially in partially substituted derivatives but these variations can be even more revealing as (2.8) demonstrates for the compound 1,1-dimethyl hexafluorocyclotetraphosphazene, 1,1-N$_4$P$_4$F$_6$Me$_2$.[9]

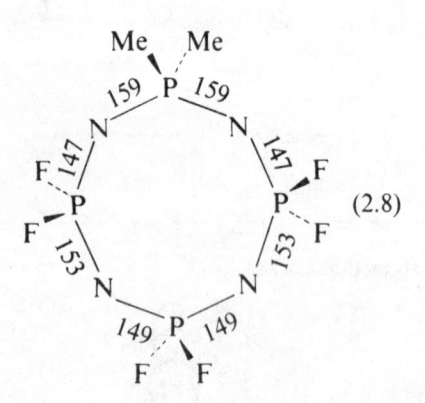

(2.8)

In 1,1-dimethyl hexafluorocyclotetraphosphazene the two methyl groups have disrupted the delocalized bonding which in N$_4$P$_4$F$_8$ makes all ring bonds equal. If the methyl groups were affecting the framework by an inductive effect this would work its way through the ring bonds becoming weaker the further the bond from the perturbed atom. An alternation of the effect, and the fact that it is transmitted through four bonds is strong evidence for disruption of a delocalized π system.

Another type of compound in which $2p\pi$–$3d\pi$ bonding plays a part is the ylids and these will be dealt with in Chapter 7.

In PX$_4^+$ cations there appears no reason for invoking 3d orbital participation and yet there is some evidence that they may be involved in stabilizing reaction intermediates and thereby dictate the course of a reaction. This would explain why the behaviour of quarternary phosphonium cations is very different from that of the corresponding quarternary ammonium cations. For example the heating of certain hydroxides produces different products in each case, compare reactions (3) and (4).

2.5 Tetracoordinate phosphorus compounds

$$R_3\overset{+}{P}-CH_2-CH_2-R^1 \quad OH^- \xrightarrow{heat} R_3PO + CH_3-CH_2R^1 \quad (3)$$

$$R_3\overset{+}{N}-CH_2-CH_2-R^1 \quad OH^- \xrightarrow{heat} R_3N + CH_2=CHR^1 + H_2O \quad (4)$$

One explanation of this difference is based on the ability of phosphorus to form a pentacoordinate intermediate and that this of necessity involves the use of 3d orbitals (2.9).

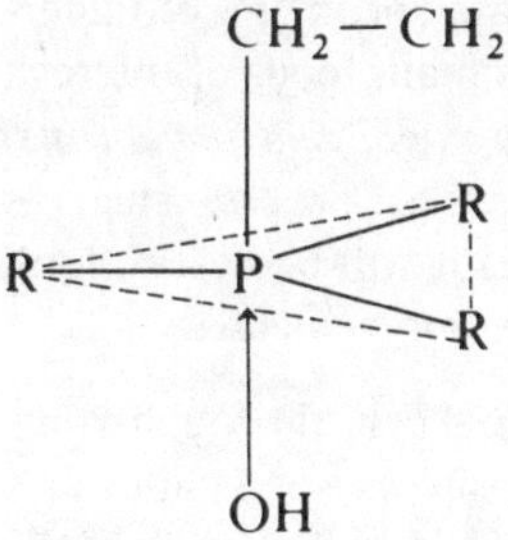

$$\begin{array}{c} CH_2-CH_2-R^1 \\ | \\ R{=\!=\!=}P{<}^R_R \\ | \\ OH \end{array} \quad (2.9)$$

It will be seen however in the next section of this chapter that there is no need to call upon 3d orbitals in the formation of such an intermediate. The above reactions (3) and (4) are no real proof of 3d involvement in PX_4^+.

Slightly more convincing proof comes from the activation by R_3P^+- of an adjacent double bond thereby enhancing its susceptibility to nucleophilic attack as in (5) where N represents a nucleophile and the product is an ylid. The corresponding vinylam-

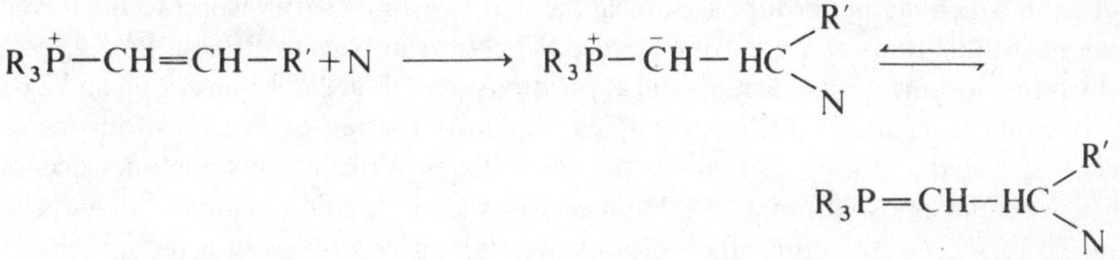

$$R_3\overset{+}{P}-CH=CH-R'+N \longrightarrow R_3\overset{+}{P}-\bar{C}H-HC{<}^{R'}_{N} \rightleftharpoons$$

$$R_3P=CH-HC{<}^{R'}_{N} \quad (5)$$

monium salts are inert to nucleophilic attack so the activation of the vinyl group is not purely inductive. Again to invoke 3d orbitals in the vinyl phosphonium compound itself seems to be using them rather as a panacea, although in the products of this reaction which are ylids these are almost certainly involved; this however is not the same thing.

2.5.2 Isomers of tetracoordinate phosphorus compounds

Optical isomers are better known for tetra- than for tricoordinate phosphorus, the reason being that they are stable to interconversion and relatively easy to prepare. The inversion of a tetrahedral arrangement requires either a bond breaking and remaking mechanism or the molecule to pass through a planar configuration on its route to the inverted form. The barrier to this process has been estimated to be in excess of 500 kJ mol^{-1} (120 kcal mole^{-1}) so that the former mechanism seems more likely, but this also represents a high energy barrier route. For these reasons optically active tetracoordinate phosphorus compounds have a stability which the tricoordinate ones lack, and they have therefore served as very useful tools in the elucidation of reaction mechanisms, as will be seen.

2.6 Pentacoordinate phosphorus compounds

Some of the most interesting facets of phosphorus chemistry are met with in pentacoordinate compounds and in pentacoordinate intermediates which are increasingly postulated in many reactions. There are two important issues which arise with the pentacoordinate state. The first is the part which 3d orbitals play in the construction of the all-important σ framework of five covalent bonds surrounding an atom. The second is the interconversion of isomers; in tri- and tetracoordination this was either simple to visualize or did not even occur and neither case presented problems of understanding. For pentacoordination this second point becomes the main focus of interest. Collectively the processes of interconversion are known as *polytopal rearrangements*, and individually according to names descriptive of a particular process such as pseudorotation. However, before dealing with these two themes it will be best to look at the physical properties of some pentacoordinate phosphorus molecules.

Relatively few pentacoordinate phosphorus compounds have been thoroughly investigated structurally. What information has been obtained shows such molecules generally to have the *tbp* configuration in which the axial bonds are longer than the equatorial ones. In some spirocyclic oxyphosphoranes an *spy* structure has been found[10]. Figure 2.6 illustrates a representative selection of structures of molecules with a pentacoordinate phosphorus atom. The compounds PF_5 (Fig. 2.6a), $MePF_4$ (Fig. 2.6b), and Me_2PF_3 (Fig. 2.6c) show the effect on bond lengths and bond angles wrought by introducing methyl groups. These changes can be explained in terms of VSEPR theory if the bonding pair of the P—Me bond were nearer the phosphorus than the P—F bonding pair. Although positional isomers are possible for $MePF_4$ and Me_2PF_3 in which methyl groups are apical these are unstable with respect to the forms shown in the diagrams and are not observed. The same applies to $PFCl_4$, PF_2Cl_3 etc. in which the fluorine atoms occupy apical positions – see Fig. 2.7. Figures 2.6d and 2.6e also illustrate other points such as the apical–equatorial siting of five and four-membered ring systems. Figure 2.6f shows the *spy* configuration which sometimes occurs with spirocyclic phosphoranes. As Holmes has shown[10], this arrangement may be preferred to the *tbp* structure if conditions are right. However even a bicyclic phosphorane may still prefer a *tbp* configuration if there is an imbalance of electronic or other effects, e.g. (2.10) is *tbp*[11].

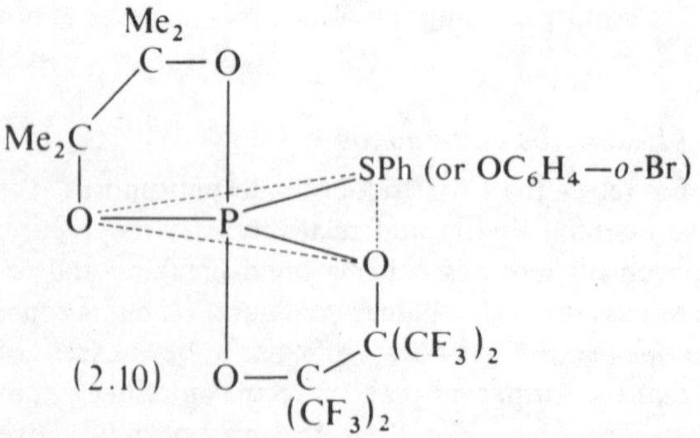

The mixed fluoride/chloride pentacoordinate phosphorus derivatives have been investigated by the measuring of their electric dipole moments and these and their structures

2.6 Pentacoordinate phosphorus compounds

Figure 2.6 The structures of pentacoordinate phosphorus molecules.

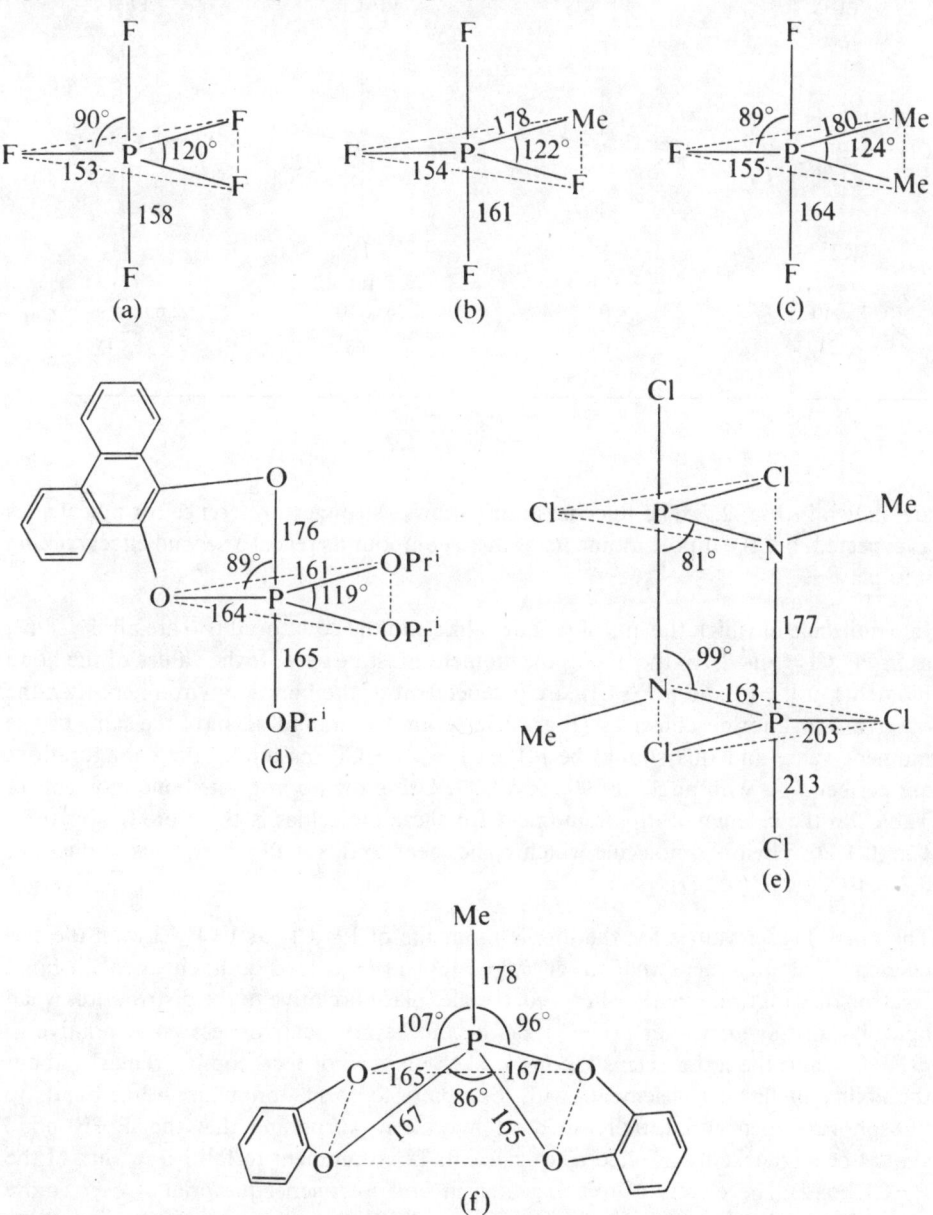

(f)

Refs: (a) L. C. Hoskins, *J. Chem. Phys.*, **42**, 2631 (1965); (b) (c) L. S. Bartell and K. W. Hansen, *Inorg. Chem.*, **4**, 1777 (1965); (d) W. C. Hamilton, S. J. LaPlaca, F. Ramirez and C. P. Smith, *J. Amer. Chem. Soc.*, **89**, 2268 (1967); (e) L. G. Hoard and R. A. Jacobson, *J. Chem. Soc. A*, 1203 (1966); (f) H. Wunderlich, D. Mootz, R. Schmutzler and M. Wieber, *Z. Naturforsch.*, **29B**, 32 (1974).

Figure 2.7 Configurations and electric dipole moments (p) of the mixed phosphorus(V)fluoride/chlorides.

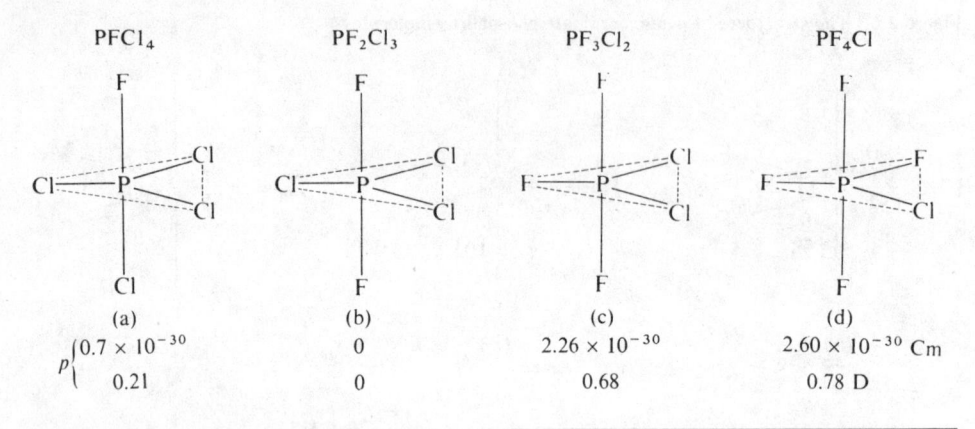

	PFCl$_4$	PF$_2$Cl$_3$	PF$_3$Cl$_2$	PF$_4$Cl
	(a)	(b)	(c)	(d)
p	0.7×10^{-30}	0	2.26×10^{-30}	2.60×10^{-30} Cm
	0.21	0	0.68	0.78 D

are shown in Fig. 2.7. The fluorine atoms show a stronger preference for apical sites as expected, but the dipole moments of these compounds reveal a second effect coming into play.

In a molecule in which the apical pair are alike and the equatorial trio are all the same, as in PF$_2$Cl$_3$ (Fig. 2.7b) the net dipole moment must be zero. If the values of the bond moments, μ(P—F) and μ(P—Cl), are independent of the bonds' environment then the other mixed fluoride chlorides (Figs. 2.7a, c and d) should all have the same dipole moment value and this should be μ(P—F) − μ(P—Cl), assuming the configurations are perfect *tbp*'s with angles of 90° and 120°. Using the appropriate bond moments of Table 2.6 the calculated dipole moment for these molecules is therefore 0.3×10^{-30} Cm (0.1 D). The only molecule which comes near to this is PFCl$_4$ with its moment of 0.7×10^{-30} Cm (0.2 D).

The much higher values for the dipole moments of PF$_3$Cl$_2$ and PF$_4$Cl than the calculated value must mean that either μ(P—F), or μ(P—Cl), or both, change their bond electron distribution greatly when equatorially placed relative to the distribution when apically positioned. Either μ(P—F) has become more polar or less polar relative to μ(P—Cl), and the latter seems the more likely in view of forthcoming remarks about the ability of first row elements with lone pairs to form donor $2p\pi \rightarrow 3d\pi$ bonds to phosphorus when equatorially sited. It may seem surprising that the P—F bond should be so markedly affected as to cause its bond moment to fall below that of the P—Cl bond. The effects of placing oxygen and nitrogen equatorial is even more marked; the phosphoryl link when it finds itself in a pentacoordinate situation clings tenaciously to an equatorial position for these very reasons. Donor $2p\pi \rightarrow 3d\pi$ bonding lies behind the observed apicophilicities of several groups, see *p.* 57.

Whatever changes have taken place in the P—F bond have done so in both PF$_3$Cl$_2$ and PF$_4$Cl since their dipole moments are very nearly the same; the discrepancy can probably be accounted for by bond-angle divergence from the true *tbp* angles.

2.6 Pentacoordinate phosphorus compounds

Thermochemical data for pentacoordinate phosphorus compounds is sparse. The enthalpies of formation, $\Delta H°_f(PX_5, g)_{298}$, are known for PF_5, -1596 kJ mol^{-1} (-381.4 kcal mole^{-1}) and PCl_5, -343 kJ mol^{-1} (-81.9 kcal mole^{-1}); PBr_5 is dissociated into PBr_3 and Br_2 in the vapour. Mean bond enthalpies can be calculated using these values and give $E(P-F)$, 465 kJ mol^{-1} (111 kcal mole^{-1}) and $E(P-Cl)$, 257 kJ mol^{-1} (61.4 kcal mole^{-1}), values which are significantly lower than those obtained from the trivalent compounds $E(P-F, PF_3)$, 490 kJ mol^{-1} (117 kcal mole^{-1}) and $E(P-Cl, PCl_3)$, 319 kJ mol^{-1} (76.2 kcal mole^{-1}). To speak of mean bond enthalpies in pentacoordinate phosphorus molecules is to deny a difference in energy between the longer apical and the shorter equatorial bonds.

Earlier in this chapter it was shown (Fig. 2.2) that the enthalpies of phosphorus–halogen bonds vary linearly with bond length, the enthalpy increasing by 3.3 kJ mol^{-1} per pm decrease in bond length (0.8 kcal mole^{-1} per 0.01 Å). Taking this as a general trend it means that in PCl_5, where the apical bonds are 207 pm and the equatorial bonds 201 pm, there should be a difference of 6×3.3 or 20 kJ mol^{-1} between the bond enthalpies. Apportioning the total bond enthalpy among the bonds of PCl_5 and taking into account this difference gives $E(P-Cl)_{ax}$, 251 kJ mol^{-1} (60.0 kcal mole^{-1}) and $E(P-Cl)_{eq}$, 271 kJ mol^{-1} (64.8 kcal mole^{-1}). The bond length of P–Cl in PCl_3 is 204 pm and $E(P-Cl, PCl_3)$ is 319 kJ mol^{-1}, which is longer yet stronger than $r(P-Cl)_{eq}$ in PCl_5, having 48 kJ mol^{-1} (11.5 kcal mole^{-1}) more enthalpy tied up in the bond. It has been suggested that the all round weakening in the bond energies of pentacoordinate phosphorus compounds is evidence of 3d involvement in σ bonding. The time has now come to grasp this particular nettle.

2.6.1 The bonding in pentacoordinate phosphorus compounds[12]

In the section on tetracoordinate phosphorus it was seen that use could be made of the 3d orbitals in the formation of π bonds, the tetracoordinate framework being composed of the single 3s and the three 3p atomic orbitals. With an arrangement of five σ bonds about phosphorus it has often been assumed that a fifth atomic orbital must be brought into play and that this would be a 3d orbital. As revealed in the spectrum of the hydrogen atom, however, the nearest orbital is not a 3d but the 4s one.* On this basis it would seem more logical for phosphorus to expand its valence shell by using the 4s orbital before calling upon its 3d orbitals. It would prefer the latter only if it could derive an even greater bonus by using these to an extent sufficient to justify the extra input of energy. Such a bonus would be the formation of stronger bonds and there are reasons for supposing that 4s would not be as suitable as 3d in bonding, (a) because of its spherical symmetry which makes it diffuse and (b) because of the need for orthogonalizing to 3s. The 3d orbitals on the other hand are spatially less diffuse and are compatible with the other valence-shell orbitals. But is this enough?

The answer to this question begs the answer to a more fundamental question – does phosphorus require the use of any but its 3s and 3p orbitals in forming a σ framework, be it tetra-, penta- or even hexacoordinate? The answer would appear to be no; at

---

* This order is almost certainly true of phosphorus since it holds for a heavier element such as potassium which has the configuration [Ar] $4s^1 3d^0$ in its ground state. The structure of the periodic table confirms this order for other elements.

least according to the molecular orbital approach. Phosphorus can manage quite well without using its 3d orbitals (or 4s for that matter). Even when 3d orbitals are brought into the picture they seem to have very little effect. There are good philosophical grounds for trying to do without 4s or 3d involvement* and we shall deal with the molecular orbital (m.o.) approach to pentacoordinate phosphorus first.

Basing their calculations on the hypothetical molecule PH_5, and using only the 3s and 3p orbitals of phosphorus and 1s of hydrogen, Hoffmann, Howell and Muetterties[13] have determined the ground state configurations for possible pentacoordinate arrangements. For PH_5 the nine atomic orbitals (3s, $3p_x$, $3p_y$, $3p_z$ and five 1s) will generate nine molecular orbitals. The maximum number of these which can be described as bonding orbitals is four (σ), with four antibonding counterparts (σ^*) leaving one which is in effect non-bonding (n.b.). Incidentally if one of the 3d orbitals of phosphorus is included this will lead to the n.b. orbital becoming a bonding orbital – it will not have much effect on the lower lying m.o.'s.

Although phosphorus can occasionally have alternative bonding arrangements such as square pyramidal (symmetry C_{4v}) it mostly prefers the more stable *tbp* structure (symmetry D_{3h}). In theory there are also two other alternatives which might just be possible and these consist of pair-trio structures (symmetry C_σ). All of these produce a m.o. picture consisting of a low-lying nodeless orbital made up of the in-phase combination of 3s with 1s functions[12]. Above this comes a group of three singly noded orbitals composed from a single 3p orbital and accompanying 1s combinations – these three m.o.'s are not far separated in energy – two are in fact degenerate. Finally there is a high-lying orbital characterized by two nodal surfaces and this is the non-bonding orbital. For the D_{3h}, *tbp*, arrangement the situation is illustrated in Fig. 2.8 with the m.o. energies. The effect of the non-bonding orbital is to lengthen the apical bonds as observed.

The participation of 3d orbitals in effect stabilizes only the $2a'_1$ orbital and then only from -11.17 eV to about -12.9 eV. The result is to shift slightly some of the electron density, which in Fig. 2.8 can be seen to be concentrated on the hydrogens of the molecule, away from the hydrogen and on to the phosphorus. So much for the contribution which 3d orbitals make to pentacoordinate bonding – their inclusion in the m.o. calculations of PH_5 has only a very small effect. Using the four 3s, $3p_x$, $3p_y$ and $3p_z$ orbitals of sulphur and the four 2s, $2p_x$, $2p_y$ and $2p_z$ orbitals of the six fluorine atoms it has also been shown that the electronic structure of SF_6 can be interpreted without recourse to 3d orbitals.

The weight of orthodoxy is now ranged behind the m.o. picture of penta- and hexa-coordinate bonding and 3d orbitals, whilst not entirely rejected, have been relegated to a trivial role. The crux of the matter is this – if the orbitals used in the formation of m.o.'s are taken to be the same as those of the free atom then the 3d orbitals are too large and therefore too diffuse to make a meaningful contribution. Only if the 3d

---

* According to William of Ockham (1300–1349) "beings must not be multiplied unnecessarily". This tenet known as Ockham's razor is generally taken to mean that an explanation should always be made in terms of as few assumptions or unknowns as possible. The fewer of these the more likely the theory is to be the correct one.

2.6 Pentacoordinate phosphorus compounds

Figure 2.8 Molecular orbital diagram of the occupied orbitals of PH$_5$ in the trigonal bipyramidal configuration[13].

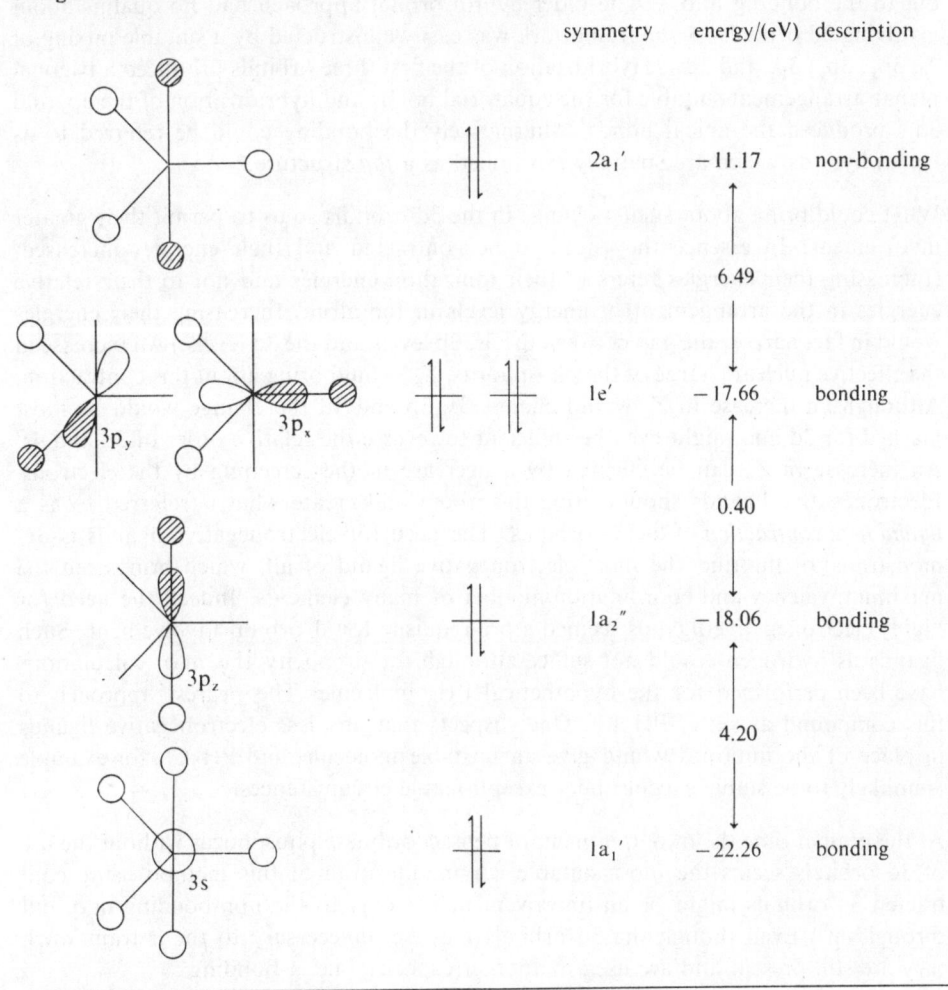

	symmetry	energy/(eV)	description
	2a$_1'$	−11.17	non-bonding
		6.49	
	1e'	−17.66	bonding
		0.40	
	1a$_2''$	−18.06	bonding
		4.20	
	1a$_1'$	−22.26	bonding

orbitals could be sharpened in some way could it become possible to justify using them in bonding.

Although it is an acceptable explanation of the bonding at pentacoordinate phosphorus, nevertheless the m.o. approach is still far from perfect. Perhaps the strongest objection to its non-inclusion of 3d orbitals is that the model for bonding which it proposes could equally well apply to another member of group M5 – nitrogen. There seems no reason why NH$_5$ should not exist especially considering that the N—H bond is the strongest single bond of any such nitrogen bond and much stronger than the P—H bond. Nor would steric overcrowding be a serious problem in NH$_5$ because the longer apical bonds could relieve this.

An alternative bonding arrangement may be possible if the 3d orbitals of the phosphorus atom in a molecular environment were contracted, and their energies brought nearer to those of 3s and 3p. The free atom 3d orbitals have only a tiny effect on the non-bonding m.o. but a modified 3d orbital may be able to contribute not only to this but to the bonding m.o.'s. The older hybrid orbital approach had no qualms about involving 3d orbitals. A *tbp* framework was easily constructed by a suitable mixing of 3s, $3p_x$, $3p_y$, $3p_z$ and $3d_{z^2}$. Hybridization of the first three orbitals produced a trigonal planar arrangement suitable for the equatorial bonds and hybridization of the $3p_z$ and $3d_{z^2}$ produced the apical bonds. Alternatively the bonding could be referred to as (sp^3d) hybrids which are spatially positioned as a *tbp* structure.*

What could bring about such a change in the 3d orbitals so as to permit their greater involvement? In essence they need to be contracted and their energies increased. (Increasing their energies refers to their ionization energies and not to their relative energies in the arrangement of energy levels in the atom. Increasing their energies would in fact narrow the gap between the 3s, 3p levels and the 3d levels.) An increase in the effective nuclear charge of the phosphorus, Z_e, would bring about this contraction. Although an increase in Z_e would change 3s, 3p and 3d the change would be most marked for 3d and might even be sufficient to reverse the relative order of 3d and 4s. An increase in Z_e can be effected by a decrease in the screening by the electrons. Electronegative ligands should bring this about and create what is referred to as a *ligand field contraction* of the 3d orbitals. The need for electronegative ligands is demonstrated by fluorine, the most electronegative ligand of all, which brings out the maximum valency and coordination number of many elements. Indeed the need for highly electronegative ligands seemed a pre-requisite for d orbital involvement. Such ligands as hydrogen would not suffice although for simplicity the m.o. calculations have been performed for the hypothetical PH_5 molecule. The nearest approach to this compound as yet is PH_2F_3. One suspects that any less electronegative ligands in place of the fluorines would give an unstable molecule, and PH_2Cl_3 for example is unlikely to be stable except under exceptionable circumstances.

At the present time the m.o. treatment of pentacoordinate phosphorus without the use of 3d orbitals seems the most suitable. (A modification of this method using contracted 3d orbitals might be an improvement not only to the non-bonding m.o. but throughout.) Even though the 3d orbitals may be unnecessary to the σ framework they are still present and are used in another capacity, i.e. π-bonding.

Until fairly recently there was no reason to postulate π-bonding in pentacoordinate phosphorus compounds since there was no phenomenon which demonstrated its possibility. As Chapter 5 will show this is no longer true. Moreover convincing evidence from n.m.r. spectroscopy has been found for such π-bonding. For example the ^{19}F n.m.r. spectrum of $PF_4(SR)$ shows the apical fluorine atoms to have distinguishable environments. This can only occur if the equatorial —SR group is restricted in its rotation about the P—S bond, preferring a configuration in which the alkyl group is not in the equatorial plane and is thus nearer one of the apical fluorine atoms. This situation would arise if the P—S bond had a π component based on donor $3p(S) \rightarrow 3d(P)$ π-bonding in the equatorial plane, Fig. 2.9a.

---

* The use of $3d_{x^2-y^2}$ instead of $3d_{z^2}$ produces (sp^3d) hybrid orbitals in the form of a square pyramid (*spy*) which is another possible arrangement of five bonds about an atom.

2.6 Pentacoordinate phosphorus compounds

Figure 2.9 Pentacoordinate phosphorus molecules displaying evidence of π-bonding.

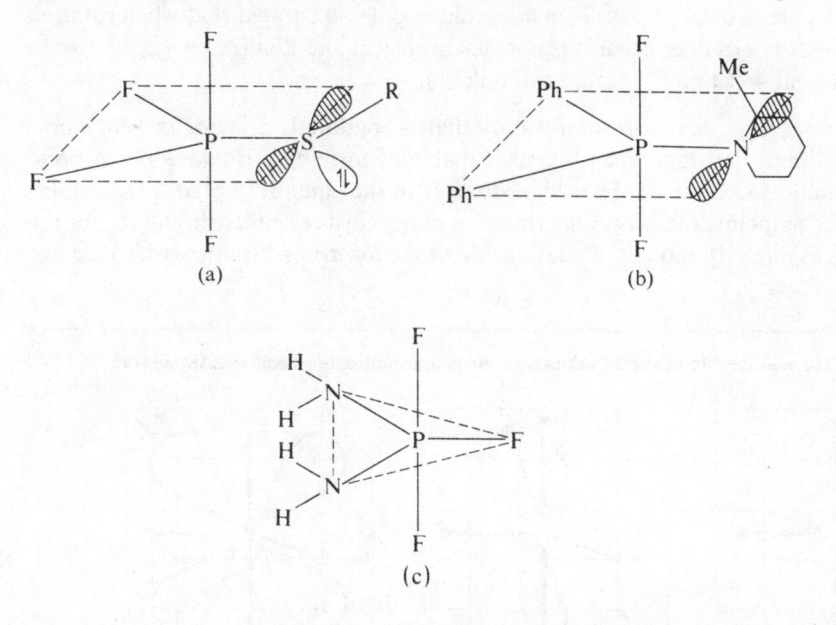

Refs: (a) S. C. Peake and R. Schmutzler, *Chem. Commun.*, 1662 (1968) and *J. Chem. Soc. A*, 1049 (1970); (b) M. J. C. Hewson, S. C. Peake and R. Schmutzler, *Chem. Commun.*, 1454 (1971); (c) E. L. Muetterties, P. Meakin and R. Hoffmann, *J. Amer. Chem. Soc.*, **94**, 5674 (1972).

The best demonstration of π-bonding in pentacoordinate phosphorus compounds comes from the barrier to rotation about the P—N bond. Such a barrier would arise from donor 2p(N)→3d(P) π-bonding although it has been suggested that in tricoordinate phosphorus–nitrogen compounds, R_2P—NR_2, the restricted rotation which the bond also shows is due to the lone pairs on P and N interacting as they appear to do in hydrazine. Although this may be the explanation for tricoordinate phosphorus derivatives it cannot be the reason for restricted rotation in pentacoordinate phosphorus compounds.

At room temperature pentacoordinate phosphorus–nitrogen compounds show unrestricted rotation about the P—N bond. Cooling to below $-40\ °C$, however, freezes out a preferred orientation of the molecule and the particular conformation can in many cases be identified. For example, the compound $Ph_2PF_2 \cdot NC_5H_4\alpha$-Me at $30\ °C$ has only one type of F shift at $+37.9$ ppm (CF_3Cl reference) in the ^{19}F n.m.r. spectrum, showing both fluorines to be equivalent and occupying the preferred apical positions. At $-70\ °C$ on the other hand two fluorine shifts at $+34.6$ and $+36.8$ ppm (CF_3Cl) can be distinguished showing their environments to be slightly different, a state of affairs consistent with structure Fig. 2.9b. In this the amine ring is constrained to the axial plane, a position in which the α-methyl group is near one of the apical fluorines. The ring is forced into this plane by the requirements of the nitrogen lone-pair in its 2p orbital to donate in the equatorial plane to an empty 3d orbital.

Work by the eminent American chemist Muetterties on the compound $PF_3(NH_2)_2$ showed that at $-40\,°C$ the conformation frozen out is that shown in Fig. 2.9c in which the H's are in axial planes. This molecule also demonstrated that when rotation about the P—N bonds does occur it is not synchronized, and that the energy barrier to rotation is about 46 kJ mol^{-1} (11.2 kcal mole^{-1}).

The above researches demonstrate not only that π-bonding is a factor in pentacoordinate compounds, but that when ligands capable of forming a single donor π bond are equatorially placed the donor orbital donates in the equatorial plane. This observation in fact pinpoints the acceptor orbital on phosphorus as either the $3d_{x^2-y^2}$ or the $3d_{xy}$ orbital. Figure 2.10 shows the relationship of the *tbp* σ-bond framework to the five

Figure 2.10 The relationship of the 3d orbitals to the pentacoordinate phosphorus framework.

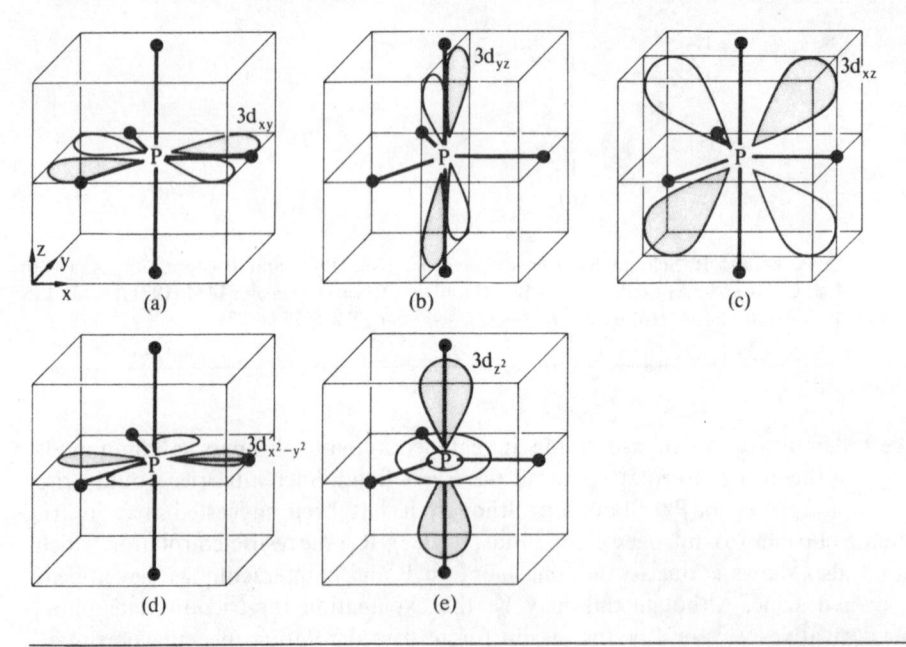

3d orbitals of phosphorus. As drawn it can be seen that the equatorial bonds interfere with those d orbitals with an xy plane component, $3d_{xy}$, $3d_{x^2-y^2}$ and $3d_{z^2}$. Depending on how the *tbp* is rotated about the z axis it is possible to minimize the coincidence between the $3d_{xy}$ orbital and one of the equatorial σ-bonds as shown in Fig. 2.10a, but this is at the expense of a partial overlap with the other two equivalent bonds. Or it is possible to minimize interference with two of these σ-bonds but at the expense of total overlap with the third, as shown for $3d_{x^2-y^2}$ in Fig. 2.10d. Possible 2p—3d π-bonds are shown for the equatorial ligands forming π-bonds in the equatorial plane which appears to be the preferred arrangement as shown by n.m.r. spectroscopy – Figs. 2.11a and 2.11b. Overlap in the apical xz plane is also shown in the Fig. 2.11c for $3d_{xz}$. For a ligand with two non-bonding electron pairs such as oxygen ligands both equatorial and apical plane π-bonding may be feasible.

2.6 Pentacoordinate phosphorus compounds

Figure 2.11 Overlap of equatorial ligand 2p and phosphorus 3d orbitals (not to scale).

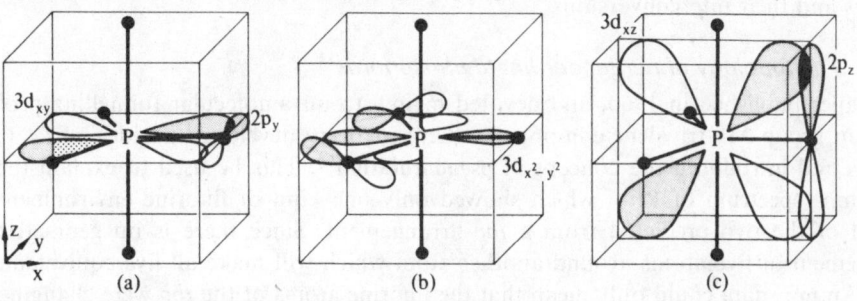

Apical ligands may also be able to form an attenuated π-bonding using the $3d_{yz}$ or $3d_{xz}$ orbitals (or both). The $3d_{z^2}$ orbital is of no use for this purpose. These ligands are further removed from the phosphorus and consequently overlap between the 2p and 3d will be less effective but may still contribute: see Fig. 2.12. Indeed it was once believed that apical ligands were in a better position for π-bonding and inspection of Figs. 2.10, 2.11 and 2.12 would seem to support this since of all the 3d orbitals only $3d_{yz}$

Figure 2.12 Overlap of apical ligand 2p and phosphorus 3d orbitals (not to scale).

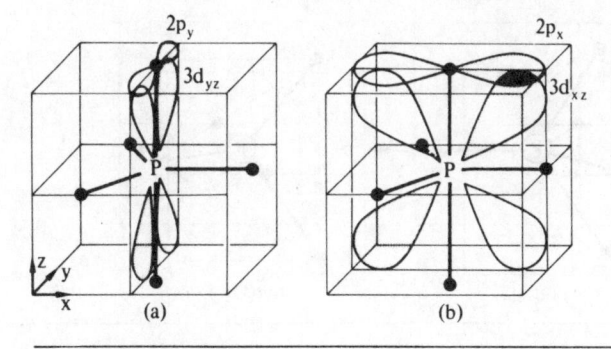

and $3d_{xz}$ which are in apical planes are entirely free of σ framework interactions. It became necessary to review this belief in the light of n.m.r. results which indicate that equatorial π-bonding is preferred.

Pi-bonding is a secondary effect however. While π-bonding may reinforce equatorial bonds more than apical bonds it is not the guiding factor which determines whether a ligand prefers to be positioned equatorially or apically in the *tbp* structure. The effect a ligand has on the σ bonds determines this; the more electron attracting a ligand is the more it will prefer the apical positions (VSEPR theory explained this in terms of reducing the Coulombic repulsions). Thus electronegative ligands should go apical and less electronegative ones go equatorial – this can be seen in structures such as Figs. 2.6b, 2.6c, 2.7a, 2.7b, 2.7c, and 2.7d. The apicophilicity (apex-seeking-ness) of ligands is a complex balance of electronegatively, π-bonding ability and possibly size with these effects in this order of importance. However before discussing the apicophilicity of ligands, which is still an unsolved problem in parts, it is necessary to discuss the geometry of the pentacoordinate state in general since this determines the isomers and their interconversions.

2.6.2 The topology of pentacoordinate phosphorus[14]

In a paper published in 1960, and devoted mainly to intramolecular tunnelling processes in group M5 trivalent compounds, Stephen Berry developed an idea of F. T. Smith's and introduced the concept of **pseudorotation**[15]. This he used to explain the ^{19}F n.m.r. spectrum of PF_5, which showed only one kind of fluorine environment instead of the two predicted from a *tbp* arrangement. Since there is no geometric arrangement of five atoms around another atom which will make all five equivalent, the ^{19}F n.m.r. data could only mean that the fluorine atoms of the *tbp* were changing place between apical and equatorial positions so rapidly that the long-exposure picture which n.m.r. produces showed an average of the two environments. Berry's achievement was to propose a credible mechanism whereby apical $\rightleftharpoons$ equatorial interchange could occur. He showed that by deforming the *tbp* structure to the square pyramidal (*spy*) configuration the apical fluorines became equivalent to two of the equatorial fluorines. Continuing the deformation produced another *tbp* arrangement in which the new apical pair were previously equatorial. Figure 2.14a illustrates the

Figure 2.14a Berry pseudorotation mechanism[15].

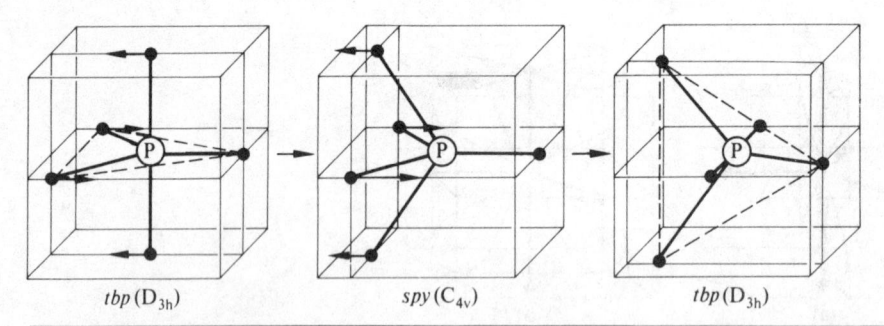

tbp (D_{3h})   *spy* (C_{4v})   *tbp* (D_{3h})

mechanism which is called pseudorotation (ψ) because after the operation the *tbp* appears to have been rotated in that the axis of the *tbp* is now at 90° to the original axis. The figure also shows how the pivot ligand remains equatorial in both *tbp*'s. By choosing a different equatorial atom to be the pivot for a subsequent ψ it can be seen that all the fluorine atoms of PF_5 can exchange between equatorial and apical environments.

Much use has been made of ψ especially in organophosphorus chemistry. Westheimer in work published in 1968 made use of it in a milestone paper[16] on the hydrolysis of phosphorus esters in which he was able to explain some results that appeared very puzzling at the time. Other chemists such as Muetterties have developed the use of ψ in general while others such as Ugi and Ramirez have proposed alternative polytopal rearrangements for the interconversion of *tbp*'s.

It can be proved that for five points on the surface of a sphere the arrangement of lowest energy is the *tbp*. Any deformation from this will be of higher energy but there

2.6 Pentacoordinate phosphorus compounds

are some alternative arrangements which are not much higher and the *spy* is one of them. Under certain circumstances phosphoranes may choose this configuration, e.g. Fig. 2.6f, although the reasons for this choice are, as yet, not understood. Another example in which *spy* is preferred over *tbp* is antimony pentaphenyl. In this case it is believed that crystal packing forces are responsible – the lattice energy of *spy* packing more than offsets the energy required to deform $SbPh_5$ from the *tbp* configuration.

There are two other pentacoordinate arrangements which have energies low enough to qualify them as possible intermediates in an isomer interconversion. These are related to one another having a pair of ligands on one side of the phosphorus atom and a trio of equivalent ligand positions on the other side. The plane encompassing the pair of ligands can be orientated in two different positions relative to the trio and these positions are referred to as eclipsed and staggered. Ugi and Ramirez[17] proposed a mechanism for the interconversion of a *tbp* in which the intermediates were of pair-trio geometry and this mechanism they called *turnstile rotation* (τ). The sequence and the two pair-trio configurations are illustrated in Fig. 2.14b. In effect the mechanism is the

Figure 2.14b Turnstile rotation mechanism.

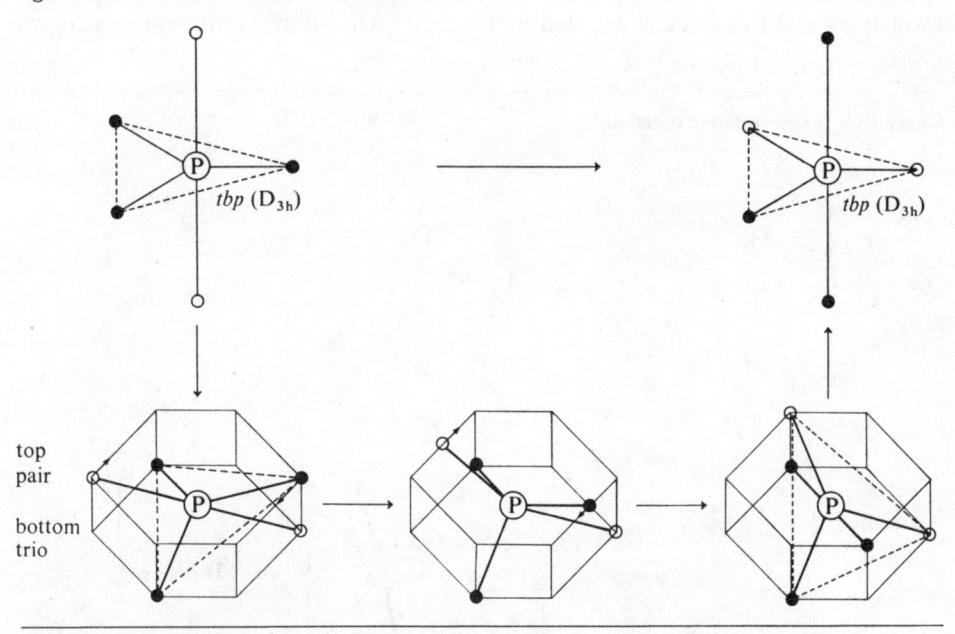

rotation of the upper pair relative to the lower trio by 60°. If the rotation were carried beyond this to 120° another configuration of the *tbp* would be produced and this is known as double turnstile rotation, τ^2. Such a simple change afforded by this double step can only be achieved by several ψ sequences, and in this respect τ has an advantage over ψ.

Which of the two polytopal rearrangement mechanisms is the more likely? This is decided by the relative energies of the intermediate states relative to the *tbp* state. In fact the *spy* configuration is of lower energy than the *pair-trio* so that pseudorotation should be the preferred route. The energy differences for the PF_5 molecule have been judged by Muetterties[13] to be about 4 kJ mol^{-1} (1 kcal mole^{-1}) for *tbp→spy* and about 40 kJ mol^{-1} (10 kcal mole^{-1}) for *tbp→pair-trio*. (Too much emphasis should not be placed on these as absolute values, but they show the order of magnitude.) It seems unnecessary then to invoke τ as a mechanism despite its simplicity although on occasion it might possibly offer an alternative route if that for ψ is restricted in some way. Nevertheless only the Berry pseudorotation will be dealt with in depth here.

2.6.3 Berry mechanism for polytopal rearrangement – pseudorotation (ψ)

Polytopal rearrangements can be classified into two categories – restricted and unrestricted – according to whether there are any restraints imposed on the rearrangement other than that of the *spy* configuration. Restrictions can be of the physical type, as when the phosphorus atom is part of a ring system, or the positional preference type, by which it is meant that certain ligands prefer particular positions and will resist a rearrangement which displaces them to a less preferred position. Most polytopal transformations are of the restricted type but to explain the phenomenon more clearly it is best to consider unrestricted ones first.

A shorthand notation is required for *tbp* structures and their pseudorotational transformations and this is demonstrated in Fig. 2.15. The atoms surrounding the phos-

Figure 2.15 Pseudorotation notation.

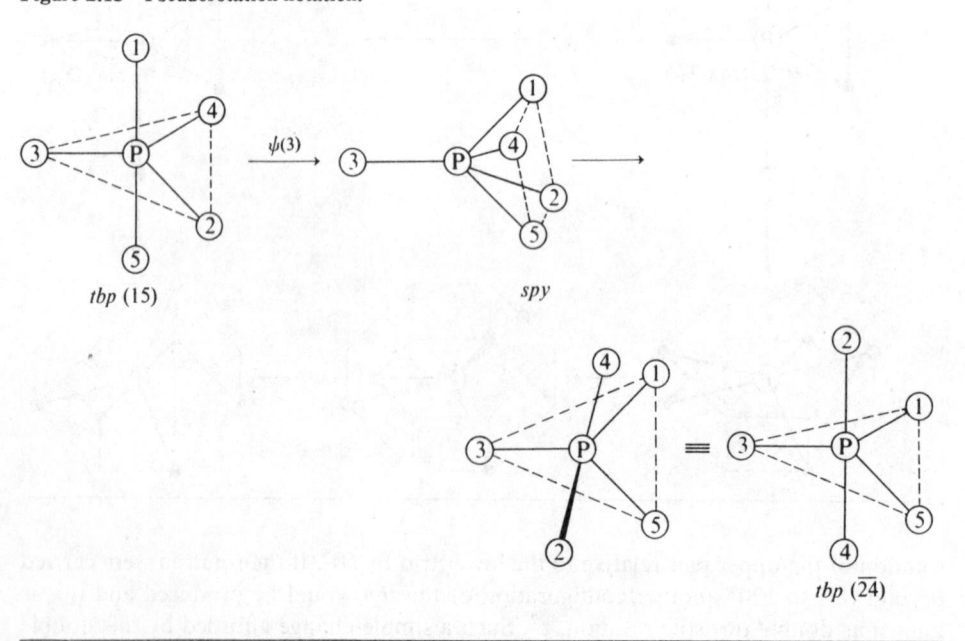

phorus have been numbered 1 to 5 and the convention is to denote a particular configuration by the numbers of the apical atoms with the *tbp* arranged so the lower apical number is uppermost. If the equatorial atoms are then arranged clockwise in ascending order the notation is not modified, but if the equatorial atoms are anticlock-

2.6 Pentacoordinate phosphorus compounds

wise in ascending order the notation has a bar over it. Thus in Fig. 2.15 the left hand *tbp* is (15) and pseudorotation about pivot ligand 3 converts it to the *tbp* on the right hand side which has the symbol $\overline{(24)}$. The complete process can be summed up neatly as (15), $\psi(3) = \overline{(24)}$.

Since there are three equatorial bonds there are three possible pivotal atoms with reference to which pseudorotation may occur and three different *tbp*'s into which the original structure can be transformed. These resulting *tbp*'s can then be further converted into others and so on. For five different ligands there are 20 possible configurations and these can be linked by 30 ψ transformations. It is not possible to represent these configurations and the transformations linking them, on a flat surface, or even on a solid surface, if we wish to avoid drawing the diagram with no pathway lines crossing. An isomer graph, as it is called, can be drawn in several ways however if the non-intersecting stipulation is borne in mind; in other words lines which cross in the diagram do not really intersect. The human eye can then easily interpret the isomer graph in Fig 2.16 in which *tbp* configurations occupy vertices of the figure, the sides represent the transformation pathways and the numbers on these are the pivotal atoms[18]. Time spent familiarizing oneself with Fig. 2.16, for example by

Figure 2.16   Graph for polytopal rearrangement of trigonal bipyramid by pseudorotation[18].

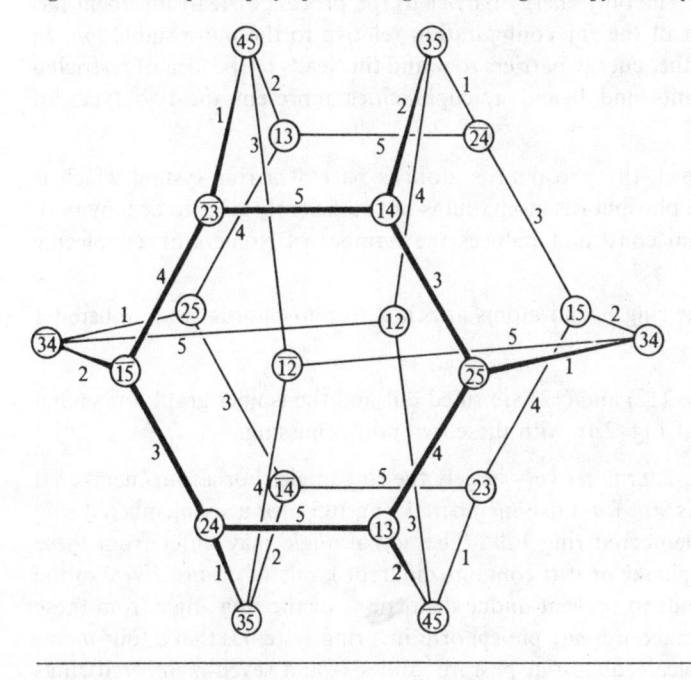

checking some of the pathways, will be amply repaid in the next section and the remainder of this book.

In addition to the general case of five ligands, all different, surrounding the central phosphorus atom there are the less complex molecules in which two or more of the ligands are the same. With only one kind of ligand, $P(L^1)_5$, as is found in PF_5 etc., pseudorotation still occurs and the ligands exchange equatorial and apical places according to the graph of Fig. 2.16. However all the points on the graph now represent identical isomers so that the isomer graph for $P(L^1)_5$ collapses to a single point. For the five other ligand mixtures – $P(L^1)_2L^2L^3L^4$ or PA_2BCD; $P(L^1)_2(L^2)_2L^3$ or PA_2B_2C; $P(L^1)_3L^2L^3$ or PA_3BC; $P(L^1)_3(L^2)_2$ or PA_3B_2; and $P(L^1)_4L^2$ or PA_4B – the isomer graphs are given in Fig. 2.17.

In the isomer graph for $P(L^1)_2L^2L^3L^4$ it is necessary to distinguish the two ligands which are the same, since ψ about these pivotal ligands when both are placed equatorially leads to different isomers. Consequently in the isomer graph they are marked as 1 and 1′ in the pathways. The big simplifying factor in this class of compounds arises from the fact that when the two like ligands are both equatorial or both apical the concept of clockwise and anticlockwise distinctions is void. Indeed (xy) and $\overline{(xy)}$ isomer pairs can only occur when all equatorial atoms are different together with both apical atoms different. This is just possible for PA_2B_2C molecules in the case of $P(AB)_{apical}(ABC)_{equatorial}$, but for PA_3BC and the rest it is not possible as Fig. 2.17 shows.

So far we have dealt with the seven cases of from one to five different, but unspecified, ligands interchanging amongst the positions of the *tbp* according to Fig. 2.16, producing isomers according to Fig. 2.16 (in the case where all ligands are different) and to Fig. 2.17 for the rest. The only energy barrier to the process of rearrangement has been assumed to be that of the *spy* configuration relative to the more stable *tbp*. In molecules there can be other energy barriers to ψ and this leads to the idea of restricted ψ; ring system constraints and ligand apicophilicities represent the two types of restriction.

Consider the case in which the phosphorus atom is part of a ring system which is flexible in its angle at the phosphorus atom, but is not sufficiently large to occupy both apical sites. This physical constraint reduces the number of isomers of a molecule

by two. If the ring ligand atoms attached to phosphorus are numbered 1 and 2 then configurations (12) and $\overline{(12)}$ are ruled out and the isomer graph for such a molecule becomes that of Fig. 2.16 with these two points missing.

The size of a planar ring determines very largely the ring angle. For a four-membered ring the preferred angle is 90°, for a five-membered ring 105°, for a six-membered ring 120°, and for a seven-membered ring 129°. The actual angle may differ from these values if the ring is not planar or if it contains different kinds of atoms. Even so the nature of a ring itself tends to prevent undue distortions of the ring angle from these values. The result for pentacoordinate phosphorus in a ring system is that a four-membered ring prefers an apical–equatorial posture, and six- and seven-membered rings an equatorial–equatorial posture – Fig. 2.18. Five-membered rings suffer a 15° distortion from the preferred angle of 105° if they are either apical–equatorial (90°) or diequatorial (120°). In this case other factors may operate to determine the configuration, e.g. $(CH_2)_4PF_3$ appears at low temperatures to be diequatorial (^{19}F n.m.r. spectroscopy)[19] a structure which allows the electronegative fluorine atoms to occupy both apical sites.

2.6 Pentacoordinate phosphorus compounds

Figure 2.17 Isomer graphs for pentacoordinate phosphorus compounds with mixed ligands.

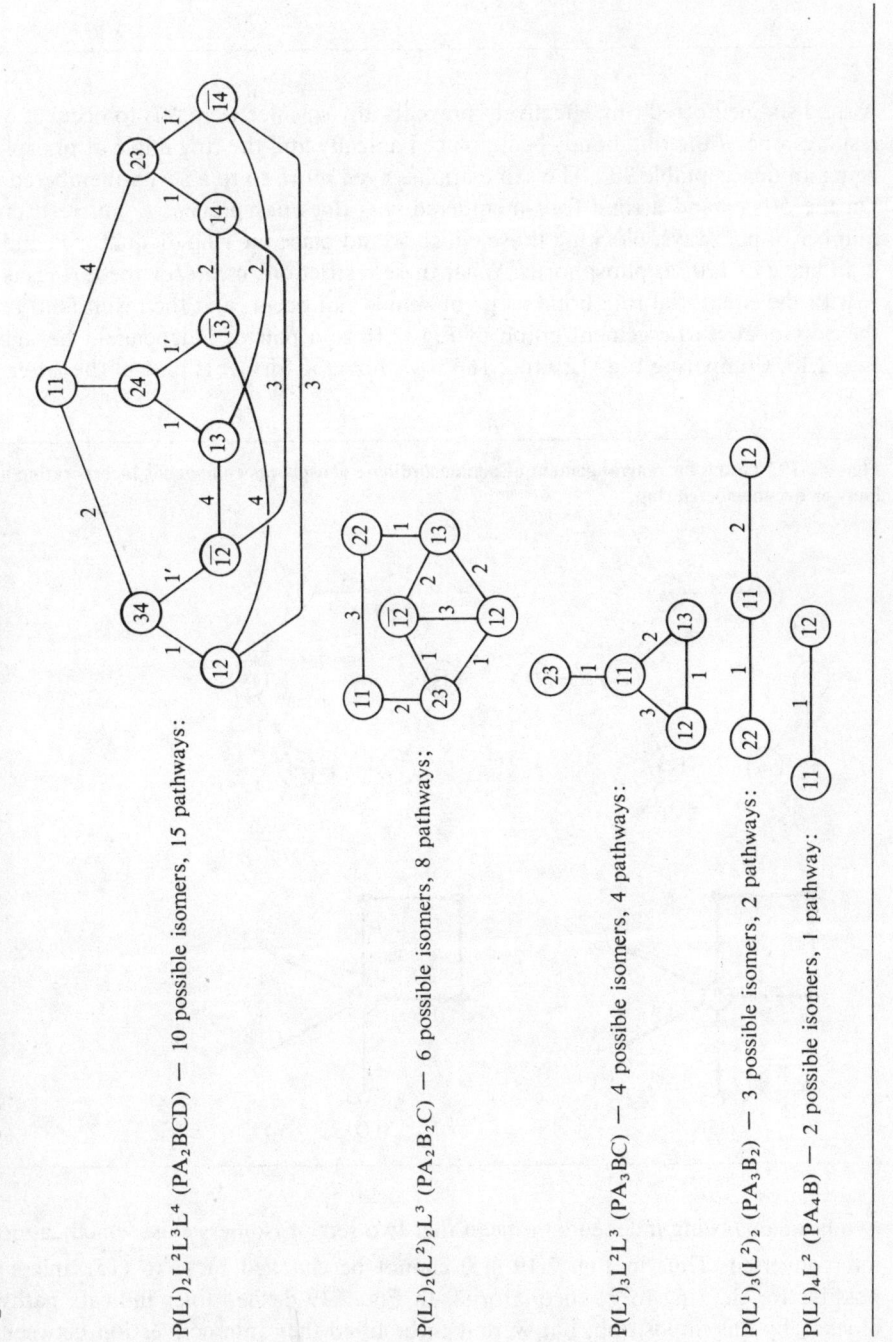

$P(L^1)_2L^2L^3L^4$ (PA_2BCD) — 10 possible isomers, 15 pathways:

$P(L^1)_2(L^2)_2L^3$ (PA_2B_2C) — 6 possible isomers, 8 pathways;

$P(L^1)_3L^2L^3$ (PA_3BC) — 4 possible isomers, 4 pathways:

$P(L^1)_3(L^2)_2$ (PA_3B_2) — 3 possible isomers, 2 pathways:

$P(L^1)_4L^2$ (PA_4B) — 2 possible isomers, 1 pathway:

Chapter 2 / Bonding at phosphorus

Figure 2.18 Preferred ring positions in pentacoordinate phosphorus compounds.

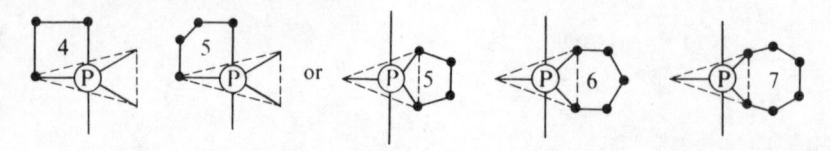

A rigid six-membered ring effectively prevents any ψ, since were this to occur it would result in one of the ring bonds being placed apically and the ring angle at phosphorus being an unacceptable 90°. The same applies even more so to a seven-membered ring. On the other hand a rigid four-membered ring does not prevent ψ but restricts the number of pathways, blocking those which would place the ring diequatorial and give it an angle of 120° at phosphorus. What these restrictions mean for these rings is that ψ with the equatorial ring bond as pivot should not occur, and the result is to reduce the polytopal rearrangement graph of Fig. 2.16 to a pair of independent hexagons – Fig. 2.19. Comparing Fig. 2.19 to 2.16 shows how the former is part of the latter. The

Figure 2.19 Graph for rearrangement of pentacoordinate phosphorus compounds incorporating a four- or five-membered ring.

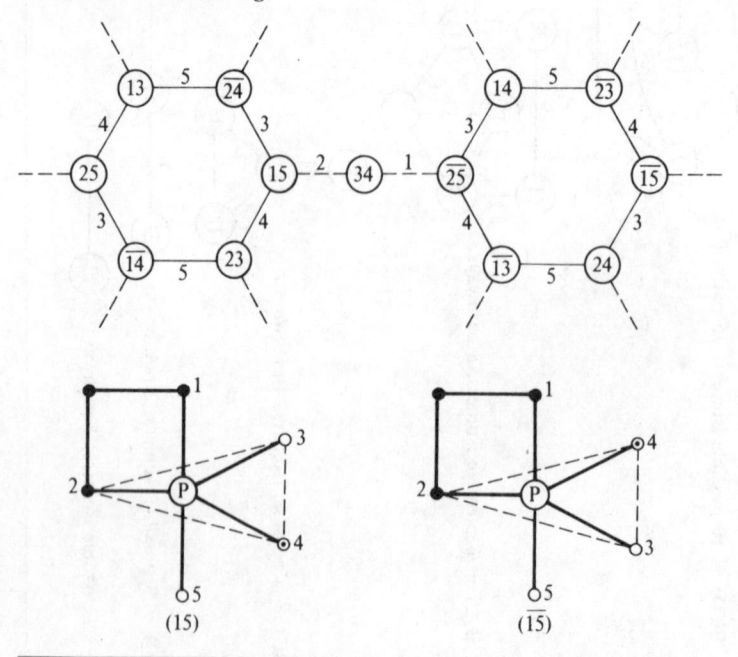

two hexagons being independent mean that two sets of isomers exist which cannot be interconverted. Thus in Fig. 2.19 (15) cannot be changed by ψ to $\overline{(15)}$ unless it is possible for the ring to go diequatorial. In Fig. 2.19 dashed lines indicate pathways blocked by this imposition, but were it to be lifted then interconversion between the two sets of isomers would become possible. In Fig. 2.19 one such bridge between the two sets is indicated.

In reality rings of atoms possess a large degree of flexibility and so must the pentacoordinate configuration. Its equatorial angle is not set at a fixed 120° any more than that of a four-membered ring is set at 90°, and a diequatorial positioning even of such a ring is feasible[20]. It should be remembered that the deforming of both systems necessary to do this will create an energy barrier, the height of which will be proportional to the strains imposed on the ring and *tbp* bonding systems. Such energy barriers may be surmounted in certain systems, especially if the ring restraint opposes the second kind of restraint, which is the desire of certain ligands to occupy apical sites in the *tbp* arrangement, i.e. apicophilicity.

2.6.4 Ligands attached to pentacoordinate phosphorus and their preferred positions

The sophistication of technique necessary to study ligand rearrangement in pentacoordinate phosphorus molecules came with the development of n.m.r. spectroscopy. Although the phosphorus nucleus, ^{31}P, has nuclear spin, $I = \frac{1}{2}$, the observation of its resonances in a magnetic field tells us only about its net electronic environment. To study the ligands attached to phosphorus in the different *tbp* positions it is necessary for them to have observable spins; for most practical purposes this has limited studies to groups containing ^{19}F and ^{1}H, although ^{13}C n.m.r. spectroscopy should soon expand the range of observable ligands.

Fluorine-19 n.m.r. spectroscopy was originally important in showing that polytopal rearrangements take place. It has also been used to show the relative effects on the rearrangements of other groups attached to the phosphorus. The method however is of limited value because of all ligands fluorine is the most apicophilic, but by measuring the temperature necessary for ψ to occur (as shown by the coalescence of the ^{19}F n.m.r. signals of the apical and equatorial fluorines) some idea can be got of the apicophilicity of ligands.

The introduction of another group in PF_5 will not affect the ψ greatly. In PF_4X the X ligand will prefer to remain equatorial and can remain so by always acting as the pivotal bond during ψ. Although some ψ pathways are blocked by this, it has no effect on the interchange of F atoms between apical and equatorial sites. However the introduction of two groups as in PF_3X_2 has a profound effect since ψ must inevitably force one of the X ligands into an apical position and at the expense of a F atom being forced equatorial. Although experimental results are scanty it would appear that ψ does not begin in PF_3Cl_2 until the temperature is raised to −80 °C. Surprisingly PF_3H_2 requires a temperature as high as −15 °C to begin the ψ process[21], and for PF_3Me_2 and $PF_3(NMe_2)_2$ ψ does not occur at room temperature nor even at 100 °C, so strong is the desire of these groups to remain equatorial. The last remarks apply also to the ring compound, $F_3P\begin{smallmatrix}CH_2-CH_2\\ CH_2-CH_2\end{smallmatrix}CH_2$ which adds a further barrier of ring angle contraint to that of the low apicophilicity of carbon ligands.

Other, smaller, rings can and do result in a *lowering* of the energy barrier to ψ since they prefer an apical–equatorial posture. Thus $F_3P\begin{smallmatrix}CH_2-CH_2\\ |\\ CH_2-CH_2\end{smallmatrix}$ shows ψ beginning when its temperature is raised to $-70\ °C$; below this the configuration is frozen out with the ring being diequatorial.* Theoretically both this and apical–equatorial are strained arrangements if the five-membered ring were exactly planar. However it can adopt the apical–equatorial position at room temperature, despite putting a fluorine atom into an unfavourable equatorial position in order to do so. Generally five-membered rings prefer to occupy apical–equatorial positions.

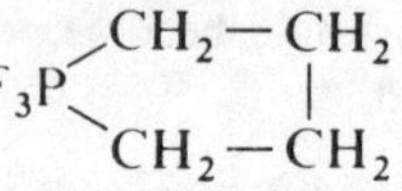

The compounds PF_2X_3 show both fluorine atoms apical at all times for X = alkyl, aryl or trifluoromethyl groups, as expected.

The studies on PF_xX_{5-x} derivatives show that the order of apicophilicity is F > Cl > H > Me ⩾ NMe₂. Other methods are necessary to confirm and extend this order. One such method, devised by DeBruin et al., is based on the conversion of [trans-$C_8H_{16}P(OMe)X]^+$ to a mixture of trans and cis products[22]. The structure of this cation and the definition of trans is (2.11)

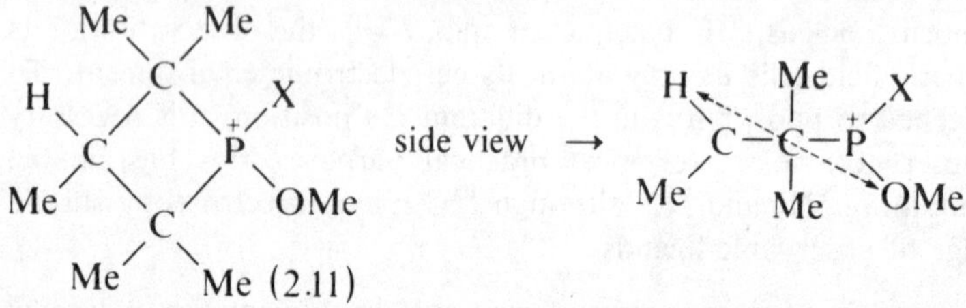

side view → (trans to H of ring) (2.11)

The ratio of cis:trans hydrolysis products from purely trans starting material is related to the apicophilicity of group X compared to that of OMe. The reaction intermediate in the hydrolysis is presumed to be the pentacoordinate compound (2.12)

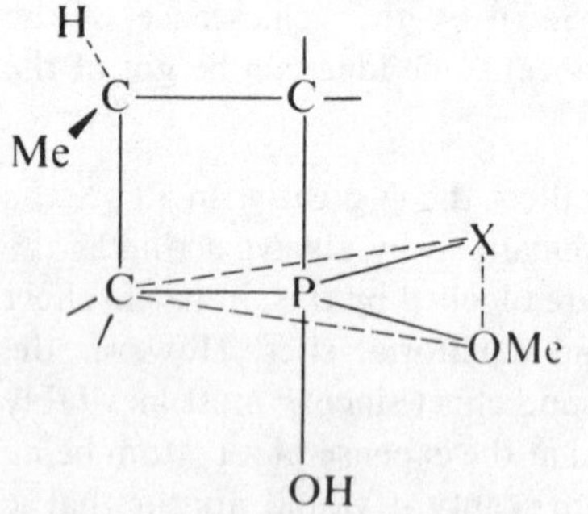

(2.12)

which can ψ with X pivotal to put OMe into the apical position from which it can leave and the product is solely trans. If on the other hand ψ(OMe) occurs because X is more apicophilic and this group leaves, the product is solely cis. (Both alternatives

---

* Even four-membered rings are forced diequatorial[20] as shown by the ¹⁹F n.m.r. spectrum of

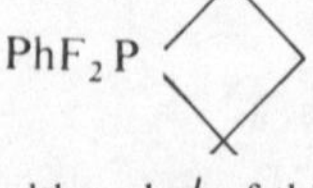

$PhF_2P\diamondsuit$

although ψ of this begins at $-95\ °C$.

2.6 Pentacoordinate phosphorus compounds

can be complicated by subsequent ψ(OH) prior to loss of X or OMe but this does not affect the *cis*: *trans* ratio.) Thus the product ratio of *cis*: *trans* reflects the competition of ψ(X): ψ(OMe) in the intermediate. The order of relative apicophilicities determined by this method was Cl ⩾ SMe > OPr^i ⩾ OMe ⩾ OEt > NMe$_2$.[22]

Other methods from which relative apicophilicities can be inferred are based on pentacoordinate compounds which incorporate the same phosphetan ring plus a 1,3,2-dioxaphospholan ring (2.13)[23]. Observation of the ^{19}F n.m.r. spectrum of the

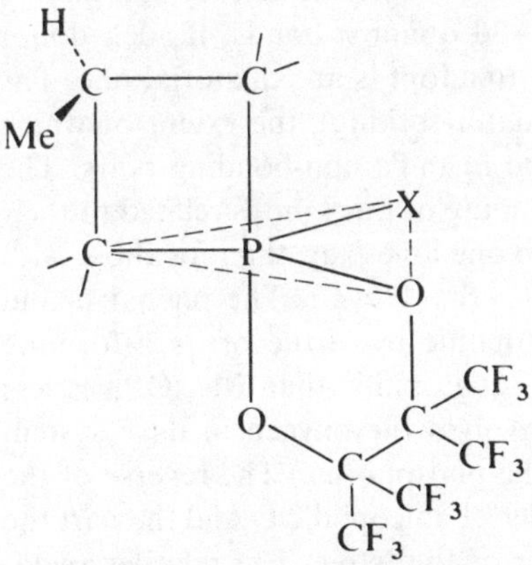

(2.13)

CF$_3$ groups over a range of temperatures led to a series of coalescence temperatures for the signals, depending upon the group X.* Free energy values $\Delta G^{\ddagger}$ for the ψ were determined and these were OPh < NMe$_2$ < Me < Ph. The reverse order should reflect the relative apicophilicities, i.e. OPh > NMe$_2$ > Me > Ph. These results may be undermined if (2.13) is not *tbp* but *spy* as are some of the other bicyclic phosphoranes; Holmes[10] suggests that a very similar compound[11] to (2.13) is in fact nearer *spy* than *tbp*.

A more theoretical approach to apicophilicity has been made by Gillespie *et al.*[14] who used the concept of binding energies (= total energy − core energy) to calculate the energies of different isomers of the *tbp*. The results indicate relative apicophilicities to be F > H > CF$_3$ > Cl > Me. The high apicophilicity of hydrogen attributed to it by this method is somewhat at variance with the observations on PF$_3$H$_2$, although others have found evidence of hydrogen's high apicophilicity. Gillespie and coworkers also calculated that the P—O$^-$ bond prefers to be equatorial and that the P—OH bond is more apicophilic.

Taking all the above information into account the overall order of relative apicophilicities of the more common ligands attached to phosphorus is F > H > CF$_3$ > OPh > Cl > SMe > OMe > NMe$_2$ > Me > Ph. What does this order reveal about phosphorus bonding? It shows that there are three factors which play a part in determining the preferred *tbp* position for a ligand – two bonding factors and a size factor.

---

* Pseudorotation is about the P—C bond as pivot. This places the four-membered ring diequatorial but this may be offset by group X being apical. Pseudorotation with P—O pivotal would place the five-membered ring diequatorial which would be less strained but this is negated by the higher apicophilicity of these particular O atoms.

The theoretical treatments of pentacoordinate phosphorus, such as the molecular orbital treatment of Muetterties, have as the keystone of their thesis, that electron density accumulates at the apical positions. Electronegative substituents should therefore prefer to occupy these regions and consequently the order of apicophilicity will reflect the order of electronegativity of the ligands. Inspection of the observed list of apicophilicities shows clearly that this is not so although there is a tendency towards this relationship.

Although the 3d orbitals are superfluous to σ-bonding requirements they can have a role in the formation of 2p→3d and possibly 3p→3d donor π-bonds. If such donor π-bonding occurs then the best position for ligand to adopt is an equatorial one. The driving force here is a positive *equatophilicity* (equator-seeking), the extent of which will increase with the π-bonding capabilities of the ligand's non-bonding pairs. The ability of ligands to undertake supplementary π-bonding of this kind is related to their Lewis basicity and is stronger for ligand atoms with one lone-pair, than for those with two lone-pairs, than those with three, i.e. $N > O > S > F > Cl$. The phenyl ligand when attached to phosphorus may also be equatophilic by virtue of its π-forming ability. (For a similar reason, though Ph is more equatophilic than Me, OPh is less equatophilic than OMe since the phenyl group involves the oxygen in its π system thus weakening the oxygen's π involvement with the phosphorus.) The reverse of the relative order of equatophilicity is of course an order of apicophilicity and the part the former plays can be judged from the overall order of the latter. The relative apicophilicities should then reflect to a good approximation the overall ($\sigma \pm \pi$) electronic effects of the various ligands.

A third effect operating to determine the best location for a ligand is its steric requirements. The larger the connecting atom or the group the more it should prefer the roomier equatorial sites, forcing the smaller atoms or groups into apical positions. This effect may serve to explain the apparently high apicophilicity of hydrogen, but in order of importance the size factor is generally regarded as third on the list.

Too much should not be read into pseudorotation and apicophilicity although they have proved useful in rationalizing many phosphorus reactions. We now turn to this area and deal with the substitution reactions of tetracoordinate phosphorus compounds in which the pentacoordinate intermediate may have a part to play.

2.6.5 *Pentacoordinate phosphorus reaction intermediates*

One of the major breakthroughs in explaining substitution at phosphorus was made by Westheimer[16]. He set himself the job of explaining two puzzling pieces of information concerning the hydrolysis of phosphate esters and found that pseudorotation provided the answer. The set of data he was attempting to explain is summarized in Fig. 2.20. Two puzzling features concern the phosphate ester hydrolysis (I): why does this reaction go much faster than the hydrolysis of the acyclic counterpart, $(MeO)_3PO$, which is not the case with reaction (III) where hydrolysis is about the same rate as that of its acyclic counterpart, $Et_2P(O)OEt$; and why does reaction (I) give such a large proportion of the product in which the ring remains intact whereas in (II) hydrolysis occurs with ring opening almost exclusively?

Westheimer was able to answer these questions by postulating pentacoordinate intermediates in which the five-membered rings took up apical–equatorial stances. These

2.6 Pentacoordinate phosphorus compounds

Figure 2.20 The acid hydrolysis of cyclic phosphorus esters.

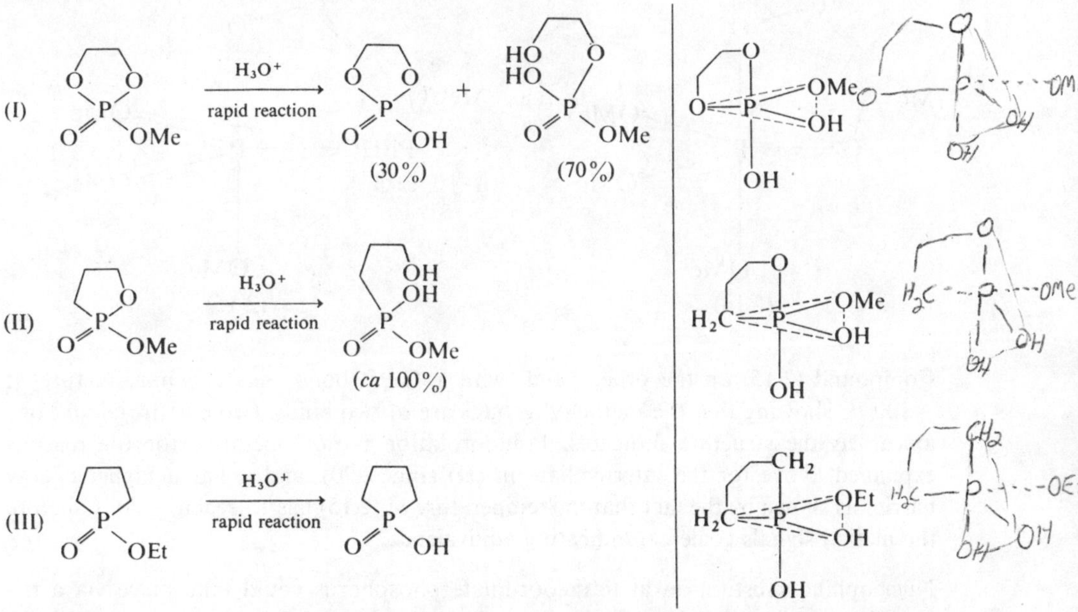

intermediates are shown in Fig 2.20 with the phosphoryl oxygen protonated and the incoming hydroxyl group apically placed.* If, as seems, likely the leaving group must depart from an apical position then intermediates in (I) and (II) can permit this step as they stand and encourage speedy departure of the apical ring oxygen by the relief of ring angle strain which it will bring. Although this step is rapid it is not so rapid as to rule out other processes and the pentacoordinate intermediate in (I) is sufficiently long-lived to permit ψ to take place. Pseudorotation of this with the equatorial OH as pivot results in the methoxy group going apical from which it can then depart, leaving a product in which the ring is preserved.

Pseudorotation, ψ(OH), of intermediate (II) is disadvantageous since it will serve to put the ring carbon apical. Nor is ψ(CH$_2$) likely as it would place the five-membered ring diequatorial. The energetically favoured outcome for (II) is cleavage of the ring P—O bond, as is found. For the intermediate in (III) ψ(OH) is again feasible, exchanging ring carbons mutually between an apical and an equatorial position, and this puts the ester group apical from which it can depart.

Westheimer received support for his theory from p.m.r. data on stable compounds which were similar to the intermediates he postulated. Although x-ray crystallography shows compound (2.14) to have the ring apical–equatorial its p.m.r. spectrum even at $-100\,°C$ shows all the methoxy groups equivalent and thus indicating that ψ is operative.

---

* The hydrolysis of these may just conceivably go via hexacoordinate intermediates[24] in which case Westheimer's explanation may be only part of the answer.

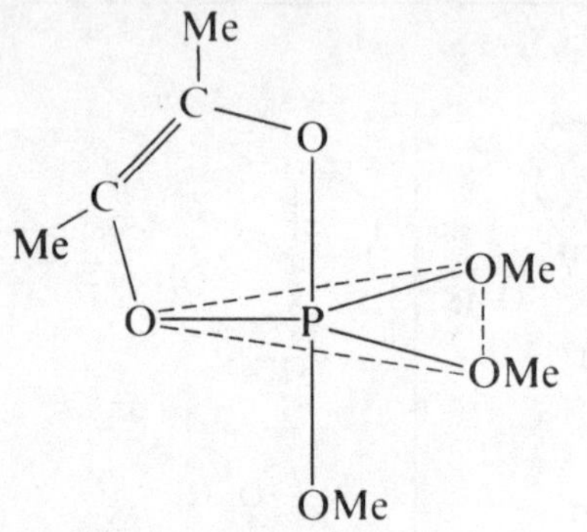

Chapter 2 / Bonding at phosphorus

(2.14) 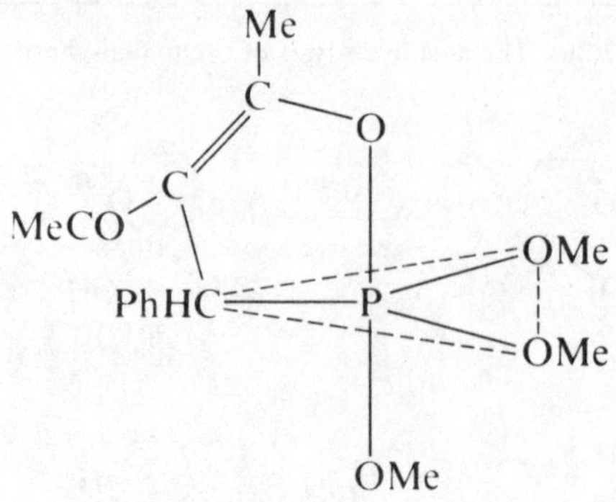   (2.15)

Compound (2.15) on the other hand, with a C—P bond, has a p.m.r. picture at −100 °C showing that the methoxyl groups are of two kinds, two equatorial and one apical, as the structure indicates. Pseudorotation is not operative for the reasons explained above for the intermediate in (II) (Fig. 2.20), and it has a higher energy barrier as shown by the fact that the temperature of (2.15) has to reach −70 °C before the methyl signals coalesce, indicating equivalence.

Nucleophilic substitution at tetracoordinate phosphorus could take place via a tricoordinate cation ($S_N1(P)$ type; ionic mechanism) (6) as well as via a pentacoordinate intermediate ($S_N2(P)$ type mechanism), (7) although the latter is the more probable.

$$S_N1(P) \quad PX_3Y \longrightarrow PX_2Y^+ + X^- \xrightarrow{Z^-} PX_2YZ \tag{6}$$

$$[Y = O, S, NR, \text{etc.}]$$

$$S_N2(P) \quad PX_3Y \xrightarrow{Z^-} [PX_3YZ]^- \longrightarrow PX_2YZ + X^- \tag{7}$$

The $S_N2(P)$ type substitution involves many energy changes, apart from that of any pseudorotation step, and these are depicted in Fig. 2.21. Often many of these changes are collected together in such loose concepts as "strong nucleophile" and "good leaving group", which is perhaps all that the present state of our knowledge will allow. Some of the steps in Fig. 2.21 involve large energy differences such as the solvation and desolvation steps, and the bond making and breaking steps. In many reactions these will balance out so that the smaller energy steps may be sufficient to tip the scales one way or the other. Pseudorotation is just one such smaller energy barrier.

So far it has been tacitly assumed that the incoming group occupies an apical position and that a leaving group departs from an apical position. The nature of the *tbp* would seem to dictate this since the apical bonds are weaker and further from the phosphorus and so should aid bond formation and bond breakage in easier stages.

When approaching a tetrahedron-shaped molecule a nucleophile can attack either the centre of a face or the midpoint along an edge. The former will produce a *tbp* structure with the incoming group as an apex; the latter type of approach will result in the incoming group being equatorial. Face attack will meet less steric interference as it

2.6 Pentacoordinate phosphorus compounds

Figure 2.21 Major energy changes involved in nucleophilic substitution at tetracoordinate phosphorus, $S_N2(P)$.

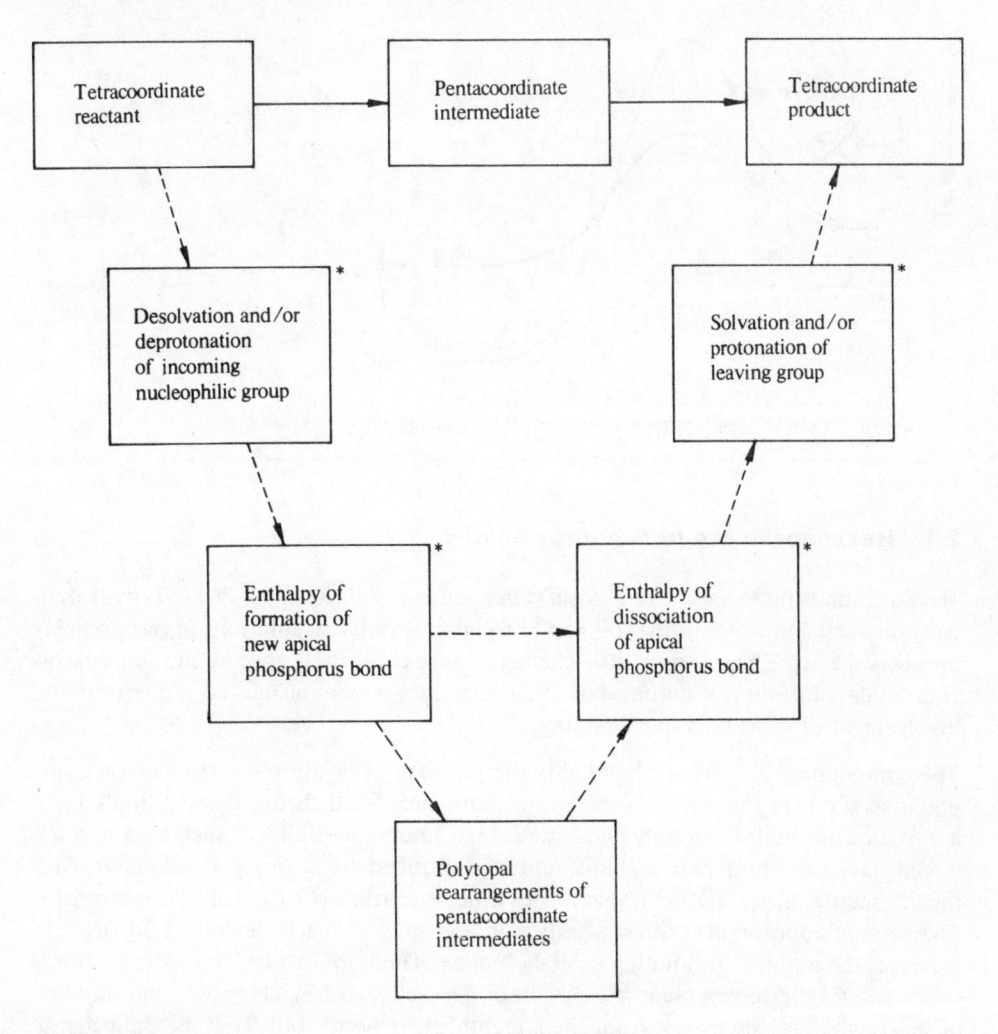

```
[Tetracoordinate reactant] → [Pentacoordinate intermediate] → [Tetracoordinate product]
        ↓*                                                            ↑*
[Desolvation and/or                                          [Solvation and/or
 deprotonation                                                protonation of
 of incoming                                                  leaving group]
 nucleophilic group]
        ↓*                                                            ↑*
[Enthalpy of                                                  [Enthalpy of
 formation of       ----→                                      dissociation
 new apical                                                    of apical
 phosphorus bond]                                              phosphorus bond]
              ↓                                              ↑
                    [Polytopal
                     rearrangements of
                     pentacoordinate
                     intermediates]
```

* Partial solvation and desolvation of phosphorus molecules and intermediates will also occur.

approaches and for this reason alone apical entry would be preferable. However once the *tbp* has been formed the apex position is the more congested, so that although the face is the easier entry route, it results in the group occupying the more hindered *tbp* position. This means that a group is easier to expel from the apical position. The energy profiles for apical vs equatorial attack and departure are shown in Fig. 2.22.

Not all chemists accept the apical substitution mechanism, nor for that matter the polytopal rearrangement of a *tbp* phosphorus molecule or intermediate[25]. Nevertheless the tide of opinion is running against them at the present time.

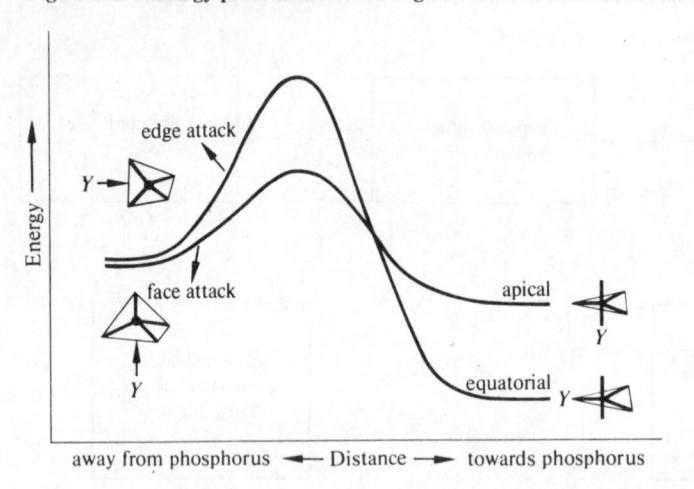

Figure 2.22 Energy profiles for face *vs* edge attack of a tetracoordinate molecule.

2.7 Hexacoordinate phosphorus anions

Hexfluorophosphate anion, PF_6^-, is a water soluble, stable anion; PCl_6^- is hydrolytically unstable but exists in the solid and probably is stable in solutions in certain polar aprotic solvents. There is very little chemistry as such of these species and other hexacoordinate phosphorus anions but their existence again brings into question the involvement of d orbitals in σ bonding.

The compound SF_6 perhaps highlights the problem even more so. How can a compound as stable as this, and it is one of the more inert of all chemical compounds, have a σ framework built from only one 3s and three 3p atomic orbitals? Such a framework would have only four m.o.'s that could be described as strongly bonding. In SF_6 these bonding m.o.'s are delocalized with a net electron drift to the fluorine atoms[12]. There are in addition also three weakly bonding m.o.'s; the inclusion of 3d orbitals increases the number of bonding orbitals to nine. The non-3d picture of the bonding seems too fragile to represent a molecule as unreactive as SF_6. However the stability of this compound stems not from the strength of its bonds but from its symmetrical octahedral structure. Indeed the bond enthalpy $E(S-F)$, SF_6 is only 328 kJ mol^{-1} (79 kcal mole^{-1}) making it exactly the same as that in SF_4 and this compound is very reactive chemically.

A similar bonding explanation holds for isoelectronic PF_6^- and other hexacoordinate phosphorus derivatives. As for 3d involvement in π bonds, this has very little relevance as yet to the debate since data is lacking on which to base proposals.

In $NaPF_6$ the anion is perfectly symmetrical with all the P—F bonds the same length, 158 pm. In the monohydrate, $NaPF_6 \cdot H_2O$, the anion is distorted by the formation of hydrogen-bonds to four of the fluorine atoms which lengthens their bonds to 173 pm. In the hexahydrate, $NaPF_6 \cdot 6H_2O$, all the bonds are hydrogen bonded and all have r(P—F) of 173 pm.

Stable salts of the anion (2.16) have been made and resolved into the enantiomers which such a system of three bidentate ligands surrounding a central atom should possess. The existence of this compound with its six phosphorus–carbon bonds seems to undermine the tenet which maintains that highly electronegative ligands are necessary for high coordination numbers and their stability.

2.8 The electronegativity (χ) of phosphorus and the electronic effects of attached groups

The word electronegativity appears in almost all chemical literature which attempts to explain the behaviour of molecules. Indeed it could be said that never have so many owed so much to so little. And it is very little. Many textbooks still quote Pauling's original table of numerical electronegativity values, in which for example phosphorus is given a value 2.2 in the mysterious units of energy$^{1/2}$, but even so most chemists in their use of χ talk only in terms of relative values assuming the order $F > O > N \simeq Cl > Br > C \simeq S \simeq I > H \simeq P$.

The concept of electronegativity is at present undergoing a fundamental reassessment[26], both in the way it is to be measured and the manner in which it is to be used. Perhaps the two most important changes which have been made are (i) that χ must be calculated in realistic units and that the units of potential, i.e. volts, are the best, and (ii) that χ of an atom varies according to its valence state. What this second point means is that χ increases with the fewer σ bonds an atom is capable of forming – another way of saying this is that χ increases with increasing s character so that $\chi(sp) > \chi(sp^2) > \chi(sp^3)$. Other important changes regarding χ are that it must be based on atomic properties, not on molecular ones such as bond moments or bond enthalpies, and moreover that in a molecule all atoms have the same electronegativity, which they attain as a result of electron redistribution. In other words χ which is quoted for atoms is translated into ionic character of bonds when those atoms are in molecules. The definition of electronegativity which best suits its modern conception is that *electronegativity is the potential of an atom in a given valence state to attract electrons to itself.*

The best χ values which meet the criteria above are the orbital electronegativities published by Hinze and Jaffé who developed Mulliken's method[27]. Unfortunately for our purpose they calculated χ values for phosphorus only in its ground state (tricoordinate, using only 3p orbitals) and its tetrahedral state, but not its pentacoordinate state. The values are quoted in Table 2.7 which includes χ values for most of the covalently bonding non-metals.

Although this table contains some surprises, especially regarding the way in which χ changes with valence state and the values of some states which become much more electronegative than fluorine in its ground state, it can be seen that almost all σ bonds to phosphorus will have the bonding pair attracted away from phosphorus and towards the other atom forming the bond. Phosphorus is more electronegative than only B(te) and As(p) when it is tricoordinate, but more electronegative than H, B, C(te), Si, As, Br and I when it is tetracoordinate.

Electronegativity differences for σ bonds may suggest the order of electron release or withdrawal which an atom attached to phosphorus may display. However it is more fruitful to seek other physical properties which will reveal an order of electron withdrawal. In theory many physical properties reveal electron-withdrawing tendencies, but few agree on a general order of electron-withdrawing power of a group with respect to phosphorus. Perhaps this is not unexpected in view of the profound bonding changes which take place depending upon whether phosphorus is tri-, tetra- or pentacoordinate. Even allowing for this a complete solution to the problem is not realized and the orders quoted in Table 2.8 for the more commonly met groups should only be taken as a rough guide.

For a particular order of groups in Table 2.8 it is difficult to assess at which point a group changes from being an overall electron-withdrawing group to an overall electron-releasing group.

Table 2.7 Orbital electronegativities[27] of the non-metals/(units of potential). Measured in electron volts but subjected to a conversion factor in order to bring the range of values into line with the Pauling range with H, *ca* 2.1 and F, *ca* 4.0. Based on H = 2.21.

M3	M4	M5	M6	M7
B	C	N	O	F
3.25, s; 1.26, p	4.84, s; 1.75, p	2.28, p	3.04, p	3.90, p
	3.29, di; 1.69π	5.07, di; 2.46π	6.60, di; 3.26π	
1.93, tr	2.75, tr; 1.68π	4.13, tr; 2.49π	5.54, tr; 3.19π	
1.81, te	2.48, te	3.68, te	4.93, te	
	Si	P	S	Cl
	3.88, s; 1.82, p	1.84, p	2.28, p	2.95, p
	2.25, te	2.79, te	3.21, te	
		As	Se	Br
		1.59, p	2.18, p	2.62, p
		2.58, te	3.07, te	
			Te	I
			2.08, p	2.52, p
			3.04, te	

di = sp, tr = sp², te = sp³

Problems

Table 2.8 Relative electron-withdrawing effects of common groups attached to phosphorus.

Tricoordinate phosphorus:
    OPh > F ≃ Cl > Br > I > H > OR(R = alkyl) > NR$_2$ > Ph ≃ Me > Et > t-Bu

Tetracoordinate phosphorus:
    F ≃ CF$_3$ > OPh > OMe ≃ Cl > Br > Et > NMe$_2$ > Ph ≃ Me > t-Bu

Pentacoordinate phosphorus:
    F > H > CF$_3$ > OPh > Cl > SMe > OMe > NMe$_2$ > Me > Ph

strongly electron withdrawing ←————————————→ electron releasing

The results are based on a variety of data although the order for the pentacoordinate case is based on apicophilicities and should be treated as the least reliable, especially as it involves other factors such as steric hindrance, and bearing in mind that the apical opportunities for π back-bonding are much less than equatorial ones as the dipole bond moments for P—F and P—Cl showed.

Problems

1 Deduce the relationship between the bond angle, α, of a trigonal pyramidal molecule PX$_3$ and the angle, β, which each bond makes with the molecule's principal axis. Calculate the bond angle, γ, at which the resultant of the three bond moments μ(P—X) is itself equal to μ(P—X).

2 What is the minimum number of pseudorotation pathways necessary to interconvert the isomers

(i)
$$\begin{array}{c} A \\ | \\ B-P \\ | \\ E \end{array} \!\!\!\! \diagdown \!\! C \atop D \quad \text{and} \quad \begin{array}{c} A \\ | \\ B-P \\ | \\ E \end{array} \!\!\!\! \diagdown \!\! D \atop C \quad \text{and (ii)} \quad \begin{array}{c} A \\ | \\ A-P \\ | \\ D \end{array} \!\!\!\! \diagdown \!\! B \atop C \quad \text{and} \quad \begin{array}{c} A \\ | \\ A-P \\ | \\ D \end{array} \!\!\!\! \diagdown \!\! C \atop B$$

3 Draw the pathway graph for pseudorotation of a pentacoordinate phosphorus molecule with two four-membered rings numbered thus

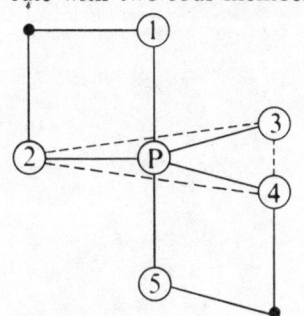

in which these cannot be placed diequatorial.

References

[1] R. J. Gillespie, *Molecular Geometry*, Van Nostrand Reinhold, London, 1972.

[2] D. E. C. Corbridge, *The Structural Chemistry of Phosphorus Compounds, Topics in Phosphorus Chemistry*, Vol 3, Interscience, New York, 1966.

[3] W. E. Dasent, *Inorganic Energetics*, Penguin, London, 1970.

[4] B. J. Walker, *Organophosphorus Chemistry*, Penguin, London, 1972.

[5] G. J. Moody and J. D. R. Thomas, *Dipole Moments in Inorganic Chemistry*, Arnold, London, 1971.

[6] P. Mauret and J-P. Fayet, *Bull. Soc. Chim. Fr.*, 2363 (1969).

[7] A. L. McClellan, *Tables of Experimental Dipole Moments*, Freeman & Co., San Francisco, 1963.

[8] M. L. Tobe, *Inorganic Reaction Mechanisms*, chap. 8, Nelson, London, 1972.

[9] W. C. Marsh, T. N. Ranganathan, J. Trotter and N. L. Paddock, *Chem. Commun.*, 815 (1970).

[10] R. R. Holmes, *J. Amer. Chem. Soc.*, **96**, 4143 (1974).

[11] J. A. Howard, D. R. Russell and S. Trippett, *Chem. Commun.*, 856 (1973).

[12] K. A. R. Mitchell, *Chem. Rev.*, **69**, 157 (1969).

[13] R. Hoffmann, J. M. Howell and E. L. Muetterties, *J. Amer. Chem. Soc.*, **94**, 3047 (1972).

[14] P. Gillespie, P. Hoffman, H. Klusacek, D. Marquarding, S. Pfohl, F. Ramirez, E. A. Tsolis and I. Ugi, *Angew. Chem., Int. Ed. Eng.*, **10**, 687 (1971).

[15] R. S. Berry, *J. Chem. Phys.*, **32**, 933 (1960).

[16] F. H. Westheimer, *Accounts Chem. Res.*, **1**, 70 (1968).

[17] I. Ugi and F. Ramirez, *Chem. Brit.*, 198 (1972).

[18] K. Mislow, *Accounts Chem. Res.*, **3**, 321 (1970).

[19] E. L. Muetterties, W. Mahler and R. Schmutzler, *Inorg. Chem.*, **2**, 613 (1963).

[20] N. J. De'Ath, D. Z. Denney and D. B. Denney, *Chem. Commun.*, 272 (1972).

[21] R. R. Holmes and C. J. Hora, *Inorg. Chem.*, **11**, 2506 (1972).

[22] K. E. DeBruin, A. G. Padilla and M. I. Campbell, *J. Amer. Chem. Soc.*, **95**, 4681 (1973).

[23] R. K. Oram and S. Trippett, *J. Chem. Soc., Perkin I*, 1300 (1973).

[24] W. C. Archie Jr. and F. H. Westheimer, *J. Amer. Chem. Soc.*, **95**, 5955 (1973).

[25] J. I. Musher, *Angew. Chem. Int. Ed. Eng.*, **8**, 54 (1969).

[26] J. E. Huheey, *Inorganic Chemistry—Principles of Structure and Reactivity*, pp 155–179, Harper & Row, New York, 1972.

[27] J. Hinze and H. H. Jaffé, *J. Amer. Chem. Soc.*, **84**, 540 (1962).

Chapter 3
³¹P n.m.r. and vibrational spectra of phosphorus compounds

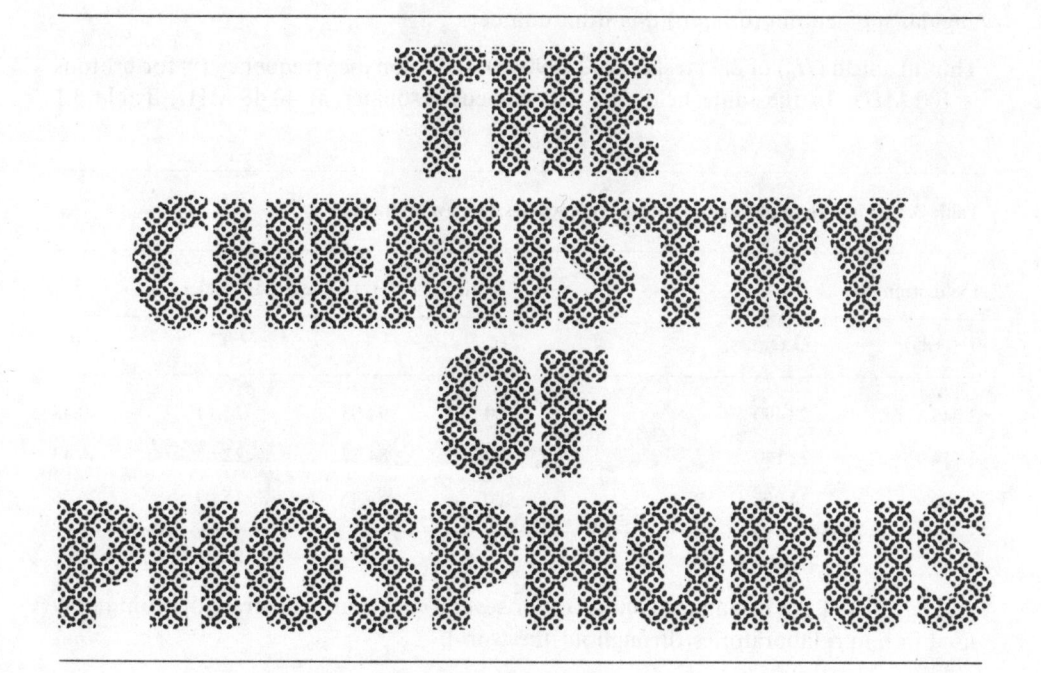

3.1 ^{31}P nuclear magnetic resonance[1-7]

The prerequisite for the observation of a nuclear magnetic resonance (n.m.r.) spectrum is that the nucleus under investigation should possess a magnetic moment. All nuclei with a spin quantum number (I) which is integral or half-integral (i.e. non-zero) fall into this category. Thus, nuclei such as ^{1}H, ^{19}F, ^{13}C and ^{31}P, all of which have $I = \frac{1}{2}$, satisfy the magnetic resonance condition and when placed in a magnetic field of suitable strength and then irradiated at the appropriate radiofrequency, they give rise to nuclear magnetic resonance signals.

The mathematical expression relating frequency (v) and the magnetic field strength (H_0) is shown in eqn. (1)

$$v = \frac{\gamma H_0}{2\pi} \qquad (1)$$

where γ is a constant known as the *magnetogyric ratio* which is a characteristic of the particular nucleus under investigation. It is defined by eqn. (2)

$$\gamma = \frac{2\pi}{h} \frac{\mu}{I} \qquad (2)$$

where h is Planck's Constant, μ is the magnetic moment of the nucleus and I is the angular momentum (or spin) quantum number.

Thus in a field (H_0) of 2.35 tesla (T) (23 490 G), the resonance frequency (v) for protons is 100 MHz. In the same field, the ^{31}P nucleus resonates at 40.48 MHz. Table 3.1

Table 3.1 Field strengths and resonance frequencies for several nuclei

Field strength		Frequency (MHz) for various nuclei			
T (tesla)	G (gauss)	^{1}H	^{19}F	^{13}C	^{31}P
2.349	23 490	100	94.08	25.14	40.48
2.114	21 140	90	84.67	22.63	36.43
1.409	14 090	60	56.45	15.09	24.28

gives details of the resonance condition for several nuclei at field strengths commonly used in n.m.r. laboratories throughout the world.

The ^{31}P nucleus occurs in 100% natural abundance (i.e. is the only stable isotope of phosphorus) and would therefore appear to be ideally suited for n.m.r. studies. However, due to a low magnetic moment, μ, the sensitivity of the ^{31}P nucleus is only 0.066 relative to that of ^{1}H. This low sensitivity has been offset to some extent by the use of large (10 or 13 mm) sample tubes and by spin-decoupling techniques which collapse (often to a single line) the complex spectra resulting from coupling with other magnetically active nuclei in the molecule. Recently the incorporation of pulsed

3.1 ^{31}P nuclear magnetic resonance

Fourier Transform techniques has increased the scope of ^{31}P n.m.r. enormously, enabling even low concentrations of molecules from biological sources to be examined. Since phosphorus plays such a critical role in many life processes (see Chapters 1 and 12), this promises to be an area of growing interest.

There are four important factors to be considered in the interpretation of an n.m.r. spectrum:

(i) the chemical shift, δ;
(ii) the coupling with other magnetically active nuclei, giving rise to characteristic coupling constants, J;
(iii) the n.m.r. peak area which is proportional to the relative amount of the active nuclei giving that particular peak, and
(iv) the line shape of the transition which is measured by the peak width at half height ($v_{1/2}$). The line shape is governed by the life-time of the excited state in accordance with Heisenberg's uncertainty principle and is characterized by a first-order rate constant T_2 which measures the time required for the spin system under investigation to re-establish its equilibrium magnetization in the direction of a radiofrequency perturbation after the perturbation has been removed. Some contribution to the line width is caused by the inhomogeneity of the applied magnetic field and further broadening of the line may occur if the presence of quadrupole moments within the molecule being examined allow more rapid return of the spin system to equilibrium. Of considerably greater chemical interest however, is the effect on the line shape of intra- or intermolecular exchange processes within the system (e.g. pseudorotation) which occur at a rate close to that associated with the particular frequency of the n.m.r. experiment.

All four factors will receive some attention in the subsequent discussion but emphasis will be placed on the use of δ and J as means of determining molecular structure in phosphorus compounds.

3.1.1 Phosphorus chemical shifts

Chemical shift (δ) is simply a measure of the position of a resonance signal relative to an arbitrary standard. The currently accepted standard in ^{31}P n.m.r. work is 85% phosphoric acid* (H$_3$PO$_4$) and the units of δ are parts per million (ppm) defined as follows:

$$\delta(\text{ppm}) = \frac{v_{\text{exp}}(\text{in Hz}) - v_{\text{H}_3\text{PO}_4}(\text{Hz})}{\text{frequency of spectrometer (MHz)}}$$

Generally speaking, the n.m.r. experiment is conducted either by varying the field strength at fixed frequency (field sweep) or by varying the frequency at fixed field (frequency sweep). In field sweep experiments, resonance signals which appear at *higher* field than H$_3$PO$_4$ (which corresponds to lower frequency signals in the frequency sweep mode) are, by convention, given *positive* δ values; conversely, signals at lower field than H$_3$PO$_4$ are given *negative* δ values. It should be noticed that this is the reverse

---

* From the practical point of view, P$_4$O$_6$ gives sharper signals than H$_3$PO$_4$ and is sometimes used as the experimental standard, particularly with older spectrometers; P$_4$O$_6$ resonates at 112 ppm lower field, relative to H$_3$PO$_4$.

Chapter 3 / ^{31}P n.m.r and vibrational spectra of phosphorus compounds

Table 3.2 ^{31}P Chemical Shifts of P(III) compounds (ppm from H$_3$PO$_4$)

Compound	δ	Phenyl analogue	δ
PH$_3$	+240		
MePH$_2$	+164	PhPH$_2$	+122
Me$_2$PH	+ 99	Ph$_2$PH	+ 41
Me$_3$P	+ 62	Ph$_3$P	+ 6
Me$_2$PCl	− 94	Ph$_2$PCl	− 81
MePCl$_2$	−191	PhPCl$_2$	−162
Me$_2$POMe	−201		
(MeO)$_3$P	−141	(PhO)$_3$P	−127
(MeS)$_3$P	−125	(PhS)$_3$P	−132
(Me$_2$N)$_3$P	−123		
Cl$_3$P	−219		
Br$_3$P	−227		
I$_3$P	−178		
F$_3$P	− 97		

Compound	δ	Compound	δ

Heterocyclic phosphines and phosphites

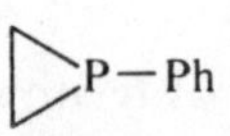

▷P—Ph	+234	⌐O\POEt / O⌐	−132
Me$_4$-cyclobutyl-P-Ph	−8.5	⌐O\POEt / O⌐ (6-ring)	−130
cyclopentyl P—Ph	+ 15		
		Me-N, N-Me cyclic PNMe$_2$	−115
cyclohexyl P—Ph	+ 23		

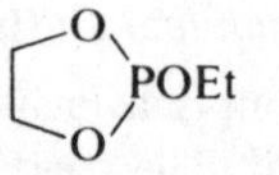

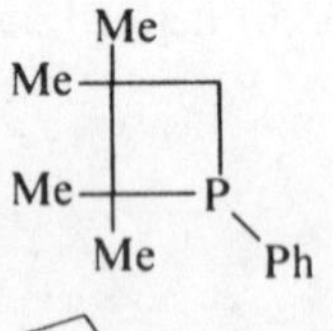

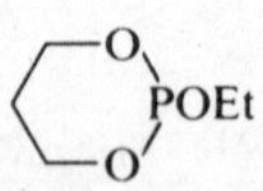

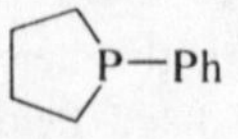

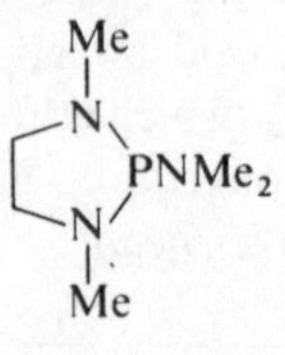

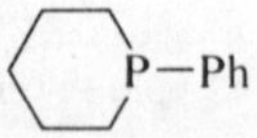

3.1 ^{31}P nuclear magnetic resonance

of the convention used in ^{1}H n.m.r. and for this and other, mathematical, reasons this convention for ^{31}P n.m.r. data has not found universal acceptance and is particularly offensive to some n.m.r. specialists. Nevertheless, most of the compilations of ^{31}P n.m.r. data use the high field $\equiv$ positive convention and it has therefore been adopted throughout this book.

As we shall see, the range of ^{31}P δ values is wide (*ca* $-$ 250 ppm to *ca* $+$ 500 ppm) and the average line width is *ca* 0.7 Hz. Since ^{31}P n.m.r. spectra often encompass a field sweep of 500 ppm (or the equivalent frequency sweep) it is not surprising that the spectra appear as very sharp lines and under these conditions, unless the *J* values exceed 0.5 ppm, even highly coupled spectra appear as singlets.

Two factors are thought to be largely responsible for the wide range of ^{31}P chemical shifts. First, electrons in the 3p and sometimes in the 3d orbitals are alleged to contribute to the shielding and second, in most of its compounds, phosphorus displays coordination numbers from three to six. The coordination number determines the shape of a given phosphorus molecule and in particular affects the electronic environment of the phosphorus atom within that molecule. Furthermore, substituents effectively protect the phosphorus atom from close-range magnetic and electronic effects due to surrounding (e.g. solvent) molecules which results in very small solvent effects on ^{31}P δ values. For instance, trimethyl phosphate, $(MeO)_3PO$, exhibits a total range of from $+1.0$ to $+2.7$ ppm in thirteen different solvents. In the usual organic solvents (except ethanol, for which $\delta = +1.0$) the range is from 1.9 to 2.1 ppm[2]. Likewise the ^{31}P δ for $Ph_3PMe^+Br^-$ ranges from -20.5 to -22.7 ppm in five different solvents with the extremes corresponding to water and dimethyl sulphoxide, respectively[2].

Typical ^{31}P δ values for tricoordinate, tetracoordinate and penta- or hexacoordinate phosphorus compounds are shown in Tables 3.2, 3.3 and 3.4 respectively. These lists are by no means exhaustive and are simply intended to show some representative values. For more comprehensive coverage, the reader is referred to the bibliography at the end of this chapter. Table 3.5 is an attempt to indicate the range of ^{31}P δ values for various types of *equivalently substituted* phosphorus molecules (e.g. PR_3). There are of course, exceptions to these ranges, and reference to Table 3.5 alone reveals a high degree of overlap between the valency states of phosphorus. Thus, caution is necessary in the use of ^{31}P δ values as diagnostic tools for structure determination.

Several rules of thumb may be deduced from the empirical data however, and these are as follows:
   (i) with $\delta < -100$ ppm, the substance is a P(III) compound bearing at least one electronegative substituent (e.g. N, O, S or halogen);
   (ii) if the compound is *known* to be *trivalent*, positive values are indicative of a phosphine, PR_3, where R = alkyl, aryl or H;
   (iii) phosphonium salts and phosphorus ylids generally fall in a narrow range of δ values from -30 to -5 ppm;
   (iv) penta- and hexacoordinate phosphorus compounds generally have *positive* δ values.

Until recently, rule (iv) was regarded as pretty well inviolate but as the range of known pentacoordinate compounds has expanded, exceptions have been reported. For

Table 3.3 ^{31}P Chemical shifts of tetracoordinate compounds (ppm from H_3PO_4)

Compound	δ	Phenyl analogue	δ
Phosphoryl compounds			
$(MeO)_3P=O$	− 2	$(PhO)_3P=O$	+18
$(EtO)_3P=O$	+ 1		
$(tBuO)_3P=O$	+ 14		
$PO_4^{3-}(3K^+)$	− 6		
$(MeO)_4\overset{+}{P}\ BF_4^-$	− 2		
$(Me_2N)_3P=O$	− 23	$(Ph_2N)_3P=O$	−2
$(EtS)_3P=O$	− 61	$(PhS)_3P=O$	−55
$Et_3P=O$	− 48	$Ph_3P=O$	−25
$nBu_3P=O$	− 43		
$Cl_3P=O$	− 3		
$Br_3P=O$	+103		
$F_3P=O$	+ 36		
Thiophosphoryl compounds			
$(MeO)_3P=S$	− 73	$(PhO)_3P=S$	−53
$(Et_2N)_3P=S$	− 77		
$(EtS)_3P=S$	− 92	$(PhS)_3P=S$	−92
$Et_3P=S$	− 54	$Ph_3P=S$	−42
$Cl_3P=S$	− 30		
$Br_3P=S$	+112		
Phosphonium salts			
$Me_4\overset{+}{P}\ Br^-$	− 25	$Ph_4\overset{+}{P}\ I^-$	−22
		$Ph_3\overset{+}{P}Me\ Br^-$	−23
		$Ph_3\overset{+}{P}CH_2Ph\ Br^-$	−23
		$Ph_3\overset{+}{P}CH_2COPh\ Br^-$	−17
		$Ph_3\overset{+}{P}CH_2CO_2Et\ Br^-$	−20
Phosphorus ylids			
$Et_3P=CH_2$	− 24	$Ph_3P=CH_2$	−20
$Et_3P=CHMe$	− 17	$Ph_3P=CHMe$	−15
$Et_3P=CHEt$	− 15	$Ph_3P=CHCOPh$	−22
		$Ph_3P=CHCOEt$	−19
		$Ph_3P=\text{cyclopentadienylidene}$ 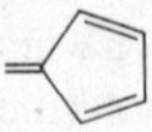	−12
Phosphazenes			
$(NPCl_2)_3$	− 20		
$(NPCl_2)_4$	+ 7		
$(NPCl_2)_5$	+ 17		
$(NPMe_2)_3$	− 31	$(NPPh_2)_3$	−14
$[NP(OMe)_2]_3$	− 22	$[NP(OPh)_2]_3$	− 9

3.1 ^{31}P nuclear magnetic resonance

Table 3.4 ^{31}P chemical shifts of pentacoordinate and hexacoordinate compounds (ppm from H_3PO_4)

Compound	δ	Compound	δ
Pentacoordinate compounds (acyclic)			
PF_5	+ 35		
PCl_5	+ 80		
PBr_5	+101	Ph_5P	+89
$MePF_4$	+ 30	$PhPF_4$	+52
$(EtO)_5P$	+ 71	$(PhO)_5P$	+86
Me_2NPMeF_3	+ 37	Me_2NPPhF_3	+53
$nBu_3P(OEt)_2$	+ 38	$Ph_3P(OEt)_2$	+54
Pentacoordinate compounds (heterocyclic)			
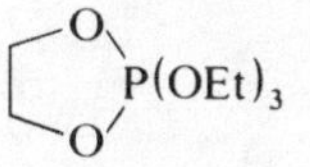 ⟨O-O⟩P(OEt)$_3$ (5-ring dioxy)	+ 52	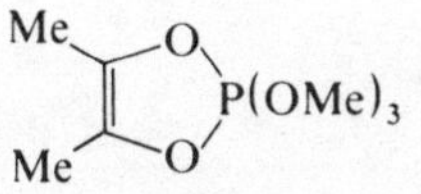 Me-C=C(Me)-O-O-P(OMe)$_3$	+49
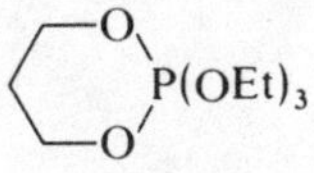 ⟨O-O⟩P(OEt)$_3$ (6-ring dioxy)	+ 72	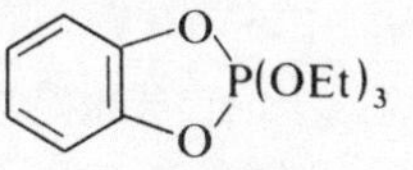 benzo-dioxy-P(OEt)$_3$	+50
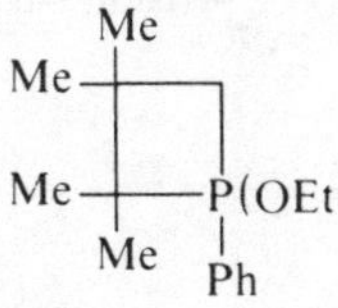 Me$_2$C-CMe$_2$-P(OEt)$_2$Ph (cyclobutane)	+ 36	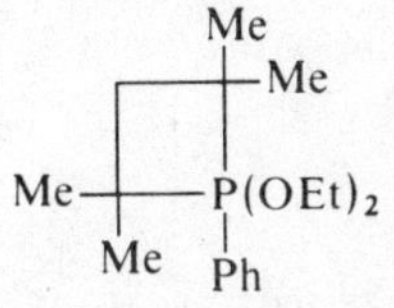 Me$_2$C-CMe$_2$-P(OEt)$_2$Ph	+22
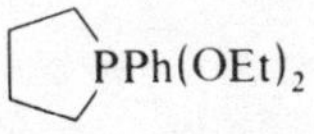 cyclopentyl-PPh(OEt)$_2$	+ 12	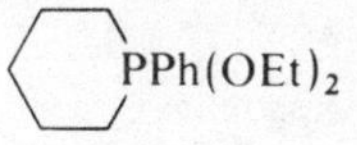 cyclohexyl-PPh(OEt)$_2$	+48
Hexacoordinate compounds			
$pClC_6H_4N_2^+$ PF_6^-	+144	PCl_4^+ PCl_6^- (α, β)	α, −96; β, +281
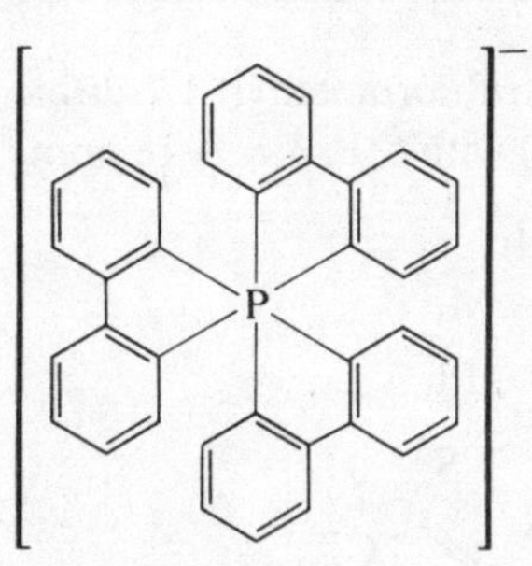 [P(biphenyl)$_3$]$^-$ Li$^+$	+181	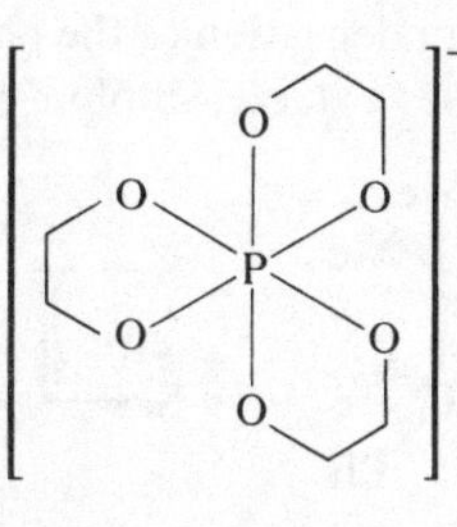 [P(OCH$_2$CH$_2$O)$_3$]$^-$ Na$^+$	+89

Chapter 3 / ^{31}P n.m.r and vibrational spectra of phosphorus compounds

Table 3.5 Range of chemical shifts for equivalently substituted phosphorus compounds

Compound	Range of δ	Overall range
Tricoordinate compounds		
(—O)$_3$P	−145 to −125	
(>N)$_3$P	−135 to −120	
(—S)$_3$P	−135 to −115	−145 to +70
(>C)$_3$P	0 to +70	
X$_3$P	−230 (Br) to −100 (F)	
Tetracoordinate compounds		
P=O		
(—O)$_3$P=O[a]	−10 to +20	
(>N)$_3$P=O	−25 to +5	
(—S)$_3$P=O	−85 to −60	−85 to +20
(>C)$_3$P=O	−50 to −20	
X$_3$P=O	−10 (Cl) to +105 (Br)	
P=S		
(—O)$_3$P=S[b]	−75 to −35	
(>N)$_3$P=S	−80 to −60	
(—S)$_3$P=S	−100 to −90	−100 to −35
(>C)$_3$P=S	−60 to −50	
X$_3$P=S	−30 (Cl) to +110 (Br)	
Phosphonium salts and phosphorus ylids	−30 to −5	
Pentacoordinate compounds		
Z$_5$P	0 to +100	

[a] Includes phosphate anions; [b] Includes thiophosphate anions.

instance, condensation of the phosphetan (3.1) with 3,4-bis(trifluoromethyl)-1,2-dithieten (3.2) gives a stable, sulphur-containing phosphorane (3.3) with ^{31}P, δ = −18 ppm.[8]

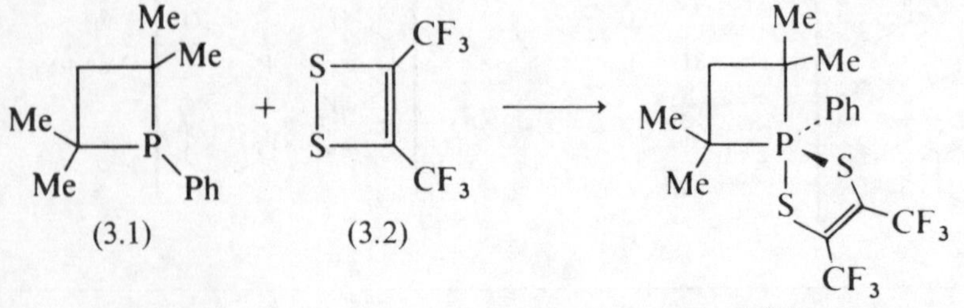

Various theoretical calculations have been carried out in attempts to predict ^{31}P chemical shifts[2]. Such calculations are complex and outside the scope of this book; suffice it to say that none, so far, has been entirely satisfactory. Likewise, attempts at empirical correlations have met with difficulty. One might expect a linear change of δ with stepwise replacement of one substituent by another (based on the δ values for equivalently substituted phosphorus compounds) and this has been shown to be valid in a few series of compounds when the range of δ values is small. Thus Grim et al.[9-12] have been able to use group contributions* from alkyl and aryl groups to predict δ values for a wide range of mixed substituent, primary (RPH$_2$), secondary (R$_2$PH) and tertiary (R$_3$P where R = alkyl or aryl) phosphines as well as tertiary (R$_3$PH$^+$) and quaternary (R$_4$P$^+$) phosphonium ions. However, for many other non-equivalently substituted phosphorus compounds, δ turns out to be *lower* (i.e. more negative) than expected.

To a first approximation it is the substituents attached *directly* to the phosphorus atom which dictate the chemical shift. The effect which a change in these substituents has on δ may be viewed in terms of (i) the change in electronegativity relative to phosphorus, (ii) the effect on bond angle and (iii) the effect on the occupation of the phosphorus dπ orbitals, when one substituent is replaced by another. It may be significant that the best empirical correlations are obtained for phosphines (R$_3$P) and phosphonium ions (R$_4$P$^+$) in which contributions to the bonding from the dπ orbitals on phosphorus are least important.

One further point is worthy of note. Reference to Table 3.3 reveals that the *charge* on the phosphorus molecule with phosphorus in a given valence state, has very little effect on the magnetic shielding at the phosphorus atom. Thus the neutral trimethyl phosphate, (MeO)$_3$PO, has a very similar δ value to that of the phosphate anion, PO$_4^{3-}$, and the tetramethoxyphosphonium ion, (MeO)$_4$P$^+$.

3.1.2 Coupling constants

When a molecule contains more than one magnetically active nucleus and the nuclei are in different chemical (and magnetic) environments, the resonance signal due to a particular nucleus may show fine structure. Thus in a phosphorus compound, individual ^{31}P peaks may be split into multiplets of smaller peaks of the same total intensity and the effect is known as "spin–spin splitting." The distance, in Hertz, between adjacent peaks in a multiplet is a measure of the strength of the interaction and is denoted by a coupling constant, J. The phenomenon is transmitted through the electronic structure of the molecule and the largest J values are observed for adjacent nuclei. Attenuation of the splitting occurs as the separation of the interacting nuclei increases and the practical limit of detection is generally for a separation of no more than three bonds.

Splitting constants between the ^{31}P nucleus and a wide variety of magnetically active nuclei are on record and numerous examples appear in Tables 3.6, 3.7 and 3.8. From the point of view of structure elucidation, spin–spin splitting (or "coupling") between the ^{31}P nucleus and ^{1}H or ^{19}F nuclei are of prime importance. With the advent of

---

* Contributions for a particular group relative to an arbitrary standard are obtained empirically, then used to calculate δ for an unknown.

Table 3.6 Spin–spin coupling constants, J_{PH}/(Hz), between ^{31}P and ^{1}H nuclei

Type of compound	J	Type of compound	J
H bonded directly to phosphorus ($^{1}J_{PH}$)			
>P—H	180–225	>P⁺—H	490–600
>P(=O)H	490–760	>P(—H)(—)	700–1000
>P(=S)H	490–650		
P—CH< coupling ($^{2}J_{PH}$)			
>P—CH₃	1–5	>P(=O)CH=	
>P(=O)CH₃	7–15		15–30
		>P(=O)CH—Ph	
>P(=S)CH₃	11–15		
>P⁺—CH₃	12–17	>P(—)—CH₃	ca. 10
P—X—CH< coupling ($^{3}J_{PH}$)			
P—OCH₃	6–15	P—NCH₃	4–17
P—SCH₃	5–20	P—C(—)—CH₃	0–2
Miscellaneous J_{PH} values			
P—NH—	13–28		
>P—C₆H₄ (phenyl)	ortho, 7–10		
	meta, 2–4		

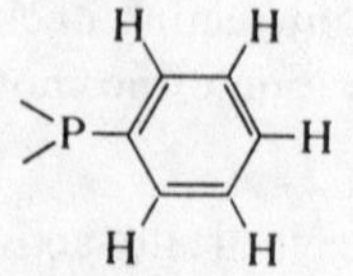

Fourier Transform spectroscopy, the natural abundance ^{13}C spectra of organophosphorus compounds are also gaining prominence and the coupling with phosphorus is in some cases (see below) proving invaluable to the assignment of stereochemical configuration. Since phosphines are commonly employed as ligands in inorganic complexes, coupling between phosphorus and certain transition metals is also important.

3.1 ^{31}P nuclear magnetic resonance

Table 3.7 Spin–spin coupling constants, J_{PF}/(Hz), between ^{31}P and ^{19}F nuclei

F bonded directly to phosphorus ($^1J_{PF}$)

Type of compound:	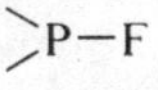 >P—F	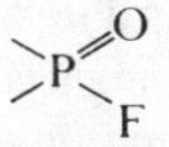 >P(=O)F	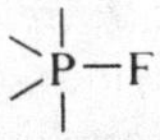 >P—F (with axial bonds)	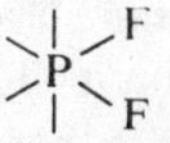 >PF$_2$ (octahedral)
J	900–1450	950–1350	600–1050 (apical)	600–700 (*trans*)
			800–1200 (equat)	800–850 (*cis*)

P—C—F ($^2J_{PCF}$) and P—C—C—F ($^3J_{PCCF}$) values

Type of compound:	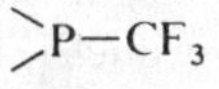 >P—CF$_3$	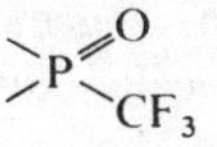 >P(=O)CF$_3$	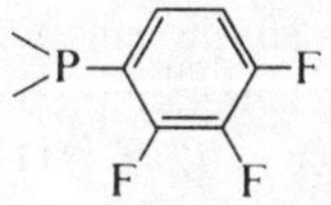 >P—C$_6$H$_2$F$_3$
J	60–90	ca 110	*ortho*, 0–60
			meta, 1–7
			para, 0–3

Table 3.8 Spin–spin coupling constants between ^{31}P and miscellaneous nuclei

Type of compound	J (Hz)	
>P—P<	220–400	($^1J_{PP}$)
>P(=O)—P(=O)<	330–500	($^1J_{PP}$)
>P(=O)—O—P(=O)<	15–20	($^2J_{POP}$)
P—^{13}C	35–250	($^1J_{PC}$)
P—C—^{13}C	0–40	($^2J_{PCC}$)
>P→M←P<	0–1100	($^2J_{PMP}$)
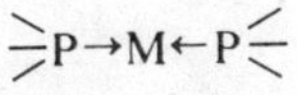 (M = transition metal)		
>P→^{63}Cu	1100–1300	($^1J_{PCu}$)
>P→^{195}Pt	2400–3900	($^1J_{PPt}$)
>P→B	50–300	($^1J_{PB}$)

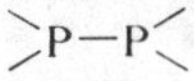

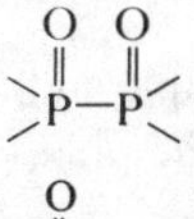

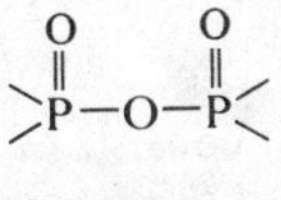

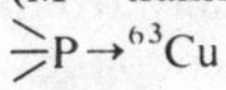

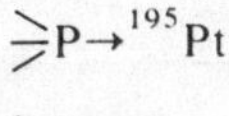

The multiplets in n.m.r. spectra due to spin–spin splitting fall into two categories: first order and second or higher order. First order spectra occur when the nuclei which are coupling with one another are separated by a chemical shift difference which is large compared to the coupling constant. Higher-order spin–spin splitting patterns are obtained when $\Delta\delta$ and J are of comparable size. Thus, most ^{31}P resonance signals exhibit first-order behaviour and this is invariably true for the coupling between ^{31}P and either ^{1}H or ^{19}F. Coupling between non-equivalent* phosphorus atoms of similar chemical shift may however, lead to higher-order spectra. A full analysis requires the use of wave mechanics but the spectra can be simulated by a computer generated fit to the experimental spectra using iterative procedures. We shall confine ourselves from hereon to a discussion of first-order spectra.

There is a simple rule governing the appearance of spin–spin multiplets. For a particular nucleus being split by n, magnetically equivalent nuclei, the observed pattern will consist of $(2nI + 1)$ lines, where I is the spin quantum number of the n nuclei effecting the splitting. Thus for ^{1}H or ^{19}F (for both of which, $I = \frac{1}{2}$) splitting a ^{31}P signal, $(n + 1)$ lines will be seen in the multiplet. A methyl group attached to phosphorus (P—CH$_3$) will therefore split the ^{31}P signal into four lines with relative intensities of 1:3:3:1. The relative intensities are given by the binomial coefficients and follow as a natural consequence of the number of ways in which n nuclei of $I = \frac{1}{2}$ may be distributed among two possible energy states in a magnetic field and the statistical probabilities of each possible distribution.

The line positions, coupling constants and signal intensities are all important features of spin–spin multiplets and can indicate the number of spin-coupled nuclei, their equivalence (or non-equivalence) and their relative positions in a molecule. Thus the

(MeO)$_2$P(H)=O   (MeO)$_2$POH   MeO-P(=O)(OMe)Me

(3.4a)   (3.4b)   (3.5)

splitting of the ^{31}P signal into a doublet with $J_{P-H} = 100$ Hz allows a simple and unambiguous assignment of the phosphonate structure (3.4a) rather than the phosphite structure (3.4b) for dimethylhydrogenphosphonate.

The first-order ^{31}P n.m.r. spectrum of dimethylmethylphosphonate (3.5) consists of twenty-eight lines due to the phosphorus signal being split into a quartet (intensity ratio, 1:3:3:1) by the CH$_3$ group attached to phosphorus, ($J_{P-C-H} = 17.3$ Hz) and each quartet being split into a septet (intensity ratio 1:6:15:20:15:6:1) by the six equivalent hydrogens of the two methoxy groups, ($J_{P-O-CH_3} = 10.9$ Hz).

Reference to Table 3.6 reveals that P—H coupling constants are diagnostic of the valency state of phosphorus, the magnitude of J_{P-H} rising quite dramatically from P(III) compounds through to P(V). The effect is not evident with P—F couplings

---

* Nuclei in a molecule which are in identical chemical environments do not couple with each other. Thus the pyrophosphate tetra anion, $^{2-}$O$_2$P(O)OP(O)O$_2^{2-}$, shows only a singlet in its ^{31}P n.m.r. spectrum.

3.1 ^{31}P nuclear magnetic resonance

(Table 3.7) but in pentacoordinate compounds, J_{P-F} is sometimes useful in deciding whether fluorine is in an apical or an equatorial situation, the latter showing the higher range of J values.

Coupling between ^{31}P and ^{13}C nuclei in some cyclic phosphines has recently been shown to be stereospecific. For instance, 1,2-dimethylphosphol-3-ene may exist as the *cis* (3.6a) and *trans* (3.6b) isomers, depending upon the configuration at phosphorus.

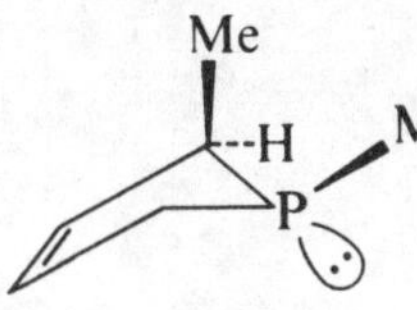

(3.6a)
cis $J_{PCCH_3} = 0$

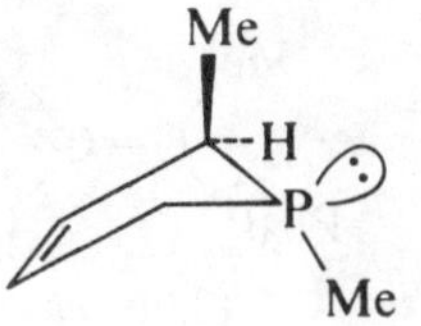

(3.6b)
trans $J_{PCCH_3} = 32$ Hz

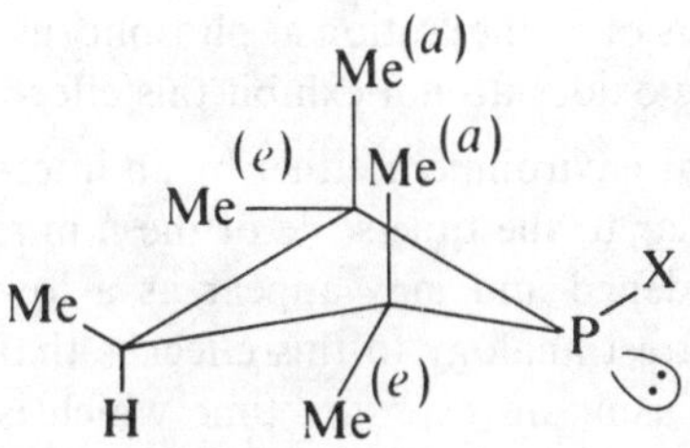

(3.7a)
cis

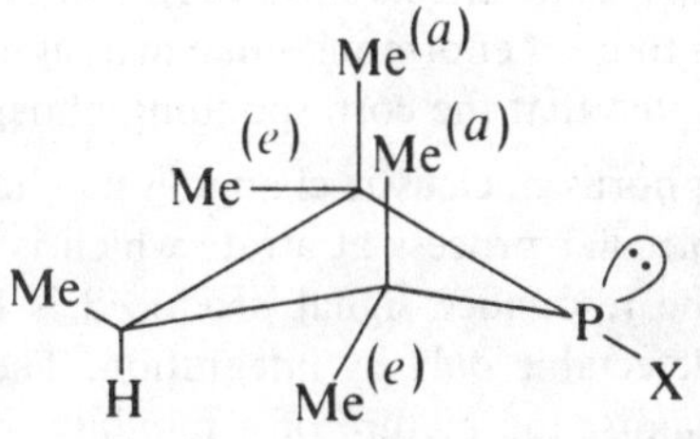

(3.7b)
trans

The ^{13}C spectrum of the *cis* isomer (3.6a) shows a singlet for the ring methyl group and a doublet for the *trans* isomer (3.6b). Likewise, the *cis*- and *trans*-pentamethylphosphetans (3.7a) and (3.7b) show stereospecific J_{PCC} values (Table 3.9), the ring

Table 3.9 J_{PCC} values (in Hz) in pentamethylphosphetans, MeHC(CMe$_2$)$_2$PX

	Isomer (3.7a)		Isomer (3.7b)	
	Axial methyls	Equatorial methyls	Axial methyls	Equatorial methyls
X = Ph	4.9	27.8	31.8	2.5
X = Me	4.3	26.9	30.5	2.1
X = Cl	2.5	33.5	37.1	0

methyl carbons which are *trans* to X showing much larger coupling constants with phosphorus[13]. Stereospecific J_{PC} values are also observed in the *cis* and *trans* forms of phosphorus ylids as evidenced by the data for (3.8) and (3.9)[14]. Similar, but less

```
     H          O⁻                    H          OEt
      \        /                       \        /
       C=C                              C=C
      /        \                       /        \
  Ph₃P⁺        OEt                 Ph₃P⁺         O⁻

    (3.8a) trans                    (3.8b) cis
    J_PC = 117                      J_PC = 130


     Me         O⁻                    Me         OEt
      \        /                       \        /
       C=C                              C=C
      /        \                       /        \
  Ph₃P⁺        OEt                 Ph₃P⁺         O⁻

    (3.9a) trans                    (3.9b) cis
    J_PC = 121                      J_PC = 128
```

dramatic stereospecific effects have been found for P—H coupling and the phenomenon promises to be of enormous value in assignments of configuration at phosphorus. It should be noted that the corresponding phosphine oxides do not exhibit this effect.

When a phosphorus nucleus is changing its chemical environment either by an inter- or an intramolecular process at a rate which is similar to the time scale of the n.m.r. experiment, the resonance signal produced is broadened and may appear as a low hump or be detectable only by integration. The nearest analogy to this effect is that of a camera taking the picture of a moving object using an exposure time which is approximately equal to the time it takes the object to cross the field of view. The result, of course, is a blurred and possibly unidentifiable image. To revert to the n.m.r. phenomenon, if the molecular process is very fast relative to the n.m.r. time-scale, an averaged (but sharp) signal will result but if the molecular process is very slow relative to the n.m.r. time-scale, signals due to both molecular species will be recorded. Hence in order to sharpen an n.m.r. signal which is broadened by exchange one may either cool the sample to slow down the molecular process and so obtain signals that are not broadened by exchange or conversely, warm the sample to accelerate the molecular process and obtain the average picture. The time-scale of the n.m.r. phenomenon is such that molecular processes with rate constants within the range from 10^{-1} to 10^5 s^{-1} may be studied.

An example of such behavior is provided by the difluorophosphorane (3.12a–d) derived from the phosphetan (3.10) and bis-trifluoromethyl peroxide (3.11)[15]. At room temperature the ^{31}P n.m.r. appears as a broad "mound" centred at ca +10 ppm At −100 °C however, the signal separates into two sharp triplets at −3 ppm, J_{PF} = 932 Hz and +30 ppm, J_{PF} = 769 Hz; integration reveals that the ratio of these two ^{31}P n.m.r. signals is 2.3 : 1.0. These observations may be explained in terms of pseudorotation (see Chapter 2) between (3.12a or 3.12b) and (3.12d) via (3.12c). At room temperature this occurs at a rate comparable with the time-scale of the n.m.r. experiment.* At low temperature, the pseudorotation process (ψ) becomes slow and the separate signals represent (3.12a/3.12b) and (3.12d) in a ratio of 2.3 : 1.0. The assign-

---

* An approximate value of k_c, the rate constant at coalescence, may be calculated from the expression, $k_c = \pi \Delta \nu / 2^{1/2}$, where $\Delta \nu$ is the maximum chemical shift difference (in Hz) between the separate signals; hence in this case, $k_c = \pi \times 1340/2^{1/2} = 3 \times 10^3$ s^{-1}.

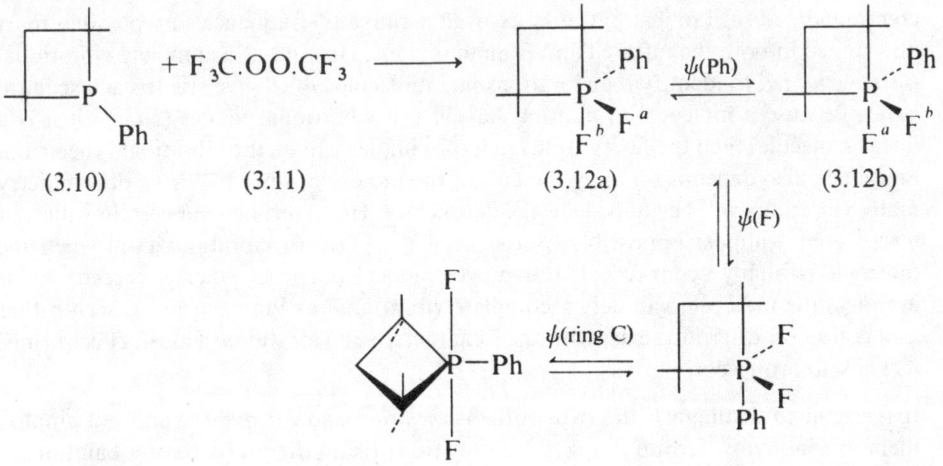

(3.10) (3.11) (3.12a) (3.12b) (3.12d) (3.12c)

ment of the signal at +30 ppm to (3.12d) is made on the basis of the *smaller* J_{PF} value for the two apical fluorines. Pseudorotamer (3.12a) undergoes an independent ψ using the phenyl group as pivot; this is a fast process and serves to equilibrate F^a and F^b and hence produce a J_{PF} value of 932 Hz which is the average of apical and equatorial P—F coupling. This interpretation is supported by ^1H and ^{19}F n.m.r. data. One is led to the conclusion that (3.12d) is only slightly less stable than (3.12a) and in view of the generalizations of Chapter 2 which referred to the extra stability afforded by an apical–equatorial disposition of small rings, this may seem surprising. Close inspection however, reveals that the diequatorial disposition of the ring in (3.12d) is compensated by *two* apical fluorines as compared with *one* apical fluorine in (3.12a) and (3.12b).

3.2 Vibrational spectra of phosphorus compounds

There are three kinds of chemist who study vibrational spectra – the preparative chemist who seeks to identify a product or by-product of his research; the spectroscopist whose aim is to probe the interactions between matter and radiation; and the physical chemist who can use absorption bands to follow the changes taking place in a reaction. We shall be concerned with the first and second types.

The preparative chemist often consults the infrared spectrum of a compound. He uses it commonly as a fingerprint to confirm his suspicions that a compound is what he thinks it is. The most demanding use to which he puts vibrational spectra is in the study of a new compound. Depending upon the nature of this he may need to delve into the far infrared, and even Raman spectroscopy, to get at the truth. However, no preparative chemist relies entirely on vibrational spectroscopy in discovering the identity of a new derivative. Indeed, other techniques such as n.m.r. spectroscopy may be far more informative, and in this event then he may consult only the 625–4000 cm^{-1} infrared spectrum. To interpret this the investigator needs only a list of infrared *correlations* which will enable him to link particular absorption bands to associated groups in the molecule.

The spectroscopist on the other hand is interested in the vibrational spectra of a compound *per se*. For his purpose, as wide a range of frequencies as possible must be scanned in both the infrared and Raman regions. He aims at a complete vibrational *assignment*, i.e. to identify the vibrations in a molecule which give rise to each spectral band. Because a molecule of n atoms has $3n - 6$ vibrational modes ($3n - 5$ if it is a linear molecule) then the fewer atoms in it the simpler will be the vibrational spectrum. Simplicity also depends on the symmetry of the molecule, since by virtue of symmetry many vibrations will be equivalent, i.e. degenerate. However, beyond certain values of n it becomes almost impossible to assign all the $(3n - 6)$ vibrations even when the molecule is highly symmetrical. For most practical purposes when n exceeds 15 or even less the molecule will defy a complete vibrational assignment and is seen rather as a collection of groups each with their characteristic vibrations. This in effect brings us back to group correlations.

It is useful to distinguish the two words *correlation* and *assignment* and not employ them as synonyms. Group frequency correlations relate the presence of a band or set of bands in the spectrum to the presence of a particular group of atoms in the molecule. A vibrational assignment identifies a band as arising from a particular vibrational mode of the molecule. Such molecular vibrations may centre round the deformation a particular bond or angle and consequently the absorption may be labelled as a stretching of this bond or a bending of that angle. Another method of notation is to label vibrations by the symmetry of the molecule when it is undergoing the vibration. Although more accurate as a system of notation this terminology is not sufficiently descriptive for most chemists, who prefer the former type of notation.

3.2.1 Group frequency correlations of phosphorus compounds

Certain correlations are invaluable in phosphorus chemistry, e.g. the groups P—CH$_3$ and P—OR can be linked to easily identified peaks which always appear in more or less the same place in the infrared spectrum. Such peaks are sharp and intense and the experienced eye can spot them and interpret them immediately. Other bonds or groups give bands in the spectrum which fall anywhere within a wide range of frequencies, are broad and of weak intensity. Group correlations are of little help in these cases and unfortunately the P—C and P—N bonds are of this type. This chapter is not the place to deal with the vagaries of infrared correlations of phosphorus compounds and the reader is referred to the works of Thomas[16] and Corbridge[17] for detailed accounts. The results of their labours however are summarized in Table 3.10 for the covalent bonds to phosphorus.

As Table 3.10 shows there is a wide range of values for some bonds, so much so that for correlation purposes they are virtually useless, especially if the range falls within the 600–1000 cm^{-1} span which tends to be heavily populated with peaks in many infrared spectra. This applies particularly to the P—C and P—N bonds. For some phosphorus bonds attempts have been made to find relationships between frequency and molecular structure, and this has been most successful in the case of the phosphoryl link.

3.2.1.1 The phosphoryl stretching vibration, νP=O

This generally appears as an intense* band centred around 1200 cm^{-1}. Since so many

---

* This band is so intense that it can be used to detect POCl$_3$ impurity in PCl$_3$ in concentrations as low as 1 part per 10^4, i.e. 0.01 %.

3.2 Vibrational spectra of phosphorus compounds

Table 3.10 Bond-stretching infrared correlations of phosphorus bonds[a] (cm^{-1})

				P—H
		Calculated value[b] →		2400
		Overall range for single bonds →		2222—2500
		Overall range for double bonds →		—

P—B	P—C	P—N	P—O	P—F
666	719	743	806	821
550—750	620—780[c]	789—1102	850—1200[d]	723—966[e]
1412—1479	1006—1389[c]	1055—1500	1097—1415	—

	P—Si	P—P	P—S	P—Cl
	434	450	481	526
	380—515	340—510	440—613[f]	400—607
			515—862[f]	—

			P—Se	P—Br
			354	421
			439—477	320—495
			421—599[f]	—

				P—I
				319
				290—350

[a] Mainly according to L. C. Thomas, ref. 16; [b] according to D. E. C. Corbridge, ref. 17; [c] a few compounds fall outside this range; [d] depends very much on the other group attached to oxygen; [e] extreme range is 500—1009 cm^{-1} if ν_sPF$_2$ and ν_{as}PF$_2$ included; [f] two bands often found within this range, particularly with P=S.

phosphorus compounds contain the P=O link a great deal of information about it has been collected and sifted. The role of νP=O in phosphorus infrared spectroscopy is rather like that of νC=O in organic and complex chemistry. The factors which have been shown to affect νP=O are as follows:

(i) the nature of the other three ligands attached to phosphorus and especially the 'electronegativity' of these;

(ii) conjugation or delocalization involving other oxygen atoms attached to phosphorus in such anions as $R_2PO_2^-$, RPO_3^{2-} and PO_4^{3-};
(iii) the involvement of the oxygen atom in hydrogen-bonding;
(iv) complex formation with the oxygen acting as donor; and
(v) solvent effects.

These are listed in order of importance from the point of view of our understanding of the effect.

(i) The position of the $\nu P{=}O$ peak can be calculated by using the empirically derived relationship (eqn. 3). This was deduced by L. C. Thomas who based it on earlier

$$\nu P{=}O \;/(\text{cm}^{-1}) = 930 + 40 \sum \pi \qquad (3)$$

suggestions that $\nu P{=}O$ was governed by the sum of the Pauling electronegativities, $\sum \chi$, of the other atoms bound to phosphorus. Although inductive effects of these atoms have an over-riding influence on $\nu P{=}O$ the Pauling electronegativities were found not to give the best results and alternative π constants* were derived for various atoms and groups from the spectra themselves. The π values for common groups are given in Table 3.11 and it can be seen that they are numerically very like Pauling's χ values.

Table 3.11 π Constants for groups attached to phosphorus[17]

H	2.5	Br	3.1	NH$_2$	1.85
F	3.9	2Br	5.5	NR$_2$	2.4
2F	8.0	3Br	8.2	CH$_3$	2.1
3F	12.0	OCH$_3$	2.9	CF$_3$	3.6
Cl	3.4	OPh	3.0	Ph	2.4
2Cl	6.3	SR	2.4	C$_6$F$_5$	2.5
3Cl	9.0	SAr	2.5	SiR$_3$	2.4

Attempts to draw up a similar equation as (3.3) for the thiophosphoryl link, $\nu P{=}S$, have met with only limited success. This bond's frequency can be related to Taft's constant, σ^*, according to eqn. (4) but the relationship holds for fewer compounds. One reason for this is that $\nu P{=}S$ tends to be coupled with other vibrations in the

$$\nu P{=}S \;/(\text{cm}^{-1}) = 599 + 14.3 \sum \sigma^\star \qquad (4)$$

molecule whereas $\nu P{=}O$ is not. Even in the anhydrides with the $\ce{>P(=O)-O-P(=O)<}$ unit the expected coupling between the P=O and P—O bonds is absent. This brings us to the second factor affecting $\nu P{=}O$.

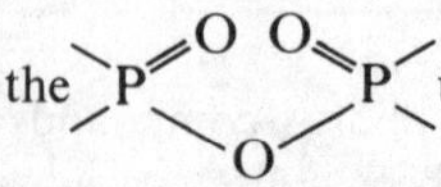

---

* This is rather an ill chosen name since it has nothing to do with π-bonding. In fact these constants probably reflect only inductive effects i.e. σ-bonding effects.

3.2 Vibrational spectra of phosphorus compounds

(ii) Further examples of the isolation of the π-bonding of the phosphoryl link comes from the infrared spectra of compounds with the C=C—P=O moiety. These show no lowering in frequency of νP=O of the kind which is found for νC=O in C=C—C=O compounds, where conjugation between the two π systems is possible. However P=O is like C=O when one or more oxygen atoms are attached to the phosphorus or carbon in that delocalization ensures equivalence in the resulting anions – i.e. $R_2PO_2^-$, RPO_3^{2-}, PO_4^{3-} and RCO_2^-, CO_3^{2-} respectively – and there is a lowering of frequency.

The spectra of the $X_2PO_2^-$ anions illustrate the effect. These show two bands of strong and comparable intensity situated in the regions 995–1164 and 1092–1323 cm^{-1} which can be attributed to the symmetric, $\nu_s PO_2^-$ and asymmetric, $\nu_{as} PO_2^-$ stretching modes. The positions of these peaks depend not only on the group X but to a small extent on the cation, as the spectra of $(RO)_2PO_2^-M^+$ (R = 2-ethylhexyl) demonstrate[18]:

$(RO)_2PO_2^-M^+$	$\nu_s PO_2^-$	$\nu_{as} PO_2^-$
$M^+ = Li^+$	1102	1196
$M^+ = Na^+$	1100	1243
$M^+ = K^+$	1098	1238

(iii) The phosphoryl stretching mode is strongly affected when the oxygen atom acts as an acceptor in the formation of a H-bond. This is the situation in phosphorus acids, $>$P(O)OH, where it is observed that νP=O is not only reduced in frequency but the band is broadened. The reductions are of the order of 50–100 cm^{-1}, e.g. vapour phase $F_2P(O)OH$ has νP=O at 1393 cm^{-1} whereas in condensed phases, where H-bonding occurs, this drops to 1335 cm^{-1}.[19] For $Me_2P(O)OH$ the fall is from 1250 to 1160 cm^{-1}.[20] H-bonding profoundly affects not only the phosphoryl link but the hydroxyl bond as well, and the whole area of H-bonding in phosphorus acids is dealt with in more detail on page 102. Infrared spectroscopy is the technique *par excellence* for probing H-bonding.

(iv) Phosphoryl compounds can act as ligands towards transition metals and they do so through the oxygen atom. Complex formation results in a lowering of νP=O as Table 3.12 shows. The table also shows the phosphoryl peak is a doublet in the complex $(Me_3PO)_2CoCl_2$ and this behaviour is not uncommon for νP=O in many compounds. There are two reasons for it; either it is due to conformers, in which case the doublet is generally poorly resolved, or it is due to Fermi resonance*, in which

---

* Fermi resonance occurs when two vibrations of a molecule accidentally have the same frequency and symmetry. The common situation is for a fundamental vibration to coincide with an overtone or combination vibration. The fundamental, which often gives rise to a strong band in the spectrum, interacts with the latter, usually a weak mode, and enhances it. The result is that the weaker band now appears as a band of comparable intensity and the two vibrations become shifted in the spectrum, one to a higher frequency the other to a low frequency. It is no longer possible to say which band is the fundamental and which is the overtone. The average of their frequencies represents the 'true' frequency.

Table 3.12 νP=O and complexes

Ligand	Free νP=O/(cm^{-1})	Complex	Complexed νP=O/(cm^{-1})		Ref.
Me$_3$PO	1174	(Me$_3$PO)$_2$CoCl$_2$	1100	1125	a
Ph$_3$PO	1195	(Ph$_3$PO)$_2$CoCl$_2$	1155		b
Cl$_3$PO	1300	(Cl$_3$PO)$_2$TiCl$_4$	1205		b
(Me$_2$N)$_3$PO	1208	{(Me$_2$N)$_3$PO}$_4$Fe(ClO$_4$)$_2$	1185		c

[a] F. A. Cotton, R. D. Barnes and E. Bannister, *J. Chem. Soc.*, 2199 (1960); [b] J. C. Sheldon and S. Y. Tyree, *J. Amer. Chem. Soc.*, **81**, 2290 (1959); [c] J. T. Donoghue and R. S. Drago, *Inorg. Chem.*, **2**, 1158 (1963).

case the peak separation may be as large as 50 cm^{-1}. Not only νP=O but several types of phosphorus bond vibrations seem prone to doublet formation and in particular νP=S. This bond gives rise to bands in one or both of two spectral regions νP=S (I) at 674–862 cm^{-1} and νP=S (II) at 515–725 cm^{-1}. The same is true, but to a lesser extent of P—S—(X) vibrations. Fermi resonance should be a randomly occurring event but the fact that particular bonds are prone to it suggests that another cause may be responsible for the doublets although at present the cause remains unknown.

Correlation theory, if theory is the right word, has made most progress in understanding the phosphoryl vibration. With other phosphorus bonds there has been less success. The P—O single bond, for example, which is found in numerous compounds, such as phosphorus acids, anhydrides and esters, has not proved amenable to a similar analysis of the factors influencing it. But some generalizations can be made. The phosphorus acids are particularly complex since the problem is exacerbated by H-bonding and νP—O—(H) probably falls in the 909–1070 cm^{-1} region. The anhydrides, X$_2$(O)P—O—P(O)X$_2$, should display ν_sPOP and ν_{as}POP vibrations but only the latter can be allotted with certainty to peaks within a range of 900–1025 cm^{-1} (mostly at 925–950 cm^{-1}). There have been suggestions that ν_sPOP is around 700 cm^{-1} but this band is not always observed in the infrared spectrum of many anhydrides. It has on occasion been detected in the Raman spectrum where its diagnosis is aided by the fact that it should be a polarized band, e.g. in C$_8$H$_{16}$(O)P—O—P(O)C$_8$H$_{16}$ (C$_8$H$_{16}$P = pentamethylphosphetan) it occurs as a strong, polarized Raman line at 611 cm^{-1} complementing the infrared and depolarized Raman lines at 952 cm^{-1} of ν_{as}POP.[21]

The P—O—C moiety gives rise to intense peaks at *ca* 1000 cm^{-1} which have proved invaluable in the analysis of infrared spectra since this grouping is extremely common in organophosphorus chemistry. For many molecules the P—O and O—C vibrations are strongly coupled and this applies particularly to aliphatic esters. The aryl esters are less coupled and the C—O bond is correlated to peaks at 1156–1242 cm^{-1} and P—O at 905–996 cm^{-1}. For aliphatic esters however, the nature of the alkyl group and the oxidation state of the phosphorus exert secondary influences on the position of the bands:

3.2 Vibrational spectra of phosphorus compounds

P(III)—O—CH$_3$	1015—1034 cm^{-1}	P(V)—O—CH$_3$	1010—1088 cm^{-1}
P(III)—O—CH$_2$R	1008—1042 cm^{-1}	P(V)—O—CH$_2$R	987—1042 cm^{-1}
P(III)—O—CHR$_2$	950—978 cm^{-1}	P(V)—O—CHR$_2$	950—1018 cm^{-1}

As these values show the sub-classification of νP—O—C$_{aliphatic}$ does not give clear cut divisions.

Such subdivisions can be found within other group correlations and this approach does represent a route to deeper truths, but along a road which traverses a maze. The alternative way to a deeper understanding is the narrow and hard pathway of deducing and attempting to assign the normal, i.e. fundamental, vibrational modes of a molecule.

3.2.2 Vibrational assignments of phosphorus molecules[22]

For a molecular vibration to interact with a photon of radiation it must be one which changes the dipole moment, p, of the molecule. If it does then radiation of the right (infrared) frequency can be absorbed by the molecule. If the vibration is so symmetrical that it causes no net change in p then it will not interact and it is said to be an infrared inactive vibrational mode. Such a mode, however, may be Raman active, the condition for this being that the vibration produces a change in the polarizability, α, of the molecule.

Raman spectroscopy involves the excitation of molecules by radiation of high frequency. The radiation which such molecules subsequently emit can be different from the exciting frequency by an amount equal to the vibrational frequencies of the molecule, and this is the origin of the Raman spectrum which is a series of bands symmetrically placed to the high and low frequency side of the exciting frequency. In addition the polarization of the emitted light can be measured by means of an analyzer and the Raman lines designated as polarized or depolarized. For symmetric vibrations of the molecule the light is polarized, symbol R(p), and for asymmetric vibrations it is depolarized, R(dp). This information can be of immense help in assigning the vibrational modes of the molecule.

The vibrations of a molecule may be very complex but they can be resolved into a set of independent motions which are the fundamental vibrations of the molecule, called the *normal modes*. One of the fascinations of vibrational spectroscopy is the detective work which leads up to pinning a particular vibration mode to a particular band in the vibrational spectra. A molecule with n atoms has $(3n - 6)$ normal modes ($3n - 5$ if it is a linear molecule). By means of mathematical group theory and character tables it is possible to deduce the degeneracy, symmetry and spectral activity of a vibration, and for this reason normal modes are often tagged with their symmetry label such as A'_1, B''_2, E_g, T_{2u} etc.*

---

* These refer to the symmetry of the molecule when deformed by the vibration. A means symmetric with respect to the principal axis of the molecule, B means asymmetric. Subscripts 1, 2 and 3 refer to secondary axes in the molecule and are used to distinguish vibrations which differ with respect to these. Superscripts ' and " are symmetry and asymmetry in a mirror plane. Subscripts u and g mean symmetry and asymmetry through a centre of inversion. If these rules permit alternative labelling the u takes precedence over 1 which takes precedence over '. A doubly degenerate mode is labelled E and a triply degenerate mode T.

The actual extent of physical change during a vibration depends on the masses of individual atoms and the need to preserve the centre of mass throughout the motion. The direction in which the atoms move is often shown in diagrams of fundamental modes by means of an arrow, and in particular cases the extent of the motion is indicated by the size of the arrow. In many molecules vibrations tend to affect mainly one bond or set of like bonds (i.e. a group) with the rest of the molecule moving only marginally in order to preserve the centre of gravity. Vibrations can be loosely classed as stretching when they affect bonds most and bending when they deform bond angles. A molecule with n atoms has $(n-1)$ bonds (n if it is cyclic) and hence $(n-1)$ vibrations which are essentially stretching. The remaining $(2n-5)$ fundamental modes ($2n-6$ if cyclic) are therefore essentially bending in character.* Many chemists prefer this semi-descriptive approach to that of the visually sterile but more accurate symmetry approach. In the following diagrams both notations are used.

Tricoordinate PX_3 phosphorus molecules are pyramidal and thus have C_{3v} symmetry. Because of this some of their fundamental modes are degenerate and only four vibrations are observed, $2A_1 + 2E$. All are infrared (i.r.) and Raman active with the A_1 modes polarized. Figure 3.1 illustrates these modes and, for a selection of molecules with this structure, lists the frequencies in the spectra assigned to them.

Upon the introduction of a second type of ligand to give PX_2Y the symmetry falls to C_σ (C_s) in which all that remains is the plane of symmetry bisecting the XPX angle. The result of this is to resolve the two degenerate E mode of C_{3v}, and the spectrum of PX_2Y molecules should show bonds for all six normal modes, $4A' + 2A''$, all i.r. and Raman active, with the $4A'$ lines polarized. How the vibrations are to be described is somewhat subjective as the information in Fig. 3.1 shows. Clearly one of the stretching modes is very different to the other two, which in the molecules listed appear near together. It is therefore logical to associate the different band with the odd bond, νPY, and the pair of peaks with the symmetric and asymmetric vibrations of the PX_2 group, $\nu_s PX_2$ and $\nu_{as} PX_2$.

For the bending modes of PX_2Y Fig. 3.1 shows that an analogous clear cut division into PY and PX_2 bending modes is not possible. For PHF_2 for example, there are two high frequency bands at 1008 and 1016 which can only come about if the P—H bond is involved in both vibrations. These modes, labelled ρPY and $\delta_{as} PXY$ have similar pairs of values in the compounds listed while the remaining band keeps to a fairly narrow range of values, 367–412 cm^{-1} in the PYF_2 compounds, showing it to be reasonably independent of the Y group, and so it can be fairly confidently labelled δPF_2.

When all the atoms are different, PXYZ, all symmetry disappears and all the six modes are i.r. and Raman active. The stretching vibrations should approximate to νPX, νPY and νPZ but the bending modes would be complex combinations involving all three ligand atoms. At the present time such simple molecules as PHFCl, PFClBr etc. defy isolation and so spectroscopic data is unavailable.

---

* The symbols used in this notation for vibrational modes are ν_s, symmetric stretching; ν_{as}, asymmetric stretching; δ, deformation (scissors or flapping); ρ, rocking; τ, twisting; ω, wagging; π, out-of-plane bending.

Figure 3.1 Tricoordinate phosphorus vibrational modes and assignments/(cm^{-1}).

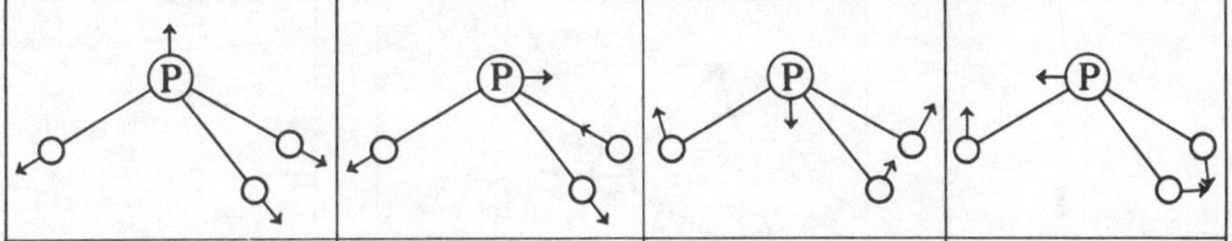

Essentially stretching modes		Essentially bending modes	
A$_1$, ν_sPX$_3$ i.r. + R(p)	E, ν_{as}PX$_3$ i.r. + R(dp)	A$_1$, δ_sPX$_3$ i.r. + R(p)	E, δ_{as}PX$_3$ i.r. + R(dp)

Pyramidal PX$_3$ molecules, symmetry C$_{3v}$

PH$_3$	2323	2328	991 (990,992*)	1121	a
PD$_3$	1683	1689	728	805	b
PF$_3$	892	860	487	344	c
PCl$_3$	504	482	252	198	d
PBr$_3$	392	392	161	116	e
PI$_3$	303	325	111	79	f

Pyramidal PX$_2$Y molecules, symmetry C$_\sigma$.

	A′, νPY i.r. + R(p)	A′, ν_sPX$_2$ i.r. + R(p)	A″, ν_{as}PX$_2$ i.r. + R(dp)	A′, ρPY i.r. + R(p)	A′, δPX$_2$ i.r. + R(p)	A″, δ_{as}PXY i.r. + R(dp)	
PHF$_2$	2240	851	838	1008	367	1016	g
PClF$_2$	544	864	853	302	412	259	h
PBrF$_2$	459	858	849	233	391	212	i
PMeF$_2$†	700	806	864	483	402	335	j

* Splitting due to inversion of molecule, † Only the CPF$_2$ framework is considered

[a] V. M. McConaghie and H. H. Nielsen, *Proc. Nat. Acad. Sci. US*, **34**, 455 (1968); [b] W. M. Ward, *Diss. Abs.*, **18**, 1823 (1958); [c] H. S. Gutowsky and A. D. Liehr, *J. Chem. Phys.*, **20**, 1652 (1952); M. K. Wilson and S. R. Palo, *J. Chem. Phys.*, **20**, 1716 (1952); **21**, 1426 (1953); [d] S. G. Frankiss and F. A. Miller, *Spectrochim. Acta*, **21**, 1235 (1965); [e] P. W. Davis and R. A. Oetjen, *J. Mol. Spectrosc.*, **2**, 253 (1958); [f] R. E. Stoup and R. A. Oetjen, *J. Chem. Phys.*, **21**, 2092 (1953); [g] R. W. Rudolf and R. W. Parry, *Inorg. Chem.*, **4**, 1339 (1965); [h] R. W. Rudolf, R. C. Taylor and R. W. Parry, *J. Amer. Chem. Soc.*, **88**, 3279 (1966); [i] A. Muller, E. Niecke and O. Glemser, *Z. Anorg. Chem.*, **350**, 256 (1967); [j] E. Griffiths, *Spectrochim. Acta*, **21**, 1135 (1965); F. Seel, K. Rudolf and R. Budenz, *Z. Anorg. Chem.*, **341**, 196 (1965);

Figure 3.2 Tetracoordinate phosphorus vibrational modes and assignments/(cm^{-1}).

	Essentially stretching modes		Essentially bending modes	
	A_1, ν_sPX$_4$ R(p)	T_2, ν_{as}PX$_4$ i.r. + R(dp)	E, δ_sPX$_4$ R(dp)	T_2, δ_{as}PX$_4$ i.r. + R(dp)
PH$_4^+$	2304	2370	1040	930	a
PCl$_4^+$	458	658	171	251	b, c
PMe$_4^+$	652	782	170	282	b
PO$_4^{3-}$	970	1080	358	500	d

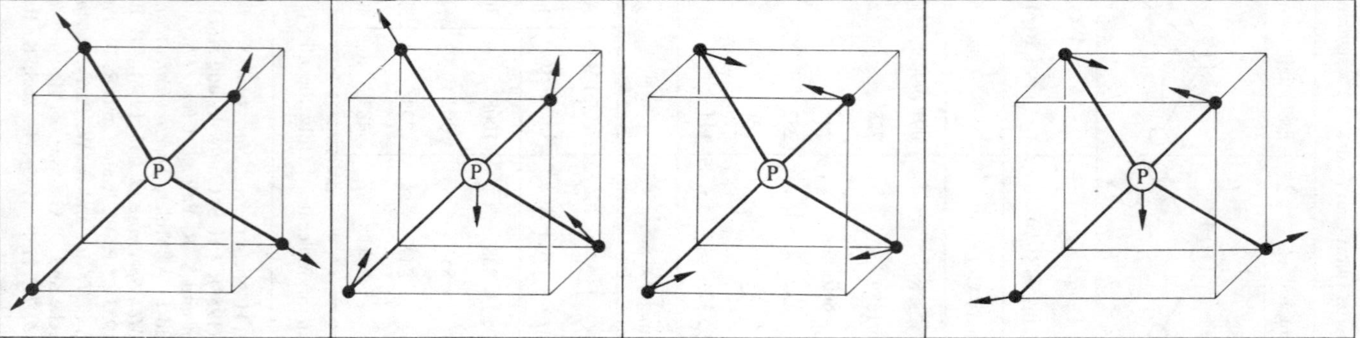

Tetrahedral PX$_4^{\pm}$ molecular ions, symmetry T$_d$

Tetrahedral PYX$_3$ molecules, symmetry C$_{3v}$

3.2 Vibrational spectra of phosphorus compounds

	$A_1, \nu_s PY$ i.r.+R(p)	$A_1, \nu_s PX_3$ i.r.+R(p)	$E, \nu_{as} PX_3$ i.r.+R(dp)	$E, \rho PY$ i.r.+R(dp)	$A_1, \delta_s PX_3$ i.r.+R(p)	$E, \delta_{as} PX_3$ i.r.+R(dp)	
POCl$_3$	1290	486	588	337	267	193	e
PSF$_3$	695	981	945	276	440	402	f
POMe$_3$	1228	671	756	—	256	331	g
HPO$_3^{2-}$	2315	979	1085	1027	567	465	h

Tetrahedral PX$_2$Y$_2^-$ molecular ions, symmetry C$_{2v}$*

	$A_1, \nu_s PY_2$ i.r.+R(p)	$A_1, \nu_s PX_2$ i.r.+R(p)	$B_1, \nu_{as} PY_2$ i.r.+R(dp)	$B_2, \nu_{as} PX_2$ i.r.+R(dp)	$A_1, \delta PY_2$ i.r.+R(p)	$A_2, \tau PX_2$ R(dp)	$A_1, \delta PX_2$ i.r.+R(p)	$B_1, \omega PX_2$ i.r.+R(dp)	$B_2, \rho PX_2$ i.r.+R(dp)	
PH$_2$O$_2^-$	2363	1042	2314	1180	469	924	1160	1086	811	h
PF$_2$O$_2^-$	834	1145	857	1311	535	286	481	512	481	i
PCl$_2$O$_2^-$	590	1147	708	1259	418	334	350	518	—	j

Tetrahedral PYX$_2$Z molecules, symmetry C$_\sigma$

	$A', \nu PY$ i.r.+R(p)	$A', \nu_s PX_2$ i.r.+R(p)	$A', \nu PZ$ i.r.+R(p)	$A'', \nu_{as} PX_2$ i.r.+R(dp)	$A', \delta PYZ$ i.r.+R(p)	$A'', \tau PX_2$ i.r.+R(dp)	$A', \delta PX_2$ i.r.+R(p)	$A', \omega PX_2$ i.r.+R(dp)	$A'', \rho PX_2$ i.r.+R(dp)	
POCl$_2$F	1358	546	907	626	374	247	330	382	205	k
POCl$_2$Br	1285	545	432	580	242	161	172	285	327	
POCl$_2$Me	1297	407	757	552	349	226	200	285	325	m

* The descriptive nomenclature is based on the assumption that mass of X < mass of Y and that mirror plane bisects XPX angle.

Refs: [a] T. C. Waddington and F. Klanberg, *J. Chem. Soc.*, 2339 (1960); [b] R. Baumgartner, W. Sawodny and J. Goubeau, *Z. Anorg. Chem.*, **333**, 171 (1964); [c] G. L. Carlson, *Spectrochim. Acta*, **19**, 1291 (1963); [d] Landolt-Bernstein, *Phys. Chem. Tabellen*, 1951; [e] M. L. Delwaulle and F. Francois, *J. Chim. Phys.*, **46**, 87 (1949); [f] J. R. Durig and J. W. Clark, *J. Chem. Phys.*, **46**, 3057 (1967); [g] L. W. Daasch and D. C. Smith, *J. Chem. Phys.*, **19**, 22 (1951); J. Goubeau and W. Berger, *Z. Anorg. Chem.*, **304**, 147 (1960); [h] M. Tsuboi, *J. Amer. Chem. Soc.*, **79**, 1351 (1957); [i] K. Buhler and W. Bues, *Z. Anorg. Chem.*, **308**, 62 (1961); [j] A. Muller and K. Dehnicke, *Z. Anorg. Chem.*, **350**, 231 (1967); [k] A. Muller, O. Glemser and E. Niecke, *Z. Anorg. Chem.*, **347**, 275 (1966); [l] J. E. Griffiths, *Spectrochim. Acta*, **24A**, 303 (1968); [m] J. Quinchon, M. Lesech and T. E. Gryzkiewier, *Bull. Soc. Chim. Fr.*, 735 (1961).

Tetracoordinate phosphorus compounds are predominant although only a few of them have tetrahedral T_d symmetry, and these must perforce be ionic such as PH_4^+ or PO_4^{3-}. The $3n - 6$, i.e. 9, normal modes give four observeable vibrational frequencies of types $A_1 + E + 2T_2$. Only the two triply-degenerate ones are i.r. active although all are Raman active with A_1 polarized. Diagrams of the vibrations are shown in Fig. 3.2 together with descriptions of the modes and the assignments of the peaks of a few simple ions; for PMe_4^+ only the PC_4 framework is considered.

The replacement of one of the atoms of a PX_4 arrangement to give PX_3Y, lowers the symmetry to C_{3v} and lifts the restriction that the species must be ionic, although some are, such as the phosphite anion HPO_3^{2-}. The more usual types of compound of this class are the phosphoryl and thiophosphoryl halides. What the lowering of symmetry means in terms of the spectra is the partial resolution of the triply degenerate modes so there appear six peaks, $3A_1 + 3E$, all i.r. and Raman active with the A_1 modes polarized. The correlation between T_d and C_{3v} is shown in Fig. 3.2 together with some assigned spectral bands for typical molecules and ions. (Again $POMe_3$ is treated as POC_3 for the assignments in Fig. 3.2.)

Further replacement of a second X atom in PX_3Y by another Y atom gives $PX_2Y_2^-$ (which must be ionic) and lowers the symmetry still more to C_{2v}. Figure 3.2 shows the relationship to C_{3v} and how the remaining degeneracies are now removed to give nine observeable bands, $4A_1 + A_2 + 2B_1 + 2B_2$, of which all but A_2 are i.r. active and all are Raman active. If, however, the second atom is different to Y so that we have PX_2YZ then the only symmetry element remaining is a mirror plane and the symmetry of the molecule is C_σ; these have $6A' + 3A''$ modes all of which are both i.r. and Raman active. A few molecules of this kind are quoted in Fig. 3.2. The descriptive tags given to the vibrations should now be treated with caution and really devised to fit the occasion. Although PXYZA type molecules are known all symmetry has gone and little is gained by a study of them in this context.

The spectra of those simple pentacoordinate phosphorus compounds which have been analyzed have shown the molecules to be *tbp* in structure, and to be the isomer predicted from apicophilicities. Figure 3.3 contains a detailed list of correlations for the vibrational modes of these molecules with the symmetries D_{3h} (PX_5 or PX_3Y_2 with both Y atoms apical), C_{3v} (PX_4Y with Y apical) and C_{2v} (PX_4Y with Y equatorial, or PX_2Y_3 with both X atoms equatorial). The arrangement PX_2Y_3 with one X apical and one equatorial would have symmetry C_σ, but apicophilicity theory excludes this arrangement if X and Y are free agents. On the other hand if X_2 were a four-membered ring and so physically constrained to an apical–equatorial configuration then this structure would be possible, although whether an assignment of the fundamental modes of the PX_2Y_3 framework would be feasible is another matter.

In this chapter we have restricted our discussion of vibrational assignments to simple compounds containing only one phosphorus centre. However, more complicated molecules can also be dealt with and this has been particularly successful with the halide derivatives of the cyclotriphosphazenes (chapter 10), e.g. the *gem* isomers of $N_3P_3F_{6-x}Cl_x$ have been fully analyzed[23].

3.2.3 Hydrogen-bonding[24,25]

For the X···H—O hydrogen bond there are three kinds of H-bond vibration frequencies in the 400–4000 cm^{-1} i.r. spectrum. In order of decreasing frequency these are:

3.2 Vibrational spectra of phosphorus compounds

the O—H stretching mode, v_s; the in-plane X···H—O bending mode, v_b; and the out-of-plane X···H—O bending mode, v_t. Carboxylic acid dimers (3.13) have been most studied and they have an obvious relationship to phosphorus acids some of which appear to be dimeric (3.14)[26]. For acetic acid (3.13, R=CH$_3$) the bands occur at

$$R-C\underset{O-H\cdots O}{\overset{O\cdots H-O}{\diagdown\diagup}}C-R \qquad \underset{R}{\overset{R}{\diagdown}}P\underset{O-H\cdots O}{\overset{O\cdots H-O}{\diagdown\diagup}}P\underset{R}{\overset{R}{\diagup}}$$

(3.13)   (3.14)

3150 cm^{-1} (v_s), 1295 cm^{-1} (v_b) and 935 cm^{-1} (v_t); these are all very broad, this being their most characteristic feature[27]. The monomer, vOH, vibration for acetic acid is at 3620 cm^{-1} which means that on H-bond formation this vibration suffers a decrease in frequency, ΔvOH, of several hundred wavenumbers. Infrared changes, ΔvOH, are often used as a measure of the enthalpy of the H-bond formed.

With phosphorus acids up to six regions of i.r. activity have been correlated with the PO$_2$H grouping[16,17]. These are
(i) 3000–2525 cm^{-1} and (ii) 2400–2000 cm^{-1}, although (ii) is not always present.*
Band (i) can be identified as v_s. Dilution of the acids does not affect the spectrum in this region, as it does with carboxylic acids when the acid monomer peak grows with dilution. This prompted some researchers to suggest that in phosphorus acids the

H-bonding may be intramolecular, i.e. $\diagdown P\underset{O}{\overset{O\cdots}{\diagdown\diagup}}H$ but the real reason appears to

be that the dimers of type (3.14) are much more strongly associated than those of type (3.13)[28].
(iii) 1900–1600 cm^{-1}, and since these bands are unaffected by deuteration, but disappear in R$_2$P(S)OH, it seems likely that they represent a combination of vP=O and the P—O band which occurs ca 500 cm^{-1} (see (vi)).
(iv) 1400–1200 cm^{-1}, which corresponds to v_b (the band is absent in neutral salts). This band has been least studied because it is often not very pronounced in the spectrum. For example it was not clearly observed in (EtO)EtP(O)OH[16] but on deuteration a peak appeared at 968 cm^{-1} due to v_bO···D—O which would correspond to a proton band at ca 1280 cm^{-1}.
(v) 1030–820 cm^{-1}, which should be v_t but some 'correlationists' attribute it to vP—O—(H). It is said to be like vP=O in that it is sensitive to changes at phosphorus and obeys the eqn. (5)[16]

$$v\text{P—O} = 650 + 40 \sum \pi \qquad (5)$$

although not so well as the phosphoryl link obeys eqn. (3).
(vi) 540–450 cm^{-1}, which is probably a deformation mode of P—O—(H).
Finally there is one small matter in which infrared spectroscopy may have a decisive role and that is in the question of hydrogen-bonding by P—H. There as yet is no

---

* The appearance of two bands is thought to be the result of quantum mechanical tunnelling.

Figure 3.3 Pentacoordinate phosphorus vibrational modes and assignments/(cm^{-1}).

Essentially stretching modes.

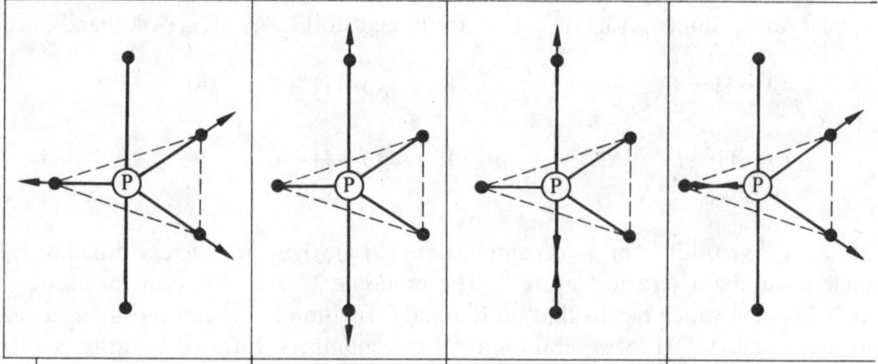

Trigonal bipyramidal PX$_5$ molecules, symmetry D$_{3h}$

	A$_1'$, ν_sPX$_3$eq R(p)	A$_1'$, ν_sPX$_2$ap R(p)	A$_2''$, ν_{as}PX$_2$ap i.r.	E', ν_{as}PX$_3$eq i.r. + R(dp)
PF$_5$	817	640	945	1026
PCl$_5$	393	282	448	581

Trigonal bipyramidal PX$_4$Y molecules, Y apical, symmetry C$_{3v}$

	A$_1$, ν_sPX$_3$eq i.r. + R(p)	A$_1$, νPYap i.r. + R(p)	A$_1$, νPXap i.r. + R(p)	E, ν_{as}PX$_3$eq i.r. + R(dp)
PCl$_4$F	422	778	388	601

Trigonal bipyramidal PX$_3$Y$_2$ molecules, Y's apical, symmetry D$_{3h}$

	A$_1'$, ν_sPX$_3$eq R(p)	A$_1'$, ν_sPY$_2$ap R(p)	A$_2''$, ν_{as}PY$_2$ap i.r.	E', ν_{as}PX$_3$eq i.r. + R(dp)
PCl$_3$F$_2$	387	633	867	625

Trigonal bipyramidal PX$_2$Y$_3$ molecules, 2Y's apical, one Y equatorial, symmetry C$_{2v}$

	A$_1$, νPYeq i.r. + R(p)	A$_1$, ν_sPY$_2$ap i.r. + R(p)	B$_1$, ν_{as}PY$_2$ap i.r. + R(dp)	A$_1$, ν_sPX$_2$eq i.r. + R(p)	B$_2$, ν_{as}PX$_2$eq i.r. + R(dp)
PCl$_2$F$_3$	893	665	925	488	500
PH$_2$F$_3$	879	614	825	2482	2549

Trigonal bipyramidal PXY$_4$ molecules, X equatorial, symmetry C$_{2v}$

	A$_1$, νPXeq i.r. + R(p)	A$_1$, ν_sPY$_2$ap i.r. + R(p)	B$_1$, ν_{as}PY$_2$ap i.r. + R(dp)	A$_1$, ν_sPY$_2$eq i.r. + R(p)	B$_2$, ν_{as}PY$_2$eq i.r. + R(dp)
PHF$_4$	2482	629	795	882	1024

Figure 3.3 (*continued*)

Essentially bending modes

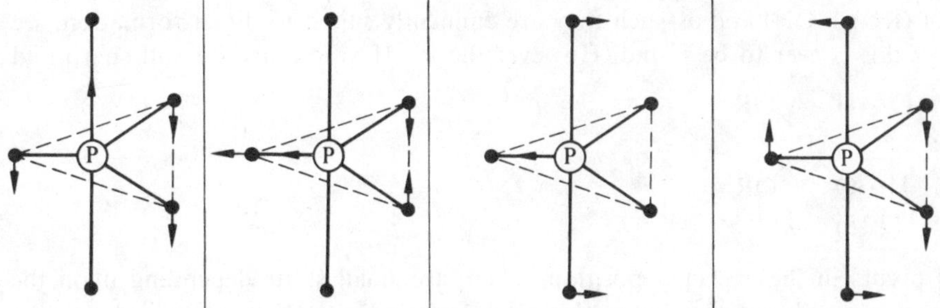

A_2'', πPX_3eq	E', δPX_3eq	E', δPX_2ap	E'', ρPX_2ap	
i.r.	i.r. + R(*dp*)	i.r. + R(*dp*)	R(*dp*)	
576	533	126	514	*a*
300	273	100	261	*b*

A_1, πPX_3eq	E, δPX_3eq	E, δPXYap	E, ρPX_4F	
i.r. + R(*p*)	i.r. + R(*dp*)	i.r. + R(*dp*)	i.r. + R(*dp*)	
265	339	110	297	*a*

A_2'', πPX_3eq	E', δPX_3eq	E', δPY_2ap	E'', ρPY_2ap	
i.r.	i.r. + R(*dp*)	i.r. + R(*dp*)	R(*dp*)	
328	404	122	357	*a*

B_1, πPYeq	A_1, δPX_2eq	B_2, ρPX_2eq	A_1, δPY_2ap	B_2, ρPY_2ap	A_2, τPY_2ap	B_1, ωPY_2ap	
i.r. + R(*dp*)	i.r. + R(*p*)	i.r. + R(*dp*)	i.r. + R(*p*)	i.r. + R(*dp*)	R(*dp*)	i.r. + R(*dp*)	
338	407	427	124	124	368	368	*a*
338	1005	1233	—	308(?)	377	472	*c*

B_1, πPXeq	A_1, δPY_2eq	B_2, ρPY_2eq	A_1, δPY_2ap	B_2, ρPY_2ap	A_2, τPY_2ap	B_1, ωPY_2ap	
i.r. + R(*dp*)	i.r. + R(*p*)	i.r. + R(*dp*)	i.r. + R(*p*)	i.r. + R(*dp*)	R(*dp*)	i.r. + R(*dp*)	
1528	525	650	200	317	336	537	*c*

[a] J. E. Griffiths, R. P. Carter and R. R. Holmes, *J. Chem. Phys.*, **41**, 863 (1964); [b] I. R. Beattie, K. Livingston and T. Gilson, *J. Chem. Soc. A*, 1, (1968); [c] R. R. Holmes and C. J. Hora Jr, *Inorg. Chem.*, **11**, 2506 (1972), reassignments of some of the bands of PHF$_4$ and PH$_2$F$_3$, especially the latter, have been made in the light of other values in this table.

conclusive evidence that P—H···X bonds can form although a neutron diffratcion study of PH$_4$I showed that some weak P—H···I bonds were present[29]. Infrared evidence is even flimsier. Dialkylphosphonates have the structure (RO)$_2$P(O)H and not (RO)$_2$P(OH) and as such they are eminently suited to dimer formation, see (3.15), if this is ever to be found. However the P—H vibrations are still sharp and

$$\begin{array}{c} RO \quad O \cdots H \quad OR \\ \diagdown P \diagup \quad \diagdown P \diagup \\ RO \quad H \cdots O \quad OR \end{array}$$
(3.15)

vP=O occurs in the expected position. There are small shifts depending upon the other groups on phosphorus but nothing definitely indicates H-bonding[30].

In non-H-bonding or weak H-bonding solvents there is no shift in vP—H, but in acetone and pyridine small shifts (<20 cm^{-1}) were observed suggestive of H-bonding via P—H···O=C< and P—H···N< respectively[31].

For H-bonding to occur there must be a sufficiently strong dipole associated with the P—H bond to make the proton attractive to a centre of high electron density such as a lone pair. Ordinarily this bond is not sufficiently polar even with strongly electron-withdrawing groups attached to phosphorus.

Problems

1 On the basis of the information provided, suggest a structure for each of the following compounds:

(a) C$_3$H$_9$O$_3$P, δ^{31}P, −140 ppm; 10 line pattern (decet), J_{PH} = 9.5 Hz.
(b) C$_2$F$_6$PBr, δ^{31}P, −33.7 septet; J_{PF} = 80.6 Hz.
(c) CH$_3$PSBr$_2$, δ^{31}P, −203.5, septet; J_{PH} = 7 Hz.
(d) CH$_2$PCl$_3$, δ^{31}P, −159, triplet; J_{PH} = 16 Hz.
(e) C$_2$H$_6$NF$_2$P, δ^{31}P, −144; triplet of septets, J_{PF} = 1200 Hz and J_{PH} = 9.3 Hz.
(f) C$_3$OF$_9$P, δ^{31}P, −2.3; decet, J_{PF} = 113.4 Hz.
(g) C$_6$H$_5$O$_2$F$_2$P, δ^{31}P, +27; triplet, J_{PF} = 1030 Hz.
(h) C$_2$H$_7$O$_3$P, δ^{31}P, −11; doublet of septets, J_{PH} = 704 Hz and J_{PH} = 14 Hz.
(i) C$_4$H$_{11}$O$_2$PS, δ^{31}P, −69; doublet of quintets, J_{PH} = 640 Hz and J_{PH} = 11 Hz.
(j) C$_4$H$_{12}$P$_2$S$_2$, δ^{31}P, −35; septet, J_{PH} = 19 Hz.
(k) C$_{19}$H$_{18}$PBr, δ^{31}P, −22.7: quartet, J_{PH} = 15 Hz.
(l) C$_3$H$_{10}$PCl, δ^{31}P, +2.8; doublet of decets, J_{PH} = 495 Hz and J_{PH} = 8 Hz.
(m) C$_{36}$H$_{30}$N$_3$O$_6$P$_3$, δ^{31}P, −9.0; singlet[1] H n.m.r. shows a signal at ca 7.3 ppm from TMS.
(n) CF$_7$P, δ^{31}P, +66.4; quintet of quartets, J_{PF} = 1103 Hz and J_{PF} = 172 Hz.
(o) C$_8$H$_{11}$NF$_3$P, δ^{31}P, +53.0; doublet of triplets, J_{PF} = 955 Hz and J_{PF} = 820 Hz.
^{1}H n.m.r. shows: δ = 7.3 (5H) and $\delta \approx$ 3 (6H, doublet)
(p) C$_{11}$H$_{15}$O$_5$P, δ^{31}P, +45; doublet of decets, J_{PH} = 32 Hz and J_{PH} = 13 Hz.
^{1}H n.m.r. shows: δ 7.3 (5H); δ 7.0 (doublet, 1H) and δ 3.6 (doublet, 9H).
(q) C$_6$H$_4$N$_2$F$_6$PCl, δ^{31}P, +144; septet, J_{PF} = 707 Hz. (^{1}H n.m.r. shows: signal ca δ = 7.3).

Problems

2 The ^{19}F n.m.r. spectrum of (1) shows a doublet at 70.4 ppm relative to CCl_3F (J_{PF} = 1018 Hz) whereas that of (2) shows a doublet at 21.7 ppm (J_{PF} = 829 Hz). Assuming

(1)   (2)

trigonal bipyramidal structures, what is the most likely orientation of the fluorine atoms in each of these molecules? Provide an explanation of any difference in structure between the two molecules.

3 When S-allyl methylphenylphosphinothioate (3) is heated at 200°C, the P—CH$_3$ resonance gradually changes from δ 1.37 (J_{P-CH_3} = 7.5 Hz) to δ 1.49 (J_{P-CH_3} = 13.0 Hz). Suggest an explanation for these results.

(3)

4 Explain the change in ^{31}P chemical shift in the following series of phosphorus ylids:

Ylid:	Et$_3$P=CH$_2$	Et$_3$P=CHMe	Et$_3$P=CHEt	Et$_3$P=CHPr^n
δ ^{31}P (H$_3$PO$_4$)	−23.6	−16.9	−14.8	−14.6

5 The phosphoryl stretching band of (EtO)$_3$PO occurs at 1277 cm^{-1}. Use equation (3.3) and the information of Table 3.11 to calculate νP=O for the compounds (EtO)$_2$P(O)NMe$_2$, (Me$_2$N)$_2$P(O)Cl and (EtO)P(O)Cl$_2$.

6 Identify the phosphorus compounds shown by the following stylized infrared spectra:

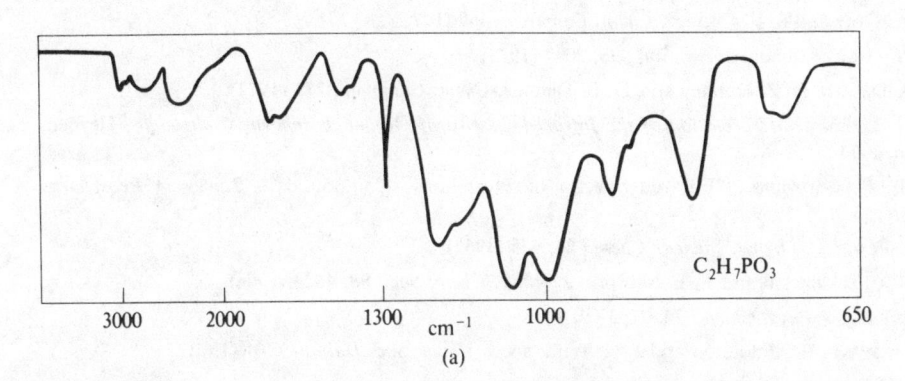

C$_2$H$_7$PO$_3$

(a)

(b)

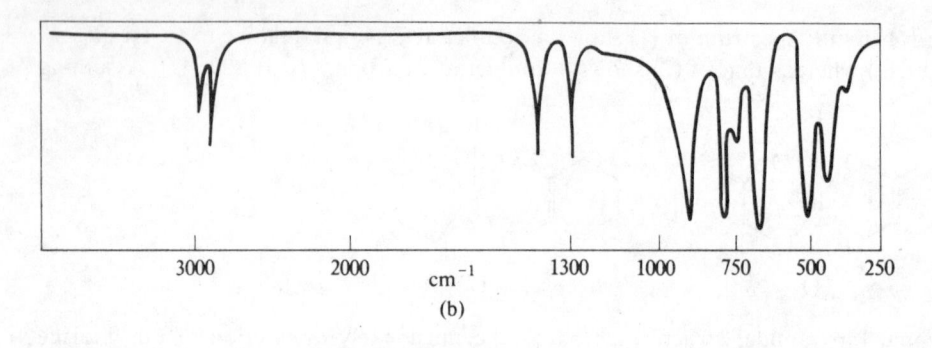

References

[1] J. R. Van Wazer, *Topics in Phosphorus Chemistry*, Vol 5, Wiley Interscience, New York, 1967. A comprehensive survey of ^{31}P n.m.r. including principles, techniques and a compilation of ^{31}P n.m.r. data covering the literature up to 1966.

[2] J. R. Van Wazer, *Determination of Organic Structures by Physical Methods*, Vol. 4, Chapter 7, Academic Press, New York, 1971. A discussion of ^{31}P chemical shifts and the coupling of phosphorus with other magnetically active nuclei.

[3] G. M. Kosolapoff and L. Maier (editors), *Organic Phosphorus Compounds*, Vols. 1–6, Wiley Interscience, New York, 1972/73. ^{31}P n.m.r. data on a wide variety of phosphorus compounds quoted where available; covers the literature up to *ca* 1971.

[4] J. A. Pople, W. G. Schneider and H. J. Bernstein, *High-Resolution Nuclear Magnetic Resonance*, McGraw-Hill, New York, 1959. A detailed and comprehensive coverage of the theory and practice of n.m.r. spectroscopy.

[5] G. Binsch, *Topics in Stereochemistry*, Vol. 3, p. 97, Wiley Interscience, New York, 1968.

[6] F. A. L. Anet and R. Anet, *Determination of Organic Structures by Physical Methods*, Vol 3, Chapter 7, Academic Press, New York, 1971.

[7] N. M. Sergeev, *Russ. Chem. Rev. (Eng. Trans.)*, **42**, 339 (1973).

Refs. 5, 6 and 7 provide a comprehensive discussion of the theory and application of dynamic n.m.r. but are devoted very largely to ^1H n.m.r. studies.

[8] N. J. De'Ath and D. B. Denney, *Chem. Commun.*, 395 (1972).

[9] S. O. Grim, W. McFarlane, E. F. Davidoff and T. J. Marks, *J. Phys. Chem.*, **70**, 581 (1966).

[10] S. O. Grim, W. McFarlane and E. F. Davidoff, *J. Org. Chem.*, **32**, 781 (1967).

[11] S. O. Grim and W. McFarlane, *Nature (London)*, **238**, 995 (1965).

[12] S. O. Grim and W. McFarlane, *Can. J. Chem.*, **46**, 2071 (1968).

[13] G. A. Gray and S. E. Cremer, *Chem. Commun.*, 367 (1972).

[14] G. A. Gray, *J. Amer. Chem. Soc.*, **95**, 5092 (1973).

[15] N. J. De'Ath, D. Z. Denney and D. B. Denney, *Chem. Commun.*, 272 (1972).

[16] L. C. Thomas, *Interpretation of the Infrared Spectra of Organophosphorus Compounds*, Heyden, London, 1974.

[17] D. E. C. Corbridge, "Infrared Spectra of Phosphorus Compounds", *Topics in Phosphorus Chemistry*, Vol. 6, 1969, pp 235–365.

[18] J. R. Ferraro, *J. Inorg. Nuclear Chem.*, **24**, 475 (1962).

[19] S. M. Chackalackal and F. E. Stafford, *J. Amer. Chem. Soc.*, **88**, 4823 (1966).

[20] S. T. King, *J. Phys. Chem.*, **74**, 2133 (1970).

[21] J. Emsley, T. B. Middleton and J. K. Williams, *J. Chem. Soc. Dalton*, 2701 (1973).

References

[22] K. Nakamoto, *Infrared Spectra of Inorganic and Coordination Compounds*, John Wiley & Sons, New York, 1963.

[23] J. Emsley, *J. Chem. Soc. A*, 109 (1970).

[24] G. C. Pimental and A. L. McLellan, *The Hydrogen Bond*, Freeman, San Francisco, 1960.

[25] S. N. Vinogradov and R. H. Linnell, *Hydrogen Bonding*, Van Nostrand Reinhold Co., New York, 1971.

[26] A. B. Burg and J. E. Griffiths, *J. Amer. Chem. Soc.*, **83**, 4333 (1961).

[27] J. Emsley, *J. Chem. Soc. A*, 2702 (1971).

[28] D. F. Peppard, J. R. Ferraro and G. W. Mason, *J. Inorg. Nuclear Chem.*, **16**, 246 (1961); and **7**, 231 (1968).

[29] A. Sequeira and W. C. Hamilton, *J. Chem. Phys.*, **47**, 1818 (1967).

[30] See ref. 16, p 64.

[31] J. G. David and H. E. Hallam, *J. Chem. Soc. A*, 1103 (1966).

Chapter 4
Tricoordinate organophosphorus chemistry

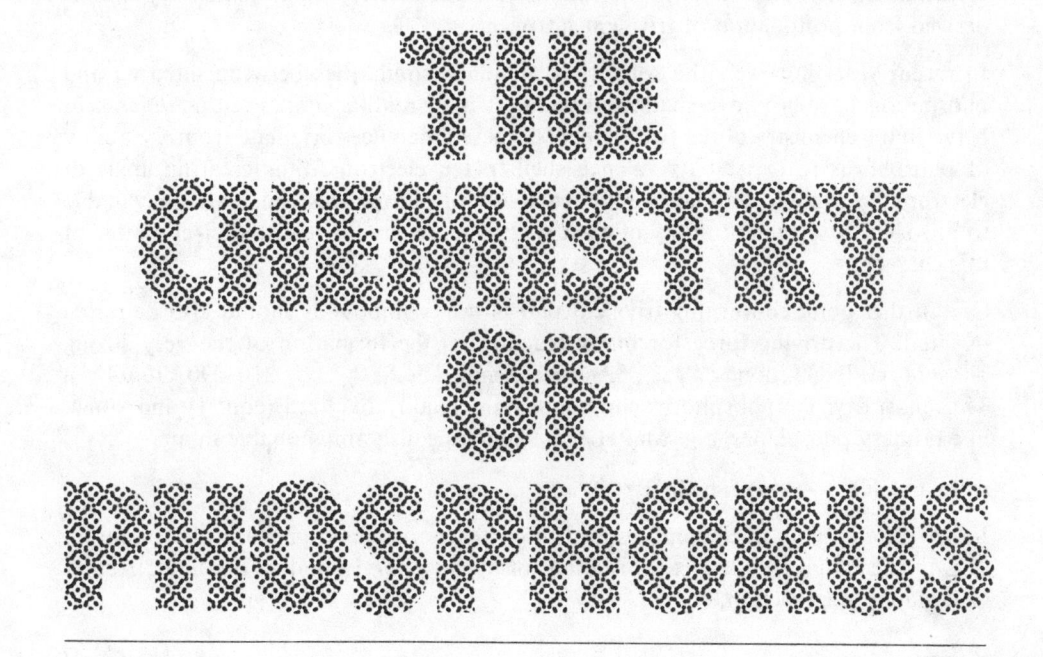

Serious interest in derivatives of the alchemist's mysterious 'coldfire'— phosphorus, began towards the end of the 18th century, when in 1783 Gengembre reported the preparation of a spontaneously inflammable gas from the reaction of phosphorus with caustic potash. The gas was, of course, phosphine (PH_3), the parent of the enormous range of trivalent phosphorus compounds now recorded in the chemical literature[1,2]. The nomenclature of such compounds is complex and notoriously difficult to assimilate but in an attempt to ease the burden, a summary is presented in the Appendix which should enable names to be derived for all the formulae of phosphorus compounds discussed in this chapter.

The fundamental characteristic of all trivalent phosphorus compounds is the presence of a lone pair of electrons on the phosphorus atom. As one might anticipate, this gives rise to a series of reactions which, to a large extent, parallel the chemistry of ammonia and trivalent compounds of nitrogen. Thus by virtue of the lone pair, trivalent phosphorus compounds behave both as *nucleophiles* and as *bases*. The greater size and lower electronegativity of phosphorus however, lead to higher polarizability and hence to higher nucleophilic reactivity than analogous nitrogen compounds. A lower degree of steric hindrance in the transition state may also be an important factor in enhancing the nucleophilic reactivity of phosphorus. On the other hand, phosphine and mono- or dialkylphosphines are weak bases in comparison with their nitrogen counterparts (see Chapter 2). This is probably due to a combination of (i) the lone-pair orbital on phosphorus having a high degree of s character (the lone pair on N is in an sp^3 orbital and therefore more available to a proton) and (ii) solvation effects which enhance the stability of the relatively small quaternary cations derived from protonation of trivalent nitrogen.

In recent years however, the somewhat mundane similarities between nitrogen and phosphorus have been overshadowed by the striking and far more exciting *differences* between the chemistry of the two elements. These differences originate from the ability of phosphorus to expand its valence shell to ten electrons, thus creating areas of electrophilic, biphilic and dienophilic reactivity for phosphorus which are unavailable to nitrogen. A systematic study of each type of reactivity forms the subject matter of this chapter.

One further point concerning trivalent phosphorus compounds should also be borne in mind. The driving force for many reactions is the formation of the very strong phosphoryl, P=O, bond ($D_{P=O}$, 523–631 kJ mol^{-1}; c.f. $D_{N^+-O^-}$ 210–290 kJ mol^{-1}). Thus most trivalent phosphorus compounds are readily oxidized (eqn. 1) and some, like primary phosphines, e.g. $MePH_2$, are spontaneously inflammable in air.

$$n\text{-}Bu_3P + [O] \longrightarrow n\text{-}Bu_3P{=}O \tag{1}$$

Furthermore, many nucleophilic substitution reactions which produce phosphonium salts in the initial step lead to formation of a phosphoryl bond in the final, irreversible step of the reaction, e.g.,

$$Et_2POEt + EtI \longrightarrow Et_3\overset{+}{P}OEt\ I^- \longrightarrow Et_3P{=}O + EtI$$

The high affinity of phosphorus for oxygen is also thought to be a manifestation of valence shell expansion, in this case by back donation of a lone pair on oxygen to vacant d-orbitals on phosphorus ($2p\pi$–$3d\pi$ bonding, see Chapter 2) so that the phosphoryl bond is best represented as a resonance hybrid of ionic and pentacovalent structures:

$$\overset{+}{R_3P}-\overset{-}{O} \longleftrightarrow R_3P=O$$

In any event the strength of the phosphoryl bond dictates the course of an enormous number of reactions throughout phosphorus chemistry[3-5] and the formation of such a bond should always be regarded as a likely final step in any reaction sequence involving phosphorus–oxygen bonds.

Methods for the preparation of trivalent phosphorus compounds are thoroughly documented[1,2,6,7] but the subsequent discussion of numerous reactions, particularly those of the phosphorus halides, will reveal synthetic sequences which ultimately lead to a wide variety of cyclic and acyclic phosphines, phosphinites, phosphonites and phosphites.

4.1 Nucleophilic reactivity

4.1.1 Displacements at saturated carbon[1,8,9]

Phosphorus nucleophiles may be broadly classified into (a) phosphide anions, derived by removal of a proton from primary or secondary phosphines and (b) neutral compounds such as trialkyl- or triarylphosphines and trialkyl or triaryl phosphite esters.

Primary and secondary phosphines are very weak acids (e.g. $PhPH_2$, $pK_a = 24$) but use of a strong base (e.g. sodium in liquid ammonia or n-butyl lithium) produces the phosphide anions. These are extremely powerful nucleophiles, capable of displacing halide ion from alkyl halides and of opening an epoxide ring; even unactivated aryl halides suffer halogen displacement. Phosphide anions are frequently employed in the synthesis of phosphorus heterocycles and examples of all these reactions appear in Table 4.1. Anions derived from phosphinous, phosphonous or phosphorous acids (i.e. anions of the type $R_2P^-(O)$ where R = alkyl, aryl or alkoxy) also fall into class (a) and although they are formally ambident nucleophiles, reaction occurs almost invariably at phosphorus with the formation of a stable phosphoryl group, thus:

$$(EtO)_2\bar{P}=O \ Na^+ \updownarrow (EtO)_2P-\bar{O} \ Na^+ \quad + MeI \longrightarrow (EtO)_2P\overset{O}{\underset{Me}{\nwarrow}} + NaI \qquad (2)$$

Neutral compounds (class (b) nucleophiles) range from phosphines to phosphites and quaternization of trialkyl- or triarylphosphines by alkyl halides with the formation of a phosphonium salt[8] (4.1) is directly analogous to the quaternization of amines (the Menschutkin reaction).

$$R_3P + R^1CH_2X \longrightarrow R_3\overset{+}{P}CH_2R^1 \ \bar{X}$$
$$(4.1)$$

Table 4.1 Nucleophilic displacements by phosphide anions at saturated carbon (THF≡tetrahydrofuran)

$$Ph_2P^- Na^+ + CH_3(CH_2)_3Br \xrightarrow{THF\ room\ temp.} Ph_2P(CH_2)_3CH_3 + NaBr$$
$$(61\%)$$

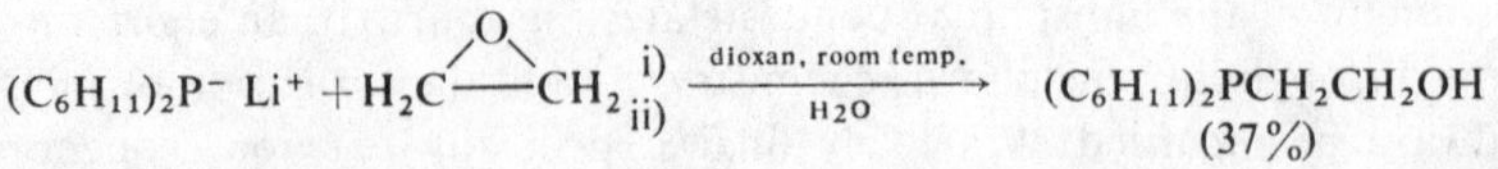

$$(C_6H_{11})_2P^- Li^+ + H_2C\underset{}{\overset{O}{-}}CH_2 \xrightarrow[ii)\ H_2O]{i)\ dioxan,\ room\ temp.} (C_6H_{11})_2PCH_2CH_2OH$$
$$(37\%)$$

$$Ph_2P^- Li^+ + p\text{-}MeC_6H_4X \xrightarrow{THF\ room\ temp.} p\text{-}MeC_6H_4Ph_2 + LiX$$
$$X = Br, I$$

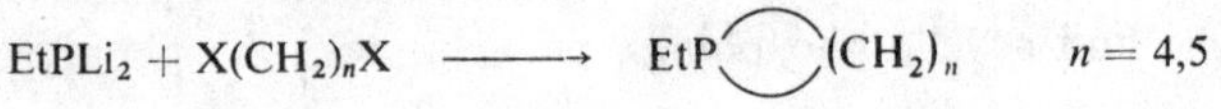

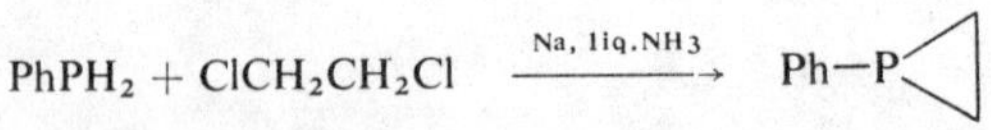

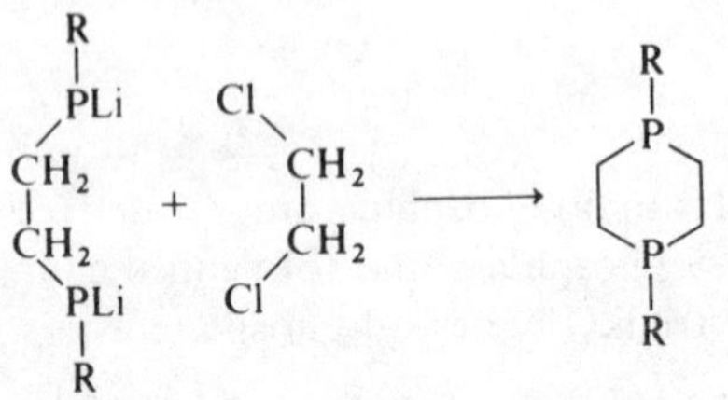

$$EtPLi_2 + X(CH_2)_nX \longrightarrow EtP\underset{}{\bigcirc}(CH_2)_n \quad n = 4, 5$$

$$PhPH_2 + ClCH_2CH_2Cl \xrightarrow{Na,\ liq.NH_3} Ph\text{-}P\triangleleft$$

The reaction, which may be carried out in a variety of organic solvents (benzene or acetonitrile being common examples), occurs readily with primary and secondary alkyl halides and shows the typical nucleophilic reactivity sequence, X = I > Br > Cl. Tertiary halides usually do not react although the corresponding phosphonium salts may sometimes be obtained by an acid-catalyzed displacement from tertiary alcohols, presumably by an S_N1 mechanism.

$$R_3P + R_3^1C\text{-}OH \xrightarrow{HBr} R_3\overset{+}{P}\text{-}CR_3^1\ Br^- + H_2O$$

All the available evidence for primary and secondary halides suggests an S_N2 mechanism. Thus displacement at a chiral carbon occurs with inversion of configuration, consistent with a bimolecular transition state.

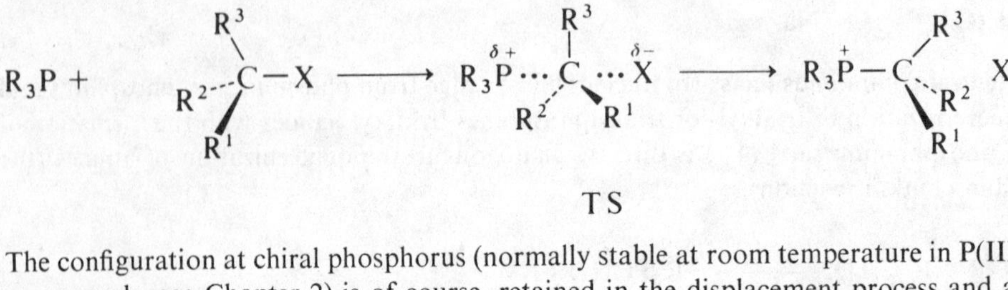

TS

The configuration at chiral phosphorus (normally stable at room temperature in P(III) compounds, see Chapter 2) is of course, retained in the displacement process and as

4.1 Nucleophilic reactivity

we shall see later, this is often an important step in relating configuration during stereochemical studies of reactions at phosphorus[10]. As early as 1934, kinetic studies established second-order behaviour and showed that phosphorus was a far more powerful nucleophile than nitrogen.

$$PhZEt_2 + EtI \longrightarrow Ph\overset{+}{Z}Et_3 \; \overset{-}{I} \quad Z = N,P; \text{ relative rates, } P:N = 520:1$$

This is exemplified further by the exclusive reaction at phosphorus between methyl iodide and the phosphorus amide (4.2)

$$Me_2PNMe_2 + MeI \longrightarrow Me_3\overset{+}{P}NMe_2 \; \overset{-}{I} \qquad (4.2)$$

The quaternization of diethylphenylphosphine, $PhPEt_2$, with ethyl iodide in acetone is accelerated by electron-donating groups in the phenyl ring. The small Hammett ρ value for this reaction (-1.01) may be compared with a ρ value of -2.77 for the quaternization of N,N-diethylaniline, $PhNEt_2$, and is a reflection of the lack of conjugation of the lone pair on phosphorus with the phenyl ring.

Because of the larger size of the phosphorus atom, steric factors are less important than in quaternization at nitrogen. Thus, for phosphines, the rate of nucleophilic displacement of halide increases in the order $PH_3 < RPH_2 < R_2PH < R_3P$ which parallels an increase in basicity whereas with amines (all of similar pK_a) a reverse rate order is observed which is due to an increasing degree of steric hindrance as the amine changes from primary to tertiary.

Phosphines may also be quaternized by trialkoxycarbonium ions (eqn. 3)

$$R_3P + Et\overset{+}{O}C(OEt)_2 \; \overset{-}{BF_4} \longrightarrow R_3\overset{+}{P}Et \; \overset{-}{BF_4} + O=C(OEt)_2 \qquad (3)$$

and under acid conditions are sufficiently nucleophilic to displace alkoxy groups from ethers, thus:

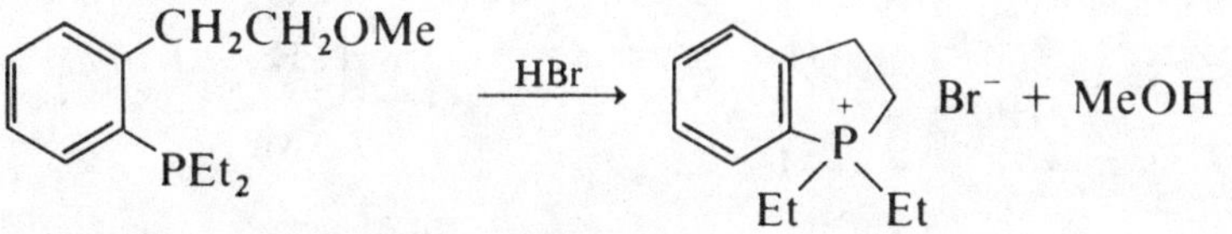

$$\underset{PEt_2}{\overset{CH_2CH_2OMe}{\text{[Ar]}}} \xrightarrow{HBr} \text{[cyclic }\overset{+}{P}Et_2\text{]} \; Br^- + MeOH \qquad (4)$$

Neutral phosphines are also capable of opening an epoxide ring. The resultant betaine which is directly analogous to the intermediate formed in the Wittig reaction (Chapter 7) collapses to form phosphine oxide and an olefin (eqn. 5).

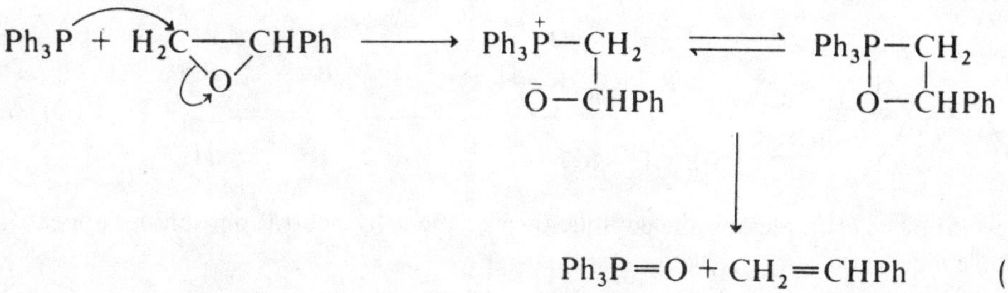

$$Ph_3P + H_2C\overset{\displaystyle\frown}{\underset{O}{\text{—}}}CHPh \longrightarrow \underset{\overset{|}{O}-CHPh}{Ph_3\overset{+}{P}-CH_2} \rightleftharpoons \underset{\overset{|}{O}-CHPh}{\overset{|}{Ph_3P-CH_2}}$$

$$\downarrow$$

$$Ph_3P=O + CH_2=CHPh \qquad (5)$$

The proposed mechanism receives support from stereochemical studies since Boskin and Denney have shown that the major product from *cis*-2,3-butene oxide is the *trans*-olefin[11]. Rotation about the 2,3 bond of (4.3) is the necessary prerequisite to collapse of the betaine (4.4).

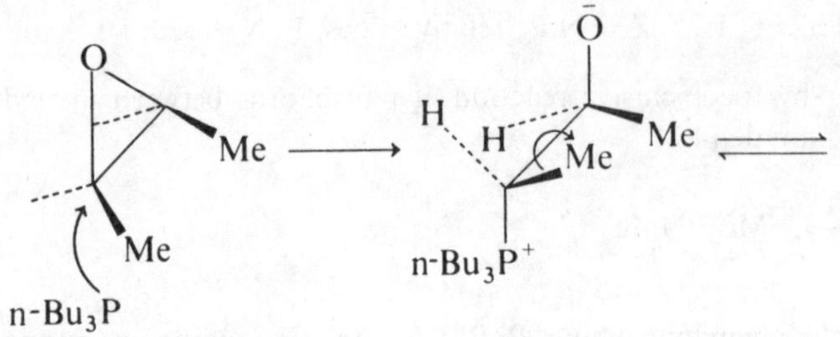

(4.3)

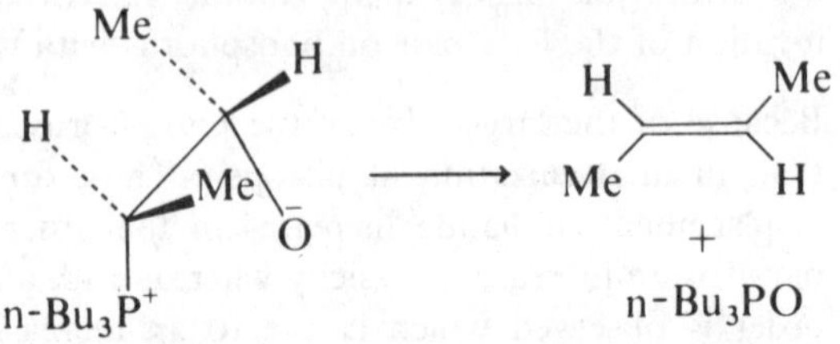

(4.4)

Small but significant amounts of the olefin of the opposite configuration may occur through attack on oxygen (see Section 4.3) or by equilibration to (4.6) through an ylid structure (4.5).

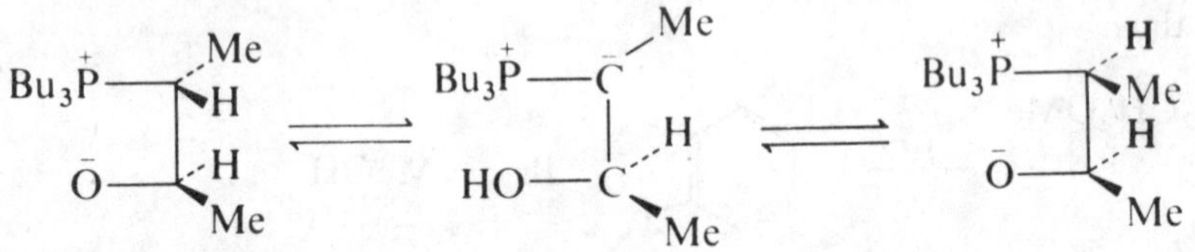

(4.4)   (4.5)   (4.6)

Recently, the deoxygenation of epoxides by lithium diphenylphosphide (in the presence of methyl iodide which quaternizes the initial product) has been shown to be stereospecific leading to an alkene of opposite configuration to that of the starting epoxide[12].

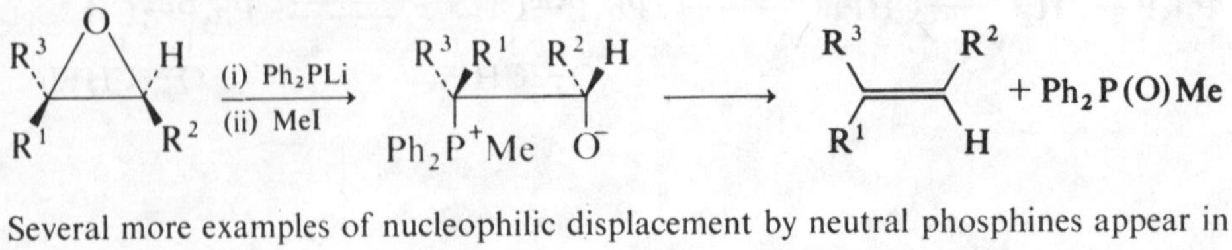

Several more examples of nucleophilic displacement by neutral phosphines appear in Table 4.2.

4.1 Nucleophilic reactivity

Table 4.2 Nucleophilic displacement by phosphines at saturated carbon.

Ph$_3$P + PhCH$_2$Br $\longrightarrow$ Ph$_3\overset{+}{P}$CH$_2$Ph $\overset{-}{B}$r

Me.Pr^n.PhP + PhCH$_2$Br $\longrightarrow$ Me.Pr^n.Ph$\overset{+}{P}$CH$_2$Ph $\overset{-}{B}$r

$[\alpha]_D^{20}$, +16.8° $\qquad\qquad\qquad\qquad$ $[\alpha]_D^{20}$, +36.8°

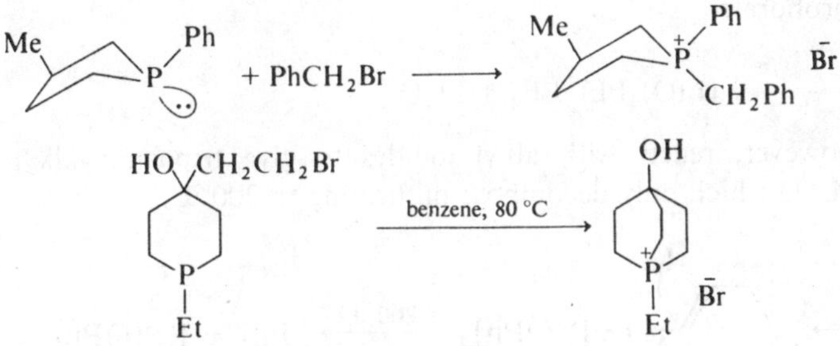

The Michaelis–Arbusov reaction[1,3,4,13,14]: When the phosphorus nucleophile contains an alkoxy group reaction with alkyl halides takes a different course. Thus triethyl phosphite reacts with methyl iodide to give diethyl methylphosphonate (4.7). This reaction, known as the Michaelis–Arbusov reaction, is quite general and may be extended to the synthesis of phosphinates (4.8) and phosphine oxides (4.9).

$$(EtO)_3P + MeI \longrightarrow (EtO)_2\overset{O}{\underset{\|}{P}}Me + EtI \quad (4.7)$$

$$(RO)_2PR^1 + R^1I \longrightarrow RO\overset{O}{\underset{\|}{P}}R^1_2 + RI \quad (4.8)$$

$$ROPR^1_2 + R^1I \longrightarrow O=PR^1_3 + RI \quad (4.9)$$

The most important feature is the formation of a carbon–phosphorus bond and this has proved useful in the synthesis of phosphonate esters which frequently form the basic molecular skeleton of insecticides and nerve gases (Chapter 12) and recently have been employed as starting materials in the Wittig reaction (Chapter 7).

The initial step, which involves formation of an alkoxy phosphonium salt (4.10), is followed by nucleophilic displacement of an alkyl group by the displaced halide ion to generate the phosphoryl bond.

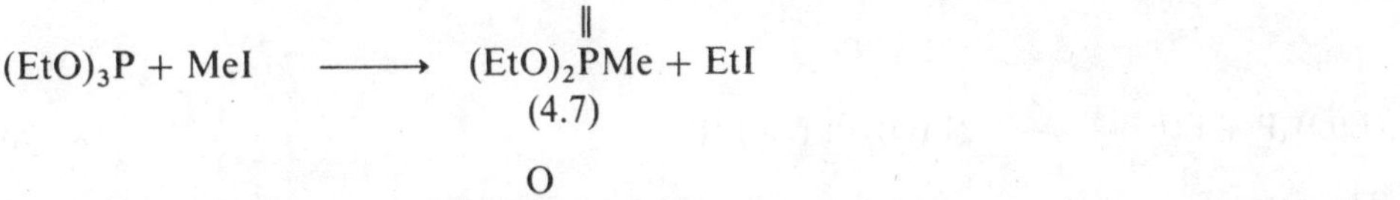

$$(RO)_3P + R^1{-}I \longrightarrow I^- \quad R{-}O{-}\overset{+}{P}(OR)_2R^1 \longrightarrow O=\underset{\underset{OR}{|}}{\overset{\overset{OR}{|}}{P}}{-}R^1 + RI$$
$$(4.10)$$

A similar but less facile reaction, occurs with trialkylthiophosphites, $(RS)_3P$, but tri-N,N-dialkylphosphoramidites, $(R_2N)_3P$, only react as far as the phosphonium salt, $(R_2N)_3PMe^+I^-$. The second, dealkylation step of the reaction is a simple bimolecular displacement on carbon as evidenced by the work of Gerrard and Green who demonstrated that the phosphite derived from (+)2-octanol reacted with ethyl iodide to give 2-iodooctane with the opposite configuration to that of the starting alcohol[15]. With alkyl phosphites, phosphonites or phosphinites isolation of the intermediate alkoxy phosphonium salts can only be achieved by using weakly nucleophilic anions such as tetrafluoroborate.

$$(EtO)_3P + Et_3\overset{+}{O}\overset{-}{BF_4} \longrightarrow (EtO)_3\overset{+}{P}Et\ \overset{-}{BF_4} + Et_2O$$

Triphenyl phosphite however, reacts with alkyl iodides to give triphenoxyalkylphosphonium iodides (4.11) which only decompose on heating to 200 °C.

$$(PhO)_3P + RI \longrightarrow \underset{R}{\underset{|}{Ph\text{-}O\text{-}\overset{+}{P}(OPh)_2}}\ I^- \xrightarrow{200\ °C} PhI + R\overset{O}{\overset{\|}{P}}(OPh)_2$$

(4.11)

This of course, parallels the resistance of the aryl halides to nucleophilic addition–elimination reactions.

The question of the rate-determining step in the normal Michaelis–Arbusov reaction was in dispute for many years but was finally resolved by Asknes and Asknes in 1964 by a quantitative study of the reaction of eqn. (6)[16]

$$(EtO)_3P + EtI \longrightarrow (EtO)_2\overset{O}{\overset{\|}{P}}Et + EtI \qquad (6)$$

The rate of product formation was followed by infrared spectroscopy and the study showed:

(a) ethyl iodide was not consumed during the reaction;
(b) the rate $\propto$ [EtI];
(c) the rate was not enhanced by increasing [I$^-$];
(d) the reaction was much faster in acetonitrile (relative permittivity, $\varepsilon_r = 36$) than in benzene ($\varepsilon_r = 2.3$)

These results are uniquely consistent with a rate-limiting first step. The transition state (TS) leading to formation of the phosphonium salt would be stabilized by polar solvents and if the second step had been rate determining, increase in the concentration of iodide ion would have enhanced the rate. Furthermore, the results are consistent with earlier relative rate studies which revealed rate sequences of the type, $Et_2POEt > EtP(OEt)_2 > (EtO)_3P$ for a given alkyl halide and for a given phosphite, relative rates in the order, MeI > EtI > i-PrI. It should be noted that in the Asknes system, ethyl iodide is both reactant and product which serves to illustrate the fact that catalytic quantities of ethyl iodide are sufficient to convert triethyl phosphite into diethyl ethylphosphonate.

4.1 Nucleophilic reactivity

Dialkyl phosphites exist mainly in the phosphoryl form (4.13) and are therefore unreactive in the Michaelis–Arbusov rearrangement.

$$(RO)_2POH \rightleftharpoons (RO)_2P(=O)H$$

(4.12) (4.13)

The anions however, which are readily available as sodium salts or which may be generated in situ from the neutral compounds and tertiary amines, react readily with alkyl halides to give high yields of phosphonates (eqn. 7)[17,18]

$$(RO)_2\bar{P}=O\ Na^+ + R^1I \longrightarrow (RO)_2\overset{O}{\underset{\|}{P}}R^1 \qquad (7)$$

This (the Michaelis–Becker reaction) is frequently the preferred method for the synthesis of phosphonates since no dealkylation step is involved and the opportunity to form mixtures of phosphonates (when R and R^1 are different) is minimized.

4.1.2 Displacements at halogen[19]

Contrary to popular belief, nucleophilic attack at a halogen atom is a common occurence in organic chemistry. One is accustomed to viewing bromine as an electrophile in additions to olefins or bromination of phenol but by the same token, the reactions may be regarded as nucleophilic attack on bromine just as base-catalyzed bromination of a ketone involves attack on the halogen by an enolate anion, e.g. eqn. (8)

$$\begin{array}{c} CH_3.CO.CH_2^- \\ \updownarrow \\ CH_3.C(O^-)=CH_2 \end{array} + Br_2 \longrightarrow CH_3.CO.CH_2Br + Br^- \qquad (8)$$

Since the halogen atom is transferred with only a sextet of electrons, reactions of this type are often said to involve 'positive' halogen and typical sources of such positive halogen are halogen molecules, hypohalites and N-haloamides (e.g. N-bromosuccinimide). Attack at halogen by P(III) compounds is particularly common due to (a) the high polarizability of phosphorus which effectively stabilizes the positive charge generated on the halogen in the transition state and (b) the high energy of phosphorus halogen bonds.* It is often difficult however to establish attack at halogen because the phosphorus–halogen bond, formed initially, is so susceptible to nucleophilic displacement of halogen. Thus for a molecule R—X (where R may be a group

---

* The energy of formation of the P—Cl bond in PCl_3 is 319 kJ mol^{-1} (76 kcal mol^{-1}) whereas the corresponding energy of the N—Cl bond in NCl_3 is only 193 kJ mol^{-1} (46 kcal mol^{-1}). The energy of the P—Cl bond in $POCl_3$ is estimated at 347 kJ mol^{-1} (83 kcal mol^{-1}) which suggests that the energies of the P—X bonds in the quaternary salts generated by nucleophilic attack at halogen are not very different from those in the neutral, P(III), molecules. See Chapter 2 for further details.

in which carbon, nitrogen, oxygen or sulphur is bound to the halogen, X) nucleophilic displacement may occur in one of two ways, A or B.

$$R^1_3P + R-X \longrightarrow R^1_3\overset{+}{P}R \quad X^- \qquad A$$

$$R^1_3P + X-R \longrightarrow R^1_3\overset{+}{P}X \quad R^- \qquad B$$
$$(4.14)$$

$$R^- + R^1_3\overset{+}{P}-X \longrightarrow R\overset{+}{P}R^1_3 \quad X^- \qquad C$$

Unfortunately in path B, the displaced anion, R^-, is almost invariably sufficiently nucleophilic to effect displacement of halogen (reaction C) and so generate the product of reaction A. This means that attack on halogen must be detected either directly by trapping the halophosphonium salt or the anion of (4.14) or indirectly by stereochemical studies at phosphorus. It is clear that reactions A and B would proceed with retention of configuration at phosphorus whereas reaction C, if it proceeds by an S_N2 displacement at phosphorus (or its equivalent, see Chapter 7) would lead to inversion, giving overall inversion for the combination of B and C. Examples of such detection procedures appear later but with respect to stereochemical studies, it is necessary to bear in mind that phosphonium salts may be in equilibrium with pentacovalent structures (4.15) which may afford racemization either by destroying the chirality at phosphorus or by ligand reorganization (e.g. by pseudorotation, Chapter 2).

$$R_3\overset{+}{P}R^1 \; X^- \;\rightleftharpoons\; R_3PR^1X$$
$$(4.15)$$

The ease of nucleophilic attack at halogen decreases in the sequence, $I > Br > Cl$ which is the reverse of the ionization potentials of the halogen atoms ($I < Br < Cl$) and is presumably a reflection of the ability of the halogen atoms to bear a positive charge in the TS of the displacement process. The ionization potentials of atoms within a given group of the Periodic Table decrease with increasing atomic size and therefore, other things being equal, one might predict that phosphorus nucleophiles would attack the halogen of oxygen–halogen bonds more selectively than the halogen of sulphur–halogen bonds. Although there is some evidence for this generalization, it is by no means a hard and fast rule.

Finally it should be noted that the *rate* at which a group will be displaced from a particular halogen atom will depend upon the acidity of its conjugate acid which is a measure of leaving-group ability. In other words, when R of reaction (B) is stabilized by resonance or inductive electron withdrawal, the rate of attack at halogen will be enhanced. Bearing these points in mind, let us consider the various kinds of halogen centre.

4.1 Nucleophilic reactivity

4.1.2.1 Halogen–halogen bonds

The simplest halogenating agents are halogen molecules which react vigorously with tertiary phosphines to give dihalophosphoranes, e.g.

$$Ph_3P + Br_2 \longrightarrow Ph_3\overset{+}{P}Br\ Br^- \rightleftharpoons Ph_3PBr_2$$

$$PhPCl_2 + Cl_2 \longrightarrow PhPCl_4$$

Trialkyl phosphites (or mixed aryl alkyl phosphites) undergo a variation of the Arbusov reaction to provide a useful synthesis of phosphoryl chlorides (4.16). Phosphoryl bromides and iodides may be formed in solution but decompose on distillation.

$$(EtO)_3P + Cl_2 \longrightarrow [(EtO)_3\overset{+}{P}Cl\ Cl^-] \longrightarrow (EtO)_2\overset{O}{\underset{\|}{P}}Cl + EtCl \quad (4.16)$$

As observed in the Arbusov reaction, optically active tri-2-octyl phosphite gave 2-chlorooctane with inversion of configuration[20]. This prompted the proposal of initial nucleophilic attack to form an intermediate chlorotrialkoxyphosphonium salt followed by an S_N2 displacement by chloride ion.

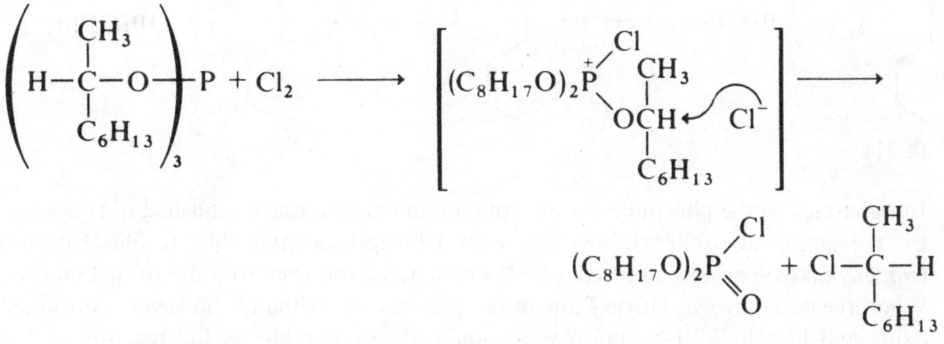

$$\left(\begin{array}{c} CH_3 \\ | \\ H-C-O \\ | \\ C_6H_{13} \end{array} \right)_3 P + Cl_2 \longrightarrow \left[(C_8H_{17}O)_2\overset{+}{P} \begin{array}{c} Cl \\ CH_3 \\ OCH \\ | \\ C_6H_{13} \end{array} Cl^- \right] \longrightarrow$$

$$(C_8H_{17}O)_2P \begin{array}{c} Cl \\ \| \\ O \end{array} + Cl-\begin{array}{c} CH_3 \\ | \\ C-H \\ | \\ C_6H_{13} \end{array}$$

In the general scheme of things however, it does not preclude a reaction via a molecular (or free radical) pathway to give a pentacovalent structure which subsequently ionizes to phosphonium ion and hence to products (eqn. 9).

$$(RO)_3P + X_2 \longrightarrow (RO)_3PX_2 \rightleftharpoons (RO)_3\overset{+}{P}X\ X^- \longrightarrow (RO)_2P(O)X + RX \quad (9)$$

Obviously the same mechanistic possibility applies to the reactions of phosphines with halogens.

The bond between two phosphorus atoms is readily cleaved by halogens. Thus tetraphenylbiphosphine (4.17) reacts with bromine to give diphenylphosphinous bromide (4.18)[21].

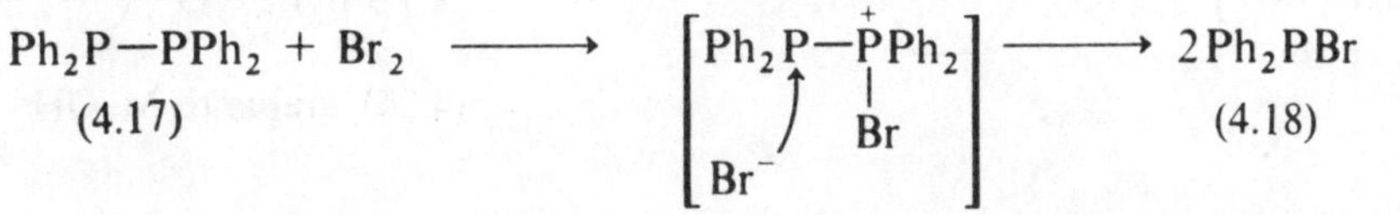

$$Ph_2P-PPh_2 + Br_2 \longrightarrow \left[\begin{array}{c} Ph_2P-\overset{+}{P}Ph_2 \\ | \\ Br \\ Br^- \end{array} \right] \longrightarrow 2Ph_2PBr$$
$$(4.17) \hspace{5cm} (4.18)$$

Triaryl phosphites also combine with halogens but the intermediate apparently reacts with a second mole of triaryl phosphite to give disproportionation products (4.19) and (4.20).

$$(PhO)_3P + X_2 \longrightarrow [(PhO)_3PX_2] \xrightarrow{(PhO)_3P} (PhO)_2PX + (PhO)_4PX$$
$$\qquad\qquad\qquad\qquad\qquad\qquad\qquad\qquad (4.19) \qquad\quad (4.20)$$

4.1.2.2 Halogen–oxygen bonds: hypochlorites

Tertiary phosphines and trialkyl- or triarylphosphites are readily oxidized to phosphine oxides and phosphates by alkyl hypochlorites, e.g. (eqns. 10 and 11).

$$Ph_3P + (CH_3)_3COCl \longrightarrow Ph_3PO + (CH_3)_3CCl \qquad (10)$$

$$(PhO)_3P + EtOCl \longrightarrow (PhO)_3PO + EtCl \qquad (11)$$

Once again the final step in the reaction appears to be an S_N2 displacement by halide ion on an intermediate alkoxyphosphonium salt since reaction of triphenylphosphite with optically active tetrahydrolinalyl hypochlorite (4.21) gives tetrahydrolinalyl chloride with inversion of configuration[22].

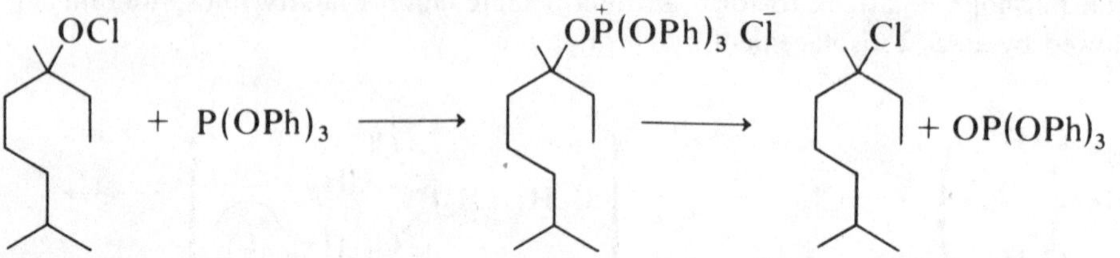

Initial attack of the phosphite (or phosphine) on halogen was established in two ways. First, reaction of triphenylphosphine with a bridgehead hypochlorite (4.22) gave a triphenylalkoxyphosphonium salt (4.24) which was stable even to water or methanol[22]. When the reaction was carried out *in the presence* of methanol however, phosphine oxide and bicyclo-2,2,1-heptanol were obtained, presumably by the reaction of the intermediate (4.23) with methanol and subsequent demethylation.

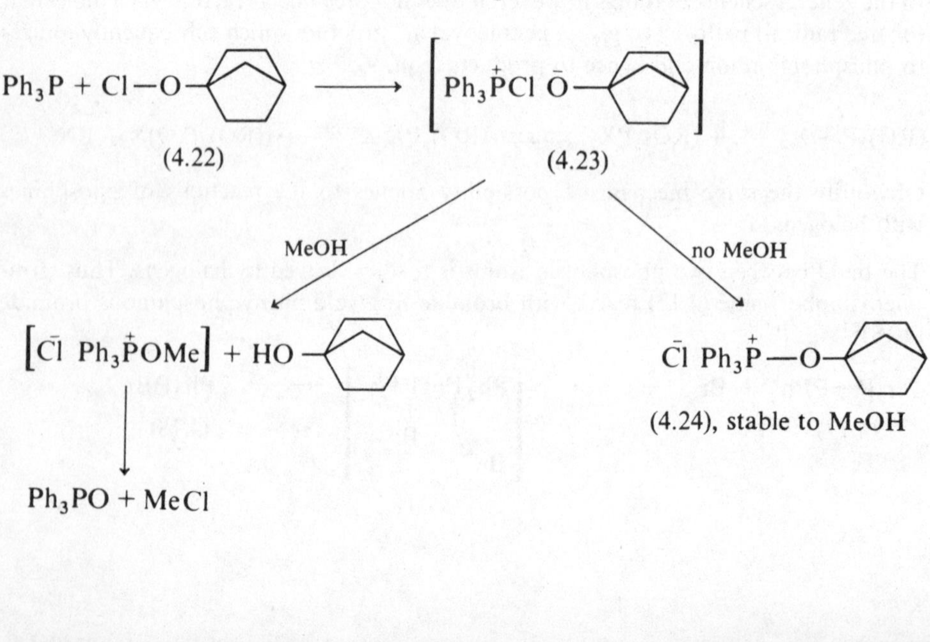

4.1 Nucleophilic reactivity

Secondly, Denney and Hanifin investigated the reaction of an optically active phosphine (4.25) with *tert*-butyl hypochlorite[23]. The presence of methanol led to a phosphine oxide (4.26) with predominant inversion at phosphorus. This was rationalized as shown below.

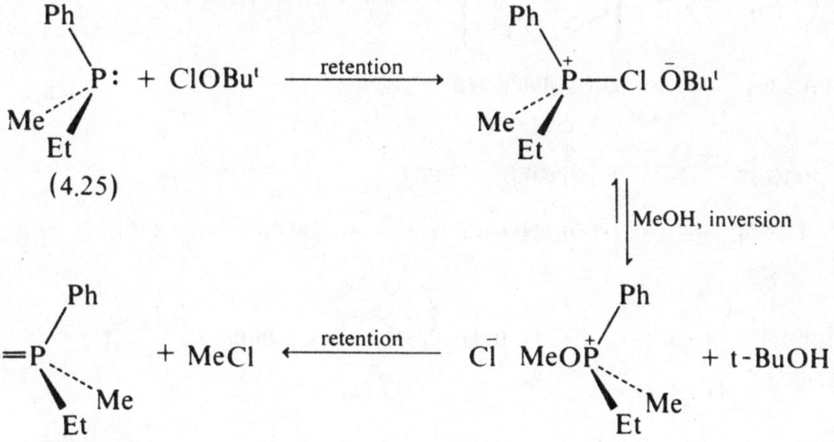

$$\underset{\underset{Et}{Me}}{Ph}\!\!\diagdown\!\!P: \;+\; ClOBu^t \;\xrightarrow{\text{retention}}\; \underset{\underset{Et}{Me}}{Ph}\!\!\diagdown\!\!\overset{+}{P}\!\!-\!\!Cl\;\;\bar{O}Bu^t$$

(4.25)

$$\Bigg\updownarrow \text{MeOH, inversion}$$

$$O\!=\!\underset{\underset{Et}{Me}}{\overset{Ph}{P}} \;+\; MeCl \;\xleftarrow{\text{retention}}\; \bar{Cl}\;\;MeO\overset{+}{\underset{\underset{Et}{Me}}{P}}\!\!\!\overset{Ph}{\diagup} \;+\; t\text{-}BuOH$$

(4.26)

The difficulty encountered in studying these systems is illustrated by the fact that the above reaction, when carried out in pentane, gives totally racemic oxide, presumably via racemization of a pentacovalent intermediate. Further examples of nucleophilic attack at halogen by trivalent phosphorus compounds (including halogen–nitrogen and halogen–sulphur bonds) are presented in Table 4.3.

4.1.2.3 Carbon–halogen bonds

Triphenylphosphine reacts with carbon tetrabromide to give a dibromomethylene ylid (4.27) and dibromotriphenylphosphorane (4.28) and the most likely mechanistic pathway is outlined below[24].

$$Ph_3P \;+\; Br\!-\!CBr_3 \;\longrightarrow\; [Ph_3\overset{+}{P}Br\;\;\bar{C}Br_3] \;\longrightarrow\; Ph_3\overset{+}{P}\!-\!CBr_3\;\;Br^-$$

$$Ph_3P \;+\; Br\!\frown\!\overset{\frown}{C}Br_2\!-\!\overset{+}{P}Ph_3\;\;\bar{Br} \;\longrightarrow\; Ph_3PBr_2 \;+\; Br_2\bar{C}\!-\!\overset{+}{P}Ph_3$$

$$\qquad\qquad\qquad\qquad\qquad\qquad\qquad (4.28)\qquad\quad (4.27)$$

The reaction with carbon tetrachloride is less facile but follows a similar course[25]. As expected, trialkyl phosphites react with carbon tetrachloride in an Arbusov-type reaction to give high yields of trichloromethyl phosphonates, (4.29).

$$(EtO)_3P \;+\; CCl_4 \;\xrightarrow{\text{heat}}\; (EtO)_2\overset{\overset{O}{\|}}{P}CCl_3 \;+\; EtCl$$

(4.29)

Table 4.3 Nucleophilic attack of trivalent phosphorus compounds at halogen.

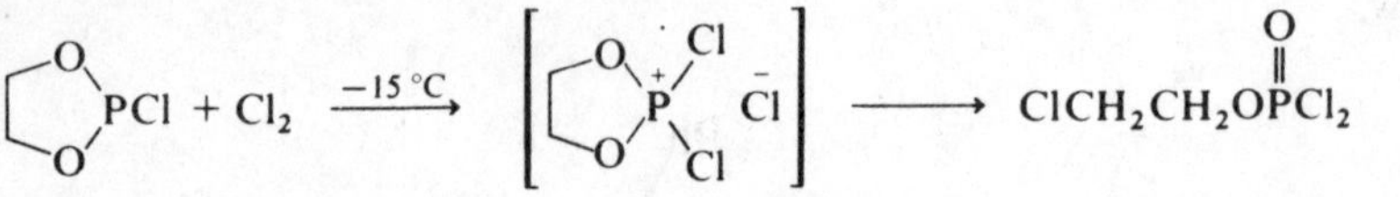

$$\underset{O}{\overset{O}{\diagdown}}PCl + Cl_2 \xrightarrow{-15\,°C} \left[\underset{O}{\overset{O}{\diagdown}}\overset{+}{P}\underset{Cl}{\overset{Cl}{\diagdown}}\right] \longrightarrow ClCH_2CH_2O\overset{O}{\overset{\|}{P}}Cl_2$$

$(C_4H_9)_2PH + Br_2 \longrightarrow (C_4H_9)_2PBr + HBr$

$Et_2NCl + (EtO)_3P \longrightarrow Et_2N\overset{O}{\overset{\|}{P}}(OEt)_2 + EtCl$ †

$(PhO)_3P + Et_2NCl \longrightarrow (PhO)_3\overset{+}{P}NEt_2 \; Cl^- \xrightarrow{heat} (PhO)_3P=NEt + EtCl$ †

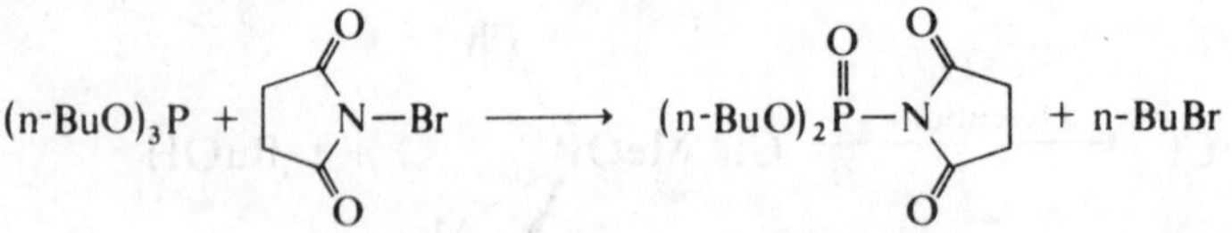

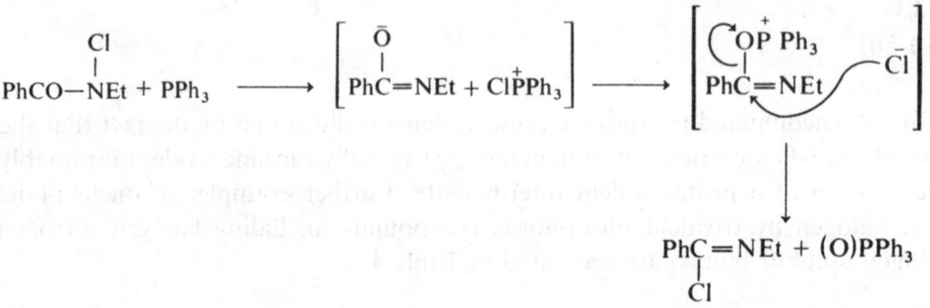

$(n\text{-}BuO)_3P + \underset{O}{\overset{O}{\diagup}}\!\!\!\!\diagdown N\text{-}Br \longrightarrow (n\text{-}BuO)_2\overset{O}{\overset{\|}{P}}-N\underset{O}{\overset{O}{\diagup}}\!\!\!\!\diagdown + n\text{-}BuBr$ †

$$\underset{PhCO-NEt}{\overset{Cl}{|}} + PPh_3 \longrightarrow \left[\underset{PhC=NEt}{\overset{\bar{O}}{|}} + Cl\overset{+}{P}Ph_3\right] \longrightarrow \left[\underset{PhC=NEt}{\overset{OPPh_3}{\frown}} \; Cl^-\right]$$

$$\downarrow$$

$$\underset{Cl}{\overset{|}{PhC}}=NEt + (O)PPh_3 \quad †$$

$(EtO)_3P + MeSCl \longrightarrow (EtO)_2\overset{O}{\overset{\|}{P}}SMe + EtCl$ †

$(RO)_3P + Cl\text{-}SO_2\text{-}Cl \longrightarrow [(RO)_3\overset{+}{P}Cl + SO_2\uparrow + Cl^-] \longrightarrow (RO)_2\overset{O}{\overset{\|}{P}}Cl + RCl$ †

$$R_2P\underset{SCl}{\overset{\diagup O}{\diagdown}} + :PR^1_3$$

$$\longrightarrow \left[R_2P\underset{S}{\overset{\diagup O}{\diagdown}}\cdots Cl\overset{+}{P}R^1_3\right] \longrightarrow$$

$$\longrightarrow \left[R_2P\underset{SPR^1_3}{\overset{\diagup O}{\diagdown}}\overset{+}{} \; Cl^-\right] \longrightarrow R_2\overset{O}{\overset{\|}{P}}Cl + S=PR^1_3$$

$$\longrightarrow \left[R_2P\underset{O-PR^1_3}{\overset{\diagup S}{\diagdown}}\overset{+}{} \; Cl^-\right] \longrightarrow R_2\overset{S}{\overset{\|}{P}}Cl + O=PR^1_3$$ †

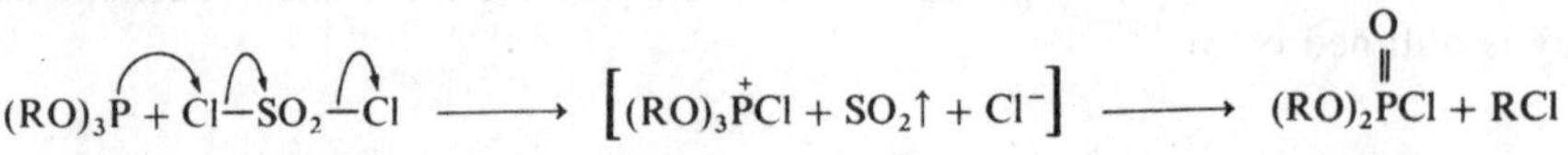

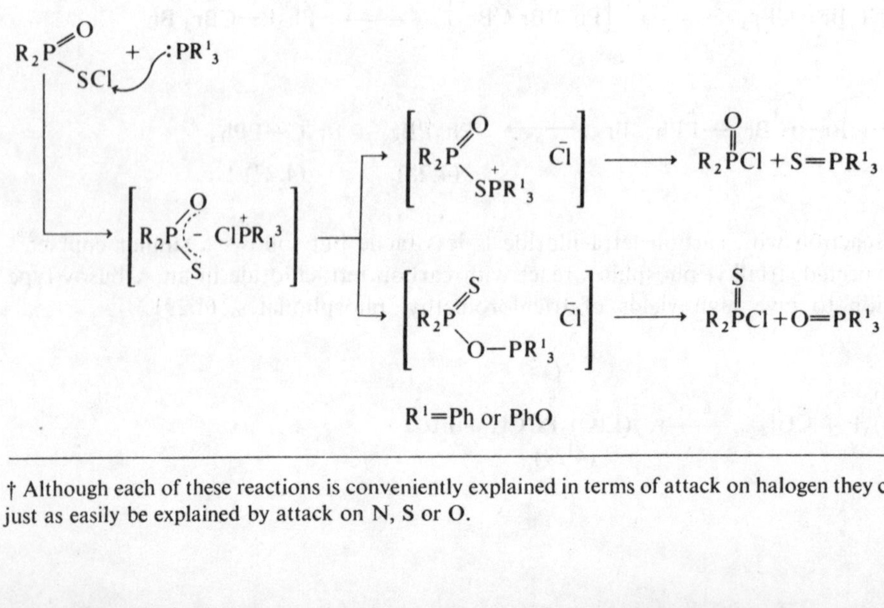

$R^1 = Ph$ or PhO

---

† Although each of these reactions is conveniently explained in terms of attack on halogen they could just as easily be explained by attack on N, S or O.

4.1 Nucleophilic reactivity

Kinetic studies with a variety of phosphites[26] revealed a rate sequence similar to that observed in the Arbusov reaction, i.e.

$(i\text{-PrO})_3P > (EtO)_3P > (MeO)_3P > (EtO)_2POPh > EtOP(OPh)_2$

which suggests rate-limiting nucleophilic attack by phosphorus, probably on chlorine.*

The mechanistic picture is by no means clear however, since evidence for a radical mechanism (catalysis by light and peroxides; inhibition by hydroquinone) has been reported for the reactions of polyhalomethanes with trialkyl phosphites. It seems likely that both ionic and free radical mechanisms may operate under the appropriate conditions since excellent yields of trichloromethylphosphonates may be obtained from reactions conducted in the dark.

Nucleophilic attack on α-haloketones affords one of the more interesting sequences of reactions at carbon–halogen bonds. Attack at halogen produces a resonance stabilized enolate anion (4.30) which is an ambident nucleophile and may attack the phosphonium ion via oxygen to give an enol-phosphonium salt (4.31) or via carbon to give a β-ketophosphonium salt (4.32)

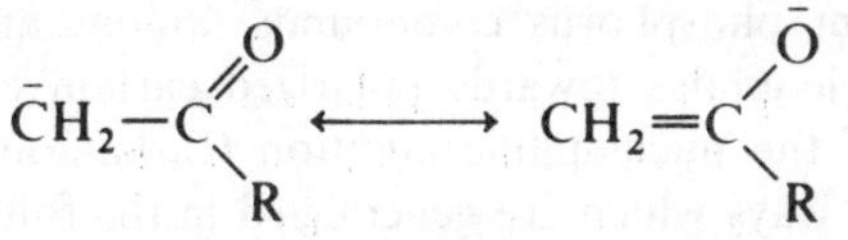

$$\bar{C}H_2-C\overset{O}{\underset{R}{\diagdown}} \longleftrightarrow CH_2=C\overset{\bar{O}}{\underset{R}{\diagdown}}$$

(4.30)

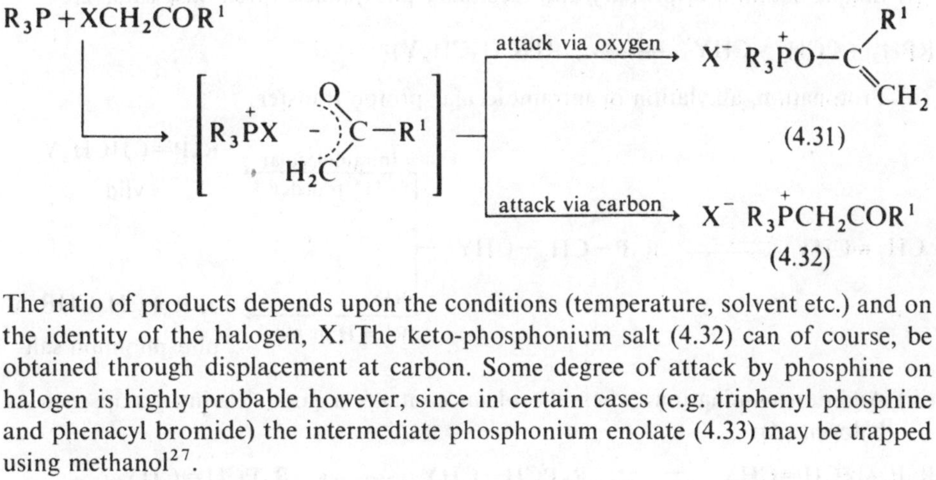

$R_3P + XCH_2COR^1$

$\longrightarrow \left[R_3\overset{+}{P}X \quad \overset{O}{\underset{H_2C}{\diagdown}}C-R^1 \right]$

attack via oxygen $\longrightarrow X^- \; R_3\overset{+}{P}O-C\overset{R^1}{\underset{CH_2}{\diagdown}}$ (4.31)

attack via carbon $\longrightarrow X^- \; R_3\overset{+}{P}CH_2COR^1$ (4.32)

The ratio of products depends upon the conditions (temperature, solvent etc.) and on the identity of the halogen, X. The keto-phosphonium salt (4.32) can of course, be obtained through displacement at carbon. Some degree of attack by phosphine on halogen is highly probable however, since in certain cases (e.g. triphenyl phosphine and phenacyl bromide) the intermediate phosphonium enolate (4.33) may be trapped using methanol[27].

$Ph_3P + BrCH_2COPh \longrightarrow [Ph_3\overset{+}{P}Br \; \bar{C}H_2COPh] \longrightarrow Ph_3\overset{+}{P}CH_2COPh \; Br^-$

(4.33) (4.34)

$\downarrow$ MeOH

$Ph_3PO + MeBr \longleftarrow [Ph_3\overset{+}{P}OMe \; \bar{B}r] + CH_3COPh$

---

* Carbon tetrabromide reacts much more vigorously with alkyl phosphites to give a complex mixture of products.

On the other hand, phenacyl chloride reacts with triphenyl phosphine to give the keto-phosphonium salt (4.32, R = R^1 = Ph; X = Cl)[27]. In this case the salt must be formed by direct nucleophilic displacement on the α-carbon since almost no change in yield occurs in the presence of methanol. Thus two compounds, differing only in the nature of the halogen, react with the same reagent under the same conditions to give analogous products (equivalent to 4.32 where R = R^1 = Ph and X = Br or Cl) but *by different mechanisms*. This is an excellent illustration of the caution to be applied to mechanistic interpretation in this area.

With α-haloesters, α-halonitriles and α-haloamides both mechanisms (i.e. attack at carbon and attack at halogen) probably still operate but the exclusive products in these cases are those analogous to (4.32). Presumably the charge density at oxygen in halo-esters and halo-amides and at nitrogen in halo-nitriles is not sufficient to afford the equivalent of (4.31).

4.1.3 Attack at unsaturated centres[1,3-6,28]

4.1.3.1 Carbon–carbon double and triple bonds

Both classes of trivalent phosphorus compounds (anions and neutral P(III) compounds) behave as nucleophiles towards polarized carbon–carbon multiple bonds. The initial products of the nucleophilic addition (carbanions or betaines) may be isolated in a number of ways which are generalized in the following reactions:

(i) Simple addition of primary and secondary phosphines, often base-catalyzed:

$$RPH_2 + 2CH_2=CHY \xrightarrow{\text{base}} RP(CH_2CH_2Y)_2$$

(ii) Protonation, alkylation or intramolecular proton transfer:

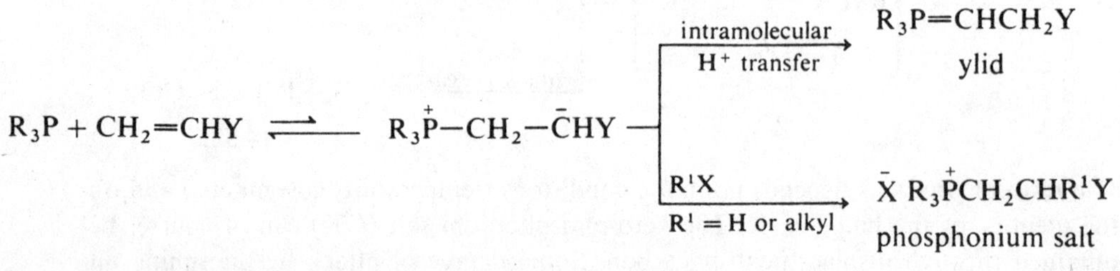

$$R_3P + CH_2=CHY \rightleftharpoons R_3\overset{+}{P}-CH_2-\overset{-}{C}HY \begin{cases} \xrightarrow{\text{intramolecular } H^+ \text{ transfer}} R_3P=CHCH_2Y \quad \text{ylid} \\ \xrightarrow[R^1 = H \text{ or alkyl}]{R^1X} \overset{-}{X} R_3\overset{+}{P}CH_2CHR^1Y \quad \text{phosphonium salt} \end{cases}$$

(iii) Addition–elimination, when the olefin contains a good leaving group such as halogen (Z):

$$R_3P + ZCH=CHY \rightleftharpoons R_3\overset{+}{P}CH-\overset{-}{C}HY \longrightarrow R_3\overset{+}{P}CH=CHY \ Z^-$$
$$\quad\quad\quad\quad\quad\quad\quad\quad\quad\quad\quad\quad |$$
$$\quad\quad\quad\quad\quad\quad\quad\quad\quad\quad\quad\quad Z$$

(iv) De-alkylation, which occurs when the nucleophile contains a P—OR bond:

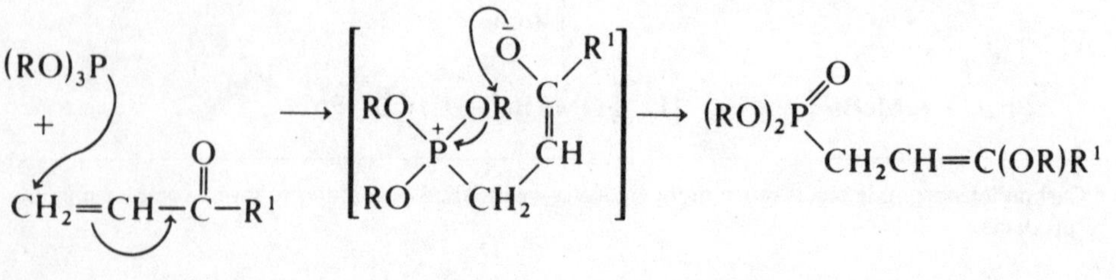

$$(RO)_3P + CH_2=CH-\overset{O}{\overset{\|}{C}}-R^1 \longrightarrow \left[\begin{array}{c} RO \quad OR \quad \overset{O^-}{\underset{\|}{C}}\overset{R^1}{\underset{CH}{}} \\ \overset{+}{P} \\ RO \quad CH_2 \end{array} \right] \longrightarrow (RO)_2P\overset{\displaystyle O}{\underset{\displaystyle CH_2CH=C(OR)R^1}{\diagdown}}$$

4.1 Nucleophilic reactivity

With neutral phosphines and phosphites, the double bond must be activated by electron withdrawal by Y before any significant reaction occurs and typically in the above reactions, Y=CN, —COR^1, —CO$_2$R^1 or —CONH$_2$. Specific examples of each type of reaction appear in Table 4.4 but beyond this, several reactions of this class are worthy of particular attention.

Synthesis of an asymmetric phosphine: The base-catalyzed addition of phenyl phosphine to acrylonitrile is the starting point for a convenient synthesis of asymmetric phosphines since bis-β-cyanoethylphenylphosphine (4.35) may be quaternized to (4.36) and a new phosphine (4.37) generated by base-catalyzed elimination. Repetition of this sequence leads to the required phosphine (4.38)[29].

$$PhPCl_2 \xrightarrow{LiAlH_4} PhPH_2 \xrightarrow{2CH_2=CHCN} PhP(CH_2CH_2CN)_2$$
$$(4.35)$$

$$\downarrow EtBr$$

$$PhEtPCH_2CH_2CN \xleftarrow{MeO^-/MeOH} PhEtP(CH_2CH_2CN)_2 \; \overset{+}{} \; Br^-$$
$$(4.37) \hspace{4cm} (4.36)$$

$$\downarrow nPrBr$$

$$PhEtPrPCH_2CH_2CN \; \overset{+}{} \; Br^- \xrightarrow{MeO^-/MeOH} PhEtPr^nP$$
$$\hspace{6cm} (4.38)$$

Betaine formation and subsequent reactions: Addition of neutral P(III) compounds to activated olefins is reversible as exemplified by the formation of betaines from tri-n-butylphosphine and *p*-substituted benzylidenemalononitriles for which Rappoport showed that the equilibrium constants varies from 1.74×10^4 for *p*-nitro to 190 for the *p*-methoxy compound[30].

$$n\text{-}Bu_3P + XC_6H_4CH=C(CN)_2 \rightleftharpoons n\text{-}Bu_3\overset{+}{P}CHC_6H_4X-\overset{-}{C}(CN)_2$$

With aryl phosphines betaine formation is not detectable but in the presence of aqueous acid, a relatively rapid hydrolysis of one cyano group occurs, presumably via protonation of the betaine[31].

$$Ph_3P + XC_6H_4CH=C(CN)_2 \rightleftharpoons Ph_3\overset{+}{P}CH.C_6H_4X-\overset{-}{C}(CN)_2$$

$$\downarrow HCl$$

$$Ph_3\overset{+}{P}CH.C_6H_4X-\overset{|}{C}HCONH_2 \xleftarrow{H_2O} Ph_3\overset{+}{P}CH.C_6H_4X-\overset{|}{C}=C=\overset{-}{N}H \; Cl^-$$
$$\hspace{2cm} CN \hspace{6cm} CN$$

Activated acetylenes: The reactions with activated acetylenes have also received considerable attention in recent years. A 1:1 adduct may be trapped as a betaine with sulphur dioxide (4.40) whereas a second mole of phosphine gives a bis-ylid (4.41)[32].

Table 4.4 Attack of phosphorus nucleophiles at carbon–carbon multiple bonds

(i) $R_2\bar{P}Li^+ + PhCH=CHPh \longrightarrow R_2PCHPh.\bar{C}HPh\ Li^+ \xrightarrow{H^+} R_2PCHPhCH_2Ph$

(i) 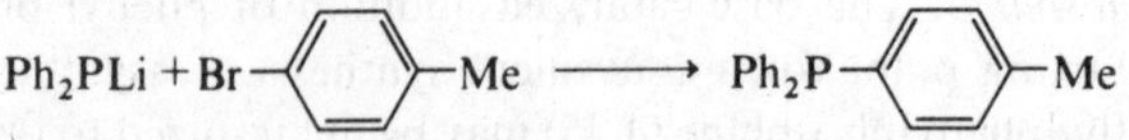 $Ph_2PLi + Br\text{-}C_6H_4\text{-}Me \longrightarrow Ph_2P\text{-}C_6H_4\text{-}Me$

(ii) $n\text{-}Bu_3P + XC_6H_4CH=C(CN)_2 \xrightarrow{HBr} n\text{-}Bu_3\overset{+}{P}CH(C_6H_4X)\text{—}CH(CN)_2\ \bar{B}r$

(ii) 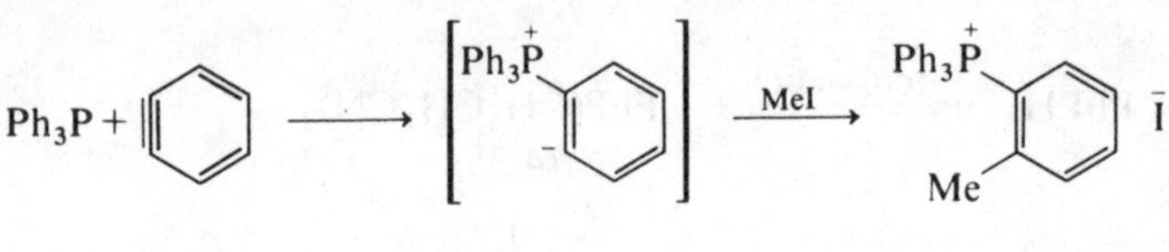 $Ph_3P +$ benzyne $\longrightarrow [Ph_3\overset{+}{P}\text{-}C_6H_4^-] \xrightarrow{MeI} Ph_3\overset{+}{P}\text{-}C_6H_4\text{-}Me\ \bar{I}$

(ii) 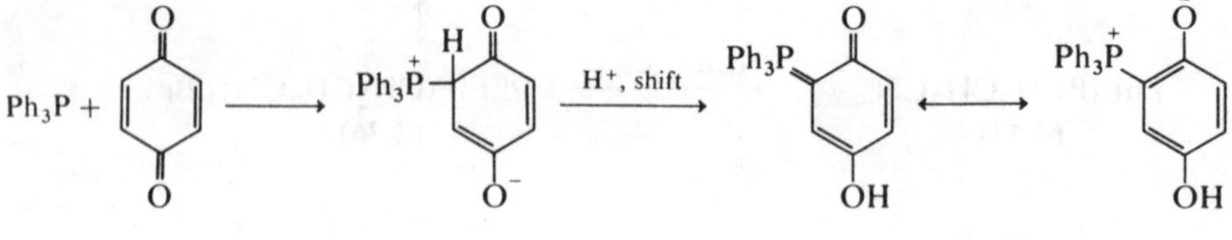 $Ph_3P +$ benzoquinone $\longrightarrow Ph_3\overset{+}{P}\text{-}(C_6H_4O^-)(=O)H \xrightarrow{H^+, shift} Ph_3P=C_6H_3(OH)(=O) \longleftrightarrow Ph_3\overset{+}{P}\text{-}C_6H_3(OH)(\bar{O})$

(iii) $Et_2PH + ClCH=CHCN \xrightarrow{Et_3N} [Et_2PCHCl\text{—}\bar{C}HCN] \longrightarrow Et_2PCH=CHCN$

(iii) $Li^+Ph_2\bar{P} + ClC\equiv CPh \longrightarrow [Ph_2P\text{—}CCl=\bar{C}Ph\ Li^+] \longrightarrow Ph_2P\text{—}C\equiv CPh + LiCl$

(iv) $(MeO)_3P + CH_2=CHCHO \longrightarrow [(MeO)_3\overset{+}{P}CH_2.CH=CH\text{-}\bar{O}] \longrightarrow$

$(MeO)_2\overset{O}{\underset{\|}{P}}CH_2.CH=CHOMe$

(v) 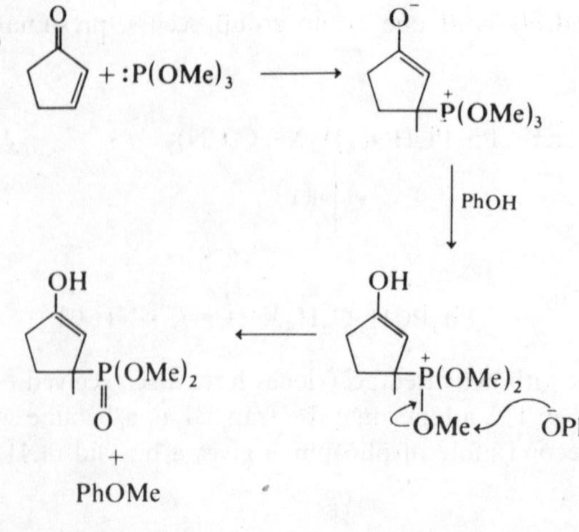 cyclopentenone $+ :P(OMe)_3 \longrightarrow$ cyclopentenolate-$\overset{+}{P}(OMe)_3$

$\downarrow PhOH$

$\text{OH-cyclopentenyl-}\overset{+}{P}(OMe)_2 (\text{OMe, OPh}) \longrightarrow \text{OH-cyclopentenyl-}P(=O)(OMe)_2 + PhOMe$

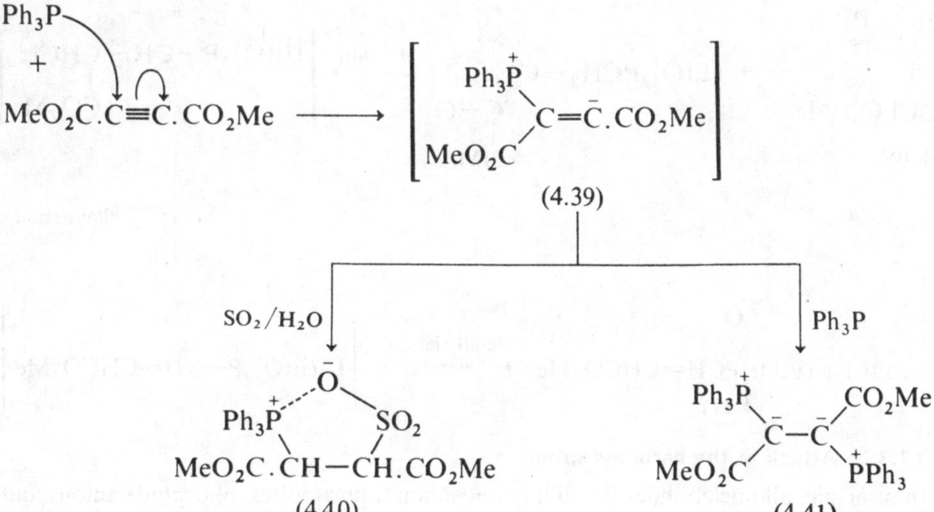

4.1 Nucleophilic reactivity

$$Ph_3P + MeO_2C.C{\equiv}C.CO_2Me \longrightarrow \left[\begin{array}{c} Ph_3\overset{+}{P} \\ \diagdown \\ MeO_2C \end{array} C{=}\overset{-}{C}.CO_2Me \right]$$

(4.39)

(4.40): $Ph_3\overset{+}{P}$—O$^-$—SO$_2$ ring with MeO$_2$C.CH—CH.CO$_2$Me (via SO$_2$/H$_2$O)

(4.41): $Ph_3\overset{+}{P}$\\C=C//CO$_2$Me, MeO$_2$C / \overset{+}{P}Ph$_3$ (via Ph$_3$P)

A 1:2 ratio of phosphine to acetylene react in refluxing ether to give the compound (4.43) derived from phenyl migration within an intermediate phosphorane (4.42)[33,34].

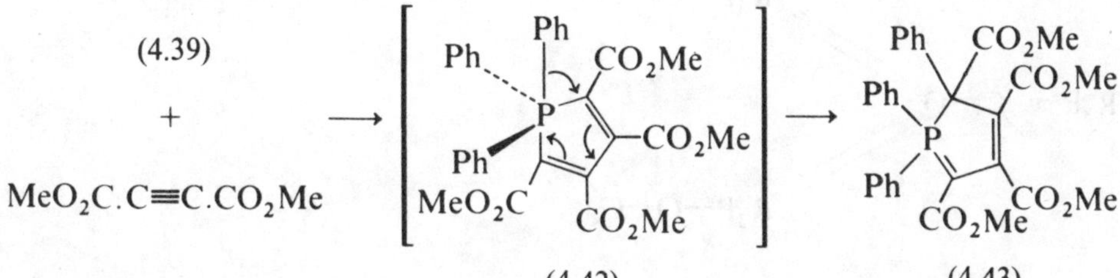

(4.39) + MeO$_2$C.C≡C.CO$_2$Me ⟶ (4.42) ⟶ (4.43)

Addition to benzyne (see Table 4.4): Addition of a phosphine with at least one alkyl substituent to benzyne also gives rise to an ylid, this time by shift of a proton[35]:

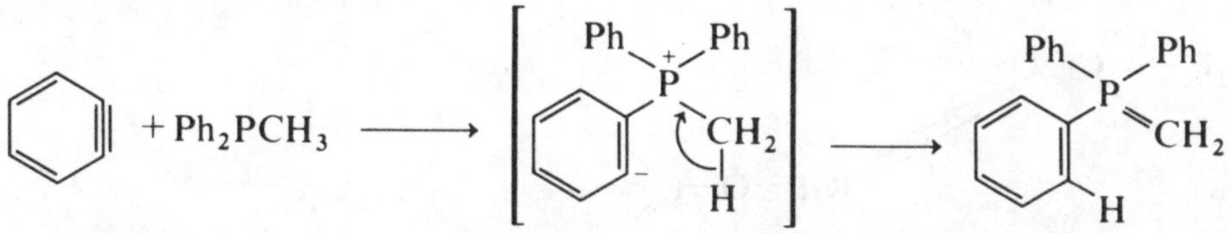

benzene + Ph$_2$PCH$_3$ ⟶ [intermediate] ⟶ product

Intramolecular rearrangement: An interesting example of de-alkylation involves the intramolecular rearrangement of diethyl 1-methylallyl phosphite (4.44) to the phosphonate (4.45)[36,37].

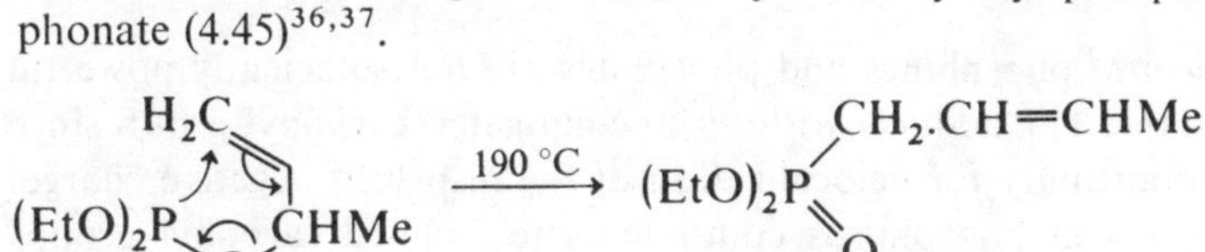

(EtO)$_2$P–O–CHMe(CH=CH$_2$) —190 °C→ (EtO)$_2$P(=O)CH$_2$.CH=CHMe

(4.44)   (4.45)

A composite reaction of types (ii), (iii) and (iv): Finally, proton shift, elimination and de-alkylation are elegantly demonstrated within one example in the case of the reaction of triethyl phosphite with α-chloromethyl acrylate (4.46) which yields diethyl β-carbomethoxyvinylphosphonate (4.47) probably by the following route[38].

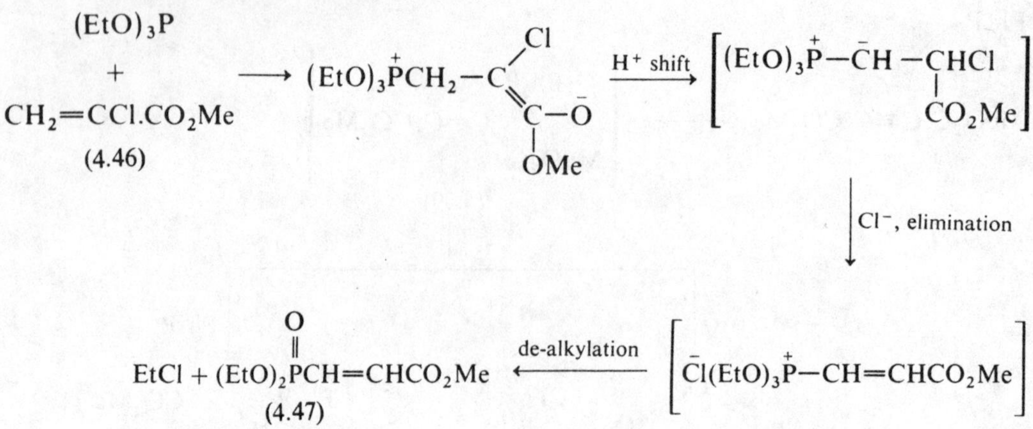

$$(EtO)_3P + CH_2=CCl.CO_2Me \longrightarrow (EtO)_3\overset{+}{P}CH_2-C\underset{OMe}{\overset{Cl}{\underset{\|}{\overset{\diagup}{C}}}}\overset{}{\underset{}{C-\bar{O}}} \xrightarrow{H^+ \text{ shift}} \left[(EtO)_3\overset{+}{P}-\bar{C}H-\underset{CO_2Me}{CHCl} \right]$$
(4.46)

$$\downarrow Cl^-, \text{ elimination}$$

$$EtCl + (EtO)_2\overset{O}{\overset{\|}{P}}CH=CHCO_2Me \xleftarrow{\text{de-alkylation}} \left[\bar{C}l(EtO)_3\overset{+}{P}-CH=CHCO_2Me \right]$$
(4.47)

4.1.3.2 Attack at the carbonyl group

In principle, all nucleophiles (including phosphines, phosphites, phosphide anions and dialkylphosphonite anions) may attack the carbonyl group at either carbon or oxygen.

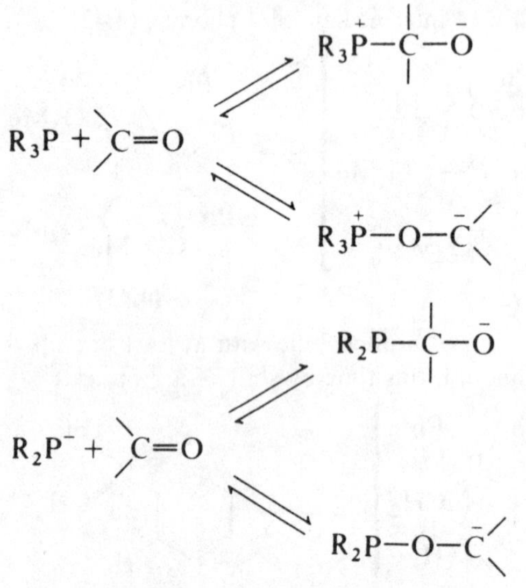

$$R_3P + \overset{}{\underset{}{\diagdown}}C=O \rightleftharpoons \begin{cases} R_3\overset{+}{P}-\overset{|}{\underset{|}{C}}-\bar{O} \\ R_3\overset{+}{P}-O-\bar{C}\overset{}{\underset{}{\diagdown}} \end{cases}$$

$$R_2P^- + \overset{}{\underset{}{\diagdown}}C=O \rightleftharpoons \begin{cases} R_2P-\overset{|}{\underset{|}{C}}-\bar{O} \\ R_2P-O-\bar{C}\overset{}{\underset{}{\diagdown}} \end{cases}$$

R = H, alkyl, aryl, alkoxy or aryloxy

In practice however, *neutral* phosphines and phosphites are not sufficiently powerful nucleophiles to form stable 1:1 adducts with an unconjugated carbonyl group since the latter affords no opportunity for delocalization of the displaced negative charge. Additions of neutral trivalent phosphorus compounds to isolated carbonyl groups are therefore highly reversible with the equilibrium lying well to the left; reaction is only observed (i) with phosphines containing a P—H bond where the hydrogen may be transferred to the anion (most readily achieved under acid-catalyzed conditions); (ii) with carbonyl compounds bearing a good leaving group, e.g. halogen; (iii) with trialkyl phosphites where de-alkylation of the intermediate betaine affords a phosphoryl group and (iv) where attack of the betaine on a second molecule of the carbonyl

4.1 Nucleophilic reactivity

compound generates a stable pentacovalent compound. The last will be discussed later but examples of the first three classes appear in Table 4.5.

Most reactions of this kind proceed via attack on carbonyl carbon as expected in view of the relative electronegativities of oxygen and carbon. The exceptions to this generalization include α-diketones and ortho-quinones (see Section 4.4) and cases where the carbanion is stabilized by resonance. For example triphenylphosphine reacts with 2,5-diphenyl-3,4-dicyano-cyclopentadienone (4.48) to give a stable betaine (4.49).

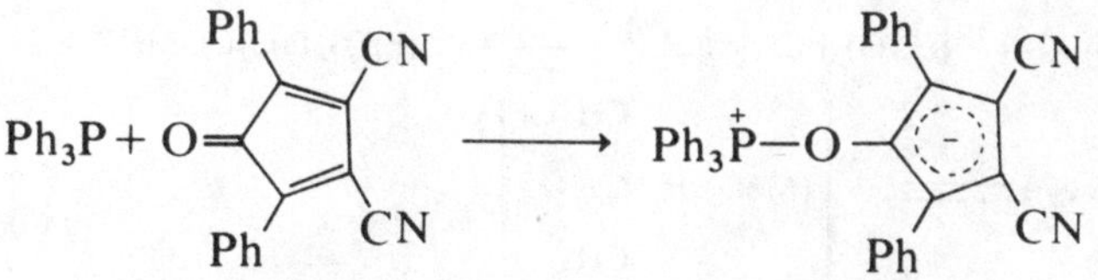

$$Ph_3P + (4.48) \longrightarrow (4.49)$$

Likewise, initial attack on oxygen followed by de-alkylation is the most reasonable explanation for the product of the reaction between triethyl phosphite and p-benzoquinone (eqn. 12)[39].

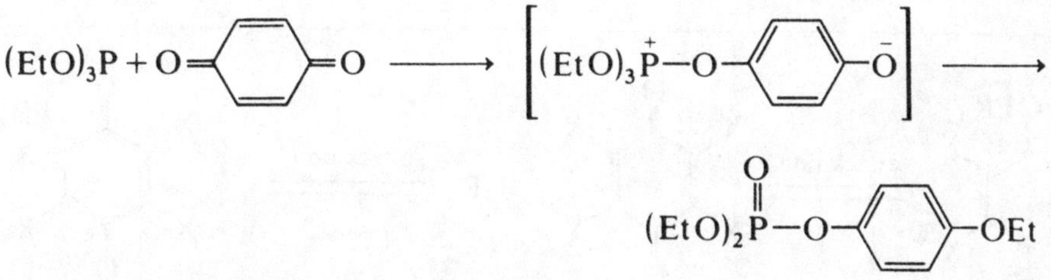

$$(EtO)_3P + O=\!\!\langle\ \rangle\!\!=O \longrightarrow [(EtO)_3\overset{+}{P}-O-\!\!\langle\ \rangle\!\!-\overset{-}{O}] \longrightarrow (EtO)_2\overset{O}{\underset{\|}{P}}-O-\!\!\langle\ \rangle\!\!-OEt \quad (12)$$

This is in marked contrast to the reaction of p-benzoquinone with triphenylphosphine where the final product is derived from addition to carbon (see Table 4.4)[40]. On the other hand, triphenylphosphine reacts with chloranil to give (4.51) the formation of which is most easily rationalized by a pathway involving (4.50) as an intermediate derived by initial attack on oxygen[40,41]. (See Chapter 9 for a more detailed discussion of this reaction.)

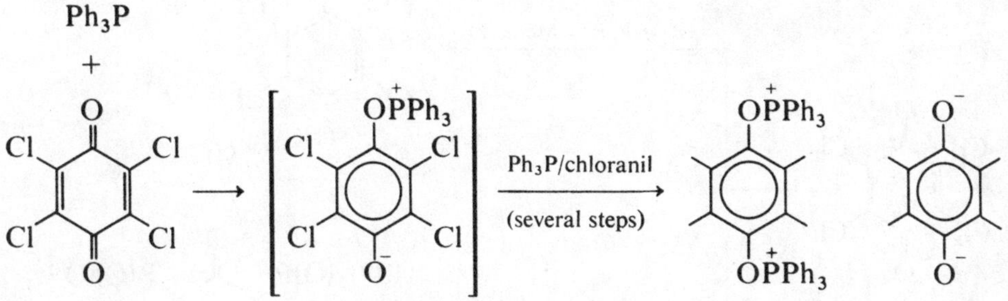

(4.50)   (4.51)

There appears to be a very complicated balance of steric and electronic factors controlling the course of each of the reactions with p-benzoquinones and the extent of the initial equilibrium to (4.52) and (4.53).

Table 4.5 Additions of neutral P(III) compounds to unconjugated carbonyl compounds

Type	Reaction
(i)	$PH_3 + 3CH_2O \xrightarrow{100°C,\ pressure} P(CH_2OH)_3$
(i)	$PH_3 + 4MeCHO \xrightarrow{H-X} (MeCHOH)_4\overset{+}{P}\ \overset{-}{X}$
(ii)	$(EtO)_3P + MeCOCl \longrightarrow [(EtO)_3\overset{+}{P}CO.Me\ \overset{-}{Cl}] \longrightarrow (EtO)_2P(O)COMe + EtCl$
(iii)	$(MeO)_3P + ClCH_2COCH_3 \rightleftharpoons \begin{bmatrix} CH_2Cl \\ | \\ (MeO)_3\overset{+}{P}-C-\overset{-}{O} \\ | \\ CH_3 \end{bmatrix}$

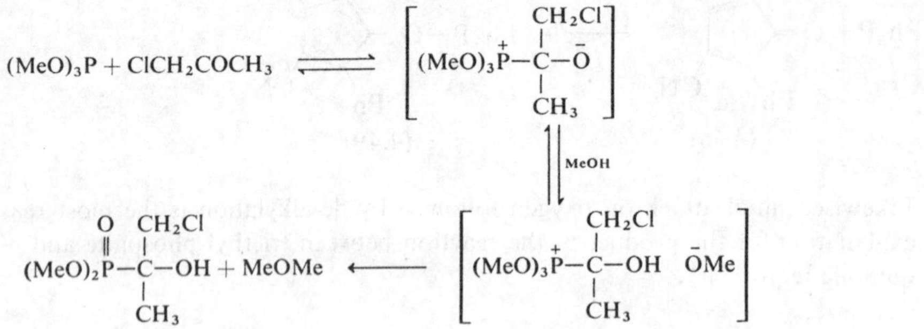

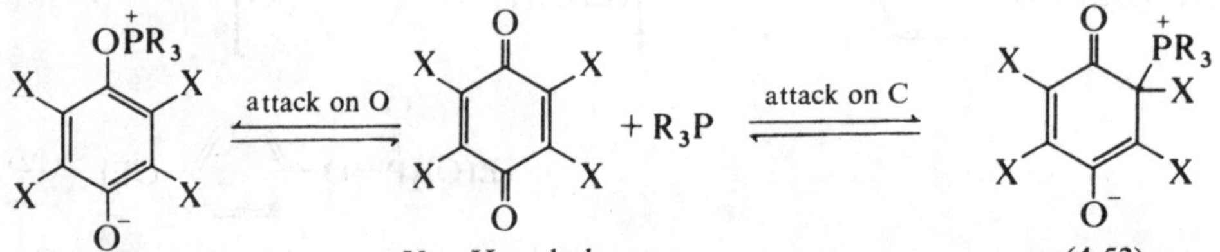

$\Updownarrow$ MeOH

$(MeO)_2\overset{O}{\overset{\|}{P}}-\underset{CH_3}{\overset{CH_2Cl}{\underset{|}{C}}}-OH + MeOMe \longleftarrow \begin{bmatrix} CH_2Cl \\ | \\ (MeO)_3\overset{+}{P}-C-OH\quad \overset{-}{O}Me \\ | \\ CH_3 \end{bmatrix}$

Thus trimethyl or triethyl phosphite and chloranil react by almost exclusive attack on oxygen to give (4.54) but tri-isopropyl phosphite gives the tetrakis phosphonate (4.55) via a Michael addition to ring carbon.

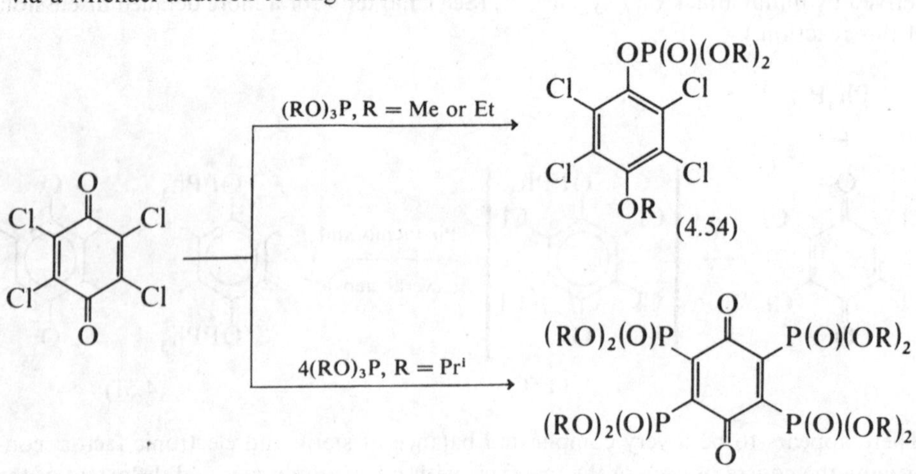

4.1 Nucleophilic reactivity

Presumably (in the tri-isopropyl case), the intermediate leading to (4.54) is too stable to permit de-alkylation and the route to (4.55) involving de-alkylation by halide ion is favoured energetically.

Some phosphines of enhanced nucleophilicity are capable of addition to the carbon of a carbonyl group to form stable betaines. Thus, Mark has characterized the crystalline 1:1 addition compounds obtained from hexamethylphosphorus triamide and aliphatic aldehydes as internal phosphonium alkoxides (eqn. 13)

$$(Me_2N)_3P + RCHO \rightleftharpoons (Me_2N)_3\overset{+}{P}-CHR-\overset{-}{O} \qquad (13)$$

Phosphines are also capable of nucleophilic addition to carbon disulphide to give red crystalline complexes with a zwitterionic structure (4.56) which may be dissociated by heat and may therefore be useful for purifying phosphines.

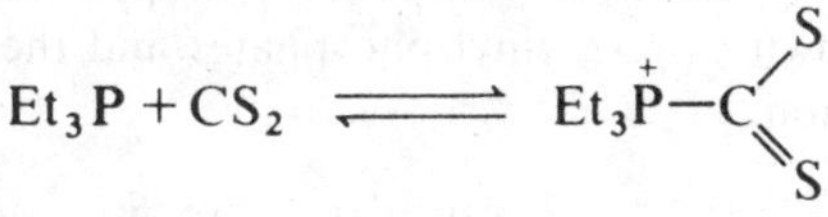

$$Et_3P + CS_2 \rightleftharpoons Et_3\overset{+}{P}-C\overset{\overset{-}{S}}{\underset{S}{\diagdown}}$$

(4.56)

Phosphide and phosphite anions are of course, very much more powerful nucleophiles and add quite readily to simple carbonyl compounds. Once again the products are isolated either by protonation, alkylation or elimination of a good leaving group (see Table 4.6).

The Perkow Reaction[1,3,4,13,14]: Trialkyl phosphites react with α-halocarbonyl compounds to give either an α-ketophosphonate (4.57) or a vinyl phosphate (4.58) or a mixture of both.

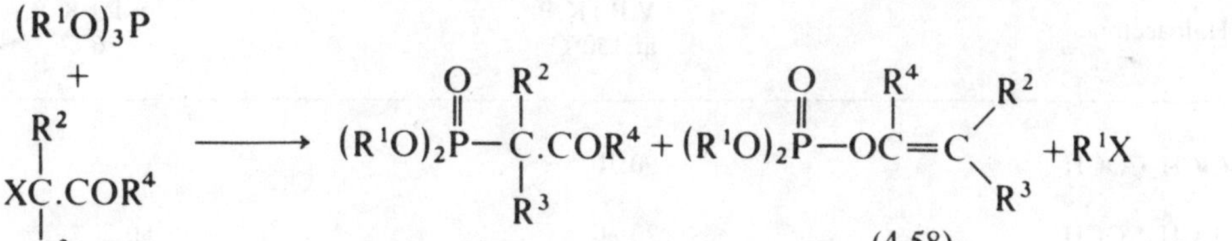

$$\begin{array}{c}(R^1O)_3P \\ + \\ R^2 \\ | \\ XC.COR^4 \\ | \\ R^3\end{array} \longrightarrow (R^1O)_2\overset{O}{\overset{\|}{P}}-\overset{R^2}{\underset{R^3}{\overset{|}{C}}}.COR^4 + (R^1O)_2\overset{O}{\overset{\|}{P}}-OC=C\overset{R^2}{\underset{R^3}{\diagup}} + R^1X$$

(4.57)   (4.58)

The formation of a vinyl phosphate was discovered in 1952 by Perkow from the reaction of triethyl phosphite with chloral (eqn. 14).

$$(EtO)_3P + Cl_3CHO \longrightarrow (EtO)_2\overset{O}{\overset{\|}{P}}-OCH=CCl_2 \qquad (14)$$

Table 4.6 Nucleophilic additions of phosphide and phosphonite anions to carbonyl compounds

$$\text{Ph}_2\bar{\text{P}}\ \text{M}^+ + \text{PhCHO} \longrightarrow \text{Ph}_2\overset{+}{\text{P}}\text{CHPh}-\text{O}^- \xrightarrow{\text{MeI}} \text{Ph}_2\text{PCHPhOMe}$$

$$(\text{EtO})_2\text{P(O)H} + \text{RCHO} \xrightarrow{\text{conc. OH}^-} (\text{EtO})_2\text{PCHOH.R}$$

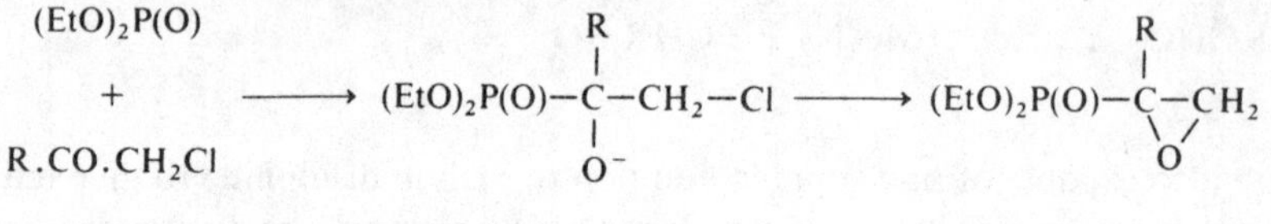

$$(\text{EtO})_2\bar{\text{P}}(\text{O}) + \text{R.CO.CH}_2\text{Cl} \longrightarrow (\text{EtO})_2\text{P(O)}-\underset{\text{O}^-}{\overset{\text{R}}{\text{C}}}-\text{CH}_2-\text{Cl} \longrightarrow (\text{EtO})_2\text{P(O)}-\underset{\text{O}}{\overset{\text{R}}{\text{C}}}-\text{CH}_2$$

Since that time the reactions of a wide variety of trivalent phosphorus compounds with α-halocarbonyl compounds have been shown to give vinyl phosphates and the reaction is now referred to as the Perkow reaction.

Several factors (including solvent, temperature and the nature of the halogen) influence the product ratio of ketophosphonate: vinyl phosphate. Low temperatures and a more electronegative halogen favour the vinyl phosphate and this is exemplified for haloacetones in Table 4.7[42,43]. Substituents on the α-carbon, especially electron-withdrawing groups, also favour the Perkow reaction.

Table 4.7 Ratio of vinyl phosphate (V.P.) to ketophosphonate (K.P.) in the reaction of haloacetones with triethyl phosphite

Haloacetone	V.P.: K.P. at 150°C	V.P.: K.P. at 36°C
ClCH$_2$COCH$_3$	90:10	—
BrCH$_2$COCH$_3$	20:80	80:20
ICH$_2$COCH$_3$	—	10:90

The mechanism of this reaction has intrigued investigators for over 20 years and is still not completely resolved. The difficulty lies in the fact that there are four possible sites for nucleophilic attack at an α-haloaldehyde or ketone (XCH$_2$.CO.R^1 where R^1 = H, alkyl or aryl). These may be listed as follows:

4.1 Nucleophilic reactivity

(i) *Attack at the α-carbon*

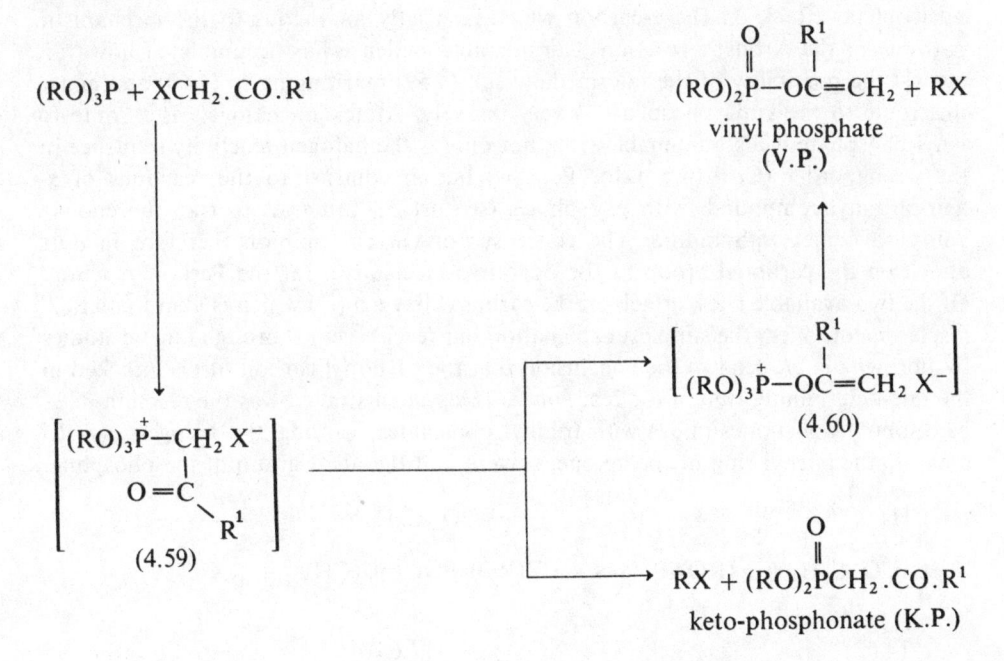

$$(RO)_3P + XCH_2.CO.R^1$$

$$\downarrow$$

$$\left[\begin{array}{c} (RO)_3\overset{+}{P}-CH_2\ X^- \\ | \\ O=C \\ \diagdown R^1 \end{array} \right]$$
(4.59)

$$\longrightarrow \left[(RO)_3\overset{+}{P}-O\underset{|}{\overset{R^1}{C}}=CH_2\ X^- \right]$$
(4.60)

$$\uparrow$$

$$(RO)_2\overset{\overset{O}{\|}}{P}-O\underset{|}{\overset{R^1}{C}}=CH_2 + RX$$
vinyl phosphate (V.P.)

$$\longrightarrow RX + (RO)_2\overset{\overset{O}{\|}}{P}CH_2.CO.R^1$$
keto-phosphonate (K.P.)

(ii) *Attack on halogen*

$$(RO)_3P + X-CH_2.CO.R^1$$

$$\downarrow$$

$$\left[(RO)_3\overset{+}{P}X + \overset{-}{C}H_2.CO.R^1 \longleftrightarrow CH_2=C\underset{R^1}{\overset{\overset{-}{O}}{\diagup}} \right]$$

Attack by C⁻ (4.59) → K.P.

Attack by O⁻ (4.60) → V.P.

(iii) *Attack on carbonyl carbon*

$$(RO)_3P + XCH_2.CO.R^1 \longrightarrow (RO)_3\overset{+}{P}-\underset{\underset{O^-}{|}}{\overset{\overset{R^1}{|}}{C}}-CH_2X \longrightarrow (4.60) \longrightarrow V.P.$$
(4.61)

(iv) *Attack on carbonyl oxygen*

$$(RO)_3P + O=C\underset{CH_2-X}{\overset{R^1}{\diagup}} \longrightarrow \left[(RO)_3\overset{+}{P}O\underset{|}{\overset{R^1}{C}}=CH_2\ X^- \right] \longrightarrow V.P.$$
(4.60)

It seems highly likely that formation of the keto-phosphonate product occurs via nucleophilic attack on the α-carbon which is strictly analogous to the mechanistic pathway of the Arbusov reaction. Considerable evidence has accumulated however, against the possibility of the intermediate salt (4.59) rearranging to (4.60) and hence this route to the vinyl phosphate is very unlikely. Attack on halogen as a route to vinyl phosphate is also untenable since not only is the halogen reactivity sequence in the wrong order (Cl > Br > I for Perkow) but in contrast to the reactions of α-halocarbonyl compounds with phosphines (see p 125.) attempts to trap the enolate anion have met with failure. The consensus of current opinion therefore favours attack on the carbonyl group as the operative mechanism for the Perkow reaction. Of the two available sites, attack on the carbonyl oxygen (iv) which is akin to an S_N2' displacement, offers the simplest explanation but recent, very thorough kinetic studies by Borowitz et al., lead to the conclusion that the carbonyl carbon (iii) is attacked in the rate-determining step of the reaction[44]. The system studied was the reaction of α-haloisobutyrophenones (4.62) with trialkyl phosphites, varying the halogen, substituents in the phenyl ring of the ketone, solvent and the alkyl group of the phosphite.

$$\begin{array}{c} CH_3 \\ | \\ X-C.CO.C_6H_4Y + (RO)_3P \\ | \\ CH_3 \\ (4.62) \end{array} \longrightarrow \begin{array}{c} O\ \ C_6H_4Y \\ \|\ \ | \\ (RO)_2POC=C(CH_3)_2 + RX \\ \\ (4.63) \end{array}$$

The experimental observations may be summarized briefly as follows:

(i) the reactions were first order in (4.62) and phosphite;
(ii) with R = Et, varying Y gave a Hammett ρ value of 1.89 with a confidence level of 99.9% using σ values and inferior correlations using $σ^-$ or mixtures of σ and $σ^-$;*
(iii) a rate enhancement of ca 2 was found on changing from benzene ($ε_r = 2.3$) to benzonitrile ($ε_r = 25.6$) as solvent;
(iv) for Y = H, (k_2 for R = Et)/(k_2 for R = i-Pr) = 2.73;†
(v) for several Y, k_2(X= Cl) > k_2(X= Br); e.g. Y= H, k_2(Cl)/k_2(Br) = 1.37;†
(vi) for X = Br, Y = H in benzene, E_a = 55.6 kJ mol^{-1} (13.3 kcal mole^{-1}), $\Delta S^\ddagger$ at 44.9 °C = −171.5 J mole^{-1} K^{-1} (−41 cal mole^{-1} K^{-1});
(vii) the reactions were powerfully catalyzed by carboxylic acids such as acetic acid.

The negative entropy of activation is indicative of a polar and highly ordered TS and is similar to that observed by Speziale for the rate-determining nucleophilic addition

---

* Hammett ρ values are simply a measure of the *sensitivity* of a reaction to substituent effects; values generally vary from ca 0.2 (low sensitivity) to ca 10 (high sensitivity) and may be positive or negative. Positive ρ values correspond to reactions accelerated by *electron-withdrawing groups*. There are various kinds of substituent parameters (σ) but for substituents on phenyl rings these may be broadly classified into σ values (non-conjugative) and $σ^+$ or $σ^-$ values (conjugative substituent constants). For further details the reader is referred to the review by C. D. Ritchie and W. F. Sager, *Progress in Physical Organic Chemistry*, Vol. 2, Academic Press, New York, 1965, p. 323.
† k_2 refers to the second-order rate coefficient derived from the kinetic studies and *not* to the k_2 of the mechanistic scheme represented by eqn. (16).

4.1 Nucleophilic reactivity

of triphenylphosphonium carbomethoxymethylide (4.64) to the carbonyl group of substituted benzaldehydes[45].

$$\begin{array}{c}Ph_3P=CHCO_2Me \\ (4.64) \\ + \\ XC_6H_4CHO\end{array} \longrightarrow \left[\begin{array}{c}\overset{+}{Ph_3P}-CHCO_2Me \\ | \\ \overset{-}{O}-CHC_6H_4Y\end{array}\right] \longrightarrow \begin{array}{c}Ph_3P(O) \\ + \\ YC_6H_4CH=CHCO_2Me\end{array}$$

Incidentally, the ylid reaction has a ρ value (in benzene) of $+2.7$ with rates also correlating with the σ values of Y.

The solvent effect in the Perkow reaction, although modest, is also indicative of polar character in the TS. The correlation with σ serves to dispel the notion of rate-determining attack on oxygen since if this were the case, the developing negative charge on carbon would be subject to resonance interaction with substituents in the phenyl ring which should have given rise to a correlation with σ^- (eqn. 15).

$$(RO)_3P + O=\overset{|}{\underset{|}{C}}-\!\!\!\!\!\bigcirc\!\!\!\!\!-Y \longrightarrow (RO)_3\overset{+}{P}-O-\overset{|}{\underset{|}{\overset{-}{C}}}-\!\!\!\!\!\bigcirc\!\!\!\!\!-Y \qquad (15)$$

The higher reactivity of the chloroketones over the bromoketones is also inconsistent with carbon–halogen cleavage being involved in the rate-determining step. Catalysis by acid is frequently observed in nucleophilic additions to carbonyl groups and finally the rate decrease in changing from triethyl to tri-isopropyl phosphite, which is qualitatively similar to that observed by Ogata during the addition of trialkyl phosphites to benzil (see Section 4.4), may be ascribed to a steric effect on the nucleophilic addition.

Whether nucleophilic addition to the carbonyl group is reversible or not, is not known. It seems likely however, that if it is reversible, the backward step (k_{-1}) is slow compared to the rearrangement (k_2) and can therefore be ignored (eqn. 16).

$$\begin{array}{c}(RO)_3P \\ + \\ \overset{|}{\underset{|}{X-C-COR^1}}\end{array} \underset{k_{-1}}{\overset{k_1}{\rightleftarrows}} \left[\begin{array}{c}\overset{\bar{O}}{|} \\ (RO)_3\overset{+}{P}-\overset{|}{\underset{R^1}{C}}-\overset{|}{\underset{|}{C}}-X\end{array}\right] \overset{k_2}{\longrightarrow} \left[\begin{array}{c}R^1 \\ | \\ (RO)_3\overset{+}{P}OC=C\diagup \quad X^- \\ \diagdown\end{array}\right] \qquad (16)$$

The concept of rate-determining nucleophilic addition to carbonyl carbon is also supported by the work of Denney et al. who studied the reaction of a series of acyclic and cyclic phosphites (4.65–4.71) with chloral and ω,ω,ω-trichloroacetophenone ($Cl_3C.CO.Ph$)[46].

(CH₃O)₃P

(4.65)

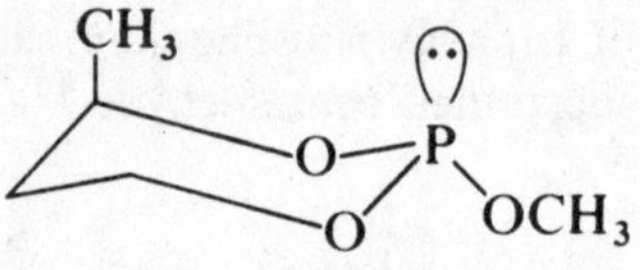

(4.66)

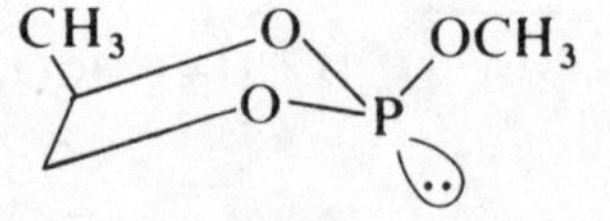

(4.67)

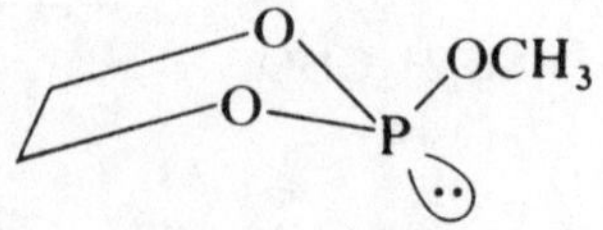

(4.68)

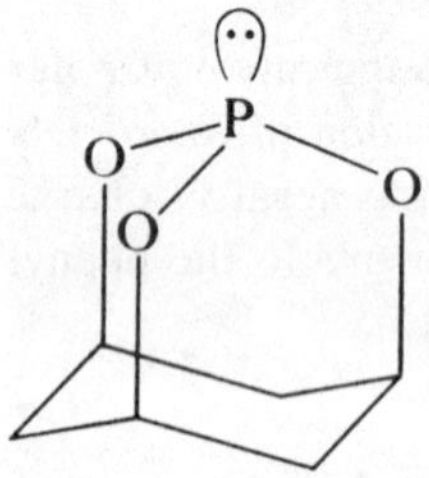

(4.69)

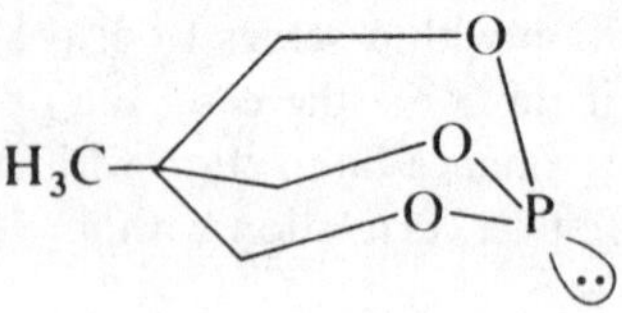

(4.70)

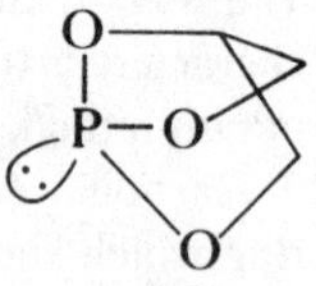

(4.71)

Rates for the Perkow reaction were in the sequence (4.65) ⩾ (4.66) > (4.67) > (4.68) > (4.69) > (4.70) > (4.71) which is identical to the sequences observed for reactions of the same phosphites with dibenzoyl peroxide (see Section 4.3 later) and alkyl halides. Reaction of (4.66) with chloral was stereospecific and occurred with retention of configuration at phosphorus.*

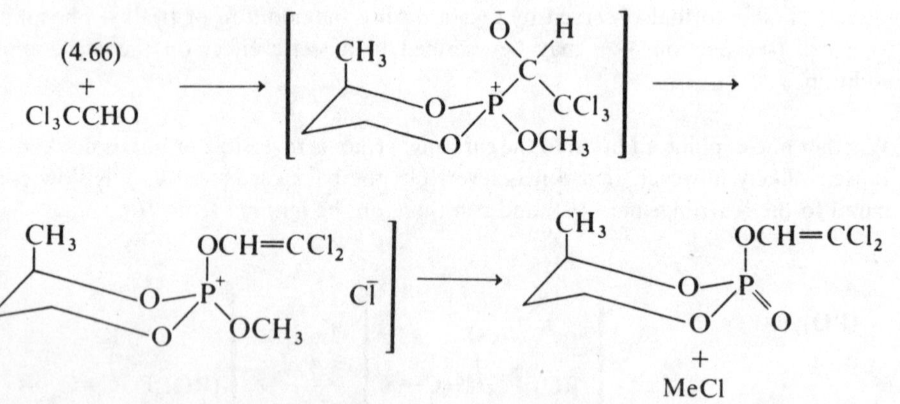

(4.66) + Cl₃CCHO ⟶ [intermediate] ⟶ [intermediate] $\overset{\bar{Cl}}{\longrightarrow}$ product + MeCl

Although the stereochemical evidence does not preclude attack on oxygen (iv) it certainly serves as another nail in the coffin of displacement on halogen (ii) for which inversion would be expected.

---

* The reaction of (4.66) with dibenzoyl peroxide is also very nearly stereospecific and leads to phosphates with predominant retention of configuration.

4.1 Nucleophilic reactivity

Where geometrical isomerism about the double bond of the product is possible, the Perkow reaction is frequently stereoselective. Thus Phosdrin (4.72) prepared from the reaction of trimethyl phosphite with α-chloroacetoacetate has a 4:1 predominance of the *cis*-crotonate structure (4.72a)[47,48].

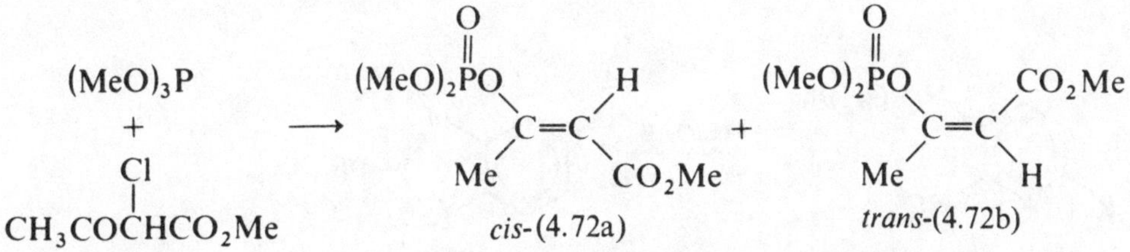

This may be rationalized by application of Cram's rule of asymmetric induction which assumes approach of the nucleophile to the less hindered side of the carbonyl group.

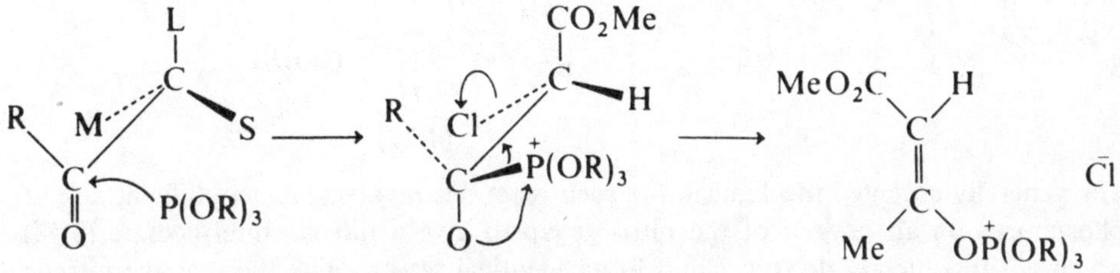

L = large (CO_2Me); M = medium (Cl); S = small (H)

4.1.3.3 Attack at nitro and nitroso groups[49,50]

The deoxygenation of nitro and nitroso compounds by phosphites and (less readily) by phosphines has attracted considerable attention since it offers a synthetic route to a variety of heterocyclic molecules. Thus *o*-nitrobiphenyl reacts with trialkyl phosphites to give carbazole (4.74) via the nitrene (4.73).

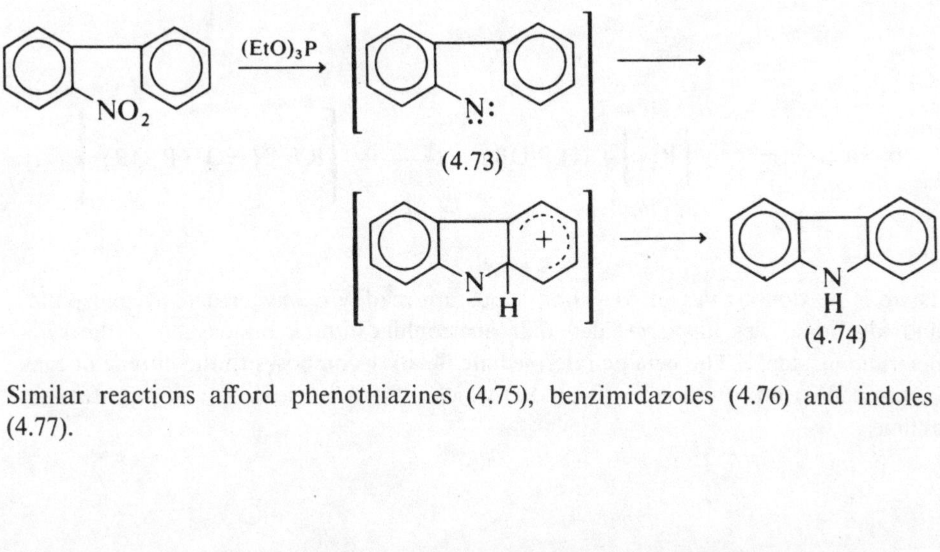

Similar reactions afford phenothiazines (4.75), benzimidazoles (4.76) and indoles (4.77).

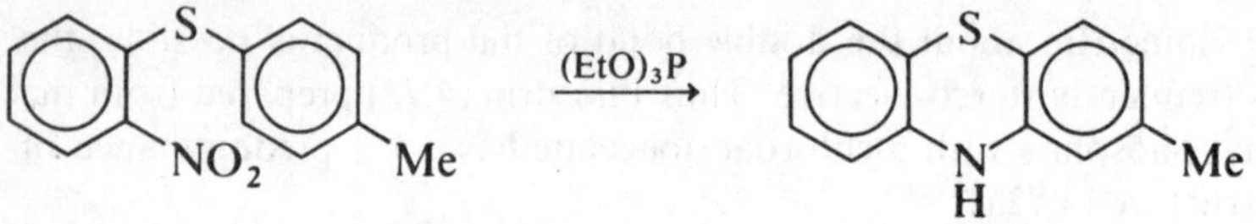

(4.75)

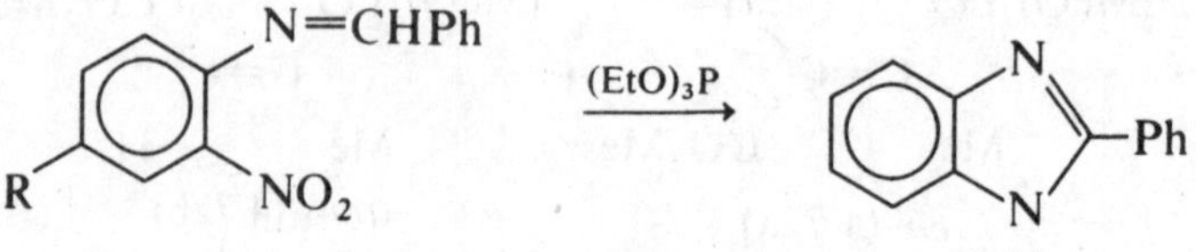

(4.76)

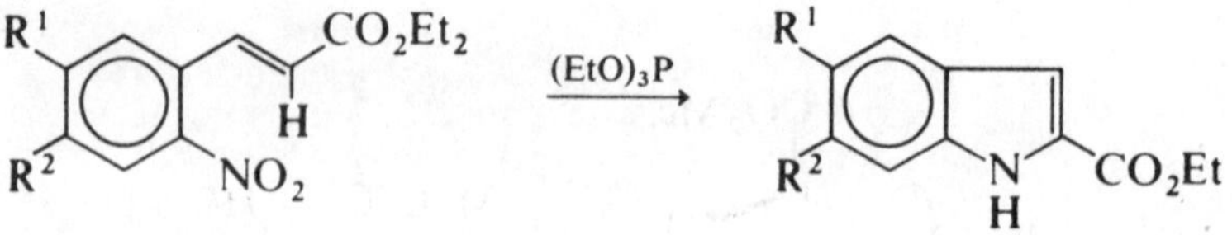

(4.77)

The generally accepted mechanism for such reactions involves nucleophilic attack of phosphorus on an oxygen of the nitro group to give a nitroso intermediate (4.78) which is subsequently deoxygenated in an identical way to give the reactive nitrene intermediate (4.79).

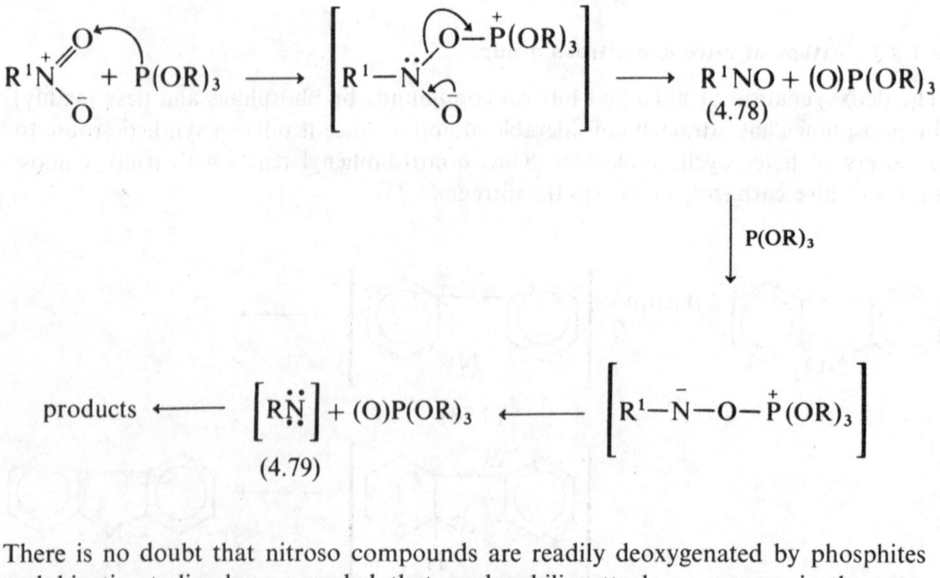

$$R^1N(O)(O) + P(OR)_3 \longrightarrow [R^1-N(O-P(OR)_3^+)(O)] \longrightarrow R^1NO + (O)P(OR)_3$$
(4.78)

$$\downarrow P(OR)_3$$

$$\text{products} \longleftarrow [R\ddot{N}] + (O)P(OR)_3 \longleftarrow [R^1-\bar{N}-O-\overset{+}{P}(OR)_3]$$
(4.79)

There is no doubt that nitroso compounds are readily deoxygenated by phosphites and kinetic studies have revealed that nucleophilic attack on oxygen is the rate-determining step[51]. The betaine intermediate (4.80) decomposes to the nitrene or may be trapped by protonation to give (4.81) which ultimately yields o- and p-substituted anilines.

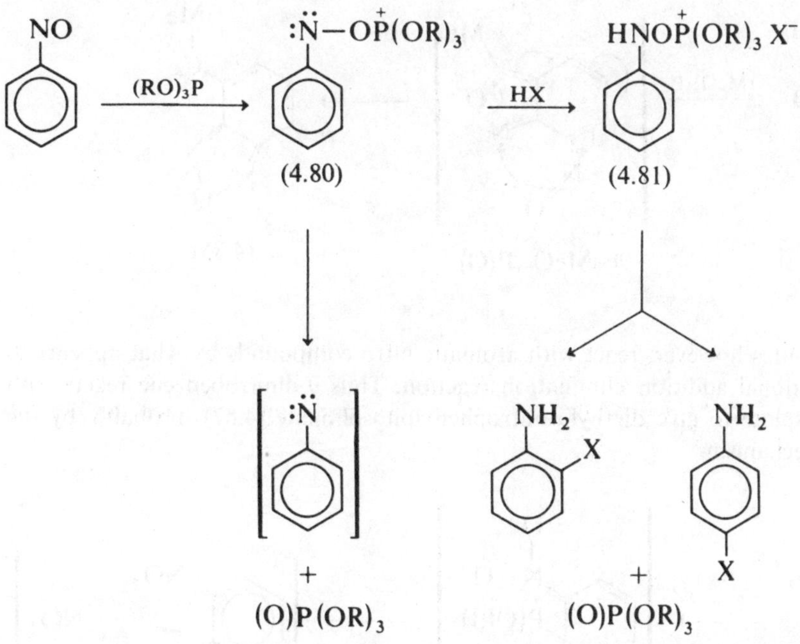

4.1 Nucleophilic reactivity

Alkyl nitroso compounds are also deoxygenated by phosphites to give N-alkylimines (4.83) by rearrangement of the nitrene (4.82)[52].

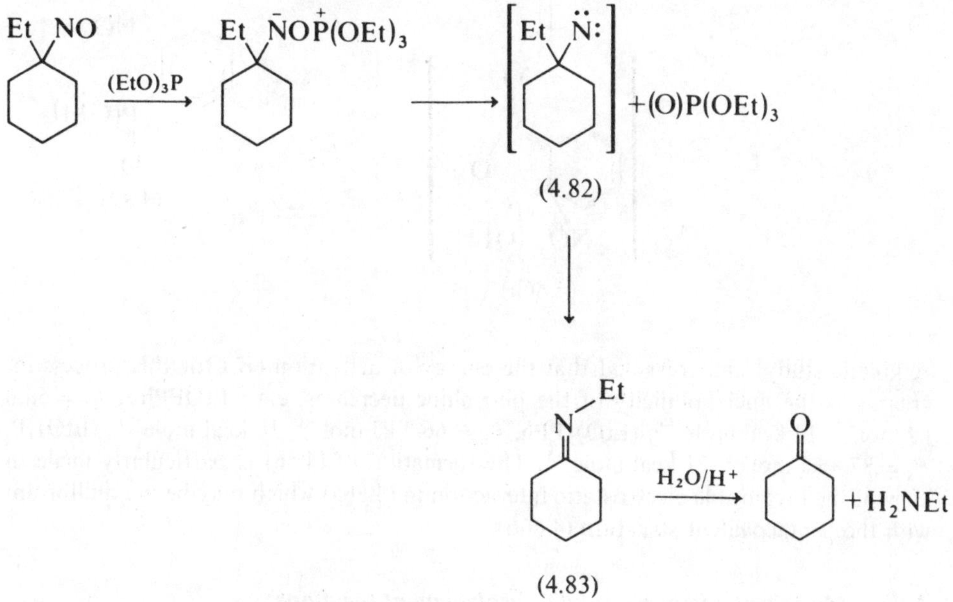

Further evidence for the intermediacy of nitroso compounds in the deoxygenation of nitro groups has been provided by the isolation of (4.85) from the deoxygenation of the nitrobenzoxazole (4.84)[53].

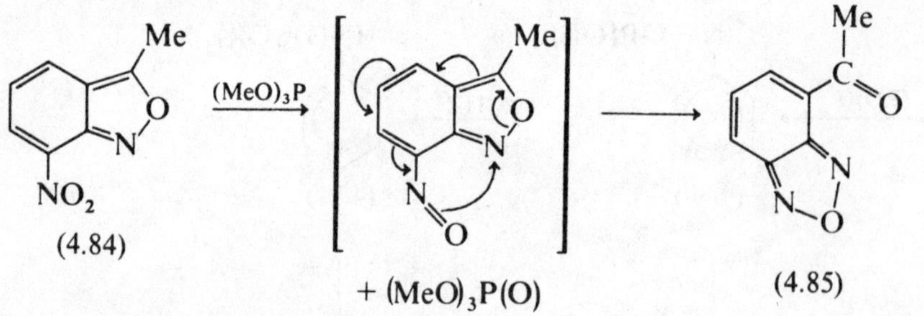

[Structures 4.84 and 4.85 with (MeO)₃P reaction giving + (MeO)₃P(O)]

Some phosphites however, react with aromatic nitro compounds by what appears to be a conventional addition–elimination reaction. Thus o-dinitrobenzene reacts with triethyl phosphite to give diethyl-o-nitrophenylphosphonate (4.87), probably by the following mechanism.

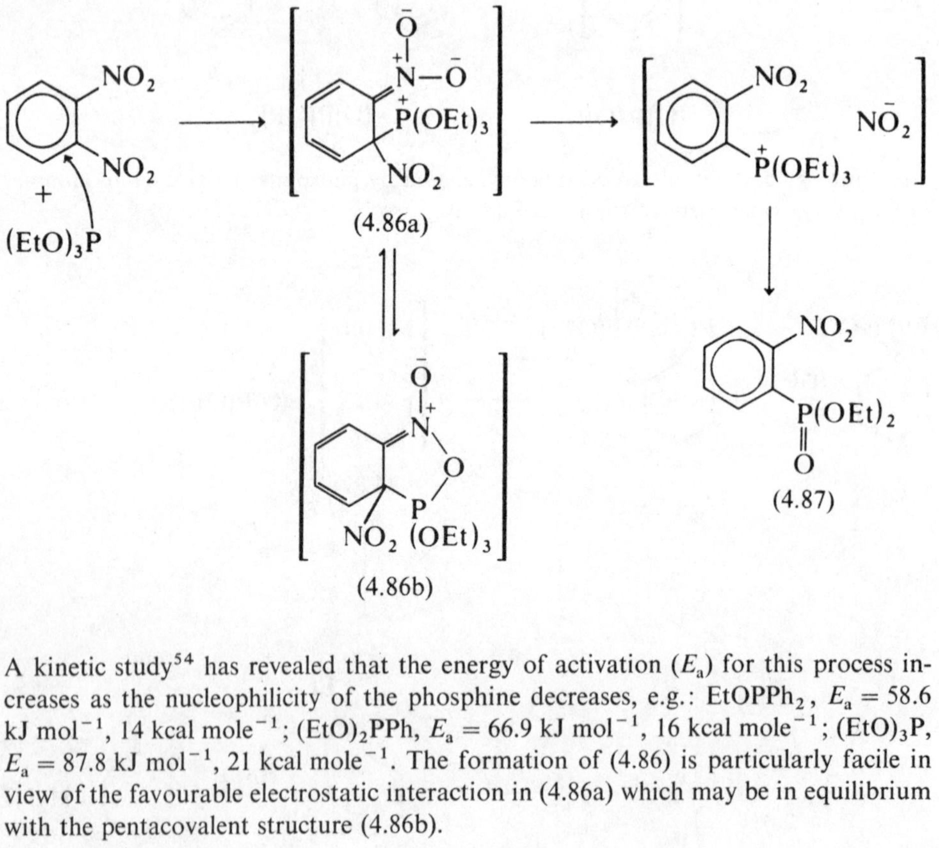

[Mechanism showing o-dinitrobenzene + (EtO)₃P → intermediates (4.86a) ⇌ (4.86b) → (4.87)]

A kinetic study[54] has revealed that the energy of activation (E_a) for this process increases as the nucleophilicity of the phosphine decreases, e.g.: EtOPPh₂, $E_a = 58.6$ kJ mol⁻¹, 14 kcal mole⁻¹; (EtO)₂PPh, $E_a = 66.9$ kJ mol⁻¹, 16 kcal mole⁻¹; (EtO)₃P, $E_a = 87.8$ kJ mol⁻¹, 21 kcal mole⁻¹. The formation of (4.86) is particularly facile in view of the favourable electrostatic interaction in (4.86a) which may be in equilibrium with the pentacovalent structure (4.86b).

4.1.4 Miscellaneous nucleophilic displacement reactions[3]

Trivalent phosphorus compounds will also react at the oxygen atom of highly polar double bonds such as the amine oxides (4.88), arsine oxides (4.89), sulphoxides (4.90) and even in some cases, the phosphoryl bond of other phosphorus compounds (4.91).

4.1 Nucleophilic reactivity

$$R_3\overset{+}{N}-\overset{-}{O} + R^1_3P \longrightarrow R_3N: + O{=}PR^1_3 \quad (4.88)$$

$$Ph_3As{=}O + Ph_3P \longrightarrow Ph_3As: + O{=}PPh_3 \quad (4.89)$$

$$Me_2S{=}O + R_3P \longrightarrow Me_2S: + O{=}PR_3 \quad (4.90)$$

$$Cl_3P{=}O + Ph_3P \longrightarrow Cl_3P: + O{=}PPh_3 \quad (4.91)$$

The reaction may also be employed to remove thiophosphoryl sulphur, e.g.

$$\underset{MePCl_2}{\overset{\overset{S}{\|}}{}} + n\text{-}Bu_3P \longrightarrow MePCl_2 + n\text{-}Bu_3P{=}S$$

Since all of these phosphoryl oxygen (or sulphur) bonds are to some extent polarized away from phosphorus (and in the case of amine oxides, totally polarized from N to O) it is difficult to accept the idea that the deoxygenation involves nucleophilic attack by phosphorus on oxygen.* Furthermore, in many cases the order of reactivity for a series of trivalent phosphorus compounds is the *reverse* of their nucleophilicity. For example, Ramirez[55] demonstrated that the ease of deoxygenation of pyridine-N-oxide decreases in the sequence, $PCl_3 > PhPCl_2 > Ph_2PCl > (PhO)_3P > (EtO)_3P > PPh_3$. Other workers have observed a similar sequence for the deoxygenation of dimethyl sulphoxide although beyond triphenylphosphine in the above sequence, reactivity increased with strongly nucleophilic compounds, i.e. $(Me_2N)_3P > n\text{-}Bu_3P > Ph_3P$.[56] One is therefore drawn to the idea of a biphilic insertion mechanism which with sulphoxides, phosphine oxides and arsine oxides may proceed via a TS described by (4.92) to an unstable intermediate (4.93) which breaks down to products.

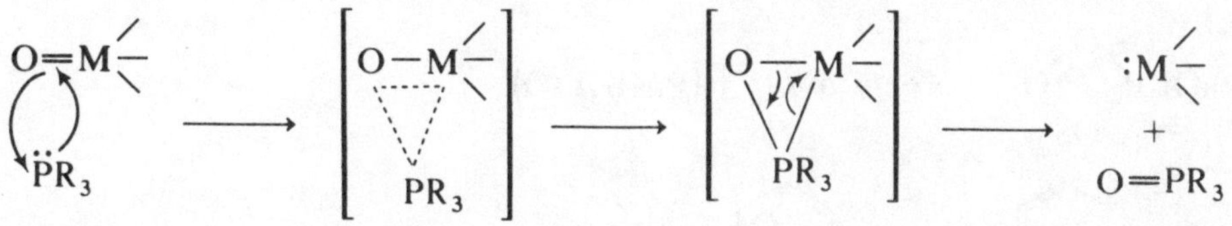

However, in order to avoid a pentacovalent nitrogen in the case of the amine oxides, it is necessary to envisage the reaction as a form of addition–elimination, steps (1) and (2) occurring almost simultaneously.

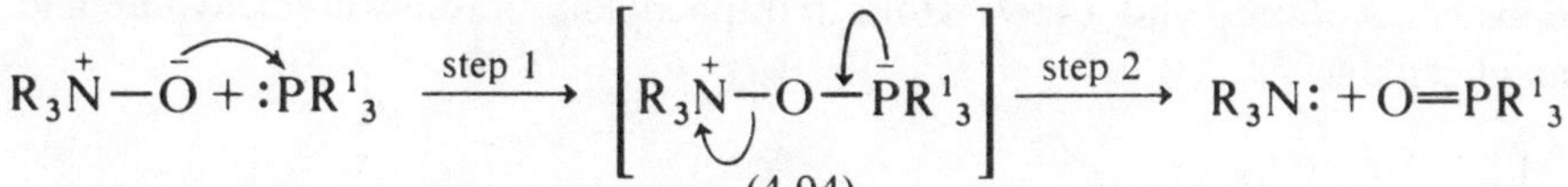

$$R_3\overset{+}{N}-\overset{-}{O} + :PR^1_3 \xrightarrow{\text{step 1}} \left[R_3\overset{+}{N}-O-PR^1_3\right] \xrightarrow{\text{step 2}} R_3N: + O{=}PR^1_3$$
$$(4.94)$$

---

* Nucleophilic attack on the positive, nitrogen end of the dipole is untenable for amine oxides since it would involve the formation of an intermediate with nitrogen in a pentacovalent state.

It is highly unlikely that an intermediate such as (4.94) would have any finite existence since in the case of $R^1 = Cl$, it would be expected to stabilize itself by elimination of chloride ion (eqn. 17).

$$[R_3\overset{+}{N}-O-\overset{-}{P}Cl_3] \longrightarrow \overset{-}{Cl}\, R_3\overset{+}{N}-OPCl_2 \tag{17}$$

We shall meet this delicate balance between nucleophilic and electrophilic characteristics of phosphorus again (Section 4.3) and come to recognize it as symptomatic of a biphilic process.

Phosphines and phosphites will also add to electrophilic oxygen and an interesting example of this involves the reaction of the bicyclic phosphite (4.95) with ozone to give the adduct (4.96) which is stable up to 0 °C.[57]

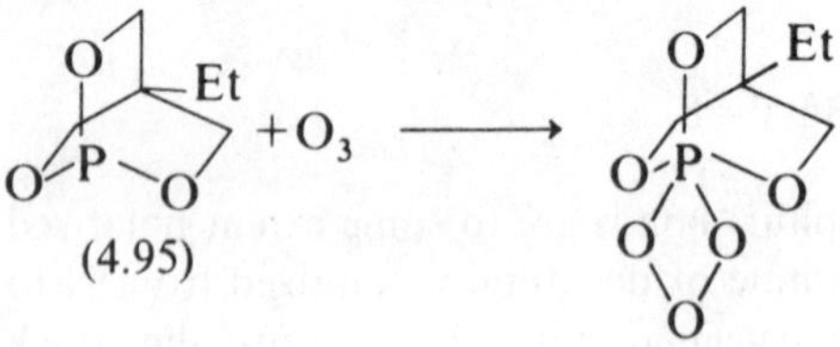

(4.95) + O$_3$ ⟶ (4.96)

Similar adducts are obtained from acyclic phosphites[58,59] but (4.97) has a half-life of

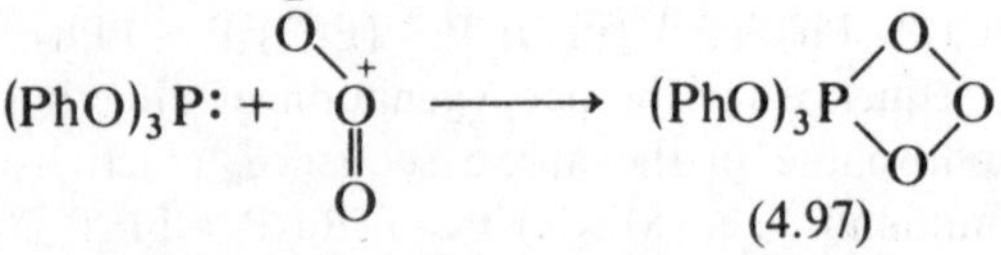

$(PhO)_3P: + \; O_3 \longrightarrow (PhO)_3P\underset{O}{\overset{O}{\diagup\diagdown}}O$

(4.97)

only 0.47 min at 10 °C (c.f. (4.96), $t_{\frac{1}{2}} = 76.2$ min at 10 °C). Both adducts decompose to give singlet oxygen which may be trapped by dienes, for example cyclohexadiene which gives norascaridole (4.98).

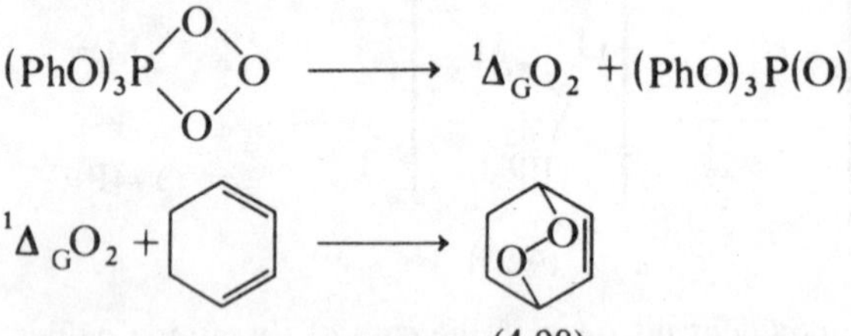

$(PhO)_3P\underset{O}{\overset{O}{\diagup\diagdown}}O \longrightarrow {}^1\Delta_G O_2 + (PhO)_3P(O)$

${}^1\Delta_G O_2 + $ ⟶

(4.98)

Finally it is worth noting that attack of trivalent phosphorus compounds on unsaturated nitrogen is also a fairly well established reaction. One example involves the formation of an imino-ylid (4.99) from tetraphenyldiazocyclopentadienylide and triphenylphosphine[60].

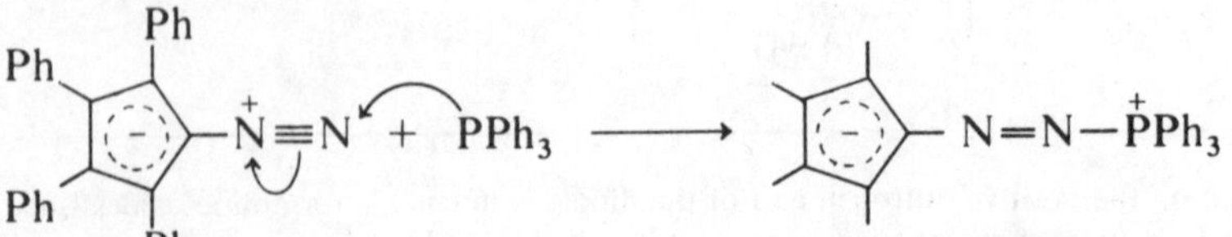

(4.99)

4.2 Electrophilic reactivity

A more exotic reaction also involves the formation of an imino-ylid (phosphazene) (4.101) this time from tetrazolopolyazines (e.g. 4.100) and triphenylphosphine, by nucleophilic attack on the triazolo ring[61].

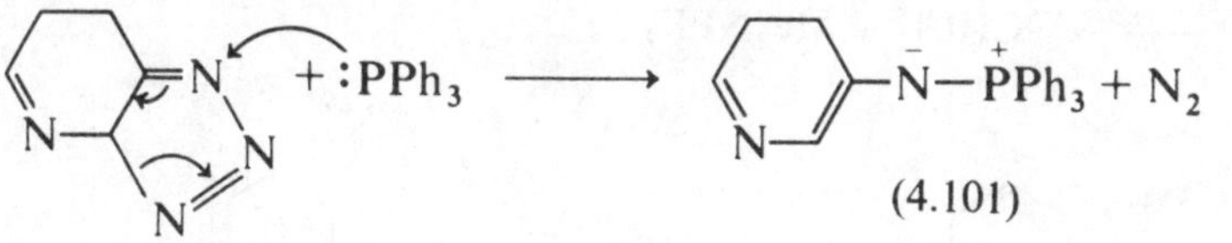

(4.100)   +:PPh$_3$ → $\overset{-}{N}-\overset{+}{P}Ph_3$ + N$_2$ (4.101)

4.2 Electrophilic reactivity[1,3,5,62]

The hydrolysis of phosphorus trichloride gives phosphorous acid (4.102) and this is perhaps the simplest example of a trivalent phosphorus compound behaving as an electrophile. The reaction also serves to highlight yet another difference between the chemistry of phosphorus and nitrogen since treatment of nitrogen trichloride with water gives ammonia and hypochlorous acid.

$$PCl_3 + 3H_2O \longrightarrow P(OH)_3 + 3HCl$$
$$(4.102)$$

$$NCl_3 + 3H_2O \longrightarrow NH_3 + 3HOCl$$

The change in behaviour has been explained in terms of electronegativity differences between phosphorus and nitrogen in comparison with chlorine, which promotes nucleophilic attack on halogen in the case of NCl$_3$ but attack on phosphorus for PCl$_3$. An equally likely explanation however, lies in the ability of phosphorus (but not nitrogen) to expand its valence shell to ten electrons and hence stabilize a transition state of type (4.103).

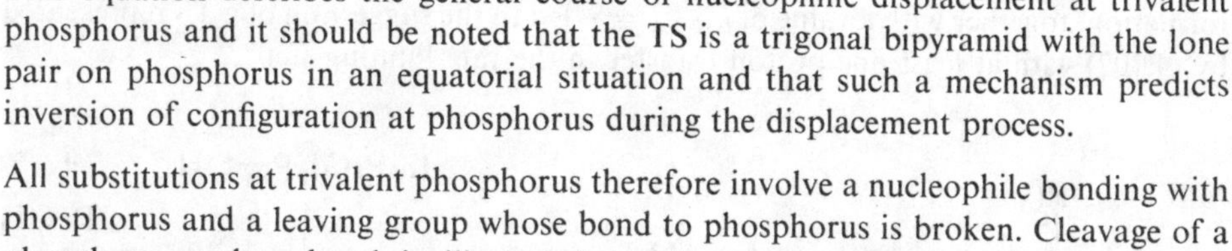

(4.103)

This equation describes the general course of nucleophilic displacement at trivalent phosphorus and it should be noted that the TS is a trigonal bipyramid with the lone pair on phosphorus in an equatorial situation and that such a mechanism predicts inversion of configuration at phosphorus during the displacement process.

All substitutions at trivalent phosphorus therefore involve a nucleophile bonding with phosphorus and a leaving group whose bond to phosphorus is broken. Cleavage of a phosphorus–carbon bond is illustrated by the base-catalyzed hydrolysis of tris-trifluoromethylphosphine (4.104) to give fluoroform. A similar example is provided by amination of (4.105). Notice that cleavage of a P—C bond normally requires stabilization of the incipient carbanion by electron-withdrawing groups. Displacement of carbon can also be achieved through attenuation of the negative charge by a good leaving group on a β-carbon (eqn. 18).

$$(CF_3)_3P \xrightarrow{H_2O/NaOH} 3CHF_3 + Na_3PO_3$$
(4.104)

$$Cl_3CPF_2 + NHMe_2 \longrightarrow CHCl_3 + Me_2NPF_2$$
(4.105)

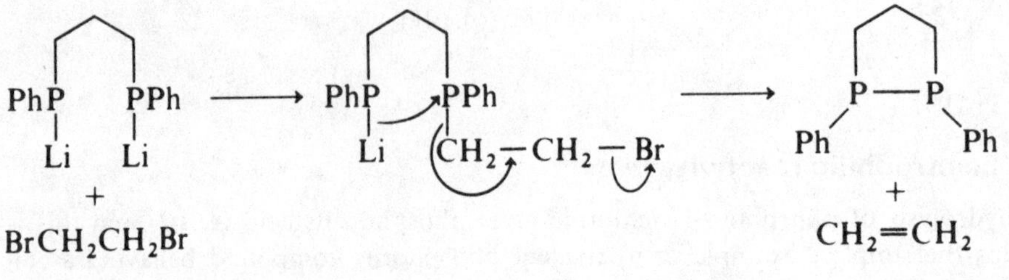

(18)

Phosphorus–nitrogen bonds may also be cleaved, particularly under acid-catalyzed conditions. Thus the rates of displacement of diethylamine from the phosphoroamidite (4.106) by phenols, increase with the acidity of the phenol used.

$$(EtO)_2PNEt_2 + ArOH \longrightarrow (EtO)_2POAr + HNEt_2$$
(4.106)

Alkoxy groups may be displaced from phosphorus either by hydrolysis or exchange reactions.

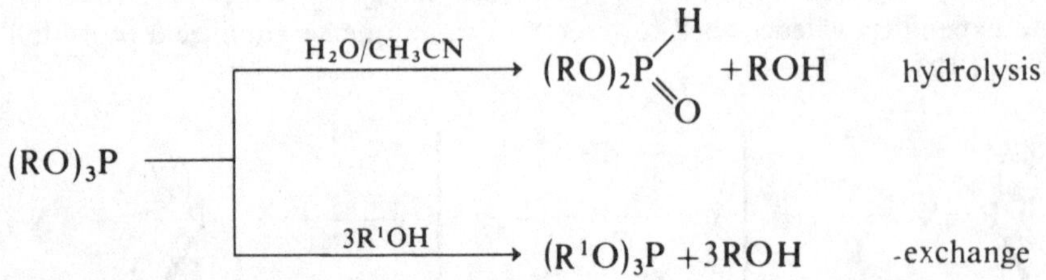

Asknes has investigated the hydrolysis of tri-n-propyl phosphite (in acetonitrile) and established an overall third-order reaction with rate $\propto [(PrO)_3P][H_2O]^2$. This information, together with a value of $k_H/k_D = 8$, led to the suggestion of a TS represented by (4.107) with at least one proton transfer in the rate-limiting step[63].

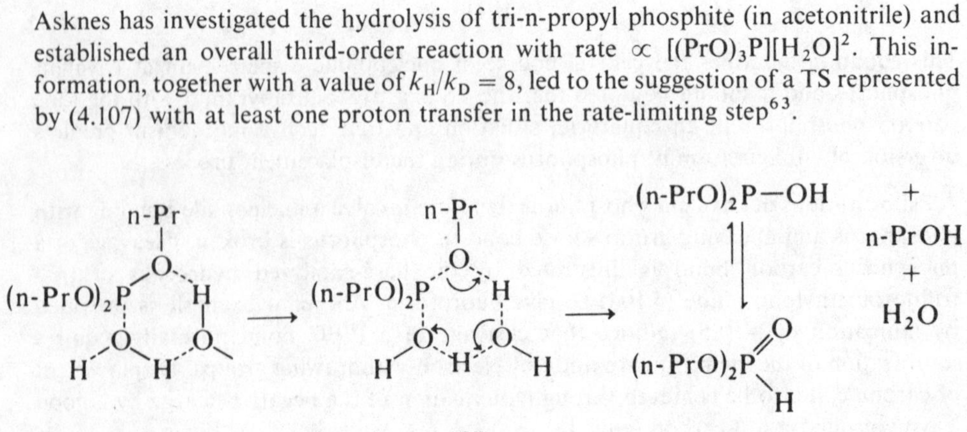

(4.107)

The exchange reaction has been exploited extensively for the synthesis of phosphites, particularly those containing five- and six-membered rings. Thus glycols give rise to 1,3,2-dioxaphospholanes (4.108 and 4.109a,b) and a trihydric alcohol such as (4.110) yields the tricylic phosphite (4.111)[64,65].

dl-glycol (4.108a) (4.108b)

meso-glycol (4.109a) (4.109b)

(4.110) (4.111)

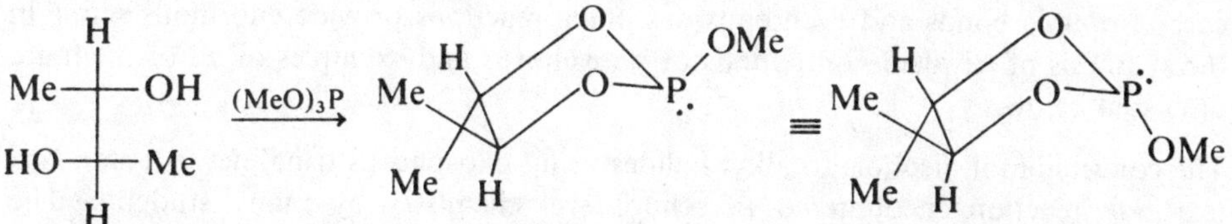

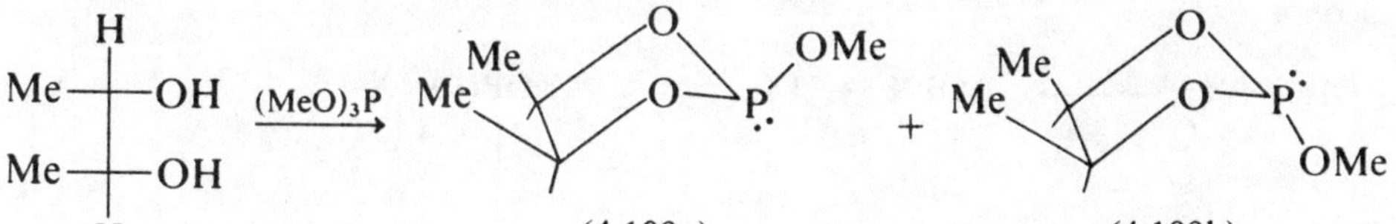

Since the cyclic phosphites are configurationally stable (high energy barrier to inversion at phosphorus), the *meso*-glycol gives *cis* and *trans* isomers (4.109a and 4.109b respectively) but the *dl*-glycol yields only one compound (4.108a), the opposite configuration at phosphorus (4.108b) being rendered superimposable on (4.108a) by inversion of the ring.

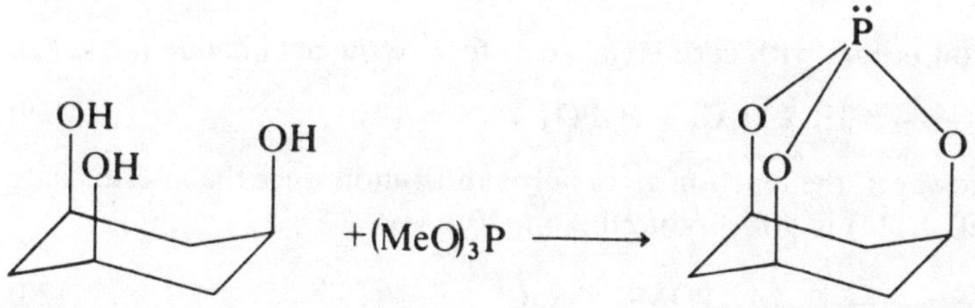

The phosphorus–oxygen bond of aryl phosphites is even more susceptible to cleavage and offers a good route for the synthesis of phosphines:

$$3PhMgBr + P(OPh)_3 \longrightarrow Ph_3P + 3PhOMgBr$$

The highest reactivity is of course afforded by the phosphorus–oxygen bonds of mixed anhydrides and facile displacement of phosphate ion from (4.112) offers a useful route to phosphoramidites (4.113)[66].

(4.112) (4.113)

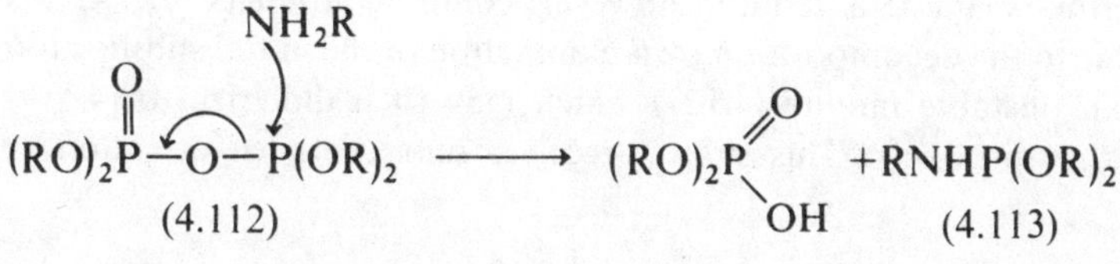

By far the most common leaving groups in substitutions at trivalent phosphorus are the halogens all of which, except fluorine, are readily displaced under the appropriate conditions by a wide variety of nucleophiles including water, alcohols and alkoxide ions, carboxylic acids, thiols, amines, phosphines, phosphite and phosphate anions, Grignard and organometallic reagents, carbanions, aromatic π-systems, carbon–carbon double bonds and even epoxides. Such reactions provide enormous scope in the synthesis of trivalent compounds of phosphorus and examples of all of them are discussed below.

The conversion of alcohols to alkyl halides using phosphorus trihalides is a standard textbook reaction encountered in school-level chemistry by many students. The mechanism involves an intermediate alkoxyphosphonium salt (4.114) which is de-alkylated to a phosphoryl halide (4.115), which then proceeds to halogenate more alcohol.

$$ROH + PX_3 \longrightarrow ROPX_2 + HX \rightleftharpoons ROPHX_2^+ \ X^-$$
$$(4.114)$$

$$O=PH(OH)_2 + 2RX \xleftarrow{2\ ROH} O=PHX_2 + RX$$
$$(4.115)$$

An analogous reaction occurs with carboxylic acids to give the acyl halide (eqn. 19).

$$3RCO_2H + PX_3 \longrightarrow 3R.CO.Cl + H_3PO_3 \tag{19}$$

With alkoxide ions however, the reaction gives only substitution since the intermediate phosphonium salt (cf. 4.114) is not produced (eqn. 20).

$$Me_2PCl + Na^+\ \bar{O}Me \longrightarrow Me_2POMe + NaCl \tag{20}$$

Thiols may be used in the same way as alcohols to generate thiophosphinites (4.116); the latter often rearrange easily to phosphine sulphides.

$$Na^+\ \bar{S}CH_2CH=CH_2 \longrightarrow Ph_2PSCH_2CH=CH_2 \xrightarrow{heat} Ph_2\overset{S}{\overset{\|}{P}}CH_2CH=CH_2$$
$$+\quad\quad\quad\quad\quad\quad\quad\quad (4.116)$$
$$Ph_2PCl$$

With secondary amines or in some instances, primary amines, aminophosphines (e.g. 4.117 and 4.118) may be prepared.

$$R_2NH + PCl_3 \longrightarrow R_2NPCl_2 + R_2\overset{+}{N}H_2\ Cl^-$$
$$(4.117)$$

$$2PhNH_2 + R_2PCl \longrightarrow R_2PNHPh + Ph\overset{+}{N}H_3\ Cl^-$$
$$(4.118)$$

As the amount of primary amine is reduced however, complex products (e.g. 4.119) are formed. This is due to the decomposition by β-elimination of the initial substitution product (4.120) to an unstable imidite (4.121) which may then dimerize to (4.119) or polymerize to other products. This subject receives more detailed attention in Chapter 10.

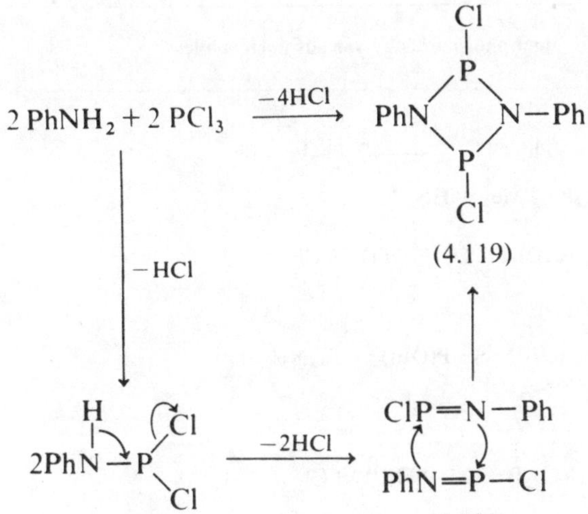

4.2 Electrophilic reactivity

$$2\,PhNH_2 + 2\,PCl_3 \xrightarrow{-4HCl} \underset{(4.119)}{\begin{array}{c}Cl\\|\\P\\PhN\diagup\quad\diagdown N-Ph\\\diagdown\quad\diagup\\P\\|\\Cl\end{array}}$$

$$\downarrow -HCl$$

$$\underset{(4.120)}{2PhN\overset{H}{-}P\overset{Cl}{\diagdown}_{Cl}} \xrightarrow{-2HCl} \underset{(4.121)}{\begin{array}{c}ClP=N-Ph\\PhN=P-Cl\end{array}}$$

With bifunctional nucleophiles (e.g. glycols, amino-alcohols) phosphorus halides will form cyclic structures, especially when five- and six-membered rings are possible products. Thus methylphosphorochloridite (4.122) reacts with 2,3-propanediol in the presence of base to give the cyclic phosphite (4.123)[64].

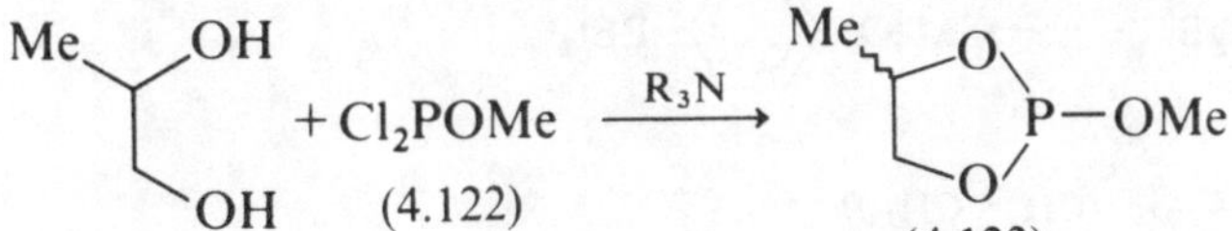

$$\underset{\text{(4.122)}}{\begin{array}{c}Me\diagup OH\\ \diagdown OH\end{array}} + Cl_2POMe \xrightarrow{R_3N} \underset{(4.123)}{\begin{array}{c}Me\\ \diagdown O\\ \diagup P-OMe\\ O\end{array}} \qquad cis:trans = 38:62$$

Likewise 2-N-phenylaminoethanol reacts with phosphorus trichloride to produce the cyclic phosphite (4.124)[67].

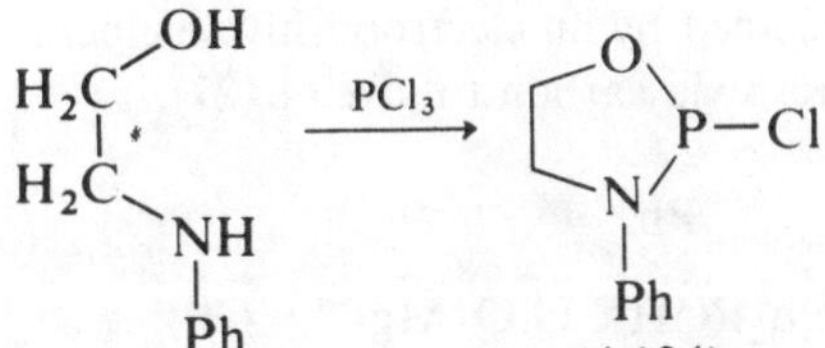

$$\begin{array}{c}H_2C\diagup OH\\ |\\ H_2C\diagdown NH\\ |\\ Ph\end{array} \xrightarrow{PCl_3} \underset{(4.124)}{\begin{array}{c}O\\ \diagup \diagdown\\ P-Cl\\ N\\ |\\ Ph\end{array}}$$

Examples of the synthetic utility of halide displacement by phosphines, phosphite and phosphate anions and organometallic reagents appear in Table 4.8.

Ambident nucleophiles generally react via the most electronegative atom of the reagent, so for example, ethyl acetoacetate gives a vinyl phosphite (4.125)[68].

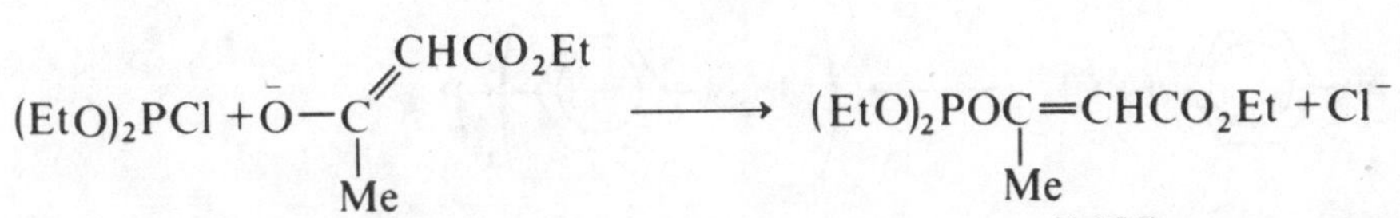

$$(EtO)_2PCl + \overset{\overset{CHCO_2Et}{\|}}{\underset{\underset{Me}{|}}{O-C}} \longrightarrow \underset{(4.125)}{(EtO)_2PO\underset{\underset{Me}{|}}{C}=CHCO_2Et} + Cl^-$$

Table 4.8 Nucleophilic displacement at trivalent phosphorus by various nucleophiles

$$Ph_2PH + Ph_2P-Cl \longrightarrow [Ph_2\overset{+}{P}H-PPh_2 \;\overset{-}{Cl}] \longrightarrow Ph_2P-PPh_2$$

$$Me_2PH + Me_2P-NMe_2 \longrightarrow Me_2P-PMe_2 + HNMe_2$$

$$(EtO)_2P-O^- + (EtO)_2P-Cl \longrightarrow (EtO)_2P-O-P(OEt)_2 + \overset{-}{Cl}$$

$$\underset{\underset{\|}{S}}{(EtO)_2P}-H + (EtO)_2P-Cl \xrightarrow{Et_3N} (EtO)_2P-S-P(OEt)_2 + Et_3\overset{+}{N}H\;\overset{-}{Cl}$$

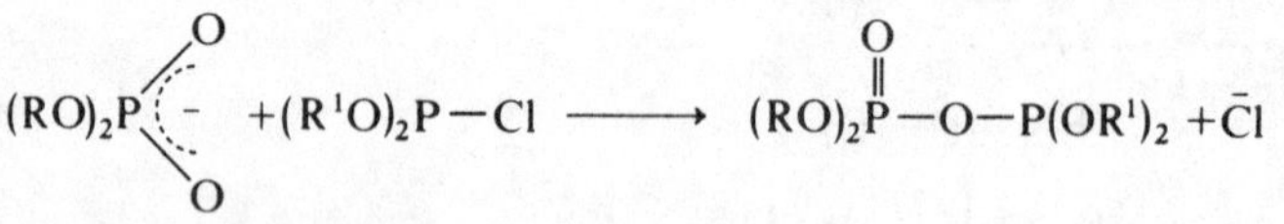

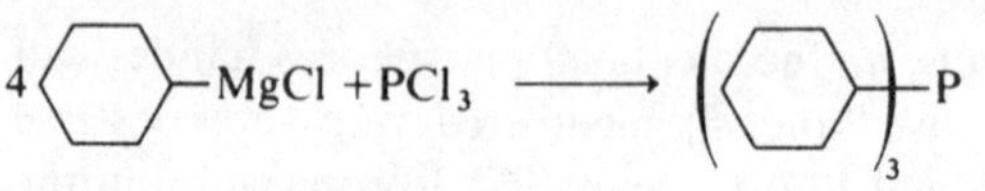

$$(RO)_2P\overset{O}{\underset{O}{\overset{\diagdown}{\diagup}}}{}^- + (R^1O)_2P-Cl \longrightarrow (RO)_2\overset{\overset{O}{\|}}{P}-O-P(OR^1)_2 + \overset{-}{Cl}$$

$$4 \langle\text{C}_6\text{H}_{11}\rangle\text{—MgCl} + PCl_3 \longrightarrow (\langle\text{C}_6\text{H}_{11}\rangle)_3\text{—P}$$

$$4\, ArMgX + PCl_3 \longrightarrow Ar_3P \quad \text{(excess Grignard reagent often necessary)}$$

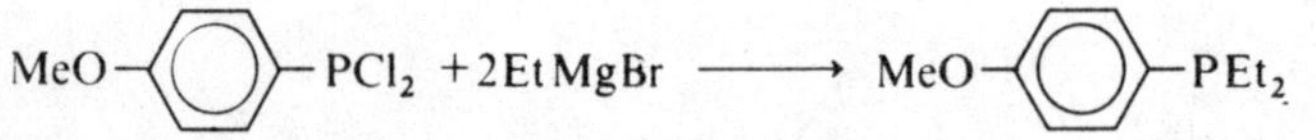

$$MeO-\langle\text{C}_6\text{H}_4\rangle-PCl_2 + 2EtMgBr \longrightarrow MeO-\langle\text{C}_6\text{H}_4\rangle-PEt_2$$

$$3\, CH_2{=}CHMgCl + PCl_3 \longrightarrow (CH_2{=}CH)_3P$$

$$3\, C_6H_5C{\equiv}CMgBr + PCl_3 \longrightarrow (C_6H_5C{\equiv}C)_3P$$

$$(C_4H_9)_2Cd + PCl_3 \longrightarrow C_4H_9PCl_2 \xrightarrow{2\,RMgX} C_4H_9PR_2$$

Exceptions to this arise when the oxyanion is shielded by an electropositive element; thus the ambident Grignard reagent (4.126) attacks via carbon to give (4.127).

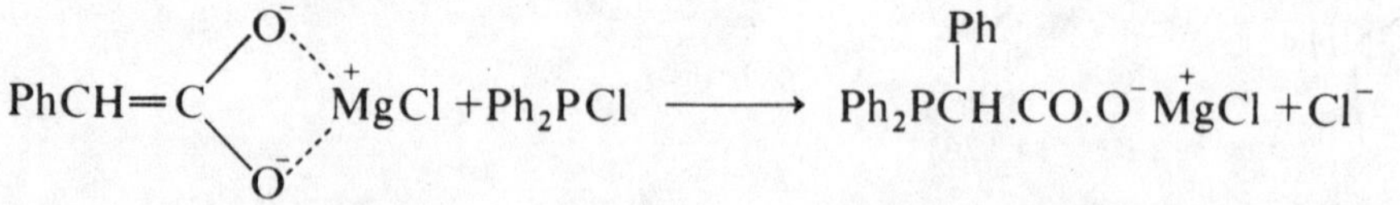

$$PhCH{=}C\underset{O\cdots}{\overset{O^-\cdots}{\diagdown\diagup}}\overset{+}{MgCl} + Ph_2PCl \longrightarrow Ph_2PCH.CO.O^-\overset{+}{MgCl} + Cl^-$$

$$(4.126) \qquad\qquad (4.127)$$

Aromatic systems activated by electron-donating groups (e.g. —NR$_2$ or —OR) react smoothly with halophosphines to give arylphosphines (e.g. 4.128) and phosphorus heterocycles such as (4.129)[69,70].

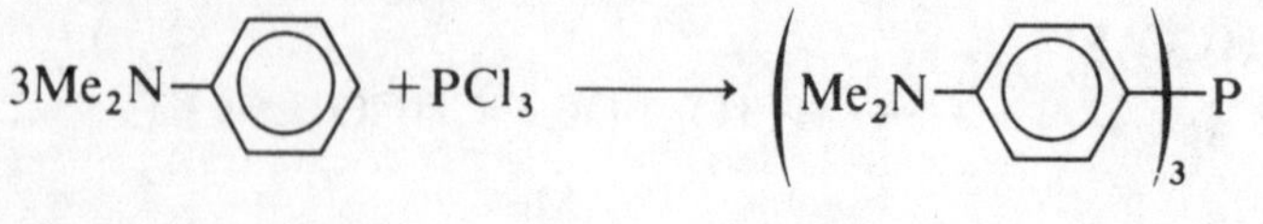

$$3Me_2N-\langle\text{C}_6\text{H}_4\rangle + PCl_3 \longrightarrow (Me_2N-\langle\text{C}_6\text{H}_4\rangle)_3-P$$

$$(4.128)$$

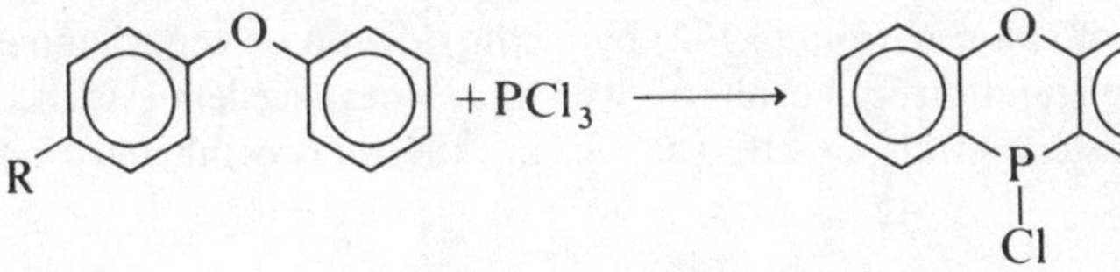

4.2 Electrophilic reactivity

(4.129)

Unactivated arenes however, require catalysis by Lewis acids; thus in the presence of aluminium chloride, benzene reacts readily to give phenylphosphonous dichloride (4.130) which is a common starting point for the synthesis of many phosphines via subsequent displacement of chloride ion by organometallic reagents.

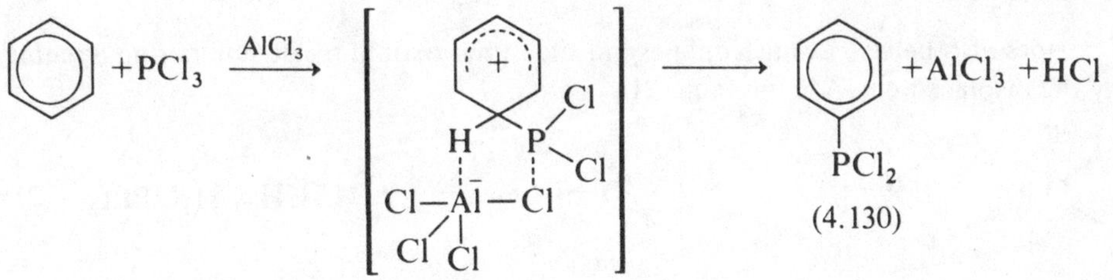

(4.130)

Olefins are also capable of reaction with phosphorus halides and one of the more unusual examples involves the formation of a four-membered ring phosphorus heterocycle (4.131)[71]. Little is known about the mechanism of this reaction but 1,2-migration of a methyl group must be involved and this suggests nucleophilic displacement of chloride by the π-bond of the olefin. The intermediate (4.131) may be hydrolyzed to give *trans*-1-chlorophosphetan 1-oxide which in turn may be reduced to *cis*- and *trans*-1-chloro-2,2,3,4,4-pentamethylphosphetan (4.132).

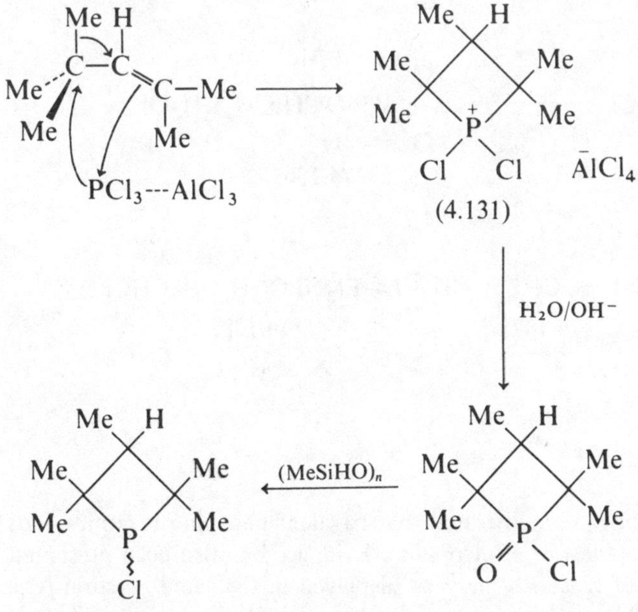

(4.131)

(4.132)

Nucleophilic displacement of chlorine from (4.132) by methoxide ion or benzylamine occurs with inversion of configuration at phosphorus which provides excellent evidence for an S_N2 TS despite the steric strain involved in placing the four-membered ring diequatorial[72].

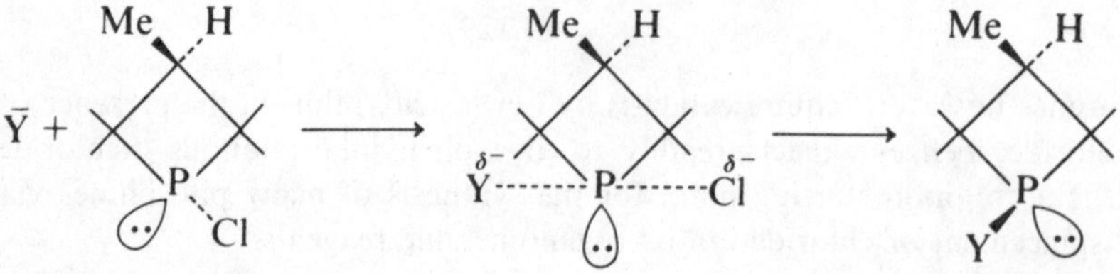

trans-(4.132) → → cis-product

Epoxides also behave as nucleophiles but the initial oxonium salt suffers ring opening by the displaced chloride ion (eqn. 21).

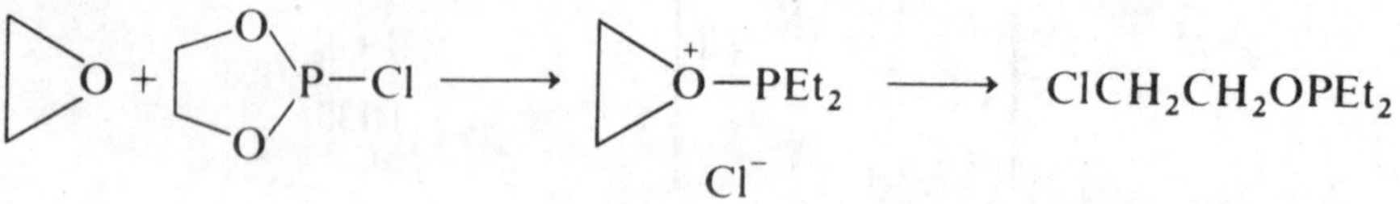

$$\triangle O + \text{(cyclic)}P-Cl \longrightarrow \triangle \overset{+}{O}-PEt_2 \quad Cl^- \longrightarrow ClCH_2CH_2OPEt_2 \quad (21)$$

Similar reactions occur with propylene oxide (4.133) and 2-methyloxetane (4.135) but these reactions are frequently complicated by ring opening on both the primary and secondary carbons to give mixtures of isomeric products (4.137 and 4.138).

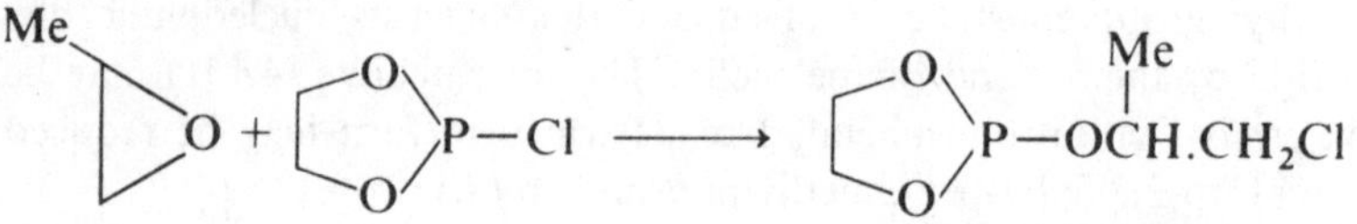

(4.133) + (cyclic)P—Cl ⟶ (cyclic)P—OCH.CH$_2$Cl (Me)

(4.134)

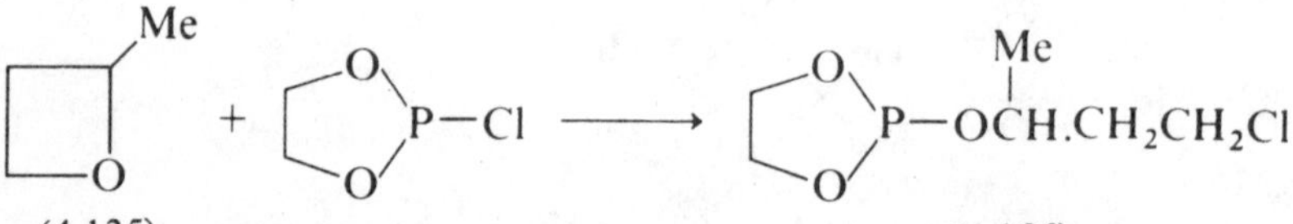

(4.135) + (cyclic)P—Cl ⟶ (cyclic)P—OCH.CH$_2$CH$_2$Cl (Me)

(4.136)

(4.135, Me-oxetane) + Et$_2$NPCl$_2$ ⟶ Et$_2$NP—OCHCH$_2$CH$_2$Cl (Cl, Me) + Et$_2$NPOCH$_2$CH$_2$CHCl (Cl, Me)

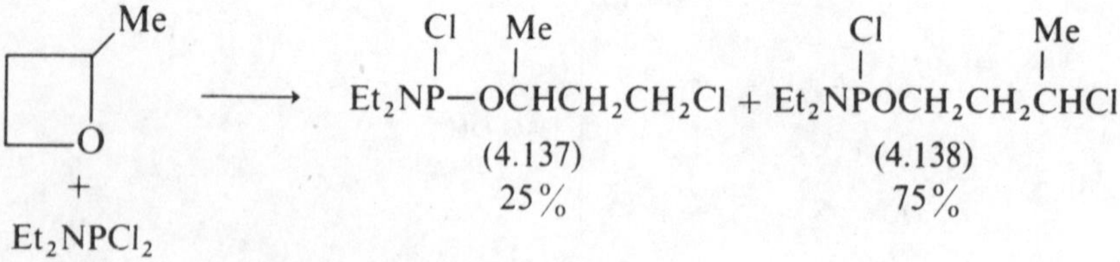

(4.137) 25%   (4.138) 75%

4.3 Biphilic reactivity[3,9]

The two previous sections have demonstrated that trivalent phosphorus compounds may behave either as nucleophiles or electrophiles. Evidence has also been presented to suggest that both types of reactivity may be displayed in the same reaction (e.g. deoxygenation of oxides). This dual function of trivalent phosphorus compounds con-

4.3 Biphilic reactivity

stitutes what is known as *biphilic reactivity* and several fascinating examples of such reactivity have appeared in recent years.

The reaction of diethyl peroxide with phosphines and phosphites in aprotic media (cyclopentane or dichloromethane) at room temperature or below affords a useful synthesis of pentacovalent phosphorus compounds[73,74] (phosphoranes) (eqn. 22)

$$R_3P + Et_2O_2 \longrightarrow R_3P(OEt)_2 \qquad (22)$$

where R = alkyl, aryl, alkoxy, aryloxy or mixtures of each. These reactions follow a reactivity sequence which is the *reverse* of that observed during nucleophilic displacement by phosphorus. Thus the phosphines (4.139–4.142) react with diethyl peroxide in a relative rate order which is almost the reverse of the order for the reactions of the same phosphines with methyl iodide[75].

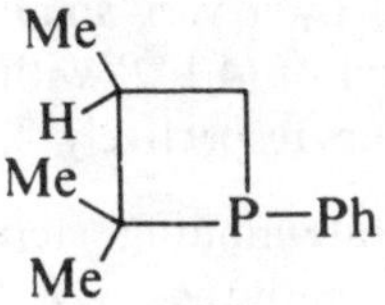

	(4.139)	(4.140)	(4.141)	(4.142)
Relative rate towards Et_2O_2:				
a) in cyclopentane	3.8	2.0	1.3	1.0
b) in CH_2Cl_2	3.1	1.7	1.6	1.0
Relative rate towards EtI:				
in CH_3CN	1.0	2.7	4.5	3.0

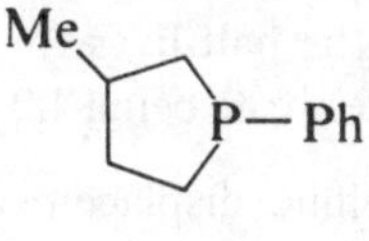

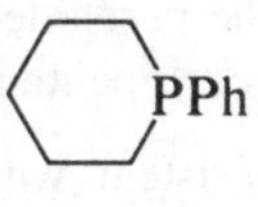

A similar reversal of relative reactivity has been observed for the phosphites (4.143–4.147) when reacted with diethyl peroxide and ethyl iodide[74]:

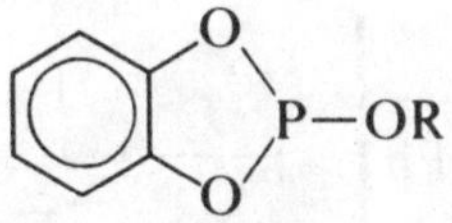

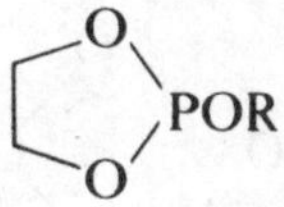

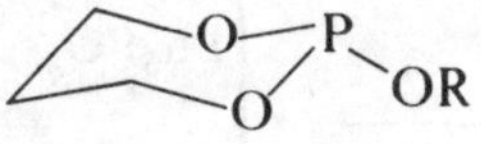

Relative rate towards:

(i) Et_2O_2: (4.143) > (4.144) > (4.145) > (4.146)
(ii) EtI   :   (4.144) < (4.145) < (4.146) < (4.147)

With diethyl peroxide the span of relative reactivity is 7:1 from (4.143) to (4.146); with ethyl iodide the corresponding span is 1:10 from (4.144) to (4.147). With ethyl iodide as substrate both series involve the formation of a TS (4.148) with tetracoordinate character at phosphorus and the rate decrease may then be explained in terms of increased strain in the TS as the ring size is decreased. The reaction with diethyl peroxide however, involves a biphilic insertion into the peroxide bond with formation of a TS with pentacovalent character (4.149) and hence the smaller rings afford stabil-

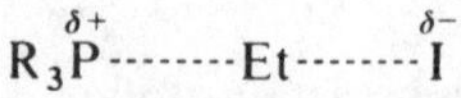

$$R_3\overset{\delta+}{P}\cdots\cdots Et\cdots\cdots\overset{\delta-}{I}$$

(4.148)

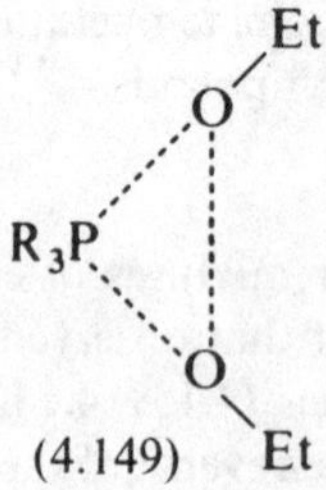

(4.149)

ization of the TS by apical–equatorial disposition of the rings in the developing trigonal bipyramid. The concept of a neutral biphilic TS is also supported by a very small solvent effect for the peroxide reaction, the half-lives for the reaction of (4.142) with diethyl peroxide in cyclopentane and acetonitrile being 4.2 and 3.7 hours respectively[75].

This is quite inconsistent with nucleophilic displacement as the rate-limiting step although it does not rule out the possibility of reaction via a radical pathway.

Reaction of trivalent phosphorus compounds with a variety of peroxides normally involves abstraction of oxygen and formation of a phosphoryl bond. Thus, hydroperoxides, peroxy-acids and diaryl or diaroyl peroxides react as follows:

$Ph_3P + ROOH \longrightarrow Ph_3P(O) + ROH$

$Ph_3P + R^1CO.OOH \longrightarrow Ph_3P(O) + R^1CO_2H$

$Ph_3P + R^1CO.OO.COR^1 \longrightarrow Ph_3P(O) + R^1CO.O.COR^1$

R = alkyl; R^1 = alkyl or aryl

The mechanisms of all these reactions appear to involve nucleophilic attack on peroxidic oxygen and this was elegantly demonstrated by a study of the reaction of triphenyl phosphine with benzoyl peroxide labelled with ^{18}O in the carbonyl group. In the resultant benzoic anhydride the ^{18}O of one carbonyl group remained intact whereas the second ^{18}O was scrambled between carbonyl and anhydride oxygens[76,77].

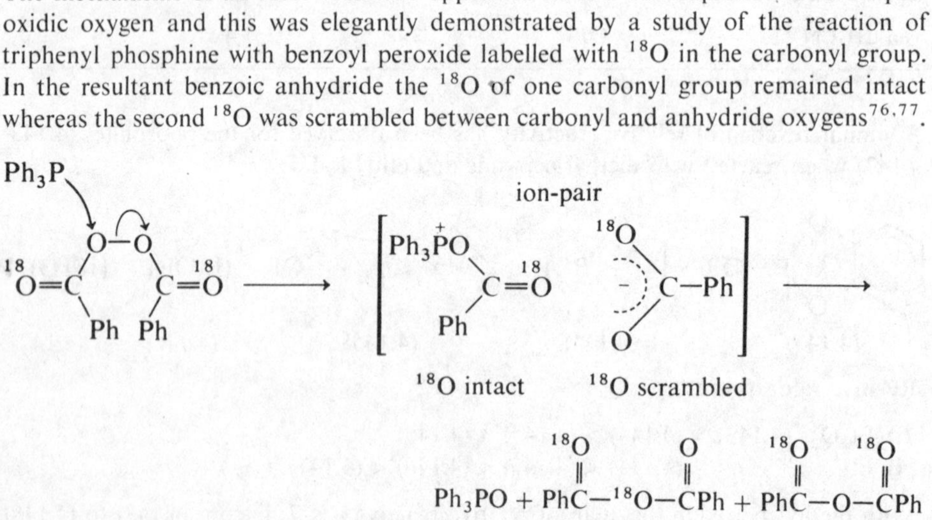

When a sufficiently polar solvent was employed (e.g. CHCl$_3$) the ion-pairs were to some extent solvent separated and in the presence of a different carboxylate ion, a mixed anhydride was produced.

4.3 Biphilic reactivity

Subsequent work with an unsymmetrical benzoyl peroxide (4.150) revealed displacement of the anion of the *weaker* acid and this curious result was explained in terms of rate-determining attack on the *more electrophilic of the peroxidic oxygens* (route 1) with a TS resembling reactants rather than products[78].

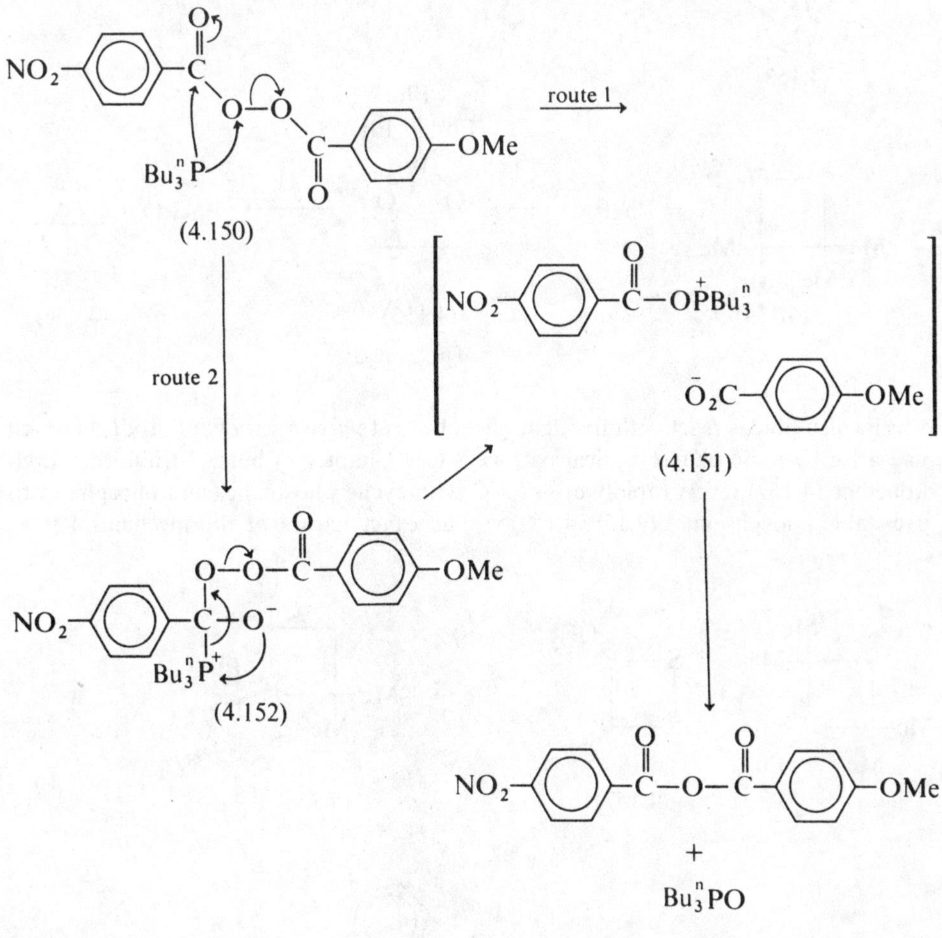

The reaction may also be explained by a change in mechanism to involve nucleophilic addition to the most electrophilic carbonyl group (route 2) followed by rearrangement of the addition product (4.152) to the ion-pair (4.151).

In any event, the concept of nucleophilic attack as the pre-eminent factor for reactions with dibenzoyl peroxide was subsequently supported by the observation of a relative rate sequence for phosphites in the order (4.146) $\geqslant$ (4.145) > (4.144), in line with the relative rates for the reaction with ethyl iodide[79].

Ozonides are also reduced by phosphines and by use of ^{18}O labelling of (4.153) it has been shown that the initial attack of phosphine is on the peroxide link[80]. Whether or not this is also a biphilic insertion is not known but, in a similar reaction of tetramethyl dioxetane (4.154) with triphenylphosphine, a stable phosphorane (4.155) is formed which decomposes smoothly to tetramethylene oxide (4.156) at 55°C.[81]

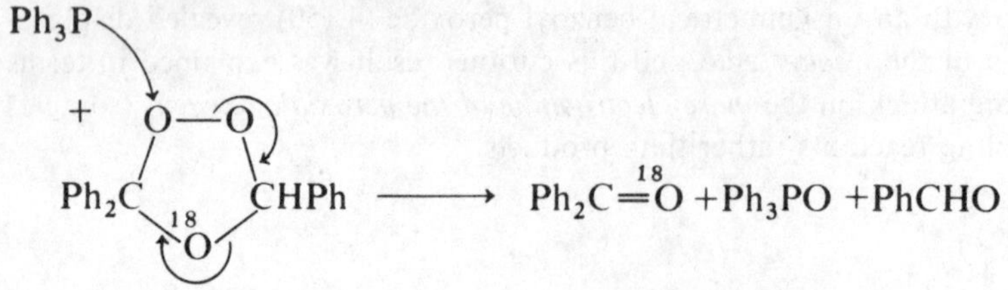

(4.153)

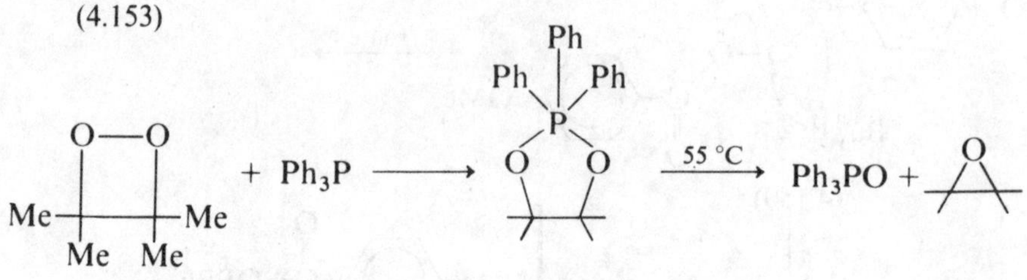

(4.154)   (4.155)   (4.156)

$\delta\ ^{31}P,\ +48.4$

Acyclic disulphides react with trivalent phosphorus to give a variety of products which may arise by ionic or free radical pathways (see Chapter 9) but 3,4-trifluoromethyl-dithietene (4.157) reacts rapidly at -78 °C with cyclic phosphines and phosphates to give stable phosphoranes (4.158, 4.159)[82]. The exact nature of the mechanism is as

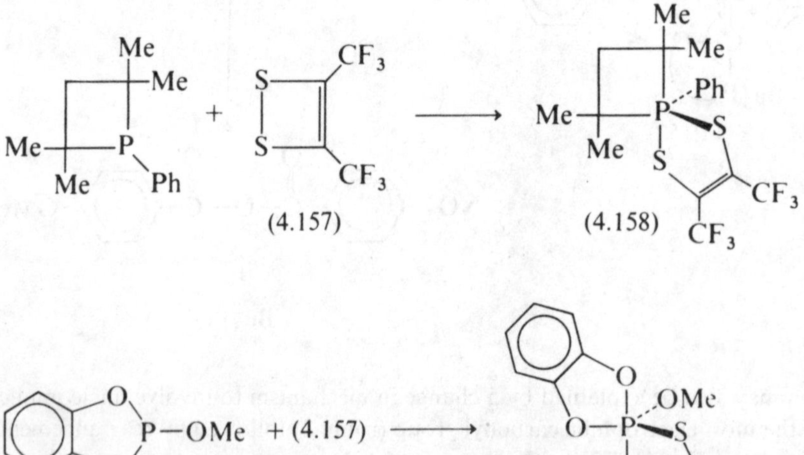

(4.157)   (4.158)

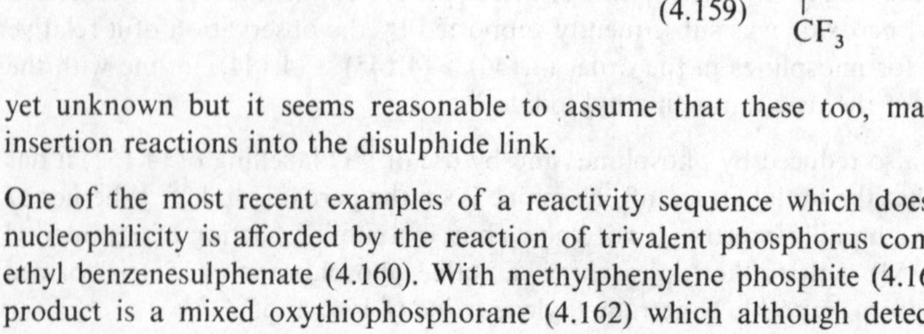

(4.159)

yet unknown but it seems reasonable to assume that these too, may be biphilic insertion reactions into the disulphide link.

One of the most recent examples of a reactivity sequence which does not parallel nucleophilicity is afforded by the reaction of trivalent phosphorus compounds with ethyl benzenesulphenate (4.160). With methylphenylene phosphite (4.161) the initial product is a mixed oxythiophosphorane (4.162) which although detectable by ^{31}P

4.4 Dienophilic reactivity

n.m.r. readily disproportionates to the oxyphosphorane (4.163) and diphenyl disulphide[83].

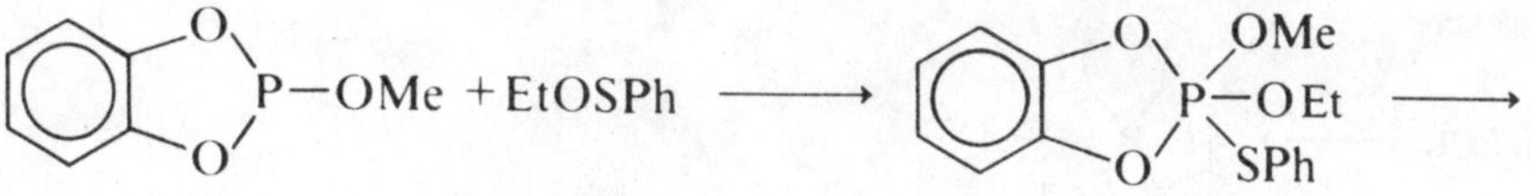

(4.161) + (4.160) EtOSPh ⟶ (4.162)   δ ^{31}P, +22

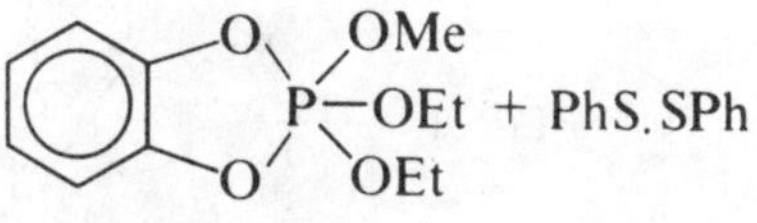

⟶ (4.163) + PhS.SPh   δ ^{31}P, +49

Similar reactions were observed with a variety of acyclic trivalent phosphorus compounds and by following changes in the ^{31}P n.m.r. spectra with time, the reactivity sequence, PhP(OEt)$_2$ > Ph$_2$POEt > (EtO)$_3$P > Ph$_3$P was established. In contrast, the reactivity sequence in a number of nucleophilic displacement reactions was Ph$_3$P ⩾ Ph$_2$POEt > PhP(OEt)$_2$ > P(OEt)$_3$. Once again, the results suggest that with (4.160) the rate-controlling step is not a normal S$_N$2 reaction but is in fact, a biphilic process.

4.4 Dienophilic reactivity[9,84,85]

4.4.1 Dienes

One of the most useful methods of preparing heterocyclic compounds with phosphorus in a five-membered ring is the McCormack synthesis which involves the reaction of trivalent phosphorus compounds with 1,3-dienes.

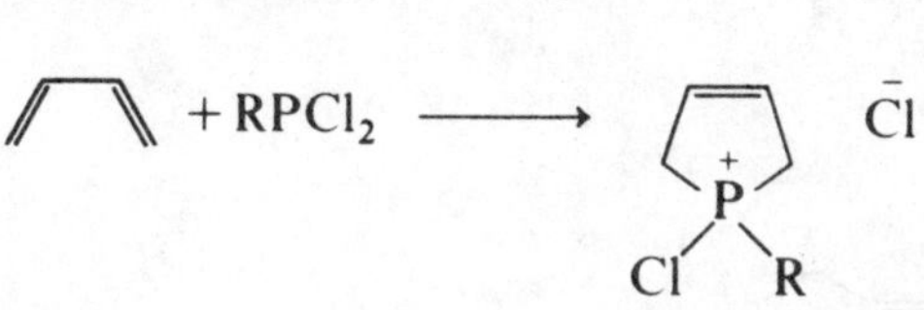

diene + RPCl$_2$ ⟶ (4.164)

The ring system produced (4.164) is known as a 3-phospholene and since its first announcement in 1953, the method has been exploited with a wide variety of dienes and phosphorus compounds (Table 4.9). Unless subsequent rearrangement reactions are possible (see later) the products are either phospholenium salts (e.g. 4.164) or phosphoranes containing a phospholene ring (e.g. 4.169). The cyclization shows all the characteristics of a Diels–Alder reaction and has been shown to be stereospecific. Reaction of *trans,trans*-hexa-2,4-diene with dimethylphosphinous chloride (4.172) gives 1,1,2,5-tetramethylphospholenium chloride (4.174) with the ring methyl groups *cis* to each other[86]. This is consistent with a concerted, disrotatory process as required by the Woodward–Hoffmann rules for a thermal (4n + 2) electron cyclo-

Table 4.9 Condensation of trivalent phosphorus compounds with dienes

(4.165)

(4.166)

(4.167)

(4.168)

(4.169)

(4.170)

(4.171)

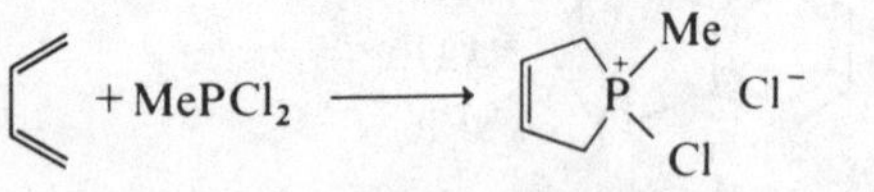

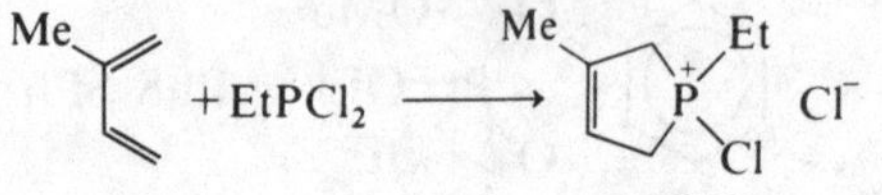

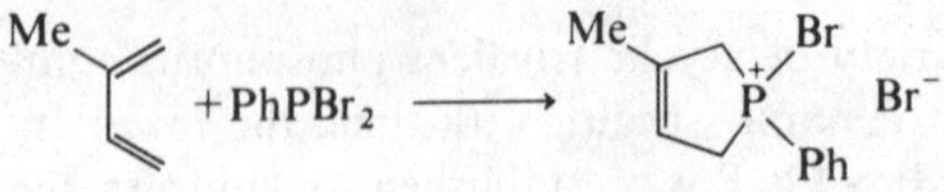

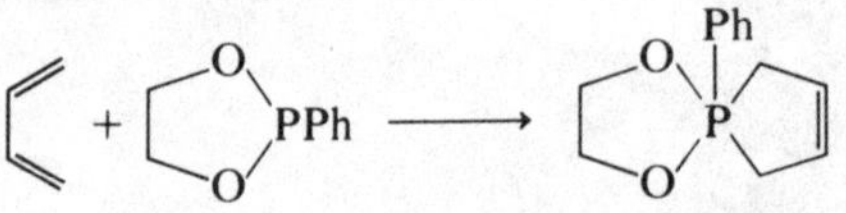

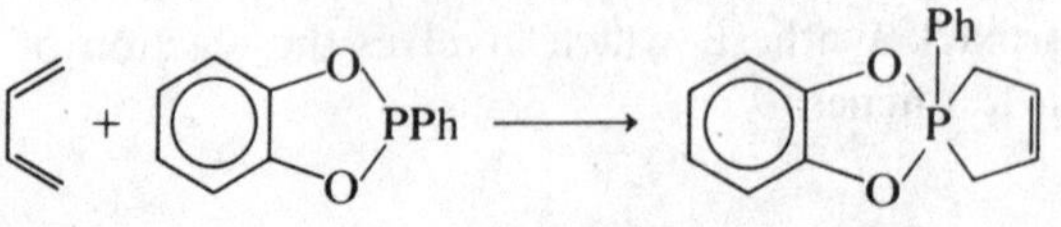

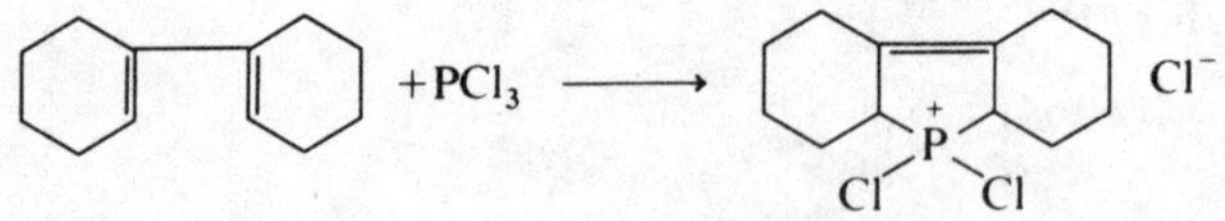

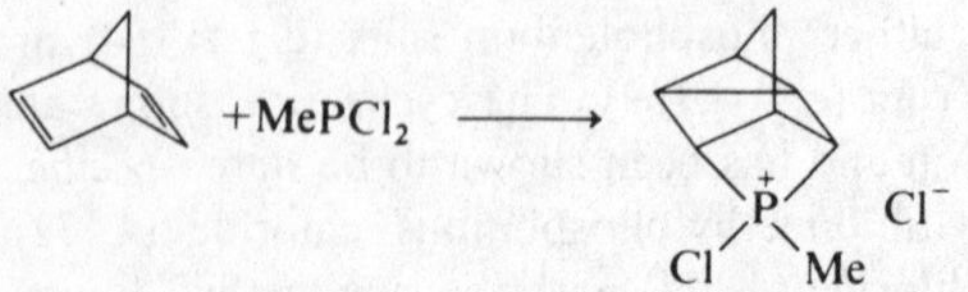

4.4 Dienophilic reactivity

addition. The TS may be envisaged as (4.173) where the filled π orbitals of the diene are interacting with the vacant d orbitals on phosphorus.

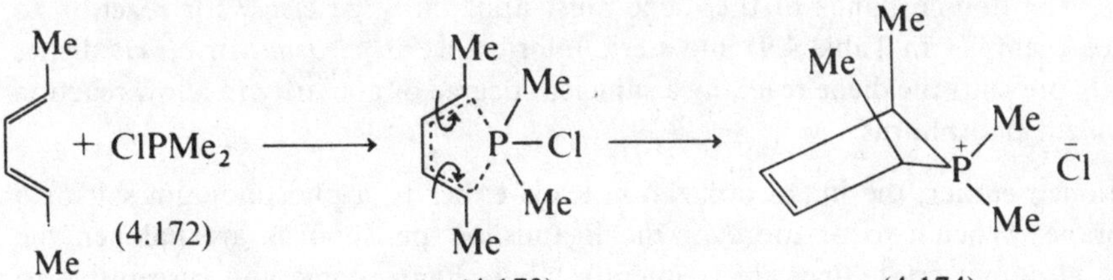

(4.172) (4.173) (4.174)

There are two alternative approaches to the interpretation of concerted reactions, viz, (i) the Dewar TS concept and (ii) the Frontier Orbital method and both of these are more readily applicable to the reactions of phosphorus compounds than the Woodward–Hoffmann rules which necessitate the construction of correlation diagrams and hence require a somewhat detailed knowledge of the orbitals involved in any chemical transformation. All three methods are discussed in Chapter 6 but it should be noted at this stage that they all predict the concerted disrotatory pathway to be "allowed" in a thermal (i.e. non-photolytic) cycloaddition of trivalent phosphorus compounds to dienes.

The stereospecificity was confirmed experimentally by the observation that certain phosphoranes, derived from *cis,cis*-1,2,5-trimethylphosphol-3-ene (4.175) and diethyl peroxide fragmented spontaneously to give *trans,trans*-2,4-hexadiene and diethylmethylphosphonite (4.176)[87].

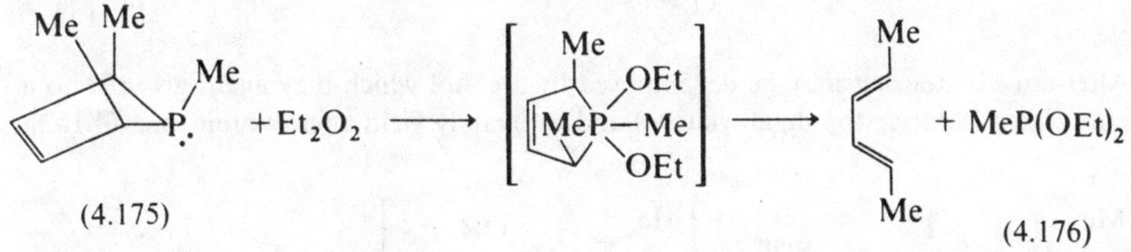

(4.175) (4.176)

This fragmentation is directly analogous to the vapour phase pyrolysis of the *cis* and *trans* isomers of 2,5-dimethyl-2,5-dihydrothiophene-1,1-dioxide (4.177 and 4.178) which at 200 °C gave *trans,trans* and *cis,trans*-2,4-hexadiene respectively with greater than 99.9% stereospecificity[88].

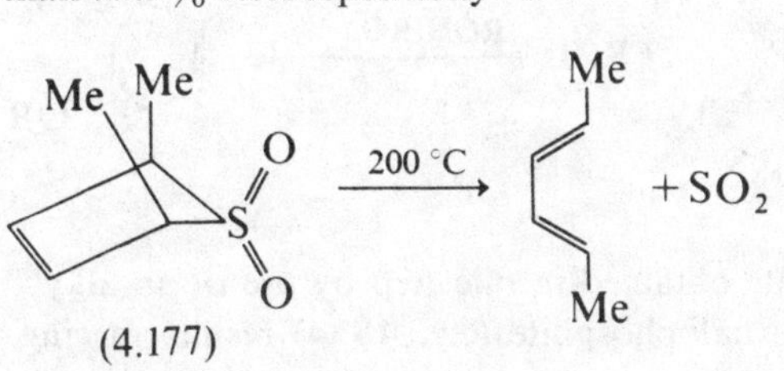

(4.177)

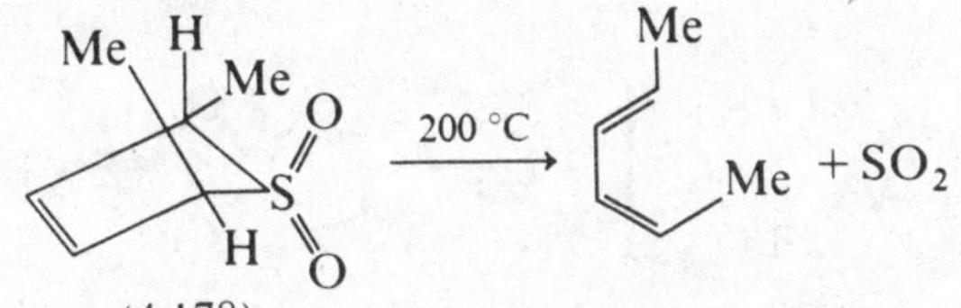

(4.178)

Unfortunately, the isomer of (4.175) with the ring methyl groups *trans* cannot be prepared by the condensation route and this reveals a limitation of the McCormack synthesis. The double bonds of the diene must apparently be *cisoid* for reaction to occur (see examples in Table 4.9) and steric interference in *cis,trans* or *cis,cis* dienes apparently prevents the diene reaching a sufficient degree of planarity to allow reaction with trivalent phosphorus.

As mentioned earlier, the initial cyclization leads either to a phosphonium salt or a phosphorane. When two or more of the ligands on phosphorus are halogen the products, although crystalline, are frequently labile, hygroscopic and susceptible to hydrolysis or rearrangement. In many cases therefore, the product is deliberately hydrolyzed to an acid or phosphine oxide in order to facilitate isolation. For example, the salt (4.179) obtained from phosphorus trichloride and isoprene may be selectively hydrolyzed to a phosphinyl chloride (4.180) or a phosphinic acid (4.181).

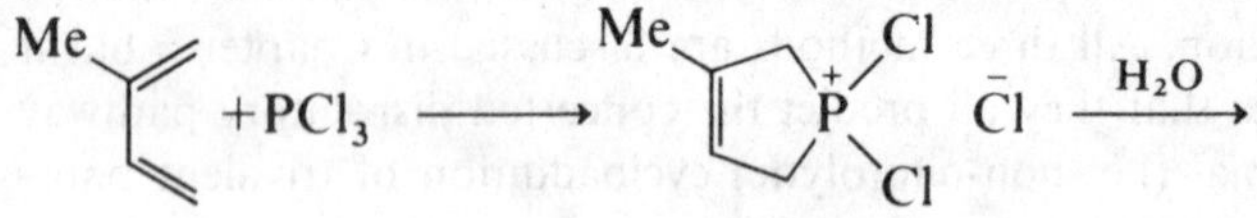

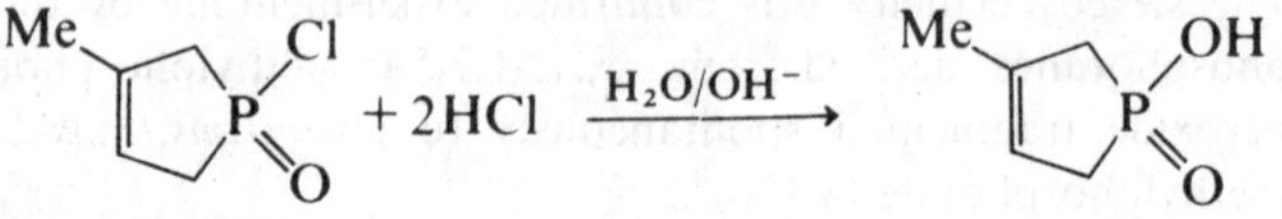

Alternatively, the salt may be decomposed by alcohol which may again give rise to a phosphoryl chloride (by de-alkylation) and ultimately yield a phosphinic ester (4.182).

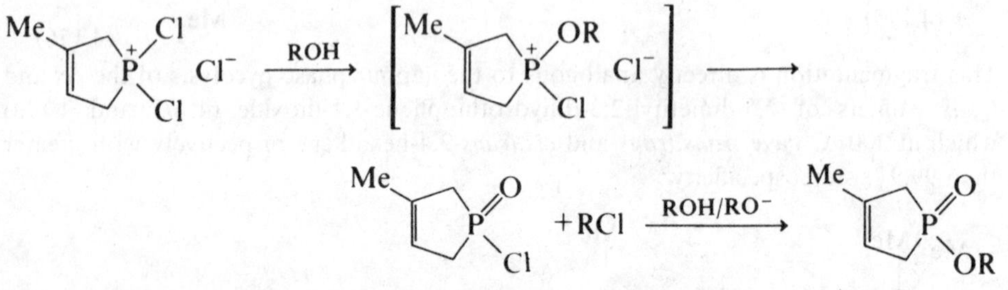

The phosphoryl chloride may of course be obtained in one step by use of an alkylphosphorous dichloride (4.183); a cyclic halophosphite (e.g. 4.184) results in ring opening to (4.185).

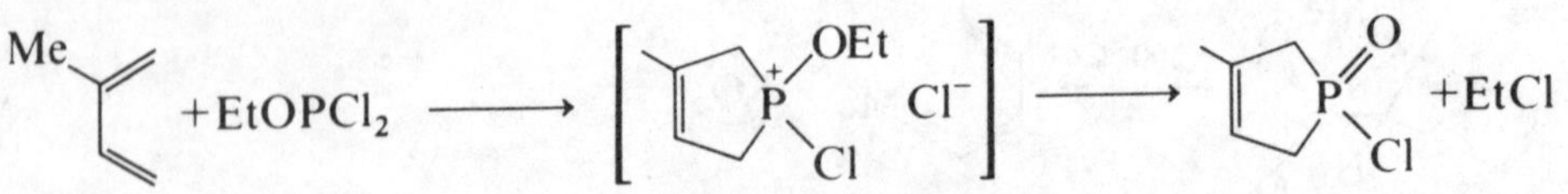

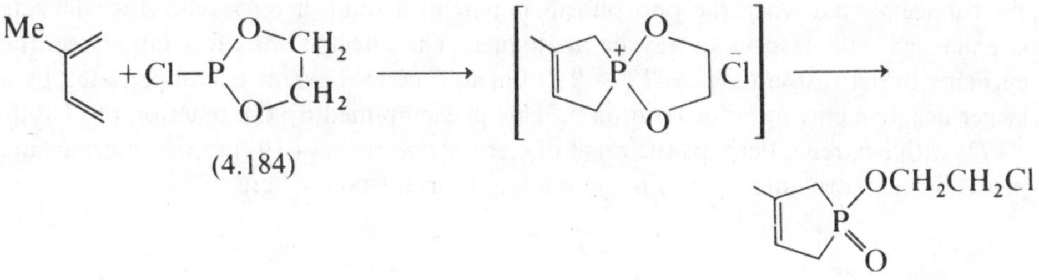

4.4 Dienophilic reactivity

(4.184)

(4.185)

In certain conditions rearrangement of the double bond from the 3- to the 2-position of the phospholene ring is observed. Thus phenylphosphonous dichloride gives mainly the 2-phospholenium salt (4.187) whereas phenylphosphonous dibromide gives the 3-isomer (4.188)[89]. It seems reasonable to assume that the higher basicity of the chloride ion is sufficient to remove a proton from the 2-position of (4.186) to give the resonance stabilized anion (4.189) which is equivalent to a phosphorus ylid and may reprotonate to (4.187).

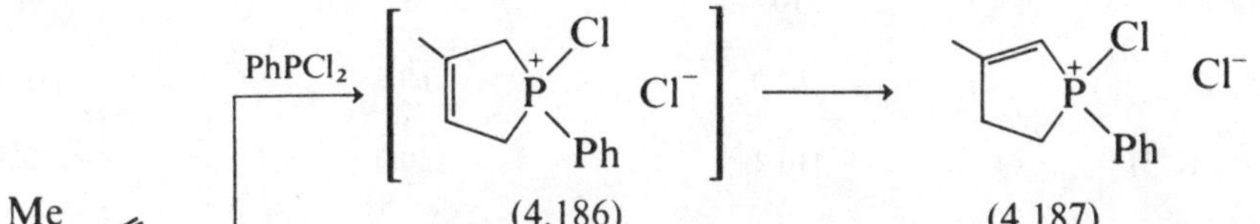

(4.186)   (4.187)

(4.188)

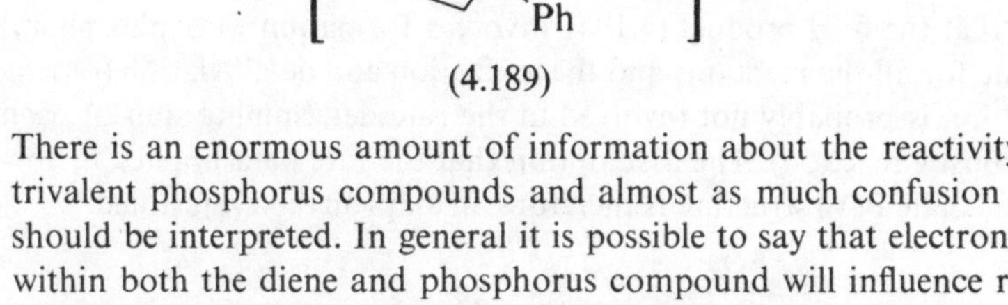

(4.186) ⟶ (4.189) ⟶ (4.187)

There is an enormous amount of information about the reactivity of dienes towards trivalent phosphorus compounds and almost as much confusion as to how the data should be interpreted. In general it is possible to say that electronic and steric factors within both the diene and phosphorus compound will influence reactivity. One must recognize however that such factors may affect both the ground state of the reactants and the transition state of the combination process. Furthermore, some reactions produce ionic, tetracovalent phosphonium salts whereas others give pentacovalent phosphoranes and the product obtained may be determined by the solvent system used. It is not surprising therefore, that a comprehensive rationalization of the facts has proved difficult. Certain broad conclusions, based mainly on relative reactivity data, may be drawn however, and these are summarized as follows.

First it seems that when the phosphorus is part of a small heterocyclic ring, the rate is enhanced with respect to acyclic analogues. The effect is manifest largely on the enthalpy of activation ($\Delta H^{\ddagger} = E_a - RT$) but to a certain extent is compensated by a larger negative entropy of activation[90]. This is exemplified by the reaction of (4.190–4.192) with isoprene. Perhaps the relief of steric strain from (4.191) to the intermediate pentacovalent structure (4.193) is the really important factor here.

	CH$_3$OPCl$_2$	(cyclic) P—Cl	(C$_2$H$_5$O)$_2$PCl
	(4.190)	(4.191)	(4.192)
Relative rate	10	8	1
E_a (kJ mol^{-1})	82.0	69.8	92.3
$\Delta S^{\ddagger}$ (J mol^{-1} K^{-1})	−116.2	−150.0	−100.3
$\Delta G^{\ddagger}$ (kJ mol^{-1})	123.4	124.2	127.5

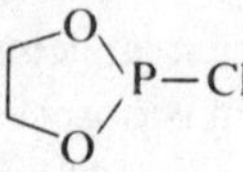

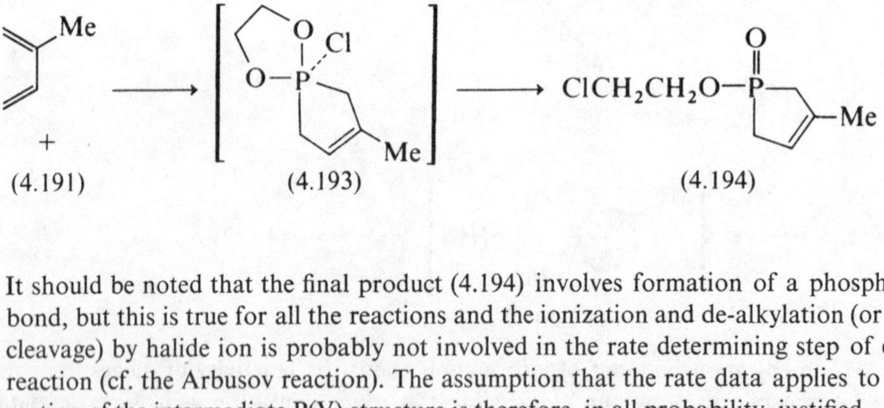

(4.191) + isoprene ⟶ [intermediate (4.193)] ⟶ ClCH$_2$CH$_2$O—P(=O)—Me (4.194)

It should be noted that the final product (4.194) involves formation of a phosphoryl bond, but this is true for all the reactions and the ionization and de-alkylation (or ring cleavage) by halide ion is probably not involved in the rate determining step of each reaction (cf. the Arbusov reaction). The assumption that the rate data applies to formation of the intermediate P(V) structure is therefore, in all probability, justified.

Secondly there is evidence to suggest that the phosphorus compounds behave as electrophiles rather than nucleophiles in the condensation process. Thus for the reactions of (4.195) with various dienes, the relative rates are in the order, isoprene (R^1 = Me, R^2 = H) > butadiene (R^1, R^2 = H) > chloroprene (R^1 = Cl, R^2 = H)[91]. With penta-1,3-diene (R^1 = H, R^2 = Me) however, the rate falls between that of butadiene and chloroprene which is presumably the result of an adverse steric effect imposed by the methyl group on a terminal carbon of the diene.

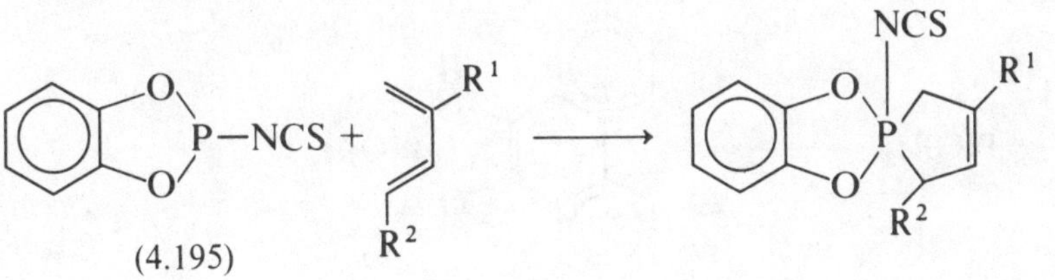

4.4 Dienophilic reactivity

[Structure (4.195) reacting with diene to form cyclic adduct]

(4.195)

Variation of the substituents on phosphorus reveals neither nucleophilic or electrophilic characteristics of the phosphorus atom as the predominant factor in determining

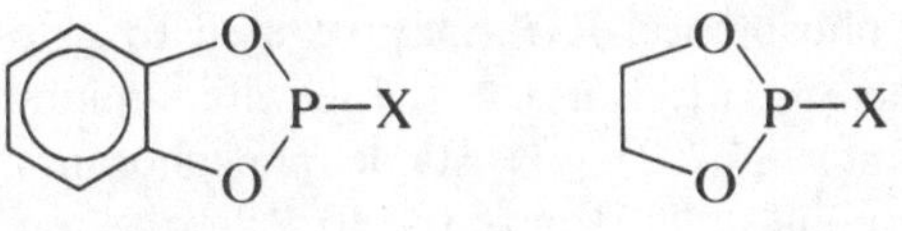

(4.196)   (4.197)

reaction rate[92,93]. Thus for the reactions of (4.196) or (4.197) with dienes, the observed relative rates *decrease* in the order, X = Ph ⩾ Br > CH$_3$, C$_2$H$_5$ > NCS > Cl > SR > F > OR > R$_2$N. There is no obvious correlation of this rate order with any empirical parameter (e.g. apicophilicity) and it seems likely that a combination of electronic and steric effects is at work on both the ground states of the reactants and transition state of the condensation. Further, careful and systematic investigation will be necessary in order to resolve this problem.

4.4.2 α-Diketones, ortho-quinones and monofunctional aldehydes and ketones[9,84,85,94,95]

One of the earliest and most useful methods of preparing pentacovalent phosphorus compounds involves the condensation of trivalent phosphorus compounds with *ortho*-quinones or α-diketones. Thus o-benzoquinone reacts with trimethyl phosphite to give the oxyphosphorane (4.198) which is a derivative of the 2,2-dihydro-1,3,2-dioxaphospholen ring system[96].

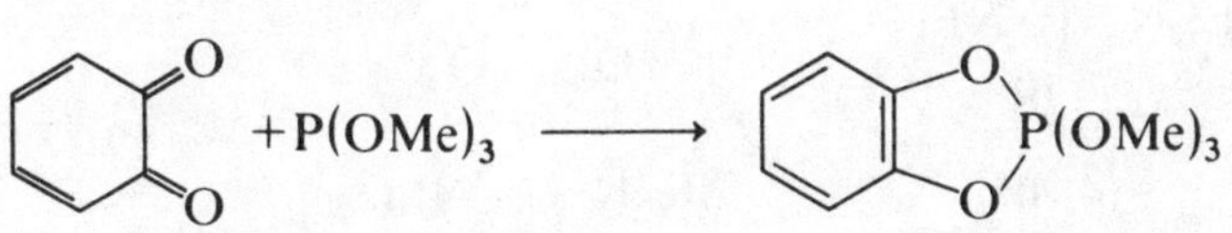

(4.198), δ ^{31}P, +50

A similar reaction occurs with α-diketones such as biacetyl or benzil (4.199)[95] and with quinone monoimines (4.200)[97].

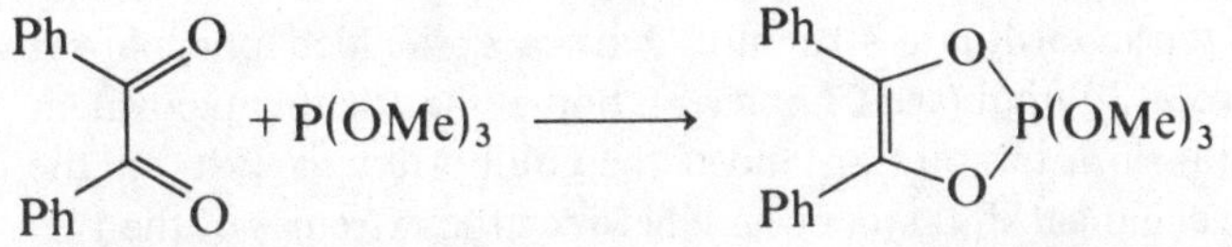

(4.199)   δ ^{31}P, +49.5

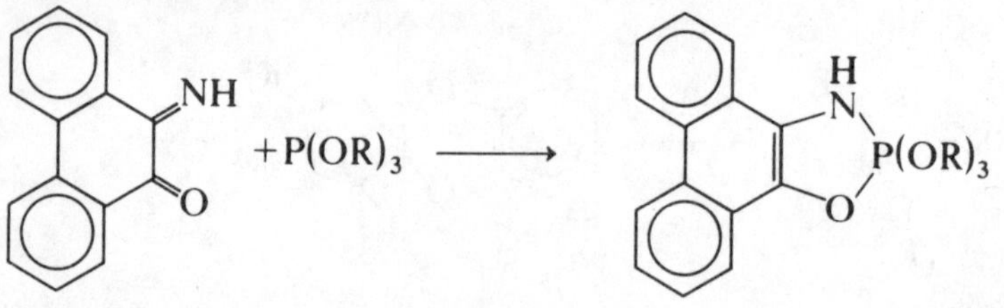

NH
+P(OR)$_3$
O
H
N
P(OR)$_3$
O

(4.200)

These and analogous reactions have been widely exploited by numerous workers but especially Ramirez and his colleagues, for the synthesis of phosphoranes. Phosphonites, $(RO)_2PR$, phosphinites, $ROPR_2$, and even phosphines, R_3P, may be used to give cyclic members of the series $(RO)_{5-n}PR_n$ where $n = 1$, 2 or 3[94,95]. Likewise, amino-phosphonites (4.201) and diamino-phosphinites (4.202) give stable phosphoranes (4.204 and 4.205 respectively) but aminophosphines (4.203) give rise to a dipolar ion (4.206) with phosphorus in a tetracoordinate state.

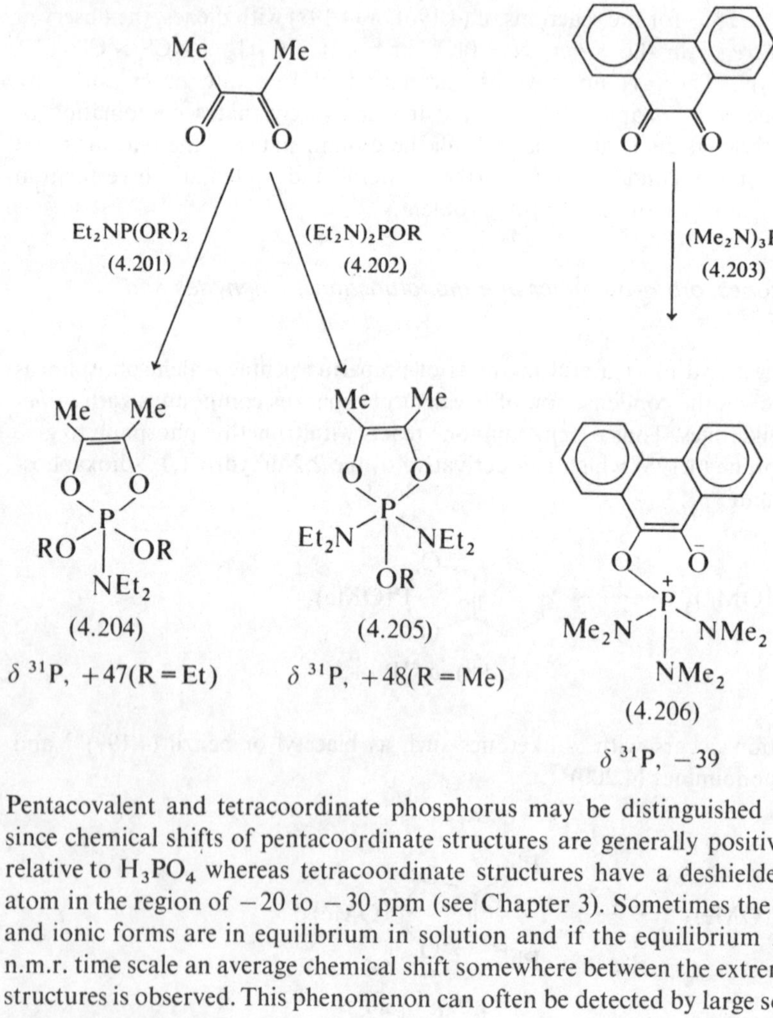

Me Me
O O
$Et_2NP(OR)_2$
(4.201)
$(Et_2N)_2POR$
(4.202)
$(Me_2N)_3P$
(4.203)
RO OR NEt$_2$
(4.204)
δ ^{31}P, +47(R = Et)
Et$_2$N NEt$_2$ OR
(4.205)
δ ^{31}P, +48(R = Me)
Me$_2$N NMe$_2$ NMe$_2$
(4.206)
δ ^{31}P, −39

Pentacovalent and tetracoordinate phosphorus may be distinguished by ^{31}P n.m.r., since chemical shifts of pentacoordinate structures are generally positive (i.e. upfield) relative to H_3PO_4 whereas tetracoordinate structures have a deshielded phosphorus atom in the region of −20 to −30 ppm (see Chapter 3). Sometimes the pentacovalent and ionic forms are in equilibrium in solution and if the equilibrium is fast on the n.m.r. time scale an average chemical shift somewhere between the extremes of the two structures is observed. This phenomenon can often be detected by large solvent shifts in

4.4 Dienophilic reactivity

the ^{31}P n.m.r. spectrum. The compound derived from tris-dimethylaminophosphine and benzil (4.207a, b) exhibits such behaviour, the non-polar solvent, hexane, stabilizing the pentacovalent structure (4.207b), and the more polar solvent shifting the equilibrium towards (4.207a)[95].

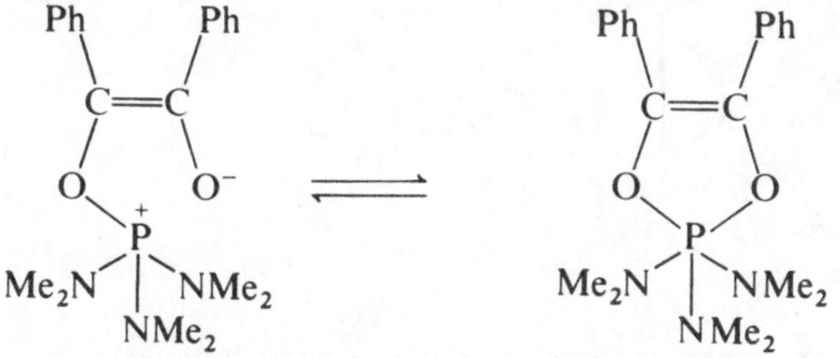

(4.207a) (4.207b)

$\delta\ ^{31}$P, +30.2 ppm in hexane (1.0 M)

$\delta\ ^{31}$P, +13.0 ppm in dichloromethane (1.0 M)

The ionic nature of (4.206) is probably due to a combination of several factors including (a) the lower electronegativity of N relative to O, (b) the availability of a lone pair on each nitrogen which helps to stabilize the positive charge on phosphorus and (c) the larger steric requirements of dialkylamino versus alkoxy groups in the crowded trigonal bipyramid structure of the pentacovalent form. That the trigonal bipyramid structure of a phosphorane may indeed be crowded was established by an x-ray analysis of the oxyphosphorane derived from tri-isopropyl phosphite and phenanthraquinone (see Chapters 2 and 6).

Incorporation of five-(or four)membered rings in the phosphorane structure apparently relieves such steric strain and this explains, at least in part, the stability of the pentacovalent forms of (4.208) and (4.209) compared to the ionic nature of (4.207) and (4.210)[95].

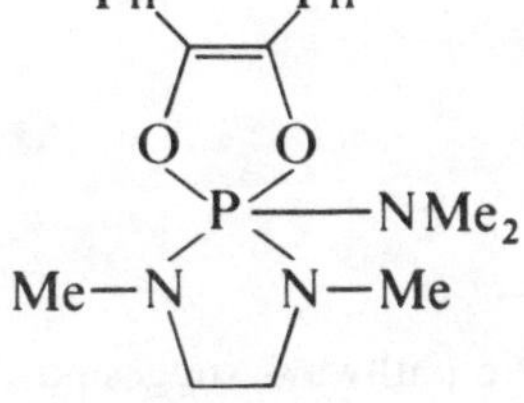

(4.208) 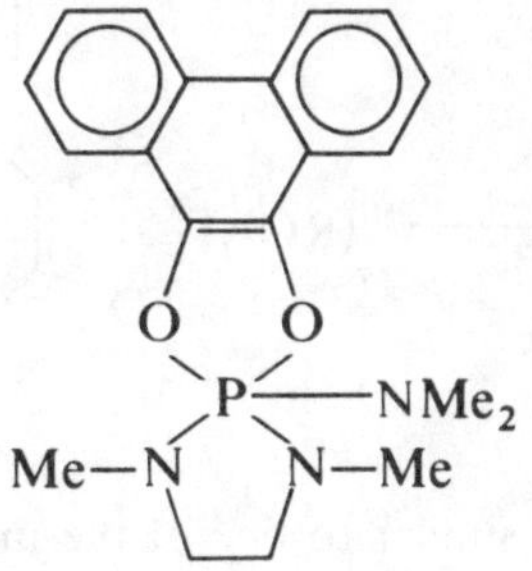   (4.209) 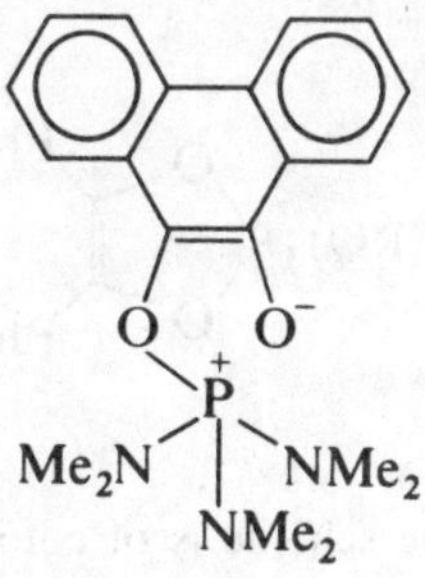   (4.210)

$\delta\ ^{31}$P, +37   +30   −39

In all the above reactions where a pentacovalent phosphorane is formed, the trivalent phosphorus compound behaves as a biphile in the sense that phosphorus both donates and receives electrons. It is therefore tempting to suggest a concerted addition to the dicarbonyl structure via a TS such as (4.211) where phosphorus is behaving like a dienophile. This may be the case for some reactions but a careful kinetic study of the addition of trialkyl phosphites to benzil by Ogata and Yamashita led to a different interpretation[98].

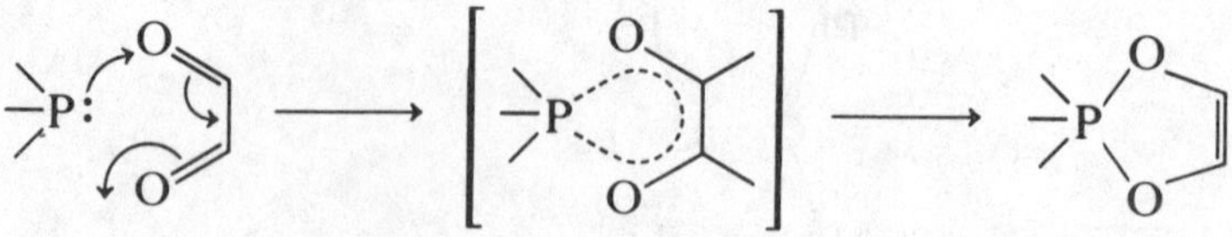

(4.211)

These workers reported that the second-order reaction with trimethyl phosphite in dioxan had activation parameters of $\Delta H^{\ddagger} = 35.1$ kJ mol^{-1} (8.4 kcal mole^{-1}) and $\Delta S^{\ddagger} = -198$ J mol^{-1} K^{-1} (-47.5 cal mole^{-1} K^{-1}). In benzene the rate constant increased linearly with low concentrations of added organic acid and showed a linear decrease with low concentrations of added base. These data are reminiscent of that associated with nucleophilic additions to carbonyl groups and led to the proposal of the following mechanistic scheme where the first step is both reversible and rate determining.

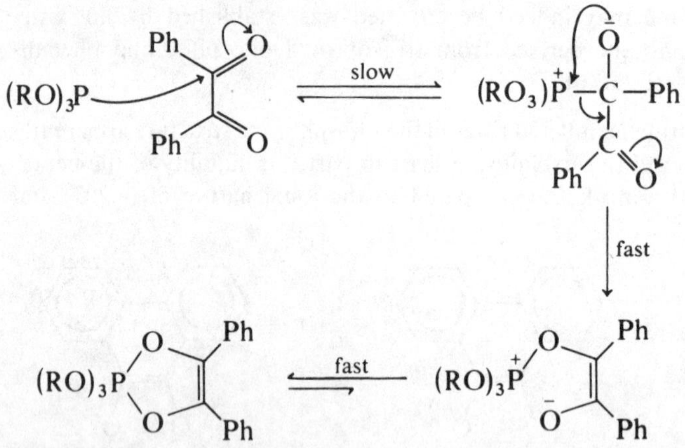

The scheme is of course, very similar to one of the mechanistic pathways suggested for the Perkow reaction.

The reactions of trivalent phosphorus compounds with α-diketones assumed much greater synthetic importance when Ramirez discovered that a second molecule of the diketone would react to give a new carbon–carbon bond. Thus biacetyl and trimethyl phosphite give the 2:1 product (4.213) as a mixture of *cis* and *trans* isomers presumably via the ionic form (4.212b) of the 1:1 adduct[99],[100] With some α-diketones and many α-keto esters the second step is faster than the first and in such cases one isolates only the 2:1 product equivalent to (4.213).

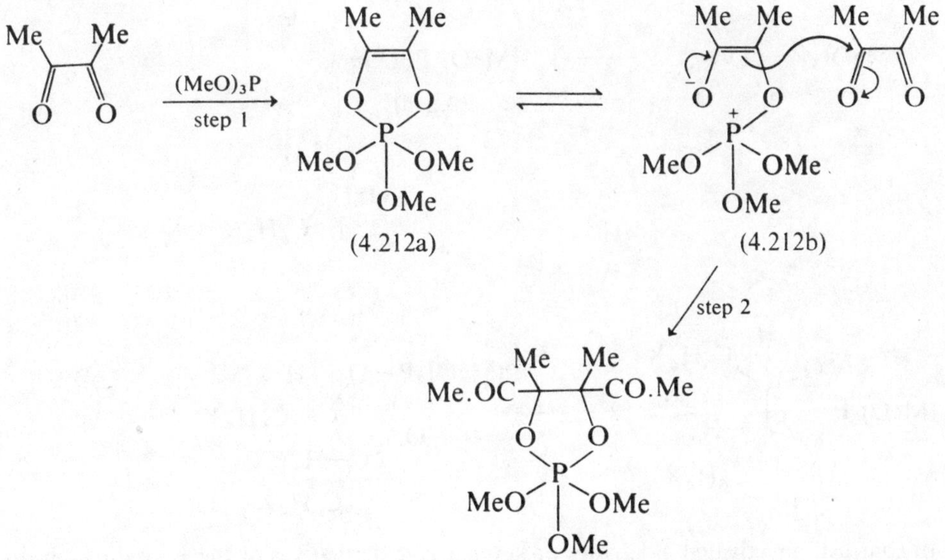

4.4 Dienophilic reactivity

(4.212a) (4.212b)

(4.213)

Ramirez et al. went on to demonstrate that the reaction of trialkyl phosphites with suitably activated monocarbonyl compounds also leads to 2:1 (aldehyde:phosphite) products which are derivatives of 1,3,2-dioxaphospholane[101,102]. Thus trimethyl phosphite and o- or p-nitrobenzaldehydes, phthalaldehydes or hexafluoroacetone yield 2:1 adducts (4.214a, b, c) and (4.215).

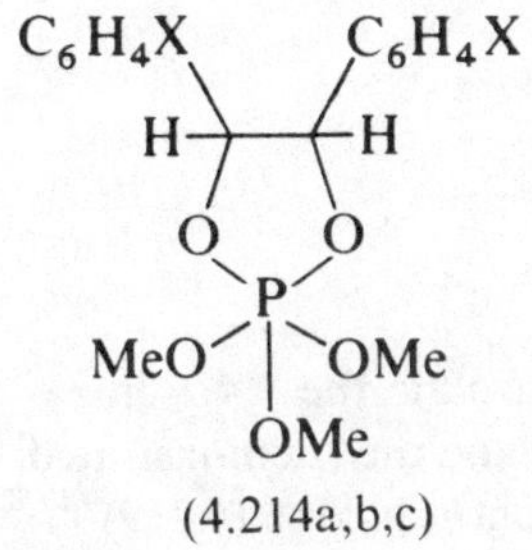

(4.214a,b,c) 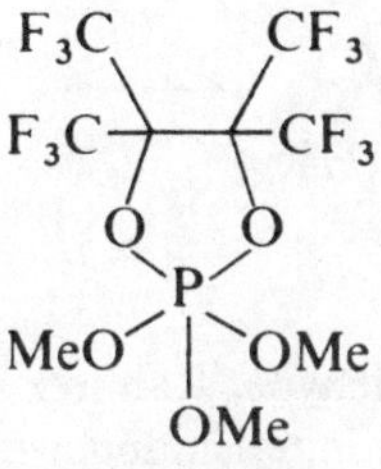   (4.215)

a, X = o-NO$_2$
b, X = p-NO$_2$ $\delta\,^{31}$P, +49 to +54 $\delta\,^{31}$P, +50.1
c, X = p-CHO

It seems likely that the initial attack by phosphite occurs on carbonyl oxygen and the negative charge, so generated, is stabilized by the electron-withdrawing substituents on carbon. The second step then involves nucleophilic addition of the carbanion (4.216) to carbonyl carbon and ring closure to the product oxyphosphorane (4.214).

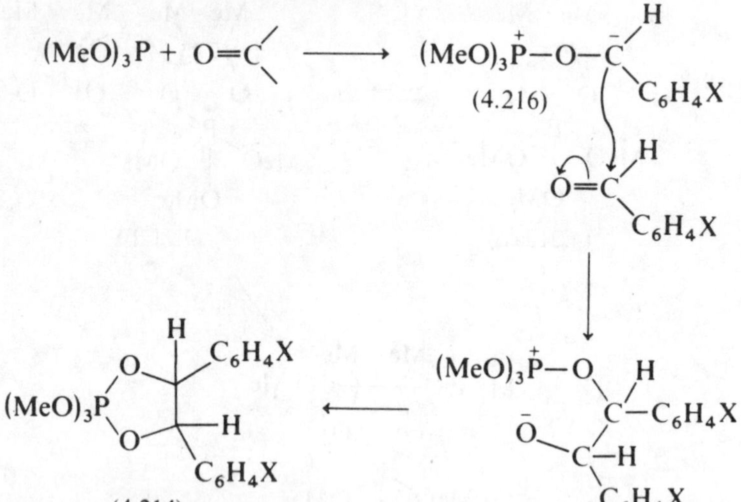

(4.216) (4.214)

In contrast, unactivated aldehydes or ketones give derivatives of the 1,4,2-dioxaphospholane system (4.217) presumably by initial attack of the phosphite on carbonyl carbon[94,95,96].

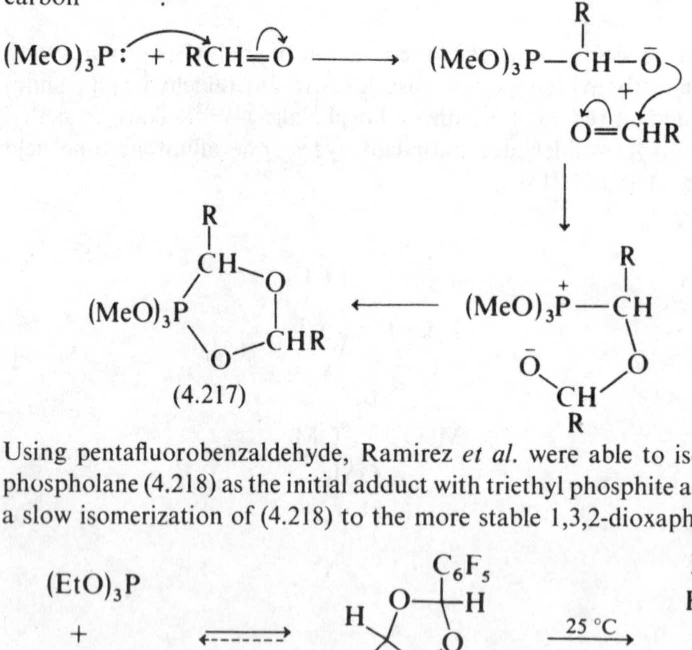

(4.217)

Using pentafluorobenzaldehyde, Ramirez *et al.* were able to isolate the 1,4,2-dioxaphospholane (4.218) as the initial adduct with triethyl phosphite and then demonstrated a slow isomerization of (4.218) to the more stable 1,3,2-dioxaphospholane (4.219)[94].

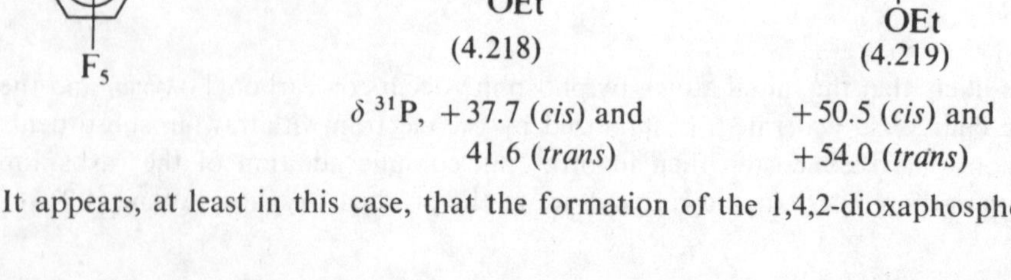

(4.218) (4.219)

δ ^{31}P, +37.7 (*cis*) and 41.6 (*trans*) +50.5 (*cis*) and +54.0 (*trans*)

It appears, at least in this case, that the formation of the 1,4,2-dioxaphospholane is

4.4 Dienophilic reactivity

reversible and the identity of the isolated product depends upon a balance of kinetic (favouring 4.218) and thermodynamic (favouring 4.219) factors.

The scope of the condensation with activated aldehydes and ketones has been extended to include phosphines. For example, triethylphosphine reacts with hexafluoroacetone to give the oxyphosphorane (4.220). On heating this undergoes a curious rearrangement to a 1,2-oxaphosphetan (4.221) which is analogous to the intermediate in a Wittig reaction (see Chapter 7) and on further heating decomposes, as expected, to a phosphinate (4.222) and an olefin (4.223)[103,104].

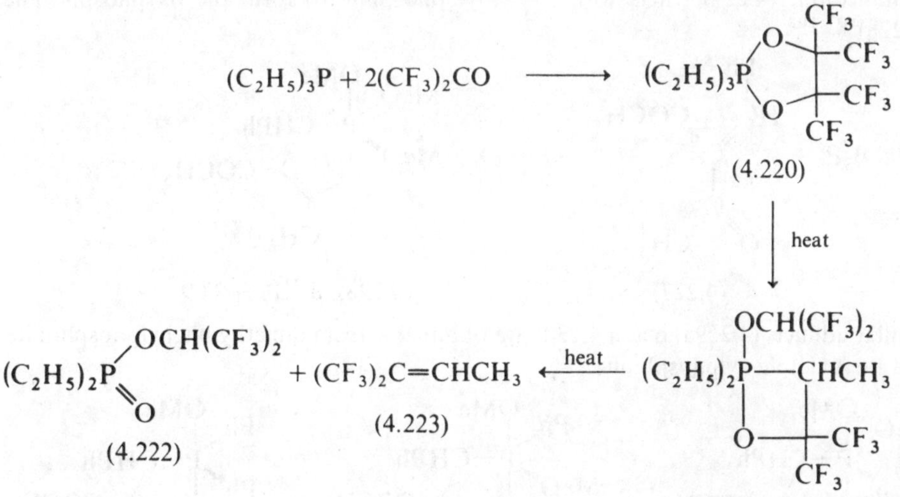

$$(C_2H_5)_3P + 2(CF_3)_2CO \longrightarrow (4.220)$$

$$(4.220) \xrightarrow{heat} (4.221)$$

$$(4.221) \xrightarrow{heat} (C_2H_5)_2P(=O)OCH(CF_3)_2 \text{ (4.222)} + (CF_3)_2C=CHCH_3 \text{ (4.223)}$$

A variety of 1-substituted phosphetans (4.224) also condense with hexafluoroacetone to give oxyphosphoranes (4.225) containing a four-membered (phosphetan) ring. By using ^{19}F n.m.r. to study the energy barrier to placing the four-membered ring in a diequatorial situation (as in 4.226) several groups of workers, but notably Trippet *et al*, have been able to provide some semi-quantitative data on the relative apicophilicity of the groups, R (see Chapters 2 and 6)[105,106].

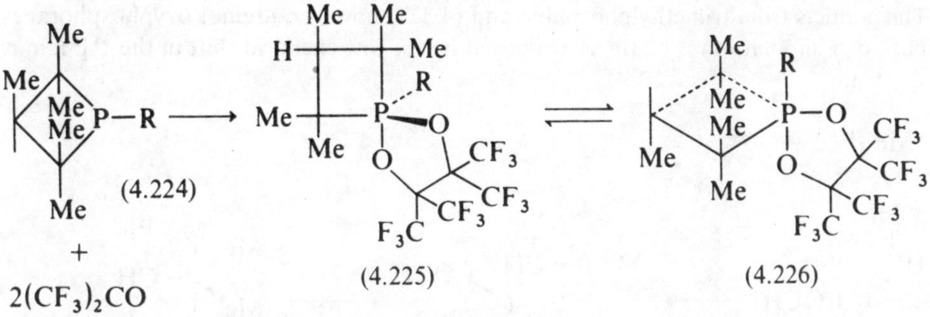

4.4.3 α,β-Unsaturated carbonyl compounds

The addition of trivalent phosphorus compounds to α,β-unsaturated carbonyl compounds sometimes involves a conventional Michael addition which may eventually produce an ylid as in the reaction between triphenylphosphine and 1,2-dibenzoylethylene (eqn. 23).

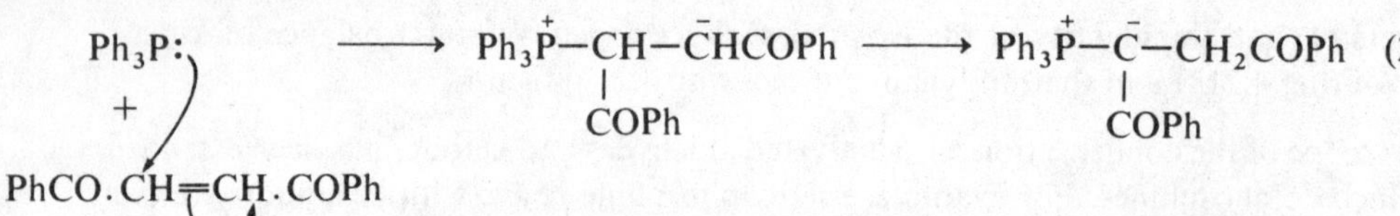

$$Ph_3P: + PhCO.CH=CH.COPh \longrightarrow Ph_3\overset{+}{P}-CH-\overset{-}{C}HCOPh \longrightarrow Ph_3\overset{+}{P}-\overset{-}{C}-CH_2COPh \quad (23)$$
$$\underset{}{} \quad\quad\quad\quad\quad\quad\quad\quad \overset{|}{COPh} \quad\quad\quad\quad \overset{|}{COPh}$$

This nucleophilic addition has already received some attention (see Section 4.1) but an alternative reaction to generate an oxyphosphorane is also possible, particularly with phosphites, phosphonites and phosphinites. For example, 3-benzylidene-2,4-pentanedione (4.227) reacts with trimethyl phosphite to form the oxyphosphorane (4.228)[107,108].

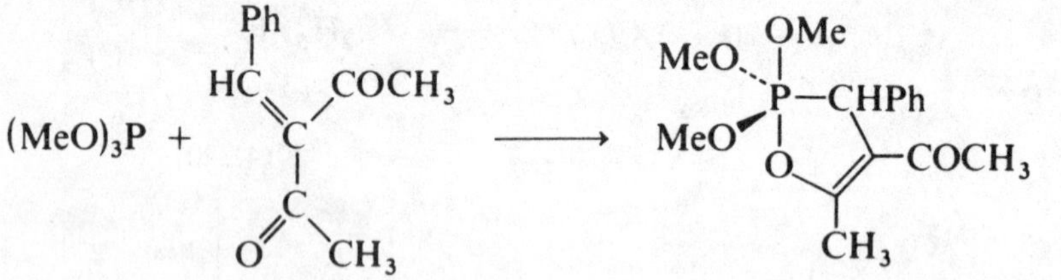

(4.227)   (4.228), $\delta\ ^{31}P$, +27.9

Similar adducts (4.229a, b and 4.230) are obtainable from dimethylphenylphosphonite and methyldiphenylphosphinite.

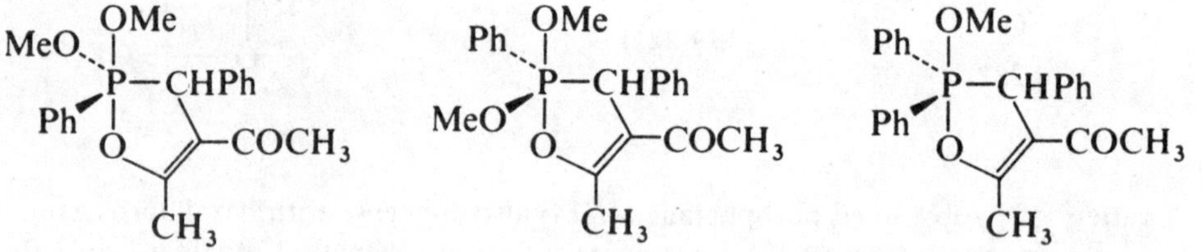

(4.229a)   (4.229b)   (4.230)

$\delta\ ^{31}P$, +16.7   +13.3   +25.7

diastereomers

The adducts from trimethylphosphine and (4.227) however, are not oxyphosphoranes but open, dipolar ions (4.231) as evidenced by the low chemical shift in the ^{31}P n.m.r. spectrum.

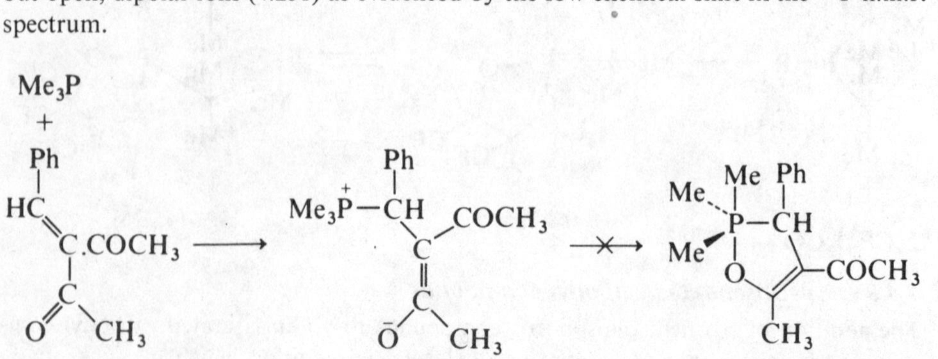

(4.231)   (4.232)

$\delta\ ^{31}P$, −10.9

In order to form a pentacovalent species (4.232) it is necessary to place a methyl group in an apical position and according to the Westheimer rules (see Chapters 2 and 6) this is energetically unfavourable.

4.5 Conclusion

The whole spectrum of reactivity for trivalent phosphorus compounds has been examined and it should be clear that the differences from nitrogen chemistry and the difficulties in interpreting the reactions of phosphorus compounds arise to a large extent from the ability of phosphorus to expand its valence shell to ten electrons. Whereas ammonia and its trivalent derivatives are essentially nucleophilic, trivalent phosphorus may display nucleophilic *and* electrophilic character. It is therefore not surprising that in certain reactions at phosphorus, particularly those leading directly from P(III) to pentacoordinate P, an intermediate situation of biphilic or dienophilic character prevails. It is obvious that in such cases, electronic factors will not have a dramatic effect on relative reactivity since increase in nucleophilicity will be countered by a corresponding decrease in electrophilicity. Thus in a biphilic (or dienophilic) reaction, rate differences over a wide range of trivalent phosphorus compounds may be small and may well be dictated by steric rather than electronic considerations. Since steric factors are much less predictable than electronic factors the interpretation of rate data becomes difficult. Nevertheless, a fairly consistent and reasonably coherent picture is emerging and it seems likely that the next few years of research in this area will enable the field to be brought into much sharper focus.

Problems

1   Inversion barriers for phosphines are normally in the region of 125 kJ mol^{-1} (30 kcal mol^{-1}). Explain why the barrier to inversion at the cyclic phosphine (1) is only 67 kJ mol^{-1} (16 kcal mol^{-1}).

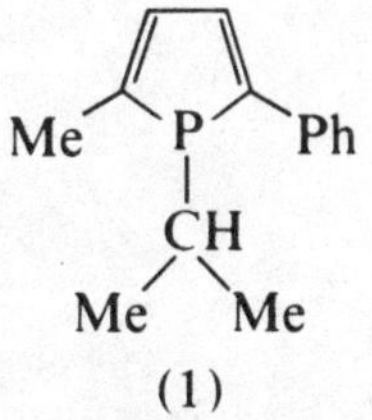

(1)

2   The driving force for many reactions in organophosphorus chemistry is the formation of the phosphoryl bond. Review the contents of Chapter 4 and give the equations of six reactions which illustrate this statement.

3   Suggest products for the following reactions:

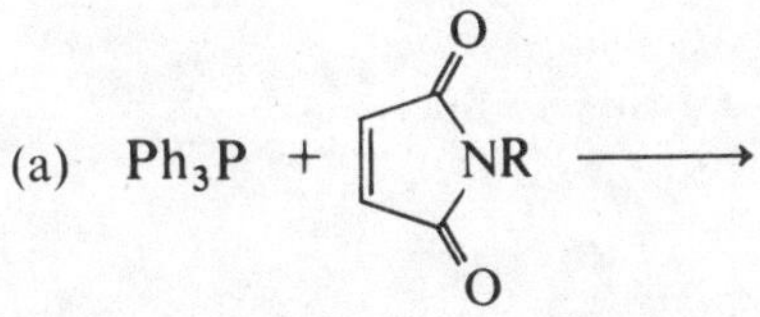

(a)   Ph$_3$P + maleimide (NR) ⟶

(b) 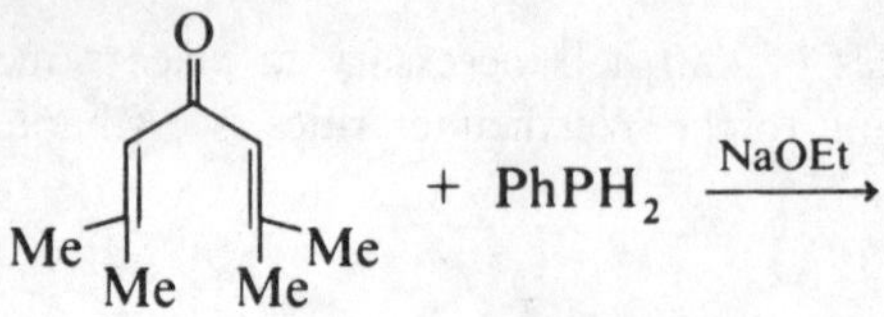 ![structure: (Me)(Me)C=CH-C(=O)-CH=C(Me)(Me)] + PhPH$_2$ $\xrightarrow{\text{NaOEt}}$

(c) ![tetrazolo-fused pyridine structure] + Ph$_3$P $\longrightarrow$

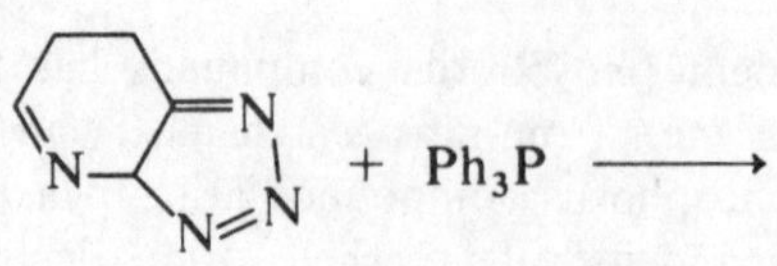

(d) 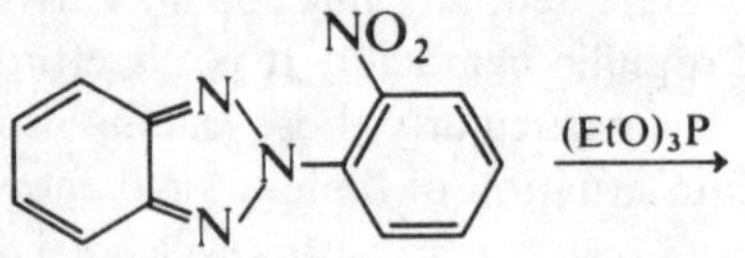 ![2-(2-nitrophenyl)benzotriazole] $\xrightarrow{\text{(EtO)}_3\text{P}}$

(e) (MeO)$_3$P + CH$_3$C(=O)-C(=CHPh)-C(=O)-CH$_3$ $\longrightarrow$

(f) (EtO)$_2$P-OCH(Me)-CH=CH$_2$ $\xrightarrow{190\,°C}$

(g) (EtO)$_2$P̄=O + RCOCH$_2$Cl $\longrightarrow$

(h) (RO)$_3$P + R$_2^1$NCl $\longrightarrow$

(i) (EtO)$_3$P + PhCO·CH=CH CCl$_3$ $\longrightarrow$

(j) 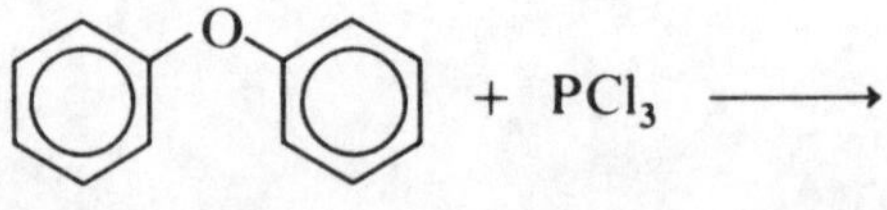 ![diphenyl ether] + PCl$_3$ $\longrightarrow$

4 Suggest mechanisms for the following reactions:

(a) 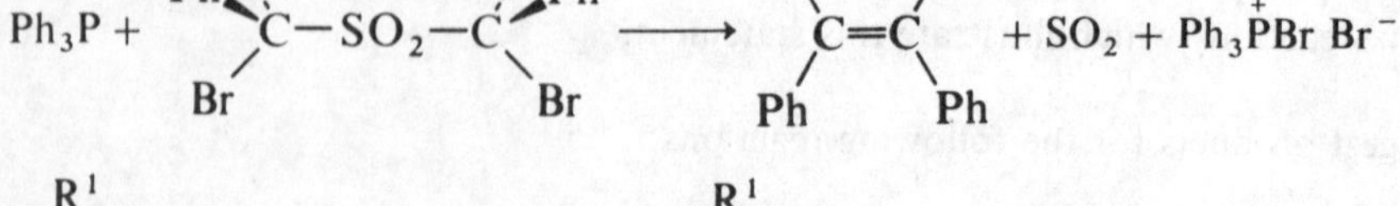 Ph$_3$P + Ph(H)(Br)C-SO$_2$-C(Br)(H)Ph $\longrightarrow$ PhHC=CHPh + SO$_2$ + Ph$_3$P̊Br Br$^-$

(b) 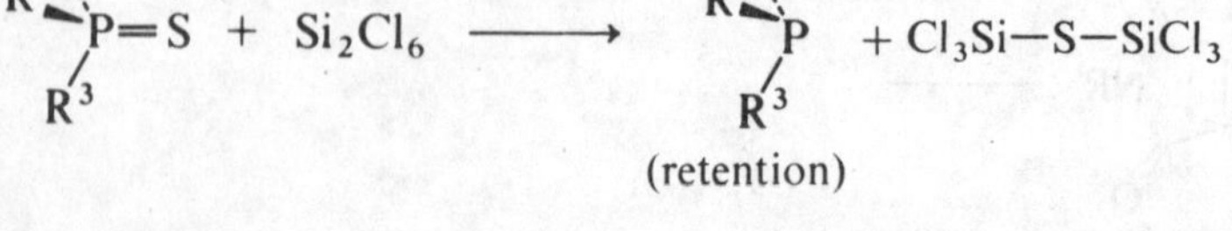 R^2(R^1)(R^3)P=S + Si$_2$Cl$_6$ $\longrightarrow$ R^2(R^1)(R^3)P + Cl$_3$Si-S-SiCl$_3$

(retention)

Problems

(c) 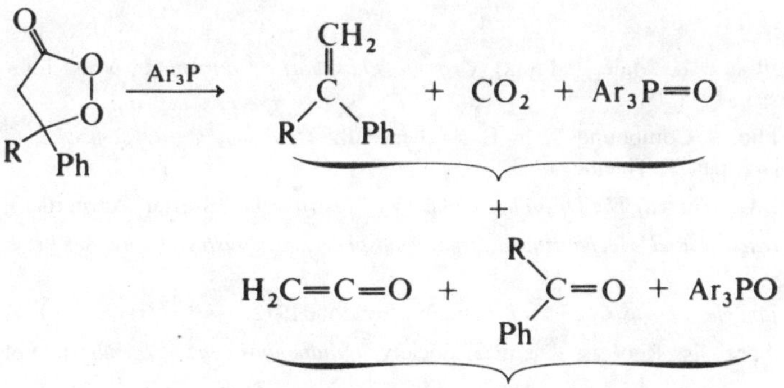
$$\underset{\underset{Ph}{R}}{\overset{O}{\underset{O}{\bigcirc}}}\xrightarrow{Ar_3P} \underset{R\ \ Ph}{\overset{CH_2}{C}} + CO_2 + Ar_3P=O$$

$$+$$

$$H_2C=C=O + \underset{Ph}{\overset{R}{C}}=O + Ar_3PO$$

(d) 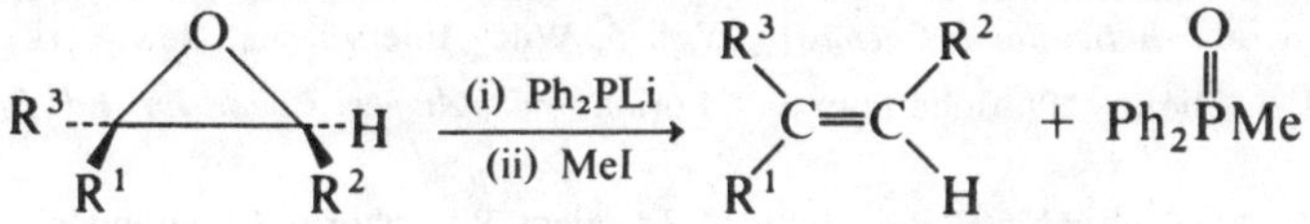 $R^3 \overset{O}{\underset{R^1\ \ R^2}{\triangle}} H \xrightarrow[\text{(ii) MeI}]{\text{(i) Ph}_2\text{PLi}} \underset{R^1}{\overset{R^3}{C}}=\underset{H}{\overset{R^2}{C}} + Ph_2\overset{O}{P}Me$

(e) $(RO)_3P + CH_2=CHCHO \longrightarrow (RO)_2\overset{O}{P}CH_2-CH=CHOR$

(f) 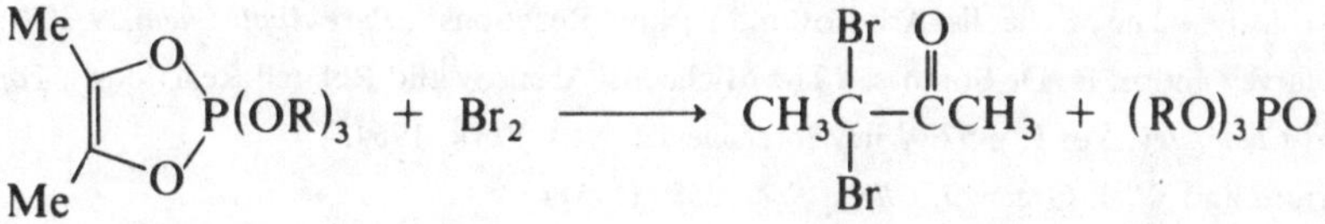 $\underset{Me}{\overset{Me}{\underset{O}{\bigcirc}}}P(OR)_3 + Br_2 \longrightarrow CH_3\underset{Br}{\overset{Br}{C}}-\overset{O}{C}CH_3 + (RO)_3PO$

(g) 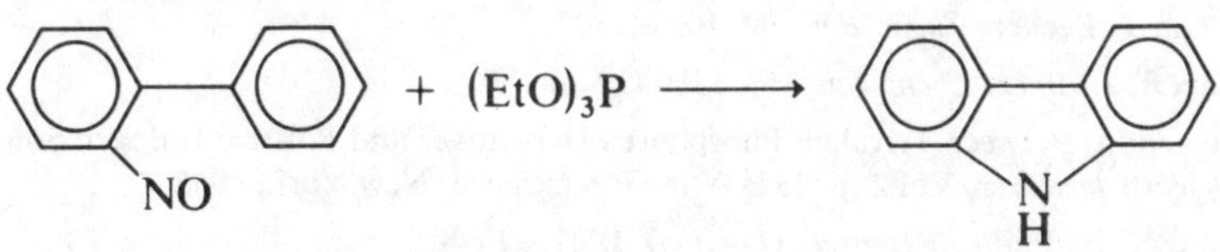 [2-nitrosobiphenyl] + (EtO)$_3$P $\longrightarrow$ [carbazole]

(h) 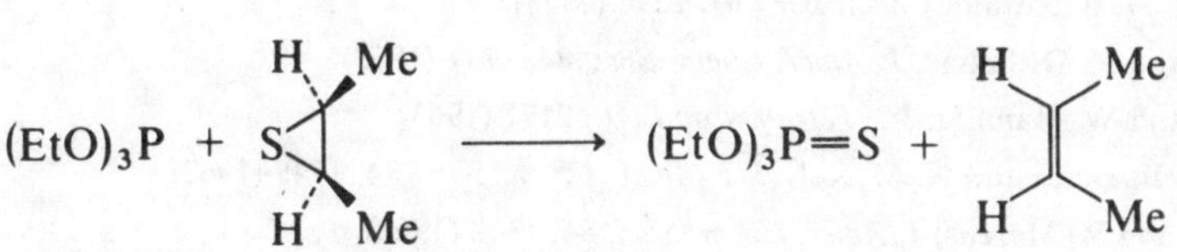 $(EtO)_3P + \underset{H\ \ Me}{\overset{H\ \ Me}{\underset{\triangle}{S}}} \longrightarrow (EtO)_3P=S + \underset{H\ \ Me}{\overset{H\ \ Me}{C=C}}$

(i) $PhCOCH_2Br + PPh_3 \xrightarrow{MeOH} PhCOCH_3 + Ph_3PO + MeBr$

(j) 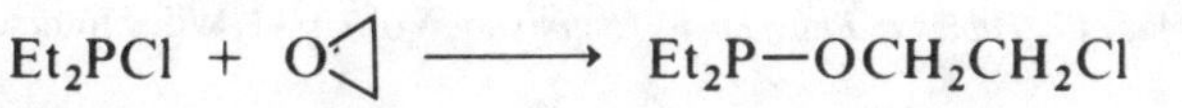 $Et_2PCl + O\triangleleft \longrightarrow Et_2P-OCH_2CH_2Cl$

(k) 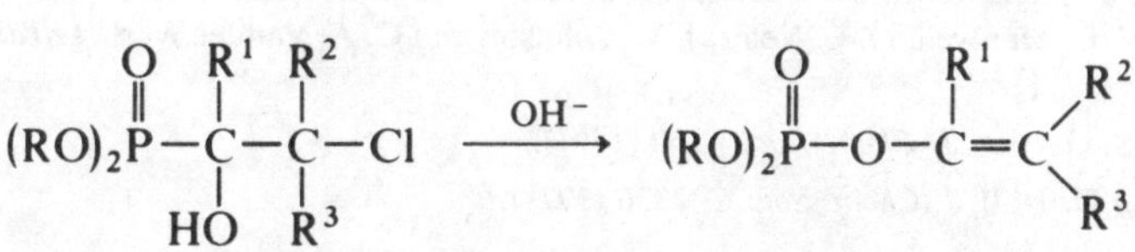 $(RO)_2\overset{O}{\underset{}{P}}-\underset{HO}{\overset{R^1}{C}}-\underset{R^3}{\overset{R^2}{C}}-Cl \xrightarrow{OH^-} (RO)_2\overset{O}{P}-O-\underset{}{\overset{R^1}{C}}=\underset{R^3}{\overset{R^2}{C}}$

References

1. G. M. Kosolapoff and L. Maier (editors), *Organic Phosphorus Compounds*, Vols. 1–6, Wiley Interscience, New York, 1972/73.
2. K. Sasse, "Phosphorus Compounds", in E. Muller (ed), *Methoden der Organischen Chemie*, (Houben Weyl), Vol 12, part 1, Thieme, 1963.
3. A. J. Kirby and S. G. Warren, *The Organic Chemistry of Phosphorus*, Elsevier, Amsterdam, 1967.
4. R. F. Hudson, *Structure and Mechanism in Organophosphorus Chemistry*, Academic Press, New York, 1965.
5. B. J. Walker, *Organophosphorus Chemistry*, Penguin, London, 1972.
6. S. Trippett (ed), Specialist Reports, Chemical Society, *Organophosphorus Chemistry*, Vols 1–5, 1969–74.
7. L. Maier, "Preparation and Properties of Primary, Secondary, and Tertiary Phosphines", in F. A. Cotton (ed) *Progress in Inorganic Chemistry*, Vol. 5, Wiley Interscience, New York, 1963.
8. H. Hoffman and H. J. Diehr, "Phosphonium Salt Formation", *Angew. Chem. Int. Ed. Eng.*, **3**, 737 (1964).
9. B. E. Ivanov and V. F. Zheltukhin, "Reactivity of Trivalent Phosphorus Compounds", *Russ. Chem. Rev.* (Eng. Trans.), **39**, (5) 358 (1970).
10. L. Horner, "Preparation and Properties of Optically Active Phosphines", *Pure Appl. Chem.*, **9**, 225 (1964).
11. M. J. Boskin and D. B. Denney, *Chem. & Ind.* (*London*), 330 (1959).
12. E. Vedejs and P. L. Fuchs, *J. Amer. Chem. Soc.*, **93**, 4070 (1971).
13. B. A. Arbusov, "The Michaelis–Arbusov and Perkow Reactions", *Pure Appl. Chem.*, **9**, 307 (1964).
14. R. G. Harvey and E. R. De Sombre, "The Michaelis–Arbusov and Related Reactions", *Topics in Phosphorus Chemistry*, Vol 1, p 57, Wiley Interscience, New York, 1964.
15. W. Gerrard and W. J. Green, *J. Chem. Soc.*, 2550 (1951).
16. G. Asknes and D. Asknes, *Acta. Chem. Scand.*, **18**, 38 (1964).
17. A. Michaelis and T. Becker, *Chem. Ber.*, **30**, 1003 (1897).
18. G. M. Kosolapoff, *J. Amer. Chem. Soc.*, **67**, 1180 (1945).
19. B. Miller, "Reactions Between Trivalent Phosphorus Derivatives and Positive Halogen Sources", *Topics in Phosphorus Chemistry*, Vol 2, p. 133, Wiley Interscience, New York, 1965.
20. W. Gerrard and N. H. Philip, *Research*, (*London*), **1**, 477 (1948).
21. W. Kuchen and H. Buchwald, *Chem. Ber.*, **91**, 2296 (1958).
22. D. B. Denney and R. Di Leone, *J. Amer. Chem. Soc.*, **84**, 4737 (1962).
23. D. B. Denney and W. Hannifin Jr., *Tetrahedron Lett.*, 2177 (1963).
24. F. Ramirez, N. B. Desai and N. McKelvie, *J. Amer. Chem. Soc.*, **84**, 1745 (1962).
25. R. Rabinowitz and R. Marcus, *J. Amer. Chem. Soc.*, **84**, 1312 (1962).
26. G. Kumai and F. M. Kharrasova, *Izv. Vyssh. Uchebn. Zaved., Khim. Khim. Tekhnol.*, **4**, 229 (1961); *Chem. Abs.*, **55**, 21762 (1961).
27. I. J. Borowitz and R. Virkhaus, *J. Amer. Chem. Soc.*, **85**, 2183 (1963).
28. M. A. Shaw and R. S. Ward, "Addition Reactions of Tertiary Phosphorus Compounds with Electrophilic Olefins and Acetylenes", *Topics in Phosphorus Chemistry*, Vol 7, p 11, Wiley Interscience, New York, 1972.
29. (a) M. Grayson, P. T. Keough and G. A. Johnson, *J. Amer. Chem. Soc.*, **81**, 4803 (1959).
    (b) D. P. Young, W. E. McEwen, D. C. Velez, J. W. Johnson and C. A. Vander Werf, *Tetrahedron Lett.*, 359 (1964).
30. Z. Rappoport and S. Gertler, *J. Chem. Soc.*, 1360 (1961).
31. R. L. Powell and C. D. Hall, *J. Chem. Soc. C*, 2336 (1971).

[32] M. A. Shaw, J. C. Tebby, R. S. Ward and D. H. Williams, *J. Chem. Soc. C*, 2795 (1968).
[33] N. E. Waite, J. C. Tebby, R. S. Ward and D. H. Williams, *J. Chem. Soc. C*, 1100 (1969).
[34] N. E. Waite, D. W. Allen and J. C. Tebby, *Phosphorus*, 1, 139 (1971).
[35] D. Seyferth and J. M. Burlitch, *J. Org. Chem.*, 28, 2463 (1963).
[36] A. N. Pudovik and I. M. Aladzhyeva, *Zh. Obshch. Khim.*, 33, 3096 (1963).
[37] A. L. Lemper and H. Tiecklemann, *Tetrahedron Lett.*, 3053 (1964).
[38] H. W. Coover, M. A. McCall and J. B. Dickey, *J. Amer. Chem. Soc.*, 79, 1963 (1957).
[39] F. Ramirez, E. H. Chen and S. Dershowitz, *J. Amer. Chem. Soc.*, 81, 4338 (1959).
[40] F. Ramirez and S. Dershowitz, *J. Amer. Chem. Soc.*, 78, 5614 (1956).
[41] E. A. C. Lucken, F. Ramirez, V. P. Catto, D. Rhum and S. Dershowitz, *Tetrahedron*, 22, 637 (1966).
[42] F. W. Lichtenthaler, *Chem. Rev.*, 61, 607 (1961).
[43] A. N. Pudovik and V. Avery'anova, *Zh. Obshch. Khim.*, 26, 1426 (1956); *Chem. Abs.*, 50, 14512 (1956).
[44] I. J. Borowitz, S. Firstenberg, G. B. Borowitz and D. Schuessler, *J. Amer. Chem. Soc.*, 94, 1623 (1972).
[45] A. J. Speziale and D. E. Bissing, *J. Amer. Chem. Soc.*, 85, 3878 (1963).
[46] D. B. Denney and F. A. Wagner Jr., *Phosphorus*, 3, 27 (1973).
[47] A. R. Stiles et al., *J. Org. Chem.*, 26, 3960 (1961).
[48] T. R. Fukoto et al., *J. Org. Chem.*, 26, 4620 (1961).
[49] J. I. G. Cadogan, *Quart. Rev., Chem. Soc.*, 16, 208 (1962).
[50] J. I. G. Cadogan, *Accounts Chem. Res.*, 5, 303 (1972).
[51] R. J. Sundberg and C. C. Lang, *J. Org. Chem.*, 36, 300 (1971).
[52] B. Sklarz and M. K. Sultan, *Tetrahedron Lett.*, 1319 (1972).
[53] A. J. Boulton, I. J. Fletcher and A. R. Katritzky, *J. Chem. Soc. C*, 1193 (1971).
[54] J. I. G. Cadogan and D. T. Eastlick, *J. Chem. Soc. B*, 1314 (1970).
[55] F. Ramirez and A. M. Aguiar, *Abstr. Amer. Chem. Soc.*, 134th meeting, Chicago, 1958.
[56] E. H. Amonoo-Neizer, S. K. Ray, R. A. Shaw and B. C. Smith, *J. Chem. Soc.*, 4296 (1965).
[57] M. E. Brennan, *Chem. Commun.*, 956 (1970).
[58] R. W. Murray and M. L. Kaplan, *J. Amer. Chem. Soc.*, 90, 4160 (1968).
[59] P. D. Bartlett and G. D. Mendenhall, *J. Amer. Chem. Soc.*, 92, 210 (1970).
[60] D. Lloyd and M. I. C. Singer, *J. Chem. Soc. C*, 2941 (1971).
[61] T. Sasaki, K. Kanematsu and M. Murata, *Tetrahedron*, 28, 2383 (1972).
[62] K. D. Berlin et al., "Nucleophilic Displacement Reactions on Phosphorus Halides and Esters by Grignard and Lithium Reagents", *Topics in Phosphorus Chemistry*, Vol. 1, p 17, Wiley Interscience, New York, 1964.
[63] G. Asknes and D. Asknes, *Acta Chem. Scand.*, 18, 1623 (1964).
[64] D. Z. Denney, G. Y. Chen and D. B. Denney, *J. Amer. Chem. Soc.*, 91, 6838 (1969).
[65] K. D. Berlin, C. Hildebrand, J. G. Verkade and O. C. Dermer, *Chem. & Ind.* (*London*), 291 (1963).
[66] J. Michalski and T. Modro, *Chem. Ber.*, 95, 1629 (1962).
[67] R. Fusco and G. M. Bertulli, *Chim. Ind.* (*Milan*)., 37, 839 (1955); *Chem. Abs.*, 53, 13182 (1959).
[68] I. F. Lutsenko and Z. S. Kraits, *Zh. Obshch. Khim.*, 32, 1663 (1962); *Chem. Abs.*, 58, 4408 (1963).
[69] M. Bouneuf, *Bull. Soc. Chim. Fr.*, 4, 1808 (1923).
[70] L. D. Freedman, G. O. Doak and J. R. Edmisten, *J. Org. Chem.*, 26, 284 (1961).
[71] J. J. McBride, E. Jungermann, J. V. Killheffer and R. J. Clutter, *J. Org. Chem.*, 27, 1833 (1962).
[72] D. J. H. Smith and S. Trippett, *Chem. Commun.*, 855 (1969).

[73] D. B. Denney, D. Z. Denney, B. C. Chang and K. L. Marsi, *J. Amer. Chem. Soc.*, **91**, 5243 (1969).

[74] D. B. Denney and D. H. Jones, *J. Amer. Chem. Soc.*, **91**, 5821 (1969).

[75] D. B. Denney, D. Z. Denney, C. D. Hall and K. L. Marsi, *J. Amer. Chem. Soc.*, **94**, 245 (1972).

[76] M. A. Greenbaum, D. B. Denney and A. K. Hoffmann, *J. Amer. Chem. Soc.*, **78**, 2563 (1956).

[77] L. Horner and W. Jurgelheit, *Justus Liebigs Ann. Chem.*, **591**, 138 (1955).

[78] D. B. Denney and M. A. Greenbaum, *J. Amer. Chem. Soc.*, **79**, 979 (1957).

[79] D. B. Denney, D. Z. Denney, S. Schutzbank and S. L. Varga, *Phosphorus*, **3**, 99 (1973).

[80] J. Carles and S. Fliszar, *Can. J. Chem.*, **47**, 1113 (1969).

[81] P. D. Bartlett, A. L. Baumstark and M. E. Landis, *J. Amer. Chem. Soc.*, **95**, 6486 (1973).

[82] N. J. De'Ath and D. B. Denney, *Chem. Commun.*, 395 (1972).

[83] L. L. Chang and D. B. Denney, *Chem. Commun.*, 84 (1974).

[84] L. D. Quin, "Trivalent Phosphorus Compounds as Dienophiles", in J. Hamer (ed), *1,4–Cycloaddition Reactions*, Academic Press, New York, 1967.

[85] K. D. Berlin and D. M. Hellwege, "Carbon–Phosphorus Heterocycles", *Topics in Phosphorus Chemistry*, Vol. 6, p 1, Wiley Interscience, New York, 1969.

[86] A. Bond, M. Green and S. C. Pearson, *J. Chem. Soc. B*, 929 (1968).

[87] C. D. Hall, J. D. Bramblett and F. S. Lin, *J. Amer. Chem. Soc.*, **94**, 9264 (1972).

[88] W. L. Mock, *J. Amer. Chem. Soc.*, **88**, 2857 (1966).

[89] L. D. Quin and T. P. Barket, *Chem. Commun.*, 914 (1967).

[90] L. I. Zubtsova, M. A. Razumova and T. V. Yakovleva, *J. Gen. Chem. USSR*, **41**, 2450 (1971).

[91] N. A. Razumova and F. V. Bagrov, *J. Gen. Chem. USSR*, **40**, 1232 (1970).

[92] N. A. Razumova, Zh. L. Evtikhov and A. A. Petrov, *J. Gen. Chem. USSR*, **39**, 1388 (1969).

[93] N. A. Razumova, F. V. Bagrov and A. A. Petrov, *J. Gen. Chem. USSR*, **39**, 2305 (1969).

[94] F. Ramirez, "Oxyphosphoranes", *Accounts Chem. Res.*, **1** 168 (1968).

[95] F. Ramirez, *Pure Applied Chem.*, **9**, 337 (1964).

[96] F. Ramirez, *Bull. Soc. Chim. Fr,.* **47**, 2443 (1966).

[97] M. M. Sidky and M. F. Zayed, *Tetrahedron Lett.*, 2313 (1971).

[98] Y. Ogata and M. Yamashita, *J. Amer. Chem. Soc.*, **92**, 4670 (1970).

[99] F. Ramirez, S. B. Bhatia, A. V. Patwardhan and C. P. Smith, *J. Org. Chem.*, **32**, 3547 and 2194 (1967).

[100] F. Ramirez, S. B. Bhatia and C. P. Smith, *J. Amer. Chem. Soc.*, **89**, 3030, 3026 (1967).

[101] F. Ramirez, S. B. Bhatia and C. P. Smith, *Tetrahedron*, **23**, 2067 (1967).

[102] F. Ramirez, M. Magabhushanam and C. P. Smith, *Tetrahedron*, **24**, 1785 (1968).

[103] F. Ramirez, C. P. Smith, J. F. Pilot and A. S. Gulati, *J. Org. Chem.*, **33**, 3787 (1968).

[104] F. Ramirez, C. P. Smith and J. F. Pilot, *J. Amer. Chem. Soc.*, **90**, 6726 (1968).

[105] R. K. Oram and S. Trippett, *Chem. Commun.*, 554 (1972).

[106] R. K. Oram and S. Trippett, *J. Chem. Soc., Perkin I*, 1300 (1973).

[107] F. Ramirez, O. P. Madan and S. R. Heller, *J. Amer. Chem. Soc.*, **87**, 731 (1965).

[108] F. Ramirez, J. F. Pilot, O. P. Madan and C. P. Smith, *J. Amer. Chem. Soc.*, **90**, 1275 (1968).

Chapter 5
Phosphorus(III) ligands in transition metal complexes

THE CHEMISTRY OF PHOSPHORUS

5.1 Introduction

The previous chapter dealt with what might be called the 'organic' chemistry of tri-coordinate phosphorus compounds; this chapter is devoted to their 'inorganic' chemistry, by which is meant their behaviour as ligands in coordination compounds.*

The phosphorus derivatives used as ligands tend to be the simple and symmetrical PX_3 kind, among which PH_3, PF_3, PCl_3, $P(OMe)_3$, $P(OEt)_3$, PR_3 (R = alkyl) and especially PPh_3 have been most used. Asymmetrical ligands such as PEt_2Ph, CF_3PF_2, Me_2NPF_2 etc. have been less employed.

The number of complexes which involve at least one phosphorus–metal bond is extremely large and growing rapidly[1]. However the focus of interest in the vast majority of them is not the phosphorus centre but the metal. Variety at phosphorus is our main aim. Ideally we should like to study changes at P whilst keeping the metal part of the molecule unchanged, but this is rarely the objective of researchers in this area and the phosphorus chemist must glean his information where he can. For our purpose we have chosen to concentrate on four types of ligand which span the range fairly representatively; these are PH_3, PF_3, $P(OR)_3$ and PR_3 (R = alkyl or phenyl). Mixed group ligands have had to be neglected, although reluctantly, since these have enormous potential in studying changes at phosphorus, but not enough has yet been done with them. Similarly we must pass over metal clusters[2] and complexes with multidentate phosphorus ligands[3,4], the reader seeking information on these is well served by reviews.

The first half of this chapter is taken up with a brief outline of known complexes of our chosen ligands and their features, keeping where possible to those complexes with a single type, or as few types of ligand as possible, most being the P(III) ligand. The second half of the chapter is devoted to the bonding between phosphorus and metal, and to the debate on the part which the P(3d) orbitals play.

To explain the particularly good bonding capabilities of PX_3 ligands it was proposed by Chatt[5] that the donor σ bond formed between the ligand and the metal was reinforced by a donor nd(M) →3d(P) π back-bond from metal to phosphorus. [The situation is similar to that of Chapter 2 for the phosphoryl link except that in the latter case the donor π is 2p(O) → 3d(P)]. Accordingly P(III) ligands are seen as members of a group of ligands, collectively called π-acceptor ligands,† which included carbonyl,

---

* The rare dicoordinate phosphorus compound mentioned in the footnote of Chapter 2, p. 36 will also act as a ligand and in a manner typical of an arene ring. The complex has the Cr atom placed centrally with respect to the phosphorin ring just as in other aromatic ring complexes, which suggests that the bonding in this ring is truly delocalized through phosphorus [H. Vahrenkamp and H. Noth, *Chem. Ber.*, **105**, 1148 (1972)].

† The π terminology may be misleading because many unsaturated organic molecules attach themselves to transition metals by donation of their π electrons; these are called π *complexes*. In this chapter the ligands we are talking about form complexes primarily by donor σ bonds, but then use empty orbitals to accept back electron density from the metal via a π system. For this reason the collective name for such complexes is π-*acceptor* or π-*acid* complexes.

5.2 Phosphine, PH$_3$, complexes

nitrosyl, isocyanide etc. These ligands of first-row elements have no 3d orbitals but are still able to relieve the metal of electrons by accepting them into unoccupied antibonding orbitals. Consequently they are efficacious in stabilizing the metal in very low oxidation states such as M(0) or even M(−I). A symbolic representation of the bonding is shown in Fig. 5.1. At one time it was believed that donor-σ and donor-π were mutually reinforcing, this being known as the synergic effect.

Figure 5.1 The bonding between π-acceptor ligands and metals.

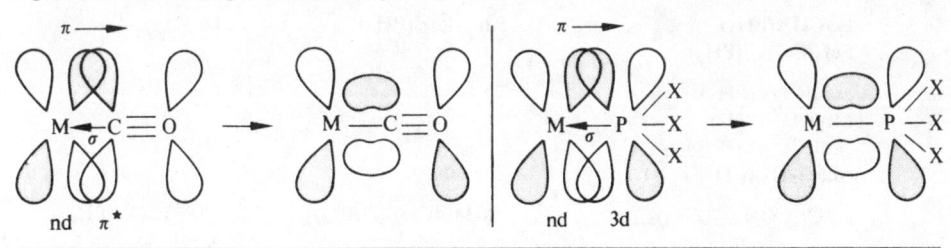

Although PF$_3$ is certainly a π-acceptor (and probably PCl$_3$ and P(OR)$_3$ as well) there are good grounds for excluding P(III) ligands with less electronegative atoms or groups such as PR$_3$ and PH$_3$. In terms of the *ligand field contraction* theory (see p. 54) only strongly electronegative atoms can sufficiently contract the 3d(P) orbitals to make them suitable for π bonding. However the use of 3d(P) orbitals seemed such a reasonable explanation of the bonding when it was first proposed that it was uncritically extended to cover all P(III) ligands. Some observations made in the early 1960's however, seemed to belie this model of the bonding. It was observed that the infrared and n.m.r. properties of these complexes could be explained in σ-bond terms alone. By 1970 it was even implied that PF$_3$ may not need the use of its 3d orbitals either since no direct proof of a π system was available. However in that year the π bonding in Ni(PF$_3$)$_4$ was incontrovertibly demonstrated by photoelectron spectroscopy[6].

5.2 Phosphine, PH$_3$, complexes

P(III) complexes are usually found with transition metals of groups to the right hand side of the Periodic Table and rarely with elements in groups T1 (Sc, Yt, La etc.), T2 (Ti, Zr, Hf) and T3 (V, Nb, Ta). In addition PH$_3$ complexes are relatively uncommon, so that it comes as something of a surprise to learn that TiCl$_4$(PH$_3$) was discovered in 1832[7]. The selection of phosphine complexes listed in Table 5.1 has been abstracted mainly from Robinson's review[8], one of the few which classifies complexes according to the nature of the P(III) ligand, and that of Chow, Levason and McAuliffe[9]. The table shows that no complex has yet been made in which PH$_3$ is the only ligand. The maximum coordination achieved by this is in the octahedral complex *cis*-Cr(CO)$_2$(PH$_3$)$_4$ prepared by the ultraviolet irradiation of a mixture of Cr(CO)$_6$ and PH$_3$.[10] This is a general method of preparation applicable to other P(III) ligands and relies upon the fact that M—CO bonds dissociate under such treatment while M—P bonds are unaffected. This photochemical substitution was also used to make Mn$_2$(CO)$_9$(PH$_3$).[11]

Table 5.1 Phosphine, PH$_3$, complexes of the transition metals

T9	CuX(PH$_3$)$_{2,1}$[a]	Ag –	Au –
T8	Ni(CO)$_3$(PH$_3$) unstable Ni(PF$_3$)$_{3,2}$(PH$_3$)$_{1,2}$	Pd(PPh$_3$)(PH$_3$)Cl$_2$ Pd$_4$(PPh$_3$)$_4$(PH$_3$)$_4$Cl$_4$[b]	Pt$_3$(PPh$_3$)$_3$(PH$_3$)$_3$I$_2$[b]
T7	Co(NO)(CO)$_2$(PH$_3$) Co(CO)$_3$(PH$_3$)$_3$[c] CoH(PF$_3$)$_3$(PH$_3$)	Rh$_6$(CO)$_8$(PH$_3$)$_8$[b]	Ir(CO)(PPh$_3$)(PH$_3$)Cl
T6	Fe(CO)$_8$(PH$_3$) FeI$_2$(CO)$_{3,2}$(PH$_3$)$_{2,3}$	Ru$_3$(CO)$_8$(PH$_3$)$_4$	Os –
T5	Mn$_2$(CO)$_9$(PH$_3$)[d] Mn(CO)$_5$(PH$_3$) MnI(CO)$_{3,4}$(PH$_3$)$_{2,1}$ Mn(C$_5$H$_5$)(CO)$_2$(PH$_3$)	Tc –	Re –
T4	cis-Cr(CO)$_{4,3,2}$(PH$_3$)$_{2,3,4}$	cis-Mo(CO)$_{4,3}$(PH$_3$)$_{2,3}$	cis-W(CO)$_4$(PH$_3$)$_2$
T3	V(C$_5$H$_5$)(CO)$_3$(PH$_3$)	Nb –	Ta –
T2	TiCl$_4$(PH$_3$)$_{1,2}$[c]	Zr –	Hf –
T1	Sc –	Yt –	La. etc. –

[a] X = Cl,Br,I; [b] cluster; [c] mentioned in text; [d] equatorial PH$_3$.

Infrared studies on νCO of carbonyl phosphine complexes indicate PH$_3$ to be a better π-acceptor ligand than PR$_3$ or PPh$_3$.[12] It can displace the latter as shown by the formation of (5.1) in eqn. (1)[13]. However it is not as good a π-acceptor as CO as evidenced

$$\text{Ir(CO)(PPh}_3)_2\text{Cl} + \text{PH}_3 \longrightarrow \text{Ir(CO)(PPh}_3)(\text{PH}_3)\text{Cl} + \text{PPh}_3 \qquad (1)$$
$$(5.1)$$

by the Co—C bond shortening from 191 pm in Co(CO)$_6$ to 184 pm in Co(CO)$_3$(PH$_3$)$_3$. Bond length as a guide to bonding has its pitfalls as we shall see, and the same applies to νCO comparisons. Generally shortening is associated with greater electron density, which in Co(CO)$_3$(PH$_3$)$_3$ implies increased π bonding between the metal and carbonyl. This can only be achieved at the expense of depleted π bonding in Co—PH$_3$. From phosphine complexes we can infer the π-acceptor abilities are CO > PH$_3$ > PR$_3 \simeq$ PPh$_3$.

5.3 Phosphorus trifluoride, PF$_3$, complexes

Unlike PH$_3$ it is possible to get transition metals fully coordinated by PF$_3$ ligands; indeed one such complex, Ni(PF$_3$)$_4$, is commercially available. The history of PF$_3$ complexes can be traced back to Moissan[15] who prepared Pt$_2$F$_4$(PF$_3$)$_2$ at the end of the last century, but as with TiCl$_4$(PH$_3$) this discovery was out of its time. Research interest only really began with the reinvestigation of Moissan's complex by Chatt[5] and with the formation of Ni(PF$_3$)$_4$ from Ni(PCl$_3$)$_4$ by Wilkinson[16] in 1950.

5.3 Phosphorus trifluoride, PF$_3$, complexes

The number of PF$_3$ complexes is now quite large and a recent review article[17] lists over 250 complexes containing this and RPF$_2$ ligands (R = H, CF$_3$, OPh, OMe, NMe$_2$, Me and Ph). This latter class holds promise for studying the finer points of electronic effects at phosphorus. Table 5.2 lists a representative cross-section of PF$_3$ and a few RPF$_2$ complexes. Most PF$_3$ ones are volatile compounds which are heat and air-stable. Although the Pt(II) complexes are water sensitive the others are only slowly hydrolyzed by water, indeed Ni(PF$_3$)$_4$ can be steam-distilled.

Table 5.2 Phosphorus trifluoride complexes of the transition metals

T9	CuCl(Me$_2$NPF$_2$)$_4$	Ag –	Au –
T8	Ni(PF$_3$)$_4$[a] b.p. 71 °C Ni(PF$_3$)$_n$(L)$_{4-n}$[b] Ni(RPF$_2$)$_4$[c]	Pd(PF$_3$)$_4$ Pd(PF$_3$)$_2$(PPh$_3$)$_2$	Pt(PF$_3$)$_4$[a] Pt(PF$_3$)$_{1,2}$(PPh$_3$)$_{4,3}$ PtCl$_2$(RPF$_2$)$_2$[c]
T7	Co$_2$(PF$_3$)$_8$ CoX(PF$_3$)$_4$[d] CoX(CO)$_{4-n}$(PF$_3$)$_n$[d]	Rh$_2$(PF$_3$)$_8$ RhX(PF$_3$)$_4$[e] Rh$_2$Cl$_2$(CO)$_{4-n}$(PF$_3$)$_n$	Ir$_2$(PF$_3$)$_8$ IrX(PF$_3$)$_4$[f]
T6	Fe(PF$_3$)$_5$[a] m.p. 45 °C FeH$_2$(PF$_3$)$_4$ Fe(CO)$_n$(PF$_3$)$_{5-n}$	Ru(PF$_3$)$_5$[a] m.p. 30 °C RuH$_2$(PF$_3$)$_4$ Ru(CO)$_n$(PF$_3$)$_{5-n}$ Ru$_3$(CO)$_6$(PF$_3$)$_6$[g]	Os(PF$_3$)$_5$[a] OsH$_2$(PF$_3$)$_4$
T5	MnH(PF$_3$)$_5$ m.p. 18.5 °C MnH(CO)$_n$(PF$_3$)$_{5-n}$ Mn$_2$(CO)$_n$(PF$_3$)$_{10-n}$ n = 1–3	Tc –	ReH(PF$_3$)$_5$ Re$_2$(PF$_3$)$_{10}$ m.p. 182 °C Re$_2$(CO)$_{10}$(PF$_3$)$_{10-n}$
T4	Cr(PF$_3$)$_6$[a] m.p. 193 °C Cr(CO)$_n$(PF$_3$)$_{6-n}$ n = 1–3 Cr(CO)$_n$(RPF$_2$)$_{6-n}$[c]	Mo(PF$_3$)$_6$[a] m.p. 196 °C Mo(CO)$_n$(PF$_3$)$_{6-n}$[a] Mo(CO)$_n$(RPF$_2$)$_{6-n}$[c]	W(PF$_3$)$_6$ m.p. 214 °C W(CO)$_{1,2}$(PF$_3$)$_{5,4}$[a] W(CO)$_n$(RPF$_2$)$_{6-n}$[c]

[a] Mentioned in text; [b] L = CO, PPh$_3$ etc; [c] R = CF$_3$, RO, R$_2$N, Me, Ph etc; [d] X = H or alkyl; [e] X = H, SnPh$_3$, X = Cl, Br, I unstable at 20 °C; [f] X = H, Cl, I; [g] cluster.

There are several ways of preparing PF$_3$ complexes. The original method of Wilkinson[16] was the fluorination of the corresponding PCl$_3$ complex and this can be achieved by means of reagents such as AsF$_3$, ZnF$_2$ and KSO$_2$F (from KF and liquid SO$_2$)[18]. The simplest method chemically is the heating of the metal and PF$_3$ at high temperatures and pressures (eqn. 2).

$$\text{Ni} + 4\text{PF}_3 \xrightarrow[70 \text{ atm}]{100 \text{ °C}} \text{Ni(PF}_3)_4 \quad (2)$$

By using the reducing properties of PF$_3$ itself it is possible to start with metal halides (eqn. 3)[19]

$$\text{PtCl}_4 + 6\text{PF}_3 \xrightarrow[\text{press.}]{\text{high}} \text{Pt(PF}_3)_4 + 2\text{PF}_3\text{Cl}_2 \quad (3)$$

although it is more general to employ another reducing agent such as copper for this purpose (eqn. 4).

$$MX_n + nPF_3 + nCu \xrightarrow[\text{press.}]{200\,°C} M(PF_3)_n + nCuX$$

$$[M = Cr, Mo, W; \quad Fe, Ru, Os; \quad Ni, Pd, Pt] \qquad (4)$$

This technique has been extensively used by Kruck et al. whose excellent review[20] on PF$_3$ complexes contains a wealth of information on preparations.

The replacement of coordinated organic ligands such as norbornadiene (eqn. 5)[21], benzene (eqn. 6)[22] or cyclopentadiene (eqn. 7)[23] is another route to synthesis. Carbonyl complexes can be partly or totally substituted by PF$_3$, the reaction being improved by the use of ultraviolet radiation (eqn. 8)[24] (eqn. 9)[25].

$$Mo(C_7H_8)(CO)_4 + 2PF_3 \longrightarrow \textit{cis-}Mo(PF_3)_2(CO)_4 + C_7H_8 \qquad (5)$$

$$Mo(C_6H_6)_2 + 6PF_3 \longrightarrow Mo(PF_3)_6 + 2C_6H_6 \qquad (6)$$

$$Ni(\pi\text{-}C_5H_5)_2 + 4PF_3 \longrightarrow Ni(PF_3)_4 + 2C_5H_5 \qquad (7)$$

$$Fe(CO)_5 + nPF_3 \xrightarrow{\text{ultraviolet}} Fe(PF_3)_n(CO)_{5-n} + nCO \qquad (8)$$

$$ReH(CO)_5 + nPF_3 \xrightarrow{\text{ultraviolet}} ReH(PF_3)_n(CO)_{5-n} + nCO \qquad (9)$$

The PF$_3$ for preparative reactions can be generated conveniently by boiling Ni(PF$_3$)$_4$ in toluene[26].

The stereochemistry of a few PF$_3$ complexes have been investigated. Ni(PF$_3$)$_4$ is tetrahedral and the PF$_3$ parameters are little changed from those of the free molecule, i.e. $r(P-F)$ is 156.1 pm[27] compared to 157.0 pm[28] and $\angle$ FPF is 99.3° compared to 97.8° in free PF$_3$. The pentacoordinate complexes, M(PF$_3$)$_5$, where M is Fe, Ru or Os, have *tbp* structures and ^{19}F and ^{31}P n.m.r. spectroscopy show all the ligand atoms equivalent, which means that they are undergoing rapid polytopal rearrangements[29]. It is because of this that isomers of mixed carbonyl PF$_3$ complexes, Fe(PF$_3$)$_n$(CO)$_{5-n}$, cannot be resolved, whereas the hexacoordinated ones of group T4 can be separated, i.e. M(PF$_3$)$_n$(CO)$_{6-n}$ where M is Cr, Mo or W.

There are two kinds of chemical reaction which PF$_3$ complexes undergo: (a) ligand displacement reactions, where the centre of attack is the metal and the M—P bond is cleaved, and (b) substitution reactions when there is nucleophilic attack at the phosphorus atom and a P—F bond is replaced.

(a) We have seen that PF$_3$ will displace other ligands, and in turn, it too can be removed by other ligands[20,30]. Triphenyl phosphine will replace one or two PF$_3$ groups (eqn. 10) and triphenylphosphite will replace all (eqn. 11)[30]. Carbon monoxide–PF$_3$ exchange has been studied for many metals (eqn. 12), and nickel and iron complexes especially.

$$Ni(PF_3)_4 + PPh_3 \longrightarrow Ni(PF_3)_{2,1}(PPh_3)_{1,2} + PF_3 \qquad (10)$$

$$Pt(PF_3)_4 + 4P(OPh)_3 \longrightarrow Ni\{P(OPh)_3\}_4 + 4PF_3 \qquad (11)$$

$$M(PF_3)_n + xCO \rightleftharpoons M(PF_3)_{n-x}(CO)_x + xPF_3 \qquad (12)$$

(b) *In situ* nucleophilic substitution of PF$_3$ can be achieved with amines, e.g. eqn. (13)[31] and alcohols, e.g. (eqn. 14)[32]. An alternative route to a single ROPF$_2$ ligand involves hydrolysis of the complex with Ba(OH)$_2$ followed by treatment with

5.4 Phosphite, P(OR)₃, complexes

$Et_3O^+BF_4^-$ (eqn. 15)[33]. The product, (5.2), is claimed to be thermally more stable than the parent complex.

$$Ni(PF_3)_4 + 2nR_2NH \longrightarrow Ni(PF_3)_{4-n}(R_2NPF_2)_n + nR_2NH_2F \quad (13)$$

$$Co(NO)(PF_3)_3 + MeOH \longrightarrow Co(NO)(PF_3)_2(MeOPF_2) + HF \quad (14)$$

$$Ni(PF_3)_4 \xrightarrow{Ba(OH)_2} Ni(PF_3)_3(PF_2O^-) \xrightarrow{Et_3O^+BF_4^-} Ni(PF_3)_3(PF_2OEt) \quad (15)$$
$$(5.2)$$

There is something special about PF_3 which makes it stand out from the other P(III) ligands. Its similarity to CO as a ligand was observed early on, and in some respects it outshines even this versatile π-acid. For example $HCo(PF_3)_4$ is stable up to 250 °C while $HCo(CO)_4$ is unstable and decomposes below room temperature. Similarly the cis-$FeX_2(PF_3)_4$(X = Cl, Br, I) products from the reaction of halogen and $Fe(PF_3)_5$ are more stable than the carbonyl analogues[34]. Even more revealing is the existence of $Pt(PF_3)_4$ whereas "$Pt(CO)_4$" does not form. These observations do not prove that PF_3 is a better π-acceptor than CO but are consistent with this suggestion. The photoelectron spectra and ionization potentials of $Cr(CO)_5(PF_3)$ do show PF_3 to be more electron withdrawing than CO, however[35], and other, less direct, evidence supports the fact that PF_3 is a stronger π-acid than CO. This will be covered in more detail in the bonding half of this chapter.

Much less work has been done on the other phosphorus trihalides as ligands. The complexes of general formula $M(CO)_5(PX_3)$, where M is Cr, Mo or W and X is Cl, Br or I, were obtained from the corresponding carbene complexes $M(CO)_5\{C(OMe)Me\}$[36]. By studying the A_1 vCO modes in the infrared spectra of these and the PH_3 and PF_3 derivatives of $W(CO)_5(PX_3)$, Fischer and Knauss[36] were able to arrange the ligands in order of net effective charge transfer from P to the metal, an order which shows that PF_3 transfers least electron density, presumably because of its ability to accept back electrons via a π system:

Charge transfer	$PH_3 > PI_3 > PBr_3 \simeq PCl_3 > PF_3$
A_1 vCO/(cm⁻¹)	2083 2087 2093 2095 2101

This inverse relationship between vCO and charge transfer of the other ligands is explained by the fact that when there is less competition from the other ligands for the π electrons then more electron density finds its way into the CO antibonding orbital. This has the effect of *weakening* the C—O bond, lowering its force constant and thereby reducing its frequency of vibration.

5.4 Phosphite, P(OR)₃, complexes

As ligands for transition metal complexes, phosphites were relative latecomers to the scene, but have made up for the delay by displaying remarkable behaviour in the ways of orthophenylation and homogeneous catalysis. So far no worker in the field has felt it timely to write a comprehensive review of the subject, presumably because this area is still rapidly expanding. However Robinson covers them in his report[8].

There are reasons to suspect that trialkyl phosphites would behave very differently to triphenyl phosphite. The σ-donor ability of the latter, as shown by bond moment data and Lewis base behaviour towards boron compounds (Chapters 2 and 11 respectively), is likely to be the poorest of any P(III) derivative, but so far no distinction has been made in its ligand behaviour as compared to the trialkyl phosphites. This suggests that what P(OPh)$_3$ loses on the σ swings it gains upon the π roundabouts. It has been proposed that electron density at the metal is greater with P(OR)$_3$ than P(OPh)$_3$,[37] but whether this is due to poorer σ donation or better π acceptance by the latter, is unclear.

Table 5.3 Phosphite complexes of the transition metals

T9	Cu{P(OMe)$_3$}$_4$ ClO$_4$[b]	Ag{P(OMe)$_3$}$_4$ ClO$_4$[b]	Au{P(OMe)$_3$}$_4$ BPh$_4$
T8	Ni{P(OR)$_3$}$_4$[a,b] Ni{P(OR)$_3$}$_{5,6}$ 2BPh$_4$	Pd{P(OEt)$_3$}$_4$ Pd{P(OEt)$_3$}$_{4,5}$ 2BPh$_4$	Pt{P(OEt)$_3$}$_4$
T7	CoH{P(OPh)$_3$}$_4$[b] Co$_4$(CO)$_{12-x}${P(OPh)$_3$}$_x$[c] Co{P(OMe)$_3$}$_5$ ClO$_4$[b] Co{P(OMe)$_3$}$_6$ 3ClO$_4$[b]	RhH{P(OPh)$_3$}$_4$ Rh$_4$(CO)$_{12-x}${P(OPh)$_3$}$_x$[c] Rh{P(OPh)$_3$}$_{4,5}$ BPh$_4$[b]	IrH{P(OPh)$_3$}$_4$[b] Ir$_4$(CO)$_{12-x}${P(OPh)$_3$}$_x$[c] Ir{P(OR)$_3$}$_5$ BPh$_4$[a,b] IrHX$_2${P(OPh)$_3$}$_3$[d]
T6	FeH$_2${P(OEt)$_3$}$_4$ Fe$_3$(CO)$_{12-x}${P(OMe)$_3$}$_x$[b,c] Fe{P(OMe)$_3$}$_6$ 2BPh$_4$	RuHCl{P(OPh)$_3$}$_4$ RuX$_2${P(OPh)$_3$}$_4$[d] Ru{P(OMe)$_3$}$_6$ 2BPh$_4$ RuX(NO){P(OPh)$_3$}$_2$[d]	OsX$_2${P(OPh)$_3$}$_2$[b,d] Os{P(OEt)$_3$}$_6$ 2BPh$_4$
T5	Mn(RCO)(CO)$_3${P(OR)$_3$}$_2$[a]	Tc –	Re –
T4	Cr(CO)$_5${P(OPh)$_3$} trans-Cr(CO)$_4${P(OPh)$_3$}$_2$	Mo(CO)$_5${P(OMe)$_3$}$_5$ Mo(C$_5$H$_5$)(CO)$_2${P(OPh)$_3$}$_2$ Mo(C$_6$H$_6$){P(OR)$_3$}$_3$[e]	W(CO){P(OMe)$_3$}$_5$

[a] R = Me,Et; [b] mentioned in text; [c] x = 1–4, cluster; [d] X = Cl,Br,I; [e] R = Me,Ph.

Table 5.3 lists a cross-section of trialkyl and triphenyl phosphite complexes. Related to the phosphite complexes are the cage phosphites (5.3) and the ligands (5.4) and (5.5) as well as the bird-cage molecule (5.6), all of which have been used as ligands. Except for (5.3), however, these are multidentate ligands and are not strictly comparable to the simple phosphites, but (5.3) is becoming important as a phosphite ligand because of its low steric requirements.

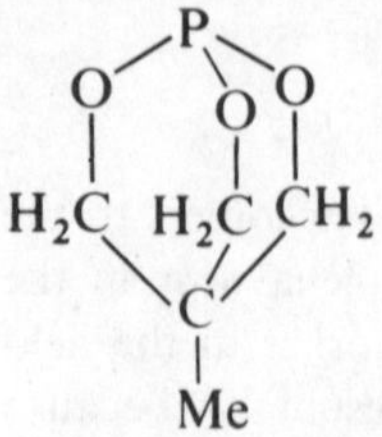

(5.3) 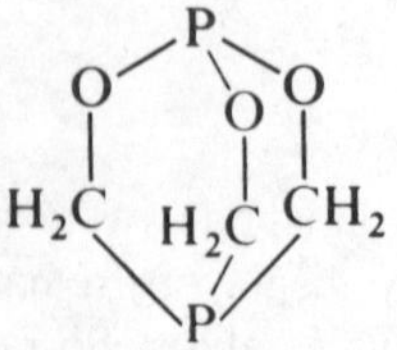   (5.4) 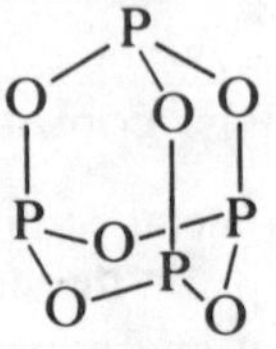   (5.5) 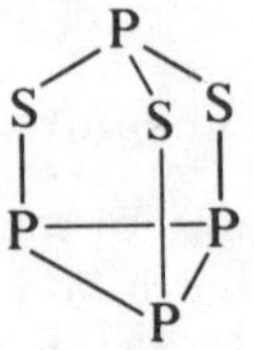   (5.6)

5.4 Phosphite, P(OR)₃, complexes

Phosphite derivatives can be prepared by several routes but generally displacement reactions are employed, particularly of carbon monoxide (eqn. 16)[38], organophosphine (eqn. 17)[39] (eqn. 18)[40] and organic substrates (eqn. 19)[41]. Another method is the alcoholysis of PF₃ complexes, e.g. (eqn. 20) which gives a quantitive yield of phosphite complex[42]. This is a promising route to phosphites and also offers an opportunity of mixed ligand derivatives, RO.PF₂ and (RO)₂PF as we have seen (eqn. 14).

$$Fe_3(CO)_{12} + P(OMe)_3 \longrightarrow Fe_3(CO)_{11,10,9}\{P(OMe)_3\}_{1,2,3} \qquad (16)$$

$$OsX_2(PPh_3)_{3,4} + P(OPh)_3 \longrightarrow OsX_2\{P(OPh)_3\}_4 \qquad X = Cl, Br, I \qquad (17)$$

$$IrH_3(PPh_3) + P(OPh)_3 \xrightarrow{heat} IrH\{P(OPh)_3\}_4 \qquad (18)$$

$$Rh_2Cl_2(C_8H_{12})_2 + P(OPh)_3 \longrightarrow Rh_2Cl_2(C_8H_{16})\{P(OPh)_3\}_2 \qquad (19)$$

$$Ni(PF_3)_4 + 4MeONa \longrightarrow Ni\{P(OMe)_3\}_4 + 4NaF \qquad (20)$$

The most notable recent advance has been in the synthesis of trimethyl and triethyl phosphite cationic complexes which Robinson et al. have shown can be isolated very effectively as their tetraphenylborate salts[43,44]. These were generally prepared by ligand exchange of PPh₃, norbornadiene etc, and Table 5.3 lists several of them.

Not all preparative reactions are ligand replacement. Some are more straightforward, e.g. AgClO₄ reacts with P(OMe)₃ to give Ag{P(OMe)₃}₄ClO₄ and Cu{P(OMe)₃}₄ClO₄ can similarly be produced. On the other hand Co(ClO₄)₂ reacts with P(OMe)₃ to give a mixture of complexes in which the Co(II) has undergone disproportionation to Co(I) and Co(III) (eqn. 21)[45].

$$2Co(ClO_4)_2 + P(OMe)_3 \longrightarrow Co\{P(OMe)_3\}_5 ClO_4 + Co\{P(OMe)_3\}_6 \, 3ClO_4 \qquad (21)$$

5.4.1 Orthophenylation of phenyl phosphites

Orthophenylation of phenyl phosphites was discovered independently by Robinson[39,46] and Knoth[47]. This is the name given to the process whereby the *ortho* hydrogen atom of one of the phenyl groups of P(OPh)₃ migrates to the metal to which the ligand is coordinated. A bond then forms between the metal and the phenyl ring at the *ortho* position, hence the name orthophenylation or orthometallation, and the process is illustrated by (eqn. 22)[47].

$$RuHCl\{P(OPh)_3\}_4 \underset{H_2}{\overset{heat\,(-H_2)}{\rightleftarrows}} \underset{O-P(OPh)_2}{\text{[C}_6H_4]}-RuCl\{P(OPh)_3\}_3 \qquad (22)$$

Triphenyl phosphite iridium(I) and iridium(III) complexes have been extensively investigated[48] and it has been shown that up to three metal-carbon bonds can be formed by successive orthophenylations of different ligands to give eventually the complex (Ir{C₆H₄O)P(OPh)₂}₃). An x-ray study of IrCl{P(OPh)₃}{(C₆H₄O)P(OPh)₂}₂ showed the arrangement at Ir to be (5.8)[49] which proved conclusively that orthophenylation had occurred, in this case with the formation of two Ir—C bonds. Orthophenylation

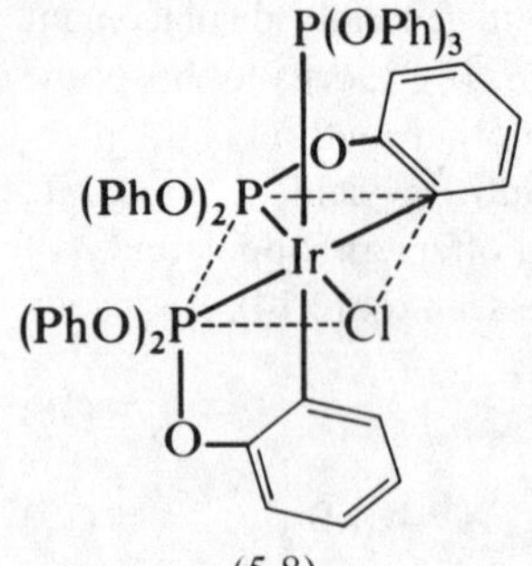

(5.8)

of Rh[47], Pd[50], Pt[50] and Os[51] complexes with P(OPh)$_3$ ligands has also been reported. As we shall see the arylphosphines also undergo orthophenylation to give a four-membered ring, the aryl phosphites are, however, better suited for orthophenylation which produces a more stable five-membered ring.

5.4.2 Homogeneous catalysis

The compound tetrakis(triethyl phosphite) nickel, Ni{P(OEt)$_3$}$_4$ can be readily protonated to form the nickel hydride NiH{P(OEt)$_3$}$_4$ and this has been intensively studied by Tolman[52]. Interest has centred on this compound because it can catalyze the coupling of 1,3-butadiene (C$_4$H$_6$) and ethylene to give hexadienes (C$_6$H$_{10}$), chiefly the 1,4-isomer in a low temperature process developed by the industrial company of du Pont. The mechanism of the reaction has been deduced by Tolman and an outline of it is shown in Fig. 5.2. Three isomers of 1,4-hexadiene are produced and so the nature of the intermediates varies slightly, but the sequence goes via (A), the active form of the catalyst, which picks up butadiene to give (B) with a Ni—C bond. This rearranges with migration of the proton from Ni to the organic moiety which becomes 1-methyl-alkyl (C$_4$H$_7$) and this forms a π-donor complex to the metal (C). This red complex

Figure 5.2 Homogeneous catalysis of hexadiene (C$_6$H$_{10}$) production (adapted from ref. 52b).

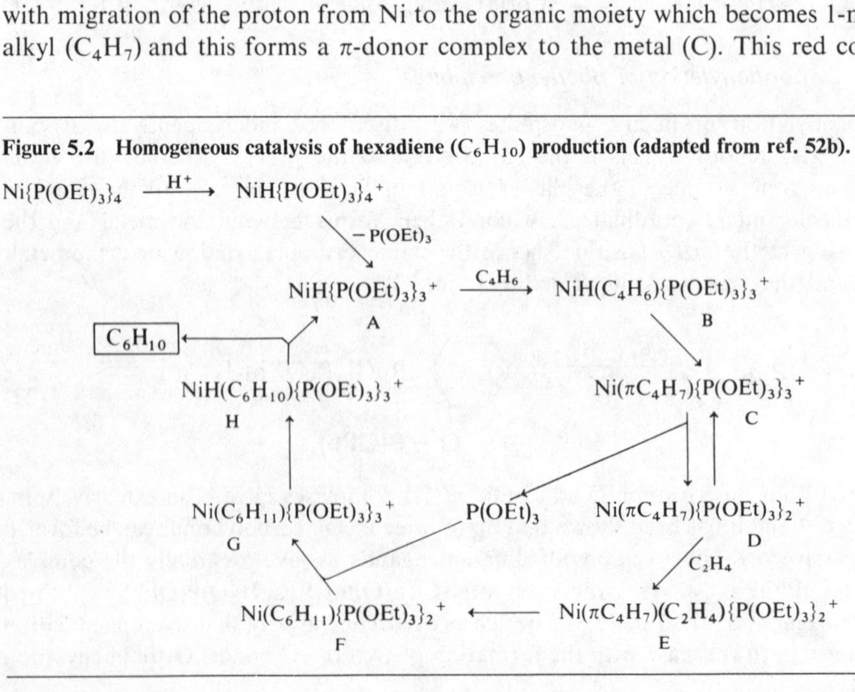

Ni{P(OEt)$_3$}$_4$ $\xrightarrow{H^+}$ NiH{P(OEt)$_3$}$_4^+$
$\searrow$ $-$P(OEt)$_3$
NiH{P(OEt)$_3$}$_3^+$ $\xrightarrow{C_4H_6}$ NiH(C$_4$H$_6$){P(OEt)$_3$}$_3^+$
A   B

C$_6$H$_{10}$ ←
NiH(C$_6$H$_{10}$){P(OEt)$_3$}$_3^+$ Ni(πC$_4$H$_7$){P(OEt)$_3$}$_3^+$
H   C

Ni(C$_6$H$_{11}$){P(OEt)$_3$}$_3^+$ P(OEt)$_3$ Ni(πC$_4$H$_7$){P(OEt)$_3$}$_2^+$
G   D
                                                    ↓ C$_2$H$_4$
Ni(C$_6$H$_{11}$){P(OEt)$_3$}$_2^+$ ← Ni(πC$_4$H$_7$)(C$_2$H$_4$){P(OEt)$_3$}$_2^+$
F   E

cation was isolated as its PF_6^- salt. (C) exists in equilibrium with (D) which is yellow, and it is form (D) which can react with ethylene to form five-coordinate (E). (At the (C)⇌(D) stage isomerization of the π alkyl can occur.) (E) rearranges to form a C—C bond between the two organic moieties (F) and triethyl phosphite re-enters the complex (G). Transfer of a hydrogen back from the organic ligand to the metal gives (H) which then produces the product hexadiene and regenerates the catalyst (A).

Another example of homogeneous catalysis by phosphites is the stereospecific polymerization of 1,3-butadiene by $Ni\{P(OPh)_3\}_4$[52c,53]. More can be expected of the phosphites in syntheses of these kinds in the future, but the key to the complexes' behaviour as a catalyst rests chiefly with the metal. For example, $NiH\{P(OEt)_3\}_4^+$ acts effectively and rapidly in the du Pont reaction but the isoelectronic complex $CoH\{P(OEt)_3\}$ is inactive. The only difference between the two is that ligand exchange on the former metal is fast but on the latter is slow. Ligand exchange is essential for a successful catalyst so that it can form a species which is temporarily unsaturated from a coordination point of view. In this state it can then attach to itself the organic reactants as substrates[54], and this is the key to the problem of what makes a viable catalyst.

Just as $ROPF_2$ and $(RO)_2PF$ bridge the gap between PF_3 and $P(OR)_3$, so phosphonic and phosphinic esters lie on the borderline between phosphites and the next class of ligand, the organophosphines. Researchers have recently turned to these, e.g. $P(OR)_2Ph$ and $P(OR)Ph_2$,[55] but progress in this area is limited as yet.

5.5 Organophosphine, PR_3, complexes[8,9,56]

There are more complexes of PPh_3 than any other phosphorus(III) ligand. There are also a substantial number of trialkylphosphine complexes. The subject has become so large that it would almost justify a whole book to itself. However, despite the mass of information, relatively little is of concern to us since in nearly all of the complexes the focus of interest is not the R_3P—M bond but some other part. The most noticeable feature of this class of complexes is the rarity of those which are coordinated purely by organophosphine ligands. Group T8 metals are exceptional in this respect. $Ni(PPh_3)_4$ is known but is substantially dissociated at room temperature to $Ni(PPh_3)_3$.[57] In solution $Pt(PPh_3)_4$ is likewise dissociated into $Pt(PPh_3)_3$ and PPh_3; $Pt(PPh_3)_2$ is also known and so are the fully substituted clusters $Pt_3(PPh_3)_6$ and $Pt_4(PPh_3)_4$.[58]

Table 5.4 lists only a limited selection of organophosphorus complexes with the emphasis on the carbonyl and hydride derivatives. Some of the hydrides mentioned undergo rapid intramolecular rearrangements, such as $WH_6(PMe_2Ph)_3$[59], $ReH_5(PPh_3)_3$[60] and $OsH_4(PR_3)_3$ where R can be a variety of alkyl and phenyl groups[61,62]. These complexes give a p.m.r. spectrum quartet, 1:3:3:1, showing that the six, five and four protons respectively are all magnetically equivalent and coupling to the three equivalent ^{31}P nuclei. Isoelectronic with these is mer-$IrH_3(PPh_3)_3$ but this has a much more complicated pattern of n.m.r. resonances[63].

There are many ways of preparing organophosphine complexes and these are generally of the ligand replacement type[56], such as the photochemical substitution of carbonyls[64].

Table 5.4 Triorganophosphine complexes of transition metals

T9	CuX(PPh$_3$)$_3$^{a} Cu(PPh$_3$)$_2$ClO$_4$	Ag{P(C$_6$H$_4$-p-Me)$_3$}$_{2,3,4}$$^+$	Au{P(C$_6$H$_4$-p-Me)$_3$}$_{2,3,4}$$^+$ Au$_{11}$(CNS)$_3$(PPh$_3$)$_7$^{b}
T8	Ni(PPh$_3$)$_{3,4}$^{c} Ni(CN)$_2$(PR$_3$)$_{2,3}$ Ni^{III}Br$_3$(PMe$_2$Ph)$_2$	Pd(CO)(PPh$_3$)$_3$	Pt(PPh$_3$)$_{2,3,4}$ Pt(CO)(PPh$_3$)$_3$ Pt(O$_2$)(PPh$_3$)$_2$
T7	CoX(PPh$_3$)$_3$^{d} CoBr$_2$(PR$_3$)$_2$ Co(CO)$_3$(PPh$_3$)$_2$$^+$ ^{e}	RhCl(PPh$_3$)$_3$ Rh(PPh$_3$)$_4$ RhX(CO)(PR$_3$)$_2$^{d} RhH(CO)(PPh$_3$)$_3$	IrCl(PPh$_3$)$_3$ IrX(CO)(PR$_3$)$_2$ IrH$_3$(PR$_3$)$_3$^{c}
T6	Fe$_3$(CO)$_{11}$(PPh$_3$)$_3$ Fe(CO)$_3$(PPh$_3$)$_2$ FeCl$_2$(PPh$_3$)$_2$ FeH$_4$(PEtPh$_2$)$_3$ FeH$_2$(PEtPh$_2$)$_3$	Ru(CO)$_{4,3}$(PPh$_3$)$_{1,2}$ RuHCl(PPh$_3$)$_3$ RuCl$_2$(PPh$_3$)$_3$ RuH$_4$(PPh$_3$)$_3$ RuH$_2$(N$_2$)(PPh$_3$)$_3$	Os(CO)$_4$(PPh$_3$) Os(CO)$_3$(PPh$_3$)$_2$^{c} Os(CO$_2$)(PPh$_3$)$_3$^{c} OsH$_4$(PR$_3$)$_3$^{c} Re^{I}Cl(CO)$_x$(PR$_3$)$_{5-x}$^{f}
T5	Mn$_2$(CO)$_9$(PR$_3$) Mn$_2$(CO)$_8$(PR$_3$)$_2$ MnH(CO)$_3$(PR$_3$)$_2$	TcBr$_4$(PPh$_3$)$_2$	Re^{III}Cl(CO)$_x$(PR$_3$)$_{4-x}$^{f} Re^{IV}Cl$_4$(PMe$_2$Ph)$_2$ Re^{V}OCl$_3$(PPh$_3$)$_2$ ReH$_5$(PR$_3$)$_3$^{c}
T4	Cr(CO)$_{6-x}$(PR$_3$)$_x$^{f} CrCl$_3$(PEt$_3$)$_3$	Mo(CO)$_{6-x}$(PR$_3$)$_x$^{f}	W(CO)$_{6-x}$(PR$_3$)$_x$ WH$_6$(PR$_3$)$_3$^{c}
T3	V(CO)$_5$(PPh$_3$)$^-$ V(C$_5$H$_5$)(CO)$_{2,3}$(PPh$_3$)$_{2,1}$ VOCl$_2$(PR$_3$)$_2$^{g,h}	Nb(CO)$_5$(PPh$_3$)$^-$ Nb(C$_5$H$_5$)(CO)$_{2,3}$(PPh$_3$)$_{2,1}$	Ta(CO)$_5$(PPh$_3$)$^-$
T2	TiCl$_4$(PR$_3$)$_{1,2}$^{g}		

^{a} X = Me,Cl,Br,I; ^{b} cluster of icosahedron of Au atoms with Au at centre; ^{c} mentioned in text; ^{d} X = Cl,Br,I; ^{e} tbp with P's apical; ^{f} x = 1,2,3; ^{g} R = alkyl or Ph; ^{h} unlike most VO compounds, which are spy with O axial, these compounds are tbp due to steric needs of PR$_3$ ligands.

Orthophenylation is inferred in certain intermediates of PPh$_3$ complexes. E.g. RuHCl(PPh$_3$)$_3$ undergoes rapid deuteration at the *ortho* position and this complex is capable of catalyzing the orthodeuteration of PPh$_3$. This shows not only that orthophenylation of intermediates is occuring but that coordinated PPh$_3$ ligands are very labile when attached to Ru.[47,65] Other orthophenylations occur with RhMe(PPh$_3$)$_3$ (eqn. 23)[66], IrCl(PPh$_3$)$_3$ (eqn. 24)[67] and *trans*-PtCl$_2$(PPhBu$^t_2$)$_2$ (eqn. 25)[68], in some

$$\text{RhMe(PPh}_3)_3 \xrightarrow{\text{heat}} \text{Rh(PPh}_3)_2(\text{C}_6\text{H}_4\text{PPh}_2) + \text{CH}_4 \tag{23}$$

$$\text{IrCl(PPh}_3)_3 \longrightarrow \text{IrHCl(PPh}_3)_2(\text{C}_6\text{H}_4\text{PPh}_2) \tag{24}$$

$$\textit{trans-}\text{PtCl}_2(\text{PPhBu}^t_2)_2 \longrightarrow \text{PtCl(PPhBu}^t_2)(\text{C}_6\text{H}_4\text{PBu}^t_2) + \text{HCl} \tag{25}$$

cases with the loss of HX from the complex.

One of the more remarkable complexes is RhCl(PPh$_3$)$_3$ which was the first practical reagent for promoting homogeneous hydrogenation of alkenes[69]. The reactions can be carried out in solution at room temperature and under a pressure of one atmosphere

5.5 Organophosphine, PR$_3$, complexes

of H$_2$. A great deal of effort has gone into the study of this reaction[70-72] and there appear to be three main steps. First (eqn. 26) the complex picks up H$_2$ and a PPh$_3$

$$RhCl(PPh_3)_3 + H_2 \rightleftharpoons RhH_2Cl(PPh_3)_2 + PPh_3 \quad (26)$$

ligand is displaced. This gives a five-coordinate Rh, although the sixth position may be occupied by a solvent molecule, especially if this is a donor solvent like acetic acid. Too strong a donor solvent such as pyridine can deactivate the catalyst because it blocks this sixth site, and prevents the attachment of the olefin which is the next step (eqn. 27). Migration of a proton from Rh to the olefin produces an alkyl

$$RhH_2Cl(PPh_3)_2 + CH_2{=}CHR \rightleftharpoons RhH_2Cl(PPh_3)_2(CH_2{=}CHR) \quad (27)$$

complex (5.9), and transfer of a second proton gives the alkane product (eqn. 28).

$$RhH_2Cl(PPh_3)_2(CH_2{=}CHR) \longrightarrow RhHCl(PPh_3)_2(CH_2CH_2R) \longrightarrow$$
$$(5.9)$$
$$RhCl(PPh_3)_2 + CH_3CH_2R \quad (28)$$

The part played by PPh$_3$ in all this is not clear, but is obviously related to its dissociability in step (26). This creates the unsaturation of coordination essential to the complex's behaviour as a catalyst. Apparently small changes can have a big effect, e.g. PPhEt$_2$ is more basic than PPh$_3$* and RhH$_2$Cl(PEt$_2$Ph) is not an effective catalyst[72].

Steric factors as well as basicity may be important. The bulkiness of the PPh$_3$ group may explain why 1-alkenes respond best to hydrogenation since these give a primary alkyl intermediate (5.9) which means a less crowded situation about Rh than if a secondary alkyl derived from a 2-alkene, say, were present. The reactions of alkynes as well as alkenes are catalyzed, e.g. Ni(CO)$_2$(PPh$_3$)$_2$ is a good agent for the cyclotrimerization of diacetylenes, R—C≡C—C≡C—R, to give 1,2,4-trialkyl-3,5,6-trialkynylbenzene[73].

Organophosphine complexes have been used as catalysts in three other kinds of alkene reaction – isomerization, hydroformylation and dismutation – enabling them to be performed under mild conditions such as room temperature and atmospheric pressure. The complexes PtCl$_2$(PPh$_3$)$_2$, PtHCl(PPh$_3$)$_2$ and PtCl(PPh$_3$)$_2$(olefin) will isomerize olefins, i.e. will change the position of the double bonds, especially in the presence of H$_2$. These complexes perform their catalytic function at about equal rates suggesting that a hydrido-Pt(II) intermediate is responsible[74].

Hydroformylation refers to the addition of H$_2$ and CO to an alkene to form an aldehyde, which may subsequently be reduced to the corresponding alcohol (eqn. 29).

$$RCH{=}CH_2 + H_2 + CO \longrightarrow RCH_2CH_2CHO \longrightarrow RCH_2CH_2CH_2OH \quad (29)$$

This reaction is catalyzed by RhH(CO)(PPh$_3$)$_3$ and this particular complex has been intensively studied and the mechanism of the reaction deduced[75]. Both RhX(CO)(PR$_3$)$_2$ and IrX(CO)(PR$_3$)$_2$ complexes have been the object of much research as catalysts[76].

---

* A study of the reaction: IrCl(CO)(PR$_3$)$_2$ + RCO$_2$H → IrHCl(CO)(RCO$_2$)(PR$_3$)$_2$ was used to show the basicity of the organophosphines and these are PMe$_3$ > PMe$_2$Ph > PMePh$_2$ > PPh$_3$ [A. J. Deeming and B. L. Shaw, *J. Chem. Soc. A*, 1802 (1969)].

Olefin dismutation is a sort of disproportionation and an example of such a reaction and its complex promoter are shown in eqn. (30)[77].

$$2CH_3CH_2CH_2CH=CH_2 \xrightarrow{MoCl_2(NO)_2(PPh_3)_2}$$
$$CH_2=CH_2 + CH_3(CH_2)_2CH=CH(CH_2)_2CH_3 \quad (30)$$

As has already been said, the chemistry of complexes with triorganophosphine ligands is very extensive and only a small portion is relevant to our discussion. A great deal is outside the sphere of influence of the phosphorus in these compounds, but some properties do depend upon the phosphorus ligand. For instance, although $NiBr_2(PEt_3)_2$ is square planar and $NiCl_2(PPh_3)_2$ is tetrahedral we may describe this as being due to the steric requirement of the ligands. However the complex $NiBr_2\{PPh_2(CH_2Ph)\}_2$ displays both a square planar trans-$NiBr_2\{PPh_2(CH_2Ph)\}_2$ and a tetrahedral $NiBr_2\{PPh_2(CH_2Ph)\}_2$ in the same crystal![78]

The structure of $Os(CO)_3(PPh_3)_2$ is a tbp with the PPh_3 groups apical[79], which suggests, if the laws of apicophilicity about Os are the same as those about phosphorus(V) discussed in Chapter 2, that PPh_3 is less effective than CO at π-bonding and is forced into the sterically more crowded apical sites. Things are not this simple, however, since $Os(CO)_2(PPh_3)_3$ has all the PPh_3 ligands equatorial[80]. This particular complex has the ability to pick up molecular oxygen to give $Os(O_2)(CO)_2(PPh_3)_2$.

Five-coordinate $[RuCl(NO)_2(PPh_3)_2]PF_6$ on the other hand does not adopt a tbp but a slightly distorted spy configuration about Ru. What is even more striking about this complex is that it contains a bent and linear nitrosyl group[81]. As (5.10) shows the axial NO group is further removed and the frequency of the NO stretching vibration is lower than the basal NO group. The former is seen to behave properly as a σ donor, whereas the latter has transferred an electron to the metal so is NO^+. As such it is isoelectronic with CO and so is behaving as a π-acceptor ligand. Hence it is linear.

$$\begin{array}{c}
\text{O} \\
\parallel \\
\nu NO \quad 1687 \text{ cm}^{-1} \\
N \\
| \, 186 \text{ pm} \\
Ru \\
Cl \diagup \quad \diagdown PPh_3 \\
174 \text{ pm} \\
Ph_3P \cdots \cdots N \\
\nu NO \quad 1847 \text{ cm}^{-1} \\
O
\end{array}$$

(5.10)

It is quite impossible to do justice, not only to the triorganophosphines, but also to the many other phosphorus ligands such as $P(NMe_2)_3$, and the phosphido ligands PR_2^- and PAr_2^- which are generally bridging ligands. The multidentate ligands also have extensive chemistries. This branch of the subject really began in 1956 when the most popular of these ligands, 1,2-bis(diphenylphosphino)ethane, $Ph_2PCH_2CH_2PPh_2$ (often called 'diphos' for short) was first synthesized; interested readers are referred to recent reviews[3,4].

5.6 Bonding between phosphorus and metals

In the first half of this chapter we have looked in a general way at the ligands PH_3, PF_3, $P(OR)_3$ and PR_3 (R = alkyl and phenyl) as complexing ligands. If they resemble any other class of ligand it would appear to be the π-acceptor group as exemplified by CO, and in terms of π-acceptor ability the order seems to be $PF_3 > CO > P(OR)_3 > PH_3 > PR_3$, with there being little distinction between alkyl and aryl derivatives. It is now time to have a closer look at the bonding between ligand and metal. This has been the subject of much controversy during the past ten years. As with so many of the bonds that phosphorus forms, the question boils down to whether it can and does utilize its 3d orbitals.

From the start it was evident that PF_3 as a ligand had a lot in common with CO, and that there were many properties of PF_3 complexes that were inconsistent with the behaviour of this ligand if its bonding were purely σ. For example, it will not form complexes with metals in high oxidation states but, like CO, it will stabilize low oxidation states. Also bond lengths show that $r(Ni-P)$ in $Ni(PF_3)_4$ is 209.9 pm which is 10% shorter than other $r(Ni-P)$ values. The early explanation of this was that PF_3, like CO, was a good π-electron acceptor using its 3d orbitals for this purpose; see Fig. 5.1.

The success of the 3d π-acceptor theory naturally encouraged others to extend it to phosphorus ligands in general and to complexes in which it perhaps was not really needed. Those who wholeheartedly supported π-acceptor bonding by P(III) ligands did not deny that it would vary from ligand to ligand, being strongest in PF_3. They sought methods of assessing π-acceptance and for this purpose used νCO frequencies in such complexes as $Mo(CO)_3P_3$[82] and $Co(CO)_3(NO)P$[83] where P refers to the phosphorus ligand. Horrocks and Taylor[83] used $\nu_s CO$, $\nu_{as} CO$ and νNO to produce the π ligand order $NO > CO > PF_3 > PCl_3 > P(OPh)_3 > P(OMe)_3 > PPh_3 > PMe_3 > PEt_3 > P(NC_5H_{10})_3$ [NC_5H_{10} = piperidine]. The rationale behind the use of νCO has been mentioned already, p. 180. It was suggested by Cotton that difference in νCO was not as reliable a guide as force constant, k, changes, and a simplified method, known now as the Cotton–Kraihanzel force field technique (CK) was advanced for their calculation[84]. This produced a slightly different order of π-acceptors to that above, the main difference being that PF_3 was shown to be a better acceptor than CO (1.32 times better in fact)[85].

This approach to the bonding, resting as it did on the assumption that all these ligands used π bonding to supplement their σ bond, first ran up against opposition in 1964 when Bigorgne suggested that the behaviour of phosphorus ligands did not require π bonding[86]. He observed the νCO values of $Ni(CO)_3P$ and $Ni(CO)_2P_2$ complexes (P = PMe_3, PEt_3, PPh_3, $P(C\equiv CPh)_3$, $P(CF_3)_3$ and PF_3) and by a judicious plot of the B_1 νCO modes of $Ni(CO)_2P_2$ he showed there to be a linear relationship between this and the Taft inductive constants (σ^*) of the groups bound to phosphorus. Because of this linear relationship he concluded that π electrons were not involved in the metal–phosphorus bonding in any of these complexes. Tolman's work on $Ni(CO)_3P$[87], to which we shall return later, covers a wider range of P ligands and a plot of his A_1 νCO values against σ^* is shown in Fig. 5.3. This shows that there is less of a linear relationship than Bigornge imagined although it does show that alkyl, aryl, alkoxy and aryloxy phosphorus ligands more or less lie on a straight line, but that PF_3,

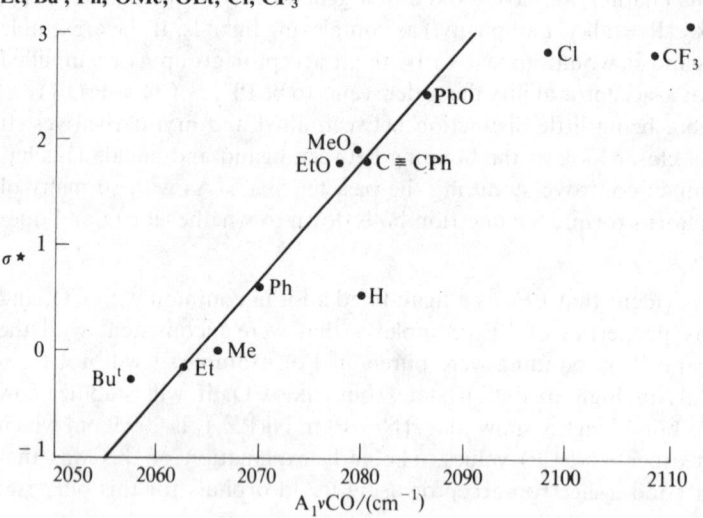

Figure 5.3 Carbonyl frequencies of Ni(CO)$_3$P vs. Taft inductive constants, σ^*; where P = H, Me, Et, Bu^t, Ph, OMe, OEt, Cl, CF$_3$

σ^* values from R. W. Taft, *Steric Effects in Organic Chem.*, in M. S. Newman (editor) 619; Wiley, New York, 1956, A$_1$ νCO values from refs. 86 and 87.

PH$_3$ and P(CF$_3$)$_3$ are way off beam. So is PCl$_3$, but not as markedly. Tolman, moreover, criticizes the idea that Taft σ^* constants are relevant since they relate to inductive effects operating through carbon which may not necessarily reflect inductive effects of atoms and groups when attached to phosphorus.

Although Bigorgne rather over-stated the anti-π case his observations point to the fact that not all P(III) ligand behaviour is in need of π-acceptor contributions to the bonding. A reassessment of his method however serves only to support such bonding for PF$_3$ and a few others.

One of the staunchest critics of π bonding has been Angelici. From 1966 he has mounted a campaign against it which has won much ground. His early foray, like that of Bigorgne, also used the same sort of νCO evidence that the supporters of π-bonding were fond of quoting. His studies were on cis-Mn(CO)$_3$(L)$_2$Br complexes[88], where L included amine ligands which cannot participate in π-acceptance. According to his claim, the pK_a of amines (taken as typical of a pure σ effect in donating to H$^+$) and νCO should vary linearly. This thesis holds for aryl and alkylamine bases, less so for pyridine bases. The relationship between pK_a and νCO he found to hold approximately for P(Bu^n)$_3$ and PEt$_2$Ph also, and so rightly concluded that these ligands did not π bond in this complex.

In another piece of work[89] Angelici reported the equilibrium constants for a complex in which an amine ligand, e.g. p-MeC$_6$H$_4$NH$_2$, was displaced by a P(III) ligand (eqn. 31). The equilibrium constants K, measured in toluene at 35.4 °C and the half

$$W(CO)_5(p\text{-}MeC_6H_4NH_2) + PX_3 \underset{}{\overset{K}{\rightleftharpoons}} W(CO)_5(PX_3) + p\text{-}MeC_6H_4NH_2 \quad (31)$$

neutralization potentials, HNP, of the phosphorus ligands are listed in Table 5.5. (The half neutralization potentials were taken to be the order of basicity of the

5.6 Bonding between phosphorus and metals

Table 5.5 Equilibrium constants for the equilibrium (31) and half neutralization potentials, (HNP)[89]

Ligand	K	HNP
P(Bu^n)$_3$	48.1	131
P(OCH$_2$)$_3$Et (cage phosphite)	37.0	665
P(OBu^n)$_3$	16.5	530
PPh$_3$	6.55	573
P(OPh)$_3$	1.16	875

ligands.) He saw these results as more in keeping with σ-bonding abilities because of the way in which they varied, except for the cage phosphite. Unfortunately his studies could not include PF$_3$. These results are less conclusive than one would like since K is so dependent on other variables as the original paper shows. They do however show that P(III) ligands bind themselves more strongly to tungsten than p-toluidine under these conditions.

N.m.r. measurements have been most quoted by those opposed to π bonding and with more success. Most of this information though is derived from complexes between phosphorus ligands and metals in higher oxidation states, where indeed it seems to be the case that only σ bonding is relevant. We shall deal with this later in the chapter under a discussion of the *trans* effect.

Before we turn to the two pieces of evidence which prove conclusively that some P—M bonds have a π component let us consider this type of bonding in general. For a comprehensive account of the bonding the reader is referred to Pidcock's review[90]. He concludes that the bonding is clear at two extremes: (i) metals in oxidation state II, or higher, form essentially pure σ bonds with organophosphines, and (ii) metals in oxidation state 0 or below form σ-with-π bonds to PF$_3$, PCl$_3$ and P(OPh)$_3$ ligands. Between these two extremes not enough information is yet available to enable definite conclusions to be arrived at. In other words, whether there is likely to be a π-acceptor bond depends not only upon the phosphorus centre but upon the metal as well. We cannot say that a particular ligand will always form a π bond whatever the state of the metal, nor that a metal in a low oxidation will always donate π electrons to a phosphorus ligand irrespective of what that ligand is. The interplay of these variables is shown diagramatically in Fig. 5.4.

Without 'electronegative' groups on phosphorus the 3d orbitals will be diffuse and so the alkylphosphines such as P(Bu^n)$_3$ may always be unsuitable for π bonding. The dependance on the metal's oxidation state was shown in studies of a series of complexes CoBr$_n$(PMePh$_2$)$_{4-n}$ $n = 0$–2 in which the oxidation state varied from Co(0) to Co(II)[91]. The p.m.r. spectrum of Co^0(PMePh$_2$)$_4$ was consistent with the 3d → 3d π bonding being the dominant spin transfer mechanism, whereas in Co^{II}Br$_2$(PMePh$_2$)$_2$

Figure 5.4 Dependence of π bonding on oxidation state of metal and 'electronegativity' of phosphorus ligands, based on tetracoordinate phosphorus, see Table 2.8 page 75.

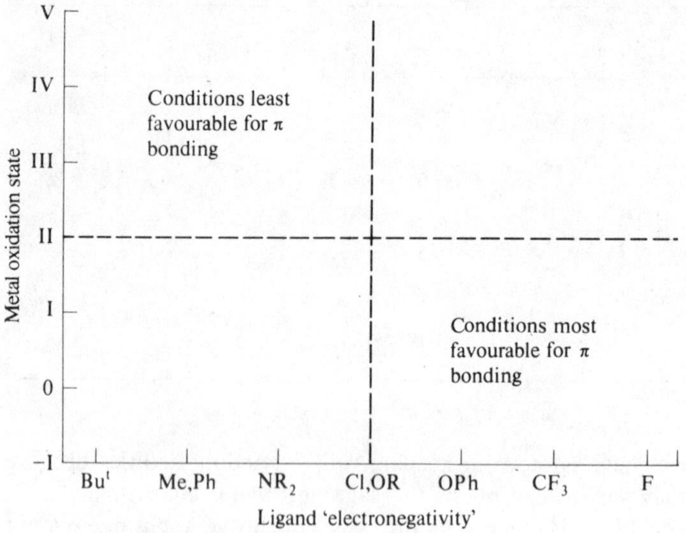
V
IV
III
II
I
0
−I
Metal oxidation state
Conditions least favourable for π bonding
Conditions most favourable for π bonding
Bu^t
Me,Ph
NR2
Cl,OR
OPh
CF3
F
Ligand 'electronegativity'

this occurs via non-orthogonality of the σ system. In $Co^{I}Br(PMePh_2)_3$ both mechanisms are operative. Although the oxidation state of the metal is the more important factor the nature of the metal also plays a part but this is less easy to judge in a comparative manner. It has been shown, however, that for carbonyls the order of donor-π ability is $W(CO)_5 > Mo(CO_5) > Cr(CO)_5 \geqslant Ni(CO)_3$.[92]

Evidence in support of π-bonding

At one end of the π-bonding spectrum is PF_3 and its complexes with zero-valent metals. An example is $Ni(PF_3)_4$, and it is for this particular complex that the best evidence of π bonding exists, which is the direct observation, by photoelectron spectroscopy, of the removal of an electron from the π system[93]. The ionization spectrum of free PF_3 shows the phosphorus lone pair to have the lowest energy within the molecule with ionization potential of 12.31 V. The complex $Ni(PF_3)_4$ shows these electrons at 13.09 V, consistent with their being the σ bond between Ni and P. In addition there now appears in the spectrum two lower energy bands at 9.55 V and 10.58 V which can only be due to electrons in π orbitals. Above 15.8 V the spectra of the free and complexed PF_3 are the same showing that the rest of the PF_3 molecule is unaffected by coordination to the metal. Moreover this research[93] indicated that the π bonds in $Ni(PF_3)_4$ were even stronger than those in $Ni(CO)_4$ as revealed by its photoelectron spectrum, thus bearing out the Cotton–Kraihanzel force constant observations.

The second piece of evidence in support of π bonding comes from bond lengths and shows that $P(OPh)_3$ can π bond. If phosphorus ligands were purely σ bonding then the more basic the ligand the stronger will be this bond. On this basis PPh_3 will be more strongly bound than the less basic $P(OPh)_3$ and consequently we would expect the M—PPh_3 distance to be shorter than the M—$P(OPh)_3$ distance. For Cr—PPh_3

5.6 Bonding between phosphorus and metals

and Cr—P(OPh)$_3$ bonds in identical environments the opposite is the case[94], proving that electron density between the Cr and P nuclei is greater in the latter, a situation which can only arise if the metal pumps electron density into a π system. Cotton and Wilkinson[95] placed a great deal of their faith in π bonding on this piece of research so it is worth looking at in more detail. Figure 5.5 contains the relevant bond lengths for the chromium carbonyl phosphorus complexes. The Cr—C bonds reveal

---

Figure 5.5 Bond length data for Cr—P complexes[94,96]/(pm).

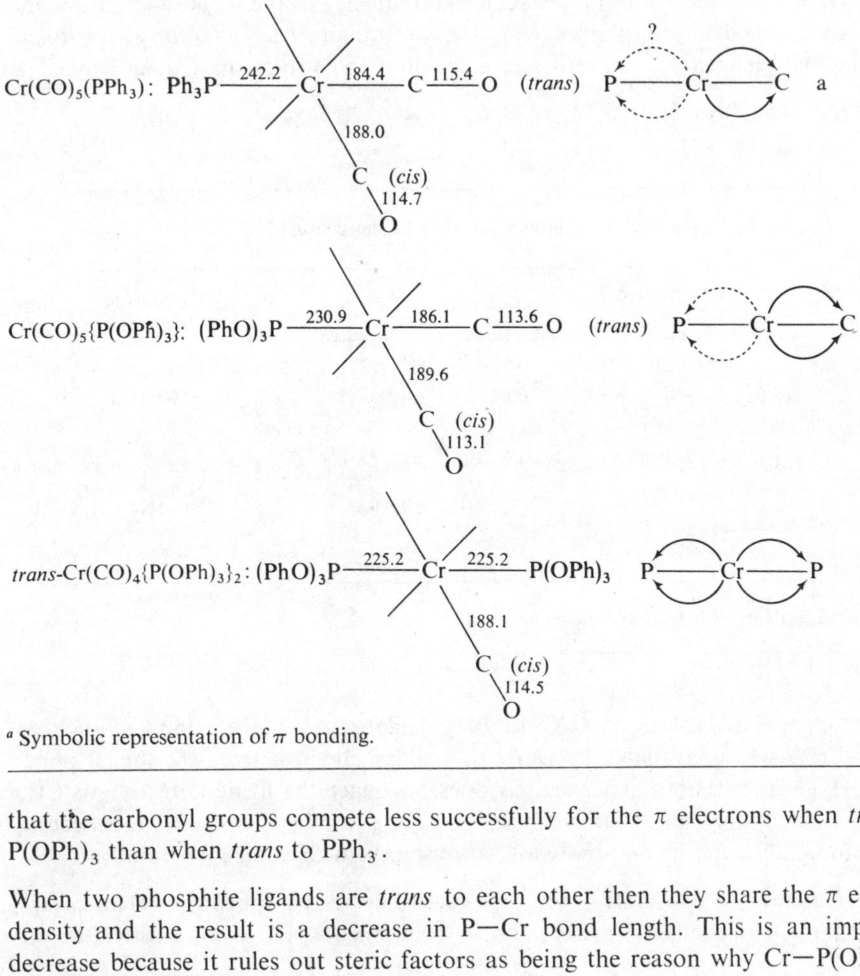

Cr(CO)$_5$(PPh$_3$): Ph$_3$P —242.2— Cr —184.4— C —115.4— O (trans) P——Cr——C a
               |188.0
               C (cis)
               ‖114.7
               O

Cr(CO)$_5${P(OPh)$_3$}: (PhO)$_3$P —230.9— Cr —186.1— C —113.6— O (trans) P——Cr——C
               |189.6
               C (cis)
               ‖113.1
               O

trans-Cr(CO)$_4${P(OPh)$_3$}$_2$: (PhO)$_3$P —225.2— Cr —225.2— P(OPh)$_3$ P——Cr——P
               |188.1
               C (cis)
               ‖114.5
               O

[a] Symbolic representation of π bonding.

---

that the carbonyl groups compete less successfully for the π electrons when *trans* to P(OPh)$_3$ than when *trans* to PPh$_3$.

When two phosphite ligands are *trans* to each other then they share the π electron density and the result is a decrease in P—Cr bond length. This is an important decrease because it rules out steric factors as being the reason why Cr—P(OPh)$_3$ is shorter than Cr—PPh$_3$; sterically the latter ligand is more hindered than the former as we shall see. The *cis* carbonyl bonds in these complexes show that the σ framework is only a little affected and that anisotropic σ effects are relatively unimportant.*

---

* If π bonding is to be considered as unimportant in these complexes then differences must be explained through variations in the σ framework. This framework, once considered isotropic about the metal, is anisotropic in metals with more than one type of ligand, but to what extent is not certain. Where there is no π bonding, however, it becomes the only vehicle for electron redistribution and the transmission of electronic effects.

The ability to form π bonds is unlikely to be an either/or property of a ligand and will depend on a variety of factors such as the oxidation state of the metal which in turn will depend on the other ligands attached to it. For a constant metal environment, however, the changes in some parameter such as νCO or k which occur when the groups attached to phosphorus change, as in $Ni(CO)_3PX_3$ complexes, may be as reliable a guide to the π bonding as we can get. Orders of π-acceptor ligands based on such information would seem to be useful rule-of-thumb indicators.

Tolman[87] studied seventy $Ni(CO)_3(PX_3)$ complexes including many with mixed ligands. He derived a relationship between the frequency of the A_1 νCO vibration and the ligands attached to phosphorus (eqn. 32), assigning to each a factor χ_X for ligand X etc. Tolman defined χ_{Bu^t} as zero and calculated the values quoted in Table 5.6.

$$A_1 \nu CO\ [Ni(CO)_3(PXYZ)]\ /(cm^{-1}) = 2056.1 + \chi_X + \chi_Y + \chi_Z \tag{32}$$

Table 5.6 Tolman's χ values (cm⁻¹) for groups attached to phosphorus[87]

Group X	χ_X	Group X	χ_X	Group X	χ_X	Group X	χ_X
CF_3	19.6	(bicyclic phosphite)—Me	10.4[a]	OEt	6.8	NC_5H_{10}[b]	2.0
F	18.2	OPh	9.7	Ph	4.3	Et	1.8
Cl	14.8	H	8.3	CH_2Ph	3.5	C_6H_{11}	0.1
C_6F_5	11.2	OMe	7.7	Me	2.6	Bu^t	0

[a] Refers to ⅓ total value for ligand; [b] piperidino.

His method enabled values of νCO to be calculated when the complex, such as $Ni(CO)_3(PH_3)$, was unavailable by using the values obtained from the mixed phosphines PH_2Ph and $PHPh_2$. The method does not place the ligands relative to CO, however. Tolman does not commit himself to saying whether his χ order is reflecting π or σ effects, although he does note a relationship to Kabachnick's σ constants.*

Graham[97] used force constants (k) in an effort to solve the difficult problem of apportioning electronic effects into σ and π components. He compared values of

---

* M. I. Kabachnik derived a set of σ constants for groups attached to phosphinic acids, $R^1R^2P(O)OH$, according to the relationship:

$$pK(R^1R^2P(O)OH) = pK(H_2P(O)OH) - \rho\sum\sigma(R)$$

where $pK(H_2P(O)OH)$ and ρ were calculated by successive approximations to be 6.13 and 1.788. The order and values of σ were CF_3, σ = +0.498; H, 0; OMe, −0.124; OEt, −0.214; OPr_n, −0.315; OBu^n, −0.411; Ph, −0.481; Me, −0.965; Et, −1.101; Bu^t, −1.546. Thus CF_3 groups withdraw electrons most strongly and Bu^t release them most strongly relative to H [M. I. Kabachnik, *Proc. Acad. Sci. USSR*, **110**, 577 (1956); translated from *Dokl. Akad. Nauk SSSR*, **110**, 393 (1956)].

5.6 Bonding between phosphorus and metals

$Mo(CO)_5(PX_3)$ complexes to $Mo(CO)_5(NH_2C_6H_{11})$ assuming that the cyclohexylamine in the latter was incapable of π acceptance. His method used $\Delta k(trans)$ and

$$\Delta k_1(trans) = \Delta\sigma + 2\Delta\pi \tag{33}$$

$$\Delta k_2(cis) = \Delta\sigma + \Delta\pi \tag{34}$$

$\Delta k(cis)$, i.e. changes in the force constants for the CO bonds *trans* and *cis* to the PX_3 group as compared to their values in the amine complex, and related them to changes in the two kinds of bonding, $\Delta\sigma$ and $\Delta\pi$ by the arbitrary eqns. (33) and (34). These values are listed in Table 5.7. Graham found little to recommend in the $\Delta\sigma$ values but

Table 5.7 Graham's $\Delta\sigma$ and $\Delta\pi$ values for phosphorus ligands relative to $C_6H_{11}NH_2$.[97]

	$\Delta\sigma$	$\Delta\pi$		$\Delta\sigma$	$\Delta\pi$
PF_3	−0.09	0.79	$P(OEt)_3$	−0.38	0.55
CO	−0.06	0.74	PBu^t_3	−0.48	0.48
PCl_3	−0.09	0.71	$P(OPh)_3$	−0.13	0.48
$P(OMe)_3$	−0.38	0.58	PPh_3	−0.15	0.27

saw his $\Delta\pi$ values as providing a meaningful order of π-acceptor abilities. This too is debatable – they may show that PF_3 is a better acceptor than CO but they also show PBu^t_3 to be comparable to $P(OPh)_3$ which is quite at variance with the facts. Graham based his calculations partly on the results of Darensbourg and Brown[98] who used force constants to calculate dipole moments. They concluded that νCO and k changes may arise from σ changes between M and C rather than from π changes because the latter are counterbalanced by π changes of the P(III) ligands. They do however conclude that π and σ effects are quite separate.

Angelici[99] studied similar compounds to Graham except that the metal was W instead of Mo, i.e. $W(CO)_5(PR_3)$ where PR_3 was PPh_3, $P(C_6H_4\text{-}p\text{-Me})_3$, $P(C_6H_4\text{-}p\text{-}OMe)$, $PPhEt_2$ and PBu^t_3. He demonstrated that νCO and k varied with pK_a for these ligands and in the same way as νCO and k varied with pK_a when the complex contained amine ligands which could not π bond. Although Angelici's phosphorus ligands were all triorganophosphines they do span a wide range of pK_a values from 2.7 (PPh_3) to 8.4 (PBu^t_3). Nevertheless these are just the kind of ligands least likely to π bond.

An uneasy truce now prevails in this area with those supporting and those opposing π-bonding each occupying certain undisputed territories, but with a large zone of no-man's land between them. The war, however, has been fought on a second front and this still goes on. The disputed area in this case being the role of d orbitals in the **trans effect**, one theory of which gives them a big part. A lot of this work has been done on Pt complexes. The study of these has been greatly facilitated by n.m.r. spectroscopy, especially since ^{195}Pt has a nuclear magnetic moment (I) of $\frac{1}{2}$, just like ^{31}P.

5.7 Trans effect and trans influence*

Ligands *trans* to each other interact in a way which *cis* ligands do not. The *trans influence* of a ligand refers to the extent to which one ligand weakens the bond of the ligand *trans* to it, and is a ground state effect. (The *trans effect* refers more specifically to the effect the ligand has on the rate of substitution of the ligand opposite itself.) The complexes most intensively studied have been the square planar ones of the group T8 metals and particularly those of platinum. Consideration of the behaviour of ligands in reactions of the type (35) have enabled them to be arranged in order of

$$\begin{array}{c} A\diagdown \diagup Cl \\ Pt \\ B\diagup \diagdown Cl \end{array} + NH_3 \longrightarrow \begin{cases} \begin{array}{c} A\diagdown \diagup Cl \\ Pt \\ B\diagup \diagdown NH_3 \end{array} & A \text{ } trans \text{ directing} > B \\ \\ \begin{array}{c} A\diagdown \diagup NH_3 \\ Pt \\ B\diagup \diagdown Cl \end{array} & B \text{ } trans \text{ directing} > A \end{cases} \quad (35)$$

trans directing ability, and such an order is $CN^- \simeq CO \simeq NO \simeq C_2H_4 > H \simeq PR_3 > CH_3 \simeq SR_2 > O_2^- \simeq I^- \simeq SCN^- > Br^- > Cl^- > C_5H_5N > RNH_2 \simeq NH_3 > OH^- > H_2O$. It is immediately obvious that the π-acceptor ligands tend to be strongly *trans* directing, which hints at a link between the two effects. Clearly π-bonding ability cannot be the only factor involved since ligands such as H and CH_3 are up among the leaders. Two principal theories have been advanced to explain this order.

Grinberg proposed that trans influence was purely electrostatic. A strong dipole interaction between a ligand and the central metal atom would tend to weaken the attachment of the ligand *trans* to it by a mis-match of dipoles (5.11). Factors favouring

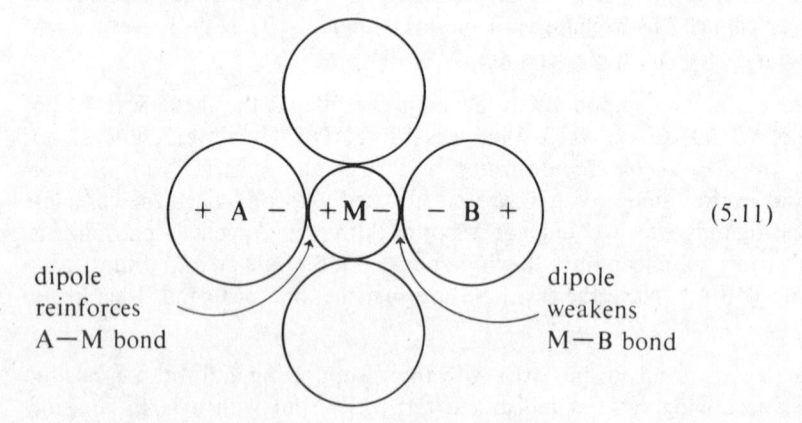

$$+ A - \quad + M - \quad - B + \qquad (5.11)$$

dipole reinforces A—M bond

dipole weakens M—B bond

a strong ligand dipole ↔ metal dipole interaction are (i) ligands with large electron clouds, e.g. $I^- > Br^- > Cl^-$, or an easily deformed electronic system such as a π

---

* For a comprehensive account of the *trans* influence the reader is directed to F. R. Hartley, *The Chemistry of Platinum and Palladium*, pp 299ff, Applied Science, London, 1973.

5.7 Trans effect and trans influence

system, e.g. C_2H_4; and (ii) large metal cation – hence *trans* influences through Pt(II) and Pd(II), which have the same radii, should be greater than through Ni(II).*

A second theory is a π-bonding theory based on *trans* ligands competing for the same metal d orbitals. Using these in π bonding at one side of the metal will reduce their availability at the opposite side. Thus a good π-acceptor ligand should have a strong *trans* influence. Symbolically this can be illustrated as (5.12). There are two d orbitals which the *trans* ligands have in common, one of which lies in the plane of the complex. As such it is common to both *cis* and *trans* ligands. The other d orbital perpendicular to this is the one which only the *trans* ligands have in common.

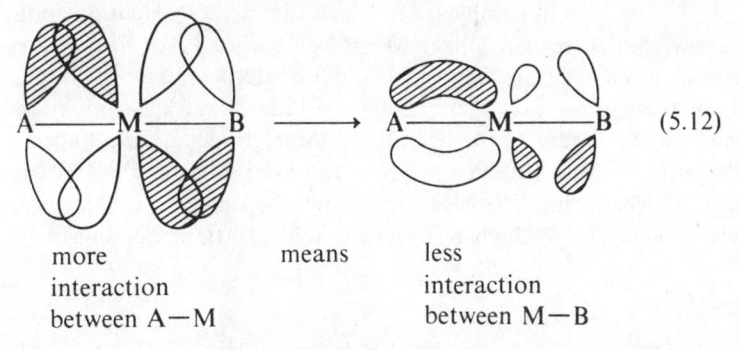

A——M——B ⟶ A——M——B   (5.12)

more interaction between A—M   means   less interaction between M—B

If there were two π-acceptor ligands in a complex then one would predict that they would orientate themselves *cis*, thereby having one each of the two out-of-plane d orbitals to themselves (in addition to sharing the in-plane one). Were they to be *trans* to each other they would compete not only for the in-plane but for the out-of-plane one as well. This sort of reasoning explains why the *cis*-$PtCl_2(PR_3)_2$ isomer is about 40 kJ mol^{-1} (10 kcal mole^{-1}) thermodynamically more stable than the *trans* isomer[100].

Bond length data, however, confronts us with evidence which suggests that π bonding may be unimportant. Consider the bond lengths in the molecules (5.13)[101], (5.14)[102] and (5.15)[103].

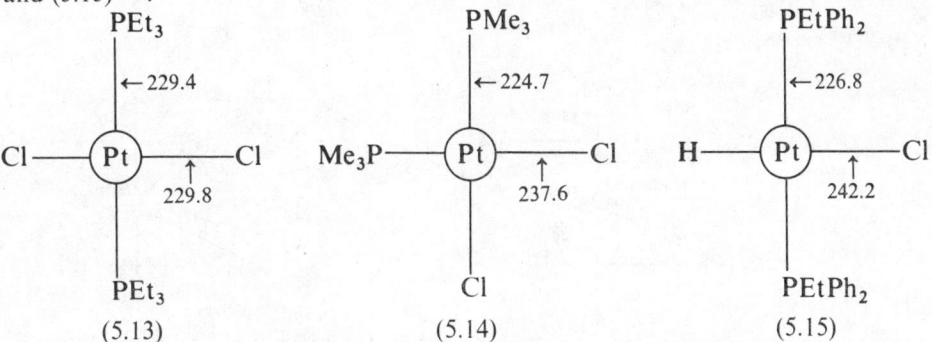

	(5.13)	(5.14)	(5.15)
top	PEt$_3$ ←229.4	PMe$_3$ ←224.7	PEtPh$_2$ ←226.8
horiz	Cl—Pt—Cl	Me$_3$P—Pt—Cl	H—Pt—Cl
↑	229.8	237.6	242.2
bottom	PEt$_3$	Cl	PEtPh$_2$

Although all the P(III) ligands are not the same they are all triorganophosphines and electronically similar. As expected, when these are *trans* (5.13) they have longer Pt—P bond lengths than when *cis* (5.14). In π terms this is because in (5.13) both are fighting

---

* Although they are the same size and have the same effective nuclear charge, Pt(II) displays a stronger *trans* influence than Pd(II). The way in which they differ is in the availability of their d orbitals, those of Pt(II) being more prominent than those of Pd(II).

for the same d orbital whereas in (5.14) each has its own. The Pt—Cl bond distances are even more informative. Chloride ligands are poor π acceptors, and do not compete effectively for d orbitals, so that when faced with a phosphine ligand *trans* to itself a Pt—Cl bond increases in length, cf. (5.13) with (5.14). So far so good, until we look at (5.15) where the Pt—Cl bond is even longer. The ligand exerting the *trans* influence in this complex is not capable of π bonding since it is simply a proton. It is '*trans*'-mitting not through a π system but through the σ framework of the molecule. This being so then the framework is also likely to be responsible for the Pt—Cl difference between (5.13) and (5.14).

Bond length ratios of M—P bonds in complexes in which the metal's oxidation state is even higher also show that there is a longer M—P bond when P(III) ligands are *trans* to each other than when they are opposite a chloride, as in (5.16)[104], (5.17), (5.18) and (5.19)[105]. In these complexes there should be little or no back donation from the metal because of the nature of the ligand and the high oxidation numbers. The ratio of M—P bond lengths (in heavy type), $R = r(P-M(P))/r(P-M(Cl))$, is not unity even in (5.16) where the difference must surely be due to σ anisotropy. In the M(III) complexes it is slightly higher suggesting that a little π bonding may be creeping in.

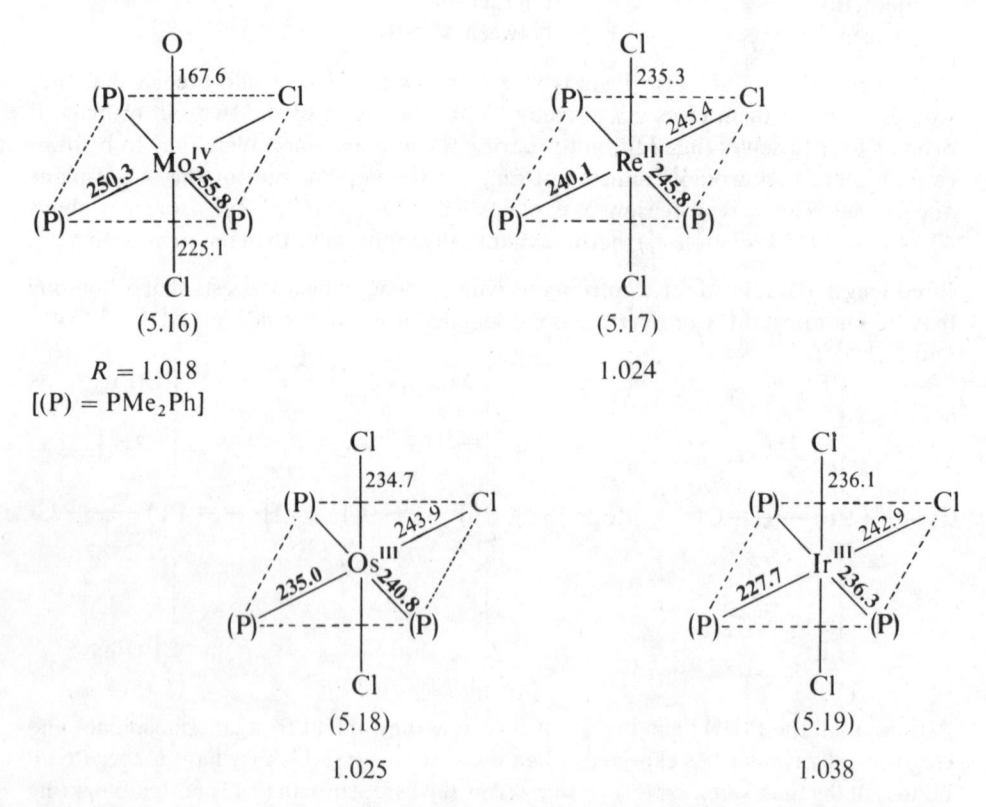

(5.16)
$R = 1.018$
[(P) = PMe$_2$Ph]

(5.17)
1.024

(5.18)
1.025

(5.19)
1.038

That the σ system is capable of acting as the agent for the *trans* influence is shown most strongly by ^{31}P n.m.r. studies. The coupling constants, $^1J_{PPt}$, are most sensitive to the s-orbital involvement in the σ-bonding and it is generally taken that the higher the value of J the greater the s-orbital contribution, and probably the more covalent

5.7 Trans effect and trans influence

is the bond (this last point is the weak link in the chain of reasoning). Consider the coupling constants for the following complexes (5.20) and (5.21)[106]. In (5.20) it would appear that the σ framework is perturbed, judging by $^1J_{PPt}$ values and that the PBu$^n_3$ *trans* to Cl is more covalently bonded than the tri-n-butylphosphines which are *trans* to each other. An explanation of (5.20) is possible in π terms – stronger π bonding strengthens the σ bond by a synergic effect (although the two systems are less interconnected than was once believed). However (5.21) seems to scotch this line of reasoning, since neither Cl nor Me is capable of significant π bonding, so both of the PEt$_3$ ligands should have almost equal opportunities for π bonding and yet in this molecule $^1J_{PPt}$ values show even more imbalance in the s character of the σ framework.

```
           PBuⁿ₃                              PEt₃
            |                                  |
         2270 Hz                            1719 Hz
            |                                  |
Buⁿ₃P ─3454 Hz─ (Pt) ─── Cl      Et₃P ─4179 Hz─ (Pt) ─── Cl
            |                                  |
           PBuⁿ₃                               Me
          (5.20)                             (5.21)
```

Again if the oxidation state of the metal atom rises then the opportunity for π bonding should be less. Thus the ratio of coupling constants for *cis* compared to *trans* complexes, which should reflect the greater opportunities for the former to π bond, will change if the oxidation state of the metal changes. Pidcock *et al*[107] have carried out such measurements for *cis*- and *trans*-Pt^{II}Cl$_2$(PBu$^n_3$)$_2$ and for *cis*- and *trans*-Pt^{IV}Cl$_4$(PBu$^n_3$)$_2$. For the Pt(II) complex the ratio $^1J_{PPt}$(cis complex)/$^1J_{PPt}$(trans complex) is 3508/2380, i.e. 1.474. In the PtCl$_4$(PBu$^n_3$)$_2$ complexes there should be a proportionately larger decrease in the *cis* $^1J_{PPt}$ if the π bonding has been considerably weakened. However the ratio turns out to be 2070/1462, i.e. 1.416 which shows only a very small change, albeit in the right direction. The conclusion to be drawn from this work is either that π contributions are insignificant or that π bonding has no synergic effect on the σ bonding, and so no effect on the coupling constants. The former choice seems the more likely explanation.

Pidcock has argued convincingly against π bonding using n.m.r. data and the reader is referred to his work for a fuller account[90,107]. On the other hand Grim has put forward π-bonding explanations which successfully rationalize a large amount of information[108]. To the observer it seems strange that the same evidence can be used to support both sides in the controversy but this is the nature of ^{31}P n.m.r. spectroscopy. So many factors are operative that one is hard put to provide any theory relating δ and J values to the bonding at phosphorus. For example the work of Meriwether and Leto[109] on Ni(0) and Ni(II) complexes with PF$_3$, PCl$_3$, P(OEt)$_3$, PEt$_3$, PPh$_3$ and PBu$^n_3$ ligands etc., led them to list the following factors which could affect the ^{31}P chemical shift: paramagnetic effects, P—Ni σ-bonding, P—Ni π-bonding electronegativity of the groups attached to phosphorus and their σ bonds to phosphorus, aromatic ring currents in PPh$_3$, bond rehybridization and steric effects. This list can obviously be shortened by sticking to one particular ligand and it is then found that a little order, but only a little, can be brought from the chaos[110,111].

The arguments over dπ bonding and the *trans* influence continue. Since triorganophosphines have been studied the most, and knowing what we do about these ligands, it would seem that π-bonding between them and metals in oxidation states of (II) and higher are likely to be very weak. However one piece of non-spectroscopic information supports π bonding in PR$_3$—Pt complexes and this comes from the behaviour of the thiocyanate when it is a ligand in such complexes. If SCN$^-$ attaches itself to a metal via the nitrogen end it cannot π bond, but if it couples via the sulphur atom it has the opportunity to π bond. In this case everything will depend upon the ligand *trans* to it, since S is only a moderate π-acceptor. If the *trans* ligand were a better π-acceptor then it will compete more successfully, and the advantage for the thiocyanate ligand of having the S attached to the atom will be lost. Consequently it will coordinate via N and indeed it does just that in Pt(PR$_3$)$_2$(NCS)$_2$ complexes. However, if the bond *trans* to the thiocyanate is non-π-bonding then it should coordinate via S and this is found in Pt(NH$_3$)$_2$(SCN)$_2$.[112]

```
        (CH₂)₃——PPh₂
              |
   Me₂N——(Pt)——SCN    (5.22)
              |
             NCS
```

Steric rather than π effects may determine how SCN$^-$ attaches itself since 2P + 2S would be more crowded than 2P + 2N atoms about Pt in Pt(PR$_3$)$_2$(NCS)$_2$. In the complex (5.22) the thiocyanates attach themselves according to the π-bonding predictions and not steric factors[113]. Nevertheless steric factors can be very important.

5.8 Steric factors and P(III) ligands

The effect of the ligand size on the geometry of a complex can be seen when one compares compound (5.16), in which there are PMe$_2$Ph ligands, with its PEt$_2$Ph counterpart. In *mer*-MoOCl$_2$(PEt$_2$Ph)$_3$ the Mo—O bond expands to 180.3 pm, which is 13 pm longer, and the Mo—Cl bond *trans* to it is 242.6 pm, which is 13 pm shorter. These bonds are both dramatically affected by the change to bulkier ligands although the P—Mo and the other Cl—Mo bond lengths suffer only minor changes, which shows the effect is not inductive but steric.

Tolman[114] in a paper devoted to a study of the competition between ligands for sites round a metal atom, studied equilibria of type (36) by ^{31}P n.m.r. spectroscopy. He

$$\text{NiP}_4 + 4\text{P}' \rightleftharpoons \text{NiP}'_4 + 4\text{P} \tag{36}$$

found the stability series for the ligands was P(OCH₂)₃CMe = P(OMe)$_3$ = P(OEt)$_3$ > P(OPh)$_3$ = PMe$_3$ > PEt$_3$ = PBu$^n_3$ = PPh$_3$ > PBu$^t_3$ = P(C$_6$H$_4$-*o*-Me)$_3$ = P(C$_6$F$_5$)$_3$. The last two ligands gave complexes that were highly dissociated in solution. This order is definitely not that expected from π bonding. Obviously another factor is

5.9 Summary

active and Tolman concluded that it was a steric factor. This led him to devise a method of measuring the steric requirements of these ligands. This he defined as the angle of the cone swept out by the van der Waals radii of the groups attached to phosphorus, with the Ni atom as the apex of the cone. He took the Ni—P bond to be a standard 228 pm (5.23). The results are given in Table 5.8, and it can be seen how they serve to explain the order he observed for NiP_4 stabilities, and this led him to state that "the property of the ligands which primarily determines the stability of the Ni(0) complexes is their size rather than their electronic character".

Table 5.8 Tolman's cone angles as a measure of the steric size of phosphorus ligands[114]

Ligand	Cone angle	Ligand	Cone angle
PH_3	87°	PEt_3	132°
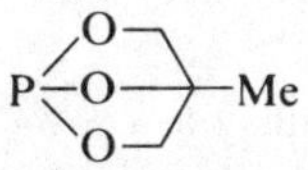 P(OCH₂)₃CMe	101°	$P(CF_3)_3$	137°
PF_3	104°	PPh_3	145°
$P(OMe)_3$	107°	$P(Pr^i)_3$	160°
$P(OEt)_3$	109°	$P(C_6H_{11})_3$	179°
PMe_3	118°	$P(Bu^t)_3$	182°
$P(OPh)_3$	121°	$P(C_6F_5)_3$	184°
PCl_3	125°	$P(C_6H_4\text{-}o\text{-Me})$	192°
PBr_3	131°		

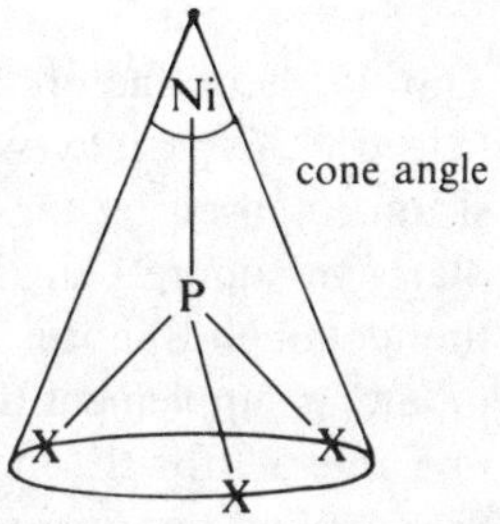

(5.23)

5.9 Summary

There are three factors which can be operative in metal complexes with P(III) ligands attached. Firstly, the σ bond. It is difficult to judge the strength of this; some relate it to the conventional basicity of the ligand[88],* others to Taft $\sigma^\star$ or Kabachnik's σ constants of the attached groups[86, 87], and yet others attempt to calculate it from force constants[97]. Another valid comparison would be to correlate σ bonding with the Lewis basicity these ligands show towards boron Lewis acids such as BH_3, or BF_3. Here steric effects and π bonding should be at a minimum and the order one gets is $P(OR)_3$ (R = alkyl) > $PR_3 \simeq PPh_3$ > $P(NMe_2)_3$ > PH_3 > PF_3 > PCl_3 > $P(OPh)_3$ (see Chapter 11 for more details).

The second factor is π bonding. Earlier theories had the σ and π reinforcing each other in the so-called synergic effect, but this seems unlikely. Again there is no clear measure of the extent to which the ligand can accept electrons from the metal in forming π bonds. Although such bonding has been proved for PF_3 complexes and

---

* PEt_3, $pK_a = 8.69$; PMe_3, 8.65; PPr^n_3, 8.64; PBu^n_3, 8.63; $P(C_5H_{11})_3$, 8.33; PPh_3, 2.73; PH_3, −14 [C. A. Streuli, *Anal. Chem.*, **32**, 985 (1960)].

P(OPh)₃ it is still a bone of contention whether the same is true of PH₃ or PR₃ ligands, and what evidence there is suggests that it is absent in the latter. In any event it will depend upon the metal's having electrons to donate. The effect which π bonding to a P(III) ligand has on other π bonds in the molecule seems the most reliable yardstick for comparing the relative merits of these ligands[83] but the method, whether based on νCO or k values, has its drawbacks.

Table 5.9 Relative σ, π and steric effects in the stability of metal–phosphorus bonds in complexes (R = n-alkyl)

σ bonding: $PBu^t_3 > P(OR)_3 > PR_3 \simeq PPh_3 > PH_3 > PF_3 > P(OPh)_3$

π bonding: $PF_3 > P(OPh)_3 > PH_3 > P(OR)_3 > PPh_3 \simeq PR_3 > PBu^t_3$

Steric effects: $PBu^t_3 > PPh_3 > P(OPh)_3 > PMe_3 > P(OR)_3 > PF_3 > PH_3$

Thirdly, steric factors have a larger effect than perhaps has been previously acknowledged[114], especially with certain very bulky ligands such as PPh₃ and PBu^{t}₃. The strain imposed by these on the metal–phosphorus bond must be quite large particularly in square planar and octahedral complexes. This strain will tend to lengthen the metal–phosphorus bond which in turn will have a disproportionately large effect on any π supplement to the bonding. The relative orders of the three effects are given in Table 5.9 for the more common ligands. Combining these effects in roughly equal proportions one arrives at the following order of all-round complexing power for P(III) ligands: $P(OR)_3 > PF_3 > PH_3 > PR_3 > P(OPh)_3 > PPh_3 \simeq PBu^t_3$.

Problems

1   Using Tolman's formula (equation 32) and χ values (Table 5.6) calculate A_1 νCO for the complexes Ni(CO)₃(PH₂Ph), Ni(CO)₃(PHPh₂), Ni(CO)₃{PMe₂(CF₃)} and Ni(CO)₃{PMe(CF₃)₂}. What would you expect the value of χ(OPr^i) to be? Guess at its numerical value and hence estimate A_1 νCO for Ni(CO)₃{P(OPr^i)₃}.

2   Draw diagrams of the nd orbitals about a metal which show that *trans* ligands are favoured in having more of them in common than *cis* ligands.

3   Why does only orthophenylation of triphenyl phosphite and phosphine ligands occur? Why is there no meta- or paraphenylation?

References

[1] See for example, *Inorganic Chemistry of the Transition Elements*, Specialist Periodical Reports, vols 1 (1972) and 2 (1973), Chemical Soc., London.

[2] R. D. Johnson, *Advan. Inorg. Chem. Radiochem.*, **13**, 487 (1970).

[3] W. Levason and C. A. McAuliffe, *Advan. Inorg. Chem. Radiochem.*, **14**, 173 (1972).

References

[4] B. Chiswell, *Transition Metal Complexes of P, As, Sb and Bi Ligands*, in C. A. McAuliffe (editor), Part 4, p 271, Macmillan, London, 1973.

[5] J. Chatt, *Nature (London)* **165**, 637 (1950); and A. A. Williams, *J. Chem. Soc.*, 3061 (1951).

[6] J. C. Green, D. I. King and J. H. D. Eland, *Chem. Commun.*, 1121 (1970).

[7] H. Rose, *Poggendorff's Annln. Phys.*, **24**, 141, 259 (1832).

[8] S. D. Robinson, *MTP Review of Science, Series One, Inorg. Chem.*, vol. 6, in M. J. Mays (editor), p 121, Butterworth, London, 1972 and *Series Two*, vol. 6, p. 71 (1975).

[9] K. K. Chow, W. Levason and C. A. McAuliffe, *Transition Metal Complexes of P, As, Sb, and Bi Ligands*, in C. A. McAuliffe (editor), Part 2, p 35, Macmillan, London, 1973.

[10] E. O. Fischer and E. Louis, *J. Organomet. Chem.*, **18**, P26 (1969).

[11] E. O. Fischer and W. A. Herrmann, *Chem. Ber.*, **105**, 286 (1972).

[12] E. O. Fischer, E. Louis and C. G. Kreiter, *Angew. Chem. Int. Ed. Eng.*, **8**, 377 (1969); and R. J. J. Schneider, *ibid.*, **7**, 136 (1968).

[13] F. Klanberg and E. L. Muetterties, *J. Amer. Chem. Soc.*, **90**, 3296 (1968).

[14] G. Huttner and S. Schelle, *J. Organomet. Chem.*, **19**, P9 (1969).

[15] H. Moissan, *Bull. Soc. Chim. Fr.*, **5**, 454 (1891); *C. R. Hebd. Seances Acad. Sci.* **102**, 763 (1886).

[16] G. Wilkinson, *Nature (London)* **168**, 514 (1951); *J. Amer. Chem. Soc.*, **73**, 5501 (1951).

[17] J. F. Nixon, *Advan. Inorg. Chem. Radiochem.*, **13**, 364 (1970).

[18] R. Schmutzler, *Advan. Chem. Ser.*, **37**, 150 (1963).

[19] T. Kruck, K. Baur and W. Lang, *Chem. Ber.*, **101**, 138 (1968).

[20] T. Kruck, *Angew. Chem. Int. Ed. Eng.*, **6**, 53 (1967).

[21] C. G. Barlow, J. F. Nixon and M. Webster, *J. Chem. Soc. A*, 2216 (1968).

[22] T. Knuck, *Z. Naturforsch.*, **19B**, 165, 670 (1964).

[23] J. F. Nixon, *J. Chem. Soc. A*, 1136 (1967) and 1089 (1969).

[24] R. J. Clark and R. I. Hoberman, *Inorg. Chem.*, **4**, 1771 (1965).

[25] W. J. Miles Jr., *Diss. Abs. (B)*, **31**, 5242 (1971).

[26] R. B. King and E. Efraty, *J. Amer. Chem. Soc.*, **93**, 5260 (1971); and **94**, 3768 (1972).

[27] A. Almeningen, B. Anderson and E. E. Astrup, *Acta Chem. Scand.*, **24**, 1579 (1970).

[28] Y. Morino, K. Kuchitzu and T. Moritani, *Inorg. Chem.*, **8**, 867 (1969).

[29] T. Kruck and A. Prasch, *Angew. Chem.*, **356**, 118 (1968).

[30] T. Kruck and K. Baur, *Chem. Ber.*, **98**, 3070 (1965); and *Z. Anorg. Allgem Chem.*, **364**, 192 (1969).

[31] T. Kruck, M. Hofler, H. Jung and H. Blume, *Angew. Chem. Int. Ed. Eng.*, **8**, 522 (1969).

[32] R. J. Clark and K. A. Morgan, *Inorg. Chim. Acta*, **2**, 93 (1968).

[33] T. Kruck, M. Hofler, K. Baur, P. Junkes and K. Glinka, *Chem. Ber.*, **101**, 3827 (1968).

[34] K. Kruck, R. Kobelt and A. Prasch, *Z. Naturforsch.*, **27B**, 344 (1972).

[35] J. Muller, K. Fenderl and B. Mertschenk, *Chem. Ber.*, **104**, 700 (1971).

[36] E. O. Fischer and L. Krauss, *Chem. Ber.*, **102**, 223 (1969).

[37] T. B. Brill, *J. Organomet. Chem.*, **40**, 373 (1972).

[38] P. J. Pollick and A. Wojcicki, *J. Organomet. Chem.*, **14**, 469 (1968).

[39] J. J. Levison and S. D. Robinson, *J. Chem. Soc. A*, 639 (1970).

[40] D. Giusto and G. Cova, *Gazz. Chim. Ital.*, **101**, 519 (1971).

[41] L. M. Haines, *Inorg. Nucl. Chem. Lett.*, **5**, 399 (1969).

[42] T. Kruck and M. Hofler, *Angew. Chem. Int. Ed. Eng.*, **6**, 563 (1967).

[43] S. D. Robinson and D. A. Couch, *Chem. Commun.*, 1508 (1971).

[44] D. A. Couch and S. D. Robinson, *Inorg. Chim. Acta*, **9**, 39 (1974).

[45] K. J. Coskran, T. J. Huttemann and J. G. Verkade, *Advan. Chem. Ser.*, **62**, 590 (1967).

[46] S. D. Robinson, *Chem. Commun.*, 521 (1968); and J. J. Levison, *ibid*, 1405 (1968). See also comments in ref. 47 regarding this work.

[47] W. H. Knoth and R. A. Schunn, *J. Amer. Chem. Soc.*, **91**, 2400 (1969); and G. W. Parshall, *ibid*, **91**, 4990 (1969).

[48] E. W. Ainscough, S. D. Robinson and J. J. Levison, *J. Chem. Soc. A*, 3413 (1971).

[49] J. M. Guss and R. Mason, *Chem. Commun.*, 58 (1971).

[50] E. W. Ainscough and S. D. Robinson, *Chem. Commun.*, 130 (1971); N. Ahmad, E. W. Ainscough, T. A. James and S. D. Robinson, *J. Chem. Soc. Dalton*, 1151 (1973).

[51] E. W. Ainscough, T. A. James, S. D. Robinson and J. N. Wingfield, *J. Organomet. Chem.*, **60**, C63 (1973).

[52] C. A. Tolman, *J. Amer. Chem. Soc.*, **92**, (a) 4217; (b) 6777; and (c) 6785 (1970).

[53] J. P. Durand, F. Dawans and Ph. Teyssie, *J. Polymer. Sci. A*, **8**, 979 (1970).

[54] J. P. Collman, *Accounts Chem. Res.*, **1**, 136 (1968).

[55] D. A. Couch and S. D. Robinson, *Inorg. Chem.*, **13**, 456 (1974); and J. N. Wingfield, *J. Chem. Soc.*, 1309 (1974).

[56] G. Booth, *Advan. Inorg. Chem. Radiochem.*, **6**, 1 (1964).

[57] C. A. Tolman, W. C. Seidel and D. H. Gerlach, *J. Amer. Chem. Soc.*, **94**, 2669 (1972).

[58] R. D. Gillard, R. Ugo, F. Caviati, S. Cenini and F. Bonati, *Chem. Commun.*, 869 (1966).

[59] J. R. Moss and B. L. Shaw, *Chem. Commun.*, 632 (1968).

[60] J. Chatt and R. S. Coffey, *Chem. Commun.*, 545 (1966).

[61] P. G. Douglas and B. L. Shaw, *J. Chem. Soc. A*, 334 (1970).

[62] G. J. Leigh, J. J. Levison and S. D. Robinson, *Chem. Commun.*, 705 (1969).

[63] J. Chatt, R. S. Coffey and B. L. Shaw, *J. Chem. Soc.*, 7391 (1965).

[64] G. Schwenzer, M. Y. Darensbourg and D. J. Darensbourg, *Inorg. Chem.*, **11**, 1967 (1972).

[65] T. Ito, S. Kitazume, A. Yamamoto and S. Ikeda, *J. Amer. Chem. Soc.*, **92**, 3011 (1970).

[66] W. Keim, *J. Organomet. Chem.*, **14**, 179 (1968); *ibid*, **19**, 161 (1969).

[67] M. A. Bennett and D. L. Milner, *J. Amer. Chem. Soc.*, **91**, 6983 (1969).

[68] A. J. Cheney, B. E. Mann, B. L. Shaw and R. M. Slade, *Chem. Commun.*, 1176 (1970).

[69] J. A. Osborn, F. H. Jardine, J. F. Young and G. Wilkinson, *J. Chem. Soc. A*, 1171 (1966).

[70] J. T. Mangue and G. Wilkinson, *J. Chem. Soc. A*, 1736 (1966).

[71] F. H. Jardine, J. A. Osborn and G. Wilkinson, *J. Chem. Soc. A*, 1574 (1967).

[72] S. Montelatici, A. van der Ent, J. A. Osborn and G. Wilkinson, *J. Chem. Soc. A*, 1054 (1968).

[73] A. J. Chalk and R. A. Jerussi, *Tetrahedron Lett.*, 61 (1972).

[74] H. A. Tayim and J. C. Bailar, *J. Amer. Chem. Soc.*, **89**, 3420 (1967).

[75] G. Yagupsky, C. K. Brown and G. Wilkinson, *J. Chem. Soc. A*, 1392 (1970).

[76] W. Strohmeier and R. Fleischmann, *Z. Naturforsch.*, **24B**, 1217 (1969) and refs. therein.

[77] E. A. Zuech, *Chem. Commun.*, 1182 (1968).

[78] B. T. Kilboum and H. M. Powell, *J. Chem. Soc. A*, 1688 (1970).

[79] J. K. Stalick and J. A. Ibers, *Inorg. Chem.*, **8**, 419 (1969).

[80] B. E. Cavit, K. R. Grundy and W. R. Roper, *Chem. Commun.*, 60 (1972).

[81] C. G. Pierpout, D. G. Van Derveer, W. Durland and R. Eisenberg, *J. Amer. Chem. Soc.*, **92**, 4760 (1970).

[82] E. W. Abel, M. A. Bennett and G. Wilkinson, *J. Chem. Soc.*, 2325 (1959).

[83] W. D. Horrocks Jr. and R. C. Taylor, *Inorg. Chem.*, **2**, 723 (1963).

References

[84] F. A. Cotton and C. S. Kraihanzel, *J. Amer. Chem. Soc.*, **84**, 4432 (1962).

[85] F. A. Cotton, *Inorg. Chem.*, **3**, 702 (1964).

[86] M. Bigorgne, *J. Inorg. Nucl. Chem.*, **26**, 107 (1964).

[87] C. A. Tolman, *J. Amer. Chem. Soc.*, **92**, 2953 (1970).

[88] R. J. Angelici, *J. Inorg. Nucl. Chem.*, **28**, 2627 (1966).

[89] R. J. Angelici and C. M. Ingemanson, *Inorg. Chem.*, **8**, 83 (1969).

[90] A. Pidcock, *Transition Metal Complexes of P, As, Sb and Bi Ligands*, in C. A. McAuliffe (editor), Part 1, Macmillan, London, 1973.

[91] G. N. LaMar, E. O. Sherman and G. A. Fuchs, *J. Coord. Chem.*, **1**, 289 (1971).

[92] O. Stelzer and R. Schmutzler, *J. Chem. Soc. A*, 2867 (1971).

[93] J. C. Green, D. I. King and J. H. D. Eland, *Chem. Commun.*, 1121 (1970).

[94] H. J. Plastas, J. M. Stewart and S. O. Grim, *J. Amer. Chem. Soc.*, **91**, 4326 (1969).

[95] F. A. Cotton and G. Wilkinson, *Advanced Inorganic Chemistry*, Third edition, p. 720, Interscience, New York, 1972.

[96] H. S. Preston, J. M. Stewart, H. J. Plastas and S. O. Grim, *Inorg. Chem.*, **11**, 161 (1972).

[97] W. A. Graham, *Inorg. Chem.*, **7**, 315 (1968).

[98] D. J. Darensbourg and T. L. Brown, *Inorg. Chem.*, **7**, 959 (1968).

[99] R. J. Angelici and M. D. Malone, *Inorg. Chem.*, **6**, 1731 (1967).

[100] J. Chatt and R. G. Wilkins, *J. Chem. Soc.*, 4300 (1952); 525 (1956).

[101] G. G. Messmer and E. L. Amma, *Inorg. Chem.*, **5**, 1775 (1966).

[102] G. G. Messner, E. L. Amma and J. A. Ibers, *Inorg. Chem.*, **6**, 725 (1967).

[103] R. Eisenberg and J. A. Ibers, *Inorg. Chem.*, **4**, 773 (1966).

[104] J. Chatt, Lj Manojovlić-Muir and K. W. Muir, *Chem. Commun.*, 655 (1971).

[105] L. Aslanov, R. Mason, A. G. Wheeler and P. O. Whimp, *Chem Commun.*, 30 (1970).

[106] J. F. Nixon and A. Pidcock, *Ann. Rev. NMR Spectrosc.*, (editor E. F. Mooney) **2**, 345 (1969).

[107] A. Pidcock, R. E. Richards and L. M. Venanzi, *J. Chem. Soc.*, 1707 (1966).

[108] S. O. Grim *et al.*, *J. Amer. Chem. Soc.*, **89**, 5573 (1967); *Inorg. Chem.*, **6**, 1133 (1967); *ibid.*, **7**, 161 (1968); *ibid.*, **8**, 1716 (1969); *Inorg. Chim. Acta*, **4**, 277 (1970).

[109] L. S. Meriwether and J. R. Leto, *J. Amer. Chem. Soc.*, **83**, 3192 (1961).

[110] S. I. Shupack and B. Wagner, *Chem. Commun.*, 547 (1966).

[111] K. J. Coskran, R. D. Bertrànd and J. G. Verkade, *J. Amer. Chem. Soc.*, **89**, 4535 (1967).

[112] A. Turco and C. Pecile, *Nature* (*London*), **191**, 66 (1961).

[113] D. W. Meek, P. E. Nicpon and V. I. Meek, *J. Amer. Chem. Soc.*, **92**, 5351 (1970).

[114] C. A. Tolman, *J. Amer. Chem. Soc.*, **92**, 2956 (1970).

Chapter 6
Pentacoordinate and hexacoordinate organophosphorus chemistry

6.1 Introduction

During the last three years the amount of published information on pentacovalent phosphorus compounds (phosphoranes) has expanded enormously[1-7]. The dramatic increase in research activity in this area undoubtedly stems from the realization that a thorough knowledge of the factors which (i) influence the stability of phosphoranes and (ii) control ligand reorganization (e.g. pseudorotation) within such molecules, is vital to the understanding of displacement reactions at tetracoordinate phosphorus (see Chapters 7 and 8).

Numerous phosphoranes have now been isolated in a pure form and electron diffraction or x-ray studies have revealed that most of them possess or approximate to, trigonal bipyramidal geometry. Details of the x-ray data for the phosphorane (6.1) derived from tri-isopropyl phosphite and phenanthraquinone appear in Chapter 2.[8] Other examples for which x-ray data are available, include the cyclic phosphoranes (6.2)[9] and (6.3)[10] and the simplest phosphorane containing five phosphorus–carbon bonds, pentaphenyl phosphorane[11] all of which are close to the trigonal bipyramid configuration.

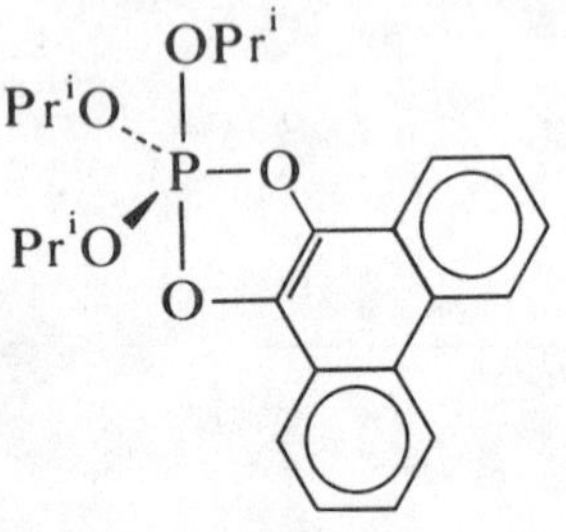

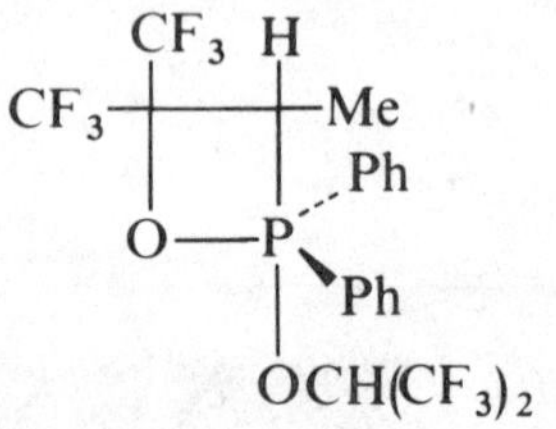

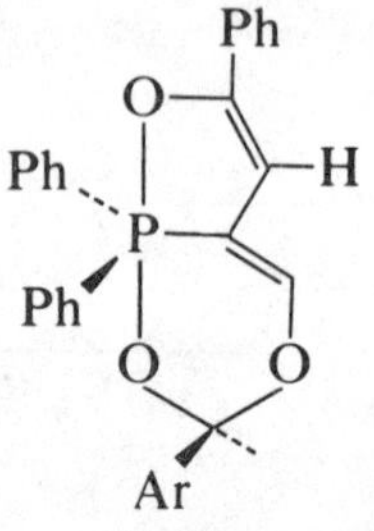

Recently however, x-ray analyses have revealed that some bicyclic phosphoranes such as (6.4) and (6.5) are much closer to the square pyramid configuration[12,13].

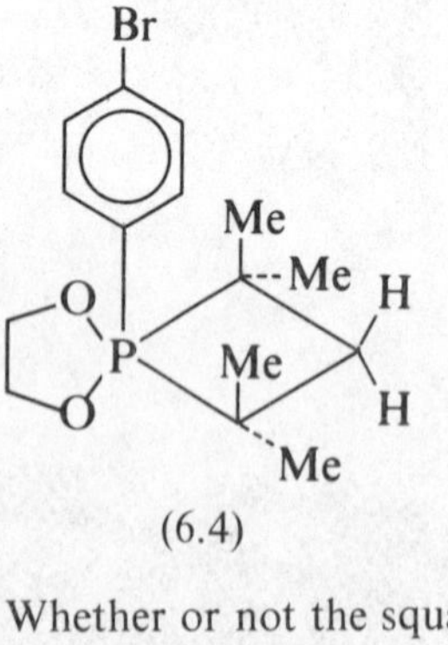

Whether or not the square pyramid is retained as the stable structure in solution is a matter for conjecture but a related compound, pentaphenylantimony, SbPh$_5$, which exists as a distorted square pyramid in the solid[14,15], shows evidence for retention of this structure in solution[16,17].

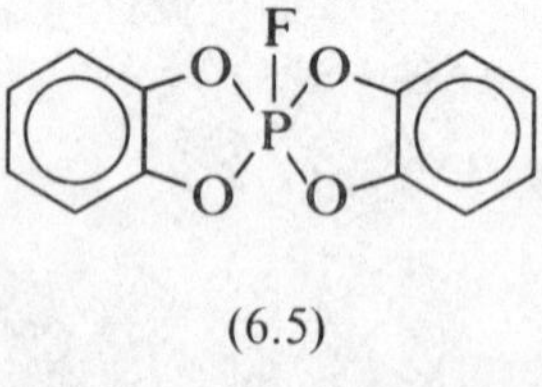

Trigonal bipyramid structures are indicated for PF$_5$ and MePF$_4$ by electron diffraction data[18], for (CF$_3$)$_2$PF$_3$ by infrared and Raman spectra[1,19] and for PCl$_2$F$_3$ by ^{35}Cl nuclear quadruple resonance spectra[1,19]. One of the latest techniques for structural

6.1 Introduction

studies is that known as x-ray photoelectron spectroscopy (ESCA) which has been used to demonstrate the existence of two groups of non-equivalent fluorine atoms in a ratio of $3:2$ in PF_5.[20]

All the data serve to reinforce the currently held views about the structures of phosphoranes in solution and these views may be summarized as follows:
(i) the most stable pentacovalent phosphorus structure is usually that of a trigonal bipyramid (for exceptions, see later);
(ii) apical bonds are longer (and weaker) than equatorial bonds;
(iii) electronegative substituents (e.g. F and —OR) prefer the apical position;
(iv) small (e.g. four- or five-membered) rings prefer an apical–equatorial disposition of the ring over a diequatorial situation.

As we have seen from Chapters 2, 3 and 4, these rules are not inviolate – it is simply a question of one structural arrangement being preferred in energetic terms, over another. Four- and five-membered rings may adopt a diequatorial disposition (B) provided that the energy barrier ($\Delta G^{\ddagger}$) between the alternative structures (A and B) is overcome.

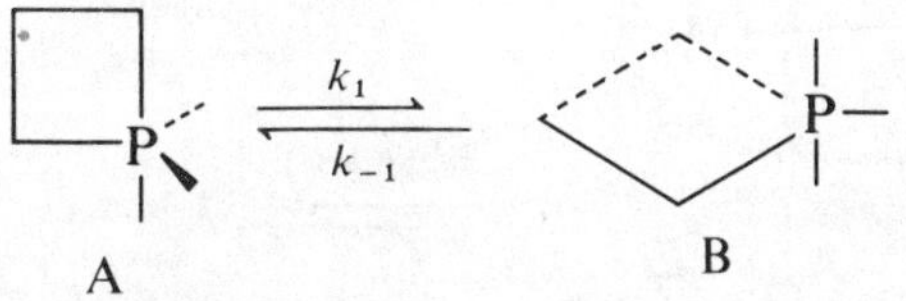

Usually, $K_1 = k_1/k_{-1} < 1$, $\Delta G° > 0$; $\Delta G^{\ddagger} > \Delta G°$

In fact there are cases in which a diequatorial disposition of the ring appears to provide the most stable structure. Thus, at low temperature, the ^{19}F n.m.r. spectrum of (6.6) indicates that the five-membered (phospholane) ring adopts the diequatorial situation and it seems likely that the extra energy required to achieve this conformation is compensated by having two electronegative fluorine atoms in apical positions[21].

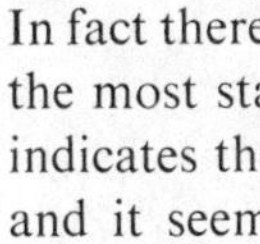

(6.6) 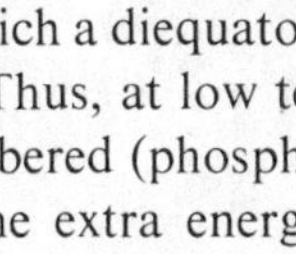   (6.7)

Likewise, as discussed in Chapter 2, the rule governing electronegative substituents is modified by the π-donor ability of such substituents. In fact, π-donors (e.g. —NR$_2$ or Ph) prefer the equatorial positions where apparently the lone pair or π-orbital of a substituent may overlap more efficiently with the d orbitals on phosphorus. Conversely π-acceptors prefer apical sites. Steric factors must also play a part in determining the preferred position of a substituent in the *tbp*. The effect of steric hindrance is very difficult to quantify but in general large groups will prefer the equatorial situations. This is simply because an equatorial substituent (R^3 in 6.7) only interacts to any significant extent with the two apical substituents (R^1 and R^2) whereas an apical substituent (R^1) suffers steric interference from *three* equatorial substituents

(R^3, R^4 and R^5). It follows that small groups will prefer the apical position and this may be the origin of the apparent high apicophilicity of hydrogen (see Chapter 2).

The question of apicophilicity is an important one since it has considerable bearing on the feasibility of ligand reorganization in phosphoranes. This in turn, may determine the stereochemical course of substitutions at tetracoordinate phosphorus when pentacovalent intermediates of finite lifetime are involved in the reactions. The details of such stereochemical studies appear in Chapters 7 and 8 but it is pertinent at this point to inquire more closely into the methods of determining apicophilicity since they are based on the now familiar concepts of pseudorotation (Chapter 2) and dynamic n.m.r. (Chapter 3). Several systems have been employed in this respect but the principles for each are similar and the one to be described is due to Trippett and his co-workers[22].

Phosphoranes of the type (6.8) were prepared from condensation of the appropriate phosphetan with hexafluoroacetone. At low temperature (room temperature or below for X = Ph, *cis* form) the ^{19}F n.m.r. spectrum showed two signals of equal intensity at +2.9 and +5.7 ppm (relative to $PhCF_3$ as internal standard) both of which showed

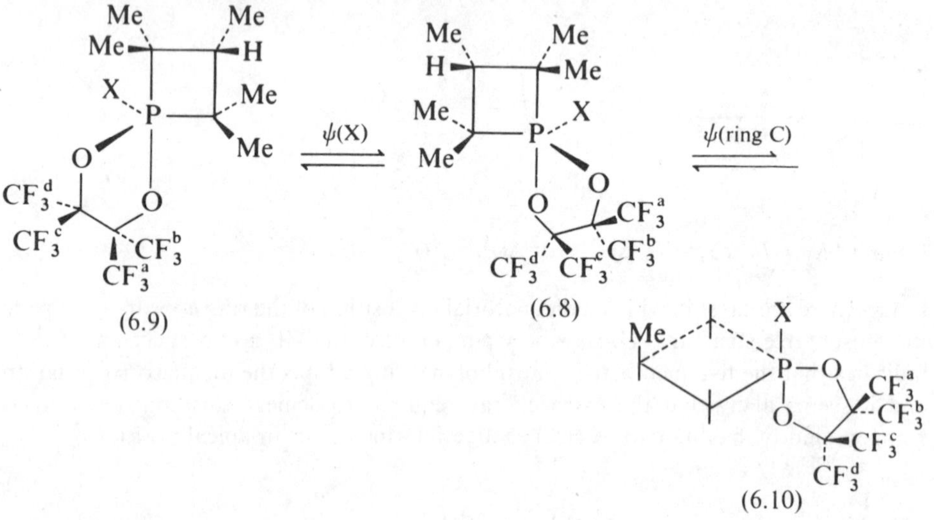

$$a \equiv c \atop b \equiv d \Big\} \text{ by } \psi(X) \qquad\qquad a \equiv b \atop c \equiv d \Big\} \text{ by } \psi(\text{ring C})$$

fine structure, presumably due to coupling with phosphorus and fluorine. Such behaviour is explained by a rapid pseudorotation with X as pivot, between (6.8) and (6.9) which maintains the four- and five-membered rings in apical–equatorial situations but serves to make CF_3^a equivalent to CF_3^c and CF_3^b equivalent to CF_3^d. The only way in which CF_3^a ($\equiv CF_3^c$) can become equivalent to CF_3^b ($\equiv CF_3^d$) is for the four-membered phosphetan ring to become diequatorial which of necessity, forces X into an apical position (6.10). At room temperature the energy barrier is sufficiently high to prevent this ψ but on raising the temperature, the two fluorine signals gradually broaden until coalescence is reached at 140 °C and beyond this temperature a singlet is observed, indicating the equivalence of all twelve fluorine atoms.

The energy barrier calculated for the interconversion of (6.8/6.9) ⇌ (6.10) for X = Ph, is 82 kJ mol^{-1} (19.6 kcal $mole^{-1}$) and energy barriers may be determined by a study

6.1 Introduction

of the variable temperature ^{19}F spectra of (6.8) for a wide variety of groups X. Thus, the relative apicophilicities of the groups, X, may be estimated*; a low $\Delta G^{\ddagger}$ indicates high apicophilicity for X and conversely a high $\Delta G^{\ddagger}$ implies low apicophilicity for X. Trippett's method established the following order of decreasing apicophilicity for X:

X   =   H > OCH(CF$_3$)$_2$ > OPh > NMe$_2$ > Me > i-Pr > CH=CMe$_2$ >
$\Delta G^{\ddagger}$ (kcal mol^{-1})   ~9   16.2   16.9   17.8   19.1

Ph(*cis*) > Ph(*trans*)
19.6   >22

It should be emphasized that the above interpretation assumes a trigonal bipyramid for the ground state geometry of these molecules. Recent work[12,13] reveals the dangers inherent in such an assumption especially for bicyclic phosphoranes such as (6.9) for which the square pyramid may be the most stable structure even in solution. If this is the case then the n.m.r. data may be providing an estimate of the energy barrier facing the interconversion of two square pyramids (6.11) and (6.13) via a trigonal bipyramid (6.12). In other words, in this case, the experimental energy barrier represents the difference in energy between (6.11) or (6.13) and (6.12) which is not a measure of the apicophilicity of X in a *tbp*.

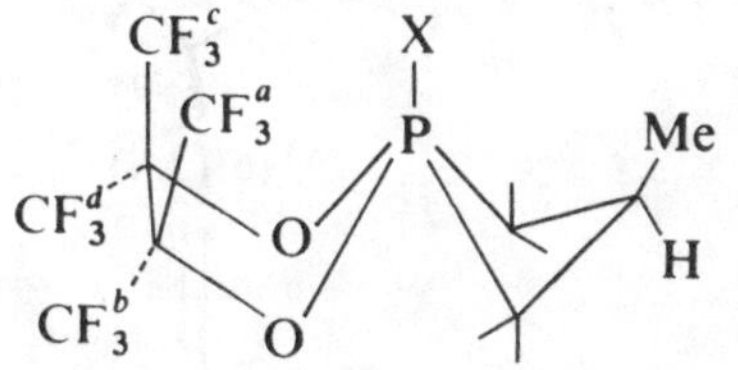

(6.11)   (6.13)

a≡c ⎫ plane of symmetry   a≡c ⎫ plane of symmetry
b≡d ⎭   b≡d ⎭

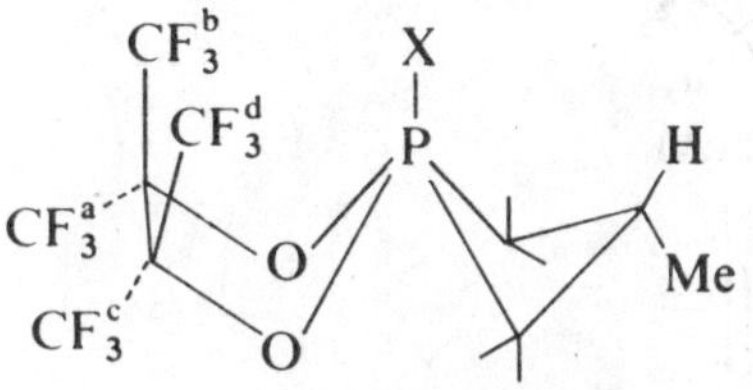

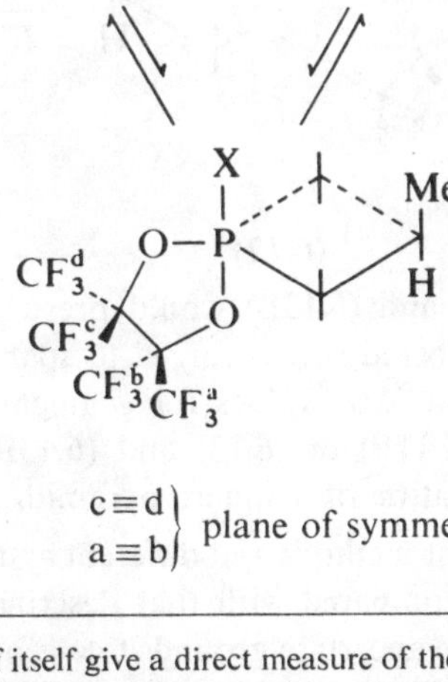

(6.12)

c≡d ⎫ plane of symmetry
a≡b ⎭

---

* This procedure does not of itself give a direct measure of the *difference* in free energy (i.e. difference in thermodynamic stability) between structures of type (6.8) and (6.10) because the calculation gives the *free energy of activation* for the pseudorotation process. However, the free energy of activation for the interconversion of structures of equivalent energy is likely to be small ($\leqslant$ 12 kJ mol^{-1}) and if one accepts the reasonable assumption that the $\Delta G^{\ddagger}$ from high to low energy species will not be greater than 12 kJ mol^{-1}, the differences in $\Delta G^{\ddagger}$ between various X represent a semi-quantitative estimate of their relative apicophilicities.

It should be noted however, that the *direct* conversion of (6.12) to (6.11) or (6.13) involves the postulate of a new type of pseudorotation process in which an *apical* group (X) in a *tbp* is the pivot and three equatorial bonds plus one apical bond in (6.12) deform simultaneously to give the two square pyramids (6.11) and (6.13). In order to avoid this situation it is necessary to invoke two other *tbp* intermediates (6.11A) and (6.13A) which would be formed via (6.11B) and (6.13B) by pseudorotation of (6.12) using either ring carbon (4) or ring carbon (2) as pivot respectively. The square

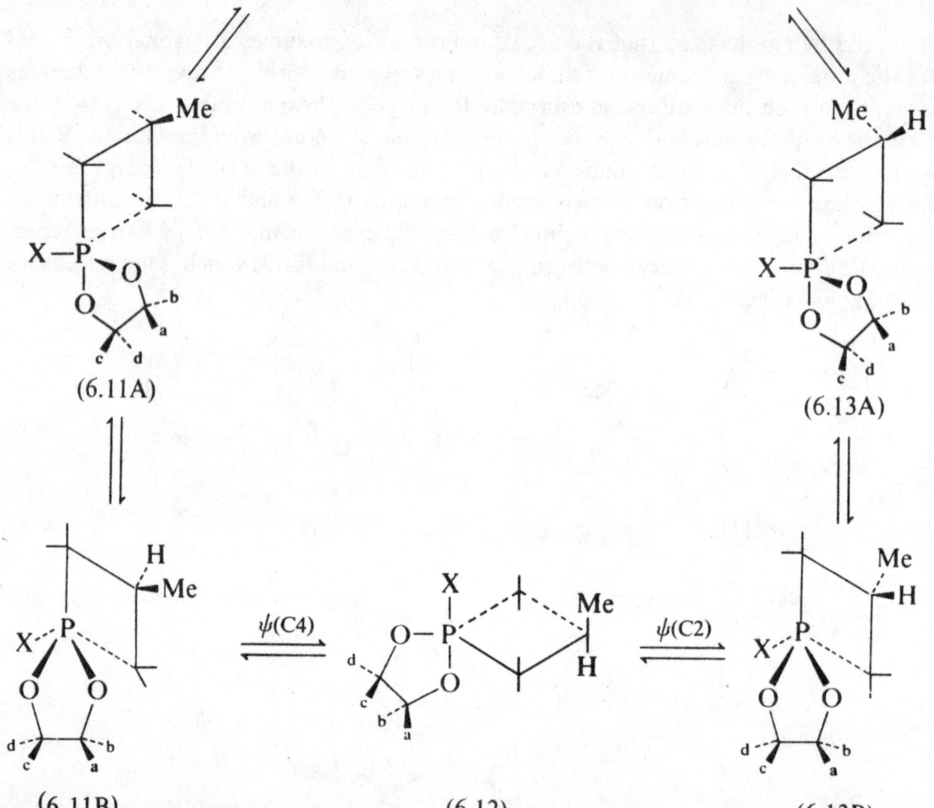

pyramidal intermediates (6.11B) and (6.13B) would presumably be of high energy since in both cases the four-membered ring is forced to span axial–basal positions of the square pyramid. In this event, $\Delta G^{\ddagger}$ values for ψ might really represent energy differences between (6.11) and (6.11B) or (6.13) and (6.13B), i.e. energy difference between X in a basal or axial position of a square pyramid.

As discussed in Chapter 2, different methods and different systems give rise to different orders of relative apicophilicity compared with that described above. In view of all the complications, in particular the uncertain ground state geometries and unquantified steric factors, a universal order of apicophilicities is probably unrealistic but a more exact solution will doubtless arrive as further effort is devoted to this area. In the interim we must use the existing empirical data with a reasonable degree of caution.

The remainder of this chapter is concerned with the synthesis of phosphoranes and aspects of their chemistry which are peculiar to the pentacovalent state.

6.2 Methods of preparing phosphoranes

Many of the preparative procedures outlined in this section, particularly those starting from P(III) compounds, have been discussed in Chapter 4. Consequently, the coverage will often be brief with the emphasis on synthetic utility and with the intention of providing a coherent picture of routes to phosphoranes.

6.2.1 From trivalent phosphorus compounds

6.2.1.1 By reaction with halogen sources.

One of the most readily available and thoroughly studied phosphoranes is pentafluorophosphorane, PF_5, which may be prepared from phosphorus trifluoride using a variety of reagents[23] including fluorine (eqn. 1) or molybdenum hexafluoride (eqn. 2).

$$PF_3 + F_2 \longrightarrow PF_5 \tag{1}$$

$$PF_3 + 2MoF_6 \longrightarrow PF_5 + 2MoF_5 \tag{2}$$

Both of these reactions and in particular the second one, for which high yields have been recorded, may be extended to alkyl- and arylphosphines to produce difluorophosphoranes (eqn. 3)[24].

$$R_3P + MoF_6 \xrightarrow[-60°C]{CH_2Cl_2} [complex] \xrightarrow{170°C} R_3PF_2 \quad R = \text{alkyl or aryl} \tag{3}$$

More unusual sources of halogen include difluorodiazirine (6.14, eqn. 4)[25], tetrafluorohydrazine (6.15, eqn. 5)[26], trifluoromethyl peroxide (6.16, eqn. 6), trifluoromethyl disulphide (6.17, eqn. 7) and trifluoromethyl hypofluorite (6.18, eqns. 8 and 9)[27,28].

$$2\,Z_3P + \begin{array}{c} F \\ F \end{array}\!\!>\!\!C\!\!<\!\!\begin{array}{c} N \\ \| \\ N \end{array} \longrightarrow Z_3PF_2 + Z_3P\!=\!N\!-\!CN \tag{4}$$

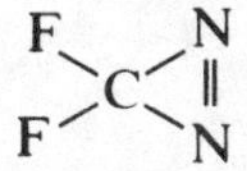

$Z = $ alkyl, aryl, $-OR$, $-SR$, $-NR_2$

$$Ph_3P + F_2N-NF_2 \longrightarrow Ph_3PF_2 \tag{5}$$ 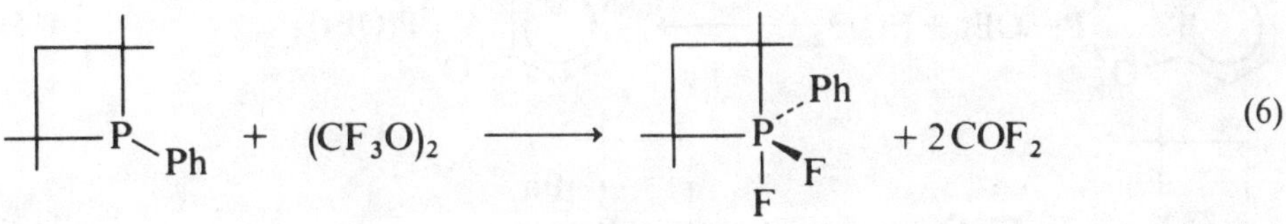
(6.15)

$$\text{[phospholane]}-P\text{-Ph} + (CF_3O)_2 \longrightarrow \text{[phospholane]}-P(Ph)(F)F + 2COF_2 \tag{6}$$
(6.16)

$$R_3P + (CF_3S)_2 \longrightarrow R_3PF_2 + 2CSF_2 \quad R = \text{n-Bu or Ph} \tag{7}$$
(6.17)

$$Z_3P + CF_3OF \longrightarrow Z_3PF_2 + COF_2 \tag{8}$$
(6.18)

$$\text{[phospholane]}-P\text{-Ph} + CF_3OF \longrightarrow \text{[phospholane]}-P(Ph)(F)F + COF_2 \tag{9}$$

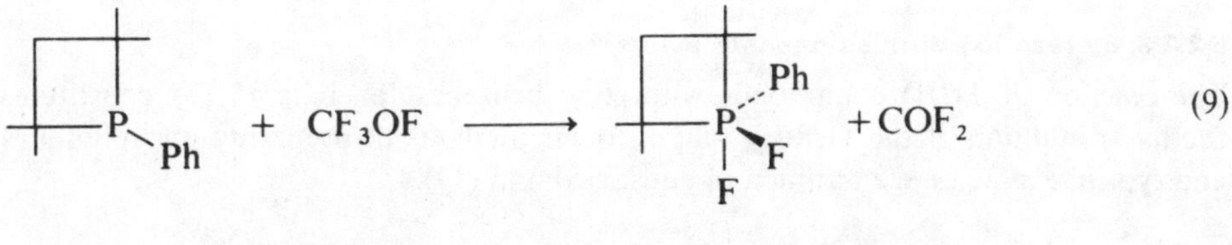

The reactions represented by eqns. (4), (8) and (9) are especially useful since they occur under mild conditions (low temperature and an organic solvent), and may be used with a wide variety of P(III) compounds.

Fluorophosphoranes constitute the most stable pentacovalent molecules containing a phosphorus–halogen bond but phosphoranes containing more than one type of halogen are also available from P(III) compounds (e.g. eqn. 10)[23,29].

$$R_2NPF_2 \xrightarrow{X_2, \text{ org. solvent, low temp.}} R_2NPF_2X_2 \qquad (10)$$

X = Cl or Br; R = Me, Et or $R_2 = -CH_2(CH_2)_3CH_2-$

Unfortunately, above 0 °C, such compounds disproportionate by an intermolecular exchange of halogen which results in a complex mixture of products (eqn. 11).

$$4Et_2NPF_2Br_2 \longrightarrow Et_2NPBrF_3 + Et_2NPF_4 + Et_2N\overset{+}{P}Br_3\ \overset{-}{F} + Et_2N\overset{+}{P}Br_3\ \overset{-}{Br} \qquad (11)$$

Similar disproportionation reactions occur with alkyldifluorodihalogenophosphoranes (RPF_2X_2, where X = Cl or Br) even below 0 °C; this makes the preparation of a pure sample of any one of such compounds extremely difficult.

6.2.1.2 By reaction with peroxides[30–33]

Typical examples of this preparative method are described by eqns. (12)–(17). Mechanistic aspects of the reaction are discussed in Chapter 4.

$$Ph_nP(OEt)_{3-n} + Et_2O_2 \longrightarrow Ph_nP(OEt)_{5-n} \quad n = 0\text{–}3 \qquad (12)$$

$$Ph_nPMe_{3-n} + Et_2O_2 \longrightarrow Ph_nPMe_{3-n}(OEt)_2 \quad n = 0\text{–}3 \qquad (13)$$

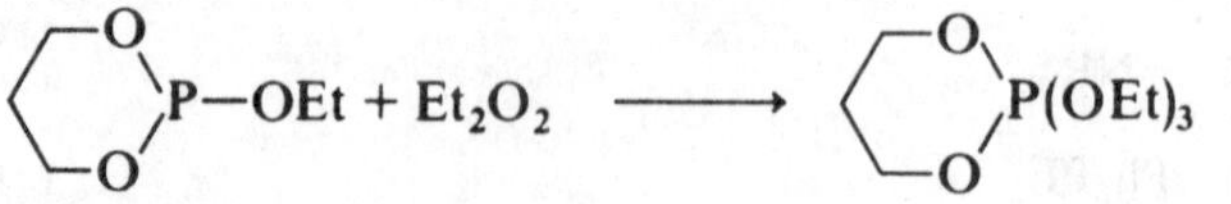

(14)

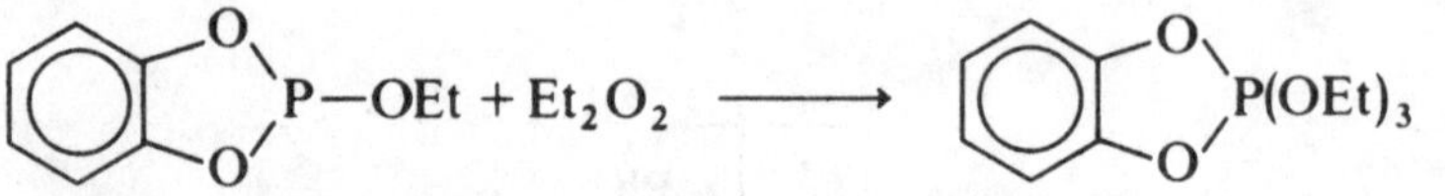

(15)

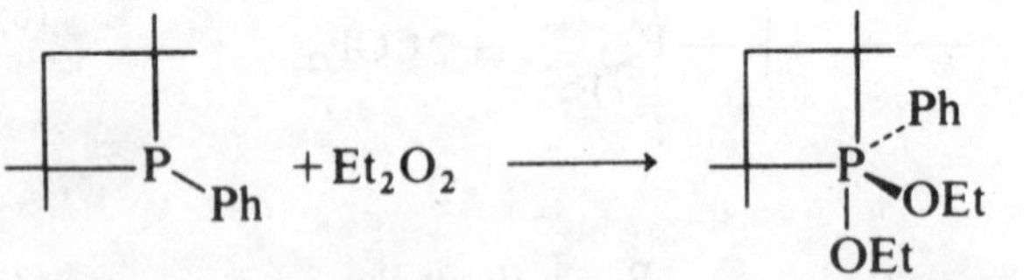

(16)

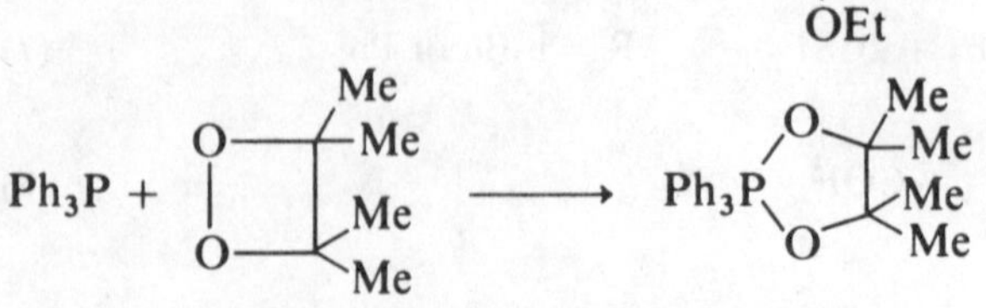

(17)

6.2.1.3 By reaction with sulphenate esters[34]

The reaction of P(III) compounds with ethyl benzenesulphenate (6.19) constitutes the latest addition of the Denney school to the methods of preparing phosphoranes and typical examples are outlined in eqns. (18) and (19).

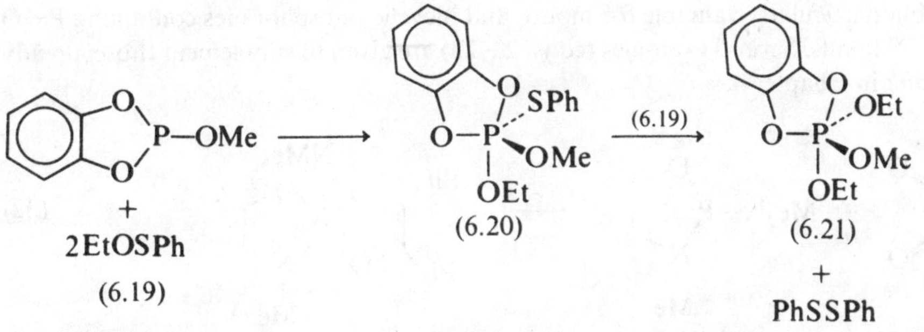

6.2 Methods of preparing phosphoranes

[Structures showing catechol-P-OMe + 2EtOSPh → (6.20) → (6.21) + PhSSPh] (18)

$$Ph_nP(OEt)_{3-n} + 2EtOSPh \longrightarrow Ph_nP(OEt)_{5-n} + PhSSPh \quad (19)$$

Unfortunately, the intermediate thiophosphorane (e.g. 6.20) is unstable and reacts with another mole of (6.19) to produce the diethoxyphosphorane (e.g. 6.21) and diphenyl disulphide probably via an ionic mechanism. This, however, carries its own bonus since the method probably represents the best route to diethoxyphosphoranes such as $Ph_3P(OEt)_2$ which is obtained in 90% yield.

6.2.1.4 By reaction with dithietens[35]

This route is particularly valuable for the preparation of phosphoranes containing P—S bonds since these molecules are generally somewhat unstable. The reaction occurs rapidly in methylene chloride at -78 °C and is especially useful for the synthesis of bicyclic(spiro)phosphoranes as exemplified by eqns. (20) and (21).

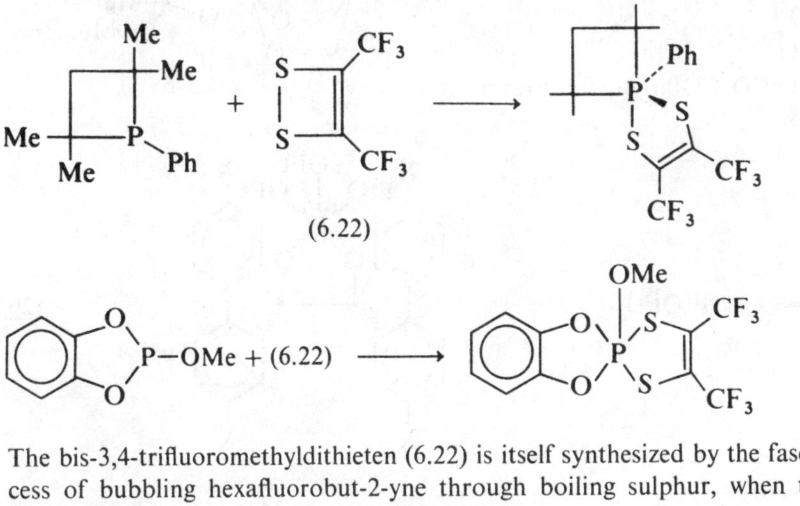

[Reaction scheme showing tetramethylphospholane + dithieten (6.22) → spirophosphorane] (20)

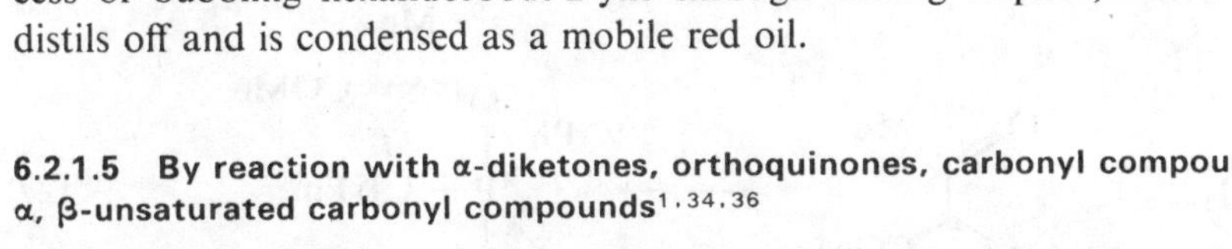

[Reaction scheme showing catechol-P-OMe + (6.22) → spirophosphorane product] (21)

The bis-3,4-trifluoromethyldithieten (6.22) is itself synthesized by the fascinating process of bubbling hexafluorobut-2-yne through boiling sulphur, when the dithieten distils off and is condensed as a mobile red oil.

6.2.1.5 By reaction with α-diketones, orthoquinones, carbonyl compounds or α, β-unsaturated carbonyl compounds[1,34,36]

The development of this preparative method is largely due to Ramirez *et al.* with subsequent modifications by a wide variety of workers. It is of general synthetic utility but particularly valuable for mono- and bicyclic phosphoranes containing P—O and P—N bonds. Several examples (eqns. 22–28) are given to supplement those already appearing in Chapter 4.

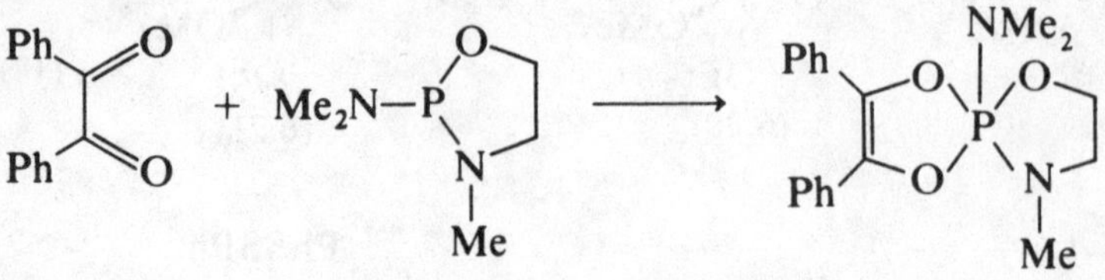

$$\text{Ph-CO-CO-Ph} + \text{Me}_2\text{N-P(OCH}_2\text{CH}_2\text{NMe)} \longrightarrow \text{bicyclic phosphorane} \quad (22)$$

$$\text{(OCH}_2\text{CH}_2\text{O)PR}^1 + 2\text{MeCO.CO}_2\text{R}^2 \longrightarrow \text{phosphorane} \quad (23)$$

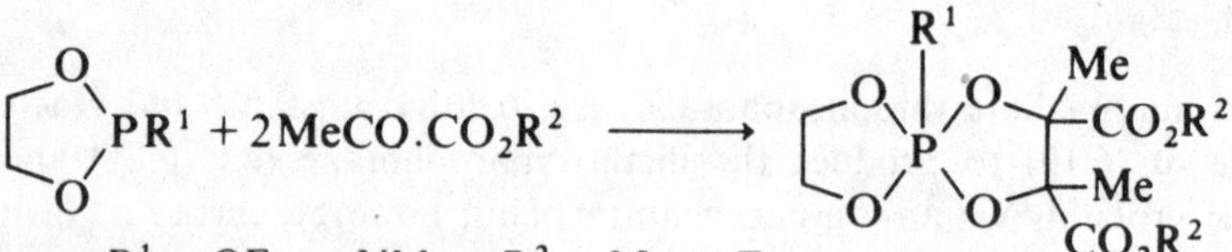

$R^1 = \text{OEt or NMe}_2$; $R^2 = \text{Me or Et}$

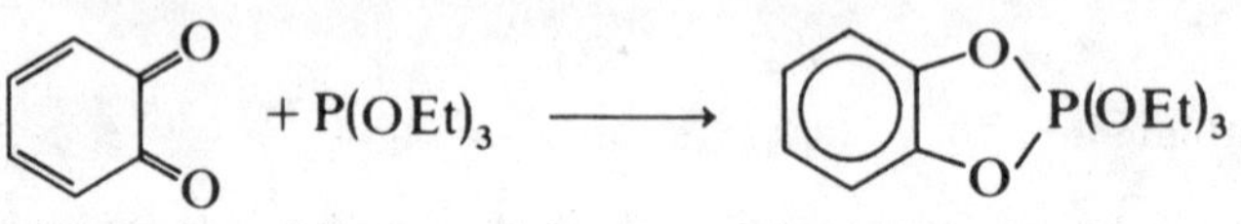

$$\text{o-benzoquinone} + \text{P(OEt)}_3 \longrightarrow \text{catechol phosphorane} \quad (24)$$

$$\text{bicyclic phosphite} + \text{MeCO.COMe} \xrightarrow{60\,°C/8\,\text{days}} \text{phosphorane} \quad (25)$$

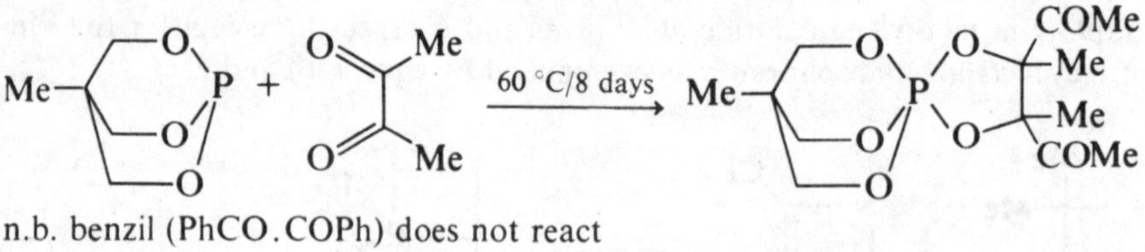

n.b. benzil (PhCO.COPh) does not react

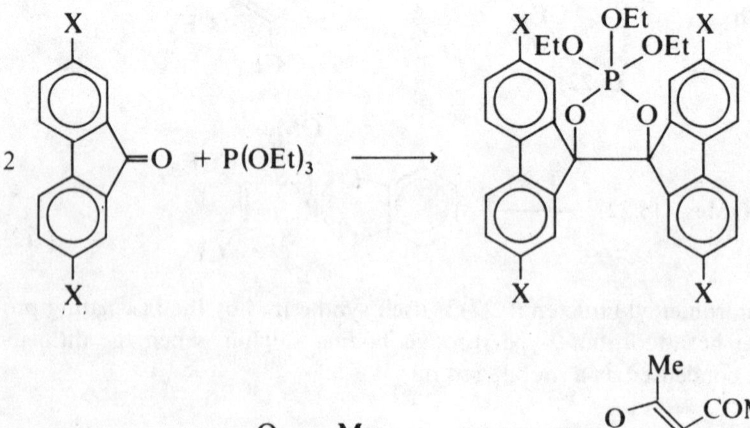

$$2\,\text{(X-substituted fluorenone)} + \text{P(OEt)}_3 \longrightarrow \text{spirophosphorane} \quad (26)$$

$$\text{Ph}_n\text{P(OMe)}_{3-n} + \text{PhCH=C(COMe)(Me)} \longrightarrow \text{phosphorane} \quad (27)$$

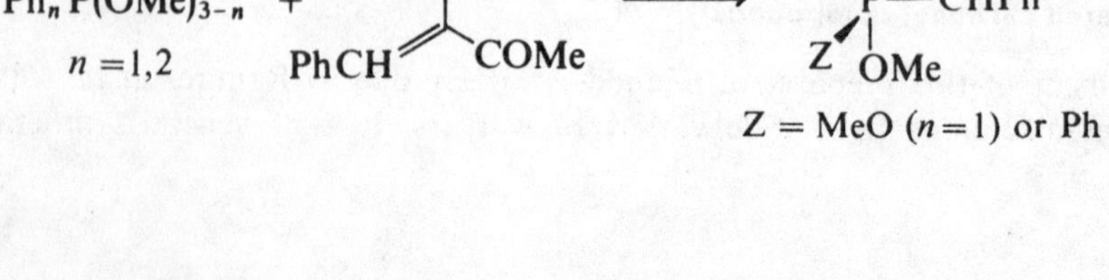

$n = 1, 2$

$Z = \text{MeO}\ (n=1)$ or $\text{Ph}\ (n=2)$

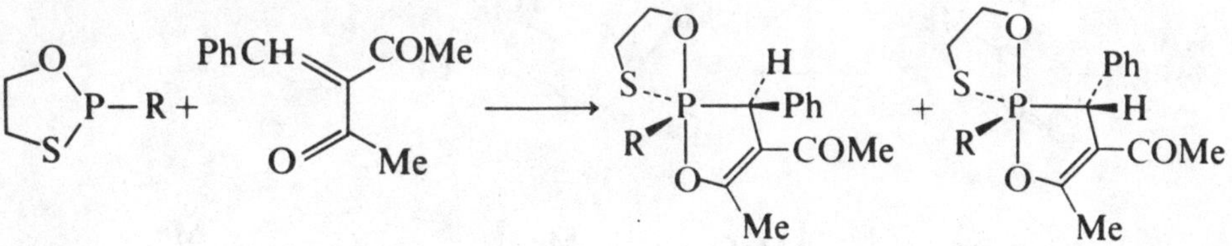

6.2 Methods of preparing phosphoranes

(28)

6.2.1.6 By reaction with dienes.[5,6,37]

The condensation of dienes with trivalent phosphorus compounds is particularly useful for the preparation of bicyclic phosphoranes (eqns. 29 and 30). A wide variety of cyclic phosphites and 1,3-dienes may be used and, as noted in Chapter 4, the con-

(29)

X = F, OMe, NMe$_2$, Me, Ph

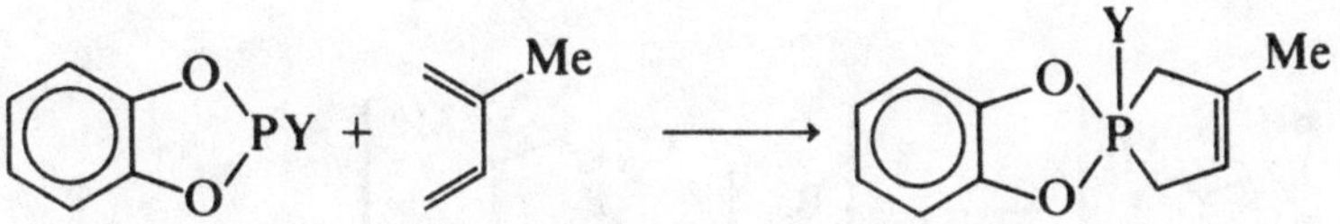

(30)

Y = Br, Cl, F, OMe, SR, NMe$_2$, NCS, OPh, Me, Ph

densation appears to be a concerted 6-electron, disrotatory process and is therefore stereospecific. Thus, *trans,trans*-hexa-2,4-diene (6.23) would give a spirophosphorane (6.24) with the methyl groups in the phospholene ring *cis* to each other.

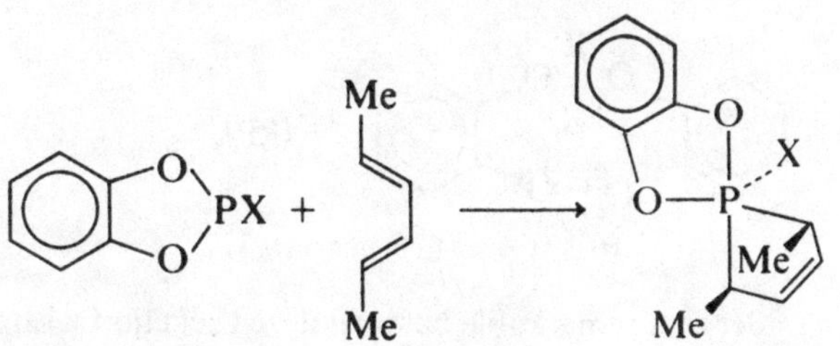

(6.23)   (6.24)

Contrary to first impressions however, the method is of somewhat limited scope. Some acyclic P(III) compounds, for example R$_3$P, R$_2$POR, RP(OR)$_2$ and P(OR)$_3$, are very unreactive towards dienes and although halophosphines (e.g. MePCl$_2$) or halophosphites (e.g. MeOPCl$_2$) will react, the products are not phosphoranes but in one case a phosphonium salt (6.25) and in the other a phosphinyl chloride (6.27) arising from rapid de-alkylation of an intermediate alkoxy phosphonium salt (6.26).

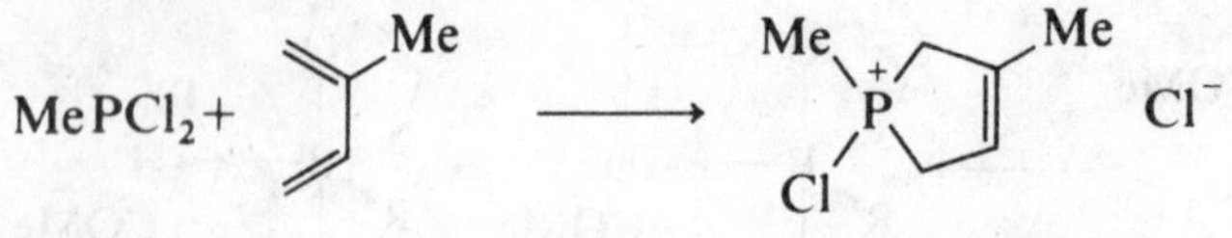

(6.25)

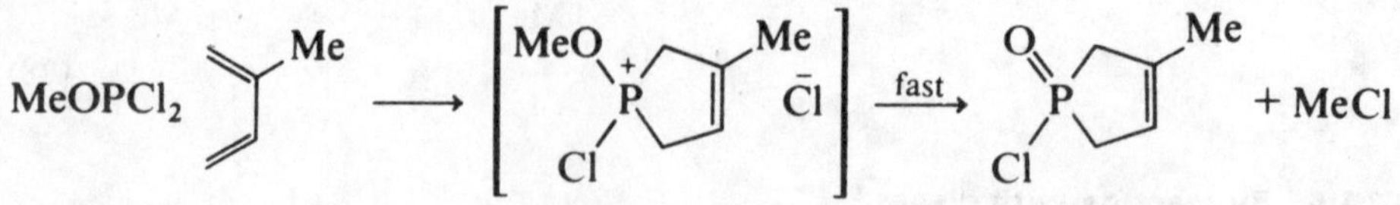

(6.26) → (6.27) + MeCl

Although such reactions represent useful syntheses of phosphorus heterocycles, any phosphoranes formed are at best in equilibrium with tetracoordinate structures. The irreversible de-alkylation step to (6.27) also highlights a limitation of the synthetic method described by eqn. (29). If the group, X, is likely to ionize (e.g. X = Cl, Br or —OPh), the phosphorane product suffers ring opening of the 1,3,2-dioxaphospholane ring to generate a cyclic phosphinate (6.28).

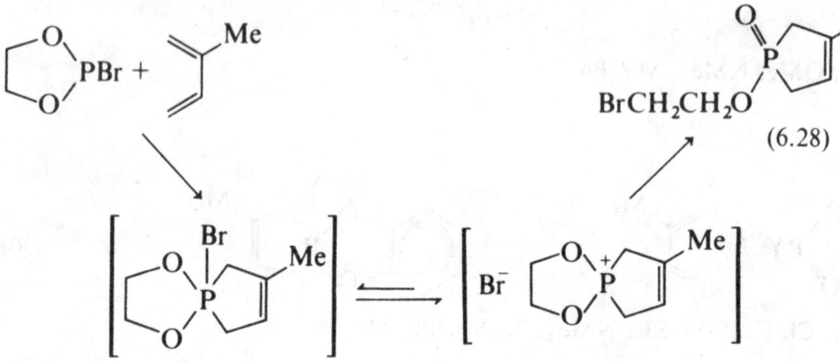

(6.28)

6.2.1.7 Miscellaneous preparations from P(III) compounds

In the presence of an acid catalyst cyclic phosphonites (e.g. 6.29) disproportionate to bicyclic phosphoranes (6.30) and cyclopolyphosphines (6.31)[38].

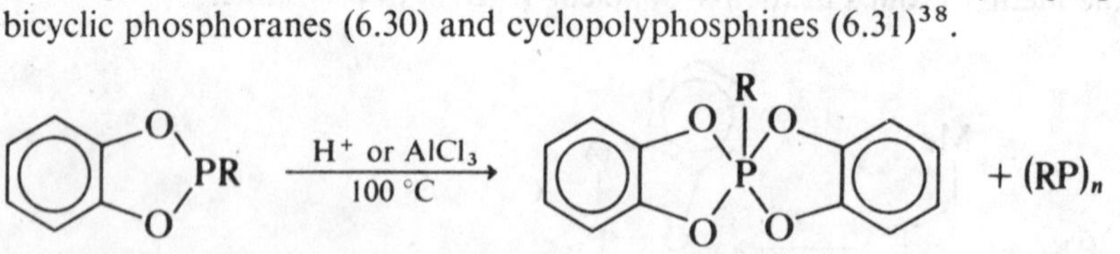

(6.29) $\xrightarrow{\text{H}^+ \text{ or AlCl}_3}_{100\,°C}$ (6.30) + (RP)$_n$ (6.31)

The same cyclic phosphonites also undergo a remarkable base-catalyzed reaction with catechol to produce phosphoranes and hydrogen[38].

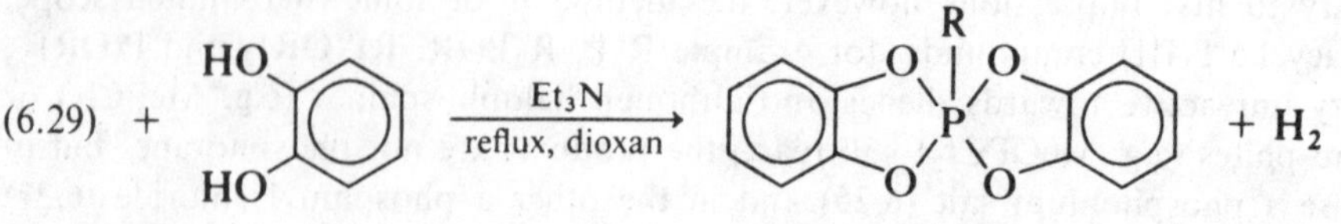

(6.29) + catechol $\xrightarrow{\text{Et}_3\text{N}}_{\text{reflux, dioxan}}$ (6.30) + H$_2$

6.2 Methods of preparing phosphoranes

It is hardly surprising therefore that the preparation of cyclic phosphonites (6.29) from catechol and phosphonous dichloride leads to a mixture of phosphonite (6.29) and phosphorane (6.30).

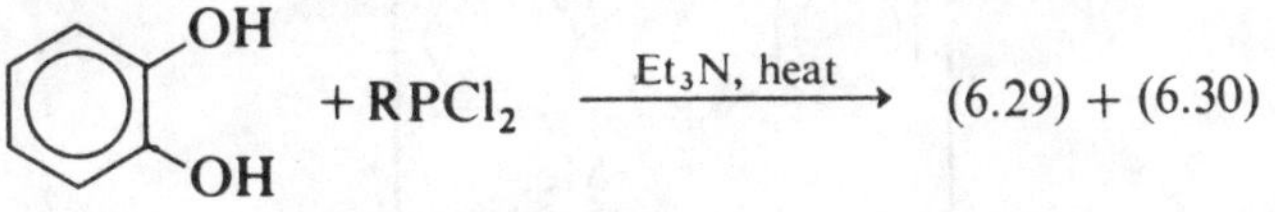

$$\text{catechol} + RPCl_2 \xrightarrow{Et_3N, \text{ heat}} (6.29) + (6.30)$$

An analogous reaction between the phosphoramidite (6.31) and catechol or pinacol gives pentacovalent structures with a P—H bond (6.32) and (6.33)[39].

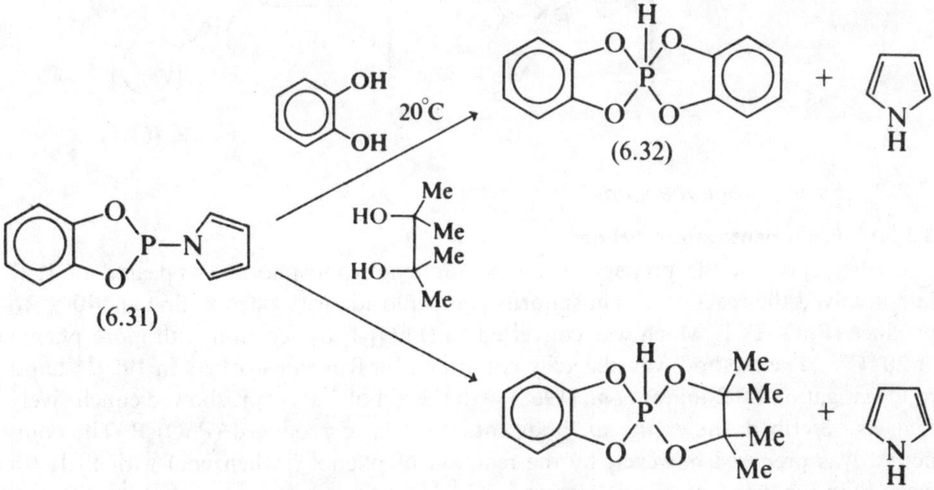

On the other hand, the phosphoramidite (6.34) gives rise to a phosphorane (6.35) which (by ^{31}P n.m.r.) is clearly in equilibrium with P(III) isomers (6.36) and (6.37)[39].

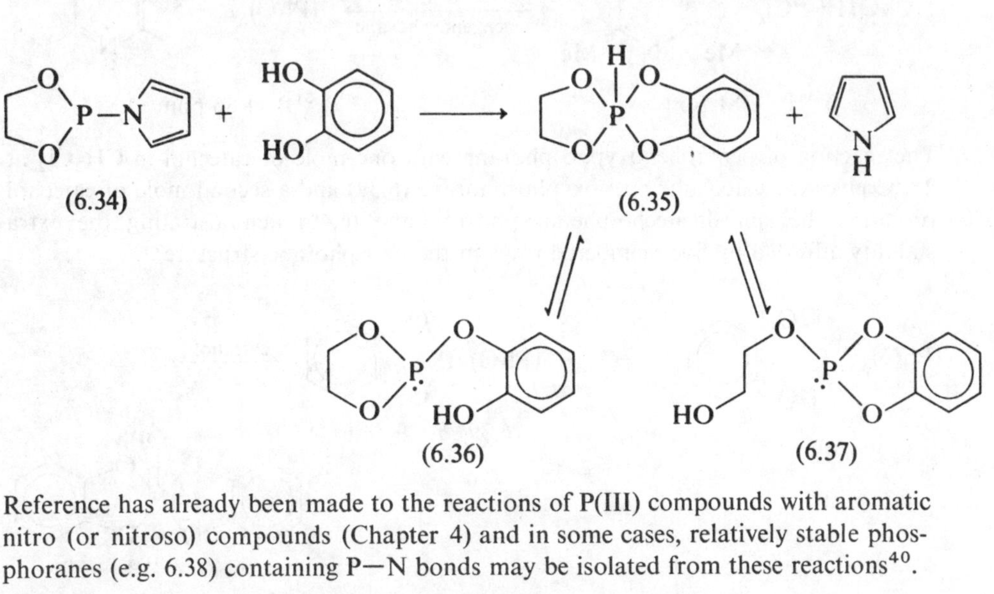

Reference has already been made to the reactions of P(III) compounds with aromatic nitro (or nitroso) compounds (Chapter 4) and in some cases, relatively stable phosphoranes (e.g. 6.38) containing P—N bonds may be isolated from these reactions[40].

(6.38)

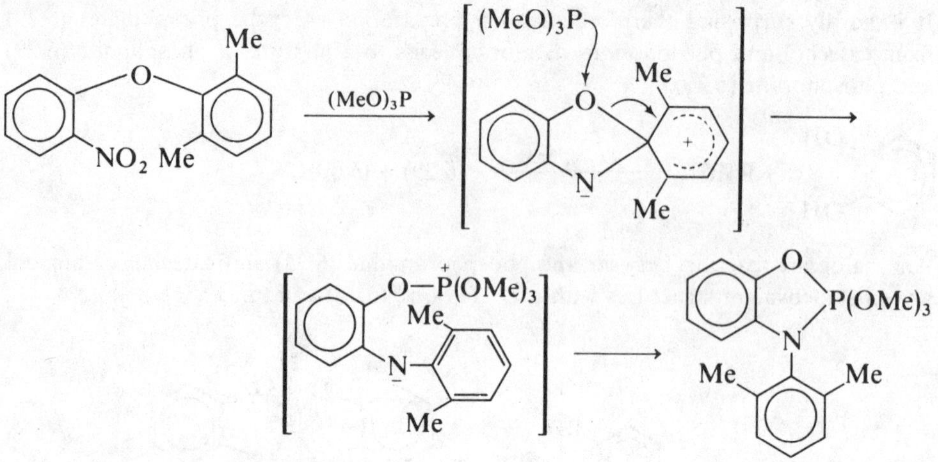

6.2.2 *By exchange reactions*

6.2.2.1 From pentavalent halides

The first report of the preparation of pentaphenoxyphosphorane appeared in 1927 and involved the reaction of phosphorus pentachloride with phenol, first at 140 °C to produce $(PhO)_3PCl_2$ which was converted to $(PhO)_5P$ by reaction with more phenol at 20 °C[41]. The method was allegedly confirmed by Russian workers in 1959[42] but a re-investigation by Ramirez *et al.* (1968) with the aid of ^{31}P n.m.r. showed conclusively that, as described, the earlier methods could not have produced $(PhO)_5P$. The compound was prepared however, by the reaction of phenol (in benzene) with PCl_5 (in hexane) in the presence of collidine at 0 °C.[4,43]

δ ^{31}P, +81 ppm δ ^{31}P, +86 ppm

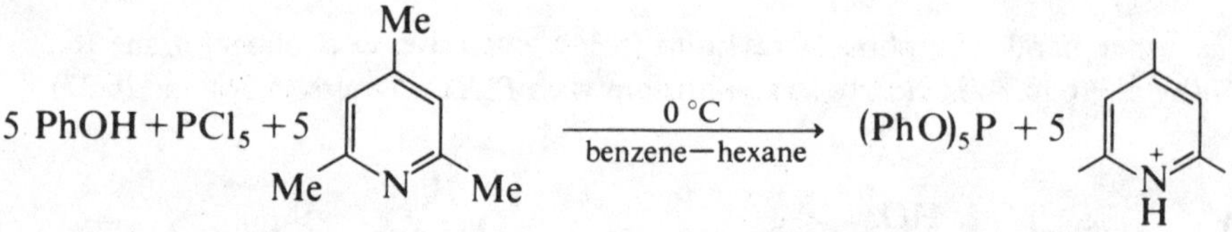

The reaction of pentaphenoxyphosphorane with one mole of catechol in CH_2Cl_2 at 25 °C gives the catecholtriphenoxyphosphorane (6.39) and a second mole of catechol produces the spirodicatecholphenoxyphosphorane (6.40) demonstrating the extra stability afforded by five-membered rings in the phosphorane structure[43].

(6.39) δ ^{31}P, +61

(6.40) δ ^{31}P, +30

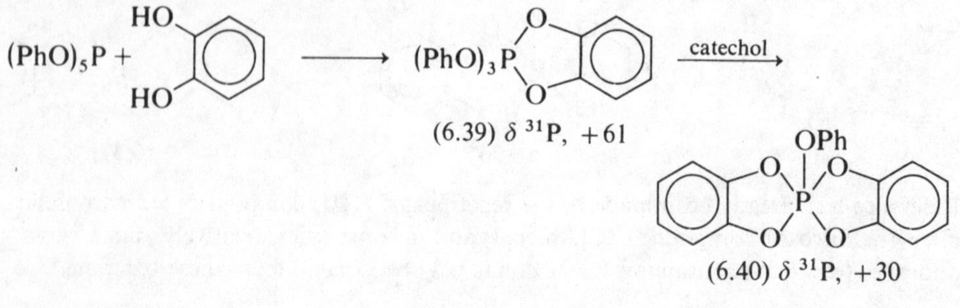

6.2 Methods of preparing phosphoranes

Phosphorus pentachloride may be converted to phosphorus pentafluoride by a variety of reagents (AgF, HF, AsF$_3$ or SbF$_5$) with antimony pentafluoride as the reagent of choice for a convenient laboratory preparation[1,2,29].

$$PCl_5 + SbF_5 \xrightarrow{\text{heat}} PF_5 + SbCl_5$$

This in turn may be converted to alkyltetrafluorophosphoranes by reaction with tetra-alkyltin (eqn. 31)[44] or to aminofluorophosphoranes by reactions with secondary amines (eqns. 32 and 33)[45].*

$$PF_5 + SnR_4 \longrightarrow RPF_4 + FSnR_3 \quad (31)$$
$$60\text{--}70\%$$

$$PF_5 + R_2NH \xrightarrow{\text{low temp.}} [PF_5 \cdot NHR_2] \xrightarrow{\text{heat}} R_2NPF_4 + HF \quad (32)$$

$$PF_5 + R_2NH \text{ (excess)} \xrightarrow{\text{heat}} F_3P(NR_2)_2 \quad (33)$$

An alternative route to aminotetrafluorophosphoranes involves the reaction of mixed halogenoaminophosphoranes with antimony pentafluoride (eqn. 34)[46].

$$R_2NPF_2 + X_2 \xrightarrow{\text{low temp.}} R_2NPF_2X_2 \xrightarrow[-30 \text{ to } +20\,^\circ\text{C}]{SbF_5} R_2NPF_4 \quad (34)$$

X = Cl or Br

The alkylfluorophosphoranes provide a starting point for the synthesis of aminoalkyl-fluorophosphoranes[47-49] (eqns. 35-37) and alkylaryloxyfluorophosphoranes (eqn. 38)[50].

$$RPF_4 + 2R^1NH_2 \longrightarrow RPF_3NHR^1 + R^1\overset{+}{N}H_3\;\overset{-}{F} \quad (35)$$

$$2RPF_4 + 2R^1_2NH \longrightarrow RPF_3NR^1_2 + R^1_2\overset{+}{N}H_2\;\overset{-}{RPF_5} \quad (36)$$

$$RPF_4 + Me_3SiNR^1_2 \longrightarrow RPF_3NR^1_2 + Me_3SiF \quad (37)$$

$$R_nPF_{5-n} + (3-n)ArOSiMe_3 \longrightarrow (ArO)_{3-n}PR_nF_2 + (3-n)FSiMe_3 \quad (38)$$

The aminofluorophosphoranes can, on the other hand, serve as starting materials for the mixed halogenophosphoranes, as evidenced by eqn. (39).

$$R_2NPF_4 \xrightarrow[X=Cl \text{ or } Br]{HX} XPF_4 \quad (39)$$

These exchange reactions are conveniently summarized in Fig. 6.1.

6.2.2.2 By exchange of alkoxy (aryloxy) groups

Exchange reactions between pentaphenoxyphosphorane and catechol (or its derivatives) has already been mentioned. The discovery of the peroxide route to oxyphos-

---

* The reaction of eqn. (33) only works well with dimethylamine, Me$_2$NH.

Figure 6.1 Synthetic routes to halogenophosphanes by exchange reactions.

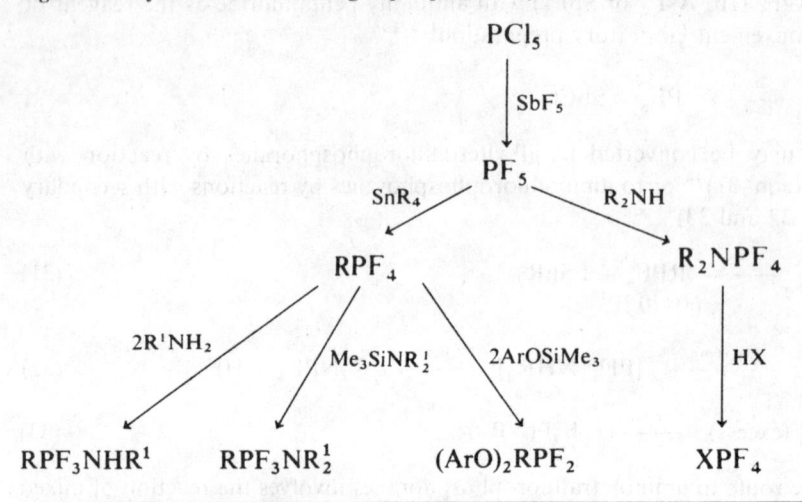

$$PCl_5 \xrightarrow{SbF_5} PF_5$$

$$PF_5 \xrightarrow{SnR_4} RPF_4 \qquad PF_5 \xrightarrow{R_2NH} R_2NPF_4$$

$$RPF_4 \xrightarrow{2R^1NH_2} RPF_3NHR^1$$
$$RPF_4 \xrightarrow{Me_3SiNR^1_2} RPF_3NR^1_2$$
$$RPF_4 \xrightarrow{2ArOSiMe_3} (ArO)_2RPF_2$$
$$R_2NPF_4 \xrightarrow{HX} XPF_4$$

---

phoranes allowed the analogous exchange of ethoxy groups to provide another source of oxyphosphoranes (eqns. 40 and 41)[51].

(EtO)$_5$P + HOCH$_2$CH$_2$OH ⟶ (EtO)$_3$P⟨O-CH$_2$-CH$_2$-O⟩ $\xrightarrow{HOCH_2CH_2OH}$ [bicyclic phosphorane] (40)

(6.41)

δ ^{31}P, +70   δ ^{31}P, +50   δ ^{31}P, +27

(EtO)$_5$P + HO—CH$_2$—CMe$_2$—CH$_2$—OH ⟶ (EtO)$_3$P⟨O-CH$_2$-CMe$_2$-CH$_2$-O⟩ $\xrightarrow{HOCH_2CMe_2CH_2OH}$ [bicyclic phosphorane] (41)

δ ^{31}P, +70

δ ^{31}P, +66

The exchange to form cyclic oxyphosphoranes only works smoothly with 1,2- and 1,3-diols since 1,4-butane diol and 1,5-pentane diol give tetrahydrofuran and tetrahydropyran respectively by a dissociation–displacement reaction (see later). Exchange reactions led to the isolation, by evaporative distillation, of the bicyclic phosphoranes (6.41, eqn. 40) and (6.42). Both were characterized by elemental analysis as well as

6.2 Methods of preparing phosphoranes

^{1}H and ^{31}P n.m.r. data and in the case of (6.42) the parent ion in the mass spectrum was found at 287 as required by the proposed structure[32].

$$Ph_2P(OEt)_3 + HN(CH_2CH_2OH)_2 \longrightarrow \underset{(6.42)}{Ph_2P(OCH_2CH_2)_2N} + 3EtOH$$

Both compounds represent early examples of phosphoranes whose structural assignment is not wholly dependent on spectroscopic data but is reinforced by the more conventional techniques of quantitative organic analysis.

6.2.2.3 By exchange of aryl groups

Pentaphenylphosphorane reacts with n-butyllithium to give triphenyl phosphonium butylide (6.44) presumably via n-butyl-tetraphenylphosphorane (6.43)[52].

$$Ph_5P + n\text{-BuLi} \longrightarrow \underset{(6.43)}{Ph_4PC_4H_9} \xrightarrow{-C_6H_6} \underset{(6.44)}{Ph_3P=CH.C_3H_7}$$

This is not a realistic method of preparing phosphoranes due to contamination of the product by ylid, but it demonstrates the exchange of aryl groups by carbanions (ex-organometallics) which is amply confirmed by the exchange of phenyl groups in PPh$_5$ by treatment with tritium labelled phenyllithium (eqn. 42)[53].

$$Ph_5M + 5Ph*Li \longrightarrow Ph_5*M + 5PhLi \qquad (42)$$

  * = tritium label; M = P, As, Sb, Bi.

6.2.3 From tetracoordinate compounds

All phosphoranes containing a phosphorus atom bound to halogen, oxygen or nitrogen may, in principle, be in equilibrium with the isomeric phosphonium salt (eqn. 43).

$$R_4PX \rightleftharpoons R_4P^+ X^- \quad X = \text{halogen, O-alkyl, O-aryl, NR}_2^1 \; (R^1 = H, \text{alkyl, aryl}) \qquad (43)$$

In practice, whenever X$^-$ is the anion of a fairly strong acid (e.g. Cl$^-$, Br$^-$, I$^-$ or PhO$^-$) the equilibrium tends to lie on the side of the phosphonium salt. As already mentioned, the pentacovalent structure is stabilized by the incorporation of small (four- or five-membered) rings and by highly electronegative substituents. Thus, when X is halogen, the use of chlorine, bromine or iodine tends to give ionic, structures, whereas fluorine is most likely to produce a pentacovalent structure. Similarly, for X = O-alkyl, a combination of the strong basicity of OR and high electronegativity of oxygen tends to promote formation of a phosphorane; in contrast, X = —OAr tends to drive the equilibrium of eqn. (43) to the right. Other factors which influence covalent *vs* ionic structure include temperature and solvent; of the latter non-polar, aprotic media favour the covalent form. As a simple example of the influence of temperature one may cite the vacuum sublimation of tetrachlorophosphonium hexafluorophosphide (6.45) which yields tetrachlorofluorophosphorane (6.46) when the sublimate is trapped at $-50\,°C^{54}$. On warming to room temperature, (6.46) dissociates to tetrachlorophosphonium fluoride (6.47) whose structure is confirmed by x-ray analysis.*

$$\overset{+}{PCl_4}\,\overset{-}{PF_6} \xrightarrow{\text{vac. sublimation}} Cl_4PF \xrightleftharpoons{\text{warm to room temp.}} \overset{+}{PCl_4}\,\overset{-}{F}$$
$$(6.45) \hspace{3cm} (6.46) \hspace{3cm} (6.47)$$

Another example of an established equilibrium situation involves the reaction of methoxide ion (in methanol) with the 2-furyltrimethylphosphonium ion (6.48, eqn. 44).

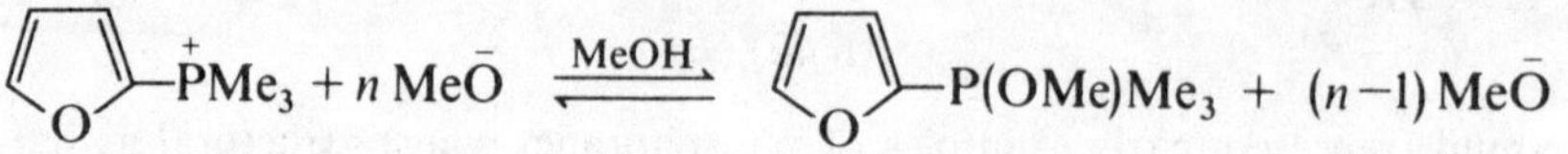

$$\text{furyl-}\overset{+}{P}Me_3 + n\,MeO^- \xrightleftharpoons{MeOH} \text{furyl-}P(OMe)Me_3 + (n-1)\,MeO^- \qquad (44)$$
$$(6.48) \hspace{4cm} (6.49)$$

With $n = 1$, the ^{31}P n.m.r. showed a signal at $+14.5$ ppm; with $n = 3$, $\delta\,^{31}P = +92$, indicating a substantial shift to the P(V) structure (6.49). At $-80\,°C$, the equilibration was slow on the n.m.r. timescale and separate signals for the two structures were observed[55].

Because of the restrictions of small rings and/or electronegative substituents, the reduction of tetracoordinate phosphorus does not provide a general route to phosphoranes. Nevertheless, some specific examples are worthy of attention.

6.2.3.1 By reduction of phosphonium salts

This may be accomplished using lithium aluminium hydride or sodium borohydride (eqn. 45)[56], phenyllithium (eqn. 46)[57] or methyllithium (eqn. 47)[58]. Notice that the phosphoranes produced all involve small rings.

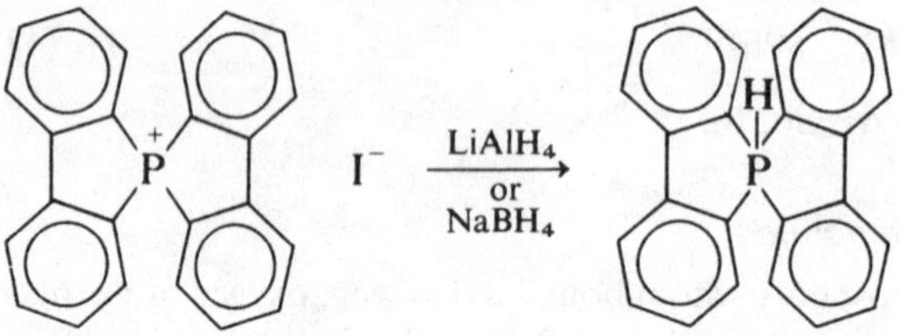

$$\text{[dibenzophospholium]}^+ I^- \xrightarrow[\text{or NaBH}_4]{\text{LiAlH}_4} \text{[dibenzophosphole-PH]} \qquad (45)$$

$$\text{[cubyl]}\overset{+}{P}Ph_2\,Br^- \xrightarrow[\text{in THF}]{PhLi} \text{[cubyl]}PPh_3 \qquad (46)$$

$$\text{[cubyl-benzophospholium]}^+ Br^- \xrightarrow{MeLi} \text{[cubyl-P(Me)(Ph)]} \qquad (47)$$

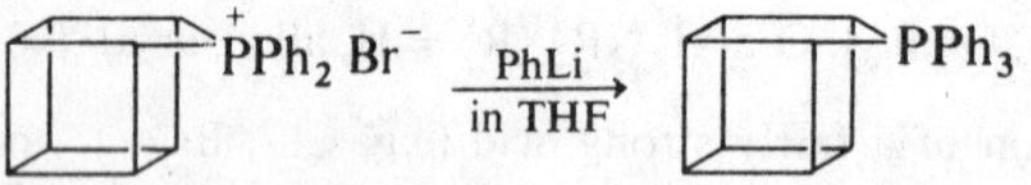

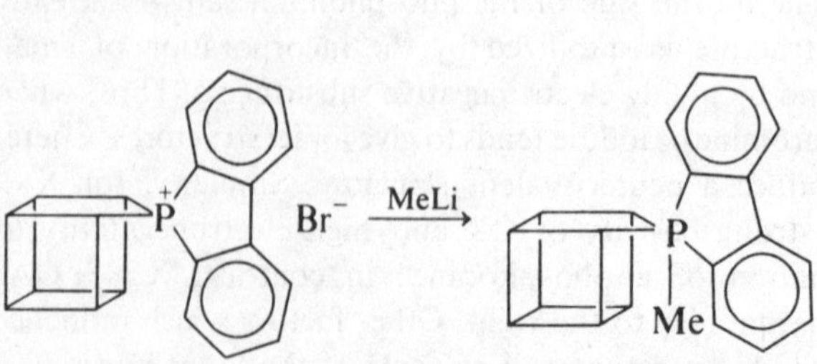

---

* Ionization of fluorine (as fluoride) rather than chlorine (as chloride) seems a little surprising but it may be connected with the symmetry of the resultant tetrachlorophosphonium ion and the effect of this on the lattice energy.

6.2.3.2 From phosphine oxides and thiophosphoryl compounds

Tristrifluoromethylphosphine oxide reacts with hexamethyldisiloxane (eqn. 48) to give a phosphorane (6.50) with two trifluoromethyl groups in apical positions[59].

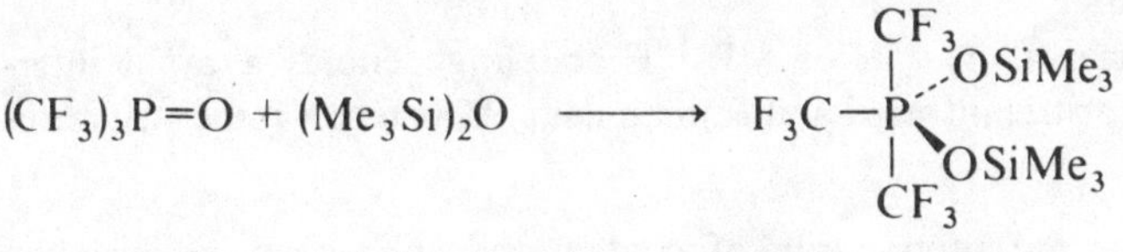

$$(CF_3)_3P=O + (Me_3Si)_2O \longrightarrow \underset{(6.50)}{F_3C-\overset{\overset{CF_3}{|}}{\underset{\underset{CF_3}{|}}{P}}\begin{smallmatrix}OSiMe_3\\OSiMe_3\end{smallmatrix}} \quad (48)$$

Thiophosphoryl (6.51) and thiophosphinyl (6.52) halides react with antimony trifluoride to give alkyltetrafluoro- and dialkyltrifluorophosphoranes (eqns. 49 and 50)[29,60]. Bis-phosphine sulphides are also capable of producing trifluorophosphoranes (eqn. 51).

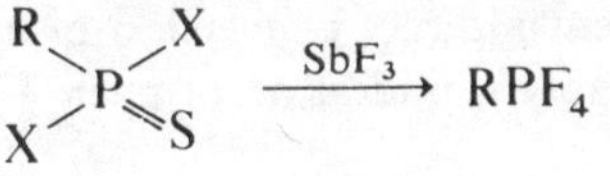

$$\underset{(6.51)}{\overset{R}{\underset{X}{>}}P\overset{X}{\underset{S}{<}}} \xrightarrow{SbF_3} RPF_4 \quad (49)$$

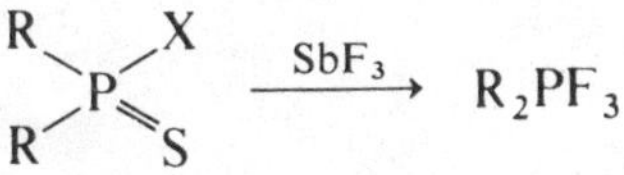

$$\underset{(6.52)}{\overset{R}{\underset{R}{>}}P\overset{X}{\underset{S}{<}}} \xrightarrow{SbF_3} R_2PF_3 \quad (50)$$

$$3 \left[\begin{matrix}P-P\\\|\\S\end{matrix}\right] + 6SbF_3 \longrightarrow 6 \left[PF_3\right] + 2Sb + 2Sb_2S_3 \quad (51)$$

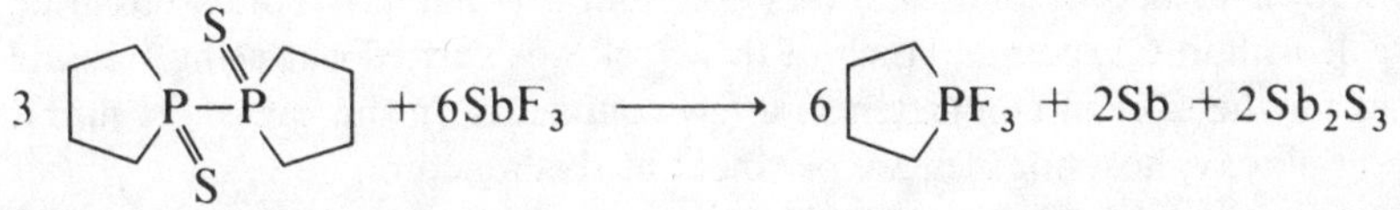

These are all examples of the use of electronegative substituents to stabilize the phosphorane structures.

Phosphine oxides may be alkylated (typically by oxonium salts) but reaction of the resultant alkoxyphosphonium salts with alkoxide ions usually regenerates the phosphine oxide (eqn. 52).

$$R_3P(O) + Et_3\overset{+}{O}\ \overset{-}{BF_4} \longrightarrow R_3\overset{+}{P}OEt\ \overset{-}{BF_4} \xrightarrow{Na^+OEt^-} R_3P(O) + Et_2O + NaBF_4 \quad (52)$$
$$(+Et_2O)$$
$$R = alkyl, aryl$$

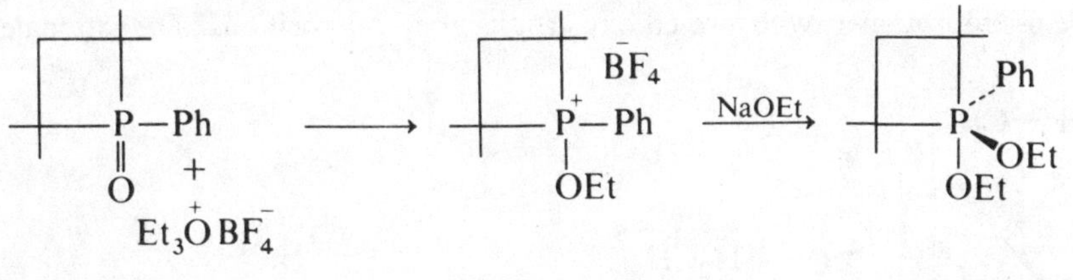

$$\left[\begin{matrix}|\\-\overset{\|}{P}-Ph\\O\end{matrix}\right] + Et_3\overset{+}{O}\overset{-}{BF_4} \longrightarrow \left[\begin{matrix}|\\-\overset{+}{P}-Ph\\OEt\end{matrix}\right]\overset{-}{BF_4} \xrightarrow{NaOEt} \left[\begin{matrix}|\\-P\overset{Ph}{\underset{OEt}{<}}\\OEt\end{matrix}\right] \quad (53)$$

If the oxide contains a small, ring, however, a substantial yield of phosphorane is obtained (eqn. 53)[32].

6.2.3.3 From phosphorus ylids

Reaction of phosphorus ylids with HF is capable of generating monofluorophosphoranes (eqn. 54)[61]. These compounds are probably in equilibrium with ionic

$$R_3P=CH_2 + HF \xrightarrow{-70\,°C} R_3P\begin{smallmatrix}F\\\\CH_3\end{smallmatrix} \rightleftharpoons R_3\overset{+}{P}CH_3\bar{F} \qquad (54)$$

R = Me, Bu or Ph

structures and the absence of $^1H-^{19}F$ or $^{31}P-^{19}F$ coupling denotes a rapid intermolecular exchange of fluorine; infrared and Raman data also suggest ionic structures for these molecules.

A convenient method for the preparation of pentaarylphosphoranes involves the reaction of aryllithium with the N-p-tosyliminoylid* (6.53, eqn. 55)[62]. Little is known

$$(PhO)_3P=NSO_2C_6H_4\text{-}p\text{-}Me + 5ArLi \longrightarrow Ar_5P \qquad (55)$$
$$(6.53) \qquad\qquad (30\text{–}49\%)$$

about the mechanism of this reaction but it does represent a direct conversion of a tetracoordinate structure to a pentacoordinate one, probably via addition of phenyllithium followed by exchange.

6.3 The chemistry of phosphoranes

6.3.1 Acyclic phosphoranes

6.3.1.1 Pseudorotation

The pseudorotation of acyclic phosphoranes containing P–halogen bonds has been adequately dealt with in Chapter 2. Much of this work was carried out using ^{19}F and ^{31}P n.m.r. and for the sake of completeness a few comments on the use of ^1H n.m.r. for the study of alkoxyphosphoranes are pertinent at this juncture.

At room temperature in deuteriochloroform, the alkoxy groups of diphenyltriethoxyphosphorane (6.54) are all equivalent, demonstrating that the three ethoxy groups undergo mutual exchange (see problem). Thus the methylene (CH$_2$) group of the ethoxy groups is seen as a quintet at 3.4 ppm with $^2J_{POCH} \approx 7$ Hz and $^1J_{H-H} \approx 7$ Hz.

$$Ph_2POEt + Et_2O_2 \longrightarrow Ph_2P(OEt)_3$$
$$(6.54)$$

At $-60\,°C$ however, two types of ethoxy group are seen with the methylene groups at $\delta = 4.39$ ppm (quintet) and $\delta = 2.83$ ppm (quintet) and in a ratio of 1:2 respectively. The low temperature n.m.r. data are consistent with a structure equivalent to (6.55), i.e. a single pseudorotamer, with two ethoxy groups in apical positions. The rationale

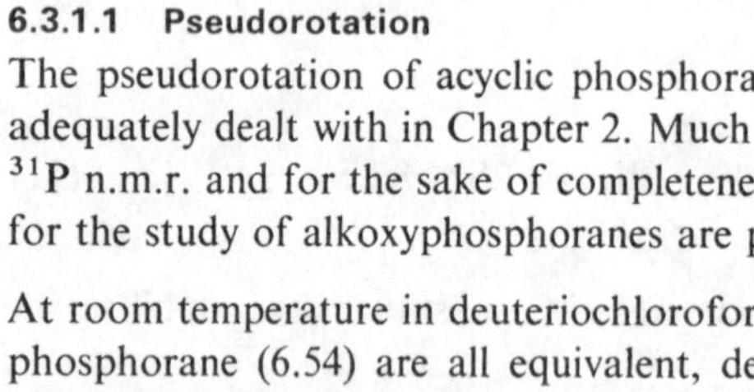

(6.55)

---

* This compound could also be named as an N-p-tosylmonophosphazene–see Chapter 10.

6.3 The chemistry of phosphoranes

offered to explain the high field shift of the apical methylenes (relative to the equatorial methylene) is that the former lies in the shielding region generated by the ring current of the equatorial phenyl groups[30].

6.3.1.2 Reactions of acyclic phosphoranes

Fluorophosphoranes are known to behave as electron acceptors and in fact, pentafluorophosphorane is classed as a powerful Lewis acid. The acceptor power of fluorophosphoranes has been established by the ^{19}F n.m.r. of their 1:1 complexes with dimethylformamide (Me_2NCHO) or pyridine and decreases in the order, PF_5 > $ArPF_4$ > (alkyl) PF_4 ≫ (alkyl)$_2PF_3$ > (alkyl)$_3PF_2$.[63] This is more or less in line with decreasing positive charge on phosphorus as the combined electronegativity of the ligands decreases.

Similar complexes are also formed with ethers and dialkyl sulphides and these may be utilized for the preparation of oxonium or sulphonium salts since on warming to room temperature, the reactions described by eqns. (56) and (57) occur[64].

$$3Me_2O \cdot PF_5 \xrightarrow{\text{room temp.}} 2Me_3\overset{+}{O}\ \overset{-}{PF_6} + O=PF_3 \qquad (56)$$

$$3Me_2S \cdot PF_5 \xrightarrow{\text{room temp.}} Me_3\overset{+}{S}\overset{-}{PF_6} + [Me_2\overset{+}{S}SMe]\ \overset{-}{PF_6} + PF_3 \qquad (57)$$

Mixed halogenoaminophosphoranes may be converted to aminophosphoryldifluorides (eqn. 58), aminodi-(alkylthio)-difluorophosphoranes (eqn. 59) or even aminodialkoxy difluorophosphoranes (eqn. 60)[29]. These rather simple reactions constitute the tip of

$$R_2NPX_2F_2 + H_2O \longrightarrow R_2NP(O)F_2 \qquad (58)$$

$$R_2NPX_2F_2 + 2EtSH \longrightarrow R_2NP(SEt)_2F_2 \qquad (59)$$

$$R_2NPX_2F_2 + 2R^1OH \xrightarrow{Et_3N} R_2NP(OR^1)_2F_2 \qquad (60)$$

a large and very complex iceberg and for further information the reader is referred to the review articles on fluorophosphoranes[1,7,23,29].

Pentaalkoxyphosphoranes, $P(OR)_5$, are ortho esters of phosphoric acid. As such they behave in exactly the same way as the ortho esters, $C(OR)_4$, of carbonic acid and readily alkylate any molecule bearing a sufficiently acidic hydrogen. For example, pentaethoxyphosphorane reacts with benzoic acid to form high yields of ethyl benzoate (eqn. 61).

$$P(OEt)_5 + PhCO_2H \longrightarrow (EtO)_3P=O + EtOH + PhCO_2Et \qquad (61)$$

Indeed, this reaction has been employed as a test for the presence of alkoxyphosphoranes in solution (eqn. 62)[65].

$$Ph_nP(OEt)_{5-n} + PhCO_2H \longrightarrow Ph_nP(O)(OEt)_{3-n} + EtOH + PhCO_2Et \qquad (62)$$
$$n = 0\text{–}3$$

Phenols (eqn. 63) and even activated methylene groups (eqn. 64) may also be alkylated[66]. The conjugate base of an activated methylene group is usually an ambident anion* and in the example shown, alkylation of oxygen predominates although some C-alkylation product is also formed.

$$\text{PhOH} + \text{Ph}_2\text{P(OEt)}_3 \longrightarrow \text{PhOEt} + \text{EtOH} + \text{Ph}_2\text{P(O)OEt} \quad (63)$$

$$\begin{array}{c} \text{MeCO.CH}_2\text{COMe} \\ + \\ \text{(EtO)}_5\text{P} \end{array} \xrightarrow{0\,°C} \left. \begin{array}{c} x\text{Me.CO.CH=C(OEt)Me} \\ + \\ (1-x)\text{MeCOCHEt.CO.Me} \end{array} \right\} + \text{EtOH} + \text{(EtO)}_3\text{PO}$$
$$1 > x > 0.5 \quad (64)$$

$$\text{(EtO)}_5\text{P} \rightleftharpoons \text{(EtO)}_4\overset{+}{\text{P}}\,\overset{-}{\text{OEt}} \quad (65)$$

The alkylation of active methylene groups is extremely facile and unique in that no acid or base catalyst is required. This may be because the reactions are auto-catalyzed by dissociation of the phosphorane (e.g. eqn. 65).

6.3.2 Cyclic phosphoranes

The chemistry of these compounds has received a great deal of attention in recent years and it would be impractical to provide a comprehensive coverage of the available information in a text of this size. We shall therefore confine ourselves to a number of examples which (a) illustrate the use of cyclic phosphoranes in studies of polytopal rearrangements (e.g. pseudorotation) and which (b) highlight some of the unusual reactions associated with these molecules.

6.3.2.1 Pseudorotation (ψ)

Reference has already been made to the apicophilicity of various groups in monocyclic phosphoranes and the influence which such groups have on the disposition (e–a or e–e) of small rings in phosphoranes (Chapters 2 and 4). A detailed discussion of the determination of relative apicophilicity by ^{19}F n.m.r. also appears earlier in this chapter.

Relatively little has been said however, about the energy barriers associated with changing a ring system in a cyclic phosphorane from an apical–equatorial situation to a diequatorial one and there is at least one very good reason for this: the energy barrier to placing a particular ring diequatorial must depend upon the nature of the remaining (exocyclic) groups attached to phosphorus. This is elegantly demonstrated by comparison of the ^{1}H n.m.r. spectra of (6.56), (6.57) and (6.58)[51]. At room temperature all the ring methyl groups are equivalent which can only be achieved by facile pseudorotations which place the six-membered rings diequatorial and in consequence, the group R in an apical position. There is no change in the spectrum of (6.56) on cooling to $-65\,°C$ but in the case of (6.57) or (6.58) the single absorption for the ring methyl groups splits into two absorptions of equal intensity. This is con-

---

* An ambident ion is one in which the charge may be located, by resonance, at two different sites in the molecule.

6.3 The chemistry of phosphoranes

sistent with an inhibition of ψ which restricts the rings to apical–equatorial situations and places R in the remaining equatorial position (6.59). The energy barrier, $\Delta G^{\ddagger}$,

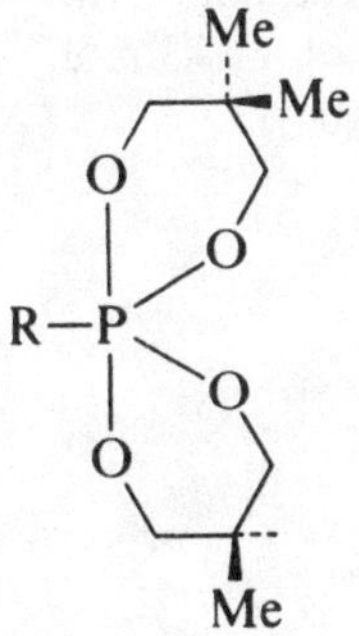

(6.56), R = OEt
(6.57), R = Ph
(6.58), R = Me

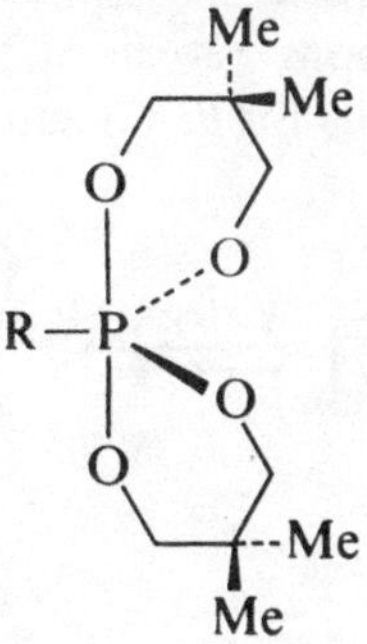

(6.59)

facing ψ with R = Ph, is calculated to be a maximum of 50 kJ mol^{-1} (12 kcal mole^{-1}) but obviously is much less for R = —OEt. This is simply a reflection of the relative apicophilicity of the phenyl and ethoxy groups with the latter showing greater affinity for the apical position. Obviously therefore, barriers to placing rings diequatorial must be estimated by using the same R group and varying the ring size whilst at the same time, not altering substitution in the rings so as to minimize steric effects. This has been achieved in a qualitative fashion by a comparison of the ^{1}H n.m.r. spectra of (6.56–6.58) with their five-membered ring analogues (6.60–6.62)[51].

At room temperature the ^{1}H n.m.r. spectra of (6.60)–(6.62) show non-equivalent ring methylene hydrogens which is exactly what would be expected if ψ was limited to apical–equatorial switching of the rings. At +172 °C, the multiplet of the ring protons for (6.60) collapses to a simple doublet. This is consistent with ψ's involving a diequatorial ring which, of necessity, would place R (=OEt) in an apical position. On heating (6.61) and (6.62) the spectra of the ring methylene hydrogens change but never become a simple doublet suggesting that the barrier to ψ effecting diequatorial disposition of the rings, is much higher in these cases. This of course, is what would be expected since phenyl and methyl groups are much less apicophilic than an ethoxy group. The overall picture suggests a much higher barrier to placing a dioxaphospholane ring diequatorial than that for placing a dioxaphosphorinane ring diequatorial.

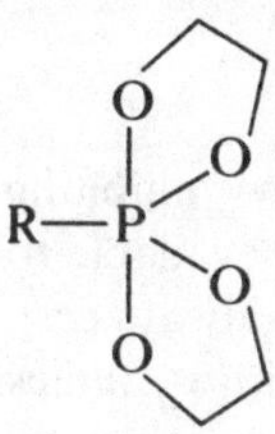

(6.60) R = OEt
(6.61) R = Ph
(6.62) R = Me

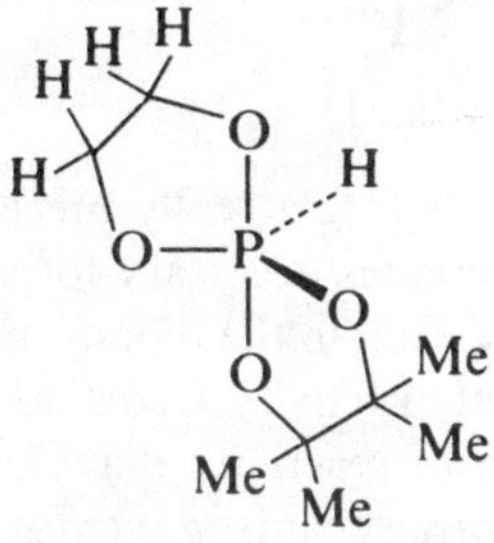

(6.63)

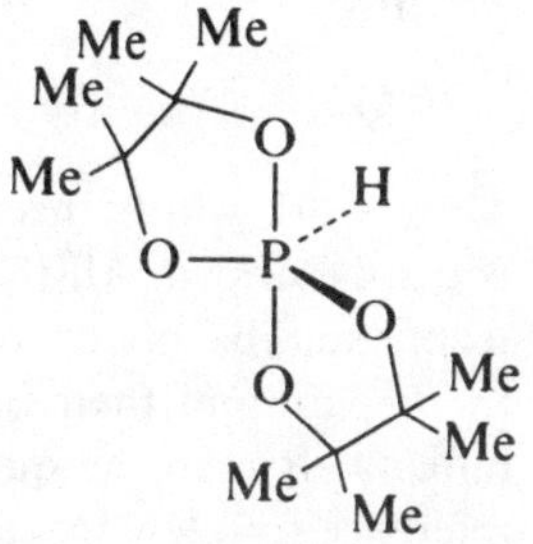

(6.64)

A more precise idea of the magnitude of the barrier to ψ is provided by the variable temperature ^{1}H n.m.r. spectra of (6.63) and (6.64)[67].

Chapter 6 / Pentacoordinate and hexacoordinate organophosphorus chemistry

Two pseudorotation processes may be identified in each case. That involving a–e switching of the rings (6.65)⇌(6.66) is rapid, even at −70 °C, but ψ (6.65)⇌(6.67) puts the 1,3,2-dioxaphospholane ring diequatorial and is slow on the n.m.r. timescale. Calculations reveal $\Delta G^{\ddagger}$ values for this "slow" process* of 66 kJ mol^{-1} (15.6 kcal mol^{-1}) for (6.63) and 77 kJ mol^{-1} (18.4 kcal mol^{-1}) for (6.64).

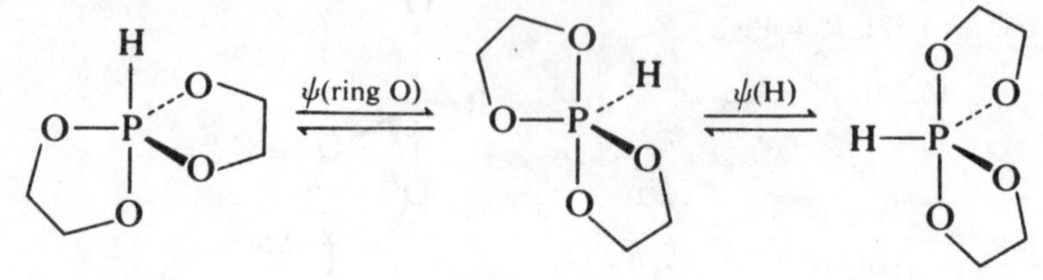

(6.67)   (6.65)   (6.66)

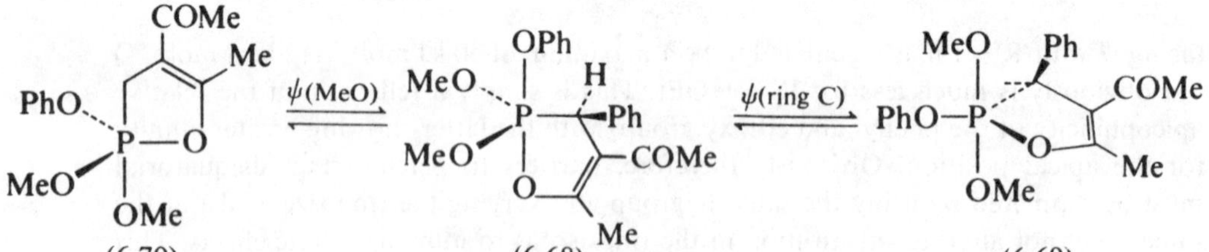

(6.70)   (6.68)   (6.69)

Another example is provided by the phosphorane (6.68). Detailed line shape analysis of the variable temperature ^{1}H n.m.r. of (6.68) shows that ψ to (6.69) has a $\Delta G^{\ddagger}$ value of ca 92 kJ mol^{-1} (22 kcal mol^{-1}). This is considerably higher than the $\Delta G^{\ddagger}$ (=55 kJ mol^{-1}) for ψ (6.68)⇌(6.70) even though the latter process puts the ring carbon into an unfavourable apical position at the expense of a phenoxy group[68]. In the case of this ring system therefore, ring strain outweighs relative apicophilicity as an inhibitor of ψ. This is in stark contrast to the cases of the fluorophosphoranes (6.6) and (6.71) in which diequatorial dispositions of carbocyclic rings appear to afford the stable structures and in which the barrier to ψ originates from the high apicophilicity of fluorine[21,28].

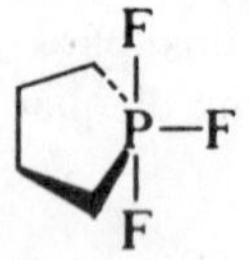

(6.6) 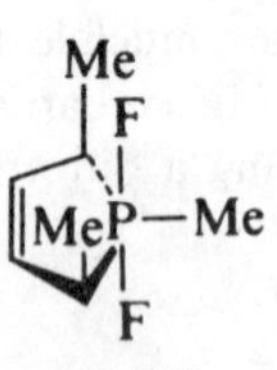   (6.71)

More work will be necessary before a really precise picture of the factors inhibiting ψ can be drawn. All the techniques necessary for such investigations are available to us but suitable phosphoranes, even in the mono- or bicyclic series are not always easy to prepare and their inherent reactivity frequently makes high temperature studies difficult. It is to the question of reactivity that we should now address ourselves by reference to a few reactions of mechanistic interest and synthetic utility.

---

* In this context "slow" is very much a relative term. An energy barrier of 15 or 18 kcal mol^{-1} is much too low to allow isolation of individual pseudorotamers at room temperature.

6.3.2.2 Reactions of cyclic phosphoranes

(i) *1,3,2-dioxaphosphoranes*: Reaction of the phosphite (6.72) with diethyl peroxide gives the phosphorane (6.73) which on heating at 120 °C for 17 h gives styrene oxide and triethyl phosphate[31]. Reaction of the tetramethyl-1,3,2-dioxaphospholane (6.74)

(6.72) (6.73)

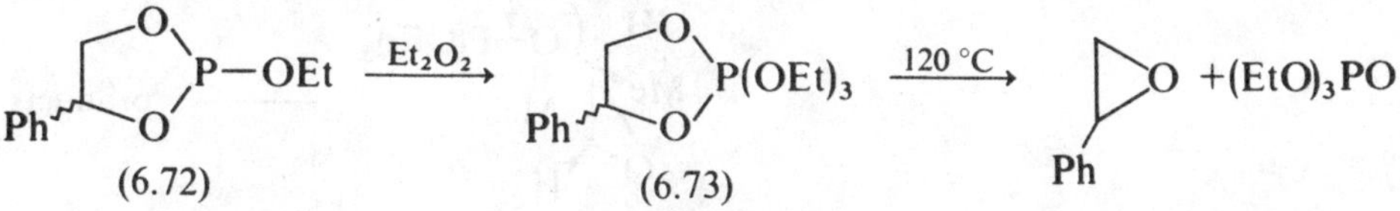

with diethyl peroxide gave an unstable phosphorane (6.75) which decomposed at room temperature to tetramethylene oxide and triethyl phosphate[31]. Both these reac-

(6.74) (6.75)

tions give some idea of the stability of phosphoranes containing a 1,3,2-dioxaphospholane ring. Formation of the epoxide is stereospecific as evidenced by the decomposition of the phosphoranes from a mixture of *dl*-phosphite (6.76a, 88 %) and *meso*-phosphite (6.76b) which gave 85 % *cis*-2-butene oxide and 15 % *trans*-2-butene oxide[31].

(6.76a) *cis*

(6.76b) *trans*

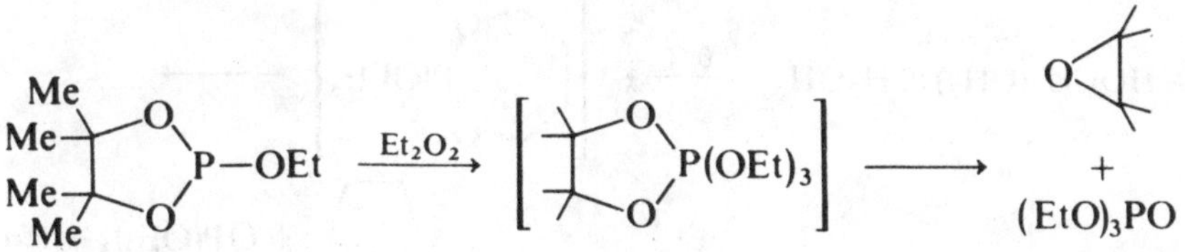

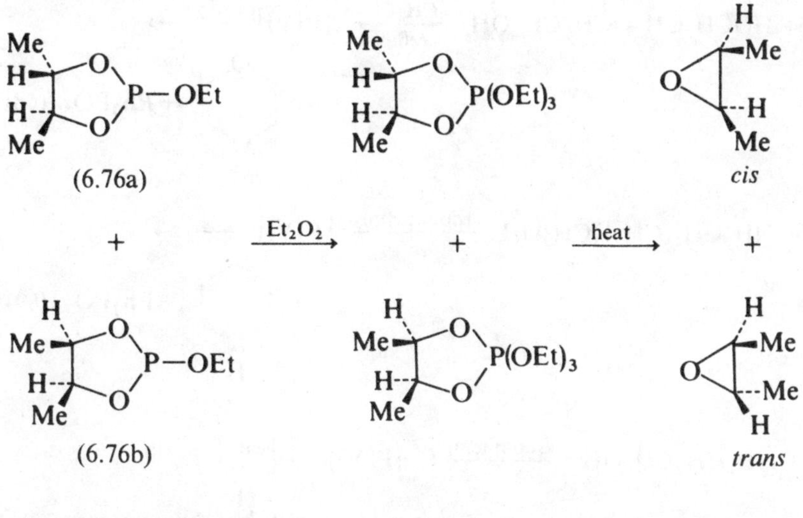

The stereochemistry also gives a clue to the mechanism which must involve dissociation of the intermediate followed by rotation about the carbon–carbon bond and back-side attack to displace phosphate, i.e. (6.77) ⇌ (6.78) ⇌ (6.79) → products.

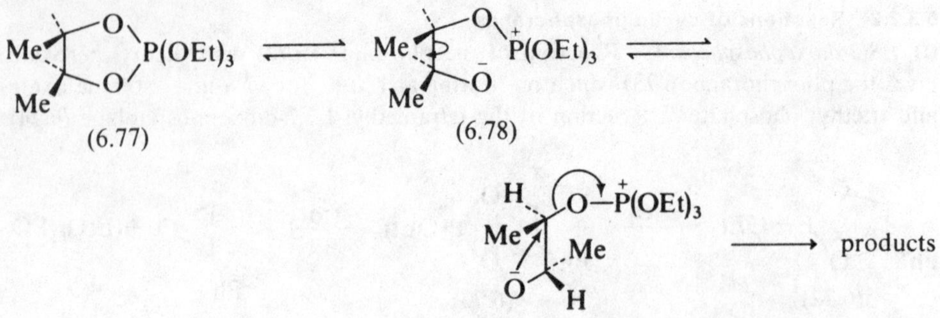

(6.77) ⇌ (6.78) ⇌

(6.79) ⟶ products

The above displacement reaction has been utilized as a general method for the synthesis of heterocyclic compounds (eqns. 66–70)[51,69].

$(EtO)_5P + HOCH_2(CH_2)_2CH_2OH \xrightarrow{0\,°C}$ [cyclic P(OEt)₃ intermediate] ⟶

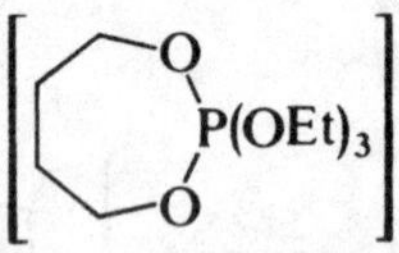

tetrahydrofuran + OP(OEt)₃ (66)
87%

$Ph_3P(OEt)_2 + HOCH_2(CH_2)_3CH_2OH \xrightarrow{room\ temp.}$ [P(V)] ⟶

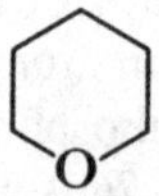

tetrahydropyran + Ph₃PO (67)
81%

$Ph_3P(OEt)_2 + HOCH_2CH_2OCH_2CH_2OH \xrightarrow[reflux]{CH_2Cl_2}$ [P(V)] ⟶

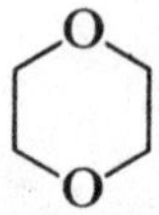

dioxane + Ph₃PO (68)
96%

$Ph_3P(OEt)_2 + NH_2CH_2(CH_2)_3CH_2OH \xrightarrow{room\ temp.}$ [P(V)] ⟶

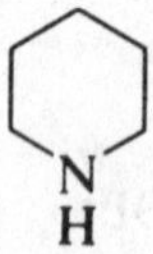

piperidine + Ph₃PO (69)
85%

$Ph_3P(OEt)_2 + NH_2CH_2CH_2OH \xrightarrow{room\ temp.}$ [P(V)] $\xrightarrow{heat}$

aziridine + Ph₃PO (70)
70%

[P(V)] = pentacovalent intermediate

6.3 The chemistry of phosphoranes

An analogous reaction is observed when the 1:1 adducts of ethylenephosphonothioites (6.80) and either phenanthraquinone or biacetyl are heated[70]. The intermediate compounds, (6.81) or (6.82) readily eliminate ethylene sulphide to form the cyclic phosphonates (6.83) and (6.84).

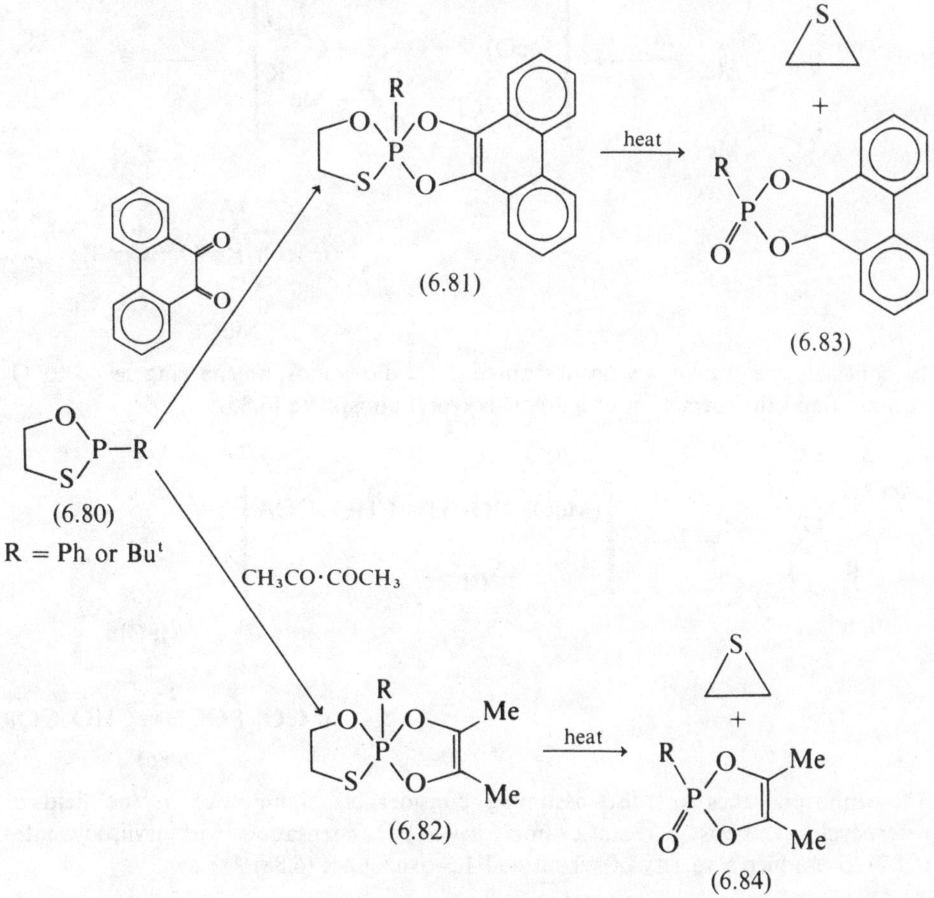

(6.80)
R = Ph or Bu^t

Phosphoranes containing a 1,3,2-dioxaphospholene ring have also found wide application in synthesis. Apart from the fact that the 1:1 adducts are capable of adding a second molecule of the diketone to form a new carbon–carbon bond (see Chapter 4), the 1:1 adducts have been employed to prepare sugar-like phosphates (eqn. 71)[71].

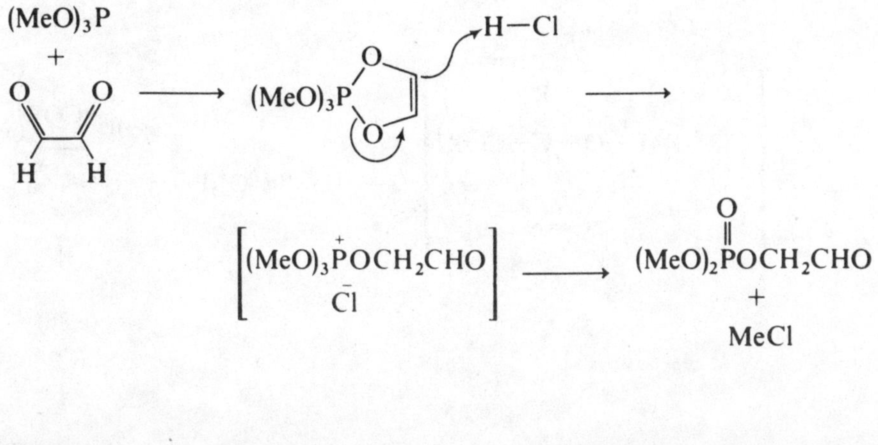

$$(MeO)_3P + OHC-CHO \longrightarrow (MeO)_3P\text{-dioxaphospholene} \xrightarrow{H-Cl} [(MeO)_3\overset{+}{P}OCH_2CHO \ \ \bar{C}l] \longrightarrow (MeO)_2\overset{O}{\overset{\|}{P}}OCH_2CHO + MeCl \quad (71)$$

Analogous phosphoranes will also attack acid chlorides to produce phosphate esters of α-hydroxy-β-diketones (6.85) by a C-acylation (eqn. 72)[72].

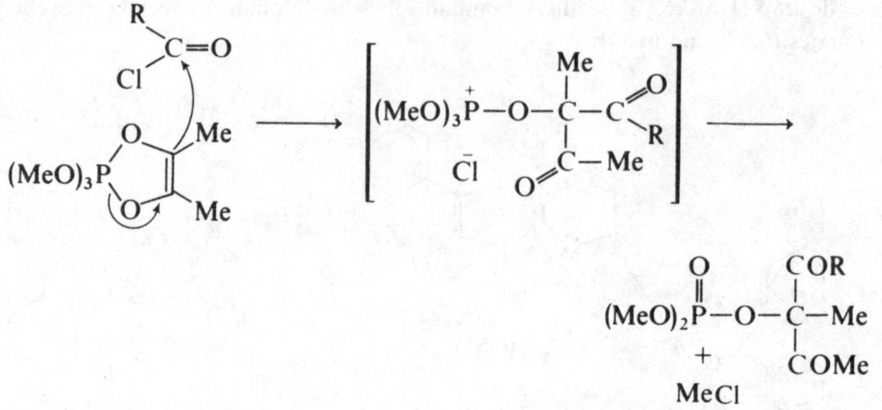

$$(MeO)_2\overset{O}{\underset{\|}{P}}-O-\underset{\underset{COMe}{|}}{\overset{\overset{COR}{|}}{C}}-Me + MeCl \quad (72)$$

In contrast, the use of an unsubstituted 1,3,2-dioxaphospholene ring leads to O-acylation and the formation of a 2-acyloxyvinyl phosphate (6.86)[73].

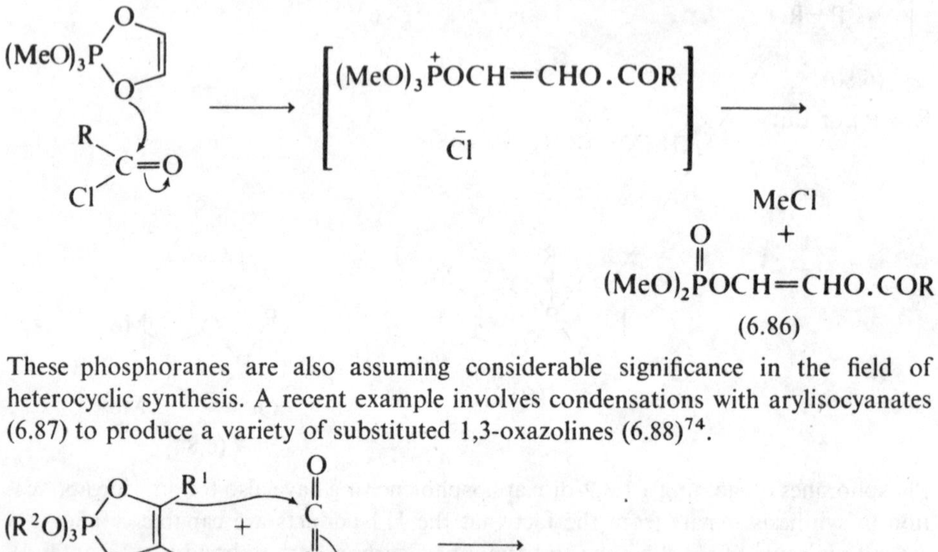

$$(MeO)_2\overset{O}{\underset{\|}{P}}OCH=CHO.COR + MeCl$$
(6.86)

These phosphoranes are also assuming considerable significance in the field of heterocyclic synthesis. A recent example involves condensations with arylisocyanates (6.87) to produce a variety of substituted 1,3-oxazolines (6.88)[74].

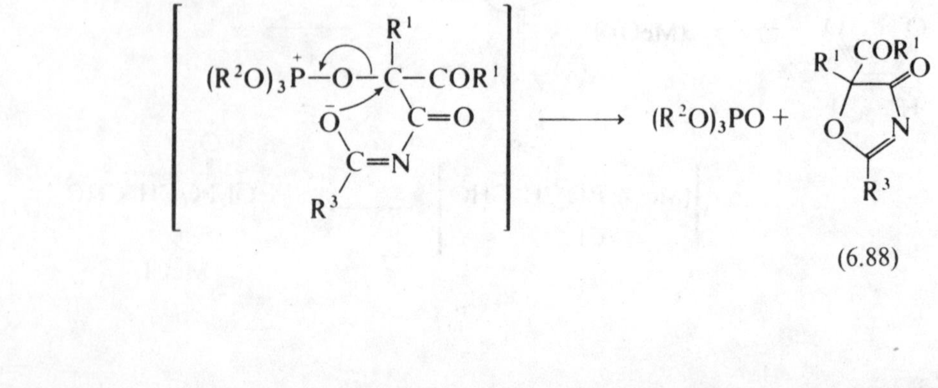

(6.87)

$\longrightarrow (R^2O)_3PO +$

(6.88)

6.3 The chemistry of phosphoranes

Addition to activated double bonds is also possible and the subsequent elimination of phosphate gives rise to a cyclopropane derivative (6.89)[75].

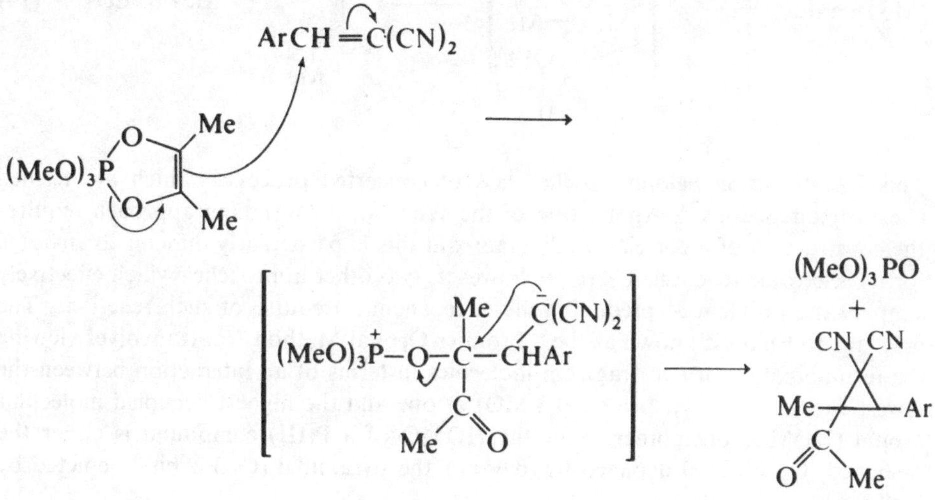

(6.89)

The possible synthetic applications are extensive and it will not have escaped the notice of the reader that the chemistry of pentaoxyphosphoranes (acyclic and cyclic) contributes a new dimension to the chemistry of the biologically indispensable phosphate esters, $(RO)_3PO$.

(ii) *The fragmentation of phosphoranes*: Two years ago, Hoffmann, Howell and Muetterties used orbital symmetry considerations to show that the concerted reaction of eqn. (73) is symmetry forbidden for apical–equatorial loss (or addition) but allowed

$$R_5P \rightleftharpoons R_3P + R_2 \qquad (73)$$

for apical–apical or equatorial–equatorial fragmentation[76]. Thus, the allowed modes for departure are represented by (6.90) and (6.91) and the forbidden mode by (6.92) even though the last is the "least motion" pathway.

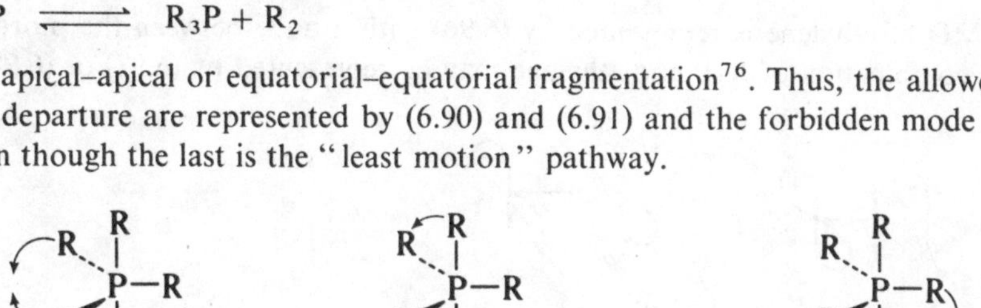

(6.90)　　　　　(6.91)　　　　　(6.92)

equatorial–equatorial　　apical–apical　　equatorial–apical (least motion)

The above fragmentation is a four-electron process and an intuitive knowledge of the Woodward–Hoffmann rules governing concerted reactions suggests that a six-electron (or in general, $4n+2$ electron) process should conform to the opposite set of rules, i.e. equatorial–apical departure should be the allowed mode and diequatorial or diapical fragmentation should be the forbidden processes. The six-electron reaction may be achieved experimentally by the fragmentation of a phosphorane (6.93) containing a five-membered phospholene ring and such a reaction has been shown to be a stereospecific, disrotatory process (eqn. 74, see Chapter 4)[77].

[Structural scheme showing fragmentation with Et₂O₂] (74)

(6.93)

This fragmentation belongs to the class of concerted processes which are termed cheletropic reactions[78]. Application of the Woodward–Hoffmann approach requires the construction of a correlation diagram and this is particularly difficult to envisage for a cheletropic reaction. There are however, two other approaches which effectively simplify the problem of predicting the stereochemical course of such reactions. The first, due to Fukui is known as the "Frontier Orbital Method"[79]. It involves viewing the recombination of two fragment molecules in terms of an interaction between the lowest vacant molecular orbital (LVMO) of one and the highest occupied molecular orbital (HOMO) of another. Now the HOMO of a P(III) compound is either the T-shaped (C_{2v}) orbital depicted by (6.94) or the pyramidal (C_{3v}) orbital depicted by (6.95).

(6.94)   (6.95)   (6.96)

The LVMO of ethylene is represented by (6.96) with a node between the p orbitals. Hence a combination of R_3P and ethylene may be represented by (6.97) or (6.98).

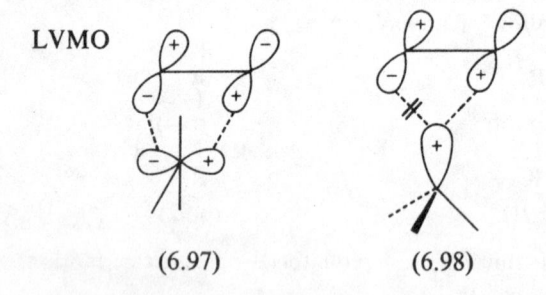

LVMO

(6.97)   (6.98)

In (6.98) which is the nearest approach to equatorial–apical entry, there is a node (//) and the process is disallowed; conversely in (6.97) there is no node, the process is allowed and this represents diequatorial entry in a four-electron process. Extending the argument to a combination of a 1,3-diene with P(III) one may envisage two situations (6.99) and (6.100). The first (6.99) which represents diequatorial entry has a node and is disallowed but the second (e–a) entry is allowed. Notice that both these arguments apply to a *disrotatory* process.

6.3 The chemistry of phosphoranes

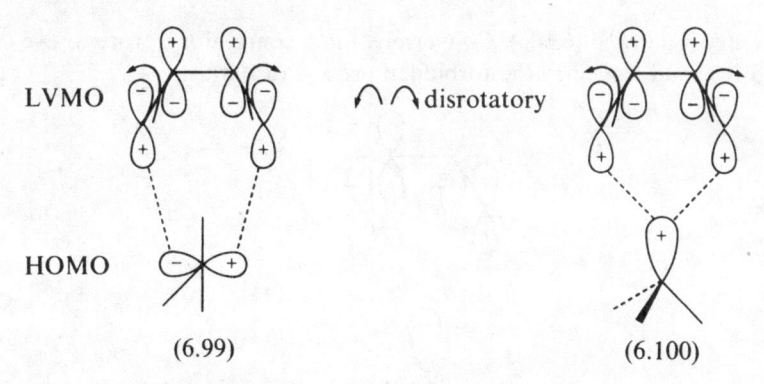

LVMO   ⌒⌒ disrotatory

HOMO

(6.99)   (6.100)

If one allows a conrotatory combination (6.101 and 6.102) the opposite conclusions may be reached, i.e. e–e entry is "allowed" and e–a entry is "forbidden".

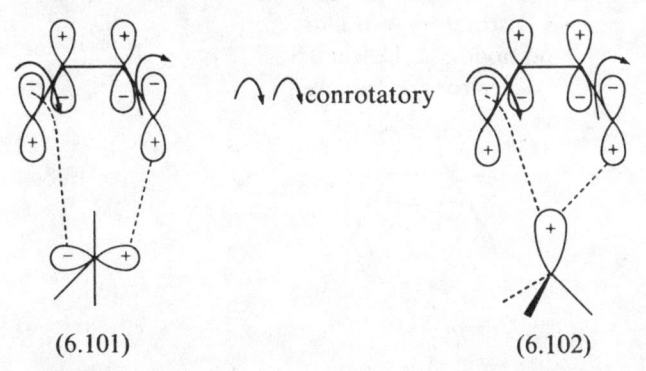

⌒⌒ conrotatory

(6.101)   (6.102)

The second approach, known as the Dewar Transition State Method[80] (after M. J. S. Dewar of the University of Texas) is in some respects, even simpler, since it assumes no knowledge of orbital energetics. It can be summed up briefly by reference to the Diels–Alder combination of ethylene with a molecule of butadiene. In the disrotatory combination (6.103) there are no nodes between the overlapping orbitals and this is known as an "aromatic" or "Dewar" transition state which is allowed for a $(4n+2)$ i.e. in this case six-electron, process. An even number of nodes also gives a Dewar TS In the conrotatory combination (6.104) there is an odd number of nodes (in fact, just one depicted by //) and this constitutes a Möbius TS which is allowed for a $4n$ electron process but, as in this case, forbidden for a $(4n+2)$ electron process.

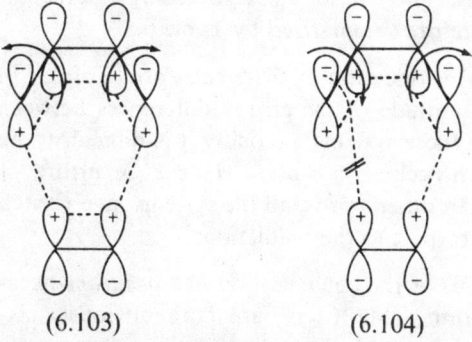

(6.103)   (6.104)

Applying this to the combination of a butadiene with a P(III) compound one sees immediately that for a disrotatory process, diequatorial entry is forbidden (6.105) but apical–equatorial entry is allowed (6.106). Conversely, for a conrotatory process, e–e entry is allowed (6.107) and e–a entry the forbidden process (6.108).

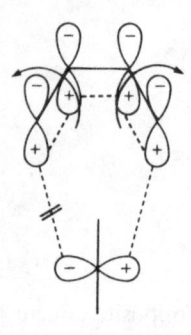

(6.105)
Disrotatory, e–e entry
One node, Möbius TS;
6 electrons, ∴ forbidden

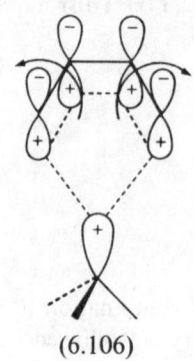

(6.106)
Disrotatory, e–a entry
no nodes, ∴ Dewar TS;
6 electrons, ∴ allowed

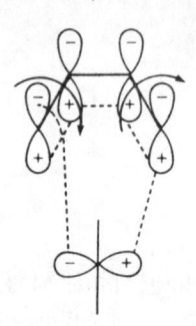

(6.107)
Conrotatory, e–e entry
No nodes, Dewar T.S.
6 electrons, ∴ allowed.

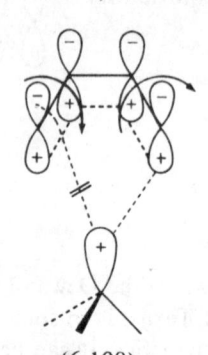

(6.108)
Conrotatory, e–a entry
One node, Möbius T.S.
6 electrons, ∴ forbidden

Exactly the opposite predictions will apply to a $4n$ electron system and although these approaches are not in any sense rigorous, since they ignore the involvement (if any) of vacant d orbitals on phosphorus, they do appear to work and to conform with the Hoffmann–Howell–Muetterties predictions. The overall rules governing fragmentations which are thermally induced are therefore summarized by Table 6.1.

The problems involved in an experimental verification of these rules are twofold. In the first place one has no idea of the magnitude of the energy differences between "allowed" and "forbidden" processes in these systems; secondly it is difficult to be certain whether an observed fragmentation occurs via e–a or via e–e departure of the non-phosphorus containing fragment. In other words, all the systems investigated so far, have been doubly degenerate with respect to the predictions.

For instance, the fragmentation of eqn. (74) or the condensation of *trans,trans*-hexa-2,4-diene with methylphosphonous dichloride (MePCl$_2$)[81] are both obviously six-electron disrotatory processes. This is consistent with the "allowed" e–a fragmentation of a $(4n + 2)$ electron system. It is well established however, that phospholene

6.3 The chemistry of phosphoranes

Table 6.1 Rules for the formation and fragmentation of phosphoranes

Stereochemistry	4n electron process		4n + 2 electron process	
	Mode of entry (or fragmentation)		Mode of entry (or fragmentation)	
	e–e (or a–a)	e–a	e–e (or a–a)	e–a
Disrotatory	allowed	forbidden	forbidden	allowed
Conrotatory	forbidden	allowed	allowed	forbidden

rings may readily adopt a diequatorial situation in a *tbp* particularly when the remaining three groups are highly electronegative. Likewise, the phosphorane (6.109) fragments via a six-electron disrotatory process to give *trans,trans*-hexa-2,4-diene[77]. The

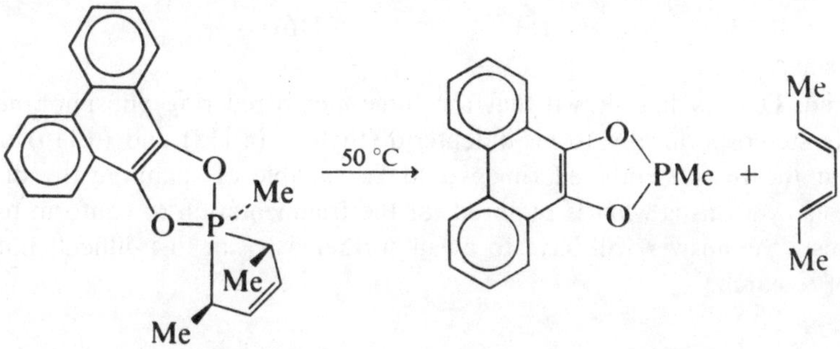

(6.109)

free energy of activation for this process is of the order of 105 kJ mol^{-1} (25 kcal mol^{-1}) which is probably in excess of that required to place the phospholene ring diequatorial, so once again confirmation of e–a departure is lacking.

On heating to 75 °C (or on irradiation) the phosphorane (6.110) fragments to (6.111) which in turn rearranges to cyclooctatetraene[57].

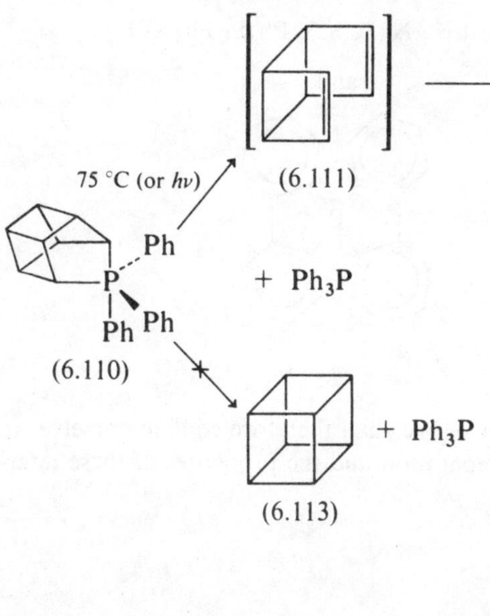

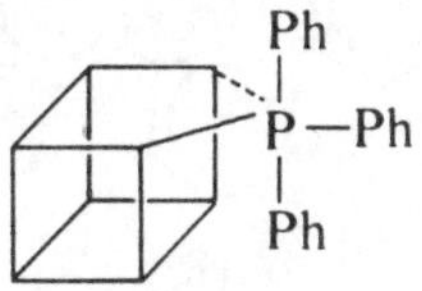

(6.110)

(6.111)

+ Ph$_3$P

(6.113)

+ Ph$_3$P

(6.112)

It is difficult to imagine (6.110) adopting the diequatorial conformation (6.112) not only because of ring strain but also as a result of severe steric crowding between one apical phenyl group and the cage structure. The elimination of tricyclooctadiene (6.111) which is a six-electron process, probably occurs therefore from the e–a conformation. A four-electron fragmentation would generate cubane (6.113) and since this is not detected, the evidence suggests the rules of Table 6.1 are correct.

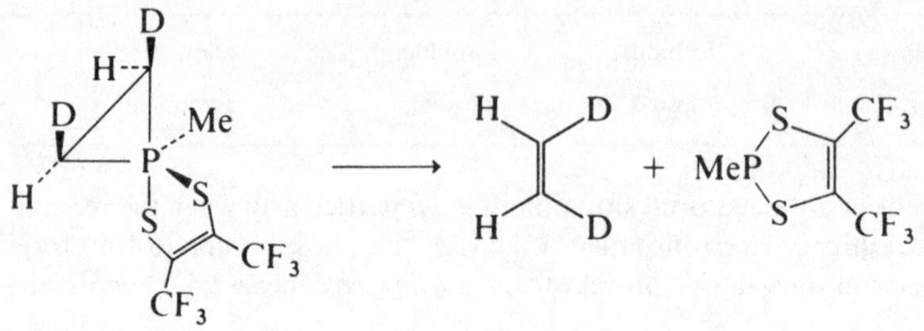

(6.114)   (6.115)   (6.116)

On the other hand, Denney has shown that the three-membered ring phosphorane (6.114) fragments stereospecifically to cis-dideuterio-ethylene (6.115) and (6.116)[82]. It is unlikely that the three-membered ring would be capable of spanning the diequatorial positions, yet this is what is required for the fragmentation to conform to the rules. Obviously the answer will have to await further work in this difficult but intriguing area of research.

6.4 Hexacoordinate phosphorus compounds[1,3]

The existence of stable compounds such as sodium hexafluorophosphate ($Na^+PF_6^-$) and tris-2,2′-biphenylenephosphate anion (6.117) poses a problem of bonding in phosphorus compounds to match that afforded by the pentacovalent structures, but

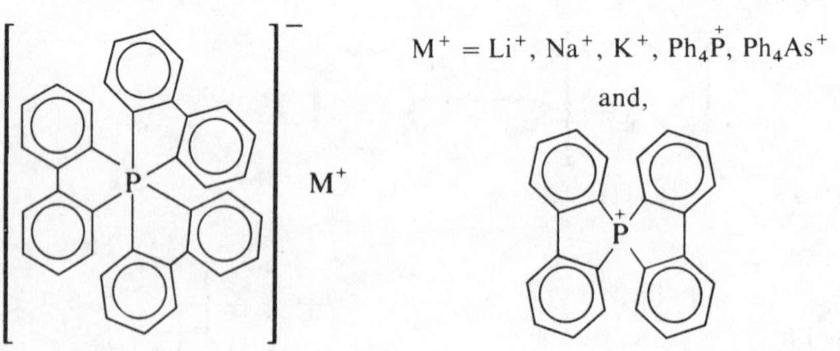

$M^+ = Li^+, Na^+, K^+, Ph_4\overset{+}{P}, Ph_4As^+$

and,

(6.117)

this problem received attention in Chapter 2. We shall therefore confine ourselves in this section to a brief discussion of the preparation and the properties of these interesting molecules.

6.4 Hexacoordinate phosphorus compounds

6.4.1 The preparation of hexacoordinate phosphorus compounds

6.4.1.1 Acyclic compounds

Direct transfer of halogen to phosphoranes may be effected in a number of ways, typified by eqn. (75)[29,48].

$$AgF + PF_5 \longrightarrow Ag^+ PF_6^- \qquad (75)$$

An indirect transfer also offers a route to hexacoordinate compounds by disproportionation of intermediate aminofluorophosphoranes (eqn. 76)[83].

$$2RPF_4 + 2R_2^1NH \longrightarrow 2[RPF_3NR_2^1] \longrightarrow RPF_5^- RPF(NR_2^1)_2^+ \qquad (76)$$

Another simple route involves the addition of HF_2^- to phosphorus trifluoride (eqn. 77) or fluorophosphines (eqn. 78)[84,85].

$$PF_3 + HF_2^- Na^+ \longrightarrow HPF_5^- Na^+ \qquad (77)$$

$$(CF_3)_nPF_{3-n} + HF_2^- Na^+ \longrightarrow (CF_3)_nPHF_{5-n}^- Na^+ \qquad (78)$$

The complexes formed between pentahalogenophosphoranes and Lewis bases may also be regarded as hexacoordinate compounds of the inner "onium-ate" type; an example involving pyridine is shown in eqn. (79)[86]. Even phosphine oxides are capable

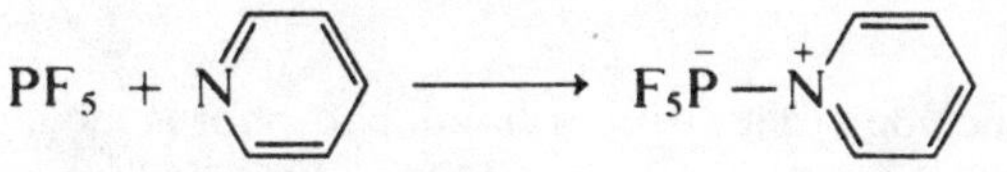

$$PF_5 + N\text{-pyridine} \longrightarrow F_5\overset{-}{P}-\overset{+}{N}\text{-pyridine} \qquad (79)$$

of forming such complexes (eqn. 80)[87] and phosphorus pentahalides will also react with

$$Ph_3P=O + PCl_5 \longrightarrow Ph_3\overset{+}{P}-O-\overset{-}{P}Cl_5 \qquad (80)$$

certain difunctional bases such as acetylacetone, phenanthroline or dimethylurea (eqn. 81)[88] with substitution of some or all of the halogen atoms and formation of another type of "onium-ate" complex (6.118).

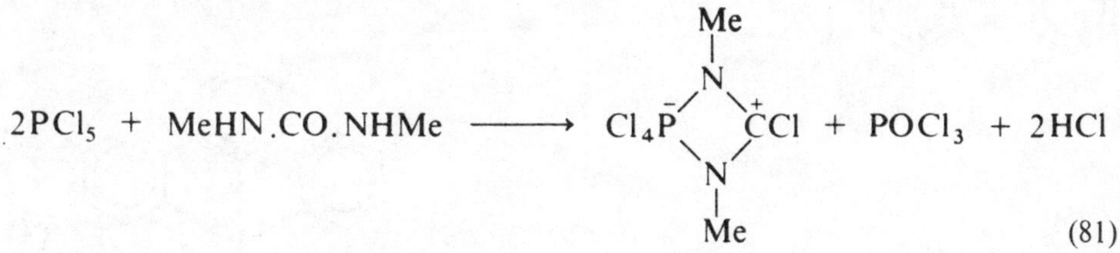

$$2PCl_5 + MeHN.CO.NHMe \longrightarrow Cl_4\overset{-}{P}\underset{\underset{Me}{N}}{\overset{\overset{Me}{N}}{\diagup\diagdown}}\overset{+}{C}Cl + POCl_3 + 2HCl \qquad (81)$$

(6.118)

6.4.1.2 Cyclic compounds

One of the early examples of such compounds was that derived from the base-catalyzed condensation of halogenocyclophosphazene (6.119) with catechol[89]. Recently an x-ray analysis has confirmed the octahedral structure of the product (6.120)[90].

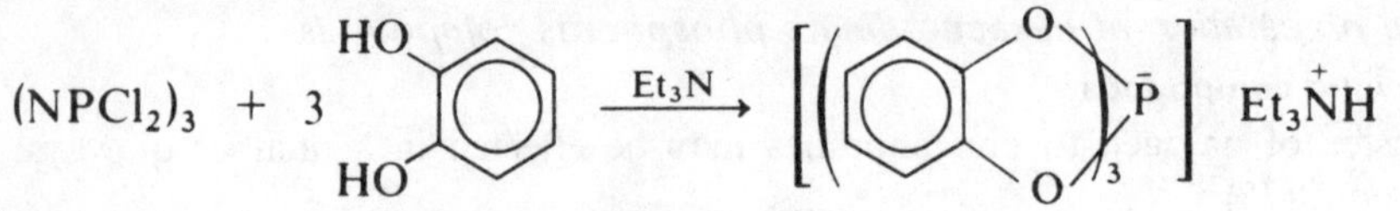

Chapter 6 / Pentacoordinate and hexacoordinate organophosphorus chemistry

$(NPCl_2)_3$ + 3 (catechol) $\xrightarrow{Et_3N}$ [spirophosphate]$^-$ $Et_3\overset{+}{N}H$

(6.119)   (6.120)

A high proportion of the work in this area is due to Hellwinkel and his collaborators. They devised the synthesis of tris-2,2′-biphenylene phosphate anion (6.117) from the reaction of bis-2,2′-dilithiobiphenyl (6.121) with either PCl_5[91] or the spirophosphonium salt (6.122)[92] and were able to demonstrate the octahedral structure of the molecule by separation of the enantiomeric forms of the chiral octahedron[92].

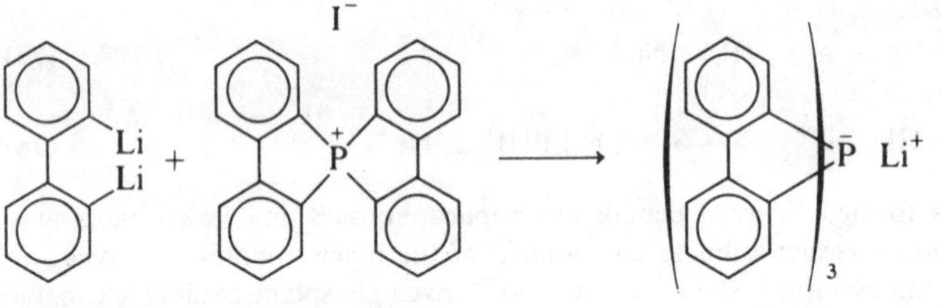

(6.121)   (6.122)   (6.117)

The method has now been extended to include cyclic six-coordinate phosphorus molecules with oxygen in the ring as exemplified by compounds (6.123) and (6.124)[93].

Denney adopted a slightly different approach in effecting a base-catalyzed reaction of

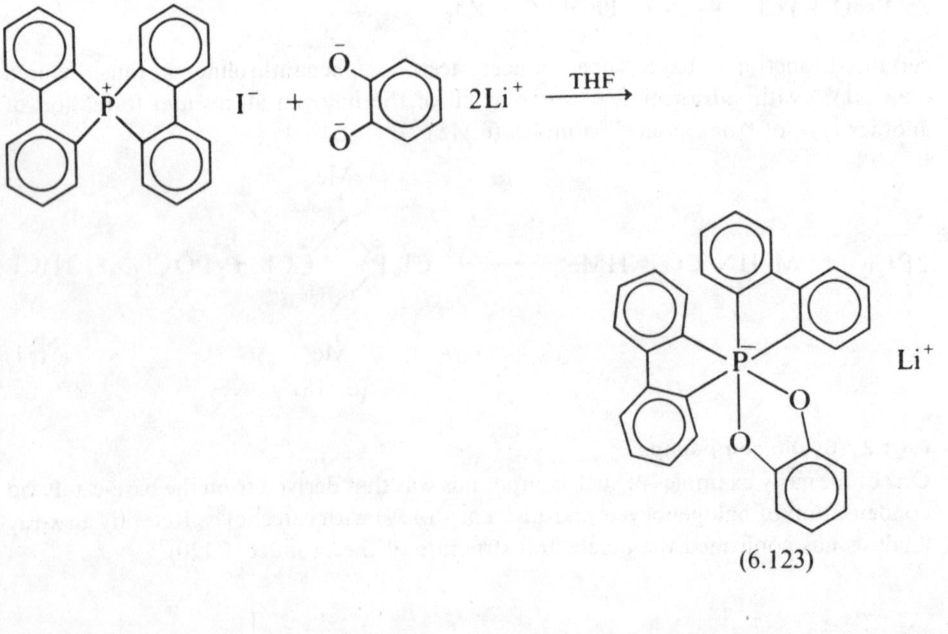

(6.123)

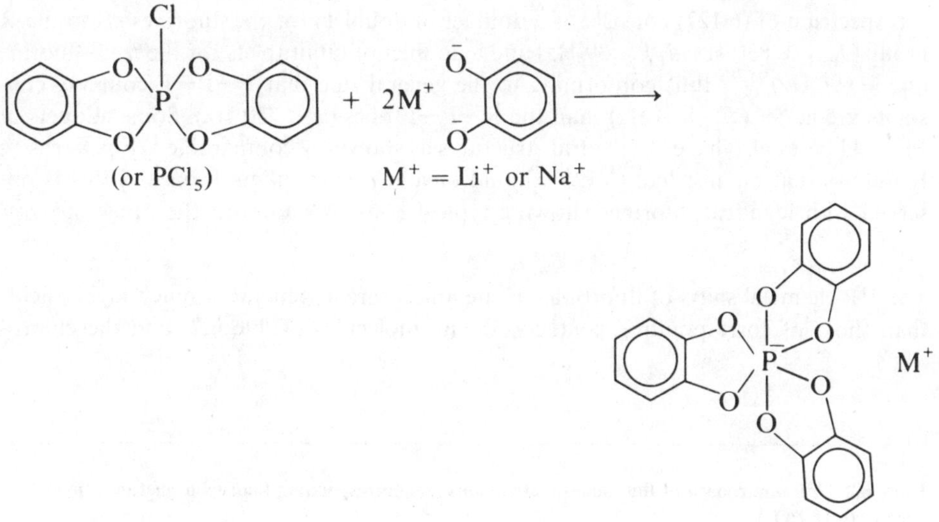

6.4 Hexacoordinate phosphorus compounds

(6.124)

the pentaoxyphosphorane (6.125) prepared by his "exchange procedure", with 1,2-ethane diol to obtain (6.126)[94].

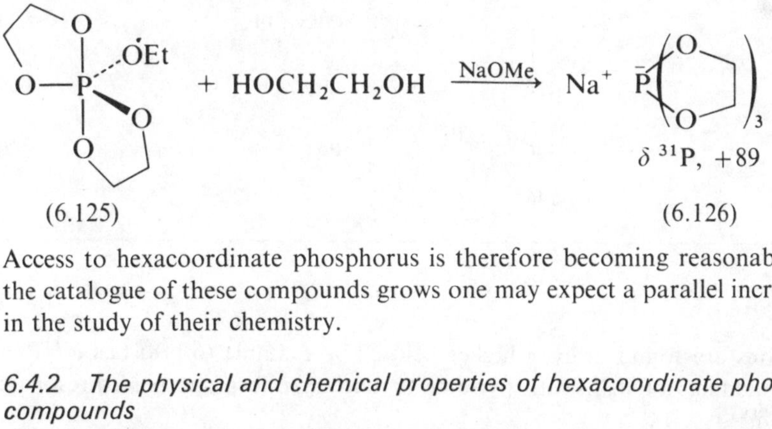

(6.125)   + HOCH$_2$CH$_2$OH $\xrightarrow{\text{NaOMe}}$ Na$^+$ (6.126)

$\delta\ ^{31}P,\ +89$

Access to hexacoordinate phosphorus is therefore becoming reasonably easy and as the catalogue of these compounds grows one may expect a parallel increase of activity in the study of their chemistry.

6.4.2 The physical and chemical properties of hexacoordinate phosphorus compounds

Most hexacoordinate phosphorus compounds are stable, crystalline solids and x-ray investigations on compounds such as (6.118)[95] and (6.120)[90] have revealed the expected octahedral structure. In the case of the former (6.118) the diazaphosphetidine ring is planar and spans adjacent positions in the octahedron with N—C ring bonds shorter than N—CH$_3$ single bonds, as required by the structure. Spectroscopic data, in particular ^{19}F n.m.r., also supports the octahedral configuration. For example, the

(6.118)   (6.127), R = Me or Ph   (6.128)

^{19}F spectrum of (6.127) consists of a doublet of doublets for the fluorines *cis* to the R group ($J_{P-F} \approx 830$ Hz; $J_{FF} \approx 39$ Hz) and a doublet of quintuplets for the *trans* fluorine ($J_{PF} = 690$ Hz)[63,96] thus conforming to the general rule that F—P—F coupling constants are larger ($J > 800$ Hz) than the R—P—F ones ($J < 730$ Hz). Some adducts of PF$_5$ and bases also have octahedral structures as shown by comparable ^{19}F patterns[97]. It follows that anions, like (6.128) should have a *trans* configuration and this is observed with identical fluorines showing typical *cis* J_{P-F} values in the range 858–898 Hz[98,99].

The ^{31}P chemical shifts of fluorophosphate anions are in general at much higher fields than those of corresponding pentacoordinate molecules (Table 6.2) and the chloro-

Table 6.2 ^{31}P n.m.r. data of fluorophosphate anions and corresponding fluorophosphoranes (ppm relative to H$_3$PO$_4$)

Anions	$\delta\ ^{31}$P	Phosphoranes	$\delta\ ^{31}$P
(alkyl) $\overline{PF_5}$	$\sim +125$	(alkyl) PF$_4$	$\sim +30$
(aryl) $\overline{PF_5}$	$\sim +136$	(aryl) PF$_4$	$\sim +50$
$\overline{PF_6}$	$+146$	PF$_5$	$+30$
$\overline{PF_5}$←CNMe	$+146$	—	

phosphate anions are found at even higher fields. For example (6.118) has $\delta\ ^{31}$P = $+200$ and the adducts of PCl$_5$ with pyridine or tertiary phosphine oxides resonate at *ca* $+300$ ppm[87].

An interesting sequence of ^{31}P δ values occurs for the spirophosphate anions, the shifts moving steadily to lower fields as the number of oxygen atoms attached to phosphorus increases (Table 6.3)[93]. Presumably, the electronegative oxygen atoms disperse the negative charge on phosphorus and so decrease the shielding. Since the compound prepared by Denney (6.126) has $\delta\ ^{31}$P = $+89$, any delocalization of charge from oxygen into the aromatic rings of (6.124) apparently has little effect on the chemical shift.

Not a great deal is known as yet about the chemistry of hexacoordinate anions. Phosphate complexes containing chlorine are sensitive to moisture but the fluorine analogues are frequently stable to water, particularly when large cations or perfluoroalkyl groups are present. Tris-(*o*-phenylenedioxy)phosphate (6.124) is stable to cold water but hydrolyses to phosphate on heating (eqn. 82)[89,93]. This immediately poses the

6.4 Hexacoordinate phosphorus compounds

Table 6.3 ^{31}P δ values of spirophosphate anions (ppm relative to H_3PO_4)

Compound	$\delta\ ^{31}P$
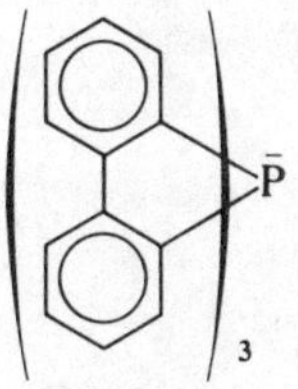 (biphenyl)$_3$P⁻	+181
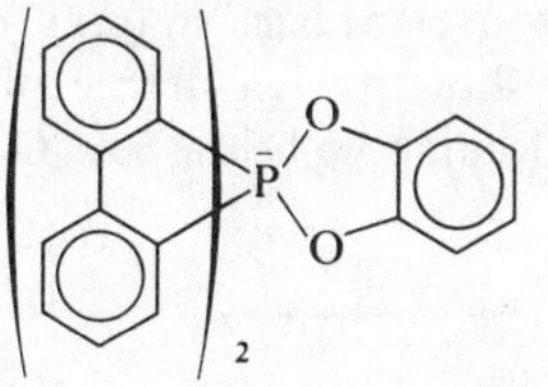 (biphenyl)$_2$P⁻(catecholate)	+147
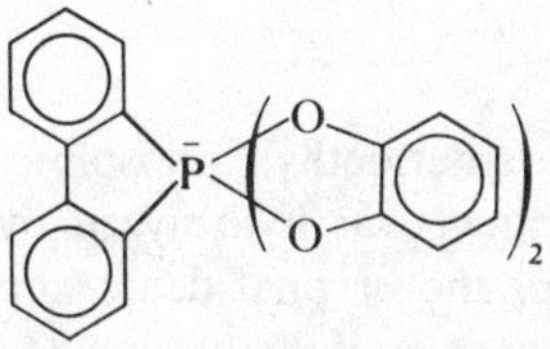 (biphenyl)P⁻(catecholate)$_2$	+106
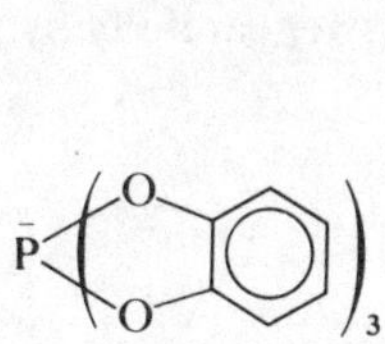 P⁻(catecholate)$_3$	+82

question of whether hexacoordinate structures may be involved in the hydrolysis of oxyphosphoranes or phosphate esters (see Chapter 8).

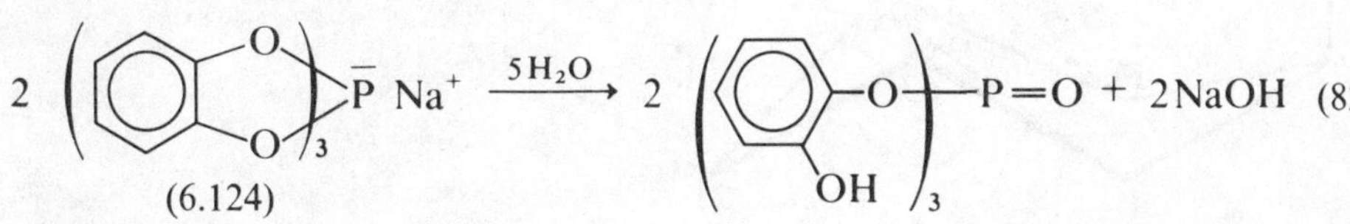

$$2\ (catecholate)_3\bar{P}\ Na^+ \xrightarrow{5H_2O} 2\ \left(\begin{array}{c}\text{C}_6\text{H}_4\text{-O}\\ \text{OH}\end{array}\right)_3 P=O\ +\ 2NaOH \quad (82)$$
(6.124)

The hexaarylphosphate ions (e.g. 6.117) are completely stable towards water and are only decomposed by strong electrophiles such as the proton or bromine (eqn. 83)[92,100].

(6.117) M⁺ $\xrightarrow{XY}$ [structure (6.130)] + MY   (83)

X = H, Br;
Y = Cl, Br   (6.130)

In the case of X = Br, the resultant phosphorane (6.130) is converted back to (6.117) by treatment with n-butyllithium. These compounds also show a high degree of thermal stability and the potassium salt of (6.117) does not decompose below 300 °C.

Problems

1 If pentaethoxyphosphorane was prepared by treating deuterioethyl phosphite, $(C_2D_5O)_3P$, with diethyl peroxide, $(C_2H_5O)_2$, and the product was hydrolyzed by water to triethyl phosphate and ethanol, what percentage of the original deuterium would you expect to find in each product? Explain your answer carefully by showing all possible products and the intermediates leading to each.

2 The topological diagram shown below is that due to Cram.

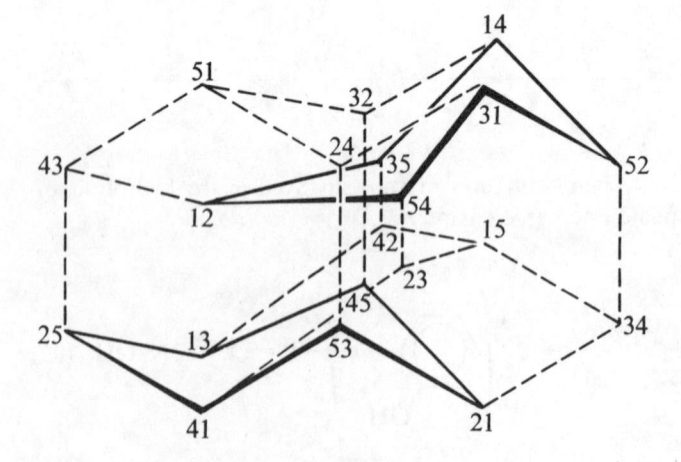

When applied to the phosphorane (1) explain what is meant by (i) each apex and (ii) each line joining any two apices.

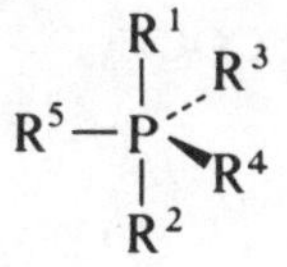

Problems   249

$$R^5-\underset{R^2}{\overset{R^1}{\underset{|}{P}}}\begin{matrix}R^3\\R^4\end{matrix}$$

(1)

$$\underset{EtO}{\overset{Ph}{\underset{MeO}{\underset{3}{\overset{4}{\underset{|}{P}}}}}}\overset{1}{\underset{CH_2}{\overset{O}{\underset{5}{\overset{Me}{\diagup}}}}}\text{COMe}$$

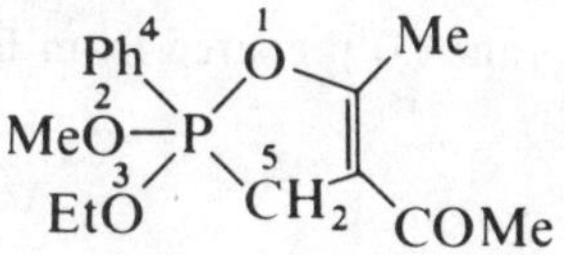

(2)

Consider the phosphorane (2) with the substituents on phosphorus numbered as shown. Assuming that the strain rule is inviolate when applied to five-membered rings, what does the topological diagram reveal about the possibility of racemization at phosphorus? Explain your reasoning.

3 Suggest mechanisms for the following reactions:

(a) [Ph₂P⁺ bridged bicyclic structure] Br⁻ $\xrightarrow{\text{PhLi,}\atop\text{THF, heat}}$ [cyclooctadiene] + Ph₃P

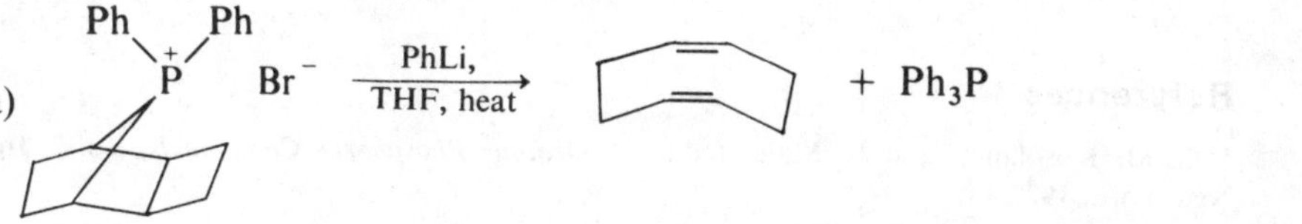

(b) [Me-substituted dioxaphospholene]Me—C(O)—O—P(OMe)₃ + RCN=C=S ⟶ [oxazoline with MeC(O), Me, S, N, R] + (MeO)₃PO

(c) (EtO)₅P + [cyclohexane-1,2-diol] ⟶ [bicyclic ether] + (EtO)₃PO

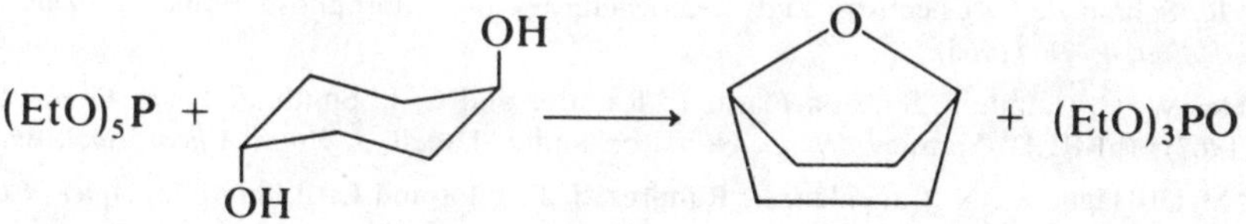

(d) [Ph,Ph-dioxaphospholene with P(OMe)₃] $\xrightarrow{\text{RSCl}}$ PhC(O)—C(Ph)(Cl)(SR) + (MeO)₃PO

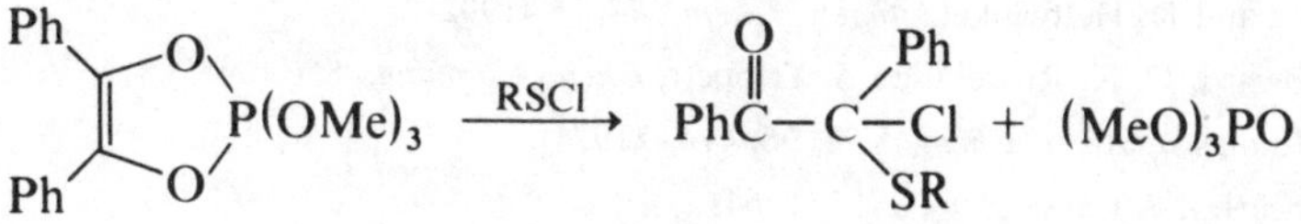

(e) [benzodioxaphosphole PR] + [catechol] $\xrightarrow{\text{Et}_3\text{N, heat}}$ [spirobis(benzodioxa)phosphorane with R] + H₂

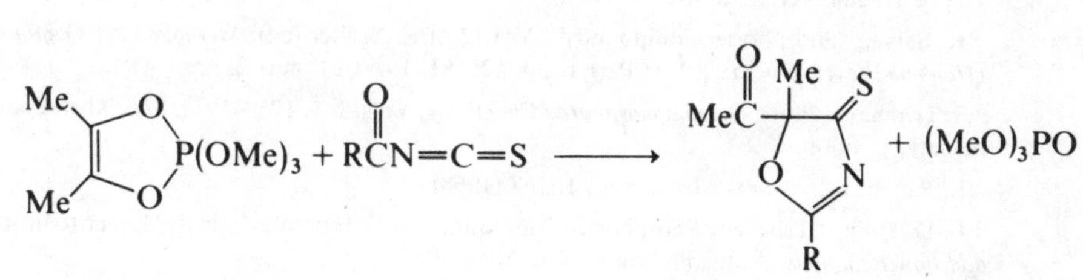

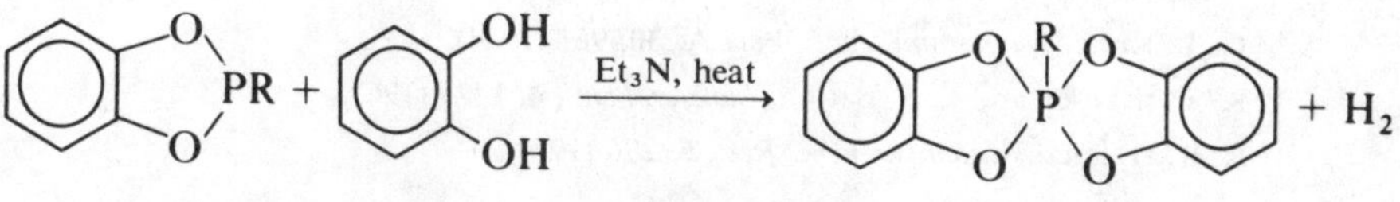

4 Consider the phosphorane (3) prepared from B_2PA and $CF_3CO.COCF_3$.

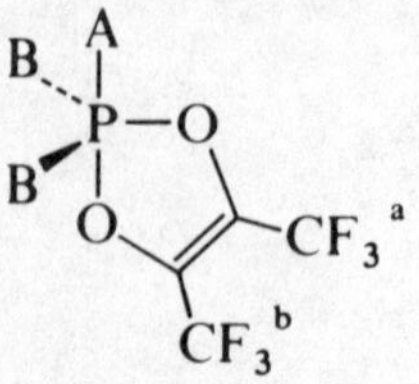

(3)

The group CF_3 (a) is not equivalent to CF_3 (b) and in this pseudorotamer the two may be distinguished by ^{19}F n.m.r. Bearing in mind the strain rule, trace the pseudorotation pathway necessary to make the CF_3 groups equivalent and therefore indistinguishable by n.m.r.

References

[1a] G. M. Kosolapoff and L. Maier (editors), *Organic Phosphorus Compounds*, Vol 3, Interscience, New York, 1972.

[1b] P. Luckenbach, *Dynamic Stereochemistry of Pentacoordinated Phosphorus and Related Elements*, Georg Thieme Verlag, Stuttgart, 1973.

[2] K. Sasse, "Phosphorus Compounds", Vol 12 of E. Müller (ed), *Methoden der Organischen Chemie (Houben-Weyl)*, Thieme, 1963; Part 1, pp. 125–81, 193–620; part 2, pp. 1–994.

[3] S. Trippett (editor), *Organophosphorus Chemistry*, Vols. 1–5, 1969–1974, Specialist reports, Chemical Society, London.

[4] F. Ramirez, *Accounts Chem. Res.*, **1**, 168 (1968).

[5] L. D. Quin, "Trivalent Phosphorus Compounds as Dienophiles", in J. Hamer (editor), *1,4–Cycloaddition Reactions*, Academic Press, New York, 1967.

[6] K. D. Berlin and D. M. Hellwege, "Carbon Phosphorus Heterocycles", *Topics in Phosphorus Chemistry*, Vol. 6, p. 1, Interscience, 1969.

[7] R. Schmutzler, "Chemistry and Stereochemistry of Fluorophosphoranes", *Angew. Chem., Int. Ed. Eng.* **4**, 496 (1965).

[8] (a) W. C. Hamilton, S. J. La Placa, F. Ramirez and C. P. Smith, *J. Amer. Chem. Soc.*, **89**, 2298 (1967); (b) R. D. Spratley, W. C. Hamilton and J. Ladell, *J. Amer. Chem. Soc.*, **89**, 2272 (1967).

[9] M-Ul-Hague, C. N. Caughlan, F. Ramirez, J. P. Pilot and C. P. Smith, *J. Amer. Chem. Soc.*, **93**, 5229 (1971).

[10] D. D. Swank, C. N. Caughlan, F. Ramirez and J. F. Pilot, *J. Amer. Chem. Soc.*, **93**, 5236 (1971).

[11] G. Wittig and D. Hellwinkel, *Angew. Chem.*, **74**, 76 (1962).

[12] J. A. Howard, D. R. Russell and S. Trippett, *Chem. Commun.*, 856 (1973).

[13] R. R. Holmes, *J. Amer. Chem. Soc.*, **96**, 4143 (1974).

[14] P. J. Wheatley, *J. Chem. Soc.*, 3718 (1964).

[15] A. L. Beauchamp, M. J. Bennett and F. A. Cotton, *J. Amer. Chem. Soc.*, **90**, 6675 (1968).

[16] I. R. Beattie, K. M. S. Livingston, G. A. Ozin and R. Sabine, *J. Chem. Soc., Dalton*, 784 (1972).

[17] G. L. Kok, *Spectrochim. Acta*, Part A, **30**, 961 (1974).

[18] K. W. Hansen and L. S. Bartell, *Inorg. Chem.*, **4**, 1775 (1965).

[19] R. R. Holmes, *Accounts Chem. Res.*, **5**, 296 (1972).

[20] R. W. Shaw Jr., T. X. Carroll and T. Darrah Thomas, *J. Amer. Chem. Soc.*, **95**, 2033 (1973).
[21] E. L. Muetterties, W. Mahler and R. Schmutzler, *Inorg. Chem.*, **2**, 613 (1963).
[22] R. K. Oram and S. Trippett, *J. Chem. Soc., Perkin I*, 1300 (1973).
[23] R. Schmutzler, *Advan. Fluorine Chemistry*, Vol. 5, p. 66, Butterworths, London, 1965.
[24] F. Mathey and J. Bensoam, *C. R. Acad. Sci. (Paris)*, **274C**, 1095 (1972).
[25] R. Mitch, *J. Amer. Chem. Soc.*, **89**, 6297 (1967).
[26] W. Firth, S. Frank, M. Garber and V. Wystrach, *Inorg. Chem.*, **4**, 365 (1965).
[27] N. J. De'Ath, D. Z. Denney and D. B. Denney, *Chem. Commun.*, 272 (1972).
[28] N. J. De'Ath, D. Z. Denney, D. B. Denney and C. D. Hall, *Phosphorus*, **3**, 205 (1974).
[29] G. I. Drozd, *Russ. Chem. Rev. (Eng. Trans.)*, **39**(1), 1 (1970).
[30] D. B. Denney, D. Z. Denney, B. C. Chang and K L. Marsi, *J. Amer. Chem. Soc.*, **91**, 5243 (1969).
[31] D. B. Denney and D. H. Jones, *J. Amer. Chem. Soc.*, **91**, 5821 (1969).
[32] D. B. Denney, D. Z. Denney, C. D. Hall and K. L. Marsi, *J. Amer. Chem. Soc.*, **94**, 245 (1972).
[33] P. D. Bartlett, A. L. Baumstark and M. E. Landis, *J. Amer. Chem. Soc.*, **95**, 6486 (1973).
[34] L. L. Chang and D. B. Denney, *Chem. Commun.*, 84 (1974).
[35] N. J. De'Ath and D. B. Denney, *Chem. Commun.*, 395 (1972).
[36] F. Ramirez, *Pure Appl. Chem.*, **9**, 337 (1964).
[37] B. E. Ivanov and V. F. Zhetukhin, "Reactivity of Trivalent Phosphorus Compounds", *Russ. Chem. Rev. (Eng. Trans.)*, **39**(5), 358 (1970).
[38] M. Wieber and W. R. Hoos, *Monatsschr.*, **101**, 776 (1970).
[39] R. Burgada and D. Bernard, *C. R. Acad. Sci. (Paris).*, **273C**, 164 (1971).
[40] J. I. G. Cadogan, D. S. B. Grace, P. K. K. Lim and B. S. Tait, *Chem. Commun.*, 520 (1972).
[41] L. Anschutz, H. Boedeker, W. Broeker and F. Wenger, *Annalen*, **454**, 71 (1927).
[42] I. N. Zhmurova and A. V. Kirsanov, *J. Gen. Chem. USSR*, **29**, 1668 (1959).
[43] (a) F. Ramirez, A. J. Bigler and C. P. Smith, *J. Amer. Chem. Soc.*, **90**, 3507 (1968); (b) F. Ramirez, A. J. Bigler and C. P. Smith, *Tetrahedron*, **24**, 5041 (1968).
[44] P. M. Treichel and R. A. Goodrich, *Inorg. Chem.*, **4**, 1424 (1965).
[45] D. Brown, G. Fraser and D. Sharp, *J. Chem. Soc. A*, 171 (1966).
[46] G. I. Drozd, M. A. Sokal'skii, V. V. Sheluchenko, *Zh. Obshch. Khim.*, **39**, 935 (1969).
[47] R. Schmutzler, *Angew. Chem.*, **76**, 570 (1964).
[48] R. Schmutzler and G. S. Reddy, *Inorg. Chem.*, **4**, 191 (1965).
[49] S. Z. Ivin, K. V. Karavanov, V. V. Lysenko and T. N. Sosina, *Zh. Obshch. Khim.*, **36**, 1246 (1966).
[50] S. C. Peake, M. Fild, M. J. C. Hewson and R. Schmutzler, *Inorg. Chem.*, **10**, 2723 (1971).
[51] B. C. Chang, W. Conrad, D. B. Denney, D. Z. Denney, R. Edelman, R. L. Powell and D. W. White, *J. Amer. Chem. Soc.*, **93**, 4004 (1971).
[52] M. Schlosser, T. Kadibelban and G. Steinhoff, *Annalen*, **743**, 25 (1971).
[53] H. Daniel and J. Paetsch, *Chem. Ber.*, **101**, 1451 (1968).
[54] L. Kolditz, *Z. Anorg. Chem.*, **286**, 307 (1956).
[55] D. W. Allen, B. G. Hutley and M. T. J. Mellor, *J. Chem. Soc. Perkin II*, 63 (1972).
[56] D. Hellwinkel, *Chem. Ber.*, **102**, 528 (1969).
[57] T. J. Katz and E. W. Turnblom, *J. Amer. Chem. Soc.*, **92**, 6701 (1970).
[58] (a) E. W. Turnblom and D. Hellwinkel, *Chem. Commun.*, 404 (1972); (b) E. W. Turnblom and T. J. Katz, *J. Amer. Chem. Soc.*, **93**, 4065 (1971).
[59] R. G. Cavell and R. D. Leary, *Chem. Commun.*, 1520 (1970).
[60] R. Schmutzler, *Angew. Chem.*, **77**, 530 (1965).

[61] H. Schmidbauer, K. H. Mitschke and J. Weidlein, *Angew. Chem., Int. Ed. Eng.*, **11**, 144 (1972).
[62] M. Schlosser, T. Kadibelban and G. Steinhoff, *Annalen*, **743**, 25 (1971).
[63] E. L. Muetterties and W. Mahler, *Inorg. Chem.*, **4**, 119 (1965).
[64] R. Goodrich and P. Treichel, *J. Amer. Chem. Soc.*, **88**, 3509 (1966).
[65] D. B. Denney and S. T. D. Gough, *J. Amer. Chem. Soc.*, **87**, 138 (1965).
[66] D. B. Denney and L. Søferstein, *J. Amer. Chem. Soc.*, **88**, 1839 (1966).
[67] D. Houalla, R. Wolf, D. Gagnaire and K. B. Robert, *Chem. Commun.*, 443 (1969).
[68] (a) D. Gorenstein and F. H. Westheimer, *J. Amer. Chem. Soc.*, **92**, 634 (1970); (b) D. Gorenstein, *J. Amer. Chem. Soc.*, **92**, 644 (1970).
[69] D. B. Denney, R. L. Powell, A. Taft and D. Twitchell, *Phosphorus*, **1**, 151 (1971).
[70] A. P. Stewart and S. Trippett, *Chem. Commun.*, 1279 (1970).
[71] F. Ramirez, S. L. Glaser, A. J. Bigler and J. F. Pilot, *J. Amer. Chem. Soc.*, **91**, 496 (1969).
[72] F. Ramirez, S. B. Bhatia, A. J. Bigler and C. P. Smith, *J. Org. Chem.*, **33**, 1192 (1968).
[73] F. Ramirez, S. L. Glaser, A. J. Bigler and J. F. Pilot, *J. Amer. Chem. Soc.*, **91**, 5966 (1969).
[74] F. Ramirez and C. D. Telefus, *J. Org. Chem.*, **34**, 376 (1969).
[75] E. Corre and A. Foucard, *Chem. Commun.*, 570 (1971).
[76] R. Hoffman, J. M. Howell and E. L. Muetterties, *J. Amer. Chem. Soc.*, **94**, 3047 (1972).
[77] C. D. Hall, J. D. Bramblett and F. S. Lin, *J. Amer. Chem. Soc.*, **94**, 9264 (1972).
[78] R. B. Woodward and R. Hoffman, *Angew. Chem., Int. Ed. Eng.*, **8**, 781 (1969).
[79] K. Fukui and H. Fujimoto, "Orbital Symmetry and Electrocyclic Rearrangements", in *Mechanisms of Molecular Migrations*, Vol 2, ed. B. S. Thyagarajan, Interscience, New York, 1969.
[80] (a) M. J. S. Dewar, *Angew. Chem. Int. Ed. Eng.*, **10**, 761 (1971); (b) H. E. Zimmerman, *Accounts Chem. Res.*, **4**, 272 (1971).
[81] A. Bond, M. Green and S. C. Pearson, *J. Chem. Soc. B*, 929 (1968).
[82] D. B. Denney and Li Shang Shih, *J. Amer. Chem. Soc.*, **96**, 317 (1974).
[83] R. Schmutzler and G. S. Reddy, *Inorg. Chem.*, **5**, 164 (1966).
[84] J. F. Nixon and J. R. Swain, *Chem. Commun.*, 997 (1968).
[85] J. F. Nixon and J. R. Swain, *Inorg. Nucl. Chem. Lett.*, **5**, 295 (1969).
[86] R. R. Holmes, W. P. Gallagher and R. P. Carter, *Inorg. Chem.*, **2**, 437 (1963).
[87] H. Binder and E. Fluck, *Z. Anorg. Allgem. Chem.*, **365**, 166 (1969).
[88] H. P. Latscha and P. B. Hormuth, *Angew. Chem., Int. Edn. Eng.*, **7**, 299 (1968).
[89] H. R. Allcock, *J. Amer. Chem. Soc.*, **85**, 4050 (1963); **86**, 2591 (1964).
[90] H. R. Allcock and E. C. Bissell, *Chem. Commun.*, 676 (1972).
[91] D. Hellwinkel, *Chem. Ber.*, **98**, 576 (1965).
[92] D. Hellwinkel, *Chem. Ber.*, **99**, 3628, 3642, 3660, 3668 (1966).
[93] D. Hellwinkel and H. J. Wilfinger, *Chem. Ber.*, **103**, 1056 (1970).
[94] B. C. Chang, D. B. Denney, R. L. Powell and D. W. White, *Chem. Commun.*, 1070 (1971).
[95] M. L. Ziegler and J. Weiss, *Angew. Chem., Int. Ed. Eng.*, **8**, 455 (1969).
[96] R. Schmutzler, *Inorg. Chem.*, **7**, 1327 (1968).
[97] F. N. Tebbe and F. L. Muetterties, *Inorg. Chem.*, **7**, 172 (1968).
[98] J. Jander, D. Börner and U. Engelhardt, *Annalen*, **726**, 19 (1969).
[99] E. O. Bishop, P. R. Carey, J. F. Nixon and J. R. Swain, *J. Chem. Soc. A*, 1074 (1970).
[100] H. J. Wilfinger, Dissertation, Heidelberg (1970).

Chapter 7
Tetracoordinate organophosphorus chemistry: Part 1
phosphonium salts and phosphorus ylids

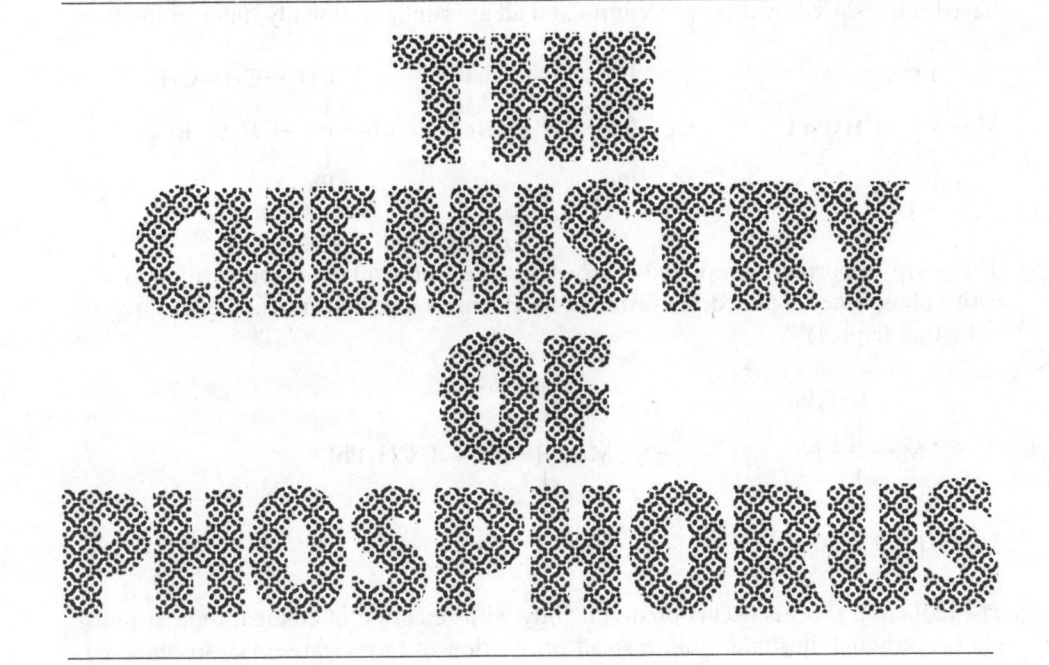

7.1 Phosphonium salts[1-7]

7.1.1 Introduction

As their name implies, phosphonium salts are derivatives of the phosphonium ion, $^+PH_4$, in which hydrogen atoms are replaced by carbon. They may be acyclic (7.1), monocyclic (7.2) or bicyclic (7.3) and in the last case phosphorus is the heteroatom of a spirophosphonium cation.

$R_n \overset{+}{P} H_{4-n}$ $(H_2C)_m \overset{+}{P} \begin{smallmatrix} R \\ R \end{smallmatrix}$

(7.1)   (7.2)   (7.3)

$n = 1-4$ $m = 0-5$

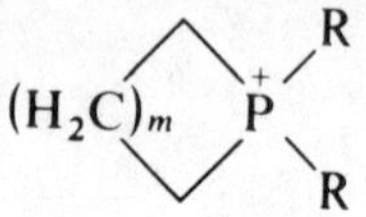

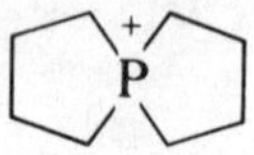

In common with the analogous ammonium ions, phosphonium ions have tetrahedral geometry and hence a phosphorus atom bound to four different substituents in a phosphonium salt, $R^1R^2R^3R^4P^+X^-$, is chiral and exists in two enantiomeric forms which may be separated by the conventional procedures of fractional crystallization using an optically active anion (e.g. $X^- = D(+)$-camphorsulphonate or $X^- = D(-)$-dibenzoylhydrogen tartrate)[8,9]. Salts such as (7.4), (7.5) and (7.6) and many others have been resolved by these procedures and all are configurationally stable in solution.

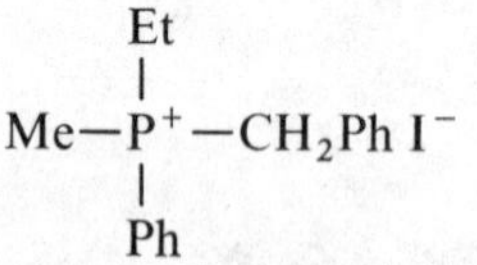

$\begin{array}{c} Et \\ | \\ Me-P^+-CH_2Ph \;\; I^- \\ | \\ Ph \end{array}$ $\begin{array}{c} Pr^n \\ | \\ Me-P^+-CH_2Ph \;\; \bar{B}r \\ | \\ Ph \end{array}$ $\begin{array}{c} CH_2-CH=CH_2 \\ | \\ Me-P^+-CH_2Ph \;\; \bar{B}r \\ | \\ Ph \end{array}$

(7.4) 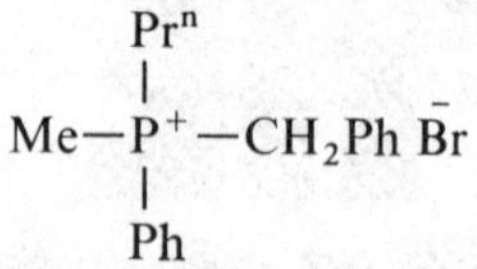   (7.5) 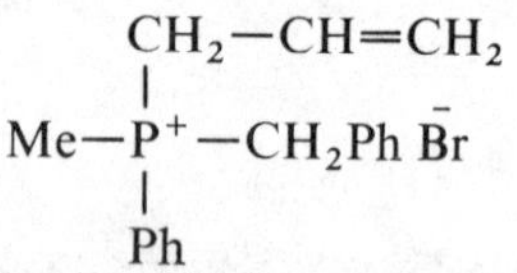   (7.6)

These are important compounds since they are frequently the source of optically active phosphines obtained, for instance, by Horner's ingenious method of electrolytic reduction (eqn. 1)[10].

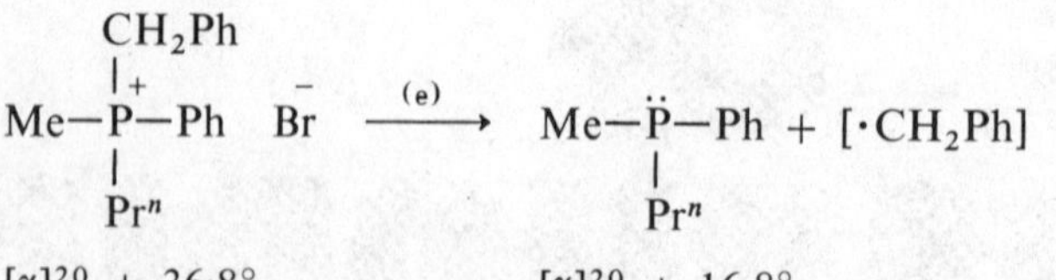

$$\begin{array}{c} CH_2Ph \\ |+ \\ Me-P-Ph \;\; \bar{B}r \\ | \\ Pr^n \end{array} \xrightarrow{(e)} \begin{array}{c} \cdot\cdot \\ Me-P-Ph \\ | \\ Pr^n \end{array} + [\cdot CH_2Ph] \qquad (1)$$

$[\alpha]_D^{20} + 36.8°$ $[\alpha]_D^{20} + 16.8°$

The reduction process occurs predominantly with retention of configuration at phosphorus although in some cases, a small proportion of racemization is also observed. Another useful method for the preparation of optically active phosphines involves the resolution of β-cyanoethylphosphonium salts as the dibenzoyl hydrogen tartrates (e.g. 7.7) and subsequent decomposition of these salts by the action of sodium methoxide in refluxing methanol to produce the optically active phosphine (7.8)[11].

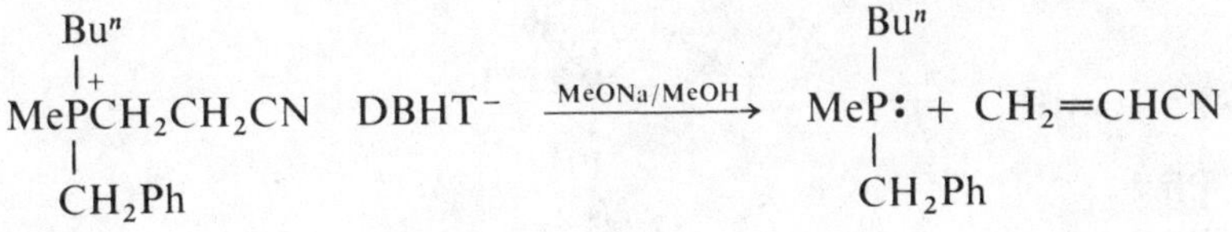

7.1 Phosphonium salts

$$\underset{(7.7)}{\underset{\underset{CH_2Ph}{|}}{\overset{\overset{Bu^n}{|}}{MePCH_2CH_2CN}}} \; DBHT^- \xrightarrow{MeONa/MeOH} \underset{(7.8)}{\underset{\underset{CH_2Ph}{|}}{\overset{\overset{Bu^n}{|}}{MeP:}}} + CH_2=CHCN$$

DBHT = dibenzoyl hydrogen tartrate.

Phosphonium salts are in general stable, crystalline solids of high melting point. As a consequence of their ionic structure they are very insoluble in non-polar organic solvents (e.g. hydrocarbons and ethers), moderately soluble in solvents of intermediate polarity (e.g. $CHCl_3$ and CH_3CN) and usually a good deal more soluble in polar solvents like methanol or ethanol. Their stability and comparitive ease of isolation make them useful as derivatives of unstable phosphines. Furthermore, the molecular weight of the resultant phosphonium salt may be determined by titration in glacial acetic acid with perchloric acid in the presence of mercuric acetate and using crystal violet as indicator (eqn. 2)[12].

$$R_4P^+Cl^- + Hg(OAc)_2 \rightleftharpoons \underset{(7.9)}{R_4\overset{+}{P}\,OAc^-} \underset{\longleftarrow}{\overset{HClO_4}{\longrightarrow}} R_4P^+ClO_4^- + HOAc \quad (2)$$

The acetate ion from the intermediate phosphonium acetate (7.9) is a strong base in glacial acetic acid and effectively neutralizes the perchloric acid. The sharp end-point is reached at a green colour between the violet of neutral crystal-violet and the yellow of the protonated form of the indicator.

The spectroscopic characteristics of phosphonium salts (particularly infrared and n.m.r.) have received attention in Chapter 3.

7.1.2 Preparation of phosphonium salts[1,2]

Most phosphonium salts (both acyclic and cyclic) are prepared by the quaternization of phosphines using the appropriate alkyl halide; eqns. (3) and (4) provide typical examples.

$$Ph_3P + PhCH_2Br \longrightarrow Ph_3\overset{+}{P}CH_2Ph \; Br^- \quad (3)$$

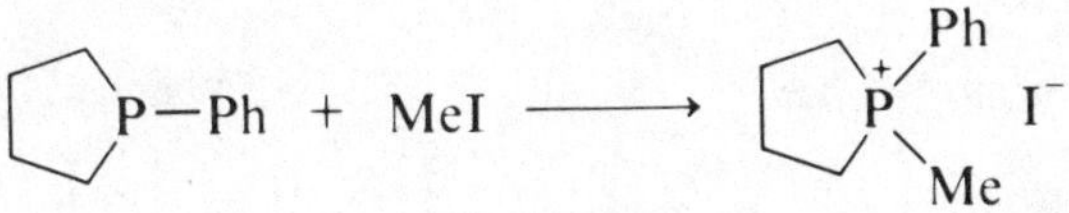

$$\underset{}{\bigcirc}P-Ph + MeI \longrightarrow \underset{}{\bigcirc}\overset{Ph}{\underset{Me}{\overset{+}{P}}} \; I^- \quad (4)$$

These nucleophilic displacement reactions have been discussed in some detail in Chapter 4. For the sake of the subsequent discussion in Section 7.1.3 however, it is worth emphasizing that quaternizations of this kind proceed with retention of configuration at phosphorus. Thus on treatment with alkyl halide, the diastereomeric phospholans (7.10) and (7.12) give two diastereomeric phosphonium salts (7.11) and (7.13) respectively.

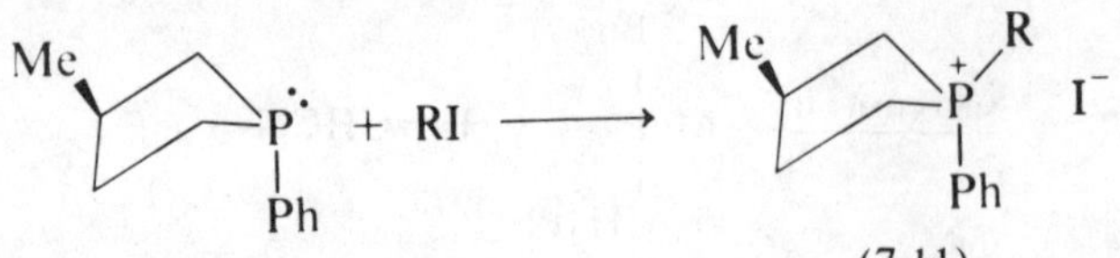

(7.10) + RI ⟶ (7.11)

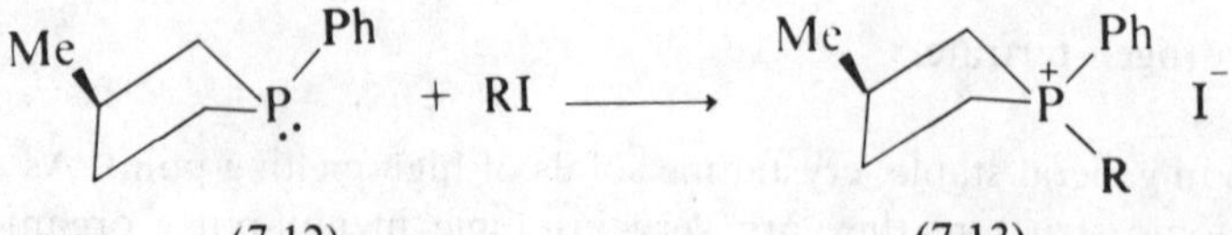

(7.12) + RI ⟶ (7.13)

The reactions are normally carried out by mixing the neat reagents and heating the mixture in a sealed tube or by heating a mixture of the reagents in a dipolar aprotic solvent such as acetonitrile from which the phosphonium salt usually crystallizes out on cooling.

Tetraarylphosphonium salts may be prepared by the complex salt method (eqn. 5)[13] which proceeds with retention of configuration at phosphorus[14] or by reaction of triarylphosphines with benzyne (eqn. 6)[15].

$$Ar_3P + Ar^1Br \xrightarrow{NiBr_2} Ar_3\overset{+}{P}Ar^1\ Br^- \quad (5)$$

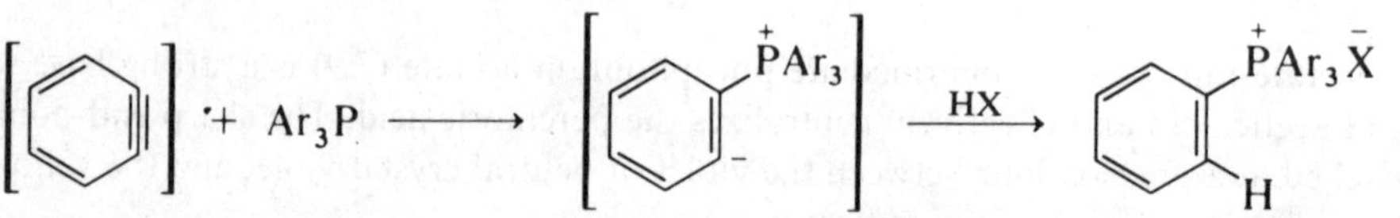

$$[benzyne] + Ar_3P \longrightarrow [C_6H_4\text{-}PAr_3]^- \xrightarrow{HX} C_6H_4(H)(PAr_3X) \quad (6)$$

Phosphonium salts may also be prepared from acid-catalyzed reactions of phosphines with alcohols or ethers (eqn. 7).

$$Ph_3P + RCH_2OR^1 \xrightarrow{HBr} Ph_3\overset{+}{P}CH_2R\ Br^- + H_2O \qquad R^1 = \text{alkyl or H} \quad (7)$$

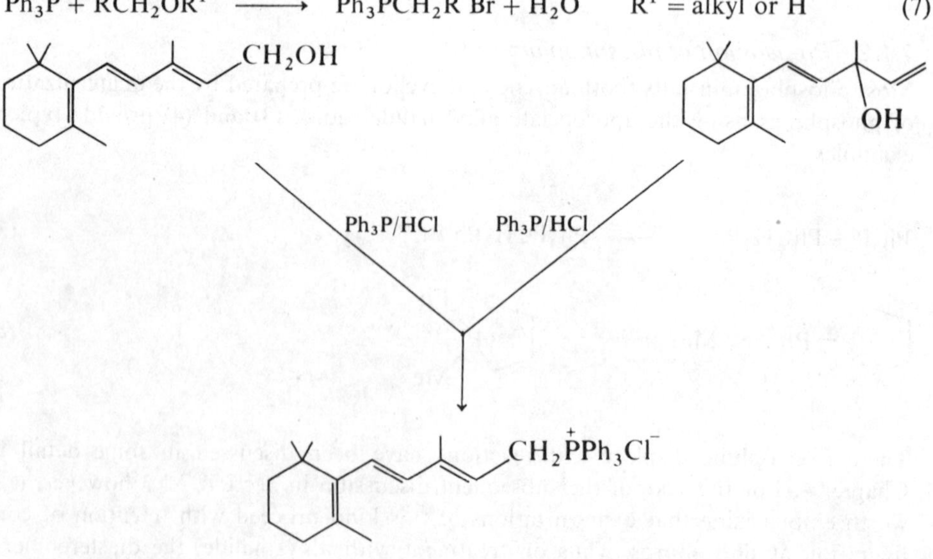

This has proved to be of value for the synthesis of compounds related to Vitamin A (see later).

7.1 Phosphonium salts

Nucleophilic addition of triphenylphosphonium halides to polyenes also provides a route to phosphonium salts which have proved valuable in the carotenoid field (eqn. 8)[16].

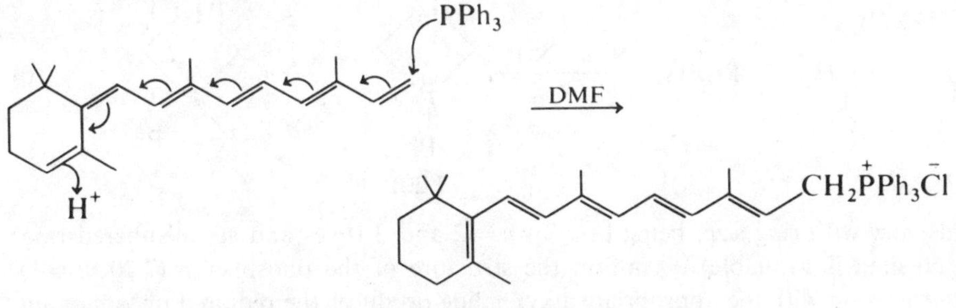

(8)

There are a number of quite specialized routes to cyclic phosphonium salts. For example, the condensation of P(III) compounds with dienes is capable of producing phospholenium salts (7.14) directly (eqn. 9)[17,18].

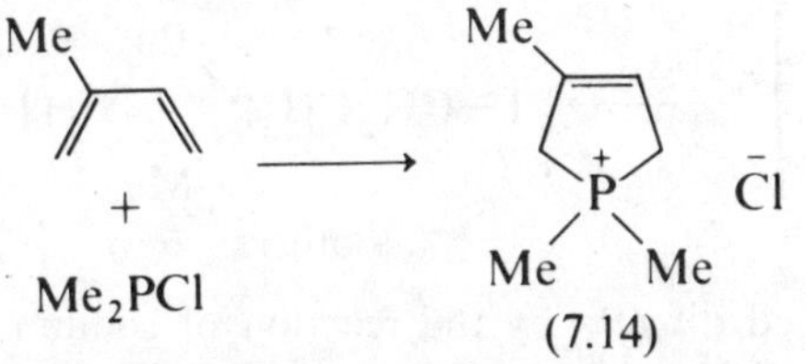

(9)

Me₂PCl   (7.14)

Highly branched olefins such as 2,4,4-trimethyl-2-pentene (7.15) will also condense with phosphonous dichlorides to give 1-chloro phosphetanium salts (7.16) which may be hydrolyzed to phosphine oxide (7.17) prior to reduction to the phosphine (7.18) and quaternization to the required phosphonium salt (7.19)[18,19].

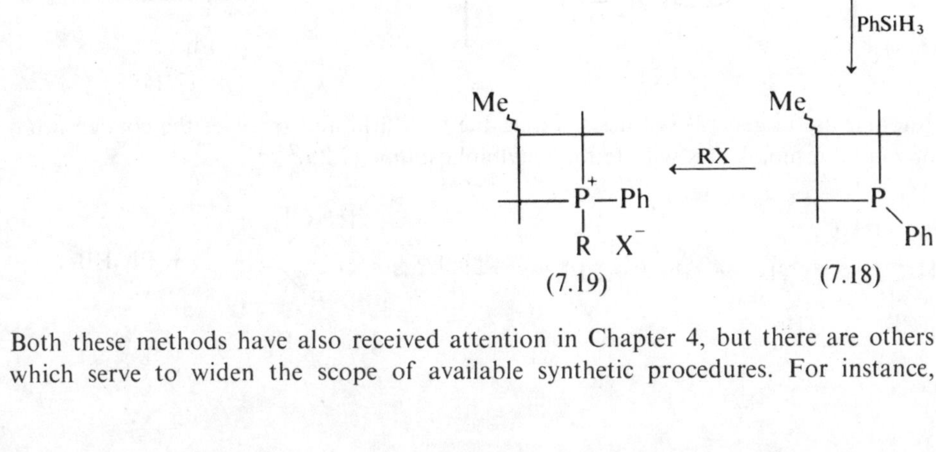

(7.15)   + PhPCl₂ $\xrightarrow{AlCl_3}$ (7.16) $\xrightarrow{H_2O}$ (7.17)

$\downarrow$ PhSiH₃

(7.19) $\xleftarrow{RX}$ (7.18)

Both these methods have also received attention in Chapter 4, but there are others which serve to widen the scope of available synthetic procedures. For instance, α,ω-dichloroalkanes condense with phenylphosphine in liquid ammonia in the presence of sodium to produce a range of phosphines (7.20) with ring sizes from 3 to 8 (eqn. 10)[20,21,22].

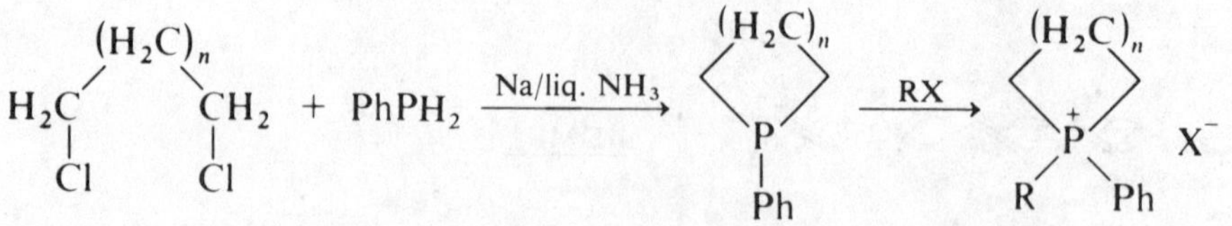

$$H_2C(Cl)-(CH_2)_n-CH_2Cl + PhPH_2 \xrightarrow{Na/liq.\ NH_3} \text{(7.20)} \xrightarrow{RX} \text{phosphonium salt} \quad X^- \quad (10)$$

$n = 0-5$

Yields vary with ring size, being best for $n = 2$ and 3 (five- and six-membered ring) and no data is available to confirm the structure of the phosphetan (7.20, $n = 1$). Quaternization with the appropriate alkyl halide produces the required phosphonium salts for $n = 2, 3, 4$ and 5 but with $n = 0$ (three-membered ring) a complex mixture of products is obtained, probably as a result of ring opening due to the strain imposed by a three-membered ring attempting to achieve a tetrahedral configuration (eqn. 11)[22].

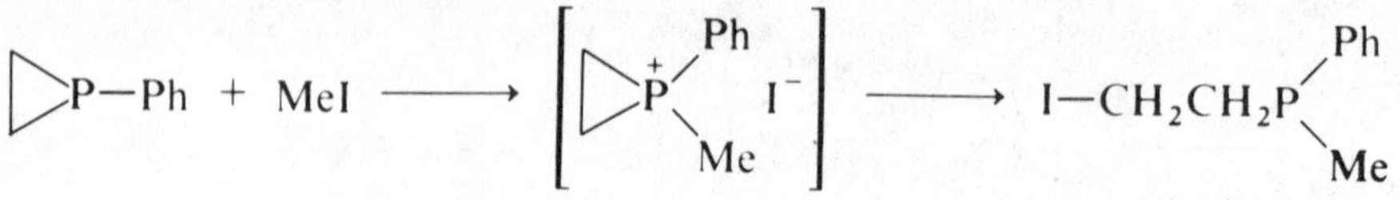

$$\triangleright P-Ph + MeI \longrightarrow [\triangleright \overset{+}{P}(Ph)(Me)]\ I^- \longrightarrow I-CH_2CH_2P(Ph)(Me) \quad (11)$$

+ others

The phosphetanium salt (7.24) can be obtained directly by the reaction of sodium diphenylphosphide (7.21) with the trichloride (7.22), the reaction proceeding via the biphosphine (7.23) which has sufficiently large groups on the β-carbon to promote intramolecular cyclization at the expense of polymerization[23].

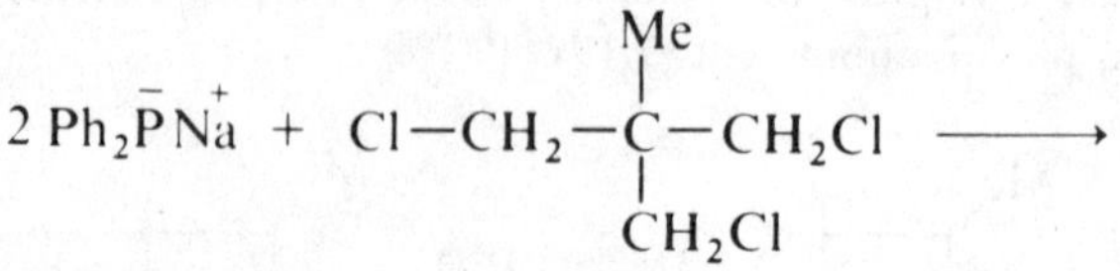

$$2\ Ph_2\overset{-}{P}\overset{+}{Na} + Cl-CH_2-\underset{CH_2Cl}{\overset{Me}{C}}-CH_2Cl \longrightarrow$$

(7.21) (7.22)

$$[Ph_2P-CH_2-\underset{CH_2-Cl}{\overset{Me}{C}}-CH_2-PPh_2] \longrightarrow Cl^- \quad Ph-\overset{+}{P}(Ph)(CH_2PPh_2)(Me)$$

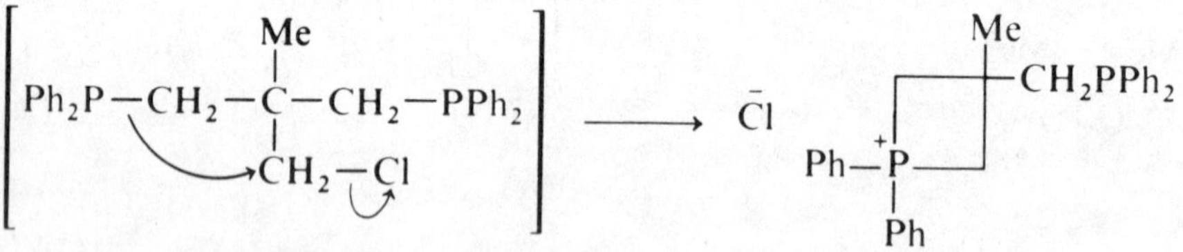

(7.23) (7.24)

Another useful general synthesis is that due to Märkl and involves the condensation of α,ω-dibromoalkanes with tetraphenylbiphosphine (7.25)[24,25].

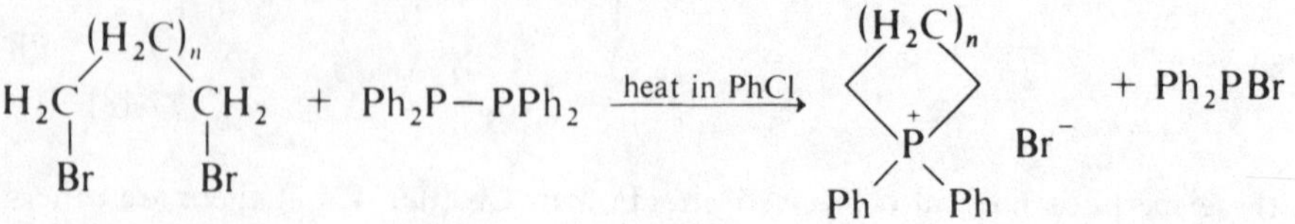

$$H_2C(Br)-(CH_2)_n-CH_2Br + Ph_2P-PPh_2 \xrightarrow{heat\ in\ PhCl} \text{cyclic}\ \overset{+}{P}(Ph)(Ph)\ Br^- + Ph_2PBr$$

(7.25)

7.1 Phosphonium salts

This works smoothly and in good yield for $n = 3$, 4 and 5, i.e. for the production of 6-, 7- and 8-membered rings.

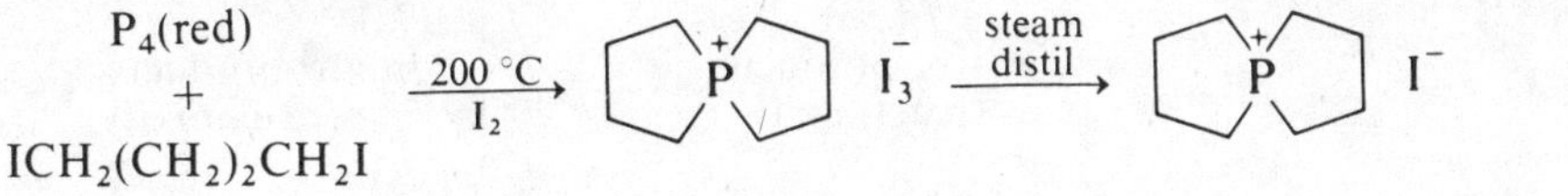

$$P_4(red) + ICH_2(CH_2)_2CH_2I \xrightarrow[I_2]{200\,°C} [\text{cyclic }\overset{+}{P}] \; I_3^- \xrightarrow{\text{steam distil}} [\text{cyclic }\overset{+}{P}] \; I^- \qquad (12)$$

Finally an interesting method for the preparation of spirophosphonium salts involves the reaction of red phosphorus with 1,4-di-iodobutane in the presence of iodine (eqn. 12)[26].

7.1.3 Reactions of phosphonium salts

7.1.3.1 General review

The important reactions of phosphonium salts are those involving nucleophiles or bases. In order to provide a general picture of reactivity one may envisage a phosphonium salt represented by (7.26) reacting with the hydroxide ion.

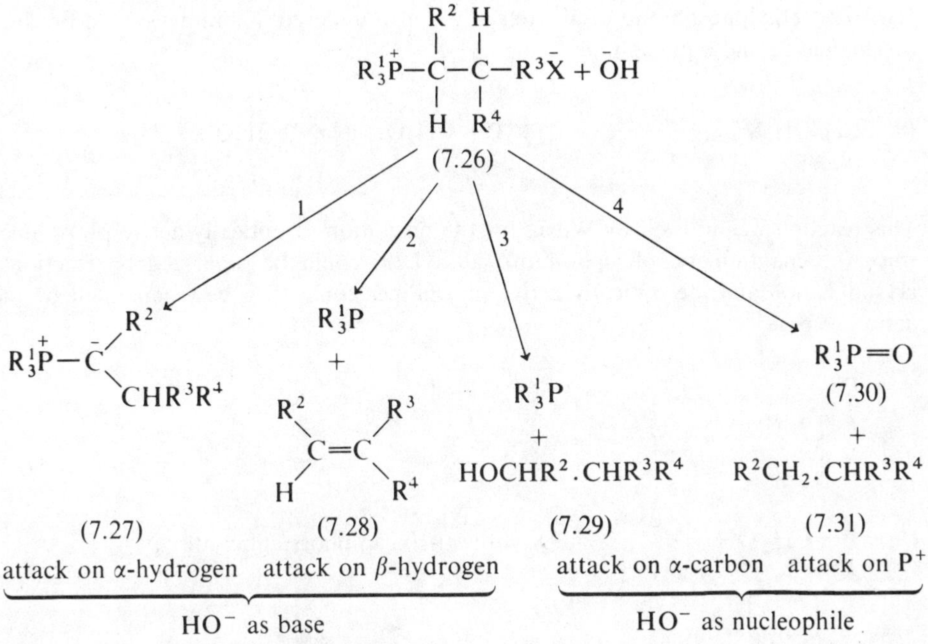

$$R^1_3\overset{+}{P}-\underset{\underset{H}{|}}{\overset{\overset{R^2}{|}}{C}}-\underset{\underset{R^4}{|}}{\overset{\overset{H}{|}}{C}}-R^3X + \bar{O}H$$
(7.26)

(7.27)	(7.28)	(7.29)	(7.30)
$R^1_3\overset{+}{P}-\bar{C}\underset{CHR^3R^4}{\overset{R^2}{\diagup}}$	R^1_3P + $\underset{H}{\overset{R^2}{\diagup}}C=C\underset{R^4}{\overset{R^3}{\diagdown}}$	R^1_3P + $HOCHR^2.CHR^3R^4$	$R^1_3P=O$ + $R^2CH_2.CHR^3R^4$ (7.31)
attack on α-hydrogen	attack on β-hydrogen	attack on α-carbon	attack on P⁺
HO⁻ as base		HO⁻ as nucleophile	

Reaction 1 involves a removal of an α-proton to form a phosphorus ylid; this will receive a more detailed treatment in Section 7.2. Reaction 2 is analogous to the Hofmann elimination of amines from quaternary ammonium salts containing alkyl groups with β-hydrogen. It is frequently only a minor pathway in the reactions of bases with phosphonium salts and only occurs to a large extent when groups R^3 and R^4 are capable of stabilizing the olefin by conjugation. At least one case of such reactivity has been quoted in terms of the elimination of phosphine from β-cyanoethylphosphonium salts (Section 7.1.1)[11]. To define the argument more clearly however, one may cite the reactions of hydroxide ion with the two phosphonium salts (7.32) and (7.35).

$$\text{Ph}_3\overset{+}{\text{P}}\text{CH}_2\text{CH}_2\text{Ph} \xrightarrow{\text{HO}^-, \text{heat}} \underset{\substack{(7.33)\\ \text{major products}\\ \text{(reaction 4)}}}{\text{Ph}_2\overset{\text{O}}{\overset{\|}{\text{P}}}\text{CH}_2\text{CH}_2\text{Ph}} + \text{PhH} + \underset{\substack{(7.34)\\ \text{minor products}\\ \text{(reaction 2)}}}{\text{CH}_2\!=\!\text{CHPh} + \text{Ph}_3\text{P}}$$
(7.32)

$$\text{Ph}_3\overset{+}{\text{P}}\text{CH}_2\text{CHPh}_2 \xrightarrow{\text{HO}^-, \text{heat}} \underset{\substack{(7.36)\\ \text{minor products}\\ \text{(reaction 4)}}}{\text{Ph}_2\overset{\text{O}}{\overset{\|}{\text{P}}}\text{CH}_2\text{CHPh}_2} + \text{PhH} + \underset{\substack{(7.37)\\ \text{major products}\\ \text{(reaction 2)}}}{\text{CH}_2\!=\!\text{CPh}_2 + \text{Ph}_3\text{P}}$$
(7.35)

In the former case, nucleophilic attack at phosphorus is the major pathway producing 2-phenylethyldiphenylphosphine oxide (7.33) with styrene as the minor product of an elimination reaction. Conversely, conjugation of the resultant olefin with *two* phenyl groups is sufficient to afford 1,1-diphenylethylene (7.37) as the major product of the reaction of OH⁻ with (7.35).

Hydroxymethylphosphonium salts (e.g. 7.38) also undergo β-eliminations to produce formaldehyde and a phosphine[27].

$$\text{Ph}_3\overset{+}{\text{P}}\text{CH}_2\text{OH} \ \text{X}^- \xrightarrow{\text{NaOH, heat}} \text{Ph}_3\text{P} + \text{CH}_2\text{O} + \text{NaX} + \text{H}_2\text{O}$$
(7.38)

This reaction was utilized by Wittig for the resolution of optically active phosphines since the enantiomeric phosphonium salts (7.39) could be separated by fractional crystallization and the optically active phosphines could then be regenerated by the action of base[28].

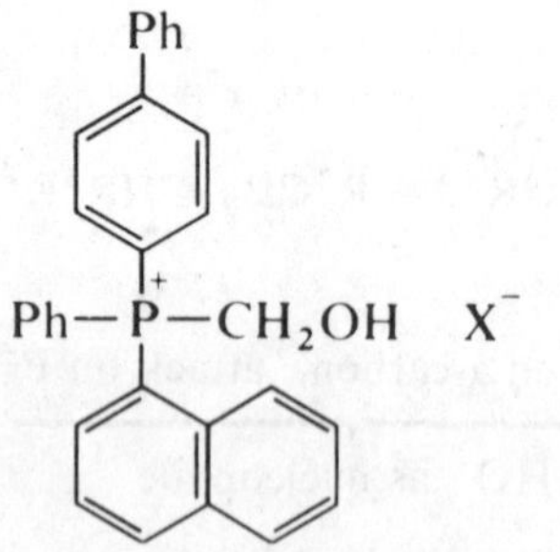

Ph—P(Ph)(naphthyl)—CH₂OH X⁻   X = D(+)-10-camphorsulphonate
(7.39)

Reaction 3 is nucleophilic attack at the α-carbon of a phosphonium salt and although it is frequently observed as a side reaction accompanying ylid formation, β-elimination or nucleophilic attack at phosphorus, it is usually the minor pathway. For example, when diphenoxymethyltriphenylphosphonium chloride (7.40) is treated with phenyl lithium, the principal product is the ylid (7.41) produced by removal of an α-proton. Some 7% of the product, however, is due to nucleophilic displacement of phenoxide ion from the α-carbon to form another phosphonium salt (7.42).

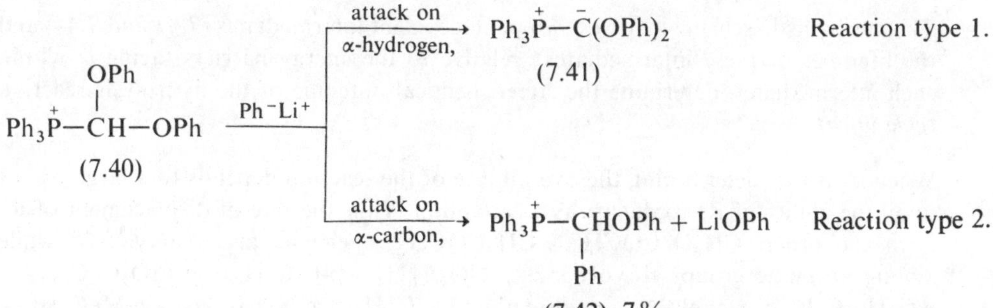

7.1 Phosphonium salts

$$Ph_3\overset{+}{P}-CH-OPh \quad \xrightarrow{Ph^-Li^+} \quad \begin{cases} \xrightarrow{\text{attack on}\\ \alpha\text{-hydrogen,}} Ph_3\overset{+}{P}-\bar{C}(OPh)_2 \quad \text{Reaction type 1.} \\ \\ \xrightarrow{\text{attack on}\\ \alpha\text{-carbon,}} Ph_3\overset{+}{P}-CHOPh + LiOPh \quad \text{Reaction type 2.} \\ \qquad\qquad\qquad | \\ \qquad\qquad\qquad Ph \end{cases}$$

(7.40) — (7.41) — (7.42), 7%

Nucleophilic attack at the positively charged phosphorus atom comprises the fourth and most commonly observed class of reactions between nucleophiles and phosphonium salts. Hydrolysis of phosphonium salts to phosphine oxides has received the most detailed attention and the remainder of this section will be devoted to an appraisal of the mechanistic aspects of this important reaction. As the subsequent discussion will show, such reactions are believed to proceed via pentacovalent intermediates and it is the lifetime of these intermediates relative to the energy barriers facing pseudorotation within them, which determine the ultimate stereochemical outcome of each reaction.

7.1.3.2. Nucleophilic attack at phosphorus: the hydrolysis of phosphonium salts to phosphine oxides[1,3–6,8,18,29,30]

Phosphonium salts hydrolyze in the presence of hydroxide ion to phosphine oxides and a hydrocarbon (eqn. 13).

$$R_3\overset{+}{P}R^1\ \bar{X} + NaOH \longrightarrow R_3P=O + R^1H + NaX \qquad (13)$$

The vast majority of these reactions obey third order kinetics, being first order in phosphonium salt and second order in hydroxide ion[31,32]. This evidence led to the proposal of the following mechanistic scheme with loss of the group, R^1, as a carbanion in the slow, rate-determining step (rate coefficient, k_3).

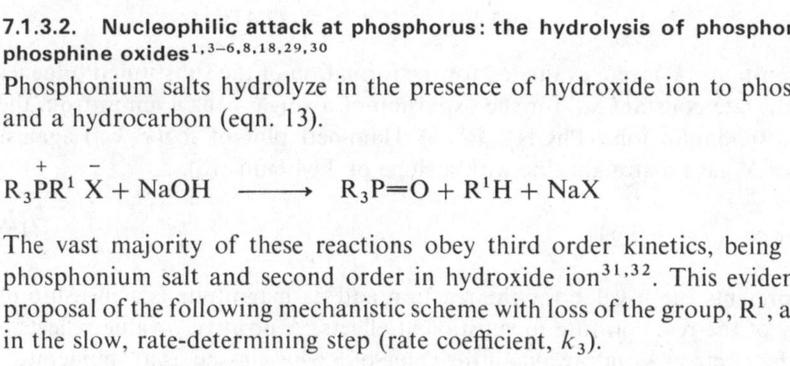

$$R_3\overset{+}{P}R^1 + \bar{O}H \underset{}{\overset{K_1}{\rightleftharpoons}} R_3P\overset{OH}{\underset{R^1}{\diagup}} \underset{}{\overset{-OH\ K_2}{\rightleftharpoons}} R_3P\overset{O^-}{\underset{R^1}{\diagup}} + H_2O$$

(7.43) — (7.44)

$$\downarrow k_3$$

$$R^1H + \bar{O}H \xleftarrow{H_2O} [\bar{R}^1] + R_3P=O$$

Thus, the overall differential rate equation for the reaction is given by eqn. (14).

$$d[R_3P=O]/dt = K_1K_2k_3[R_3\overset{+}{P}R^1][\bar{O}H]^2$$
$$= k_{obs}[R_3\overset{+}{P}R^1][\bar{O}H]^2 \qquad (14)$$

The mechanistic scheme involves two pentacovalent intermediates (7.43 and 7.44) and the lifetimes of these intermediates* relative to the energy barriers facing ψ within such intermediates, determine the stereochemical outcome of the hydrolysis reaction (*vide infra*).

When k_3 is rate determining the overall rate of the reaction depends to a large extent upon the ability of R^1 to depart as a carbanion. Thus the rate of displacement of R^1 is in the order, $CH_3O.COCH_2 \geqslant CH_3COCH_2 >$ benzyl $>$ aryl $>$ alkyl[33–36] while among aromatic groups the order is $p\text{-}NO_2C_6H_4 > p\text{-}ClC_6H_4 > p\text{-}EtO_2C.C_6H_4 > p\text{-}C_6H_5.C_6H_4 > \alpha$-naphthyl or β-naphthyl $> C_6H_5 > p\text{-}NH_2C_6H_4 > p\text{-}MeC_6H_4 > p\text{-}MeOC_6H_4$.[31,37] A detailed kinetic study of the alkaline hydrolysis of para-substituted benzyltribenzylphosphonium bromides (7.45) revealed that the rate of reaction and the yield of substituted toluene (7.46) both increased as the substituent (Y) became more electron withdrawing[38].

$(PhCH_2)_3 \overset{+}{P} CH_2$—⟨C₆H₄⟩—Y $\xrightarrow{HO^-}$ $(PhCH_2)_3 P{=}O + (PhCH_2)_2 P(O)CH_2$—⟨C₆H₄⟩—Y

(7.45)

$+$ $+$

H_3C—⟨C₆H₄⟩—Y   ⟨C₆H₅⟩—CH_3

(7.46) (7.47)

Partial rate constants (k') were evaluated for the formation of the substituted toluenes together with a rate constant k_0' for the expulsion of a single benzyl anion from the tetrabenzylphosphonium ion, $(PhCH_2)_4P^+$. A Hammett plot of $\log(k'/k_0')$ against the σ values of Y gave a straight line with a slope of 3.64 (eqn. 15).

$$\log(k'/k_0') = \sigma\rho, \quad (\rho = +3.64) \tag{15}$$

The slope represents the ρ-value for the reaction and its magnitude is a measure of the sensitivity of the reaction rate to substituent effects. A positive ρ-value reflects a rate increase by electron withdrawal and the value of 3.64 is classed as a "moderate" substituent effect. Partial rate factors ($k^\star$) for the displacement of toluene were also evaluated and were also found to increase with increasing electron withdrawal by Y. Presumably this is due to an increase in either or both of the equilibrium constants K_1 or K_2 as electron withdrawal by Y increases the concentration of the pentacovalent intermediates.

The hydrolysis of most acyclic phosphonium salts occurs with inversion† of configuration at phosphorus[39–41]. This was elegantly demonstrated by the following stereochemical cycle[42].

---

* Determined by the rate of elimination of R^1 as a carbanion, i.e. the magnitude of k_3.
† Recently, the alkaline hydrolysis of $Ph.Me.PhCH_2P^+R\ Br^-$ has been shown to occur with varying degrees of inversion depending upon the group, R.[R. Luckenbach, *Phosphorus*, 1, 223 (1972)]. Furthermore, diarylphosphonium salts, $Ph.Pr^n.MeP^+Ar\ Br^-$, where $Ar = \alpha\text{-}C_{10}H_7$, $\beta\text{-}C_{10}H_7$ or $p\text{-}Ph\text{-}C_6H_4\text{—}$, hydrolyze with loss of Ar and partial retention of configuration. It seems that nothing in chemistry is sacred!

7.1 Phosphonium salts

$$\underset{(+)}{\underset{(7.48)}{\overset{Me}{\underset{Ph}{Et\text{-}\!\!-\!\!P:}}}} \xrightarrow[\text{(retention)}]{H_2O_2} \underset{(+)}{\underset{(7.49)}{\overset{Me}{\underset{Ph}{Et\text{-}\!\!-\!\!P=O}}}}$$

via PhCHO (retention) from ylid (7.51):

$$\overset{Me}{\underset{Ph}{Et\text{-}\!\!-\!\!P=CHPh}} \quad (7.51)$$

$$\underset{(+)}{\underset{(7.50)}{\overset{Me}{\underset{Ph}{Et\text{-}\!\!-\!\!P^+\!\!-\!\!CH_2Ph}}}}\; I^- \xrightarrow[\text{inversion}]{OH^-} \underset{(-)}{\underset{(7.52)}{\overset{Me}{\underset{Ph}{O=P\text{-}\!\!-\!\!Et}}}}$$

(7.48) → (7.50) via PhCH$_2$I (retention); (7.50) → (7.51) via PhLi (retention).

Reaction of the (+)phosphine (7.48) with hydrogen peroxide gives the (+)oxide (7.49) which may also be obtained via the phosphonium salt (7.50) and phosphorus ylid (7.51) by a Wittig reaction of the ylid with benzaldehyde. All the steps to the (+)oxide involve retention of configuration at phosphorus. The oxide of opposite rotation and therefore of opposite configuration at phosphorus, is obtained by the alkaline hydrolysis of the salt (7.50) and this reaction must therefore involve inversion.

Phosphonium salts are tetrahedral (7.53) and nucleophilic attack on a tetrahedron may occur via any one of the four faces or any one of the six edges. Attack on a face represents apical entry whereas attack on an edge is equivalent to equatorial entry.

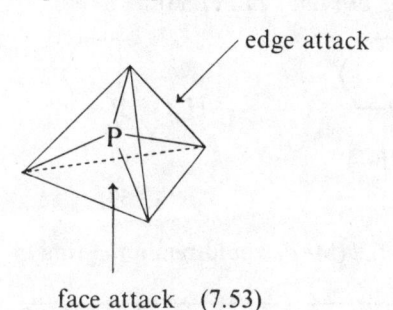

edge attack / face attack (7.53)

The most likely explanation for inversion at phosphorus is apical entry of the hydroxide ion and apical departure of the leaving group (cf. Chapter 2). Equatorial entry and departure would also lead to inversion but this is regarded as less probable[8,29]. Conversely, apical entry and equatorial departure would lead to retention of configuration at phosphorus. It has been suggested that this last pathway would violate an "extended" principle of microscopic reversibility[43,44], and although the concept is open to criticism[29], the "extended principle of microscopic reversibility" has achieved widespread acceptance as a valid argument against apical entry–equatorial departure. Thus apical entry and apical departure is frequently proposed for displacement at tetracoordinate phosphorus and at least it serves to simplify the rationalization of much of the observed stereochemistry.

With this in mind it is time to explore the facts and interpretations surrounding the base-catalyzed hydrolysis of cyclic phosphonium salts. The story begins with the observation by Asknes of a large rate enhancement for the hydrolysis of a phospholanium salt (7.54) over its six-membered ring phosphorinanium analogue (7.55) and its acyclic analogue (7.56)[4,5]. See Table 7.1.

Table 7.1 Rates of hydrolysis of cyclic phosphonium salts k_3 (l mol^{-2} min^{-2})

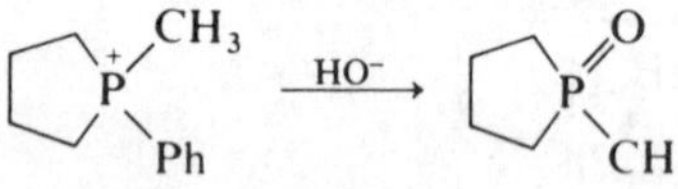 (7.54) → product	1862
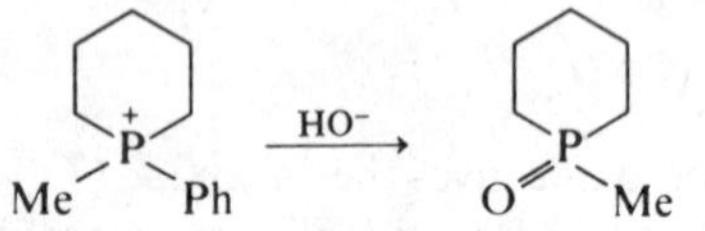 (7.55) → product	1.2
Me$_3$P$^+$Ph (7.56) →$^{HO^-}$ Me$_3$P=O	1.7

The most likely explanation for this effect is the relief of strain involved in the formation of the pentacovalent intermediate (7.57) which would increase K_1 of eqn. (14) relative to those for the pentacovalent counterparts of (7.55) and (7.56).

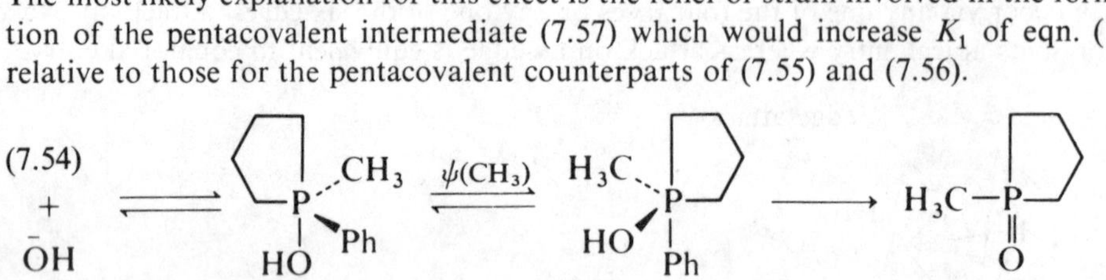

(7.54) + $\bar{O}H$ ⇌ (7.57) $\xrightarrow{\psi(CH_3)}$ (7.58) ⟶ (7.59)

One pseudorotation with the methyl group as pivot, $\psi(Me)$* would retain the ring in

---

* Actually, $\psi(Me)$ probably occurs within the oxyanion of (7.57), i.e.

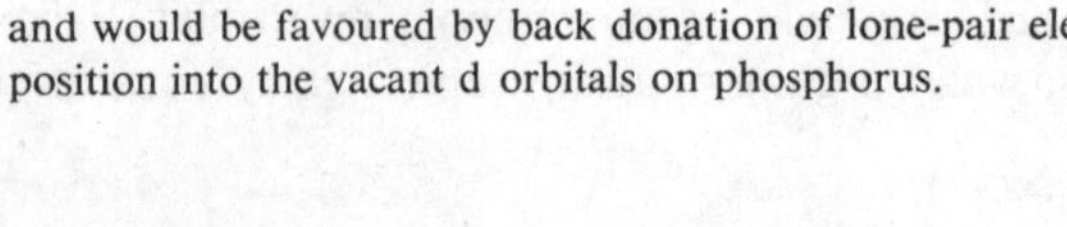

(7.57) $\xrightarrow{OH^-}$ ⇌ $\xrightarrow{\psi(CH_3)}$

This avoids placing an electronegative substituent such as OH in an equatorial position, as in (7.58) and would be favoured by back donation of lone-pair electrons from the oxyanion in an equatorial position into the vacant d orbitals on phosphorus.

7.1 Phosphonium salts

an apical–equatorial situation (7.58) but place the phenyl group in an apical position from which it would leave to form the product oxide (7.59). Notice that this would involve *retention of configuration* at phosphorus.

This stereochemical prediction was verified convincingly by Marsi through studies of the alkaline hydrolysis of the pure diastereomeric phospholanium salts, *cis*- and *trans*-(7.60)[46]. The hydrolyses were stereospecific and involved retention of configuration at phosphorus in both cases.

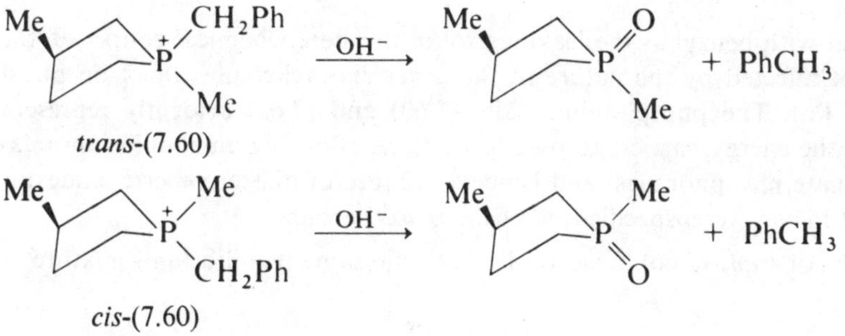

The stereochemical course for one diastereomer is outlined below.

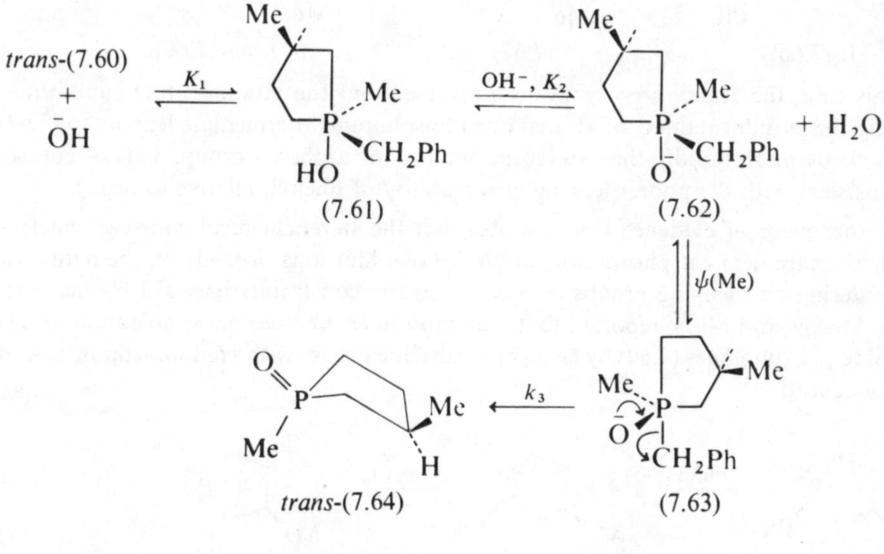

Notice that apical entry of the hydroxide ion leaves the ring in an apical–equatorial situation; removal of a proton forms (7.62) which may pseudorotate to (7.63) keeping the ring apical–equatorial and maintaining the two methyl groups *trans* to each other. Loss of the benzyl group from an apical position gives the oxide (7.64) with the methyl groups still *trans*, i.e. the configuration at phosphorus is retained.

Marsi went on to demonstrate retention of configuration in the hydrolysis of another pair of diastereomeric phospholanium salts, *cis*- and *trans*-(7.65)[47].

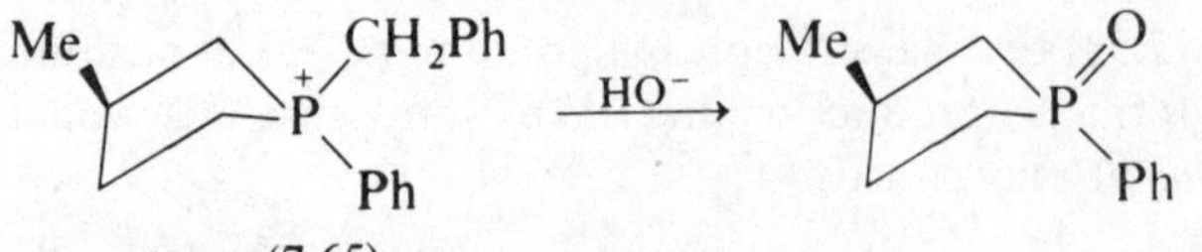

trans-(7.65)

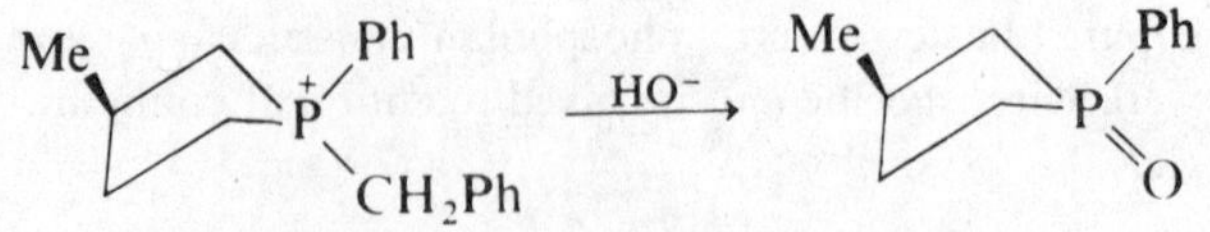

cis-(7.65)

This showed that with benzyl as the leaving group, the stereochemical course of the reaction was not affected by the nature of the other exocyclic substituent on phosphorus (Me or Ph). The phosphonium salts (7.60) and (7.65) evidently represent examples where the energy barriers to pseudorotations affording an equilibrium mixture of intermediate phosphoranes (and hence a mixture of diastereomeric oxides) is higher than that facing stereospecific loss of the benzyl group.

Hydrolysis of *cis*- or *trans*-(7.66) however, leads to the same (equilibrium) mixture of oxides (7.67)[47].

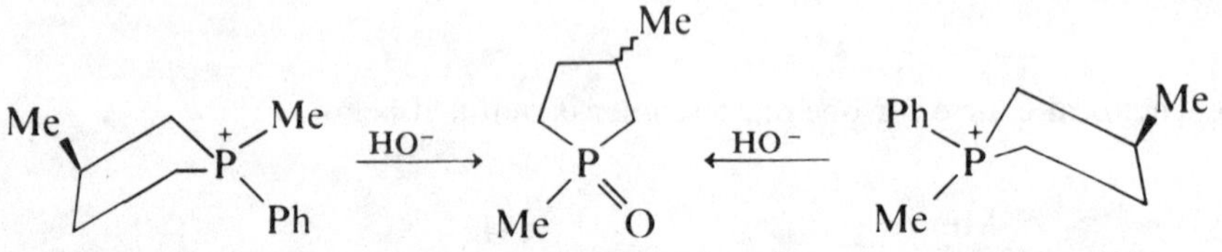

cis-(7.66) (7.67) *trans*-(7.66)

In this case, the reaction energetics are reversed and the attainment of equilibrium among the pseudorotamers of at least one phosphorane intermediate leading to (7.67) must occur more rapidly than stereospecific loss of a phenyl group. This of course, is consistent with the poorer leaving group ability of phenyl, relative to benzyl.

A further piece of evidence demonstrates that the stereochemical course of nucleophilic displacements at phosphorus in phospholanium ions depends on the nature of the entering and leaving groups as well as on the constraints imposed by the ring. Thus Mislow and Marsi reported that reduction of *cis* or *trans* phospholanium oxides (7.68) to phospholanes (7.69) by hexachlorodisilane occurs with predominant inversion of configuration[48].

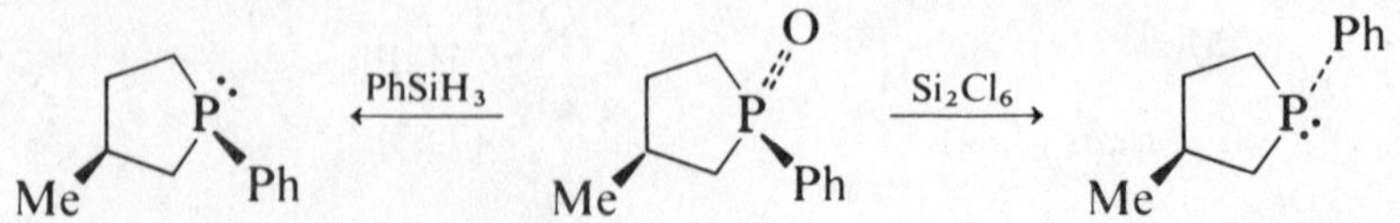

cis-(7.69) *cis*-(7.68) *trans*-(7.69)

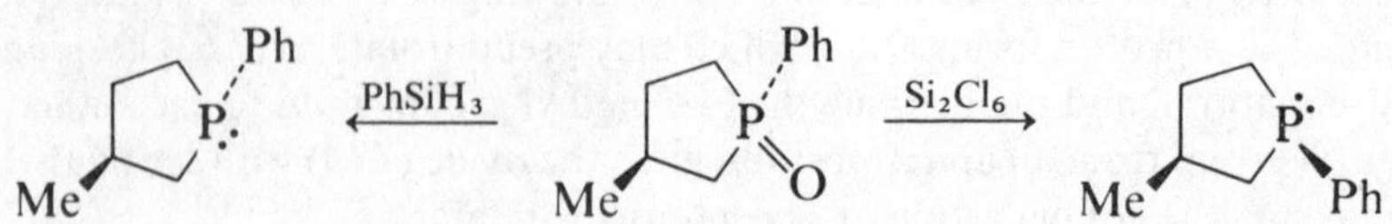

trans-(7.69) *trans*-(7.68) *cis*-(7.69)

retention inversion

7.1 Phosphonium salts

In contrast, reduction with phenylsilane occurs with retention of configuration[49]. The inversion mechanism probably involves an intermediate such as (7.70), where the strain involved in placing the ring diequatorial is compensated by the two, electronegative substituents ($SiCl_3$ and $OSiCl_3$) occupying apical positions.

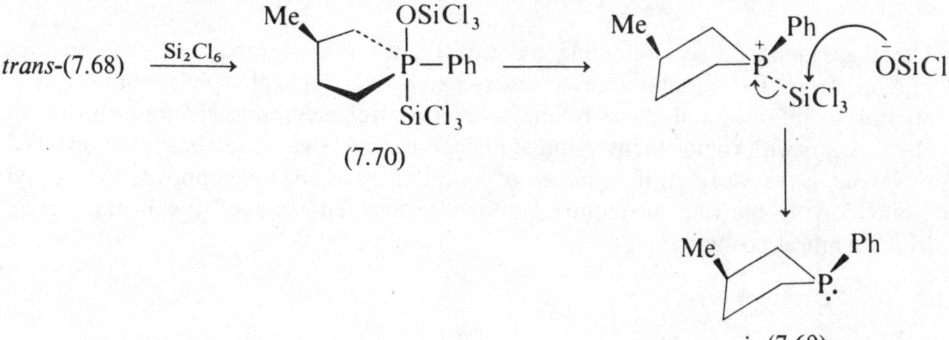

trans-(7.68) $\xrightarrow{Si_2Cl_6}$ (7.70) ⟶ ⟶ cis-(7.69)

The situation with six-membered rings is more complex. Hydrolysis of the *cis*-phosphorinanium salt (7.71) gives a mixture of 48 % *cis*-oxide (*cis*-(7.72) i.e. retention) and 52 % *trans*-oxide (*trans*-(7.72) i.e. inversion). In contrast, the *trans*-salt of (7.71) gives 22 % *cis*-oxide (inversion) and 78 % *trans*-oxide (retention)[50].

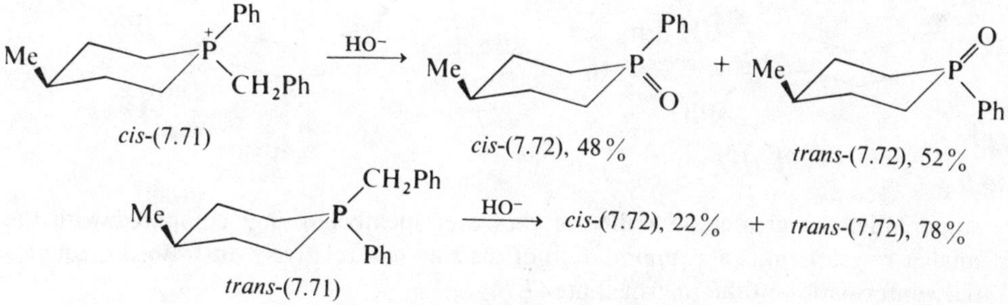

cis-(7.71) $\xrightarrow{HO^-}$ cis-(7.72), 48% + trans-(7.72), 52%

trans-(7.71) $\xrightarrow{HO^-}$ cis-(7.72), 22% + trans-(7.72), 78%

This is most easily explained by a combination of the retention mechanism as described for the phospholanium salts and a simultaneous mechanism of the McEwen type in which the more flexible six-membered ring becomes diequatorial with subsequent departure of the benzyl group from an apical position with inversion of configuration. In the McEwen mechanism the *cis* salt (7.71) would give (7.73) and the *trans* salt of (7.71) would give (7.74).

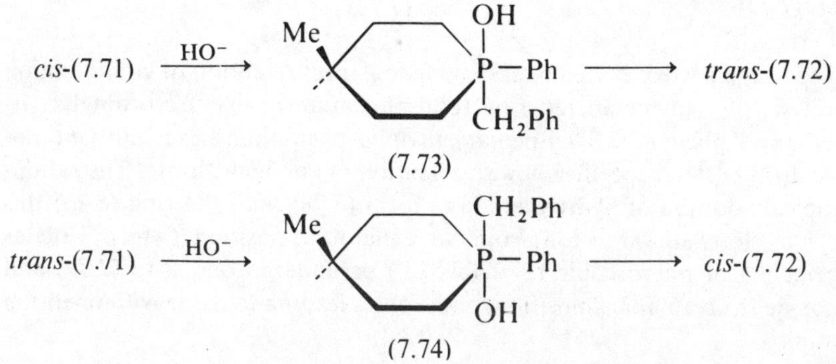

cis-(7.71) $\xrightarrow{HO^-}$ (7.73) ⟶ trans-(7.72)

trans-(7.71) $\xrightarrow{HO^-}$ (7.74) ⟶ cis-(7.72)

The two intermediates are diastereomers and are therefore of unequal ground state energies. It seems reasonable to assume that because of steric interactions between the methyl and benzyl groups, that (7.74) would be of higher energy than (7.73) and hence the degree of inversion for *trans*-(7.71) would be less than for *cis*-(7.71), as observed.

The suggestion of two competing mechanisms for phosphorinanium salts receives support from the results with a seven-membered ring (phosphepanium) salt[51]. Hydrolysis of *cis*- and *trans*-1-benzyl-4-methyl-1-phenylphosphepanium bromides (7.75) occurs with complete inversion at phosphorus and this is most easily rationalized by invoking the McEwen mechanism of apical entry at the face opposite the benzyl group, forcing the ring diequatorial and allowing departure of the benzyl group from an apical position.

cis-(7.75) *trans*-oxide

trans-(7.75) *cis*-oxide

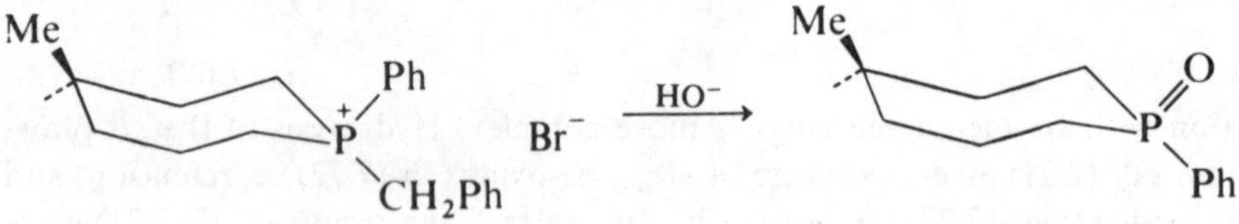

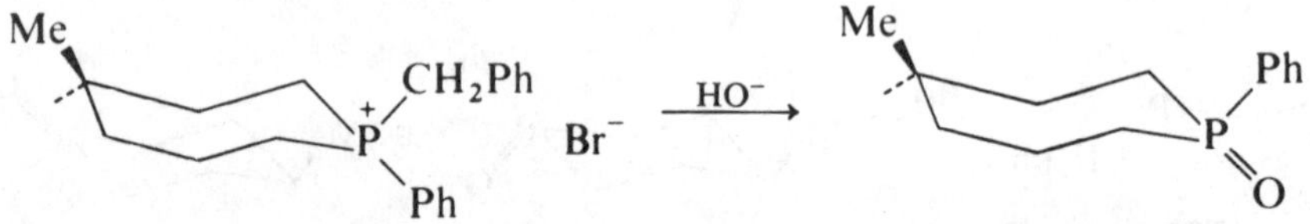

Apparently, the greater flexibility of the seven-membered ring compared with the smaller rings, permits accommodation of the ring in a relatively unstrained diequatorial conformation in the intermediate (7.76).

trans-(7.75) (7.76)

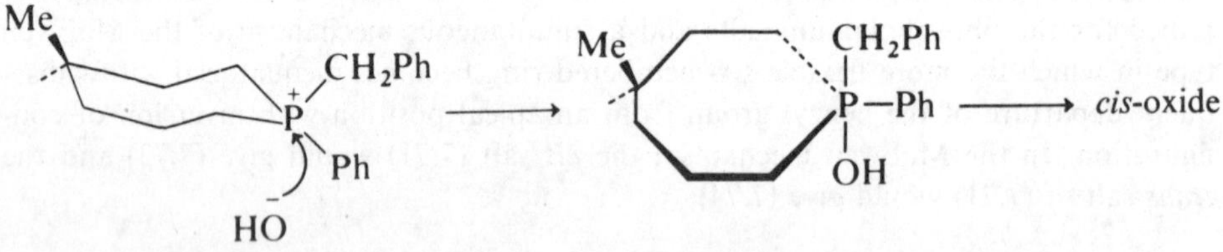

On the basis of Marsi's work it seems reasonable to expect retention of configuration for the hydrolysis of four-membered ring (phosphetanium) salts. Accordingly, *cis*- and *trans*-1-ethoxy-1-phenyl-2,2,3,4,4-pentamethylphosphetanium hexachloroantimonates (7.77) hydrolyze stereospecifically with retention of configuration[52]. The rationale involves apical addition of hydroxide ion to form (7.78) with the ring (a–e); this intermediate may lose ethoxide ion from an equatorial position (which violates "extended principle of microscopic reversibility") or undergo one ψ to (7.79) and eliminate ethoxide from an apical position, both routes leading to oxide with retention of configuration.

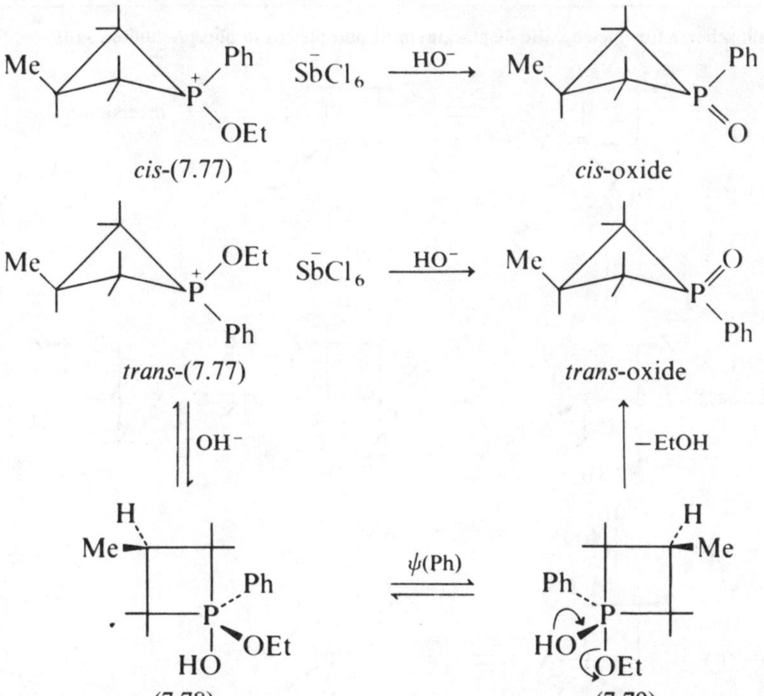

7.1 Phosphonium salts

cis-(7.77) → cis-oxide

trans-(7.77) → trans-oxide

(7.78) ⇌ ψ(Ph) (7.79)

In certain cases however, the situation is complicated by polytopal rearrangements within the phosphorane intermediates. For example, epimerization has been found to proceed faster than hydrolysis in the base-catalyzed hydrolysis of cis- and trans-1-benzyl-1-phenyl-2,2,3,4,4-pentamethylphosphetanium bromides (7.80)[53] and the product is an equilibrium mixture of oxides (7.81).

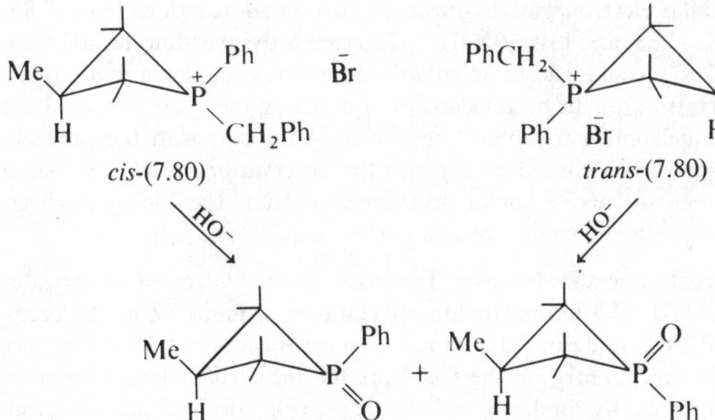

cis-(7.80)   trans-(7.80)

cis-(7.81), 10%   +   trans-(7.81), 90%

It has been suggested[54] that for any phosphetanium salt (7.82) the rate of displacement of a group (e.g. X in 7.82) will depend upon the electronegativity of that group. If one assumes that the initial nucleophilic addition gives a phosphorane with an apical–equatorial ring (7.83, Fig. 7.1) then if X is highly electronegative relative to Y, one

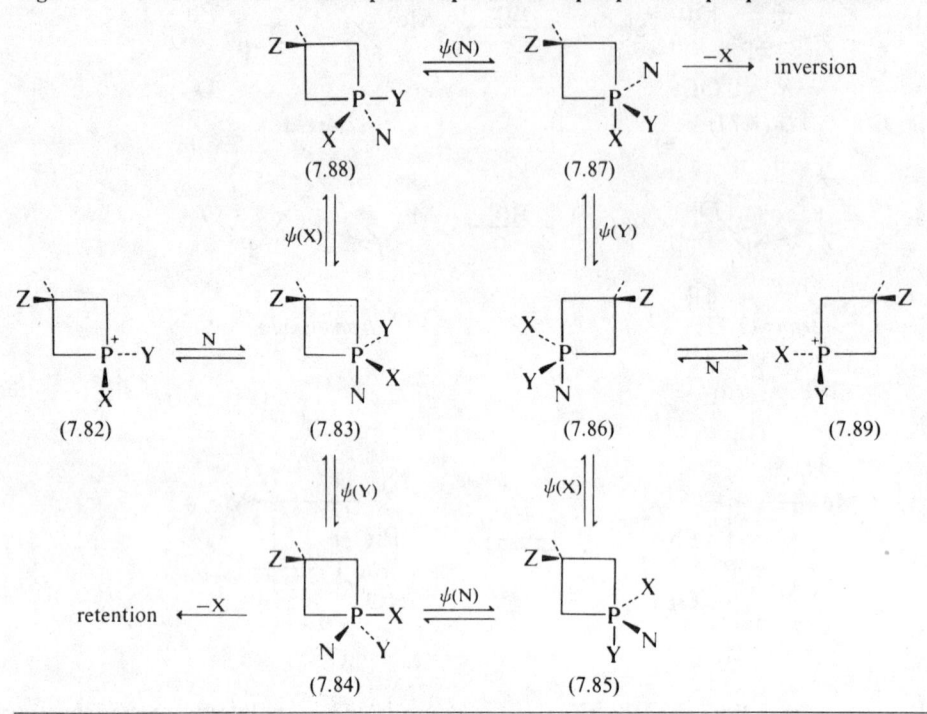

Chapter 7 / Tetracoordinate organophosphorus chemistry: Part 1

Figure 7.1 General scheme for nucleophilic displacement at phosphorus in phosphetanium salts.

ψ to (7.84) will place X in an apical position from which it can depart to give a product with retention. This is a generalized version of the result obtained with *cis*- and *trans*-(7.77).

If X and Y are of similar electronegativity however, two pseudorotations from (7.83) will produce (7.87) via (7.88) and loss of X from (7.87) leads to inversion. Equilibrium between (7.84) and (7.87) would lead to an equilibrium mixture of isomeric products. Alternatively, epimerization might be achieved by completing the cycle to give (7.86) and eliminating the nucleophile to obtain the phosphonium salt with the opposite configuration at phosphorus, (7.89). This explains the observations with (7.80) since the benzyl and phenyl groups are of similar electronegativity and the scheme has been used to rationalize other stereochemical results in this area[50,52,55,56].

There are some apparent anomalies however. For instance, the hydrolysis of optically active 1-benzyl-1-phenyl-2,2,3,3-tetramethylphosphetanium bromide (7.90) proceeds with *retention* of configuration despite the similarity in electronegativity of the benzyl and phenyl groups[57]. Apical entry on the face opposite the tertiary ring carbon to give (7.91) is the least sterically hindered pathway and ψ(Ph) to (7.92) allows apical departure of the benzyl group to give the oxide with retention of configuration. This ψ would be energetically favourable since placement of the tertiary carbon of the ring in an equatorial position would afford relief of the steric strain imposed by a tertiary carbon in an apical position. The alternative ψ(CH₂Ph) to (7.94) affords a similar release of steric strain but places the phenyl group in an apical position. Arguments have been advanced[58] to suggest that further pseudorotations to (7.93) or (7.95) which would ultimately lead to racemization via (7.96) (compare Fig. 7.1) are inhibited by the steric constraints of a tertiary carbon in the apical position. Hence the energetically favourable pathway leads to retention. Notice that the analogous set of intermediates from (7.80) have *both* tertiary ring carbons adjacent to phosphorus and all would therefore be of similar energy; the result is an equilibrium situation leading to epimerization.

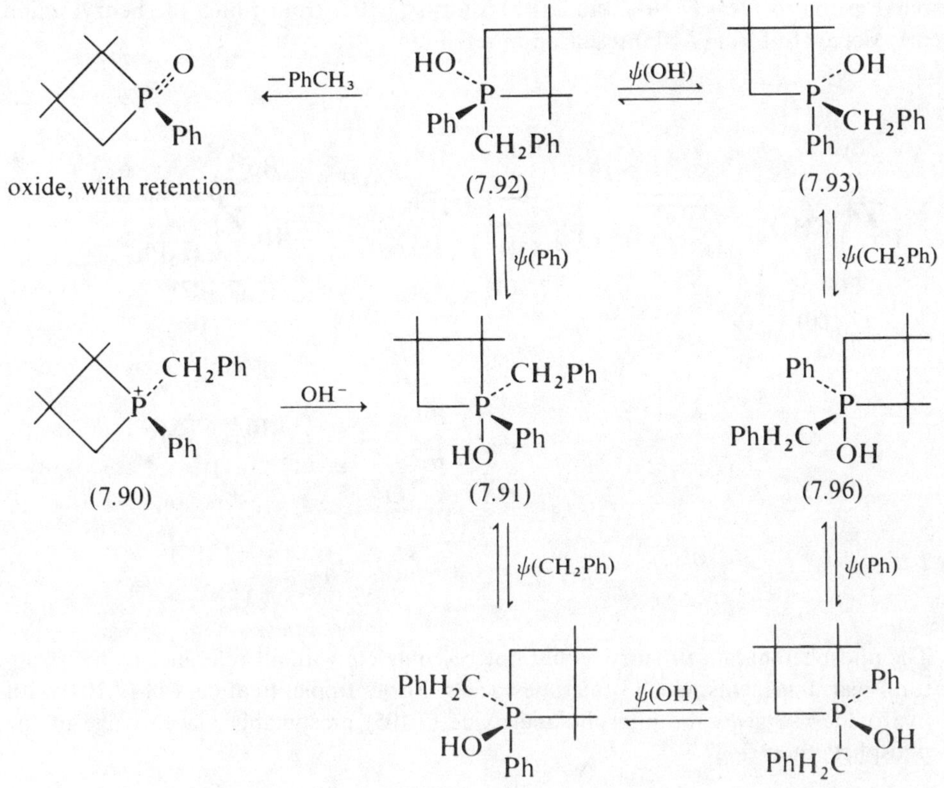

oxide, with retention (7.92) (7.93)

(7.90) (7.91) (7.96)

(7.94) (7.95)

There is no doubt that bulky substituents can influence the rate and stereochemistry of nucleophilic substitution at tetracoordinate P since Trippett has shown that (7.98) hydrolyzes about fifty times more slowly than (7.97) and that 21 % of (7.99) remained unchanged after eleven days at 100 °C in 90 % ethanolic sodium hydroxide[59].*

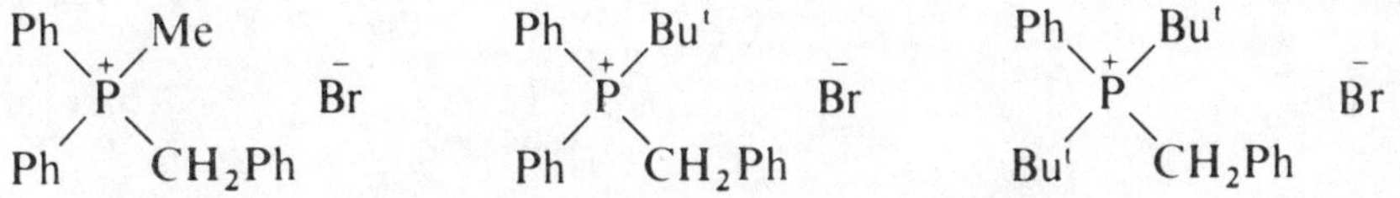

(7.97) (7.98) (7.99)

---

* In fact the major product of this reaction was that from a Hofmann elimination – benzyl-t-butylphenylphosphine, $PhCH_2.Bu^t.P.Ph$.

The extremely low rate with (7.99) may be ascribed either to the steric hindrance afforded by two t-butyl groups if the nucleophile approaches on the face opposite the benzyl group or to the steric hindrance imposed by a t-butyl group in an apical position if the nucleophile enters by the face opposite either of the t-butyl groups. The latter appears to be the preferred route since the alkaline hydrolysis of (−)benzyl-t-butylmethylphenylphosphonium iodide (7.100) occurs with predominant retention of configuration at phosphorus. Presumably, initial attack must occur opposite the t-butyl group to form (7.101) and ψ(Ph) affords (7.102) from which the benzyl anion may depart to form (7.103) resulting in retention.

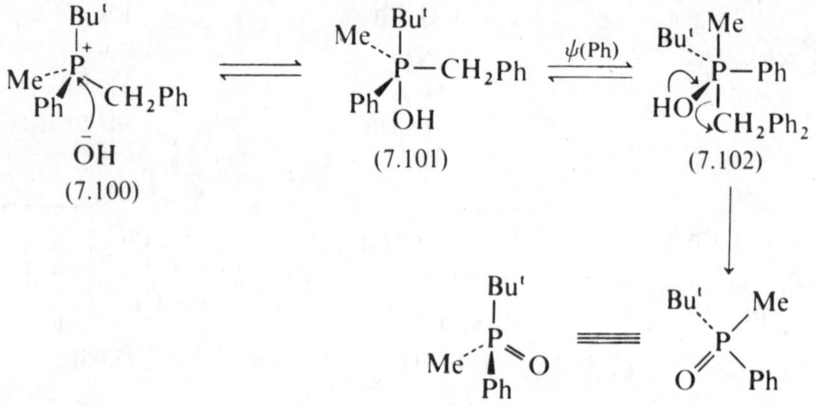

The phosphetanium salt story would not be complete without reference to the structural rearrangements which sometimes occur. For example, treatment of (7.104) with hydroxide ion gives the phospholane oxide (7.105) presumably via cleavage of the phosphetanium ring[60].

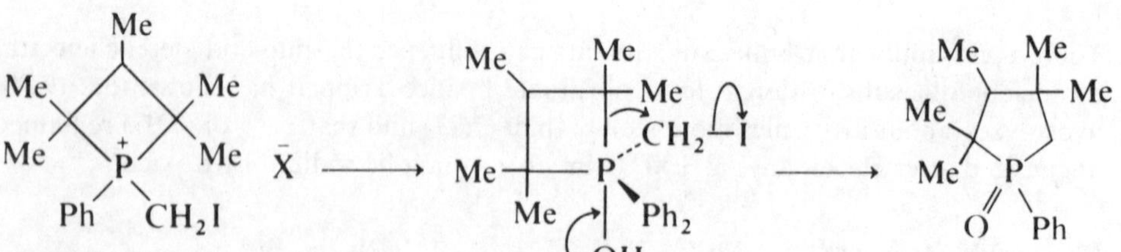

Likewise, the phosphetanium salt (7.106) rearranges to (7.107) on treatment with base[60]. It seems that methyl and phenyl are such poor leaving groups that ring cleavage and rearrangement is preferred to the normal hydrolysis pathway. The structure of the product was confirmed by deuterium labelling[61].

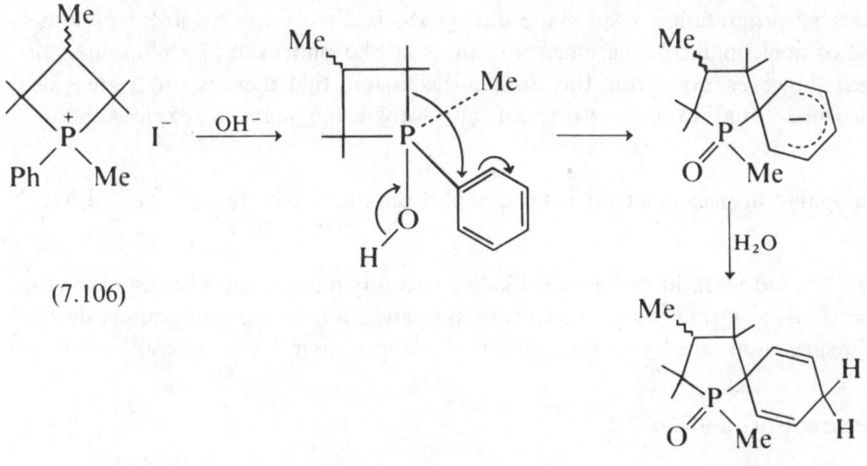

7.1 Phosphonium salts

(7.106)

(7.107)

Rearrangements are not limited to cyclic phosphonium salts and in fact the alkaline hydrolysis of triphenylvinylphosphonium bromide (7.108) gives mainly the oxide (7.109) plus minor amounts of styrene and the bis-phosphine oxide (7.110), thought to be formed by the route shown[62].

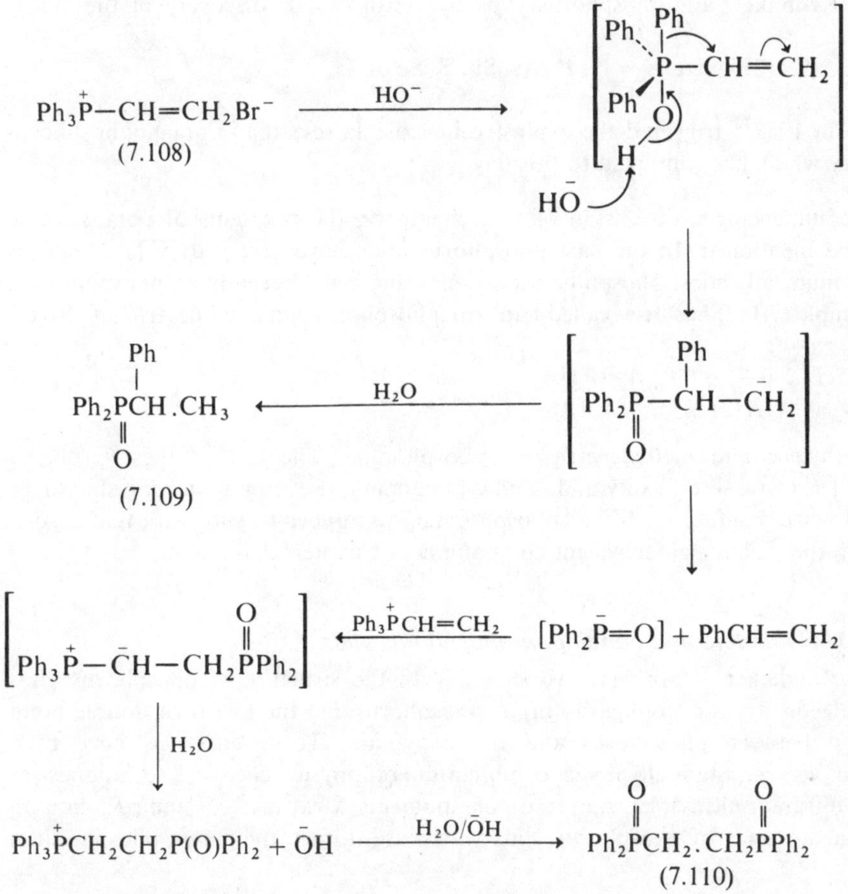

$Ph_3\overset{+}{P}-CH=CH_2 Br^-$ (7.108) $\xrightarrow{HO^-}$

$Ph_2PCH.CH_3$, with Ph and =O substituents (7.109) $\xleftarrow{H_2O}$

$[Ph_3\overset{+}{P}-\overset{-}{C}H-CH_2PPh_2=O] \xleftarrow{Ph_3\overset{+}{P}CH=CH_2} [Ph_2\overset{-}{P}=O] + PhCH=CH_2$

$\downarrow H_2O$

$Ph_3\overset{+}{P}CH_2CH_2P(O)Ph_2 + \overset{-}{O}H \xrightarrow{H_2O/\overset{-}{O}H} Ph_2PCH_2.CH_2PPh_2$ (with two =O groups)
(7.110)

Considerable progress has been made during the last ten years towards our understanding of nucleophilic displacement reactions at phosphorus in phosphonium salts. It is clear, however, even from this limited discussion, that there is still a great deal to be learned. For instance, efforts are currently being made to explore the field of nucleophilic displacement on heteroarylphosphonium salts (e.g.

(7.111)

(7.111))[63-65] and work in this area is likely to occupy phosphorus chemists for years to come. It is now time however, to turn our attention to the compounds derived from phosphonium salts by removal of an α-hydrogen, namely, the phosphorus ylids.

7.2 Phosphorus ylids[1-6,66,67]

7.2.1 Introduction

An ylid is a substance of general formula (7.112) and is composed essentially of a carbanion attached directly to a heteroatom, X. Phosphorus ylids have been known since 1894[68] but the first flurry of activity in ylid chemistry occurred in the early 1920's in Staudinger's laboratory[69] and was followed by investigations of pyridinium ylids by Krohnke[70] and phosphorus ylids by Wittig[71]. The discovery of the Wittig

$\overset{}{\underset{}{>}}\bar{C}-\overset{+}{X}$ (7.112) where X = N, P, As, Sb, S, Se or Te

reaction in 1953[72] triggered the explosive increase in research in organophosphorus chemistry which has continued to this day.

Before commencing a discussion of ylid chemistry, the problems of nomenclature should be mentioned. In the past phosphorus ylids have been variously named as phosphonium alkylides, phosphine methylenes and more recently as phosphoranes. For example (7.113) has been called triphenylphosphonium methylide, triphenylphos-

$Ph_3\overset{+}{P}-\overset{-}{CH_2} \longleftrightarrow Ph_3P=CH_2$
(7.113)

phinemethylene and methylenetriphenylphosphorane. The last of these probably presents least overall difficulty and is now in common use[5] but a warning should be sounded not to confuse the term "phosphorane" as applied to ylids with that associated with the strictly pentacovalent compounds of Chapters 2, 4 and 6.

7.2.2 The structure and bonding in phosphorus ylids

The two fundamental problems associated with the structure of phosphorus ylids were the geometry (i.e. configuration) at phosphorus and the extent of double bond formation between phosphorus and the carbanion. These questions have been answered very substantially by a combination of physico-chemical techniques including infrared, ultraviolet, n.m.r., dipole moments, x-ray analysis and pK_a data on the conjugate acids of phosphorus ylids which are in fact, the parent phosphonium salts[66].

7.2 Phosphorus ylids

Many phosphorus ylids display a band between 1200 and 1220 cm^{-1} in the infrared which has been assigned to a C=P stretching vibration[73].

A broad band between 300 and 400 nm in the ultraviolet has been ascribed to a π–π* transition of the C=P bond[74]. Tailing of this ultraviolet band into the visible is probably responsible for the yellow or orange colour of many phosphorus ylids. Examples of the ^{31}P n.m.r. chemical shifts of phosphorus ylids appear in Table 3.3 and it is sufficient to note that they all fall in the region associated with phosphonium salts at around -20 ppm relative to H$_3$PO$_4$. Dipole moments provide evidence for varying degrees of double bond character in the ylid link depending upon the ylid under study; a typical example is provided by (7.114a, b) for which the experimental moment of 23×10^{-30} Cm (7.0 D) falls exactly halfway between that calculated for the covalent (ylene) structure (7.114a, $\mu = 0$) and the ionic structure (7.114b), calculated $\mu = 47 \times 10^{-30}$ Cm (14.0 D)[75]. This suggests a 50% contribution of each canonical form to the resonance hybrid.

(7.114a) ⟷ (7.114b)

The most rigorous structural analysis however, comes from x-ray data and early examples are provided by the α-halophenacylidenetriphenylphosphoranes (7.115a, b)[76,77]. Both compounds have a planar skeleton with trigonally hybridized ylid carbons and the oxygen and phosphorus atoms *cis* to each other. The length of the ylid bond is between that of a single C—P bond (187 pm) and the double bond (167 pm). Bond angle data reveal that the phosphorus atom is tetrahedral despite the double bond character of the ylid bond (see Table 7.2).*

Finally, deuterium exchange studies leave no doubt that ylid carbanions are stabilized to varying degrees by conjugation with the heteroatom. Doering[78] made a quantitative comparison of the stability of various types of ylid by measuring the rates of alkali-catalyzed deuterium exchange of the α-hydrogen atoms of the corresponding salts, Me$_4$X$^+$ where X = N, P, As or Sb or Me$_3$Z$^+$ where Z = S or Se; these are shown in Table 7.3. The acidity of the salt and hence the stability of the ylid, is affected enormously by changes in structure and significantly, the nitrogen ylid is the least stable molecule.

The explanation lies in the nature of the bonding. This involves back donation of the negative charge from the ylid carbon into the vacant d orbitals on the heteroatom except for the nitrogen atom of nitrogen ylids which has no suitable, vacant d orbitals to accommodate the negative charge of the carbanion. This donor pπ—dπ bonding situation has been discussed in Chapter 2.

7.2.3 The preparation of phosphorus ylids[1,2]

The removal of α-hydrogen from a phosphonium salt (7.116) produces a phosphorus ylid (7.117) and this, the so-called "salt method", is by far the most common proce-

---

* A more recent x-ray analysis on pure Ph$_3$P=CH$_2$ [J. C. J. Bart, *J. Chem. Soc. C*, 350 (1969)] revealed a P—C bond length of 166 pm, again suggesting a high degree of double bond character in the ylid bond.

Table 7.2 X-ray data on (7.115a) and (7.115b).

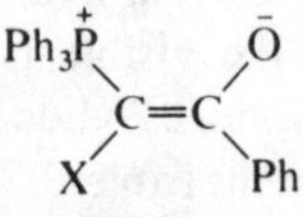

$$\underset{X}{\overset{Ph_3\overset{+}{P}}{\diagdown}}C=C\underset{Ph}{\overset{\overset{-}{O}}{\diagup}} \quad \begin{array}{l} a, X = Cl \\ b, X = I \end{array}$$

(7.115a,b)

Bond	Bond lengths (pm)	
	(7.115a)	(7.115b)
$\overset{+}{P}-\overset{-}{C}$	174	171
P—Ph	180–182	177–182
C—CO	136	135
C—O	130	128

Table 7.3 Relative rates of proton exchange of salts Me_4X^+ and Me_3Z^+

X or Z	N	P	As	Sb	S	Se
$k (\times 10^{10})$	1.8	4.3×10^6	2.5×10^5	2.4×10^4	3.6×10^7	8.1×10^5

dure for the preparation of phosphorus ylids. A variety of bases have been used for this reaction including ammonia, triethylamine, pyridine, sodium carbonate, sodium hydroxide, sodium and potassium alkoxides, sodium amide, lithium diethylamide, sodium hydride, n-butyllithium, phenyllithium, sodium acetylide and several others[1].

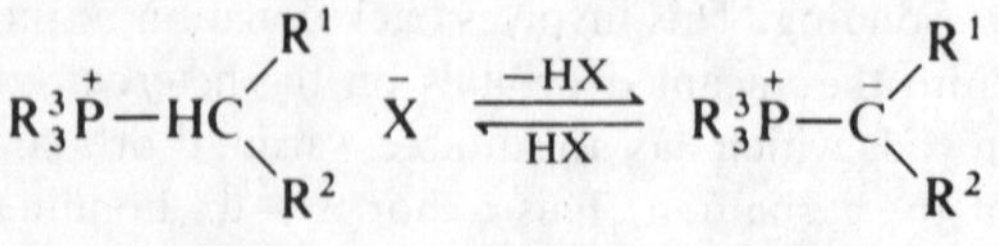

$$R_3^3\overset{+}{P}-HC\underset{R^2}{\overset{R^1}{\diagdown}}\quad \overset{-}{X} \quad \underset{HX}{\overset{-HX}{\rightleftharpoons}} \quad R_3^3\overset{+}{P}-\overset{-}{C}\underset{R^2}{\overset{R^1}{\diagdown}}$$

(7.116) (7.117)

Obviously the strength of the base required depends upon the acidity (pK_a) of the corresponding phosphonium salt and if the groups R^1 and R^2 are capable of stabilizing a negative charge by inductive electron withdrawal or resonance delocalization, relatively weak bases such as ammonia or sodium carbonate will suffice. For example, phenacylidenetriphenylphosphorane (7.118) and p-nitrophenacylidenetriphenylphosphorane (7.119) may be prepared from the corresponding phosphonium salts by treatment with sodium carbonate[79,80].

7.2 Phosphorus ylids

$$\overset{+}{Ph_3P}CH_2COPh\ \overset{-}{Br}\ \xrightarrow{aq.\ Na_2CO_3}\ \overset{+}{Ph_3P}-\overset{-}{C}HCOPh$$
pK_a, 5.5 $\qquad\qquad\qquad\qquad\qquad\qquad\qquad$ (7.118)

$$\overset{+}{Ph_3P}CH_2COC_6H_4p\text{-}NO_2\ \overset{-}{Br}\ \xrightarrow{aq.\ Na_2CO_3}\ \overset{+}{Ph_3P}-\overset{-}{C}HCOC_6H_4p\text{-}NO_2$$
pK_a, 4.2 $\qquad\qquad\qquad\qquad\qquad\qquad\qquad$ (7.119)

In contrast, if R^1 and R^2 are not capable of stabilizing the carbanion, much stronger bases are required. For example, the preparation of methylenetriphenylphosphorane (7.120) or ethylidenetriphenylphosphorane (7.121) requires the use of sodium hydride[81] or n-butyllithium[82].

$$\overset{+}{Ph_3P}CH_3\ \overset{-}{Br}\ \xrightarrow{NaH}\ \overset{+}{Ph_3P}-\overset{-}{C}H_2 + NaBr + H_2$$
$\qquad\qquad\qquad\qquad\qquad$ (7.120)

$$\overset{+}{Ph_3P}CH_2CH_3\ \overset{-}{Br}\ \xrightarrow{n\text{-}BuLi}\ \overset{+}{Ph_3P}-\overset{-}{C}HCH_3 + LiBr$$
$\qquad\qquad\qquad\qquad\qquad\qquad$ (7.121)

Incidentally, ylids and especially alkylidenetrialkylphosphoranes form stable complexes with lithium salts which dissociate only at elevated temperatures. Such complexes often interfere with the course of subsequent reactions of the ylids and in consequence, the alkyl (or aryl) lithium method must be used with caution (see later). The use of sodium amide in boiling tetrahydrofuran[81] constitutes one of the more recent methods of preparing pure alkylidenetrialkylphosphoranes.

When groups R^1 and/or R^2 are strongly electron withdrawing, the ylid carbanion is often sufficiently stabilized to be unreactive towards water. On the other hand, strongly basic ylids (i.e. those derived from weakly acidic phosphonium salts) behave like unprotected carbanions and will rapidly deprotonate water to form the corresponding phosphonium hydroxides (7.122) which in turn, decompose to phosphine oxides (7.123) and a hydrocarbon. Consequently the preparation of strongly basic ylids requires the use of aprotic or basic media such as hydrocarbons, ethers and liquid ammonia.

$$\overset{+}{R_3P}-\overset{-}{C}\underset{R^2}{\overset{R^1}{\diagup}} + H_2O \longrightarrow \overset{+}{R_3P}CH\underset{R^2}{\overset{R^1}{\diagup}}\ \overset{-}{O}H \longrightarrow R_3P=O + R^1CH_2R^2$$
$\qquad\qquad\qquad\qquad\qquad\qquad\qquad\qquad\qquad\qquad$ (7.123)
$\qquad\qquad\qquad\qquad\qquad\qquad$ (7.122)

$R^1, R^2 = H$, alkyl

It should be noted that the reactivity of strongly basic ylids towards water (and oxygen) has frequently made the isolation of these compounds difficult. This has led to a rough classification of ylids as either "stabilized" (i.e. readily isolable due to low carbanion reactivity) or "non-stabilized" (i.e. highly reactive). Although this classification is often useful it should be remembered that the choice of the dividing line is somewhat arbitrary.*

---

* Ylids such as $Ph_3P=C(CO_2Et)_2$ and $Ph_3P=CMe_2$ represent the extremes of "stabilized" and "non-stabilized" ylids, respectively. A structure such as $Ph_3P=CPh_2$ represents an ylid of intermediate stability.

Apart from the salt method there are a number of more exotic procedures for the preparation of ylids and we shall now turn our attention to a few of these.

Alkylidenephosphoranes with one or two halogen atoms on the α-carbon may be prepared from a phosphine and the corresponding carbene (eqn. 16).

$$R_3P: + :C\begin{smallmatrix}X\\Y\end{smallmatrix} \longrightarrow R_3P=C\begin{smallmatrix}X\\Y\end{smallmatrix} \qquad (16)$$

This is exemplified by the reaction of triphenylphosphine with dichlorocarbene in eqn. (17)[83,84].

$$Ph_3P + :CCl_2 \longrightarrow Ph_3P=CCl_2 \qquad (17)$$

Diazoaliphatics may also serve as the carbene source provided they are decomposed by cuprous salts (eqn. 18)[85].

$$Ph_3P + \overset{+}{N}\equiv N-\overset{-}{C}R_2 \xrightarrow{Cu^+} Ph_3P=CR_2 + N_2 \qquad (18)$$

Dihalomethylenephosphoranes are also formed from the reaction of triphenylphosphine with carbon tetrachloride or tetrabromide (eqn. 19) and this reaction may also involve carbene intermediates (see Chapter 4)[86,87].

$$2Ph_3P + CX_4 \longrightarrow Ph_3P=CX_2 + Ph_3PX_2 \quad X = Cl \text{ or } Br \qquad (19)$$

Reactions of phosphines with activated olefins are also capable of producing ylids and a typical example is provided by eqn. (20)[88].

$$Ph_3P + PhCOCH=CHCOPh \longrightarrow \left[Ph_3\overset{+}{P}-CH(COPh)-\overset{-}{C}HCOPh\right] \longrightarrow$$

$$Ph_3\overset{+}{P}-\overset{-}{C}\begin{smallmatrix}COPh\\CH_2COPh\end{smallmatrix} \qquad (20)$$

Ylids are also available from the reaction of dihalophosphoranes (the P(V) compounds) with activated methylene compounds (eqn. 21)[89] or in a few cases from the reaction of phosphines with benzyne (eqn. 22)[90].

$$Ph_3PX_2 + CH_2.CN.CO_2Me \longrightarrow Ph_3\overset{+}{P}-\overset{-}{C}(CN)CO_2Me \quad X = Cl, Br \qquad (21)$$

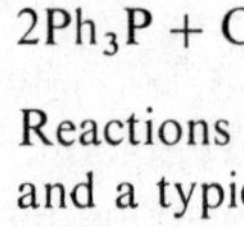

$$[C_6H_4] + Ph_2PCH_3 \longrightarrow Ph_3P=CH_2 \qquad (22)$$

A large number of ylids are obtained by the alkylation, acylation, alkoxycarbonylation and halogenation of simple alkylenephosphoranes (eqn. 23)[1]. The phosphonium salt formed in the first step is deprotonated by a second molecule of ylid in what is termed a "transylidation" reaction. Silylmethylenephosphoranes (7.124) are available by

7.2 Phosphorus ylids

this method (eqn. 24)[91a] and silyl-substituted methylenephosphoranes may be desilylated by trimethylsilanol (7.125), or methanol[91b].

$$R^1_3P=CHR^2 + R^3X \longrightarrow R^1_3\overset{+}{P}-HC\underset{R^3}{\overset{R^2}{\diagup}} \bar{X} \xrightarrow{R^1_3P=CHR^2} R^1_3P=C\underset{R^3}{\overset{R^2}{\diagup}} + R^1_3\overset{+}{P}-CH_2R^2 \quad (23)$$

R^3 = alkyl, $R\overset{O}{\overset{\|}{C}}$, $RO\overset{O}{\overset{\|}{C}}$, or halogen

$$2Ph_3P=CH_2 + Me_3SiCl \longrightarrow Ph_3\overset{+}{P}Me\ \bar{Cl} + Ph_3P=CHSiMe_3 \quad (24)$$
$$(7.124)$$

This provided the first successful route to pure alkylidenetrialkylphosphoranes (eqn. 25).

$$Me_3P=CHSiMe_3 + Me_3SiOH \longrightarrow Me_3P=CH_2 + Me_3SiOSiMe_3 \quad (25)$$
$$(7.125)$$

7.2.4 The Wittig reaction.[1,3-6,8,66,67,93-95]

7.2.4.1 Introduction

Any reaction in organic chemistry which creates a new carbon–carbon bond is likely to be of synthetic importance and the basic significance of the Wittig reaction is that it offers a route to olefins. The original paper[72] described the condensation of methylenetriphenylphosphorane with benzophenone (eqn. 26) to form 1,1-diphenylethylene and it was this report which initiated the last 20 years of research into the mechanism and synthetic utility of the reaction.

$$Ph_3P=CH_2 + PhCOPh \longrightarrow Ph_3P(O) + CH_2=CPh_2 \quad (26)$$

In fact, Wittig's reaction was not the first of its kind. As early as 1919, Staudinger described the condensation of an ylid (7.126) with phenylisocyanate to form a ketenimine (7.127)[92], and in his address to the IUPAC symposium on organophosphorus

$$Ph_3P=CPh_2 + PhNCO \longrightarrow Ph_3P(O) + PhN=C=CPh_2$$
$$(7.126) \qquad\qquad\qquad (7.127)$$

compounds in Heidelberg (May, 1964) Wittig entitled his paper, "Variatzionen zu einem Thema von Staudinger"[93], in recognition of the pioneering work of the Staudinger school. Nevertheless, the credit for the development of the reaction as an important synthetic tool undoubtedly belongs to Wittig and his collaborators and latterly to H. J. Bestmann and his group. The next section attempts to provide some idea of the scope of this olefin synthesis.

7.2.4.2 Synthetic applications of the Wittig reaction

One of the first striking examples was the preparation of methylenecyclohexane (7.128) free of its endocyclic isomer (7.129)[96].

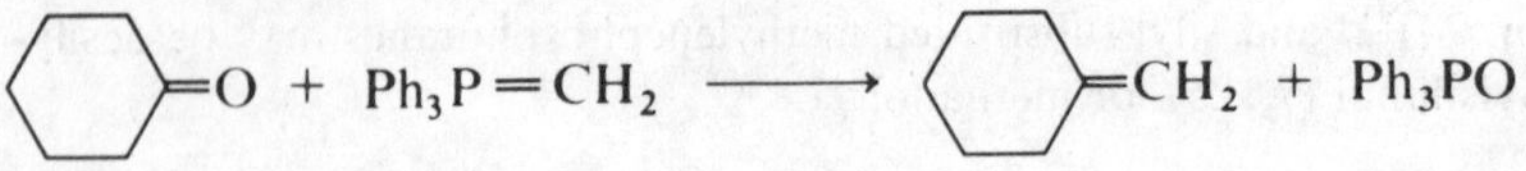

$$\text{cyclohexanone} + Ph_3P=CH_2 \longrightarrow \text{methylenecyclohexane} + Ph_3PO \tag{7.128}$$

$$\not\longrightarrow \text{1-methylcyclohexene} \tag{7.129}$$

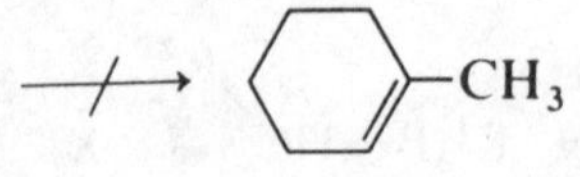

An extension of this allowed the synthesis of a derivative of methylenecyclopropene (7.130) at a time when interest in non-benzenoid aromatics was intense[97].

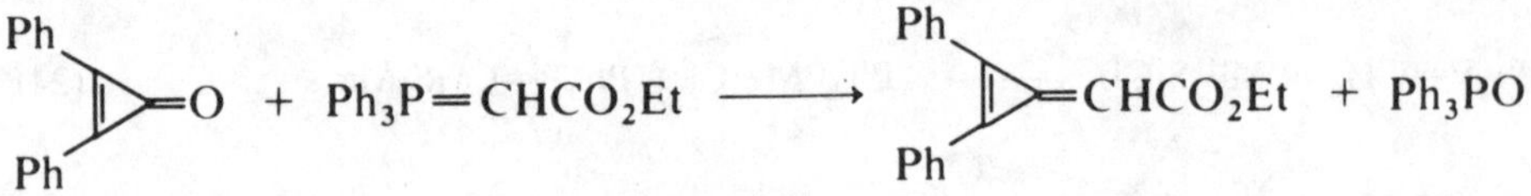

$$\underset{Ph}{\overset{Ph}{>}}\!\!\!\!\triangleright\!\!=\!\!O + Ph_3P=CHCO_2Et \longrightarrow \underset{Ph}{\overset{Ph}{>}}\!\!\!\!\triangleright\!\!=\!\!CHCO_2Et + Ph_3PO \tag{7.130}$$

The conversion of ketones into aldehydes containing one more carbon atom is illustrated by eqns. (27) and (28)[98,99] and the synthesis of a deuterium-labelled olefin is accomplished easily using a deuterio-ylid (7.131) and benzaldehyde[100].

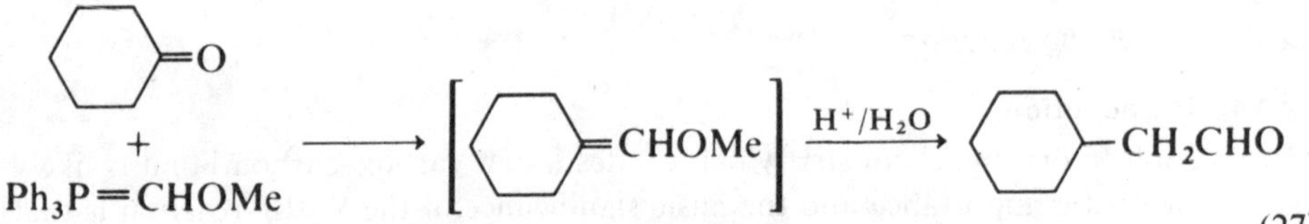

$$\text{cyclohexanone} + Ph_3P=CHOMe \longrightarrow [\text{=CHOMe}] \xrightarrow{H^+/H_2O} \text{-CH}_2CHO \tag{27}$$

$$PhCHO + Ph_3P=CHOMe \longrightarrow PhCH=CHOMe \xrightarrow{H^+/H_2O} PhCH_2CHO \tag{28}$$

$$Ph_3\overset{+}{P}-\overset{-}{C}DPh + PhCHO \longrightarrow PhCH=CDPh$$
(7.131)

A variation of the Wittig reaction using oxygen instead of a carbonyl compound allows the preparation of a 1,2-dideuteriooolefin (7.132)[100] and a similar reaction with electrophilic oxygen from a peracid also affords olefins (eqn. 29)[101].

$$(7.131) \xrightarrow{O_2} Ph_3P=O + O=CDPh \xrightarrow{Ph_3\overset{+}{P}-\overset{-}{C}DPh} Ph_3P=O + PhCD=CDPh \tag{7.132}$$

$$2Ph_3P=CHCOPh + RCO_3H \longrightarrow 2Ph_3P=O + RCO_2H + PhCOCH=CHCOPh \tag{29}$$

Condensation of a bifunctional ylid (7.133) with phthaldehyde gives a bicyclic hydrocarbon, 3,4-benzoheptatriene (7.134) in 28 % yield[102] and among the early applications to the synthesis of heterocyclic systems one may cite the preparation of 3H-pyrrolizine (7.136) in 87 % yield from vinyltriphenylphosphonium bromide (7.135) and 2-pyrrolealdehyde, presumably via the route shown below[103].

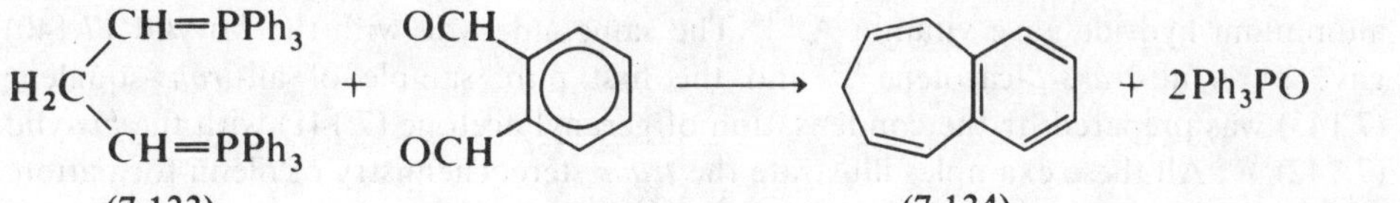

7.2 Phosphorus ylids

(7.133) + (7.134) → [product] + 2Ph$_3$PO

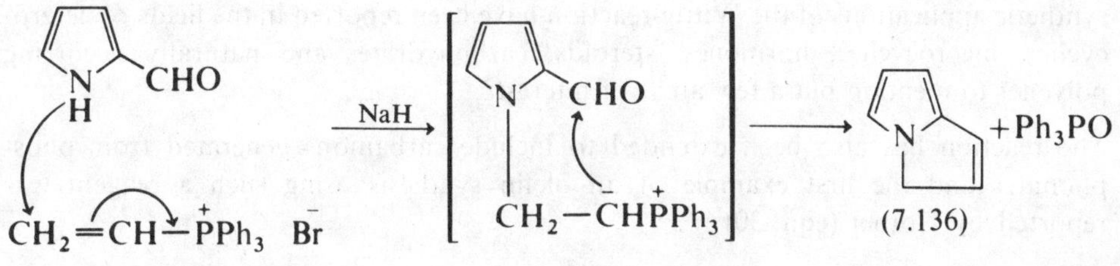

(7.135) + pyrrole-2-CHO →[NaH] [intermediate] → (7.136) + Ph$_3$PO

Applications in the field of natural products are illustrated by the examples in Table 7.4. Wittig and Pommer condensed β-ionylideneacetaldehyde (7.137) with the ester-ylid (7.138) and obtained the polyene ester (7.139) which on reduction with lithium

---

Table 7.4 Applications of the Wittig reaction to the synthesis of natural products

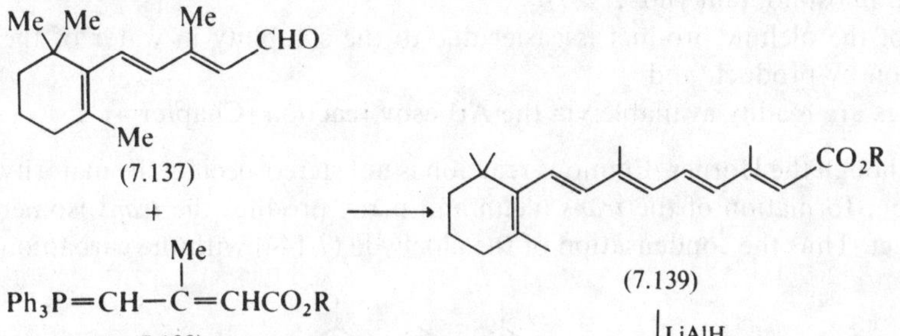

(7.137) + (7.138) Ph$_3$P=CH—C(Me)=CHCO$_2$R → (7.139) →[LiAlH$_4$] Vitamin A

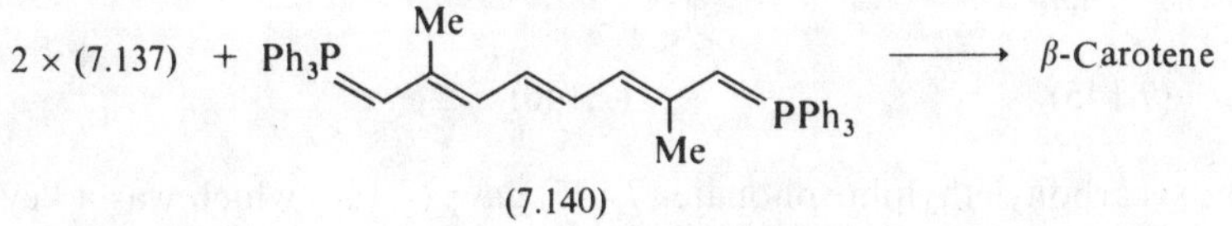

2 × (7.137) + (7.140) Ph$_3$P=CH–C(Me)=CH–CH=CH–C(Me)=CH–PPh$_3$ → β-Carotene

2 (7.141) + (7.142) Ph$_3$P=CH(CH$_2$)$_2$CH=PPh$_3$ → (7.143) trans-squalene

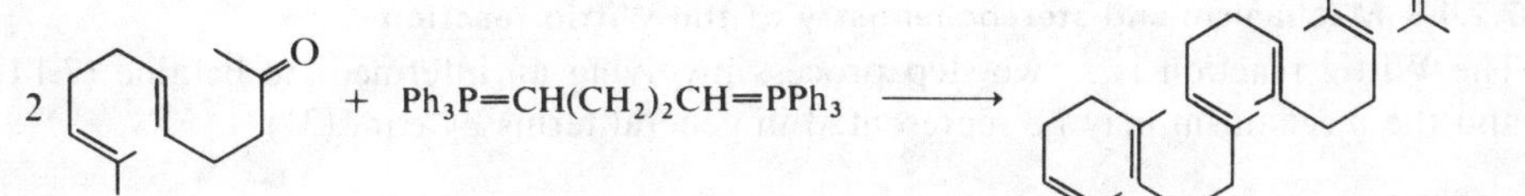

aluminium hydride gave vitamin A.[104] The same aldehyde with the *bis*-ylid (7.140) gave pure, all-*trans*-β-carotene[105] and the first pure sample of all-*trans*-squalene (7.143) was prepared by the condensation of geranyl acetone (7.141) with the *bis*-ylid (7.142)[106]. All these examples illustrate the *trans* stereochemistry of olefin formation, frequently (but not always) found in the Wittig reaction. In recent years further synthetic applications of the Wittig reaction have been reported in the fields of heterocyclics, macrocyclics, hormones, steroids, carbohydrates and naturally occurring polyenes to mention but a few areas of interest[5].

The reaction has also been extended to include carbanions generated from phosphonates and the first example of an olefin synthesis using such a reagent was reported by Horner (eqn. 30)[107].

$$(EtO)_2P(O)CH_2Ph \xrightarrow{NaNH_2} (EtO)_2P(O)\overset{-}{C}HPh \xrightarrow{Ph_2CO} Ph_2C=CHPh + (EtO)_2PO_2^- \quad (30)$$

Subsequently, Wadsworth and Emmons carried out a broad study of this reaction[108] and a recent review by Boutagy and Thomas[109] gives a comprehensive coverage of the mechanism, stereochemistry and synthetic scope of the so-called "Horner–Wittig" or "Horner–Emmons" reaction. It has a number of advantages over the conventional Wittig reaction and these may be summarized as follows:

(i) reaction occurs with a wider variety of aldehydes and ketones under relatively mild conditions due to the higher nucleophilicity of phosphonate carbanions compared to phosphonium ylids;
(ii) separation of the olefinic product is easier due to the solubility in water of the phosphate ion by-product, and
(iii) phosphonates are readily available via the Arbusov reaction (Chapter 4).

Furthermore, although the Horner–Emmons reaction is not stereospecific, the majority of reactions favour formation of the *trans* olefin and many produce the *trans* isomer as the sole product. Thus the condensation of the aldehyde (7.144) with the carbanion

(7.144)   (7.145)   (7.146)

from diethyl-1-ethoxycarbonylethylphosphonate (7.145) gave (7.146) which was a key step in the synthesis of the sesquiterpenoid, α-farnesene[110].

7.2.4.3 Mechanism and stereochemistry of the Wittig reaction

The Wittig reaction is a two-step process involving an intermediate betaine (7.147) and the mechanism may be represented in general terms by eqn. (31).

$$\underset{O=CR^3R^4}{\overset{R_3\overset{+}{P}-\overset{-}{C}R^1R^2}{+}} \underset{k_2}{\overset{k_1}{\rightleftharpoons}} \underset{(7.147)}{\overset{R_3\overset{+}{P}-CR^1R^2}{\underset{-O-CR^3R^4}{|}}} \xrightarrow{k_3} \underset{R^1R^2C=CR^3R^4}{\overset{R_3P=O}{+}} \quad (31)$$

7.2 Phosphorus ylids

In some cases the intermediate has been trapped either by protonation (eqn. 32)[111] or by complex formation with lithium salts (eqn. 33)[112] and in one case an oxyphosphorane (7.148) has been isolated and characterized by ^{31}P n.m.r.[113].

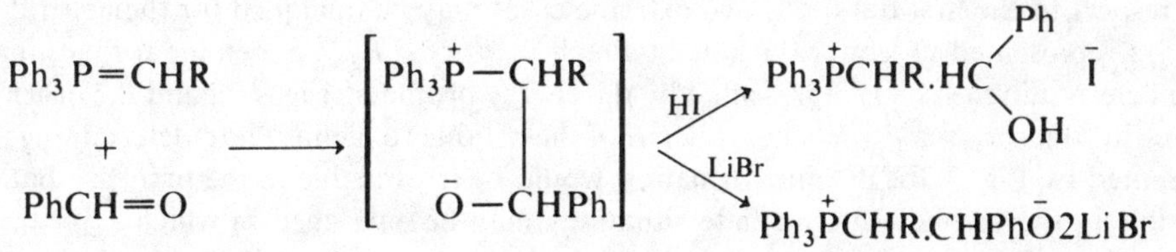

$$Ph_3P=CHR + PhCH=O \longrightarrow \left[\begin{array}{c} Ph_3\overset{+}{P}-CHR \\ | \\ \overset{-}{O}-CHPh \end{array} \right] \underset{LiBr}{\overset{HI}{\diagup\!\!\!\diagdown}} \begin{array}{l} Ph_3\overset{+}{P}CHR.HC\diagup^{Ph}_{\diagdown OH} \bar{I} \quad (32) \\ \\ Ph_3\overset{+}{P}CHR.CHPh\bar{O}2LiBr \quad (33) \end{array}$$

$$\begin{array}{c} Ph_3\overset{+}{P}-\overset{2-}{C}-\overset{+}{P}Ph_3 \\ + \\ (CF_3)_2C=O \end{array} \longrightarrow \begin{array}{c} Ph_3P-C\diagup^{PPh_3} \\ | \quad \diagdown CF_3 \\ O-C \\ \diagdown CF_3 \end{array}$$

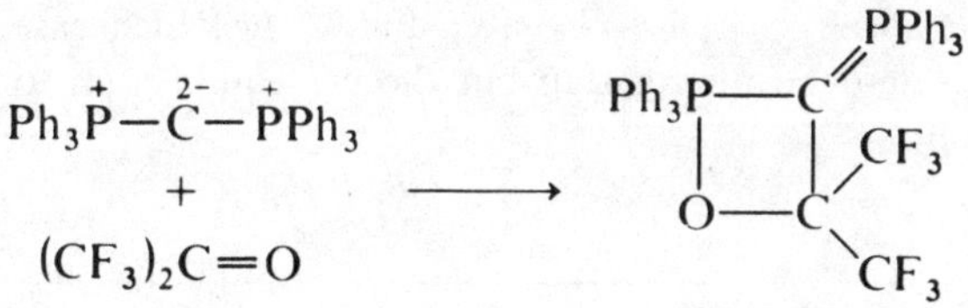

(7.148)

The second stage of the reaction is thought to proceed through a four-membered cyclic intermediate such as (7.149), cf. (7.148) and it seems reasonable to assume that the driving force for this reaction is the formation of the P=O bond.

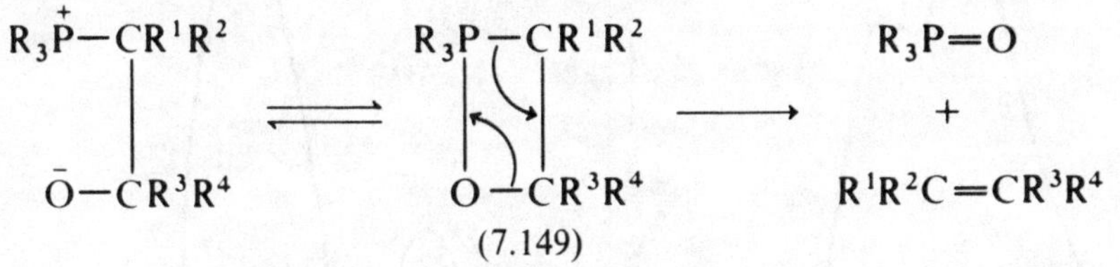

$$\begin{array}{c} R_3\overset{+}{P}-CR^1R^2 \\ | \\ \overset{-}{O}-CR^3R^4 \end{array} \rightleftharpoons \begin{array}{c} R_3P-CR^1R^2 \\ | \quad | \\ O-CR^3R^4 \end{array} \longrightarrow \begin{array}{c} R_3P=O \\ + \\ R^1R^2C=CR^3R^4 \end{array}$$

(7.149)

As required by such a four-centre rearrangement with the ring apical–equatorial, the configuration at phosphorus is retained during the reaction. This was elegantly demonstrated by the scheme outlined in Fig. 7.2[114,115].

Figure 7.2 Retention of configuration at phosphorus during a Wittig reaction.

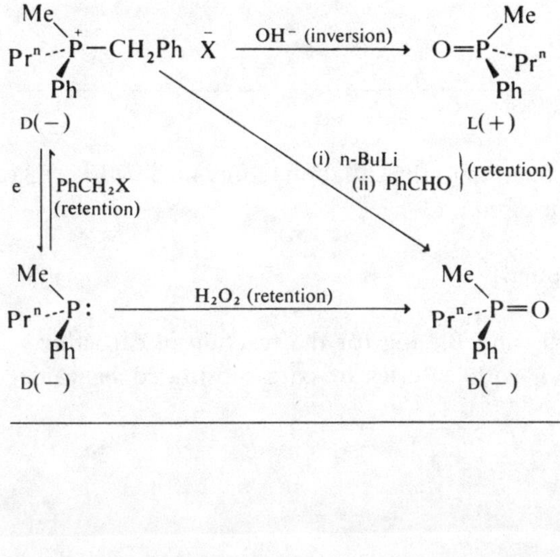

$$\begin{array}{c} \text{Me} \\ \diagdown \\ \text{Pr}^n\text{---}\overset{+}{P}-CH_2Ph \quad X^- \\ \diagup \\ \text{Ph} \\ \text{D}(-) \end{array} \xrightarrow{OH^- \text{ (inversion)}} \begin{array}{c} \text{Me} \\ \diagdown \\ O=P\text{---}\text{Pr}^n \\ \diagup \\ \text{Ph} \\ \text{L}(+) \end{array}$$

$\Updownarrow e$ | PhCH$_2$X (retention)

(i) n-BuLi
(ii) PhCHO } (retention)

$$\begin{array}{c} \text{Me} \\ \diagdown \\ \text{Pr}^n\text{---}P: \\ \diagup \\ \text{Ph} \\ \text{D}(-) \end{array} \xrightarrow{H_2O_2 \text{ (retention)}} \begin{array}{c} \text{Me} \\ \diagdown \\ \text{Pr}^n\text{---}P=O \\ \diagup \\ \text{Ph} \\ \text{D}(-) \end{array}$$

With this much established the important mechanistic questions to be answered are (i), which step of the reaction is rate determining; (ii) is betaine formation reversible and (iii) what factors control the stereochemistry of the product olefin?

With respect to the first question, two extreme cases may be imagined for the general situation represented by eqn. (31): one in which $k_1 < k_2 < k_3$, i.e. betaine formation is rate determining which is represented by the energy profile of Fig. 7.3a and the other extreme in which $k_1 > k_2 > k_3$, i.e. collapse of the betaine to olefin is rate-determining, represented by Fig. 7.3b. Betaine formation would be irreversible in the first case but reversible in the second. Intermediate situations may be envisaged in which $k_2 > k_3$ but $k_1 < k_2$ and k_3 (betaine formation remaining the single step of highest activation energy but now becoming reversible) or $k_2 > k_1$ but $k_3 < k_1$ and k_2 in which case betaine decomposition remains the rate-determining step but the pre-equilibrium to betaine lies on the side of the reactants.

Figure 7.3 Wittig reaction energy profiles.

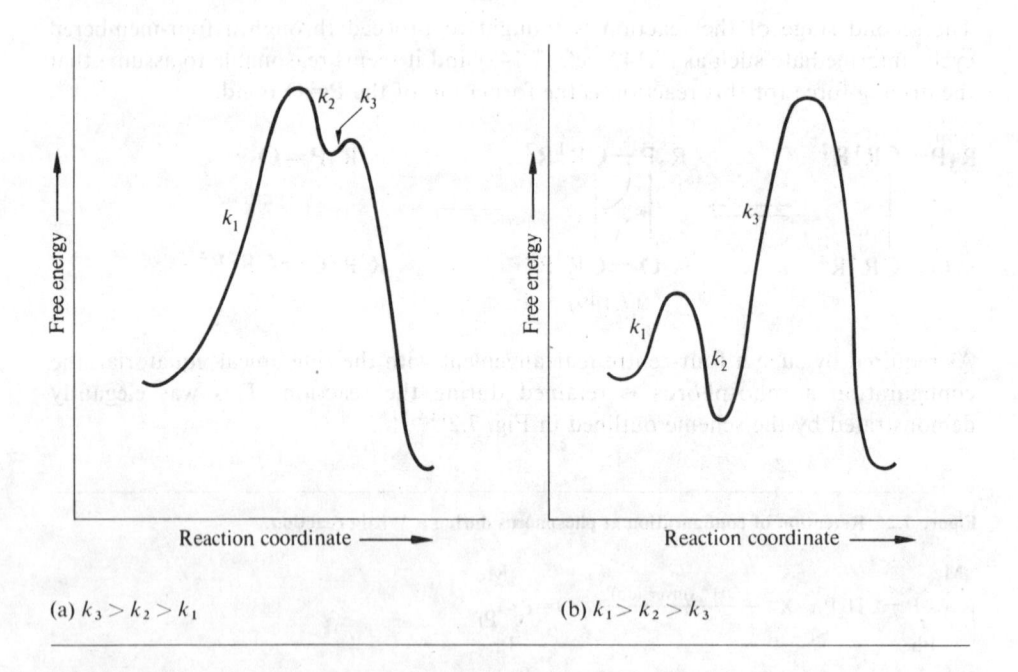

(a) $k_3 > k_2 > k_1$

(b) $k_1 > k_2 > k_3$

Application of the steady-state approximation to the situation represented by Fig. 7.3a leads to the differential rate expression of eqn. (34).

$$d[\text{olefin}]/dt = k_1[\text{ylid}][\text{carbonyl compound}] \tag{34}$$

Such a situation was found by Speziale and Bissing for the reaction of carbethoxymethylenetriphenylphosphorane (7.150) with a series of para-substituted benzaldehydes[116].

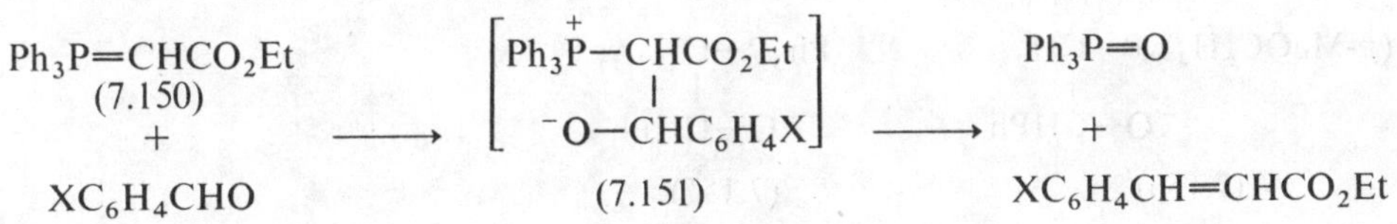

7.2 Phosphorus ylids

$$Ph_3P=CHCO_2Et \quad (7.150) \quad + \quad XC_6H_4CHO \quad \longrightarrow \quad \begin{bmatrix} Ph_3\overset{+}{P}-CHCO_2Et \\ | \\ {}^-O-CHC_6H_4X \end{bmatrix} \quad (7.151) \quad \longrightarrow \quad Ph_3P=O \quad + \quad XC_6H_4CH=CHCO_2Et$$

$$X = p\text{-}NO_2 > m\text{-}Cl > p\text{-}Cl > H > p\text{-}Me > p\text{-}MeO$$

The reaction was second order (first order in each component) and the rate of disappearance of ylid was the same as the rate of formation of olefin, consistent with eqn. (34). Furthermore, the reaction was accelerated by polar solvents (by a factor of 6 in $CHCl_3$ and ca 100 in CH_3OH compared to the rate in benzene) and the entropies of activation, $\Delta S^{\ddagger}$, were about -170 J K^{-1} mol^{-1} (-40 cal K^{-1} mol^{-1}) at 25 °C, both facts suggesting a highly polar and highly orientated transition state as required by rate-determining formation of the betaine (7.151). Finally, the second order rate constants gave a good Hammett correlation with the σ-constants of the groups, X, with a ρ-value of $+2.7$ indicating a moderately high degree of negative charge developing on carbonyl oxygen in the transition state.

However, in the case where $k_2 \geqslant k_3$ but $k_2 > k_1$ (i.e. betaine formation rate-determining, but reversible) a similar rate expression (eqn. 35) holds and the kinetic data do not suffice to distinguish the mechanistic pathways.

$$d[\text{olefin}]/dt = k_1 k_3 [\text{ylid}][\text{carbonyl compound}]/k_2 + k_3 \qquad (35)$$

In fact, the reversibility of betaine formation was demonstrated in a related system by trapping the ylid (7.154) formed by dissociation of a betaine (7.152) prepared from

---

Figure 7.4 Reversibility of betaine formation in the Wittig reaction.

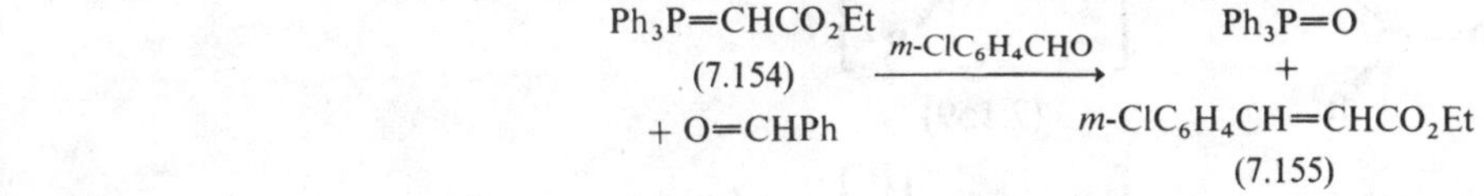

$$Ph_3P + PhHC\overset{O}{-\!\!\!\triangle\!\!\!-}CHCO_2Et \longrightarrow \begin{matrix} Ph_3\overset{+}{P}-CHCO_2Et \\ | \\ {}^-O-CHPh \\ (7.152) \end{matrix} \longrightarrow \begin{matrix} Ph_3P=O \\ + \\ PhHC=CHCO_2Et \\ (7.153) \end{matrix}$$

$$\updownarrow$$

$$\begin{matrix} Ph_3P=CHCO_2Et \\ (7.154) \\ + O=CHPh \end{matrix} \xrightarrow{m\text{-}ClC_6H_4CHO} \begin{matrix} Ph_3P=O \\ + \\ m\text{-}ClC_6H_4CH=CHCO_2Et \\ (7.155) \end{matrix}$$

---

triphenyl phosphine and an epoxide (Fig. 7.4)[116]. Isolation of ethyl *m*-chlorocinnamate (7.155) as well as ethyl cinnamate (7.153) led Speziale and Bissing to propose that *stabilized ylids* (such as 7.150 and 7.154) react with carbonyl compounds in a slow, *reversible* first step followed by rapid decomposition to olefin and phosphine oxide.

On this basis it seems intuitively reasonable that one should find betaine decomposition to be rate determining for the most reactive, i.e. least stabilized, ylids and judging by the isolation of betaines such as (7.156) and (7.157) this appears to be so[112,117].

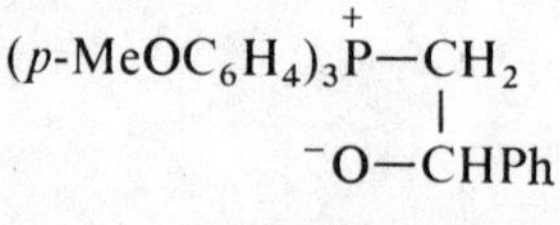

$(p\text{-MeOC}_6\text{H}_4)_3\overset{+}{\text{P}}-\text{CH}_2$ $\text{Ph}_3\overset{+}{\text{P}}-\text{CH}_2$
            | |
      $^-\text{O}-\text{CHPh}$ $^-\text{O}-\text{CHPh}$
         (7.156) 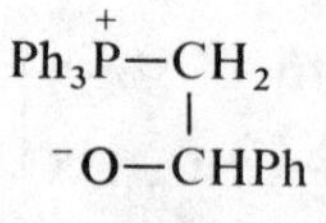   (7.157)

But reactive ylids are often generated using alkyl or aryl lithium and lithium salt complexes of the betaines are remarkably stable. This is probably the origin of the change in rate-determining step from betaine formation to betaine decomposition in the case of the non-stabilized ylids since recent competitive rate studies and stereochemical evidence suggests that in salt-free media, betaine formation between non-stabilized or partially stabilized ylids (7.158) and para-substituted benzaldehydes remains the rate-determining step[118].

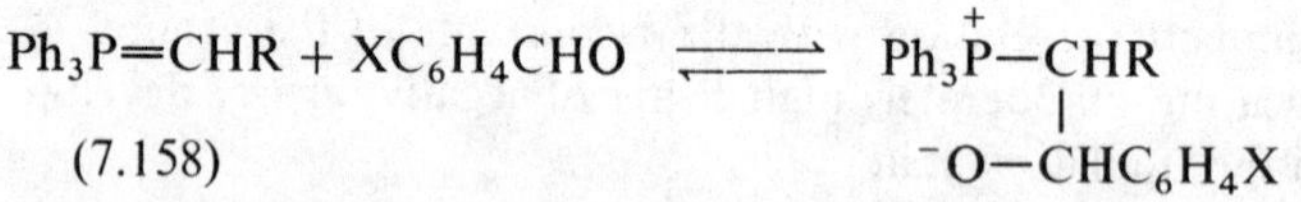

$$\text{Ph}_3\text{P}=\text{CHR} + \text{XC}_6\text{H}_4\text{CHO} \rightleftharpoons \text{Ph}_3\overset{+}{\text{P}}-\text{CHR}$$
$$\qquad\qquad\qquad\qquad\qquad\qquad\qquad\qquad |$$
$$(7.158) \qquad\qquad\qquad\qquad\qquad\qquad ^-\text{O}-\text{CHC}_6\text{H}_4\text{X}$$

$R = \text{Et or Ph}$;   Rate: $X = m\text{-Cl} > \text{H} > p\text{-Me} > p\text{-MeO}$.

The rationalization of the stereochemistry of the olefin formed in the Wittig reaction is a fascinating, but complex story. Stabilized ylids produce mostly the *trans*-olefin when reacted with aldehydes or ketones and since betaine formation is reversible (see Fig. 7.4) this is most readily explained in the following way.

Figure 7.5 General scheme for formation of *cis* and *trans* olefins.

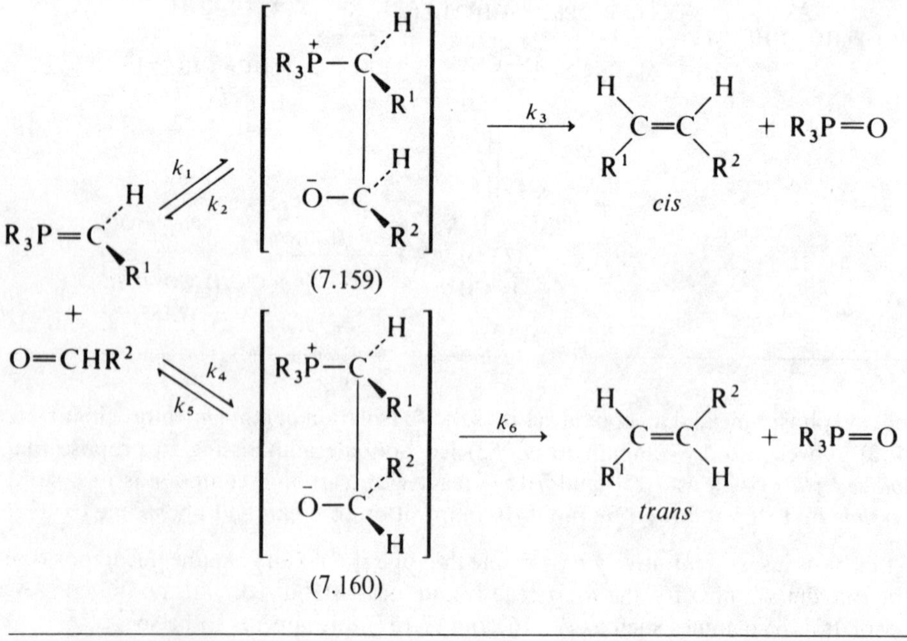

7.2 Phosphorus ylids

The general scheme for formation of *cis* and *trans* olefins is depicted in Fig. 7.5. It has been shown[119] that for all ylids (stabilized or unstabilized) the ratio of *cis/trans* olefins is given by eqn. (36) which may be transposed to eqn. (37).

$$cis/trans = k_1 k_3 (k_5 + k_6)/k_4 k_6 (k_2 + k_3) \quad (36)$$

$$cis/trans = k_1/k_4 [(k_5/k_6 + 1)/(k_2/k_3 + 1)] \quad (37)$$

From eqn. (37) it is evident that if $k_1/k_4 > 1$ and $k_5/k_6 > k_2/k_3$ the *cis/trans* ratio will be >1. Conversely, if $k_1/k_4 < 1$ and $k_5/k_6 < k_2/k_3$ the *trans* isomer will predominate. Now the ratios k_2/k_3 and k_5/k_6 may be evaluated by competitive studies similar to that outlined in Fig. 7.4 but starting with the *cis* or *trans* epoxide[116]. For $R = R^2 =$ phenyl and $R^1 = -CO_2Et$, $k_2/k_3 = 0.5$ and $k_5/k_6 = 0.15$, i.e. $k_2/k_3 > k_5/k_6$ which favours formation of the *trans*-olefin. Since in Fig. 7.5 betaine (7.159) has R^1 and R^2 eclipsed it will undoubtedly be more crowded sterically than betaine (7.160) and hence one might expect k_1 to be less than k_4, i.e. $k_1/k_4 < 1$ which would again favour formation of the *trans* isomer. The experimental *cis/trans* ratio of 16:84 confirms this expectation and by substituting the values of *cis/trans*, k_5/k_6 and k_2/k_3 in eqn. (37), one can place a numerical value on k_1/k_4 of 0.25. Thus the rate of betaine formation does have a profound effect on the *cis/trans* isomer ratio, i.e. the stability of the intermediate betaines is a major factor in determining the resultant stereochemistry of the olefin.

It appears that protonic solvents like methanol increase the proportion of *cis*-olefin from stabilized ylids (Table 7.5). This is readily explained by assuming that methanol solvates the oxy-anion of the intermediate betaine, thus reducing electrostatic interactions between P^+ and O^- and allowing the formation of alternative conformations (7.161) and (7.162), the most stable form of which (7.161) leads to the *cis*-olefin.

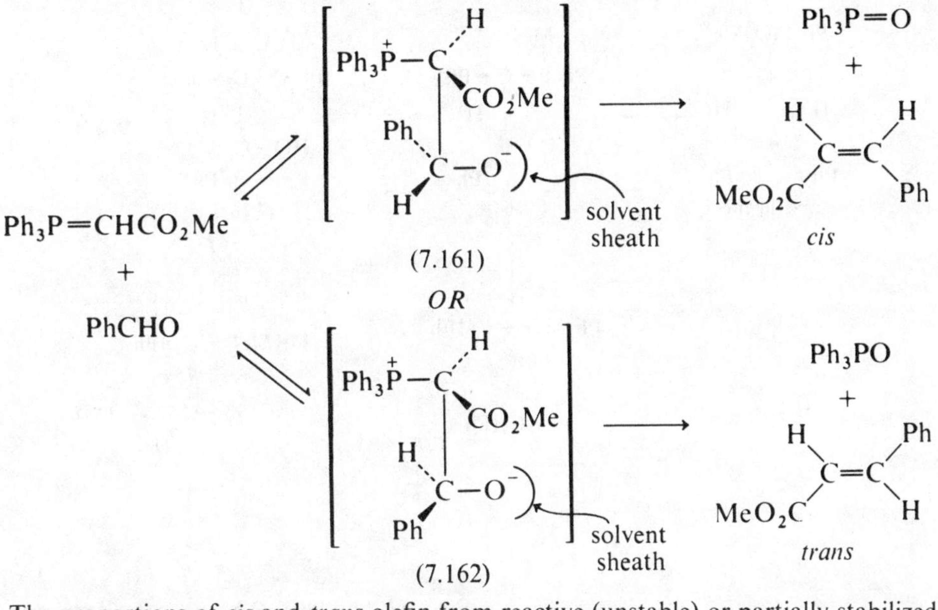

The proportions of *cis* and *trans* olefin from reactive (unstable) or partially stabilized ylids depends upon the reaction conditions such as solvent and the presence of salt. By means of the scheme outlined in Fig. 7.6, Trippett has shown that formation of

Table 7.5 Effect of solvent on *cis/trans* olefin ratios from stabilized ylids for the reaction
$Ph_3P=CHCO_2Me + MeCHO \longrightarrow Ph_3P=O + MeCH=CH\ CO_2Me$

Solvent	Overall yield	% cis	% trans
CH_2Cl_2	88	6	94
Me_2NCHO (DMF)	98	3	97
MeOH	96	38	62

the betaine is definitely reversible[119]. The starting point was the diastereomerically pure phosphonium salt (7.163) prepared as shown, which on treatment with base gave the betaine (7.164) which in turn collapsed to *cis*-stilbene or dissociated to benzylidenephosphorane (7.165) and benzaldehyde. Recombination via (7.166)

Figure 7.6 Reversibility of betaine formation with reactive ylids.

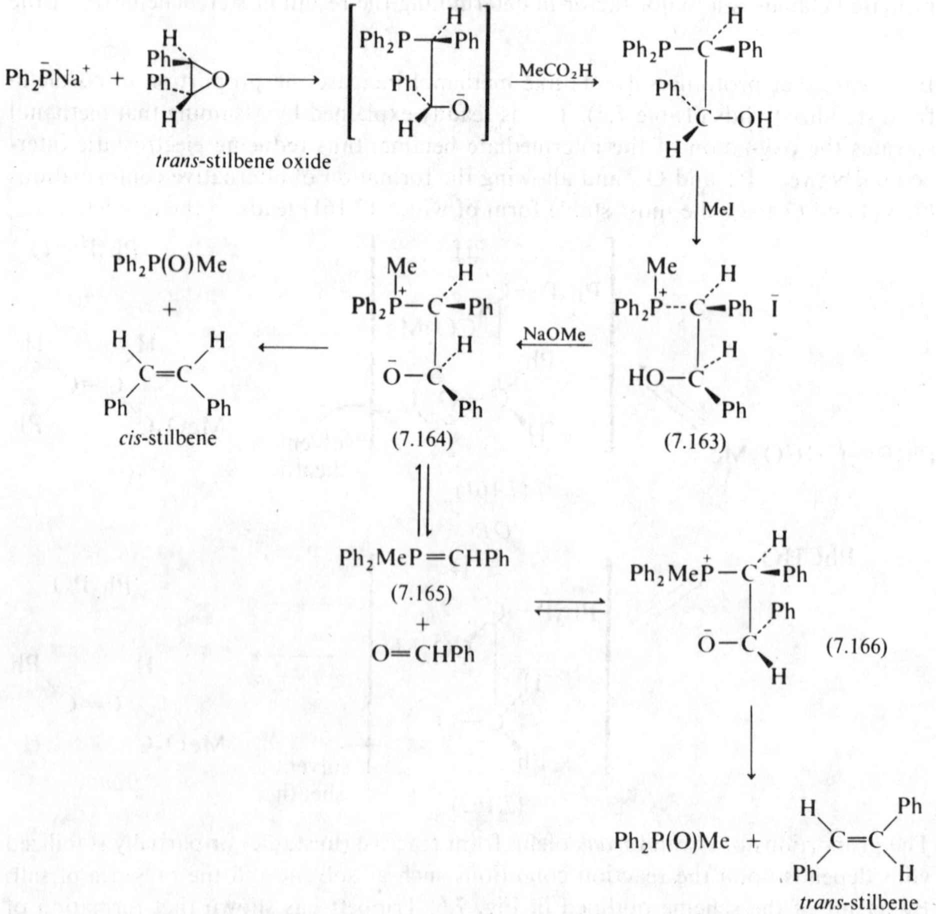

7.2 Phosphorus ylids

led to the *trans*-stilbene and in fact, *cis*- and *trans*-stilbene were obtained in almost equal proportions. Since Trippett established that betaines (7.164) and (7.166) were not directly interconvertible the results could only be explained by invoking reversibility in betaine formation. With this in mind it is possible to explore the stereochemistry of olefin formation with semi-stabilized and unstabilized ylids.

For the semi-stabilized ylid (7.165) the experimental *cis/trans* ratio in methanol is 22:78.[119] Once again this means that either $k_1 < k_4$ (if $k_5/k_6 \approx k_2/k_3$) or if $k_1 \approx k_4$ then $k_2/k_3 > k_5/k_6$.

Using the salt (7.163) from *trans*-stilbene oxide, treatment with sodium methoxide gives a 40:60 mixture of *cis:trans* olefin. From this it is possible to calculate that $k_2/k_3 = 3.25$ but since an experimental value for k_5/k_6 is not available for this system, a clear cut decision on the factors controlling the stereochemistry of the reaction is not possible.

In polar, aprotic solvents like DMF, or in non-polar solvents like benzene, *reactive* (unstabilized) ylids give mainly the *cis*-olefin. The reactions in DMF are virtually unaffected by the presence of lithium salts but in non-polar solvents the same reactions show an increase in the proportion of the *trans* isomers and a considerable dependence on the size of the anion of the salt (Table 7.6)[120].

Table 7.6 Salt effects in benzene-light petroleum at 0°C for the reaction

$Ph_3P=CHCH_3 + PhCHO \longrightarrow Ph_3PO + CH_3CH=CHPh$

Salt	Overall yield of olefin (%)	*cis/trans* ratio
None	98	87/13
LiCl	70	81/19
LiBr	68	61/39
LiI	76	58/42
LiBPh$_4$	63	50/50

It is not easy to rationalize these facts. Trippett has shown that even with very reactive ylids, betaine formation is reversible[119] and evidence has already been mentioned[118] which suggests betaine formation remains the rate-limiting step. With reference to Fig. 7.5 it seems likely that k_2 will always be greater than k_5 and hence if k_1 and k_4 are comparable, k_6 must be considerably less than k_3 to satisfy the experimental observation of a high *cis/trans* isomer ratio. Since k_6 affords the thermodynamically more stable *trans*-olefin, this seems unreasonable. The alternative hypothesis is that with very reactive ylids $k_1 > k_4$; this too seems hard to believe. The mystery remains unsolved but the answer may be connected with the nature of the transition state leading to betaine. With a highly reactive ylid, one would expect an "early" transition state for betaine formation, i.e. one in which bond formation and charge separation was slight. This might permit the formation of (7.167) or (7.169) as the initial intermediates. Since (7.167) is less sterically hindered than (7.169), the former would be more stable and would therefore be formed preferentially. Internal rotation of (7.167) would lead to (7.168) which would collapse to *cis*-olefin. By analogy, the betaine (7.169) would give the *trans*-olefin via (7.170).

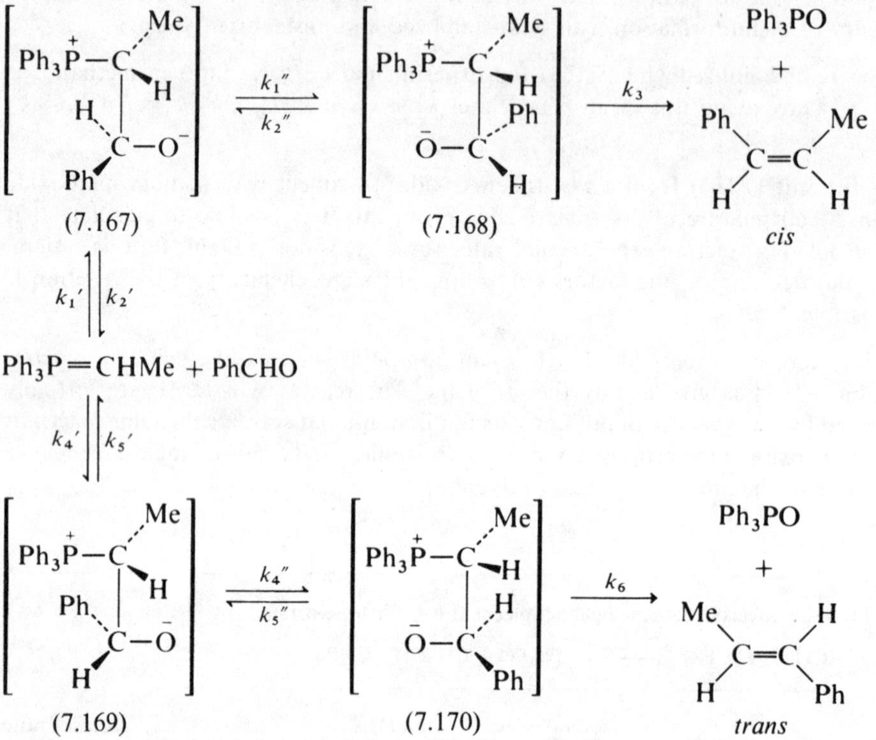

$$[Ph_3\overset{+}{P}-C(Me)(H)-C(H)(Ph)-O^-] \underset{k_2''}{\overset{k_1''}{\rightleftharpoons}} [Ph_3\overset{+}{P}-C(Me)(H)-C(Ph)(H)-\bar{O}] \overset{k_3}{\longrightarrow} Ph_3PO + \underset{cis}{Ph(H)C=C(Me)(H)}$$

(7.167) (7.168)

$k_1' \updownarrow k_2'$

$Ph_3P=CHMe + PhCHO$

$k_4' \updownarrow k_5'$

$$[Ph_3\overset{+}{P}-C(Me)(H)-C(Ph)(H)-\bar{O}] \underset{k_5''}{\overset{k_4''}{\rightleftharpoons}} [Ph_3\overset{+}{P}-C(Me)(H)-C(H)(Ph)-\bar{O}] \overset{k_6}{\longrightarrow} Ph_3PO + \underset{trans}{Me(H)C=C(H)(Ph)}$$

(7.169) (7.170)

In other words, by this mechanism $k_1' > k_4'$, i.e. $k_1'/k_4' > 1$ and this could be the origin of the predominance of *cis*-olefin. The suggestion of an "early" transition state receives some support from Bergelson's observation that the ρ value for the reactions of a highly reactive ylid with various substituted aldehydes is *ca* $+1.0$ whereas in the cases where *trans*-olefin predominates (stabilized and semi-stabilized ylids) the ρ values are both *ca* $+2.7$. The positive sign indicates that betaine formation remains the rate-limiting step but the lower numerical value implies a much lower sensitivity to substituent effects and is consistent with a low degree of bond formation and charge separation in the transition state. Whatever the reasons however, experiment indicates that highly reactive ylids give largely the *cis*-olefin in aprotic, salt-free media.

7.2.5 Miscellaneous reactions of phosphorus ylids

7.2.5.1 Hydrolysis and alcoholysis

As mentioned earlier, phosphorus ylids are hydrolyzed to phosphine oxides by the general scheme outlined in eqn. (38).

$$R_3P=CR_2^1 + H_2O \rightleftharpoons [R_3\overset{+}{P}-CHR_2^1\overset{-}{O}H] \longrightarrow R_3P=O + R_2^1CH_2$$
$$\text{or } R_2P(O)CHR_2^1 + RH \quad (38)$$

7.2 Phosphorus ylids

The ease of hydrolysis depends upon the nature of the group, R^1, and varies from almost instantaneous hydrolysis in cold water for R^1 = alkyl, to stability in hot dilute hydroxide for $R^1 = -CO_2Et$.* Also, as anticipated from our knowledge of the hydrolysis of phosphonium salts, the group which is displaced (R^- or $^-CHR^1{}_2$) is that forming the most stable carbanion. For example, the hydrolysis of methylenetriphenylphosphorane gives methyldiphenylphosphine oxide (7.171) and benzene[96].

$$Ph_3P=CH_2 + H_2O \longrightarrow Ph_2P(O)Me + PhH$$
$$(7.171)$$

A pre-equilibrium with the parent phosphonium salt is suggested by the incorporation of deuterium into the product *p*-nitrotoluene (7.173) when (7.172) is subjected to base-catalyzed alcoholysis in deuterioethanol[121].

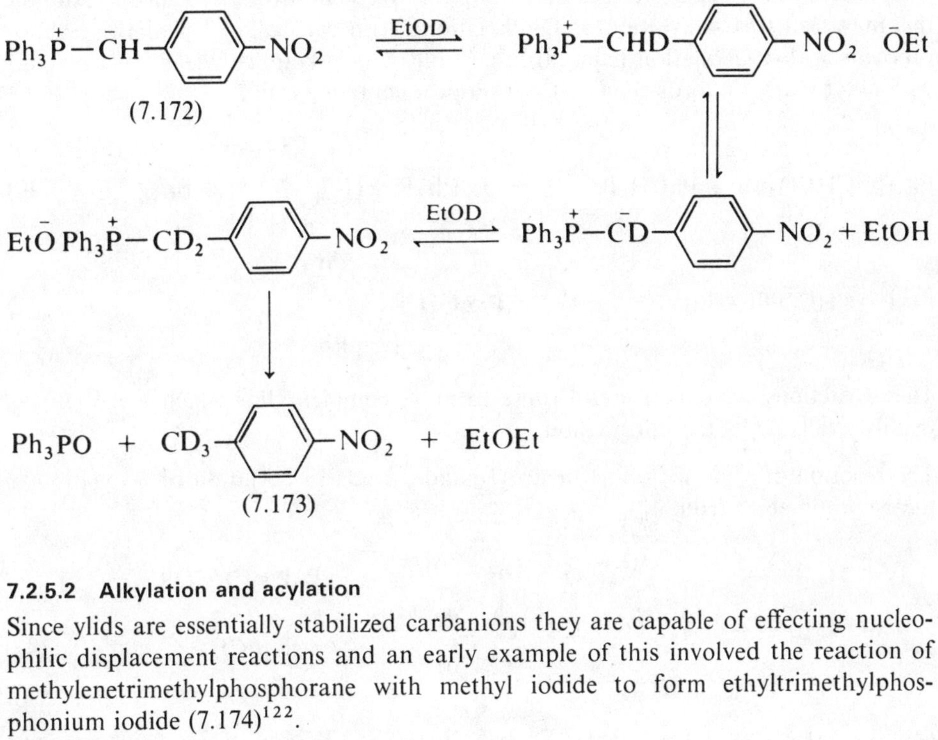

$$Ph_3\overset{+}{P}-\overset{-}{C}H-C_6H_4-NO_2 \overset{EtOD}{\rightleftharpoons} Ph_3\overset{+}{P}-CHD-C_6H_4-NO_2 \quad \overline{O}Et$$
$$(7.172)$$

$$Et\overline{O} \; Ph_3\overset{+}{P}-CD_2-C_6H_4-NO_2 \overset{EtOD}{\rightleftharpoons} Ph_3\overset{+}{P}-\overset{-}{C}D-C_6H_4-NO_2 + EtOH$$

$$Ph_3PO + CD_3-C_6H_4-NO_2 + EtOEt$$
$$(7.173)$$

7.2.5.2 Alkylation and acylation

Since ylids are essentially stabilized carbanions they are capable of effecting nucleophilic displacement reactions and an early example of this involved the reaction of methylenetrimethylphosphorane with methyl iodide to form ethyltrimethylphosphonium iodide (7.174)[122].

$$Me_3P=CH_2 + MeI \longrightarrow Me_3\overset{+}{P}-CH_2CH_3 \; \overset{-}{I}$$
$$(7.174)$$

---

* Actually, this is at first sight rather puzzling since a phosphonium ion such as $R_3P^+CHR^1{}_2$ with R^1 = alkyl would be expected to undergo base-catalyzed hydrolysis rather slowly, e.g. boiling for several hours in dilute (*ca* 1 N) sodium hydroxide. The enigma has recently been neatly resolved as a rather dramatic solvent effect. Reactive ylids are invariably prepared in aprotic media and it is the mixing of the aprotic medium with water which creates the unusual solvent condition necessary for rapid hydrolysis of the protonated ylid [A. Schnell and J. C. Tebby, *Chem. Commun.*, 134, (1975)].

A similar reaction may be effected with triethyloxonium tetrafluoroborate as alkylating agent (eqn. 39)[123].

$$Ph_3P=CHPh + Et_3O^+BF_4^- \longrightarrow \underset{\underset{Et}{|}}{Ph_3\overset{+}{P}-CHPh}\ \overset{-}{BF_4} + Et_2O \qquad (39)$$

Complications arise however, with stabilized ylids such as (7.175)

$$\underset{\underset{}{}}{Ph_3\overset{+}{P}-\overset{-}{CH}-\overset{\overset{O}{\|}}{C}R} \longleftrightarrow Ph_3\overset{+}{P}-CH=\underset{\underset{}{}}{\overset{\overset{O^-}{|}}{C}R}$$

(7.175)

since the negative charge resides on carbon and oxygen in the resonance hybrid and this may lead to C-alkylation or O-alkylation. With carboalkoxy ylids the reaction proceeds with C-alkylation (eqn. 40)[124,125] but with β-ketomethylene ylids reaction with alkyl iodides affords the O-alkylation product (eqn. 41)[126].

$$Ph_3P=CHCO_2Me + PhCH_2Br \longrightarrow Ph_3\overset{+}{P}-HC\overset{CO_2Me}{\underset{CH_2Ph}{\diagdown}}\quad Br^- \qquad (40)$$

$$Ph_3P=CHCOPh + EtI \longrightarrow Ph_3\overset{+}{P}-CH=C\overset{OEt}{\underset{Ph}{\diagdown}} \qquad (41)$$

These reactions serve as a useful route to more complex ylids which are often not readily available by the salt method.

The reaction of ylids with acyl or aroyl halides leads to acylation or aroylation of the α-carbon atom (eqn. 42).

$$\begin{matrix} R_3P=CH_2 \\ + \\ R^1COX \end{matrix} \longrightarrow \left[\underset{}{R_3\overset{+}{P}CH_2\overset{\overset{O}{\|}}{C}R^1\ X^-}\right] \xrightarrow{R_3P=CH_2} \begin{matrix} R_3P=CHCOR^1 \\ + \\ R_3\overset{+}{P}CH_3\ \overset{-}{X} \end{matrix} \qquad (42)$$

(7.176)

R^1 = alkyl or aryl

However, the intermediate phosphonium salt invariably contains a relatively acidic hydrogen and this is removed by a second mole of the starting ylid to give a mixture of acyl (or aroyl) ylid and a phosphonium salt. Thus methylenetriphenylphosphorane and benzoyl chloride give a 50:50 mixture of benzoylmethylenetriphenylphosphorane (7.177) and methyltriphenylphosphonium chloride (7.178)[127,128].

$$\begin{matrix} Ph_3P=CH_2 \\ + \\ PhCOCl \end{matrix} \longrightarrow \left[Ph_3\overset{+}{P}CH_2COPh\ \overset{-}{Cl}\right] \xrightarrow{Ph_3P=CH_2} \begin{matrix} Ph_3P=CHCOPh \\ (7.177) \\ + \\ Ph_3\overset{+}{P}CH_3\ \overset{-}{Cl} \\ (7.178) \end{matrix}$$

7.2 Phosphorus ylids

The yield of acylated ylid is therefore limited to 50% and since ylids are expensive reagents, alternative acylation procedures have been sought and found. One such procedure involves the use of carboxylate esters which acylate reactive ylids with the elimination of ethoxide (eqn. 43)[96,129,130].

$$Ph_3P=CH_2 + RCO_2Et \longrightarrow \left[Ph_3\overset{+}{P}CH_2\overset{O}{\overset{\|}{C}}R \ \overset{-}{OEt} \right] \longrightarrow Ph_3P=CHCOR + EtOH \quad (43)$$

The ethoxide ion is then sufficiently basic to remove a proton from the intermediate phosphonium salt to give a high yield of acylated ylid.

Just as in the case of alkylation, β-keto ylids may undergo acylation on carbon or oxygen. Acid anhydrides give C-acylation (7.179) but acid chlorides give O-acylation (7.180) and Chopard has suggested that O-acylation is kinetically controlled whereas C-acylation leads to the most stable product from thermodynamic control of the reaction[131].

$$Ph_3P=CHCOR \xrightarrow{(R^1CO)_2O} Ph_3P=C\underset{COR^1}{\overset{COR^1}{\diagup}} \quad \text{C-acylation; thermodynamic control} \quad (7.179)$$

$$\xrightarrow{R^1COCl} Ph_3\overset{+}{P}-CH=C\underset{O.COR^1}{\overset{R}{\diagup}}\ \overset{-}{Cl} \quad \text{O-acylation; kinetic control} \quad (7.180)$$

The acylation reactions are also useful ways of synthesizing more complex ylids which in turn offer routes to olefins (via the Wittig reaction) or ketones (via hydrolysis or reduction of the ylids).

7.2.5.3 Reduction

Both phosphonium salts and phosphorus ylids are reduced to phosphines and hydrocarbons by lithium aluminium hydride but in terms of the hydrocarbon group displaced, the reactions frequently take different courses. For example, benzyltriphenylphosphonium bromide is reduced to triphenylphosphine and toluene (eqn. 44)[132] but benzylidenetriphenylphosphorane gives benzyldiphenylphosphine and benzene (eqn. 45)[133]. Reduction of the phosphonium salts appears to follow a similar mechanism as that for base-catalyzed hydrolysis of phosphonium salts and in both cases the group forming the most stable carbanion is displaced. The ylid reductions apparently follow a different mechanism which presumably does not involve a pentacovalent intermediate. Whatever the mechanistic details are, it seems that the ylid carbon is never displaced and that reduction occurs by direct displacement of one of the remaining groups on phosphorus by the incipient hydride ion of $LiAlH_4$. Obviously, phosphonium salt reductions do not take place via the ylid.

$$Ph_3\overset{+}{P}CH_2Ph \xrightarrow{LiAlH_4} Ph_3P + CH_3Ph \quad (44)$$

$$Ph_3P=CHPh \xrightarrow{LiAlH_4} Ph_2PCH_2Ph + PhH \quad (45)$$

On the other hand, the reduction of stable ylids (e.g. 7.181) by zinc in acetic acid[134] or hydrogen over Raney nickel[135] takes place with displacement of the most electronegative group.

$$Ph_3P=CHCOPh + Zn/HOAc \longrightarrow Ph_3P + CH_3COPh$$
(7.181)

It should be apparent therefore, that reduction sometimes offers an alternative to hydrolysis for the removal from phosphorus of organic groups which have been modified by acylation or aroylation of ylids.

7.2.5.4 Reactions with carbon–carbon multiple bonds

Phosphorus ylids, like any other carbanion, will add to carbon–carbon double bonds or triple bonds provided such bonds are sufficiently activated towards nucleophilic attack by electron-withdrawing groups such as CN, —C(O)R or —CO$_2$R. Three distinct reaction pathways may be envisaged and these are generalized in Scheme 7.1. All three types of reaction have been observed and examples of each appear below.

Scheme 7.1 Possible reactions of ylids with activated olefins

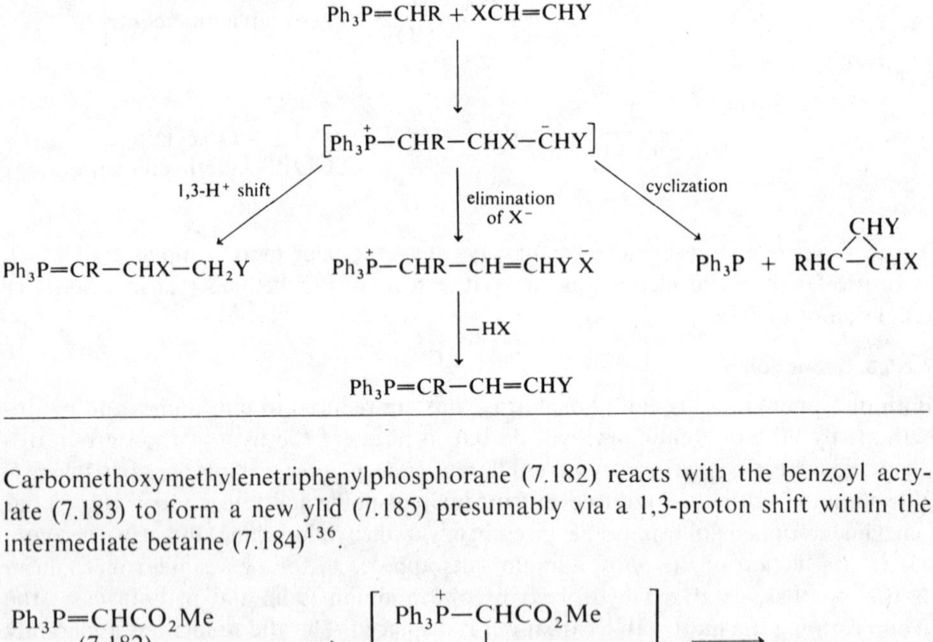

$$Ph_3P=CHR + XCH=CHY$$

$$\downarrow$$

$$[Ph_3\overset{+}{P}-CHR-CHX-\overset{-}{C}HY]$$

1,3-H$^+$ shift ↙ elimination of X$^-$ ↓ cyclization ↘

$$Ph_3P=CR-CHX-CH_2Y \qquad Ph_3\overset{+}{P}-CHR-CH=CHY\ \overset{-}{X} \qquad Ph_3P\ +\ RHC\overset{CHY}{\underset{}{-}}CHX$$

$$\downarrow -HX$$

$$Ph_3P=CR-CH=CHY$$

Carbomethoxymethylenetriphenylphosphorane (7.182) reacts with the benzoyl acrylate (7.183) to form a new ylid (7.185) presumably via a 1,3-proton shift within the intermediate betaine (7.184)[136].

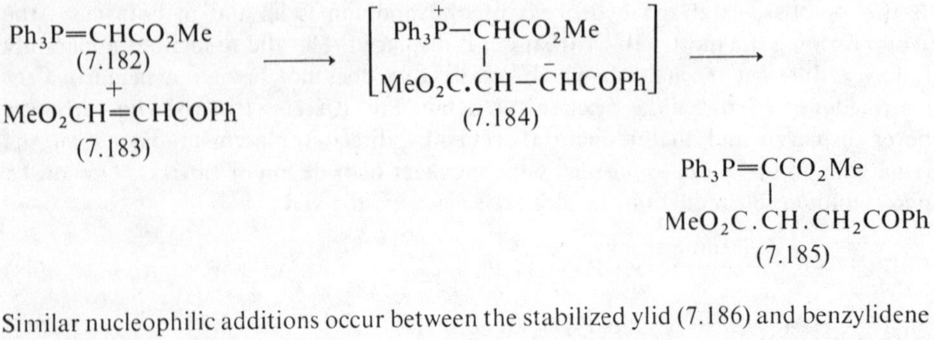

$$Ph_3P=CHCO_2Me$$
(7.182)
+
$$MeO_2CH=CHCOPh$$
(7.183)

$$\longrightarrow \begin{bmatrix} Ph_3\overset{+}{P}-CHCO_2Me \\ | \\ MeO_2C.CH-\overset{-}{C}HCOPh \end{bmatrix} \longrightarrow$$
(7.184)

$$Ph_3P=CCO_2Me$$
$$|$$
$$MeO_2C.CH.CH_2COPh$$
(7.185)

Similar nucleophilic additions occur between the stabilized ylid (7.186) and benzylidene

7.2 Phosphorus ylids

malononitrile (7.187) or tricyanovinylbenzene (7.188) except that attack is now via the 2-position of the cyclopentadiene ring rather than the ylidic carbon[137,138]. The product (7.189) from the latter substrate is capable of undergoing base-catalyzed elimination of hydrogen cyanide to form the fully conjugated product (7.190) and kinetic studies of this reaction provided one of the early, authentic examples of an E1cB elimination via ion-pairs[139].

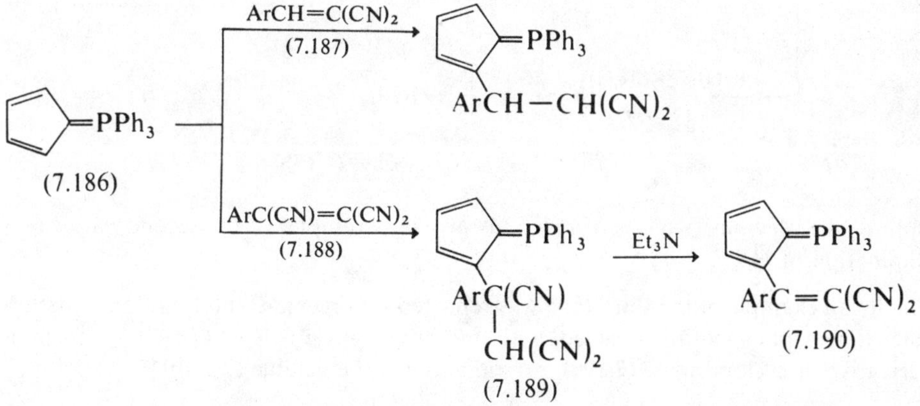

Reaction of cyanomethylenetriphenylphosphorane (7.191) with ethoxymethylenemalonate (7.192) or tetracyanoethylene (7.193) gave new ylids (7.194) and (7.195) through loss of ethanol or HCN from the intermediate betaines[140].

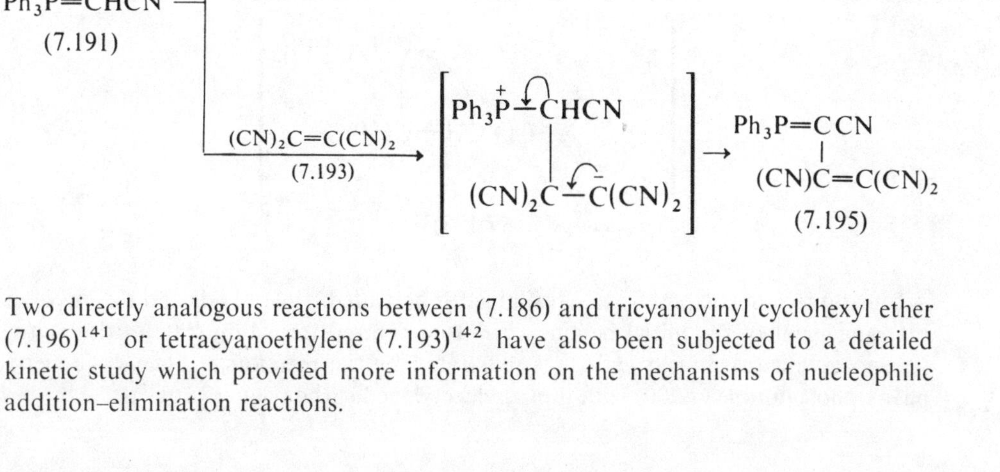

Two directly analogous reactions between (7.186) and tricyanovinyl cyclohexyl ether (7.196)[141] or tetracyanoethylene (7.193)[142] have also been subjected to a detailed kinetic study which provided more information on the mechanisms of nucleophilic addition–elimination reactions.

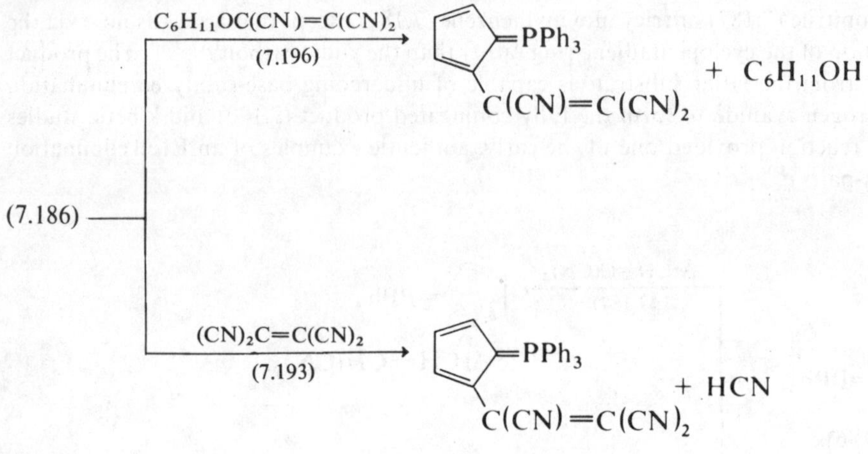

296 Chapter 7 / Tetracoordinate organophosphorus chemistry: Part 1

(7.186)

$C_6H_{11}OC(CN)=C(CN)_2$ (7.196) → [cyclopentadienylidene]=PPh$_3$ with $C(CN)=C(CN)_2$ substituent + $C_6H_{11}OH$

$(CN)_2C=C(CN)_2$ (7.193) → [cyclopentadienylidene]=PPh$_3$ with $C(CN)=C(CN)_2$ substituent + HCN

Both of the previous two sets of reactions provide examples of the second pathway – elimination of HX.

Finally an example of the third type was reported by Freeman who found that methylenetriphenylphosphorane reacted with mesitoylphenylethylene (7.197) to form a derivative of cyclopropane (7.199), presumably via the betaine (7.198)[143].

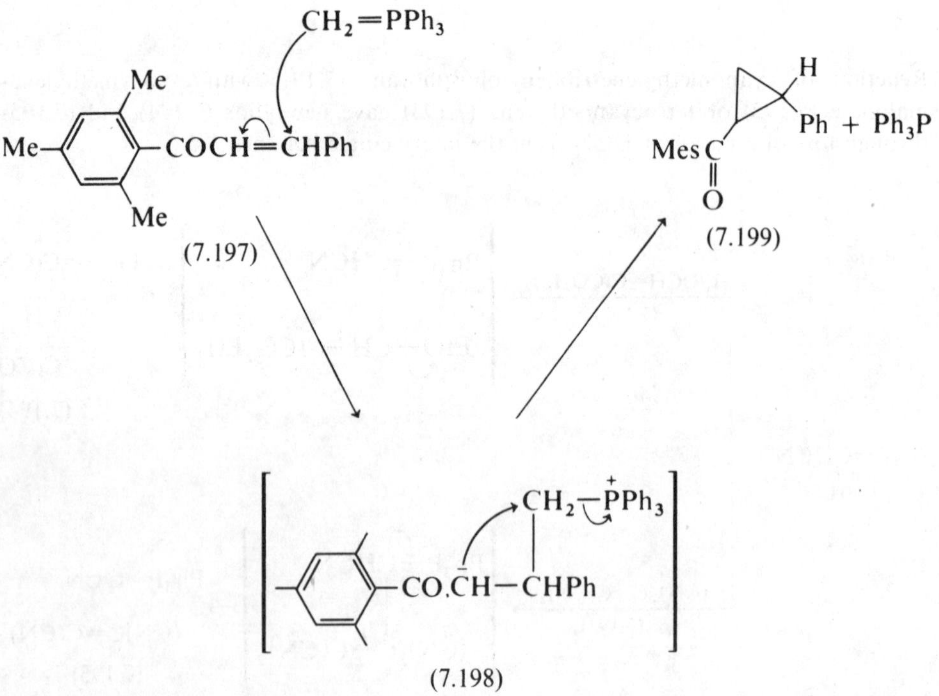

Mes–COCH=CHPh (7.197) + CH$_2$=PPh$_3$ → cyclopropane derivative MesC(=O) with H, Ph substituents (7.199) + Ph$_3$P

via betaine [–CO.ĊH–CHPh with CH$_2$–$\overset{+}{P}$Ph$_3$] (7.198)

A fourth mechanistic pathway is available for activated triple bonds and this involves rearrangement of the initial betaine via a four-centred transition state of the Wittig type to give a rearranged ylid. It is illustrated by the reaction of phenacylidenetriphenylphosphorane (7.200) with dimethylacetylenedicarboxylate (7.201)[144].

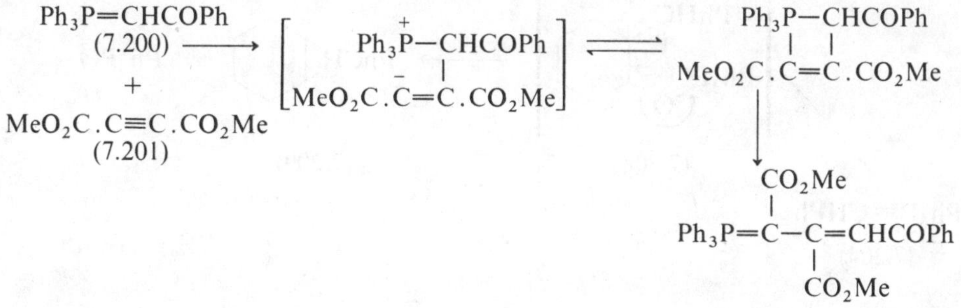

7.2 Phosphorus ylids

$$Ph_3P=CHCOPh \quad (7.200)$$
$$+$$
$$MeO_2C.C\equiv C.CO_2Me \quad (7.201)$$

$$\longrightarrow \left[\begin{array}{c} Ph_3\overset{+}{P}-CHCOPh \\ | \\ MeO_2C.\overset{-}{C}=C.CO_2Me \end{array} \right] \rightleftharpoons \begin{array}{c} Ph_3P-CHCOPh \\ | \quad | \\ MeO_2C.C=C.CO_2Me \end{array}$$

$$\downarrow$$

$$\begin{array}{c} CO_2Me \\ | \\ Ph_3P=C-C=CHCOPh \\ | \\ CO_2Me \end{array}$$

In fairness it should be added that a certain amount of controversy surrounds the structure of the products of such reactions[140,145].

7.2.5.5 Reactions with epoxides

A variety of products are formed from the reactions of ylids with epoxides[66] and a comprehensive mechanism to account for all of them has been proposed[146]. Perhaps the most interesting products, however, are the cyclopropane derivatives obtained by an intramolecular rearrangement of the initial betaines. For example, Denney et al. found that carbethoxymethylenetriphenylphosphorane (7.202) reacted with styrene oxide (7.203) to form the 1-carbethoxy-2-phenylcyclopropane (7.206) via intermediates (7.204) and (7.205)[147]; similar reactions occur with 1-octene oxide and cyclohexene oxide.

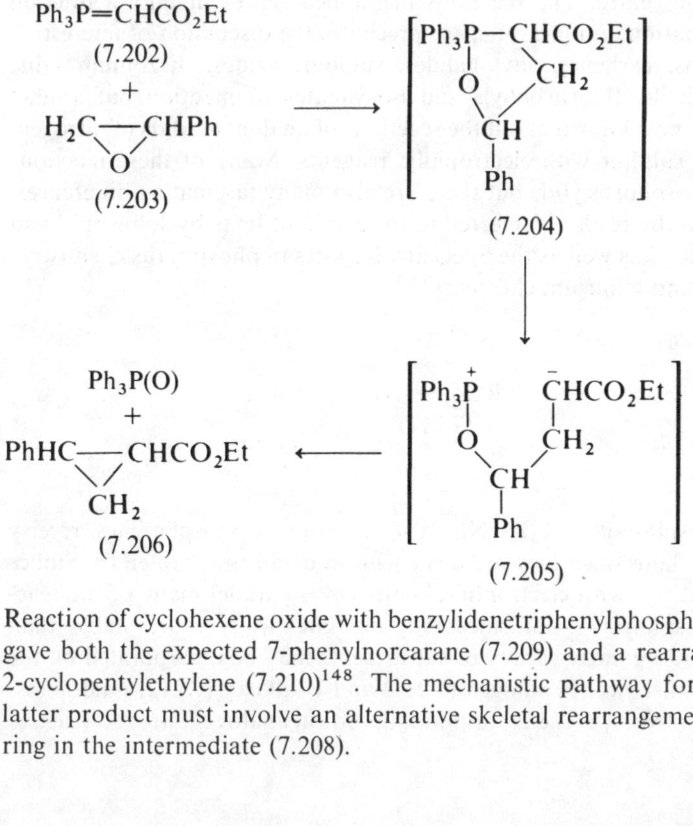

$$Ph_3P=CHCO_2Et \quad (7.202)$$
$$+$$
$$H_2C\!\!-\!\!\!\underset{O}{\diagdown\!\!\diagup}\!\!-\!\!CHPh \quad (7.203)$$

$$\longrightarrow \left[\begin{array}{c} Ph_3P-CHCO_2Et \\ | \quad \diagdown CH_2 \\ O \quad \diagup \\ \quad CH \\ \quad | \\ \quad Ph \end{array} \right] \quad (7.204)$$

$$\downarrow$$

$$Ph_3P(O)$$
$$+$$
$$PhHC\!\!-\!\!\!\underset{CH_2}{\diagdown\!\!\diagup}\!\!-\!\!CHCO_2Et \quad (7.206)$$

$$\longleftarrow \left[\begin{array}{c} Ph_3\overset{+}{P} \quad \overset{-}{C}HCO_2Et \\ \diagdown \quad \diagup \\ O \quad CH_2 \\ \diagdown \quad \diagup \\ CH \\ | \\ Ph \end{array} \right] \quad (7.205)$$

Reaction of cyclohexene oxide with benzylidenetriphenylphosphorane (7.207) however, gave both the expected 7-phenylnorcarane (7.209) and a rearranged olefin 1-phenyl-2-cyclopentylethylene (7.210)[148]. The mechanistic pathway for the formation of the latter product must involve an alternative skeletal rearrangement of the cyclohexane ring in the intermediate (7.208).

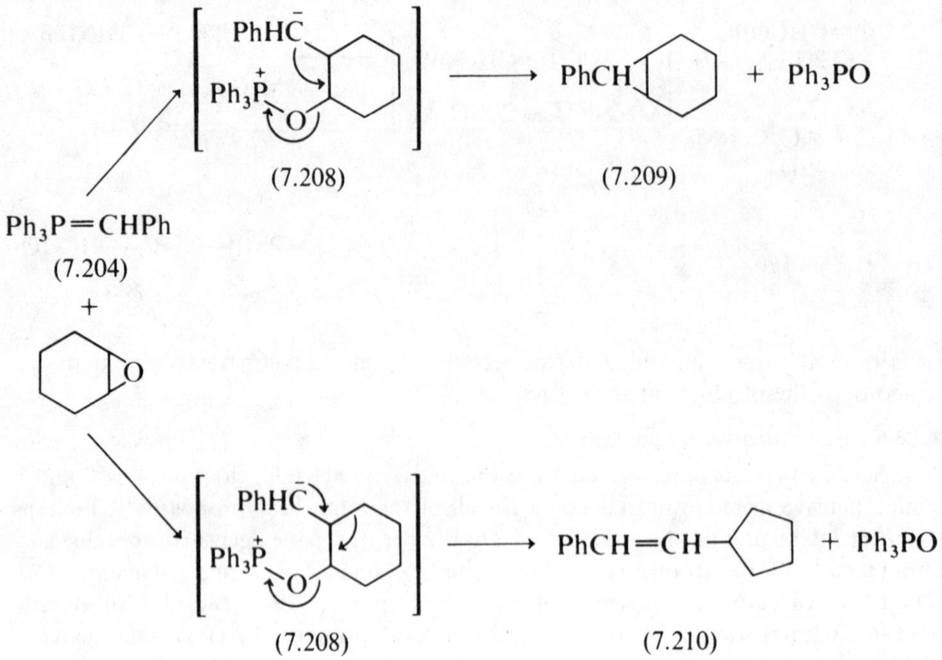

Ph₃P=CHPh
(7.204)
+
[epoxycyclohexane]

[PhHC⁻–Ph₃P⁺–O...] (7.208) ⟶ PhCH[cyclohexene] + Ph₃PO
(7.209)

[PhHC⁻–Ph₃P⁺–O...] (7.208) ⟶ PhCH=CH–[cyclopentane] + Ph₃PO
(7.210)

In conclusion it is evident that the carbanion of phosphorus ylids will react with almost any electrophilic centre. The reactions mentioned represent only a fraction of the published information and lack of space precludes the discussion of interesting reactions with halogens, carbenes, silyl halides, sulphur, azides, diazonium salts, nitroso compounds, nitrile, thiocarbonyls and isocyanates to mention but a few! Likewise, a great deal is now known about the reactions of analogous ylids of nitrogen, arsenic, antimony and sulphur with electrophilic reagents. Many of these reactions parallel those of the phosphorus ylids but there are also many fascinating differences. For further information the reader is referred to the excellent texts by Johnson[66] and by Doak and Freedman[149] as well as the Specialist Reports in phosphorus chemistry[5] and sulphur, selenium and tellurium chemistry[150].

$X_3P=NR$
(7.211), X = F, Cl, Br

$(RO)_3P=NR^1$
(7.212)

Finally, the imido phosphoranes, $R_3P=NR$, also called monophosphazenes, receive only a fleeting mention here since they are dealt with in detail in Chapter 10. Suffice it to say that their reactions with electrophilic centres also parallel many of the reactions of methylenephosphoranes but since they are chemically more stable than methylenephosphoranes, a greater variety of substituents may be incorporated on the phosphorus atom. For instance, P–halogen (7.211) and P–alkoxy (7.212) imidophosphoranes are well known whereas the corresponding carbon analogues are too reactive to permit isolation.

Problems

1 The hydrolysis of *trans*-1,3-dimethylbenzylphospholanium bromide (7.60) in aqueous alkali proceeds with retention of configuration at phosphorus. The mechanistic pathway for hydrolysis is given in the text. Devise mechanistic pathways which would lead to (i) inversion and (ii) racemization.

2 (a) Suggest mechanisms for the following reactions:

(i) [dibenzophospholium salt with R and CH₂I substituents] + OH⁻/H₂O → [dibenzophosphole oxide with R and CH₂ bridge]

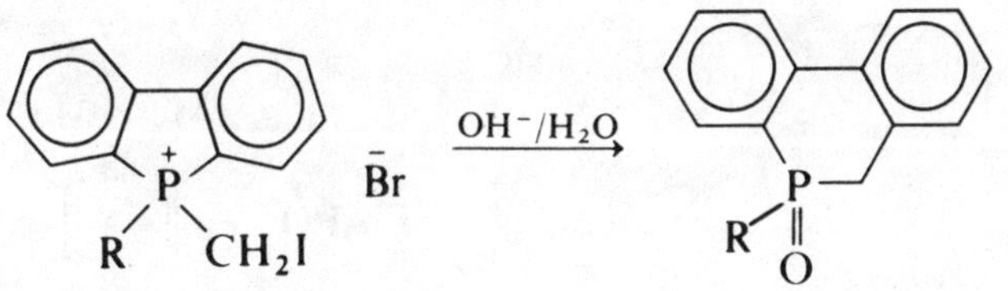

(ii) [tetramethyl phosphetanium with Ph, Me substituents and Me, H on ring] + OH⁻/H₂O → Ph-P(=O)(Me)(CMe₂CHMe₂)

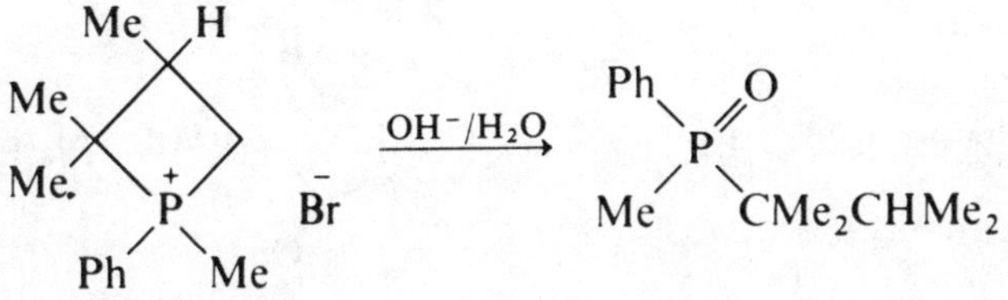

(iii) [pentamethyl phosphetanium with Ph, Me] + PhLi → Ph(Me)P-C(Me)₂-CHMe-C(Me)₂-Ph

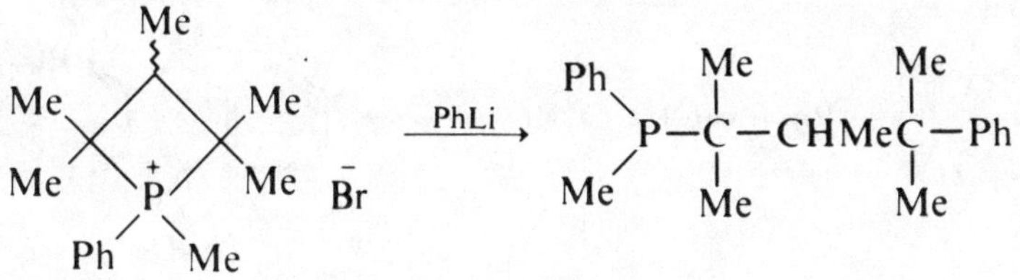

(iv) Ph₄P⁺X⁻ + LiCH=CH₂ ⟶ Ph₃P + PhCH=CH₂.

(b) Treatment of (7.106) with base gives (7.107) see page 273.

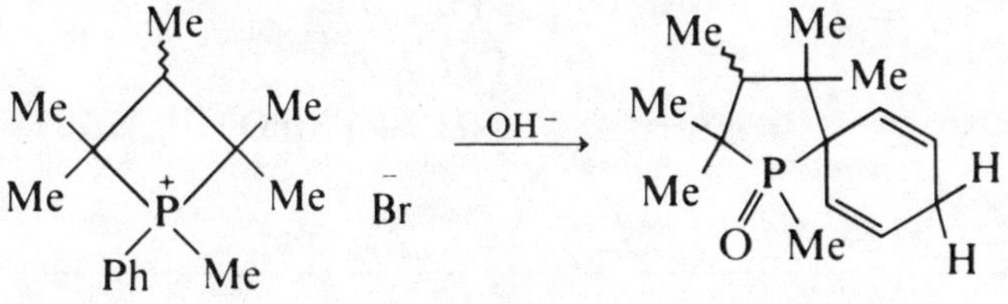

(7.106) [phosphetanium bromide] —OH⁻→ (7.107) [phosphorane structure]

Suggest an alternative structure for the rearrangement product.

3 When 0.152 g of a monobasic compound (A) containing the elements C, H and P was dissolved in glacial acetic acid, 23.3 ml of a 0.02 M solution of perchloric acid in glacial acetic acid was required to neutralize the compound to crystal violet indicator. Calculate the molecular weight of A. The ^{1}H n.m.r. spectrum of A showed multiplets at δ 7.5 (15H), 6.3 (2H) and 6.45 (2H). The ^{31}P n.m.r. showed a signal at -12 ppm (*vs* H₃PO₄). Suggest a structure for A.

4 Devise mechanisms for the following reactions:

(i) Ph₂C₃(=O) (diphenylcyclopropenone) + Ph₃P=CHCOMe $\xrightarrow{\text{reflux}}$ 4-methoxy-3-phenyl-4-methyl-2H-pyran-2-one derivative + Ph₃P

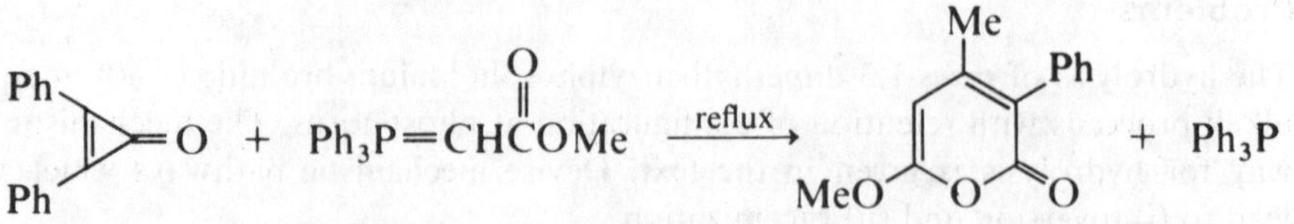

(ii) Ph₃P=CH.CH=CH₂ + MeC(Me)=C(CO₂Et)COMe $\longrightarrow$ Ph₃PO + cyclohexadiene derivative (Me, Me, CO₂Et, Me substituted)

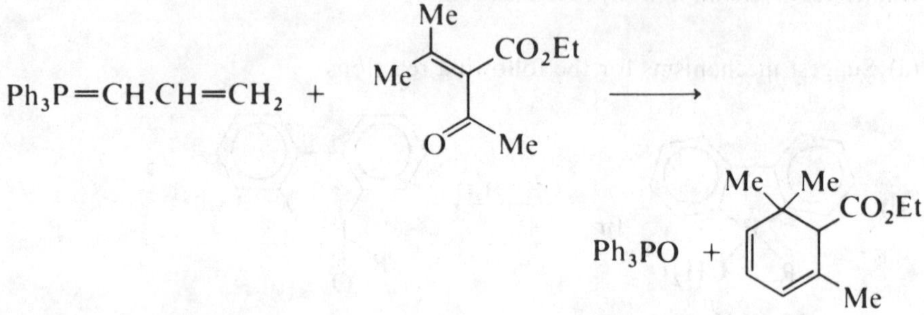

(iii) pyrrole-2-COR¹ + Ph₃P=CR²CO₂Et $\longrightarrow$ pyrrolizinone (R¹, R² substituted) + Ph₃PO

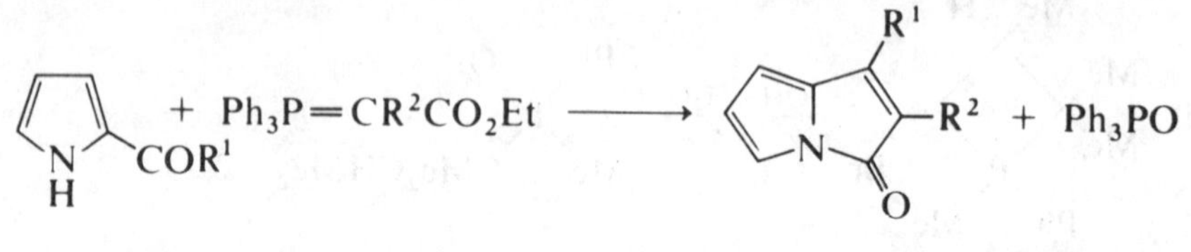

(iv) isatoic anhydride (N-R) + Ph₃P=CHCO₂Et $\longrightarrow$ 3-(triphenylphosphoranylidene)quinoline-2,4-dione (N-R) + CO₂ + EtOH

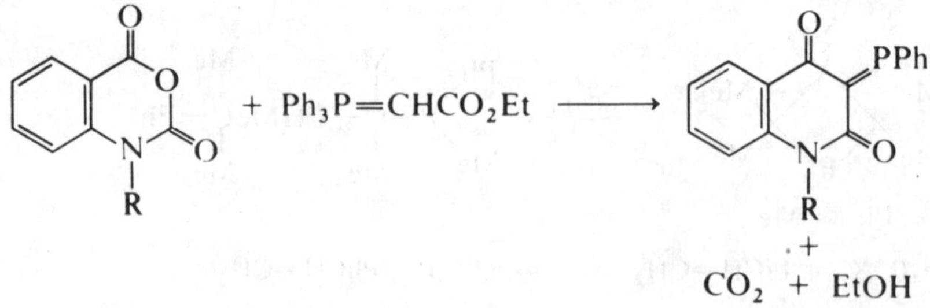

(v) PhCOCH=PPh₃ + PhN=N⁺=N⁻ $\longrightarrow$ 4-phenyl-1H-1,2,3-triazole (NH, N-Ph) + Ph₃PO

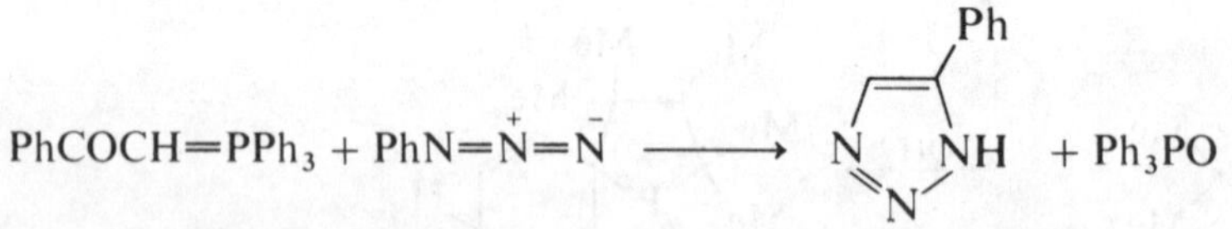

References

[1] G. M. Kosolapoff and L. Maier (editors), *Organic Phosphorus Compounds*, Vols. 2 and 3, Wiley Interscience, New York, 1972.

[2] K. Sasse, "*Phosphorus Compounds*", vol. 12 of E. Muller (ed.), *Methoden der Organischen Chemie* (Houben Weyl), Thieme, 1963, part 1, pp 79–123.

[3] A. J. Kirby and S. G. Warren, *The Organic Chemistry of Phosphorus*, Elsevier, Amsterdam, 1967.

[4] B. J. Walker, *Organophosphorus Chemistry*, Penguin, London, 1972.

[5] S. Trippett (ed.), *Organophosphorus Chemistry*, Vols 1–5, 1969–74, Specialist Reports, Chemical Society, London.

References

[6] R. F. Hudson, *Structure and Mechanism in Organophosphorus Chemistry*, Academic Press, New York, 1965.

[7] H. Hoffman and H. J. Diehr, "Phosphonium Salt Formation", *Angew. Chem., Int. Ed. Eng.*, **3**, 737 (1964).

[8] (a) W. E. McEwan, "Stereochemistry of Reactions of Organophosphorus Compounds", *Topics in Phosphorus Chemistry*, **2**, 1 (1965); (b) M. J. Gallagher and I. D. Jenkins, *Topics in Stereochem.*, **3**, 1 (1968).

[9] L. Horner, "Preparation and Properties of Optically Active Phosphines", *Pure Appl. Chem.*, **9**, 225 (1964).

[10] L. Horner, H. Fuchs, H. Winkler and A. Rapp, *Tetrahedron Lett.*, 965 (1963).

[11] D. P. Young, W. E. McEwan, D. C. Velez, J. W. Johnson and C. A. Van der Werf, *Tetrahedron Lett.*, 359 (1964).

[12] D. B. Denney and S. T. Ross, *Anal. Chem.*, **32**, 1896 (1960).

[13] L. Horner and H. M. Duda, *Tetrahedron Lett.*, 5177 (1970).

[14] L. Horner, R. Luckenbach and W.-D. Balzer, *Tetrahedron Lett.*, 3157 (1968).

[15] G. Wittig and H. Matzura, *Annalen*, **732**, 97 (1970).

[16] H. Freyschlag, H. Grassner, A. Murrenbach, H. Pommer, W. Reif and W. Sarnecki, *Angew. Chem., Int. Ed. Eng.*, **4**, 287 (1965).

[17] L. D. Quin, "Trivalent Phosphorus Compounds as Dienophiles", in J. Hamer (editor), *1,4-Cycloaddition Reactions*, Academic Press, New York, 1967.

[18] K. D. Berlin and D. M. Hellwege, "Carbon Phosphorus Heterocycles", *Topics in Phosphorus Chemistry*, Vol. 6, p. 1, Wiley Interscience, New York, 1969.

[19] S. E. Cremer and R. J. Chorvat, *J. Org. Chem.*, **32**, 4066 (1967).

[20] R. I. Wagner, U.S. Patent 3,086,053 (1963); *Chem. Abs.*, **59**, 10124 (1963).

[21] R. I. Wagner, U.S. Patent 3,105,096 (1963); *Chem. Abs.*, **60**, 5563 (1964).

[22] R. I. Wagner, L. D. Freeman, H. Goldwhite and D. G. Rowsell, *J. Amer. Chem. Soc.*, **89**, 1102 (1967).

[23] D. Berglund and D. W. Meek, *J. Amer. Chem. Soc.*, **90**, 518 (1968).

[24] G. Märkl, *Angew. Chem., Int. Ed. Eng.*, **2**, 620 (1963).

[25] K. L. Marsi, D. M. Lynch and G. D. Homer, *J. Heterocyclic Chem.*, **9**, 331 (1972).

[26] N. Ya. Derkach and A. V. Kirsanov, *Zh. Obsch. Khim.*, **38**, 331 (1968).

[27] H. Hellmann and O. Schumacher, *Angew. Chem.*, **72**, 211 (1960).

[28] G. Wittig, H. J. Cristan and H. Braun, *Angew. Chem., Int. Ed. Eng.*, **6**, 700 (1967).

[29] K. Mislow, *Accounts Chem. Res.*, **3**, 321 (1970).

[30] R. F. Hudson and C. Brown, *Accounts Chem. Res.*, **5**, 204 (1971).

[31] M. Zanger, C. A. Van der Werf and W. E. McEwen, *J. Amer. Chem. Soc.*, **81**, 3806 (1959).

[32] H. Hoffmann, *Annalen*, **634**, 1 (1960).

[33] G. Asknes, *Acta Chem. Scand.*, **15**, 438 (1961).

[34] G. Märkl, *Chem. Ber.*, **94**, 3005 (1961).

[35] S. T. D. Gough and S. Trippett, *J. Chem. Soc.*, 2333 (1962).

[36] G. W. Fenton and C. K. Ingold, *J. Chem. Soc.*, 2342 (1929).

[37] L. Horner and H. Hoffmann, *Chem. Ber.*, **91**, 52 (1958).

[38] W. E. McEwen, G. Axelrad, M. Zanger and C. Van der Werf, *J. Amer. Chem. Soc.*, **87**, 3948 (1965).

[39] K. E. Kumli, C. A. Van der Werf and W. E. McEwen, *J. Amer. Chem. Soc.*, **81**, 3805 (1959).

[40] L. Horner, A. Winkler, A. Rapp, A. Mentrup and P. Beck, *Tetrahedron Lett.*, 161 (1961).

41 L. Horner, H. Fuchs, H. Winkler and A. Rapp, *Tetrahedron Lett.*, 965 (1963).

42 A. Blade Font, C. A. Van der Werf and W. E. McEwen, *J. Amer. Chem. Soc.*, **82**, 2396 (1960).

43 F. H. Westheimer, *Accounts Chem. Res.*, **1**, 70 (1968).

44 K. E. De Bruin, K. Naumann, G. Zon and K. Mislow, *J. Amer. Chem. Soc.*, **91**, 7031 (1969).

45 C. Asknes and K. Bergesen, *Acta Chem. Scand.*, **19**, 931 (1965).

46 K. L. Marsi, *J. Amer. Chem. Soc.*, **91**, 4724 (1969).

47 K. L. Marsi, F. B. Burns and R. T. Clark, *J. Org. Chem.*, **37**, 238 (1972).

48 W. Egan, G. Chauvière, K. Mislow, R. T. Clark and K. L. Marsi, *Chem. Commun.*, 733 (1970)

49 K. L. Marsi, *J. Org. Chem.*, **39**, 265 (1974).

50 K. L. Marsi and R. T. Clark, *J. Amer. Chem. Soc.*, **92**, 3791 (1970).

51 K. L. Marsi, *J. Amer. Chem. Soc.*, **93**, 6341 (1971).

52 D. E. De Bruin, G. Zon, K. Naumann and K. Mislow, *J. Amer. Chem. Soc.*, **91**, 7027 (1969).

53 S. E. Cremer, R. J. Chorvat and B. C. Trevedi, *Chem. Commun.*, 769 (1969).

54 J. R. Corfield, N. J. De'Ath and S. Trippett, *Chem. Commun.*, 1502 (1970).

55 J. R. Corfield, M. J. P. Harger, J. R. Shutt and S. Trippett, *J. Chem. Soc. C*, 1855 (1970).

56 K. E. De Bruin and M. J. Jakobs, *Chem. Commun.*, 59 (1971).

57 J. R. Corfield, J. R. Shutt and S. Trippett, *Chem. Commun.*, 789 (1969).

58 K. E. De Bruin and K. Mislow, *J. Amer. Chem. Soc.*, **91**, 7393 (1969).

59 N. J. De'Ath and S. Trippett, *Chem. Commun.*, 172 (1969).

60 S. E. Fishwick, J. Flint, W. Hawes and S. Trippett, *Chem. Commun.*, 1113 (1967).

61 S. E. Cremer, *Chem. Commun.*, 1132 (1968).

62 J. R. Shutt and S. Trippett, *J. Chem. Soc. C*, 2038 (1969).

63 D. W. Allen and B. G. Huntley, *J. Chem. Soc., Perkin II*, 67 (1972).

64 D. W. Allen, B. G. Huntley and M. J. Miller, *J. Chem. Soc., Perkin II*, 63 (1972).

65 D. W. Allen, S. J. Grayson, I. Harness, B. G. Hutley and I. W. Mowat, *J. Chem. Soc., Perkin I*, 1912 (1973).

66 A. W. Johnson, *Ylid Chemistry*, Academic Press, New York, 1966.

67 R. F. Hudson, *Chem. Brit.*, 287 (1971).

68 A. Michaelis and H. W. Gimborn, *Chem. Ber.*, **27**, 272 (1894).

69 H. Staudinger and J. Meyer, *Helv. Chim. Acta*, **2**, 619 (1919).

70 F. Krohnke, *Chem. Ber.*, **68**, 1177 (1935).

71 G. Wittig and M. Rieber, *Annalen*, **562**, 177 (1949).

72 G. Wittig and G. Geissler, *Annalen*, **580** 44 (1953).

73 L. C. Thomas and R. A. Chittenden, *Spectrochim. Acta*, **21**, 1905 (1965).

74 A. W. Johnson, *J. Org. Chem.*, **24**, 282 (1959).

75 F. Ramirez and S. Levy, *J. Amer. Chem. Soc.*, **79**, 67 (1957).

76 A. J. Speziale and K. W. Ratts, *J. Amer. Chem. Soc.*, **87**, 5603 (1965).

77 F. S. Stephens, *J. Chem. Soc.*, 5640, 5658 (1965).

78 W. Von E. Doering and A. K. Hoffmann, *J. Amer. Chem. Soc.*, **77**, 521 (1955).

79 F. Ramirez and S. Dershowitz, *J. Org. Chem.*, **22**, 41 (1957).

80 S. Flizar, R. F. Hudson and G. Salvatori, *Helv. Chim. Acta*, **46**, 1580 (1963).

81 R. Köster, D. Simić and M. A. Grassberger, *Annalen*, **739**, 211 (1970).

82 D. D. Coffmann and C. S. Marvel, *J. Amer. Chem. Soc.*, **51**, 3496 (1929).

83 A. J. Speziale, G. T. Marco and K. W. Ratts, *J. Amer. Chem. Soc.*, **82**, 1260 (1960).

References

[84] A. J. Speziale, G. T. Marco and K. W. Ratts, *J. Amer. Chem. Soc.*, **84**, 854 (1962).

[85] G. Wittig and M. Schlosser, *Tetrahedron*, **18**, 1023 (1962).

[86] R. Rabinowitz and R. Marcus, *J. Amer. Chem. Soc.*, **84**, 1312 (1962).

[87] F. Ramirez, M. Desai and N. B. McKelvie, *J. Amer. Chem. Soc.*, **84**, 1745 (1962).

[88] F. Ramirez, O. P. Madan and C. P. Smith, *Tetrahedron*, **22**, 567 (1966).

[89] L. Horner and H. Oediger, *Chem. Ber.*, **91**, 437 (1958).

[90] D. Seyforth and J. M. Burlitch, *J. Org. Chem.*, **28**, 2463 (1963).

[91] (a) N. E. Miller, *J. Amer. Chem. Soc.*, **87**, 390 (1965); (b) H. Schmidbaur and W. Tronich, *Chem. Ber.*, **101**, 595 (1968).

[92] H. Staudinger and J. Meyer, *Helv. Chim. Acta*, **2**, 635 (1919).

[93] G. Wittig, *Pure Applied Chem.*, **9**, 245 (1964).

[94] S. Trippett, "The Wittig Reaction", *Pure Appl. Chem.*, **9**, 255 (1964).

[95] H. J. Bestmann, *Angew. Chem., Int. Ed. Eng.*, **4**, 583, 654, 830 (1965).

[96] G. Wittig and U. Schollkopf, *Chem. Ber.*, **87**, 1318 (1954).

[97] M. A. Battiste, *J. Amer. Chem. Soc.*, **86**, 942 (1964).

[98] S. G. Levine, *J. Amer. Chem. Soc.*, **80**, 6150 (1958).

[99] G. Wittig and E. Knauss, *Angew. Chem.*, **71**, 127 (1959).

[100] M. Schlosser, *Chem. Ber.*, **97**, 3219 (1964).

[101] D. B. Denney, L. S. Smith, J. Song, C. J. Rossi and C. D. Hall, *J. Org. Chem.*, **28**, 778 (1963).

[102] G. Wittig, H. Eggers and P. Duffner, *Annalen*, **619**, 10 (1958).

[103] E. E. Schweizer and K. K. Light, *J. Amer. Chem. Soc.*, **86**, 2963 (1964).

[104] G. Wittig and H. Pommer, *German Patent*, 950,552, Oct. 1956; *Chem. Abs.*, **53**, 436g,i (1959).

[105] J. D. Surmatis and A. Ofner, *J. Org. Chem.*, **26**, 1171 (1961).

[106] S. Trippett, *Chem. & Ind. (London).*, 80 (1956).

[107] L. Horner, H. Hoffmann and H. G. Wippel, *Chem. Ber.*, **91**, 61 (1958).

[108] W. S. Wadsworth and W. D. Emmons, *J. Amer. Chem. Soc.*, **83**, 1733 (1961).

[109] J. Boutagy and R. Thomas, *Chem. Revs.*, **74**, 87 (1974).

[110] O. P. Vig, R. C. Anand, G. Kad and J. M. Sehgal, *J. Indian Chem. Soc.*, **47**, 999 (1970).

[111] U. Schollkopf, *Angew. Chemie*, **71**, 260 (1959).

[112] R. F. Hudson, S. Fliszar and G. Salvadori, *Helv. Chim. Acta*, **46**, 1580 (1963).

[113] G. H. Birum and C. N. Matthews, *Chem. Commun.*, 137 (1967).

[114] W. E. Blade-Font, C. A. Van der Werf and W. E. McEwen, *J. Amer. Chem. Soc.*, **82**, 2396 (1960).

[115] L. Horner and H. Winkler, *Tetrahedron Lett.*, 3265 (1964).

[116] A. J. Speziale and D. J. Bissing, *J. Amer. Chem. Soc.*, **85**, 3878 (1963).

[117] G. Wittig, H. D. Weigman and M. Schlosser, *Chem. Ber.*, **94**, 676 (1971).

[118] L. D. Bergelson, L. I. Barsukov and M. M. Shemyakin, *J. Gen. Chem. USSR*, **38**, 810 (1968).

[119] M. E. Jones and S. Trippett, *J. Chem. Soc. C*, 1090 (1966).

[120] M. Schlosser, G. Müller and K. F. Christmann, *Angew. Chem., Int. Ed. Eng.*, **5**, 667 (1966).

[121] M. Grayson and P. T. Keough, *J. Amer. Chem. Soc.*, **82**, 3919 (1960).

[122] G. Wittig and M. Rieber, *Annalen*, **562**, 177 (1949).

[123] G. Märkl, *Tetrahedron Lett.*, 1027 (1962).

[124] H. J. Bestmann and H. Schulz, *Tetrahedron Lett.*, 5 (1960).

[125] H. J. Bestmann and H. Schulz, *Chem. Ber.*, **95**, 2921 (1962).

[126] F. Ramirez and S. Dershowitz, *J. Org. Chem.*, **22**, 41 (1957).

[127] H. J. Bestmann and B. Arnason, *Chem. Ber.*, **95**, 1513 (1962).
[128] S. T. D. Gough and S. Trippett, *Proc. Chem. Soc.*, (*London*), 302 (1961).
[129] S. Trippett and D. M. Walker, *Chem. & Ind.* (*London*), 202 (1960).
[130] H. J. Bestmann and B. Arnason, *Tetrahedron Lett.*, 455 (1961).
[131] P. A. Chopard, R. J. G. Searle and F. H. Devitt, *J. Org. Chem.*, **30**, 1015 (1965).
[132] W. J. Bailey and S. A. Buckler, *J. Amer. Chem. Soc.*, **79**, 3567 (1957).
[133] S. T. D. Gough and S. Trippett, *J. Chem. Soc.*, 4263 (1961).
[134] S. Trippett and D. Walker, *J. Chem. Soc.*, 1266 (1961).
[135] A. Schonberg, K. H. Brosowski and E. Singer, *Chem. Ber.*, **95**, 2984 (1962).
[136] H. J. Bestmann and F. Seng, *Angew. Chem.*, **74**, 154 (1962).
[137] E. Lord, M. P. Naan and C. D. Hall, *J. Chem. Soc. B*, 1401 (1970).
[138] E. Lord, M. P. Naan and C. D. Hall, *J. Chem. Soc. B*, 213 (1971).
[139] E. Lord, M. P. Naan and C. D. Hall, *J. Chem. Soc. B*, 220 (1971).
[140] S. Trippett, *J. Chem. Soc.*, 4733 (1962).
[141] C. W. Rigby, E. Lord, M. P. Naan and C. D. Hall, *J. Chem. Soc. B*, 1192 (1971).
[142] M. P. Naan, A. P. Bell and C. D. Hall, *J. Chem. Soc., Perkin II*, 1821 (1973).
[143] J. P. Freeman, *Chem. Ind.* (*London*), 1254 (1959).
[144] H. J. Bestmann and O. Rothe, *Angew. Chem.*, **76**, 569 (1964).
[145] J. B. Hendrickson, *J. Amer. Chem. Soc.*, **83**, 2018 (1961).
[146] S. Trippett, *Quart. Rev., Chem. Soc.*, **17**, 406 (1964).
[147] D. B. Denney, J. J. Vill and M. J. Boskin, *J. Amer. Chem. Soc.*, **84**, 3944 (1962).
[148] E. Zbiral, *Monatsh. Chem.*, **94**, 78 (1963).
[149] G. O. Doak and L. D. Freedman, "*Organometallic Compounds of Arsenic, Antimony and Bismuth*, Wiley Interscience, New York, 1970.
[150] D. H. Reid (ed.), *Organic compounds of sulphur, selenium and tellurium*, Specialist Reports, Chemical Soc., London, Vols 1, 2 (1970 and 1973).

Chapter 8
Tetracoordinate organophosphorus chemistry: Part 2
phosphoryl esters and related compounds

THE CHEMISTRY OF PHOSPHORUS

8.1 Introduction

This chapter focuses attention on the reactions of phosphoryl esters, i.e. compounds containing the P=O or P=S group, and in particular, attempts to present a concise view of nucleophilic substitution at phosphorus for this class of compounds.[1-8] Before embarking on the subject however, it seems desirable to review the nomenclature (to supplement the general survey of the appendix) since without a knowledge of the names of the basic structural types, one is likely to find a study of this area, particularly difficult.

Phosphorus esters, as their name implies, are simply esters of the corresponding phosphorus oxy-acid. There are however, several classes of phosphorus acid containing the phosphoryl (P=O) bond. These include phosphoric acid (8.1), phosphorous or phosphonic acid (8.2) and phosphinic acid (8.3). They are tribasic, dibasic and monobasic respectively and the pK_a values and possible oxy-esters which can be derived from them are shown in Table 8.1 which also includes alkylphosphonic acids (8.4) and mono or dialkyl phosphinic acids (8.5) and their respective esters.

The phosphonic (8.2) and phosphinic (8.3) acids are in tautomeric equilibrium with their trivalent isomers, phosphorous acid (8.6) and phosphonous acid (8.7) respectively, but owing to the strength of the phosphoryl group, the equilibrium lies on the left. Similar equilibria in the esters (e.g. 8.8 ⇌ 8.9) also lie on the side of the phosphonate structure and the simplest class of phosphorus acids, the dialkyl phosphinous acid (8.11) exists largely (>99 %) in the phosphoryl form (a dialkyl phosphine oxide, 8.10).

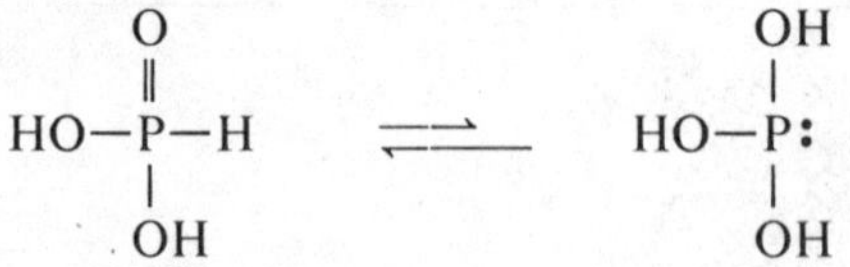

$$\begin{array}{cc} \text{O} & \text{OH} \\ \| & | \\ \text{HO—P—H} \rightleftharpoons & \text{HO—P:} \\ | & | \\ \text{OH} & \text{OH} \\ (8.2) & (8.6) \\ \text{phosphonic acid} & \text{phosphorous acid} \end{array}$$

$$\begin{array}{cc} \text{O} & \text{OH} \\ \| & | \\ \text{H—P—H} \rightleftharpoons & \text{H—P:} \\ | & | \\ \text{OH} & \text{OH} \\ (8.3) & (8.7) \\ \text{phosphinic acid} & \text{phosphonous acid} \end{array}$$

$$\begin{array}{cc} \text{O} & \text{OH} \\ \| & | \\ \text{RO—P—H} \rightleftharpoons & \text{RO—P:} \\ | & | \\ \text{OH} & \text{OH} \\ (8.8) & (8.9) \\ \text{O-alkyl phosphonate} & \text{monoalkyl phosphite} \end{array}$$

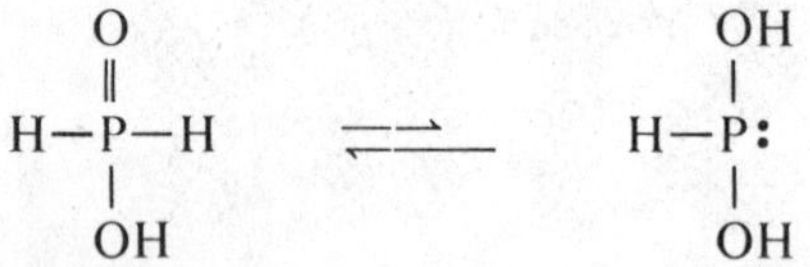

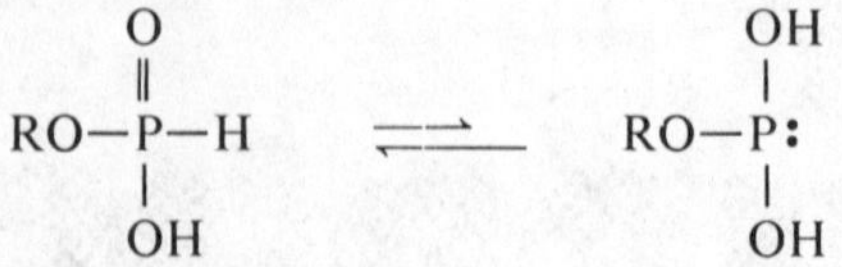

$$\begin{array}{cc} \text{O} & \text{OH} \\ \| & | \\ \text{R—P—H} \rightleftharpoons & \text{R—P:} \\ | & | \\ \text{R} & \text{R} \\ (8.10) & (8.11) \\ \text{dialkylphosphine oxide (99\%)} & \text{dialkylphosphinous acid (1\%)} \end{array}$$

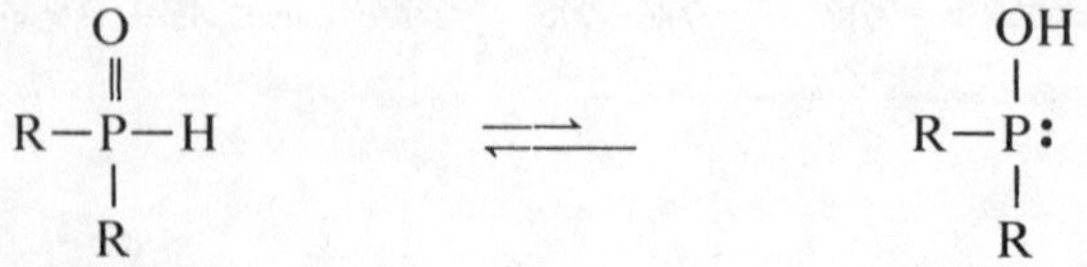

8.1 Introduction

Table 8.1 Phosphorus oxy-acids

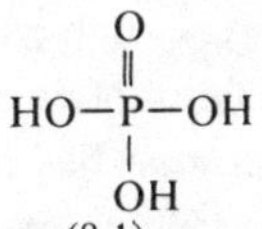 $\begin{array}{c} O \\ \parallel \\ HO-P-OH \\ \mid \\ OH \end{array}$ (8.1) phosphoric acid pK_1 2.1 pK_2 7.1 pK_3 ca 12	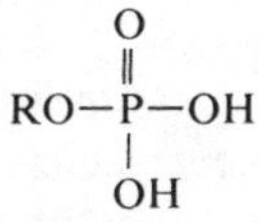 $\begin{array}{c} O \\ \parallel \\ RO-P-OH \\ \mid \\ OH \end{array}$ alkyl phosphate	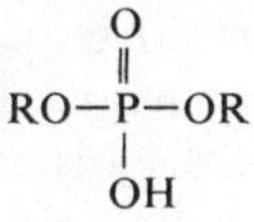 $\begin{array}{c} O \\ \parallel \\ RO-P-OR \\ \mid \\ OH \end{array}$ dialkyl phosphate	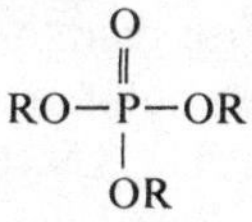 $\begin{array}{c} O \\ \parallel \\ RO-P-OR \\ \mid \\ OR \end{array}$ trialkyl phosphate
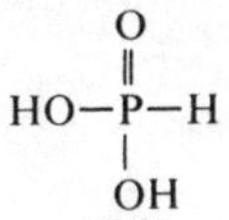 $\begin{array}{c} O \\ \parallel \\ HO-P-H \\ \mid \\ OH \end{array}$ (8.2) phosphonic acid pK_1 1.3 pK_2 6.7	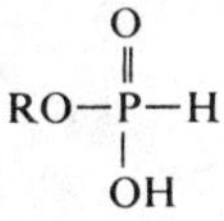 $\begin{array}{c} O \\ \parallel \\ RO-P-H \\ \mid \\ OH \end{array}$ O-alkyl phosphonate (monoalkyl phosphite)	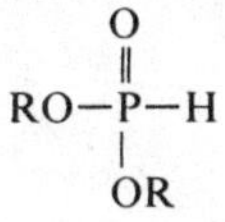 $\begin{array}{c} O \\ \parallel \\ RO-P-H \\ \mid \\ OR \end{array}$ O,O-dialkyl phosphonate (dialkyl phosphite)	
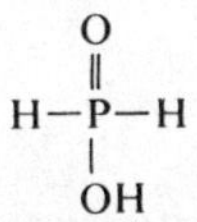 $\begin{array}{c} O \\ \parallel \\ H-P-H \\ \mid \\ OH \end{array}$ (8.3) phosphinic acid pK_1 1.1	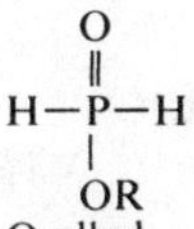 $\begin{array}{c} O \\ \parallel \\ H-P-H \\ \mid \\ OR \end{array}$ O-alkyl phosphinate (monoalkyl-phosphonite)		
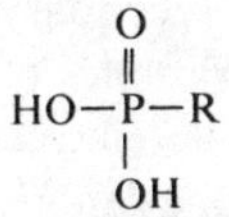 $\begin{array}{c} O \\ \parallel \\ HO-P-R \\ \mid \\ OH \end{array}$ (8.4) alkyl phosphonic acid pK_1 2.3 (R = Me) pK_2 7.9 (R = Me)	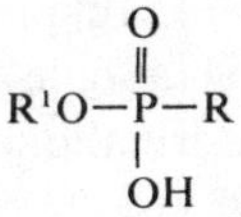 $\begin{array}{c} O \\ \parallel \\ R^1O-P-R \\ \mid \\ OH \end{array}$ O-alkyl alkyl-phosphonate	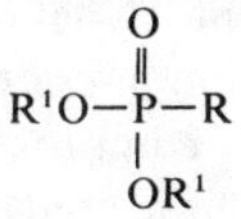 $\begin{array}{c} O \\ \parallel \\ R^1O-P-R \\ \mid \\ OR^1 \end{array}$ O,O-dialkyl alkyl-phosphonate	
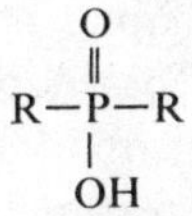 $\begin{array}{c} O \\ \parallel \\ R-P-R \\ \mid \\ OH \end{array}$ (8.5) dialkyl-phosphinic acid pK_1 3.1 (R = Me)	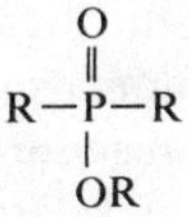 $\begin{array}{c} O \\ \parallel \\ R-P-R \\ \mid \\ OR \end{array}$ O-alkyldialkyl-phosphinate		

Notice that for acids containing a P=O group the names end with the suffix "ic" whereas the trivalent isomers have the suffix "ous". The corresponding esters are "ates" and "ites" respectively. Furthermore, compounds with three hydroxy or alkoxy groups on phosphorus bear the generic name "phosphor", e.g. phosphoric (8.1) or phosphorous (8.6), compounds with two hydroxy groups and one hydrogen or alkyl/aryl group attached to phosphorus are named "phosphon", e.g. phosphonic acid (8.2), and compounds with only one hydroxy or alkoxy group but two hydrogens or hydrocarbon groups attached to phosphorus are termed "phosphin", e.g. phosphinic acid (8.3).

Of course there are many derivatives of these acids involving other elements such as nitrogen, sulphur and the halogens and the possible permutation of formulae involving just P, H, C, N, O, S and halogen is enormous. Nevertheless, with the above basic rules in mind one can name most derivatives of the acids of phosphorus. A few examples (8.12–8.15) are shown below and throughout the chapter new names are accompanied by the appropriate formula.

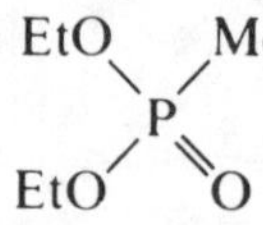

(8.12) O,O-diethyl methylphosphonate

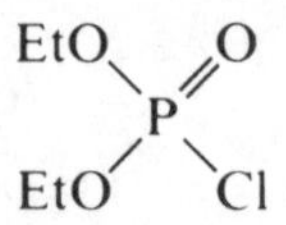

(8.13) diethyl phosphorochloridate

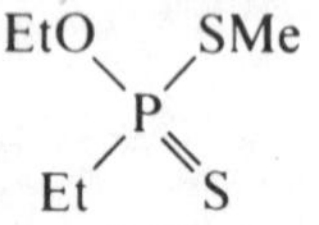

(8.14) O-ethyl-S-methyl ethylphosphonodithioate

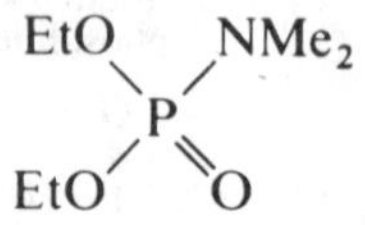

(8.15) diethyl-N,N-dimethylphosphoramidate

Phosphorus acids are often prepared by the hydrolysis of phosphorus halides and similar nucleophilic displacement of halide by alcohols or primary and secondary amines leads to oxy-esters and amido-esters respectively. Some common examples of such reactions appear in Table 8.2 but for detailed information the reader is referred to the specialist texts dealing with the preparative aspects of the subject[1,2]. One can see that by an appropriate choice of the starting halide, the whole range of oxy-acids from dialkylphosphine oxides (phosphinous acids) to phosphoric and polyphosphoric acids may be prepared; an analogous range of oxy-esters is also obtainable. All of these reactions are examples of nucleophilic displacement at phosphorus and those involving displacement at trivalent phosphorus have been discussed in Chapter 4. But compounds containing the phosphoryl (P=O) or thiophosphoryl (P=S) group are tetracoordinate and tetrahedral (see Chapter 2) and reactions with nucleophiles may therefore be expected to display different mechanistic and stereochemical features. Since a knowledge of such reactions is vital to an understanding of the chemistry of the biologically indispensable phosphate esters, this field of study has received considerable attention and it is now time to present some of the conclusions.

8.2 Nucleophilic displacement at phosphorus: acyclic compounds

Nucleophilic displacement at phosphorus is generalized by eqn. (1)

$$Y^- + \underset{B}{\overset{Z}{\underset{\|}{A-P-X}}} \longrightarrow Y-\underset{B}{\overset{Z}{\underset{\|}{P-A}}} + X^- \tag{1}$$

8.2 Nucleophilic displacement at phosphorus: acyclic compounds

Table 8.2 The hydrolysis, alcoholysis and amidolysis of phosphorus halides

Phosphorus halide	With H$_2$O	With R^{1}OH	With R$_2^1$NH
P(III)			
R$_2$PCl	R$_2$P(=O)H	R$_2$P(OR^1)	R$_2$PNR$_2^1$
RPCl$_2$	RPH(OH) (=O)	RP(OR^1)$_2$	RP(NR$_2^1$)$_2$
PCl$_3$	HP(OH)$_2$ (=O)	P(OR^1)$_3$	P(NR$_2^1$)$_3$
P(V)			
R$_2$PCl$_3$	R$_2$POH (=O)	R$_2$P(OR^1) (=O)	R$_2\overset{+}{P}$(NR$_2^1$)$_2$ Cl$^-$
RPCl$_4$	RP(OH)$_2$ (=O)	RP(OR^1)$_2$ (=O)	R$\overset{+}{P}$(NR$_2^1$)$_3$ Cl$^-$
PCl$_5$	O=P(OH)$_3$	O=P(OR^1)$_3$	(R$_2^1$N)$_4\overset{+}{P}$ Cl$^-$
Phosphoryl halides			
ROP(=O)Cl$_2$	ROP(=O)(OH)–O–P(=O)(OH)OR dialkyldiphosphate		
(RO)$_2$P(=O)Cl	(RO)$_2$P(=O)–O–P(=O)(OR)$_2$ tetraalkyldiphosphate^a		

^{a} These compounds are extremely poisonous – see Chapter 12.

where Y$^-$ is the nucleophile, X is the leaving group, A and B are hydrogen, alkyl, aryl, hydroxy, alkoxy, aryloxy or thio-analogues of the last three and Z is commonly O, S or N. Typical examples of such nucleophilic displacements are shown in eqns. (2)–(4).

$$Z = O, \quad PhNH_2 + (CH_3)_2P(=O)Cl \longrightarrow PhNHP(=O)(CH_3)_2 \qquad (2)$$

$$Z = S, \quad EtNH_2 + (EtO)_2P(=S)Cl \longrightarrow EtNHP(=S)(OEt)_2 \qquad (3)$$

T.C.O.P.—L

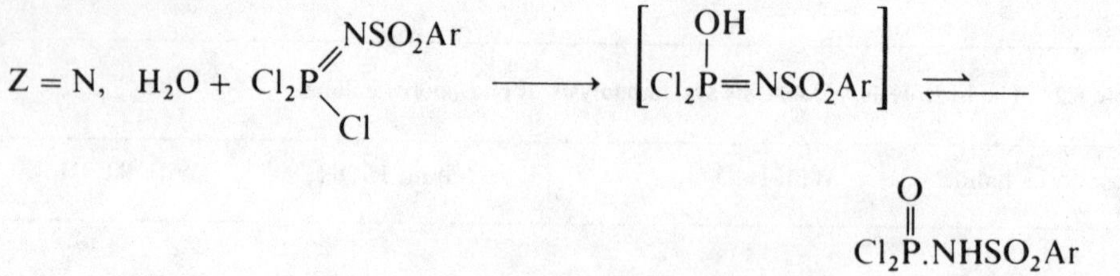

$$Z = N, \quad H_2O + Cl_2P\begin{matrix}NSO_2Ar\\ \\Cl\end{matrix} \longrightarrow \left[Cl_2\overset{OH}{\underset{|}{P}}=NSO_2Ar\right] \rightleftharpoons$$

$$Cl_2\overset{O}{\underset{\|}{P}}.NHSO_2Ar \quad (4)$$

An example of the importance of nucleophilic attack at phosphorus in a biological process is provided by the phosphorylation of an amino acid (8.16) by adenosine triphosphate (ATP, 8.17) which occurs during protein synthesis.

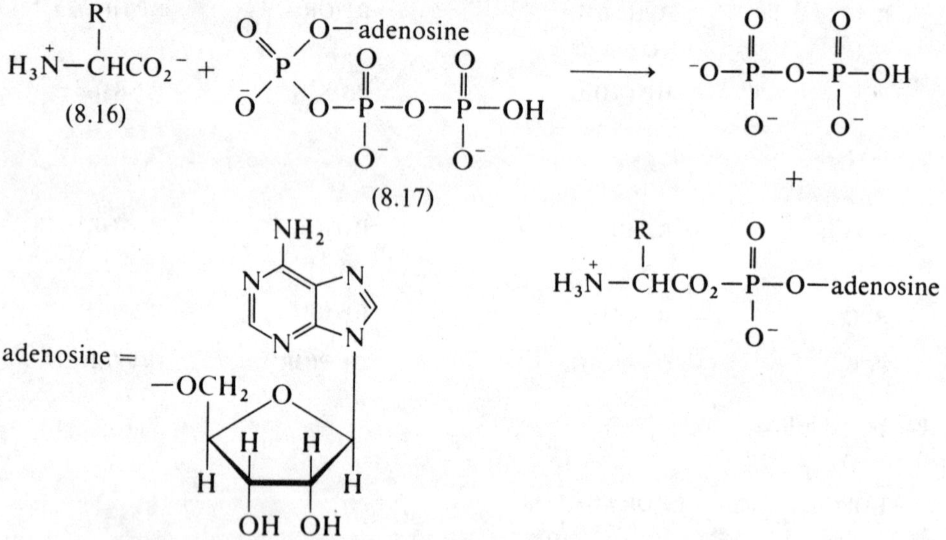

$$\underset{(8.16)}{H_3\overset{+}{N}-\underset{\underset{R}{|}}{C}HCO_2^-} + \underset{(8.17)}{\overset{O}{\underset{O^-}{\overset{\|}{P}}}\underset{}{\overset{O-adenosine}{\diagdown}}\overset{O}{\underset{O^-}{\overset{\|}{P}}}-O-\overset{O}{\underset{O^-}{\overset{\|}{P}}}-OH} \longrightarrow \ ^-O-\overset{O}{\underset{\underset{O^-}{|}}{\overset{\|}{P}}}-O-\overset{O}{\underset{\underset{O^-}{|}}{\overset{\|}{P}}}-OH$$

$$+$$

$$H_3\overset{+}{N}-\underset{\underset{R}{|}}{C}HCO_2-\overset{O}{\underset{\underset{O^-}{|}}{\overset{\|}{P}}}-O-adenosine$$

adenosine = [adenosine structure with adenine base attached to ribose via −OCH₂− group; ribose bears OH OH groups]

There are three possible mechanisms for such nucleophilic substitutions, viz: (i) $S_N1(P)$, (ii) addition–elimination, and (iii) $S_N2(P)$. Each will receive some attention in the subsequent discussion but a few general points should be noted before the detailed study of each mechanistic class.

It is important to remember that in compounds of the type, ABXP=Z, extensive conjugation of the type (8.18 ↔ 8.19) does not occur. This is in contrast to the *planar* carbonyl compounds where lone pairs adjacent to the carbonyl group are extensively delocalized (e.g. 8.20 ↔ 8.21).

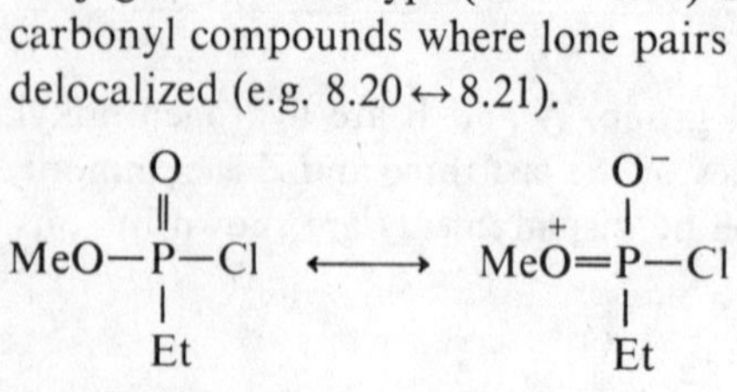

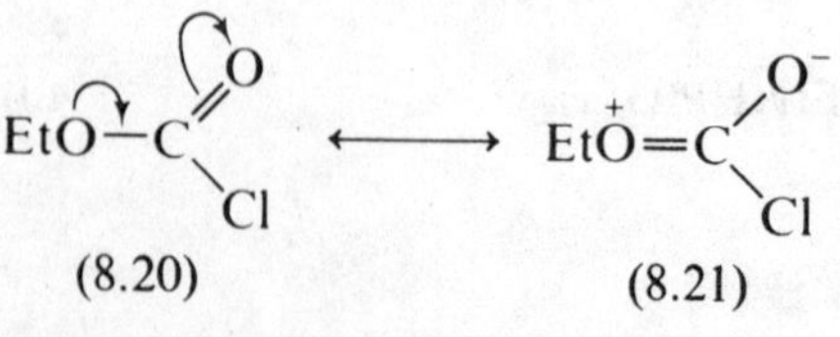

$$\underset{(8.18)}{MeO-\underset{\underset{Et}{|}}{\overset{\overset{O}{\|}}{P}}-Cl} \longleftrightarrow \underset{(8.19)}{MeO\overset{+}{=}\underset{\underset{Et}{|}}{\overset{\overset{O^-}{|}}{P}}-Cl}$$

$$\underset{(8.20)}{EtO-\underset{\underset{Cl}{\diagdown}}{\overset{\overset{O}{\diagup}}{C}}} \longleftrightarrow \underset{(8.21)}{Et\overset{+}{O}=\underset{\underset{Cl}{\diagdown}}{\overset{\overset{O^-}{\diagup}}{C}}}$$

8.2 Nucleophilic displacement at phosphorus: acyclic compounds

The result is that minor structural differences have very large effects on the rates of hydrolysis of acyl compounds but only small effects on the analogous phosphoryl compounds (Table 8.3). In fact nucleophilic attack at a phosphoryl centre shows a

Table 8.3 Relative rates of hydrolysis of acyl and phosphoryl compounds

Compound:	MeO.COCl	MeCOCl	MeO—P(=O)(Et)—Cl	Et—P(=O)(Et)—Cl
Rate of hydrolysis:	1	10^4	1	15

much greater dependence on the strength of the covalent bond between the leaving group and phosphorus than does attack at a carbonyl centre; in the latter case, covalent bond formation with the incoming nucleophile controls reactivity (see later).

8.2.1 $S_N1(P)$ mechanism (elimination–addition)[9]

The $S_N1(P)$ mechanism is generalized by eqn. (5) and is directly analogous to the S_N1 mechanism of carbon chemistry. Intermediates of type (8.22) – metaphosphates* – are extremely reactive and have never been isolated as monomers.

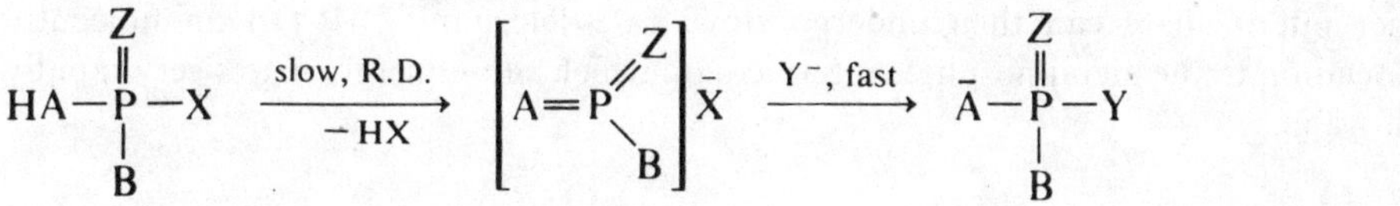

$$HA-\underset{B}{\overset{\overset{Z}{\|}}{P}}-X \xrightarrow[-HX]{\text{slow, R.D.}} \left[A=\underset{B}{\overset{Z}{P}}\right]^- X \xrightarrow{Y^-, \text{ fast}} \bar{A}-\underset{B}{\overset{\overset{Z}{\|}}{P}}-Y \quad (5)$$

$$(8.22)$$

Examples include the hypothetical metaphosphoric acid (8.23) and the metaphosphorthioimidate (8.24), which is known as the dimer (8.25)[10].

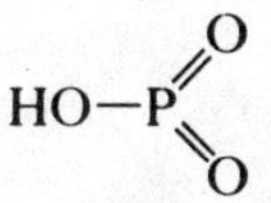

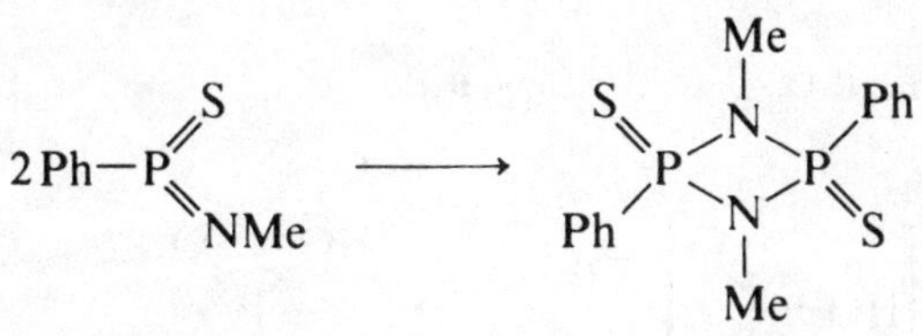

(8.23) HO—P(=O)(=O)   (8.24) 2 Ph—P(=S)(NMe) ⟶   (8.25)

The metaphosphates are unstable but their existence as transient intermediates has been deduced from kinetic and stereochemical evidence and one of the best established examples involves the hydrolysis of monoesters of phosphoric acid.

Monoalkyl phosphates, $ROP(O)(OH)_2$, show a rate maximum for hydrolysis at pH 4

---

* Not to be confused with the other class of metaphosphates of formula $(PO_3^-)_n$ which are cyclic polyphosphates.

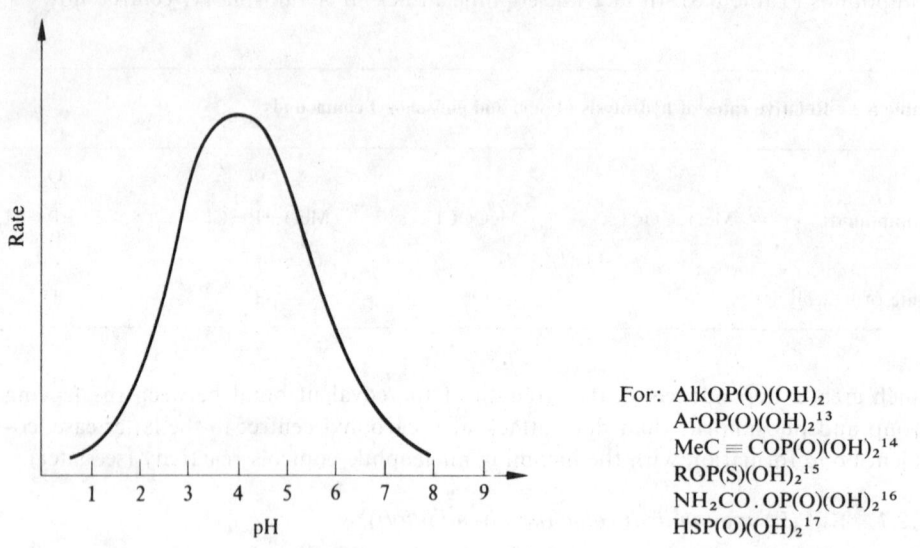

Figure 8.1 pH–rate profile for the hydrolysis of ROP(O)(OH)$_2$.

For: AlkOP(O)(OH)$_2$
ArOP(O)(OH)$_2$[13]
MeC≡COP(O)(OH)$_2$[14]
ROP(S)(OH)$_2$[15]
NH$_2$CO.OP(O)(OH)$_2$[16]
HSP(O)(OH)$_2$[17]

and the pH–rate profile is described by a bell-shaped curve (Fig. 8.1)[11,12]. Other monoesters show the same profile[13-17] and the rate maximum corresponds to the maximum concentration of the mono anion (8.26) or its kinetically equivalent ion-pair (8.27). Either intermediate can then undergo slow, rate-determining (R.D.) unimolecular dissociation to the metaphosphate intermediate which subsequently reacts very rapidly with water.

$$\underset{(8.26)}{\text{RO}-\underset{\underset{\text{OH}}{|}}{\overset{\overset{\text{O}}{\|}}{\text{P}}}-\bar{\text{O}}} \qquad \underset{(8.27)}{\text{RO}-\underset{\underset{\text{OH}}{|}}{\overset{\overset{\text{O}}{\|}}{\text{P}}}-\bar{\text{O}}\cdots\text{H}_3\overset{+}{\text{O}}}$$

slow, R.D.   slow, R.D.

$$\left[\text{HO}-\text{P}\underset{\text{O}}{\overset{\text{O}}{\lessgtr}}\right] \xrightarrow{\text{H}_2\text{O, fast}} \underset{\text{HO}}{\overset{\text{HO}}{\diagdown}}\text{P}\underset{\text{OH}}{\overset{\text{O}}{\diagup}}$$

Support for the mechanism is available from a considerable body of ancillary evidence which may be summarized as follows:

(a) When the hydrolysis is conducted in ^{18}O-labelled water, one atom of ^{18}O is incorporated into the phosphoric acid product and none into the alcohol; this demonstrates exclusive P—O bond cleavage[11,12].

(b) Optically active alcohols, e.g. (+)2-butanol, are displaced with retention of optical activity and configuration at carbon, again showing exclusive P—OR bond cleavage[12].

8.2 Nucleophilic displacement at phosphorus: acyclic compounds

(c) The entropies of activation are often slightly positive; e.g. PhOP(O)(OH)$_2$ has $\Delta S^\ddagger = +3.8$ J K^{-1} mol^{-1} (+0.9 cal K^{-1} mol^{-1}) and α-naphthyl—OP(O)(OH)$_2$ has $\Delta S^\ddagger = +17.2$ J K^{-1} mol^{-1} (+4.1 cal K^{-1} mol^{-1})[12,18]. This suggests dissociation, i.e. an increase in the degree of disorder, in the rate-determining step.

(d) The product distribution in aqueous methanol is a precise reflection of the mole fraction of methanol in the solvent. This is indicative of a highly reactive, non-discriminating intermediate such as metaphosphoric acid (e.g. eqn. 6)[19].

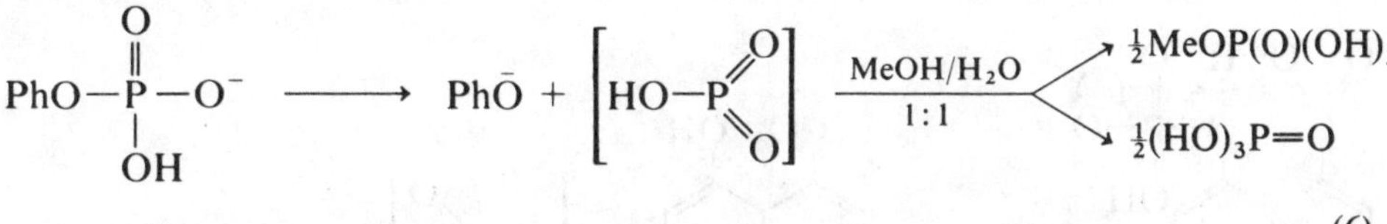

$$PhO-P(O)(OH)-O^- \longrightarrow PhO^- + [HO-PO_2] \xrightarrow{MeOH/H_2O \; 1:1} \begin{array}{l} \tfrac{1}{2}MeOP(O)(OH)_2 \\ \tfrac{1}{2}(HO)_3P=O \end{array}$$

(6)

In fact, the true intermediate is probably the metaphosphate anion (8.28) derived through departure of the leaving group as the protonated species.

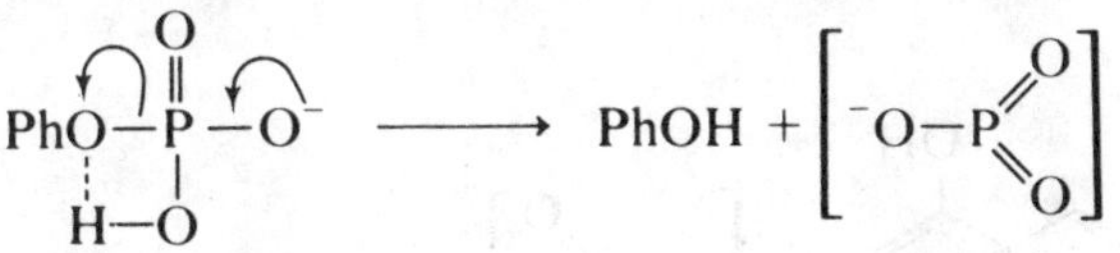

$$PhO-P(O)(O^-)\cdots H-O \longrightarrow PhOH + [^-O-PO_2]$$

(8.28)

Evidence to support this suggestion is provided by the hydrolysis of a series of XP(O)(OH)$_2$ compounds for which a plot of log (rate) vs pK_a of HX gives a straight line of slope -0.27.[20] The plot is another linear free energy relationship and just like the ρ value of Hammett plots, the slope measures the sensitivity of the reaction to changes in X. Obviously the rate varies with the nature of X but the sensitivity to changes in X is low and this is best explained as a unimolecular decomposition of the monoanion with concerted (route A) or pre-equilibrium (route B) protonation of X (Scheme 8.1).

Scheme 8.1 Mechanistic pathways for the decomposition of phosphate monoanions

$$X-P(O)(O^-)\cdots H-O \xrightarrow[\text{route A}]{k_2, \text{ slow, R.D.}} XH + [^-O-PO_2] \xrightarrow[\text{fast}]{H_2O} H_2PO_4^-$$

$$\downarrow k_1, \text{ fast}$$

$$HX^+-P(O)(O^-)-O^- \xrightarrow[\text{route B}]{k_2', \text{ slow, R.D.}}$$

This proposal is substantiated to some extent by the observation that for X = MeO, $k_{H_2O}/k_{D_2O} = 0.87$ whereas for X = 2,4-dinitrophenoxy, $k_{H_2O}/k_{D_2O} = 1.45$.[20] Thus as the leaving group ability of X increases, the slow step is facilitated, which serves to increase the relative importance of k_1 in route B or the relative importance of the proton transfer step in the synchronous route, A.*

The proton can also be supplied by other functional groups within the monoester. Thus 8-carboxy-1-naphthyl phosphate (8.29) hydrolyzes with $\Delta S^{\ddagger} = +44.5$ J K^{-1} mol^{-1} (+10.6 cal K^{-1} mol^{-1}) in comparison with $\Delta S^{\ddagger} = 17.2$ J K^{-1} mol^{-1} (+4.1 cal K^{-1} mol^{-1}) for 1-naphthyl phosphate (8.30)[13]. The difference is probably due to the less stringent geometrical requirements of intramolecular proton transfer in (8.29).

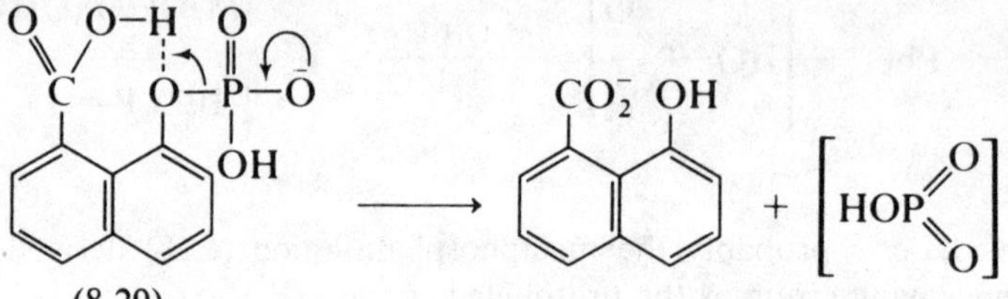

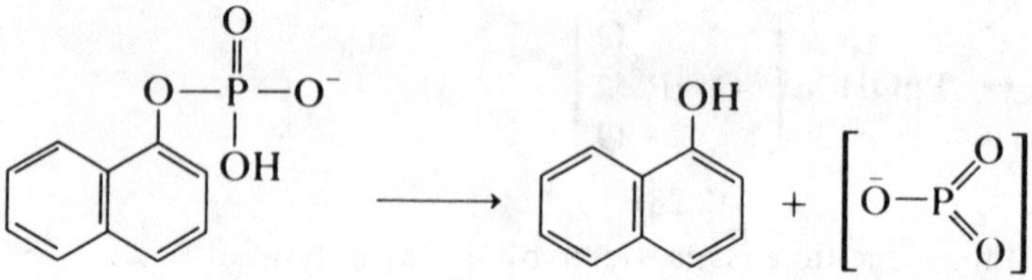

Similar kinetic studies, including entropies of activation, and solvent deuterium isotope effects, on the hydrolysis of phosphoric acid monoester dianions (8.31) suggest that the reactions also proceed by an $S_N1(P)$ elimination–addition mechanism via the

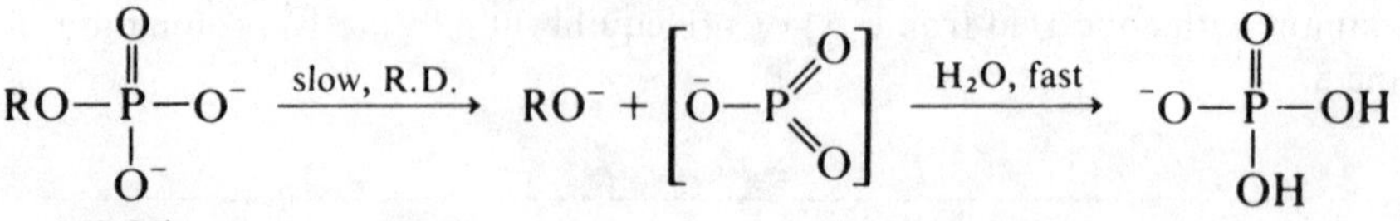

metaphosphate ion[9,20]. Here of course, there is no possibility of synchronous proton transfer to RO$^-$ unless R contains an acidic hydrogen, as in (8.29).

There are a number of other interesting examples where metaphosphates are thought to be essential intermediates on the mechanistic pathway. For instance, β-chloro-

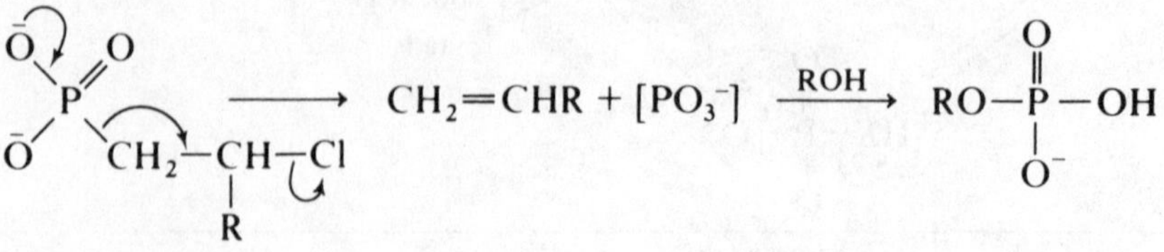

---

* In general, with a reversible proton transfer prior to the rate-determining step, $k_{H_2O}/k_{D_2O} < 1$ but if proton transfer is involved in the rate-determining step, $k_{H_2O}/k_{D_2O} > 1$.

8.2 Nucleophilic displacement at phosphorus: acyclic compounds

alkylphosphonates (8.32) decompose under alkaline conditions and phosphorylate alcohols, probably via metaphosphate[21].

This reaction has been shown to be a stereospecific *trans*-elimination as evidenced by the alkaline hydrolysis of stereochemically pure (8.33)[22].

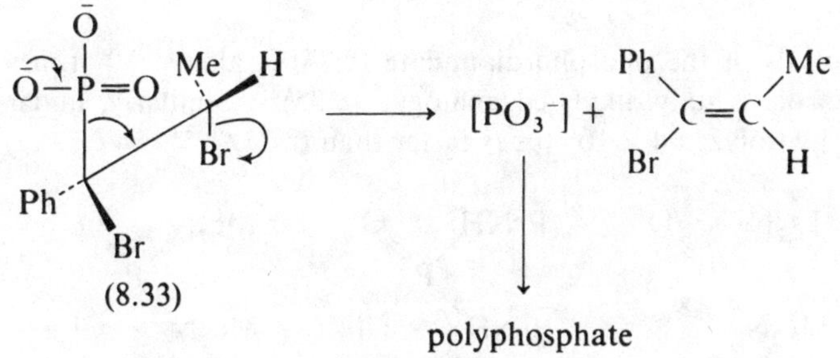

$$\text{(8.33)} \longrightarrow [PO_3^-] + \underset{Br}{\overset{Ph}{>}}C=C\underset{H}{\overset{Me}{<}}$$

$$\downarrow$$

polyphosphate

The pH–rate profiles for the hydrolysis of monoalkyl pyrophosphates also show maxima about pH 4, have low values of $\Delta S^{\ddagger}$ (*ca* 0) and are catalyzed by metals[23,24]. This information suggests an $S_N1(P)$ mechanism (eqn. 7).

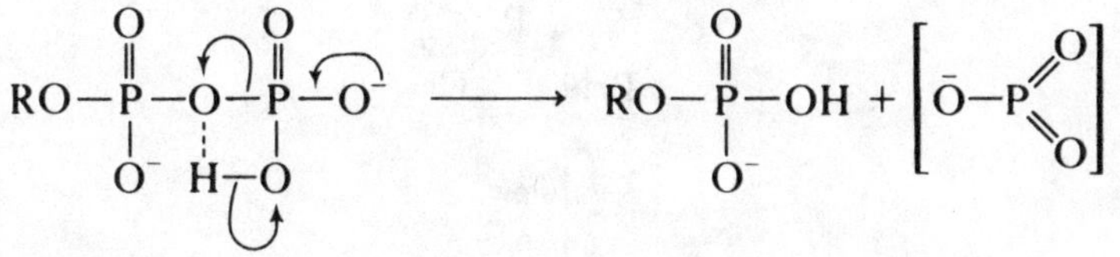

$$RO-\overset{O}{\underset{O^-}{\overset{\|}{P}}}-O-\overset{O}{\overset{\|}{P}}-O^- \quad H-O \longrightarrow RO-\overset{O}{\underset{O^-}{\overset{\|}{P}}}-OH + \left[\bar{O}-P\overset{O}{\underset{O}{<}}\right] \quad (7)$$

Hydrolysis of ATP is also catalyzed by various metals especially divalent ones; ^{18}O is incorporated into the inorganic phosphate produced and the rate maximum is at pH 5.7[25]. The $S_N1(P)$ mechanism outlined in eqn. (8) is consistent with the facts, coordination of the metal ion with the phosphate anion assisting departure of the leaving group ADP.

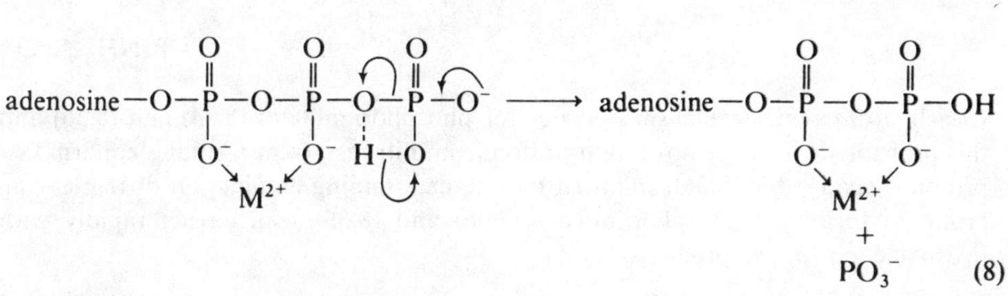

$$\text{adenosine}-O-\overset{O}{\underset{O^-}{\overset{\|}{P}}}-O-\overset{O}{\underset{O^-\searrow M^{2+}}{\overset{\|}{P}}}-O-\overset{O}{\overset{\|}{P}}-O^- \quad H-O \longrightarrow \text{adenosine}-O-\overset{O}{\underset{O^-}{\overset{\|}{P}}}-O-\overset{O}{\underset{O^-\searrow M^{2+}}{\overset{\|}{P}}}-OH$$
$$+ PO_3^- \quad (8)$$

Evidence has also accumulated for an $S_N1(P)$ mechanism in the base-catalyzed hydrolysis of N-substituted phosphoramidates, $(HO)_2P(O)NHR$. One atom of ^{18}O is incorporated in the phosphate product on hydrolysis in $H_2^{18}O$ and with $R = CO_2Et$, a rate maximum is observed at pH 4.[26] Values of $\Delta S^{\ddagger}$ are also positive, e.g. for $R = SO_2Ph$, $\Delta S^{\ddagger} = 2.5$ J K^{-1} mol^{-1} (0.6 cal K^{-1} mol^{-1}) and for $R = CO_2Et$, $\Delta S^{\ddagger} = 50.5$ J K^{-1} mol^{-1} (14.4 cal K^{-1} mol^{-1}) [27,28], and these facts may be rationalized by eqn. (9).

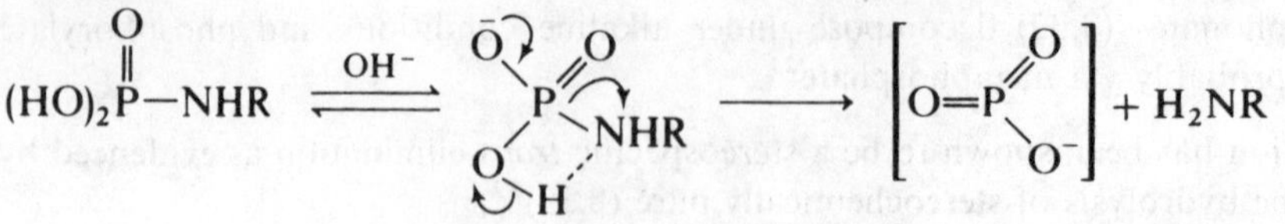

$$(HO)_2\overset{\displaystyle O}{\underset{\|}{P}}-NHR \underset{}{\overset{OH^-}{\rightleftharpoons}} \begin{bmatrix} \bar{O} & O \\ \diagdown \| \\ P \\ \diagup \diagdown \\ O & NHR \\ \diagdown H \end{bmatrix} \longrightarrow \begin{bmatrix} O \\ \| \\ O=P \\ | \\ O^- \end{bmatrix} + H_2NR \qquad (9)$$

The base-catalyzed hydrolysis of the phosphordiamidate (8.34) is almost 10^4 times faster than the hydrolysis of its fully alkylated analogue (8.35)[29]. Similarly, under basic conditions (8.36) is hydrolyzed 4×10^6 times faster than (8.37)[30,31].

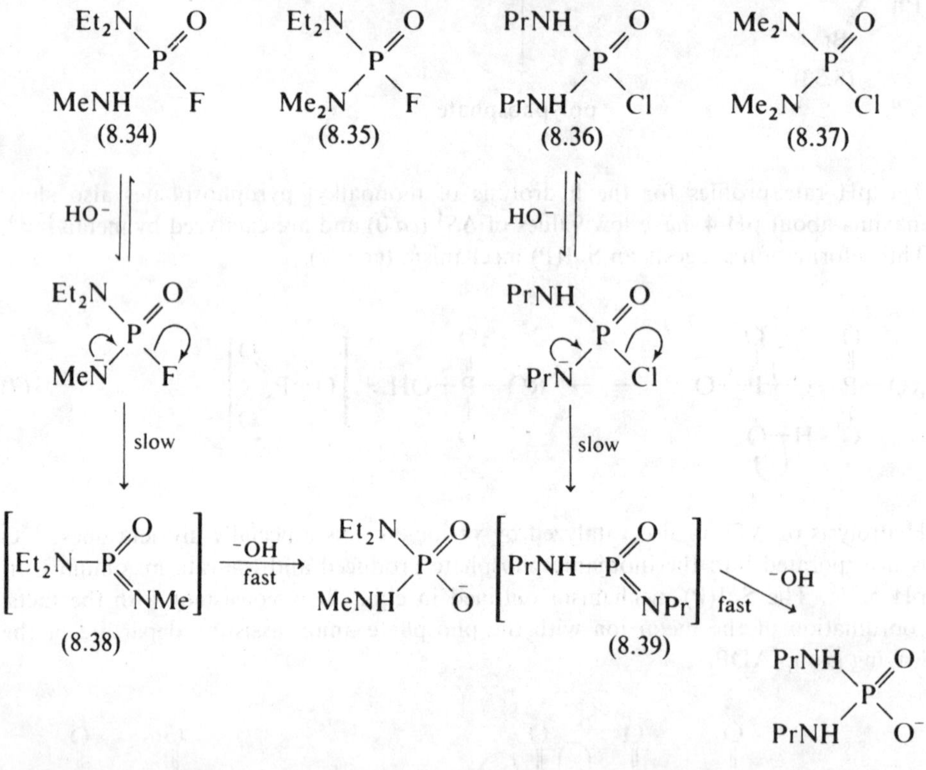

| Et$_2$N O
 \\//
 P
 / \\
 MeNH F
 (8.34) | Et$_2$N O
 \\//
 P
 / \\
 Me$_2$N F
 (8.35) | PrNH O
 \\//
 P
 / \\
 PrNH Cl
 (8.36) | Me$_2$N O
 \\//
 P
 / \\
 Me$_2$N Cl
 (8.37) |

HO$^-$ ⇅ (for 8.34)   HO$^-$ ⇅ (for 8.36)

Et$_2$N O   PrNH O
 \\//   \\//
  P   P
 / \\   / \\
MeN̄ F   PrN̄ Cl

↓ slow   ↓ slow

$\begin{bmatrix} Et_2N-P \overset{O}{\diagdown} \\ \diagdown NMe \end{bmatrix} \xrightarrow[fast]{^-OH}$ Et$_2$N O / P \\ MeNH Ō $\begin{bmatrix} PrNH-P \overset{O}{\diagdown} \\ \diagdown NPr \end{bmatrix} \xrightarrow[fast]{^-OH}$ PrNH O / P \\ PrNH Ō

(8.38)   (8.39)

Clearly a different mechanism operates for phosphoramidates (8.34) and (8.36) and this probably involves removal of a proton from nitrogen (n.m.r. studies confirm fast proton exchange for (8.36)) followed by rate-determining elimination of the leaving group to form metaphosphorimidates (8.38) and (8.39) which react rapidly with hydroxide ion to give products.

Stereochemical evidence has been published to support this interpretation. The neutral hydrolysis of optically active methyl-N-cyclohexyl phosphoramidothioic chloride (8.40) in aqueous dimethoxyethane occurs with inversion of configuration (presumably by an $S_N2(P)$ mechanism, see later) but hydrolysis under alkaline conditions occurs 10^4–10^5 times faster and yields a racemic product. The inference is that under alkaline conditions, a planar metathiophosphorimidate intermediate (8.41) is formed, which gives rise to racemic products[32].

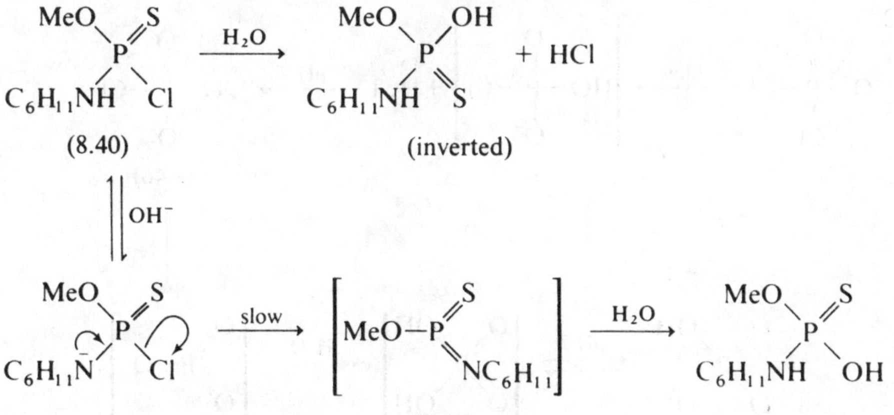

8.2 Nucleophilic displacement at phosphorus: acyclic compounds

[Structures (8.40) reacting with H₂O to give inverted product + HCl, with OH⁻ equilibrium leading through slow step to intermediate (8.41) then H₂O to racemic product]

Alkaline hydrolysis of the *p*-nitrophenyl ester analogue (8.42) occurs with extensive, but not complete, racemization[33] which suggests either a mixed S_N1 and S_N2 mechanism or alternatively a non-planar intermediate. In fact in the alkaline hydrolysis of (8.43) which apparently occurs via the anion (8.44), the intermediate (8.45) is thought to be non-planar[34].

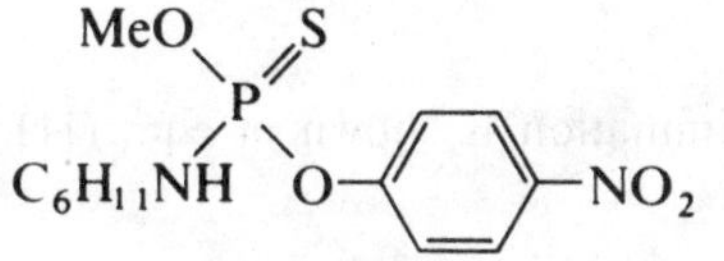

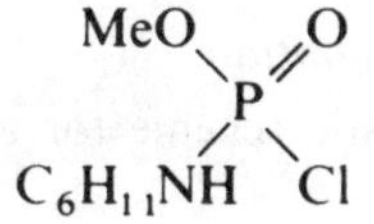

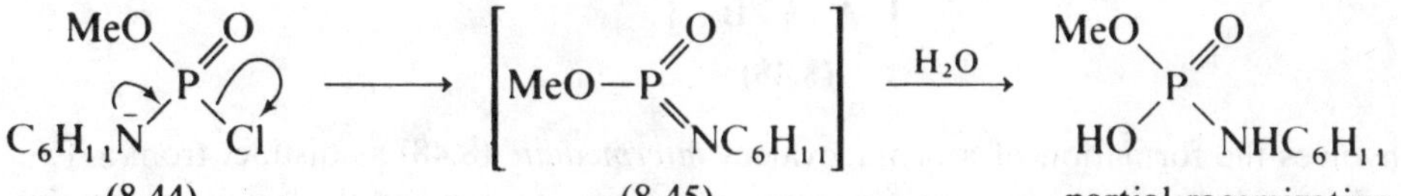

[Structures (8.42) and (8.43)]

[Structure (8.44) → intermediate (8.45) → H₂O → partial racemization product]

The structural criteria for the $S_N1(P)$ mechanism are therefore firmly established. The molecule under attack must have acidic hydrogens and must contain a good leaving group (e.g. Cl) or receive assistance in the bond breaking process by protonation of the leaving group (e.g. 8.29) or by electrophilic catalysis (e.g. eqn. 8).

The importance of the leaving group is illustrated by the hydrolysis of phosphorus oxychloride which at pH 7 is a two stage process (eqn. 10)[35,36].

$$POCl_3 + H_2O \xrightarrow{fast} {}^-O-PCl_2(=O) \xrightarrow[2H_2O]{slow} H_3PO_4 + 2HCl \qquad (10)$$

In the slow step, two chloride ions are removed almost simultaneously; in other words, the hydrolysis of $(HO)_2P(O)Cl$ is very fast. The most likely explanation involves a metaphosphate intermediate (8.47) formed by fast elimination of chloride ion from (8.46).

$$\text{H}_2\text{O} \diagdown \underset{\underset{\text{Cl}}{|}}{\overset{\overset{\text{O}}{\|}}{\text{O}-\text{P}-\text{Cl}}} \xrightarrow{\text{slow}} \left[\underset{\underset{\text{O}^-}{|}}{\overset{\overset{\text{O}}{\|}}{\text{HO}-\text{P}-\text{Cl}}} \right] + \text{HCl} \xrightarrow{\text{pH 7}} \underset{\underset{\text{O}^-}{|}}{\overset{\overset{\text{O}}{\|}}{{}^-\text{O}-\text{P}-\text{Cl}}} \quad (8.46)$$

$$\downarrow \text{very fast}$$

$$\underset{{}^-\text{O}}{\overset{\text{O}}{\diagdown}}\text{P}\underset{\text{O}^-}{\overset{\text{OH}}{\diagup}} \rightleftharpoons \left[\underset{\bar{\text{O}}}{\overset{\text{O}}{\diagdown}}\text{P}\underset{\text{OH}}{\overset{\text{OH}}{\diagup}} \right] \xleftarrow{\text{H}_2\text{O}} \left[\underset{\text{O}}{\overset{\text{O}}{\diagdown}}\text{P}-\text{O}^- \right] \quad (8.47)$$

This is in marked contrast to the hydrolysis of POF_3 where $(\text{HO})_2\text{P(O)F}$ hydrolyzes more slowly than $(\text{HO})\text{P(O)F}_2$[37] and probably by an $\text{S}_\text{N}2(\text{P})$ mechanism. The difference is attributable to the leaving group ability of chlorine which is superior to that of fluorine and hence promotes formation of the metaphosphate as the energetically favourable pathway.

8.2.2 Addition–elimination

The general mechanistic scheme for addition–elimination is shown in eqn. (11). It

$$\text{Y}^- + \underset{\underset{\text{B}}{\diagup}\text{X}}{\overset{\overset{\text{Z}}{\|}}{\text{A}\cdots\text{P}}} \rightleftharpoons \left[\underset{\text{A} \quad \text{B}}{\overset{\overset{\text{Z}^-}{|}}{\text{Y}-\text{P}-\text{X}}} \right] \longrightarrow \underset{\underset{\text{B}}{\diagup}}{\overset{\overset{\text{Z}}{\|}}{\underset{\text{Y}}{\diagup}\text{P}\cdots\text{A}}} + \text{X}^- \quad (11)$$

(8.48)

involves the formation of a pentacovalent *intermediate* (8.48) as distinct from a T.S. and provided that the nucleophile enters an apical position and the leaving group (X) departs from an apical position, the mechanism leads to inversion at phosphorus.*

Thus, if alkaline hydrolysis of diethyl phosphorochloridate (8.49) occurred by addition–elimination, the mechanism would be adequately represented by eqn. (12).

$$\text{HO}^- + \underset{\underset{\text{EtO}}{\diagup}\text{Cl}}{\overset{\overset{\text{O}}{\|}}{\text{EtO}\cdots\text{P}}} \rightleftharpoons \underset{\text{EtO} \quad \text{OEt}}{\overset{\overset{\text{O}^-}{|}}{\text{HO}-\text{P}-\text{Cl}}} \longrightarrow \text{HOP(O)(OEt)}_2 + \text{Cl}^- \quad (12)$$

(8.49) \qquad (8.50)

---

* Equatorial entry and departure would have the same stereochemical consequence (inversion); apical entry and equatorial departure *prior* to pseudorotation would lead to retention. On the other hand, if ψ is facile compared to the energetics of P—X bond cleavage, one would expect extensive racemization.

8.2 Nucleophilic displacement at phosphorus: acyclic compounds

By analogy with what we know about the hydrolysis of phosphonium salts (Chapter 7), the mechanistic scheme of eqn. (12) would seem to be a reasonable guess. In fact there is very little evidence to suggest that intermediates of type (8.48) or (8.50) have any finite existence in nucleophilic displacements on acyclic phosphorus esters. (The cyclic phosphorus esters are a different story but we shall come to that later.) In most cases moreover, the operation of this mechanism can be disproved.

In this respect one of the most damning lines of evidence involves exchange studies of reactions carried out in $H_2^{18}O$. It is well known that during the hydrolysis of carbonyl compounds, exchange occurs to incorporate two atoms of ^{18}O in the product acid (Scheme 8.2)[38].

Scheme 8.2 ^{18}O exchange during hydrolysis of carbonyl compounds.

$$R-\overset{O}{\underset{\|}{C}}-X + {}^{18}OH^- \rightleftharpoons R-\underset{\underset{{}^{18}OH}{|}}{\overset{\overset{O^-}{|}}{C}}-X \longrightarrow R-\overset{O}{\underset{\|}{C}}-{}^{18}OH + X^-$$

$$\Updownarrow$$

$$R-\underset{{}^{18}O}{\overset{O}{\underset{\|}{C}}}-X + OH^- \rightleftharpoons R-\underset{\underset{{}^{18}O^-}{|}}{\overset{\overset{OH}{|}}{C}}-X$$

$$\Updownarrow {}^{18}OH^-$$

$$R-\underset{\underset{{}^{18}O^-}{|}}{\overset{\overset{{}^{18}OH}{|}}{C}}-X \longrightarrow R-\underset{{}^{18}O}{\overset{O}{\underset{\|}{C}}}-{}^{18}OH + X^-$$

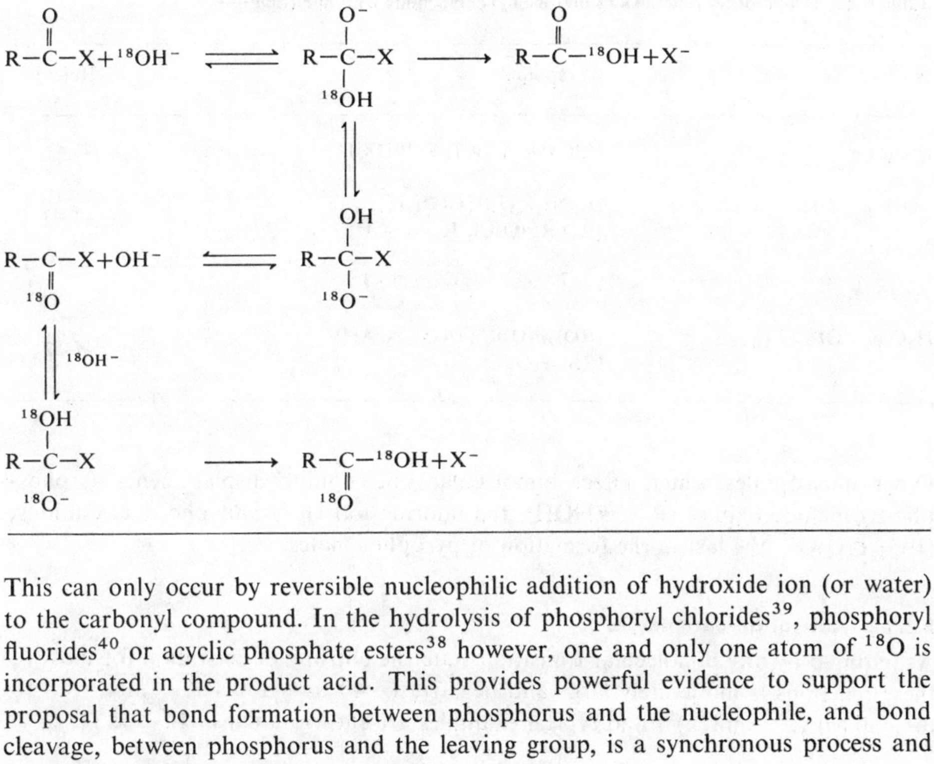

This can only occur by reversible nucleophilic addition of hydroxide ion (or water) to the carbonyl compound. In the hydrolysis of phosphoryl chlorides[39], phosphoryl fluorides[40] or acyclic phosphate esters[38] however, one and only one atom of ^{18}O is incorporated in the product acid. This provides powerful evidence to support the proposal that bond formation between phosphorus and the nucleophile, and bond cleavage, between phosphorus and the leaving group, is a synchronous process and leads us to a discussion of the $S_N2(P)$ mechanism.

8.2.3 $S_N2(P)$ mechanism

This mechanism is directly analogous to nucleophilic displacement at carbon and views the *tbp* of eqn. (13) as a *transition state* rather than an intermediate.

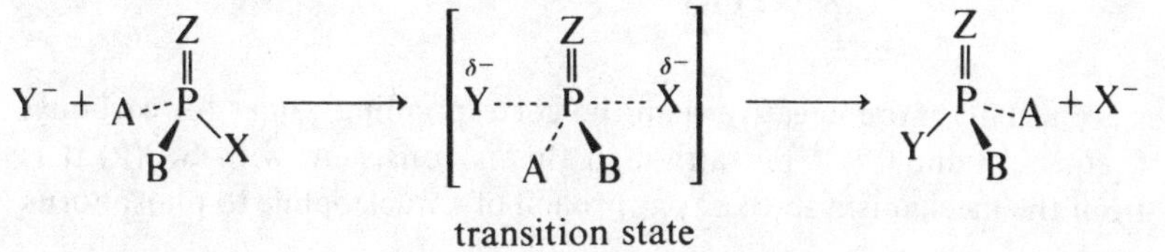

$$Y^- + A\cdots\underset{B}{\overset{\overset{Z}{\|}}{P}}\diagdown_X \longrightarrow \left[Y\overset{\delta-}{\cdots\cdots}\underset{\underset{B}{\diagup}\overset{A}{}}{\overset{\overset{Z}{\|}}{P}}\overset{\delta-}{\cdots\cdots}X\right] \longrightarrow \underset{Y}{\overset{}{}}\underset{B}{\overset{\overset{Z}{\|}}{P}}\cdots A + X^- \qquad (13)$$

$$\text{transition state}$$

Many lines of evidence have been developed to support this mechanism rather than addition–elimination and Kirby and Warren provide a detailed and exemplary review of the subject[3]. Here we shall only cover the main lines of evidence which are taken as diagnostic of the $S_N2(P)$ process.

8.2.3.1 Kinetic order

Most reactions of the esters of phosphoric, phosphonic and phosphinic acids (or their thioanalogues) with nucleophiles are first-order in the phosphorus compound and first-order in nucleophile (Table 8.4)[3,9]. This, of course, is a prerequisite of the $S_N2(P)$ mechanism but does not exclude addition–elimination.

Table 8.4 Second-order reactions of phosphoryl compounds with nucleophiles

Nucleophile	Compounds	Reference
amines	(EtO)MeP(O)Cl; R$_2$P(O)Cl	41,42
$^-$OH or $^-$OPh	(R$_2$N)$_2$P(O)F; (RO)$_2$P(O)Cl (RO)RP(O)Cl; R$_2$P(O)Cl	43–47
H$_2$O or $^-$OH	(RO)$_2$P(O)R^1; (RO)$_2$POAr; (RS)$_2$P(O)R [O(S) ‖]	47–52

Other nucleophiles which effect bimolecular nucleophilic displacements at phosphorus include oximes (R$_2$C=NOH), the fluoride ion (F$^-$) and phosphate anions, (RO)$_2$P(O)O$^-$, the last in the formation of pyrophosphates.

8.2.3.2 Activation parameters

As required by any bimolecular transition state the entropy of activation for most of these reactions is moderately large and negative (-40 to -130 J K^{-1} mol^{-1}; -10 to -30 cal K^{-1} mol^{-1})[53] and typical examples are shown in eqns. (14) and (15).

$$(\text{EtO})_2\text{P(O)Cl} + \text{H}_2\text{O} \longrightarrow (\text{EtO})_2\text{P(O)OH} \quad \Delta S^\ddagger = -92 \text{ J K}^{-1}\text{ mol}^{-1} \quad (14)$$

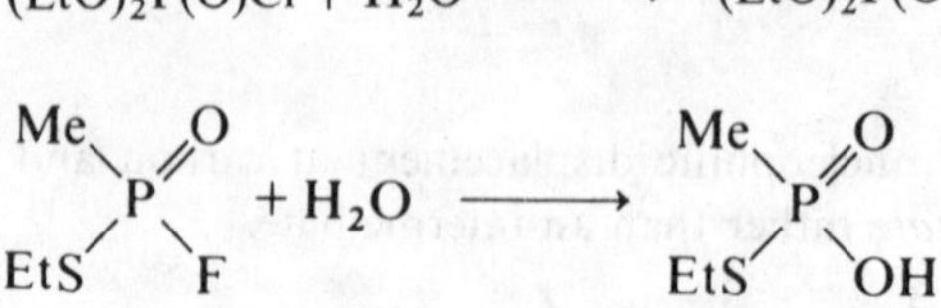

$$\underset{\text{EtS}}{\overset{\text{Me}}{>}}\text{P}\underset{\text{F}}{\overset{\text{O}}{<}} + \text{H}_2\text{O} \longrightarrow \underset{\text{EtS}}{\overset{\text{Me}}{>}}\text{P}\underset{\text{OH}}{\overset{\text{O}}{<}} \quad \Delta S^\ddagger = -92 \text{ J K}^{-1}\text{ mol}^{-1} \quad (15)$$

The $\Delta S^\ddagger$ values are usually more negative than the corresponding values for carbonyl compounds (cf. eqns. 16 and 17)[54] but although this is consistent with $S_N2(P)$ it is hardly diagnostic of the mechanism since any approach of a nucleophile to phosphorus

8.2 Nucleophilic displacement at phosphorus: acyclic compounds

would be expected to create a higher degree of order than addition to the trigonal carbonyl and this would be reflected in the value of $\Delta S^{\ddagger}$ by either mechanism, e.g.

$$CH_3COEt + H_2O \longrightarrow CH_3-\underset{\underset{^+OH_2}{|}}{\overset{\overset{O^-}{|}}{C}}-OEt \qquad \Delta S^{\ddagger} = -88 \text{ J K}^{-1} \text{ mol}^{-1} \qquad (16)$$

$$\underset{\underset{OEt}{|}}{\overset{\overset{OEt}{|}}{CH_3P}}=O + H_2O \longrightarrow CH_3-\underset{\underset{^+OH_2}{|}}{\overset{\overset{OEt}{|}}{P}}\overset{O^-}{\underset{OEt}{\diagdown}} \qquad \Delta S^{\ddagger} = -143 \text{ J K}^{-1} \text{ mol}^{-1} \qquad (17)$$

8.2.3.3 Isotopic labelling

Isotopic labelling has been used to establish P—O bond cleavage in the neutral or alkaline hydrolysis of phosphorus esters. But more interesting from a mechanistic point of view is that only *one* atom of ^{18}O is found in the product (eqn. 18)[55].

$$H^{18}O^- + MeO\cdots\overset{\overset{O}{\|}}{\underset{MeO}{P}}\diagup_{OMe} \longrightarrow H^{18}O-\overset{\overset{O}{\|}}{\underset{OMe}{P}}-OMe + MeOH \qquad (18)$$

Under acidic conditions the reaction proceeds via nucleophilic attack of water on carbon (eqn. 19)[55] and no ^{18}O is incorporated in the dimethyl phosphate product.

$$H_2^{18}O + CH_3-\overset{+}{\underset{\underset{H}{|}}{O}}-P(O)(OCH_3)_2 \longrightarrow H^+ + H^{18}OCH_3 + HOP(O)(OCH_3)_2 \quad (19)$$

8.2.3.4 Kinetic isotope effects

Kinetic isotope effects may also be employed to detect a bimolecular reaction. Thus for a general base-catalyzed reaction (described by eqn. 20, where B may be water) the value of k_{H_2O}/k_{D_2O} is expected to be *ca* 2;[56] the value for the hydrolysis of $(Me_2N)_2P(O)Cl$ is 1.34.[29] On the other hand specific acid catalysis is expected to show an inverse solvent isotope effect $(k_{D_2O}/k_{H_2O} \approx 2)$[57] and this is found for the acid-catalyzed hydrolysis of phosphoramidic acid and derivatives (eqn. 21)[58,59].

$$B + H-\underset{\underset{H}{|}}{O} + R_2\overset{\overset{Z}{\|}}{P}-X \longrightarrow B\overset{+}{H} + HOPR_2 + X^- \qquad (20)$$

$$(HO)_2P(O)NHR + H^+ \rightleftharpoons (HO)_2\overset{H_2O}{\overset{\downarrow}{P(O)\overset{+}{N}H_2R}} \longrightarrow (HO)_3P(O) + H^+ + NH_2R \qquad (21)$$

For $R = H$, $k_{D_2O}/k_{H_2O} = 1.4$; for $R = COPh$, $k_{D_2O}/k_{H_2O} = 1.3$.

8.2.3.5 Stereochemistry

The transition state as shown in eqn. (13) involves apical entry and apical departure which would, of course, lead to inversion at phosphorus. Alternative transition states might include equatorial entry (attack on the edge of the tetrahedron) and equatorial departure (8.49) again leading to *inversion* or apical entry and equatorial departure (8.50), or its converse, which would lead to retention.

$$\text{inversion} \longleftarrow \quad Z=P \begin{smallmatrix} A \\ | \\ B \end{smallmatrix} \begin{smallmatrix} X^{\delta-} \\ Y^{\delta-} \end{smallmatrix} \qquad\qquad B-P \begin{smallmatrix} A \\ | \\ Y^{\delta-} \end{smallmatrix} \begin{smallmatrix} Z \\ X^{\delta-} \end{smallmatrix} \longrightarrow \text{retention}$$

(8.49)   (8.50)

Now (8.50) may be regarded as analogous to front-side displacement on carbon which has not been observed in a synchronous process even under the most favourable circumstances[60]. The situation may be different in phosphorus chemistry where vacant d orbitals are potentially available for coordination with the nucleophile. Nevertheless it seems reasonable to assume that where retention is observed, the most likely explanation is the formation of a pentacoordinate *intermediate* which after one ψ may eject the leaving group, X, from an apical position.

Where racemization is observed, it may be the result of several pseudorotations of a relatively stable phosphorane or in cases where protons can be removed from the substrate may be due to a mechanism involving a planar or rapidly inverting metaphosphate intermediate.

The vast majority of second-order nucleophilic displacements at phosphorus proceed with inversion of configuration at phosphorus. Hudson demonstrated this very elegantly by the equilibration of optically active, ^{14}C labelled O-methyl methylphenylphosphinate (8.51) with its unlabelled analogue (8.52) in methanol[61]. The rate of racemization of (8.51) was exactly twice the rate of exchange of methoxide and this can only be explained by a bimolecular displacement with complete inversion.

(8.51)   (8.52)

Thiol esters, (RO)R^{1}P(O)SR^2 and phosphonochloridates, (RO)R^{1}P(O)Cl also react with inversion[3]. To be certain of a change in configuration it is necessary to construct a Walden cycle and a typical example of such a cycle appears in Scheme 8.3, where all the reactions at the chiral phosphorus involve inversion[62].

Many other facets of these reactions including solvent effects, steric hindrance, inductive and mesomeric effects of substituents on phosphorus and product composition, i.e. discrimination by mixtures of nucleophiles, have been studied[3] and they all tend to substantiate the S$_N$2(P) mechanism.

It should be remembered, however, that distinction between the S$_N$2(P) and addition–elimination mechanisms is a matter of finely balanced energetics. The only really

8.2 Nucleophilic displacement at phosphorus: acyclic compounds

convincing evidence for $S_N2(P)$ is the lack of isotopic (^{18}O) exchange and inversion of configuration at phosphorus. If an intermediate phosphorane has any finite existence one might expect different stereochemical consequences in some cases, through pseudo-rotation. This may be the explanation for some recent and apparently anomalous stereochemical results. For instance, Mislow *et al.* found that the optically active

Scheme 8.3 Walden cycle to establish inversion at phosphorus.

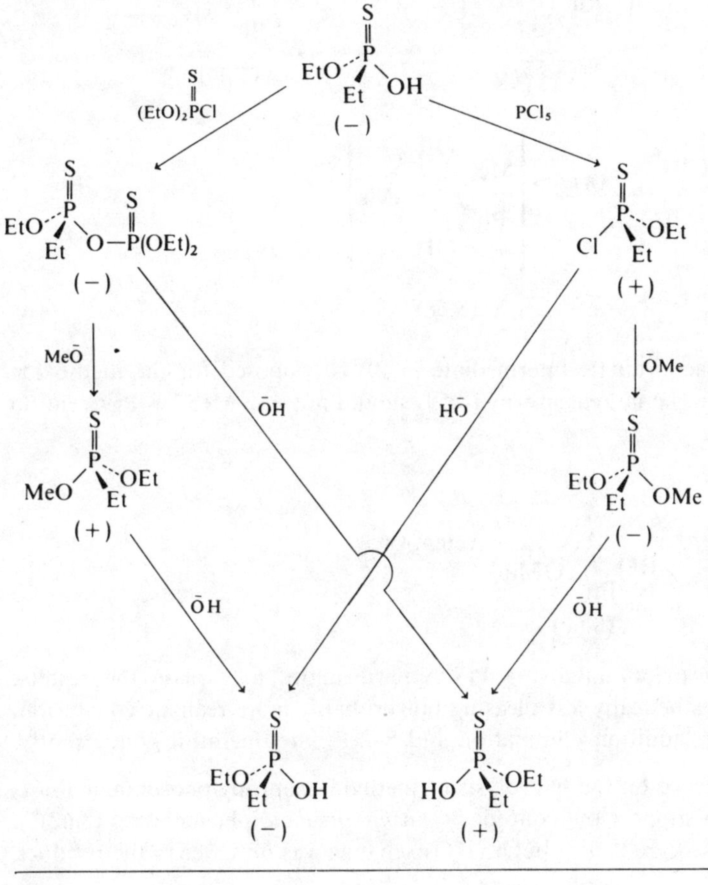

phosphonothiolate (8.53) reacted with methoxide ion with inversion of configuration – $S_N2(P)$ – but nucleophilic displacement by the methyl anion (from methylmagnesium bromide) occurred with retention[63,64]. The alkaline hydrolysis of the phosphonium salt (8.57) also occurs with retention[65] to give the same phosphonate (8.56) as that derived from (8.53). The two observations of retention are most easily explained by invoking two pentacoordinate intermediates (8.55) and (8.58) which lose methane thiol either from an equatorial position or, after one ψ, from an apical position.

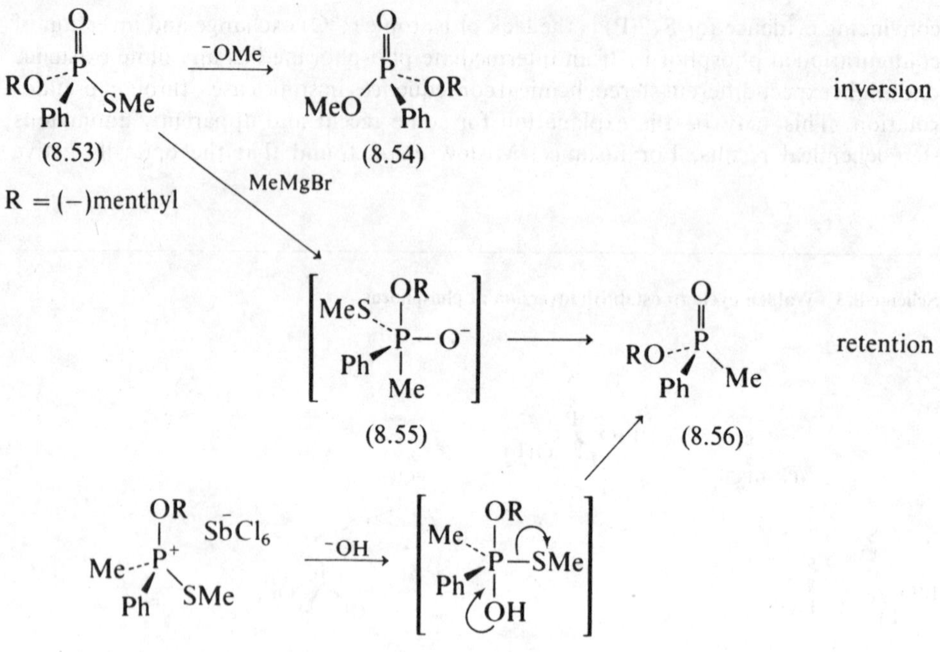

[Structures 8.53 → 8.54 with "inversion"; MeMgBr path to 8.55 → 8.56 with "retention"; 8.57 → 8.58]

If an analogous pentacoordinate intermediate (8.59) is proposed for the methoxide reaction, there seems to be no reason why (8.59) should not lose MeS⁻ with retention like the others.

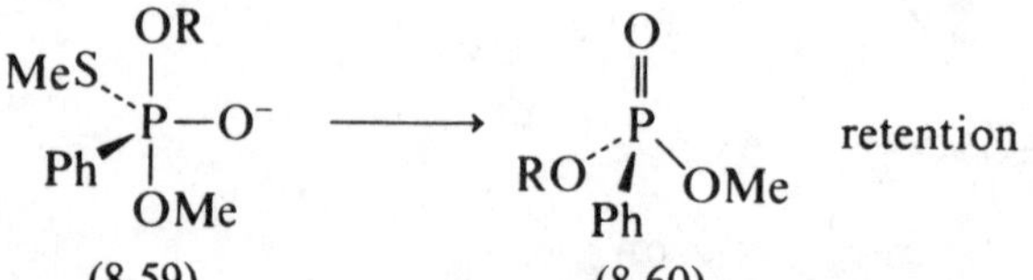

[Structures 8.59 → 8.60 with "retention"]

There is no unifying theory involving P(V) intermediates to explain the results. One is left with the aesthetically less pleasing but probably more realistic conclusion, that both mechanisms, addition–elimination and $S_N2(P)$, are operating concurrently.

Recently, kinetic evidence on the hydrolysis of methyldi-isopropylphosphinate (8.61) has been presented to support the concept of an *intermediate* phosphorane (8.62)[66]. Haake *et al.* found $k_1 \approx k_2 \approx 0.1 k_{-1}$ but no ^{18}O exchange was observed in the product.

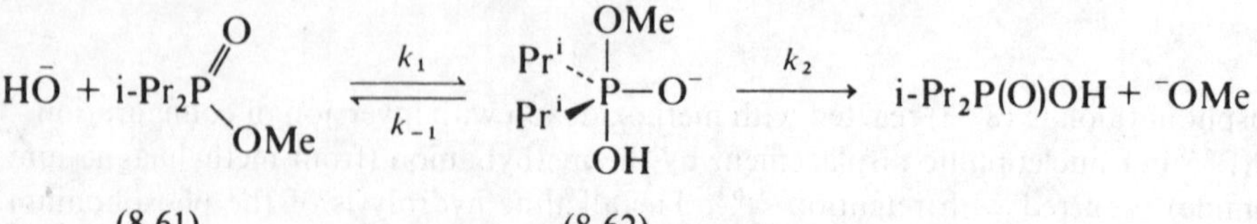

$$HO^- + i\text{-}Pr_2P(O)OMe \underset{k_{-1}}{\overset{k_1}{\rightleftharpoons}} [Pr^i_2P(OMe)(O^-)(OH)] \xrightarrow{k_2} i\text{-}Pr_2P(O)OH + {}^-OMe$$

(8.61) (8.62)

This latter fact may be explained if one assumed ^{18}O exchange would involve unfavourable pseudorotations, e.g. (8.62) → (8.63) by ψ(i-Pr) which would place an isopropyl group apical at the expense of a methoxy group; if one also assumes that HO⁻

8.2 Nucleophilic displacement at phosphorus: acyclic compounds

may only be lost from an apical position (see later) then (8.63) may be the necessary intermediate for exchange via proton transfer to (8.64) and apical loss of HO$^-$.

$$\underset{(8.62)}{\overset{Pr^i}{\underset{Pr^i}{\diagdown}}\overset{OMe}{\underset{|}{P}}\overset{}{\underset{OH}{-}}O^-} \quad \underset{\psi(i-Pr)}{\rightleftharpoons} \quad \underset{(8.63)}{Pr^i-\overset{Pr^i}{\underset{|}{\overset{|}{P}}}\overset{OH}{\underset{O^-}{\diagup\diagdown}}{OMe}} \quad \underset{}{\overset{H^+ \text{ transfer}}{\rightleftharpoons}} \quad \underset{(8.64)}{Pr^i-\overset{Pr^i}{\underset{|}{\overset{|}{P}}}\overset{\bar{O}}{\underset{OH}{\diagup\diagdown}}{OMe}} \quad \longrightarrow \text{ exchange}$$

8.2.4 Catalysis

Nucleophilic displacements may be catalyzed in one of four ways, viz: nucleophilic catalysis, general base catalysis, acid catalysis and electrophilic catalysis. Examples of each are presented below.

8.2.4.1 Nucleophilic catalysis

This may be represented by the general eqn. (22).

$$\underset{\text{(catalyst)}}{W^-} \overset{Z}{\underset{B}{\underset{|}{\overset{||}{A--P}}}}{\underset{X}{\diagdown}} \quad \underset{}{\overset{k_1}{\rightleftharpoons}} \quad \overset{Y^- \text{ (nucleophile)}}{\underset{B}{\underset{|}{\overset{||}{\underset{W}{P}-A}}}} \quad \underset{}{\overset{k_2}{\rightleftharpoons}} \quad \underset{B}{\underset{|}{\overset{||}{\overset{Z}{A--P}}}}{\underset{Y}{\diagdown}} + W^- \qquad (22)$$

For W to behave as a catalyst, k_1 and k_2 must be faster than reaction of Y$^-$ with the substrate alone, i.e. $k_1, k_2 > k_Y$. The phenomenon is often difficult to distinguish from general base catalysis but an established example is provided by the imidazole-catalyzed alcoholysis of the diphosphate (8.65)[67].

$$(PhCH_2O)_2\overset{O}{\overset{||}{P}}-O-\overset{O}{\overset{||}{P}}(OCH_2Ph)_2 \longrightarrow (PhCH_2O)_2\overset{O}{\overset{||}{P}}-\overset{+}{N}\diagdown\diagup NH$$

(8.65)

$$Pr^nOH \qquad + \bar{O}-\overset{O}{\overset{||}{P}}(OCH_2Ph)_2$$

$$\downarrow$$

$$(PhCH_2O)_2\overset{O}{\overset{||}{P}}OPr^n + N\diagdown\diagup NH + H^+$$

Many linear polyphosphates suffer this type of catalytic decomposition (eqn. 23) but the catalysis is often inhibited by steric hindrance.

$$\underset{\text{(ATP)}}{\text{adenosine triphosphate}} \quad \underset{\underset{\text{pyridine}}{\text{catalyzed by}}}{\overset{H_2O}{\longrightarrow}} \quad \underset{\text{(ADP)}}{\text{adenosine diphosphate}} + HO-\overset{\overset{O}{||}}{\underset{\underset{O^-}{|}}{P}}-O^- \qquad (23)$$

For example, β- and δ-methylpyridines (picolines) catalyze the decomposition of ATP but α-picoline is ineffective as a nucleophilic catalyst.

Sometimes the intermediate breaks down in another way and an interesting example of this is afforded by the hydrolysis of Sarin (8.66, see Chapter 12) when catalyzed by oximes (8.67)[68].

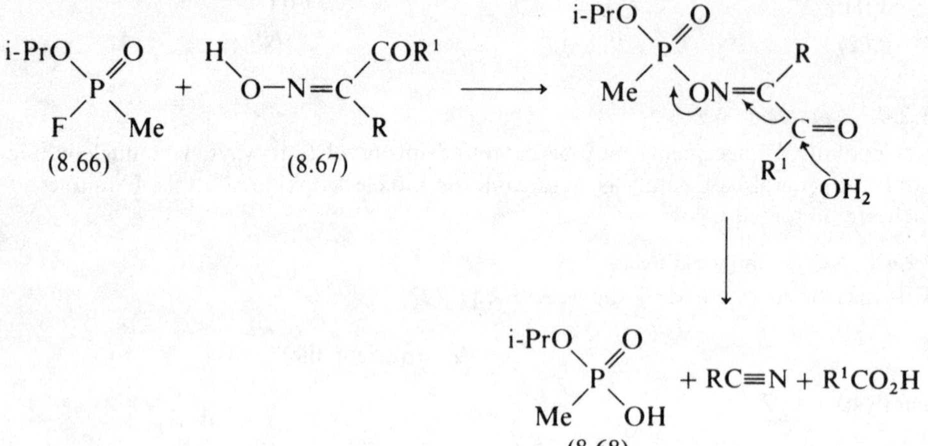

$$\text{(8.66)} + \text{(8.67)} \longrightarrow \text{intermediate} \longrightarrow \text{(8.68)} + RC\equiv N + R^1CO_2H$$

8.2.4.2 General base catalysis

This is represented by the general scheme of eqn. (24).

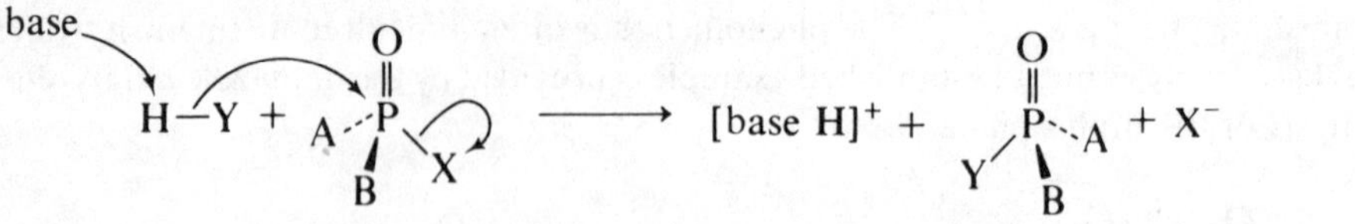

$$H-Y + \underset{B}{\overset{O}{\underset{\|}{A-P-X}}} \longrightarrow [\text{base H}]^+ + \underset{B}{\overset{O}{\underset{\|}{Y-P-A}}} + X^- \quad (24)$$

It is distinguished from nucleophilic catalysis by a kinetic deuterium isotope effect, $k_H/k_D \geqslant 2$ and by a lower sensitivity to steric hindrance.

For example, the alcoholysis of diphosphates or phosphorochloridates is catalyzed by the sterically hindered 2,6-dimethyl pyridine (eqn. 25) and this reaction has $k_H/k_D = 3.4$, indicating a proton transfer in the rate-determining step of the reaction[67].

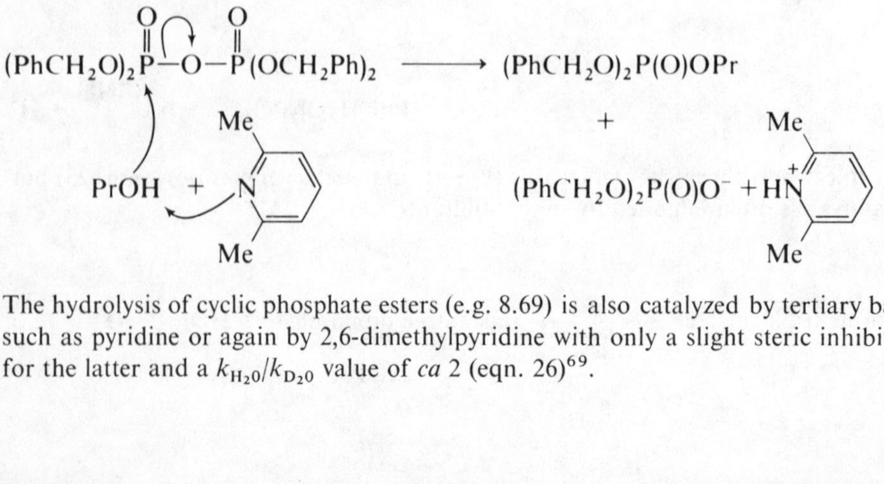

$$(PhCH_2O)_2\overset{O}{\underset{\|}{P}}-O-\overset{O}{\underset{\|}{P}}(OCH_2Ph)_2 \longrightarrow (PhCH_2O)_2P(O)OPr$$
$$+ PrOH + \text{2,6-lutidine} \qquad\qquad (PhCH_2O)_2P(O)O^- + H\overset{+}{N}\text{-lutidinium} \quad (25)$$

The hydrolysis of cyclic phosphate esters (e.g. 8.69) is also catalyzed by tertiary bases such as pyridine or again by 2,6-dimethylpyridine with only a slight steric inhibition for the latter and a k_{H_2O}/k_{D_2O} value of ca 2 (eqn. 26)[69].

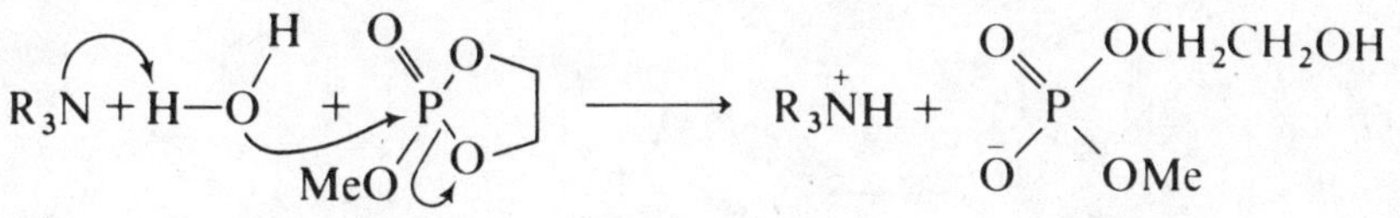

8.2 Nucleophilic displacement at phosphorus: acyclic compounds 327

$$R_3N + H-O-H + \underset{MeO}{\overset{O}{\underset{||}{P}}}\overset{O}{\underset{O}{\big\langle}} \longrightarrow R_3\overset{+}{N}H + \underset{\bar{O}}{\overset{O}{\underset{||}{P}}}\overset{OCH_2CH_2OH}{\underset{OMe}{}} \quad (26)$$

(8.69)

8.2.4.3 Acid catalysis

Acid catalysis is often associated with an inverse solvent isotope effect, i.e. $k_{D_2O}/k_{H_2O} > 1$. This has been mentioned already for the specific acid-catalyzed hydrolysis of phosphoramidic acid, $(HO)_2P(O)NHR$[58,59] and there are some interesting examples of intramolecular, general acid catalysis. One such is the displacement of fluoride from Sarin by phenate ions; the reaction with catechol monoanion (eqn. 27) is faster than the reaction with sodium phenate, presumably due to hydrogen bonding of the remaining phenolic proton of catechol with either fluorine or the phosphoryl oxygen of the substrate[70].

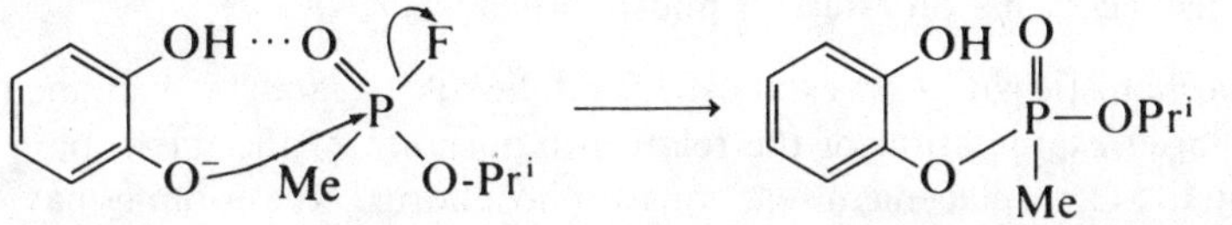

$$\text{catechol-P(Me)(O-Pr}^i\text{)F} \longrightarrow \text{catechol-P(Me)(O)-OPr}^i \quad (27)$$

Another example involves the hydrolysis of the carboxyphenylphosphonate esters (8.70a, b). The *ortho* isomer is extremely susceptible to hydrolysis and reacts at least 10^8 times faster than the para isomer by intramolecular acid catalysis acting either on phosphoryl oxygen or on the leaving group (OEt)[71].

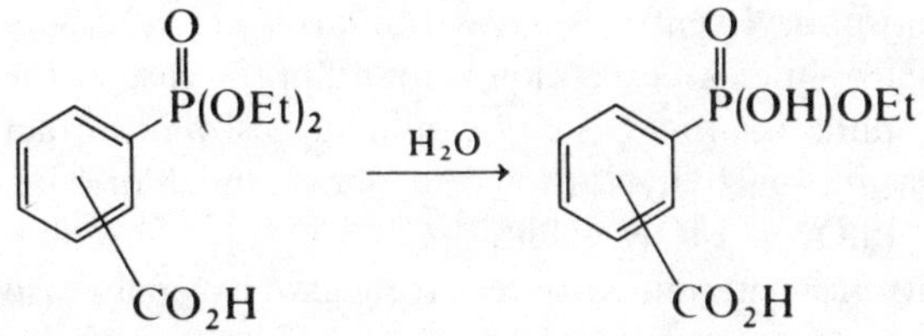

$$\text{Ar-P(OEt)}_2 \text{ (with CO}_2\text{H)} \xrightarrow{H_2O} \text{Ar-P(OH)OEt (with CO}_2\text{H)}$$

(8.70a,b)

a = *o*-CO₂H
b = *p*-CO₂H (d)

8.2.4.4 Electrophilic catalysis

It is well known that certain metals, e.g. Cu(II), U(VI), Zr(VI), Mo(VI) and the lanthanides, are capable of catalyzing nucleophilic displacement reactions at phosphorus, presumably by coordination of the metal with the phosphoryl (P=O) group and/or, the leaving group (X). An example is provided by the enhanced rate of hydrolysis of Sarin in the presence of Cu^{2+} complexes of dipyridyl (8.71), probably via (8.72)[72].

328   Chapter 8 / Tetracoordinate organophosphorus chemistry: part 2

[Structure 8.71: i-PrO, Me, F substituents on P=O] + Cu²⁺ + bipyridine ⇌ [Structure 8.72: Cu²⁺ coordinated complex]

(8.71)   (8.72)

↓ H₂O

[Structure: i-PrO, Me, OH on P=O] + HF + (8.71)

Gels of lanthanum hydroxide also catalyze the hydrolysis of phosphate ester[73,74] and recent kinetic studies have probed the details of electrophilic catalysis by lanthanides during nucleophilic displacements on esters of phosphoric acids[75].

With the variety of mechanistic pathways established and the role of catalysts examined it is time to make a brief examination of the relative importance of the nucleophile and the leaving group in S_N2 displacement reactions at phosphorus. Nucleophiles have been described in terms of three parameters, viz. (i) basicity, (ii) polarizability and (iii) the α-effect[76]. Classification of an electrophile depends upon which of the nucleophilic parameters is most important in the transition state of the reaction. Thus, tetragonal carbon, as exemplified by alkyl halides, falls into the class of electrophiles reacting with class (ii) nucleophiles since it is most readily attacked by polarizable nucleophiles such as RS⁻ or I⁻. On the other hand, the reactivity of carbonyl compounds as electrophiles depends upon the basicity of the nucleophile, i.e. class (i).

In general, a phosphoryl or thiophosphoryl centre behaves similarly, i.e. reactivity depends upon the basicity of the nucleophile. An exception is found in the case of the fluoride ion which is a better nucleophile than its basicity would suggest and in fact it heads the following list of decreasing reactivities towards a phosphinochloridate, $R_2P(O)Cl$ for which, F⁻ > HO⁻ > PhO⁻ > EtOH > PhS⁻, $CH_3CO_2^-$ [77]. This suggests that size and/or apicophilicity (as determined by electronegativity) may also play a very significant part in determining reactivity towards phosphoryl phosphorus.

The above order is illustrated further by phosphorus nucleophiles which are ambident ions. Thus the phosphinothioate anion (8.73) reacts via the more basic oxygen rather

$R_2P(=S)O^-$ + $R_2P(=S)Cl$ ⟶ $R_2P(=S)-O-P(=S)R_2$

(8.73)   (8.74)

than sulphur in forming the pyrophosphinodithioate product (8.74)[78]. Nucleophiles such as NH₂OH, in which an atom adjacent to the nucleophilic atom has one or more lone pairs exhibit enhanced nucleophilic reactivity and this phenomenon is known as the α-effect[79]. Its origins are still a matter of controversy but empirical results leave

8.3 Nucleophilic displacement at phosphorus: cyclic phosphorus esters

no doubt of its existence. For instance, Sarin undergoes nucleophilic displacement of fluoride by hydroxylamine some 1.5×10^4 times faster than nucleophilic displacement by water; moreover the relative rates of displacement of fluoride from Sarin by $^-$OH and $^-$OOH are $1:10^3$.[80] There are numerous other examples of a similar nature and the α-effect is probably responsible, at least in part, for the high nucleophilicity of oxime anions, $R_2C=NO^-$, towards phosphorus.

Bond breaking between substrate and the leaving group is much more important in displacements at phosphorus than in displacements at the carbonyl group. This is demonstrated by the observation that nucleophilic attack on carboxylic anhydrides (eqn. 28) occurs at the more electrophilic carbonyl (cf. Chapter 4, phosphines–diaroyl peroxides) whereas attack at pyrophosphates occurs at the less electrophilic phosphorus with displacement of the better leaving group, the anion of the stronger acid (eqn. 29)[81].

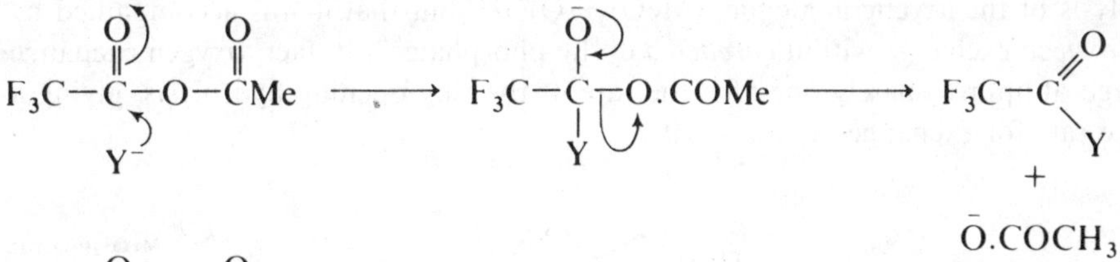

$$F_3C-\underset{Y^-}{\overset{\overset{O^-}{\|}}{C}}-O-\overset{O}{\overset{\|}{C}}Me \longrightarrow F_3C-\underset{Y}{\overset{\overset{O^-}{|}}{C}}-O.COMe \longrightarrow F_3C-C\overset{O}{\underset{Y}{\diagdown}} \quad (28)$$

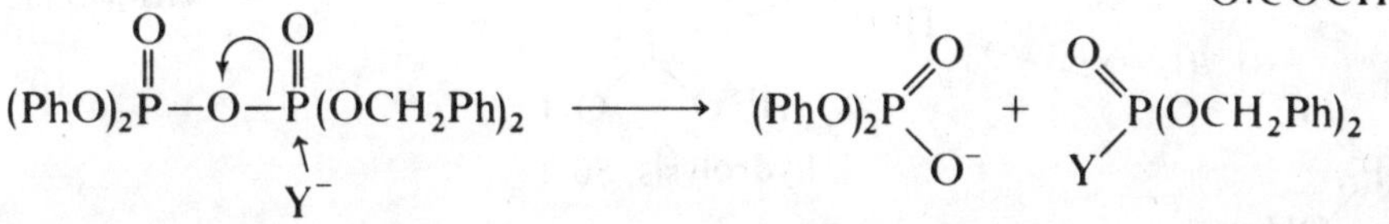

$$+ \quad \bar{O}.COCH_3$$

$$(PhO)_2\overset{O}{\overset{\|}{P}}-O-\overset{O}{\overset{\|}{\underset{\uparrow}{P}}}(OCH_2Ph)_2 \longrightarrow (PhO)_2P\overset{O}{\underset{O^-}{\diagdown}} + \overset{O}{\underset{Y}{\diagdown}}P(OCH_2Ph)_2 \quad (29)$$
$$Y^-$$

This is essentially the reason why nucleophilic attack at phosphorus is susceptible to general acid and electrophilic catalysis, because the bond breaking process assumes considerable importance in the transition state and acid or electrophilic catalysis is effective in weakening the bond between phosphorus and the leaving group.

8.3 Nucleophilic displacement at phosphorus: cyclic phosphorus esters

It is now well known that the hydrolyses in acid or base of five-membered cyclic esters of phosphoric or phosphonic acids proceed millions of times faster than the hydrolyses of their acyclic analogues[8,82,83]. For example, methyl ethylene phosphate (8.75) hydrolyzes in acid with 70 % ring opening (to 8.76) and 30 % exocyclic cleavage (to 8.77) both reactions occurring at rates over a million times faster than that for the hydrolysis of trimethyl phosphate.

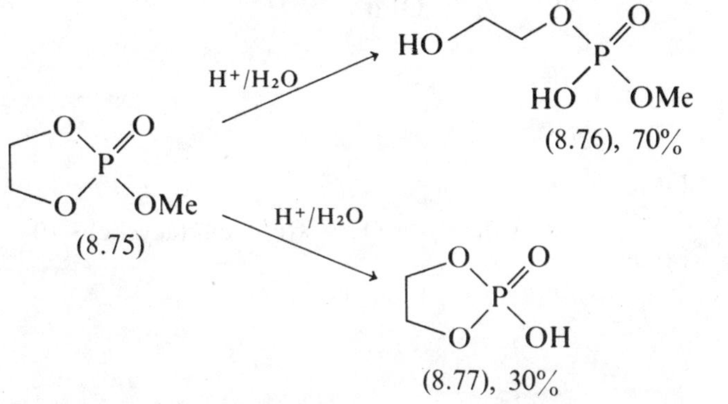

$k/k_{(MeO)_3PO}$

(8.76), 70% 2×10^6

(8.77), 30% 1×10^6

The rate ratio for exocyclic hydrolysis (to 8.77) *vs* hydrolysis of (MeO)$_3$PO corresponds to a difference in free energy of activation, $\Delta(\Delta G^{\ddagger})$, for the two reactions of *ca* 36 kJ mol^{-1} (8.5 kcal mol^{-1}). By contrast, six- and seven-membered ring cyclic phosphates (8.78 and 8.79) hydrolyze at "normal" rates, i.e. the rates found for acyclic analogues[84,85].

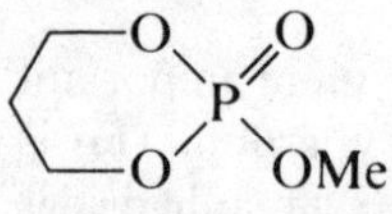

(8.78) 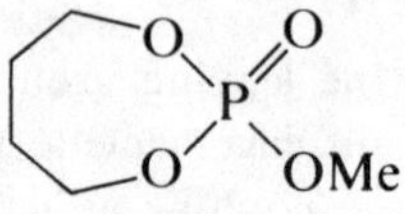   (8.79)

A key discovery in providing an explanation for these results was that hydrolysis of hydrogen ethylene phosphate (8.80) in acid or base was not only 10^8 times faster than hydrolysis of the acyclic analogue, (MeO)$_2$P(O)OH, but that it was accompanied by rapid oxygen exchange within *unreacted* cyclic phosphate[86]. In fact, oxygen exchange occurred at approximately one fifth the rate of the ring opening hydrolysis, giving a relative rate for exchange of *ca* 2 × 10^7.

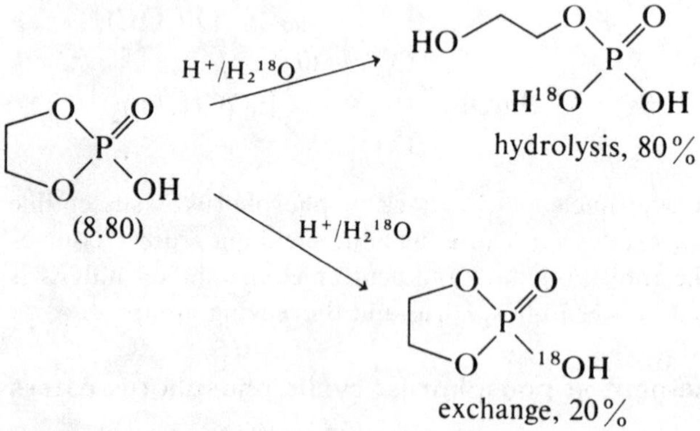

$k/k_{(MeO)_2P(O)OH}$

hydrolysis, 80%   10^8

exchange, 20%   2 × 10^7

Similar rate enhancements for hydrolysis where the ring is preserved are observed for (8.81)[8] and (8.82)[87].

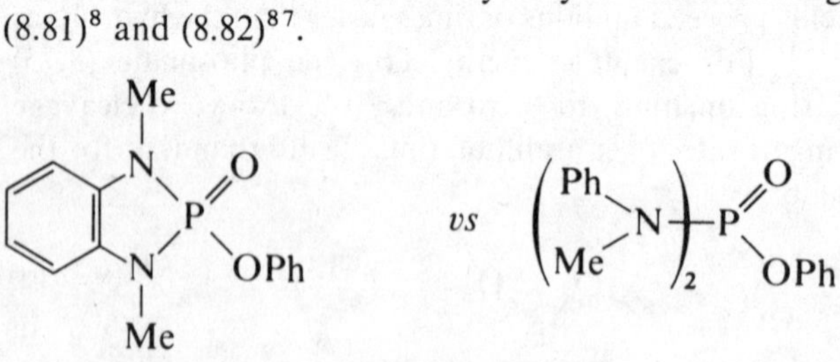

(8.81)   *vs* k(cyclic)/k(acyclic) = 10^6

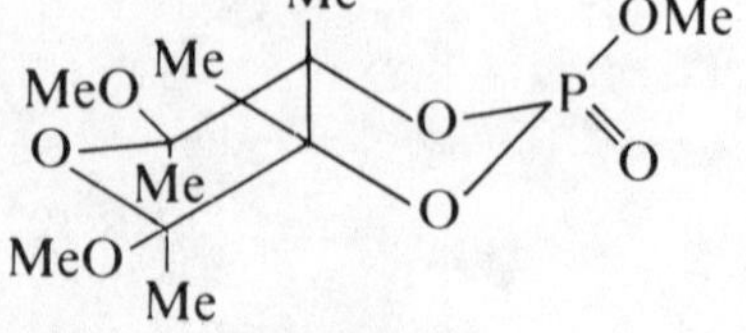

(8.82)   *vs* (MeO)$_3$P=O k(cyclic)/k(acyclic) ~ 10^6

8.3 Nucleophilic displacement at phosphorus: cyclic phosphorus esters

In cyclic phosphonates (e.g. 8.83) the rate enhancement (in acid or base) is preserved but hydrolysis occurs with almost exclusive ring cleavage to (8.84)[88]. The product ratio (8.84:8.85) corresponds to a difference in $\Delta G^{\ddagger}$ for the two hydrolytic routes of ca 17 kJ mol^{-1} (4 kcal mol^{-1}).

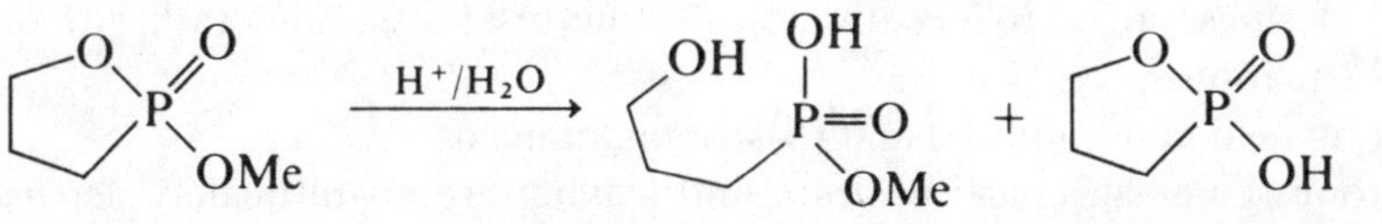

(8.83) →[H$^+$/H$_2$O] (8.84) >99.8% + (8.85) <0.2%

On the other hand, the rates of hydrolysis of cyclic phosphinates (8.86–8.88) which of necessity occurs *external* to the ring, are only slightly greater than the rates for their acyclic analogues (8.89–8.91)[88,89].

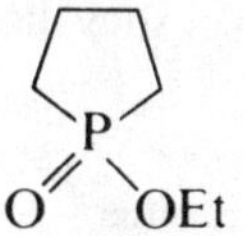

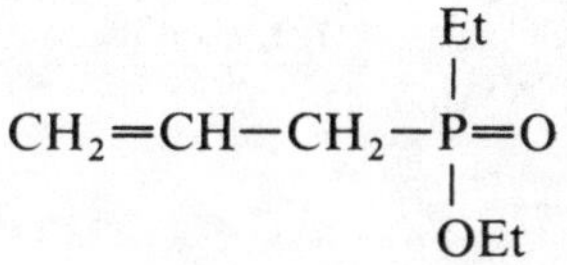

(8.86) 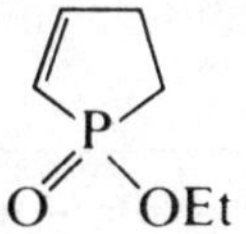   (8.87) 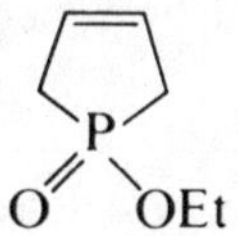   (8.88)

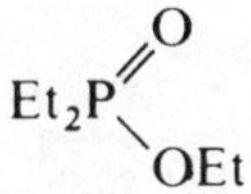

(8.89) Et$_2$P(=O)OEt

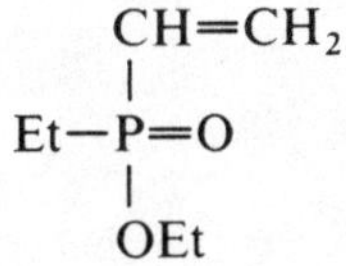

(8.90) Et–P(=O)(OEt)(CH=CH$_2$)

(8.91) CH$_2$=CH–CH$_2$–P(=O)(OEt)(Et)

Two questions arise from the above data. First, what is the explanation for the enhanced rate of hydrolysis of five-membered cyclic phosphate and phosphonate esters and second, how can the observed product ratios of ring cleavage to exocyclic cleavage be rationalized?

The chief driving force for enhanced rates of ring cleavage is probably ring strain. Calculations on methyl ethylene phosphate (8.75)[90] predicted an O—P—O ring angle of 99° which was confirmed by x-ray analysis[91], and this obviously constitutes a strained situation.* Hydrolysis with ring cleavage would relieve such strain and this may account for the fact that the enthalpy of hydrolysis $\Delta H°$ of (8.75) exceeds that of trimethyl phosphate by 21–25 kJ mol^{-1} (5–6 kcal mol^{-1}).[92] Since $\Delta(\Delta G^{\ddagger})$ for the hydrolysis of (8.75) vs (MeO)$_3$PO is ca 38 kJ mol^{-1} (8.5 kcal mol^{-1}) and ΔE_a for the same two reactions is ca 33 kJ mol^{-1} (7.5 kcal mol^{-1}) the thermochemical energy difference of 21–25 kJ mol^{-1} may account for most, but not all, of the enhanced reaction rate. But this explanation does *not* account for the enhanced rate of exocyclic hydrolysis, i.e. cleavage of P—OMe in (8.75) nor does it account for the high rate of ^{18}O exchange in (8.80). Thus the crucial question is how strain, or any other feature of the ring structure can accelerate hydrolysis *without* ring opening. The answer, as proposed by Westheimer[8], lies in the *intermediacy* of pentacoordinate structures which can undergo pseudorotation (ψ), i.e. the hydrolysis proceeds by an addition–

---

* The optimum O—P—O angle for minimum strain would be ca 108°.

elimination mechanism in certain cyclic systems. The rate acceleration of cyclic esters may then be rationalized by making four assumptions about the intermediate viz:

(i) it is energetically unfavourable for carbon atoms to occupy apical sites at the expense of oxygen or other electronegative atoms;
(ii) five-membered rings prefer to occupy apical–equatorial situations rather than diequatorial situations*;
(iii) ψ is facile, provided conditions (i) and (ii) are maintained;
(iv) attacking groups enter an apical position and leaving groups ultimately depart from an apical position.

Notice that assumptions (i), (ii) and (iii) are borne out by n.m.r. studies on stable phosphoranes (see Chapters 2 and 6).

On this basis, the hydrolysis (including cyclic and exocyclic cleavage) of (8.75) may be envisaged by Scheme 8.4. According to this scheme the enhanced rate of exocyclic

---

Scheme 8.4 The hydrolysis of methyl ethylene phosphate by the addition–elimination mechanism

---

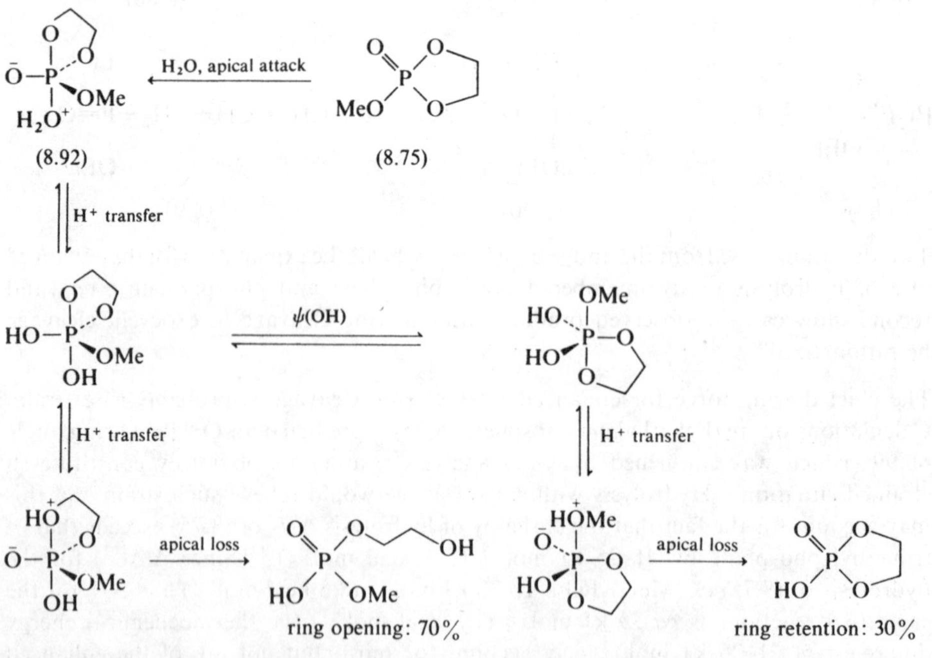

cleavage must arise to some extent from the relief of strain in the five-membered ring in changing from the tetracoordinate (8.75) to the pentacoordinate state (8.92) with the

---

* Calculations[90] also suggest that minimization of strain energy for a possible *tbp* intermediate occurs on forming the intermediate with a 90° angle at P (i.e. ring in a–e conformation) and a large increase in strain occurs on forming the intermediate with a 120° angle at P (i.e. ring in e–e conformation).

8.3 Nucleophilic displacement at phosphorus: cyclic phosphorus esters

ring apical–equatorial. The facile ψ(OH) then places the MeO group apical from which it can depart as MeOH.

The concept of apical attack and apical departure is in effect an extension of the principle of microscopic reversibility. This "extended" principle can be justified if one makes two assumptions, namely:
(a) a single mechanism (i.e. apical attack and apical departure) accounts for ring opening, exocyclic cleavage and exchange;
(b) EtO or MeO may be substituted for OH in the arguments, i.e. MeOH, EtOH and H_2O are so alike as leaving groups that they will all depart from the same geometrical situations.

With these two assumptions in mind it is possible to examine the other two mechanistic sequences of equatorial attack and departure or apical attack and equatorial departure (or its converse). Both, in fact, are inconsistent with the chemistry of the cyclic phosphonate (8.83). Equatorial attack and departure, although conforming to the principle of microscopic reversibility, would presumably involve an intermediate (or transition state) of type (8.93) where the ring oxygen is frozen in an apical position. This would not allow ring cleavage which is contrary to the experiment. Similarly, apical entry and equatorial departure would involve an intermediate (or transition state) of type (8.94) which again would not permit ring cleavage.

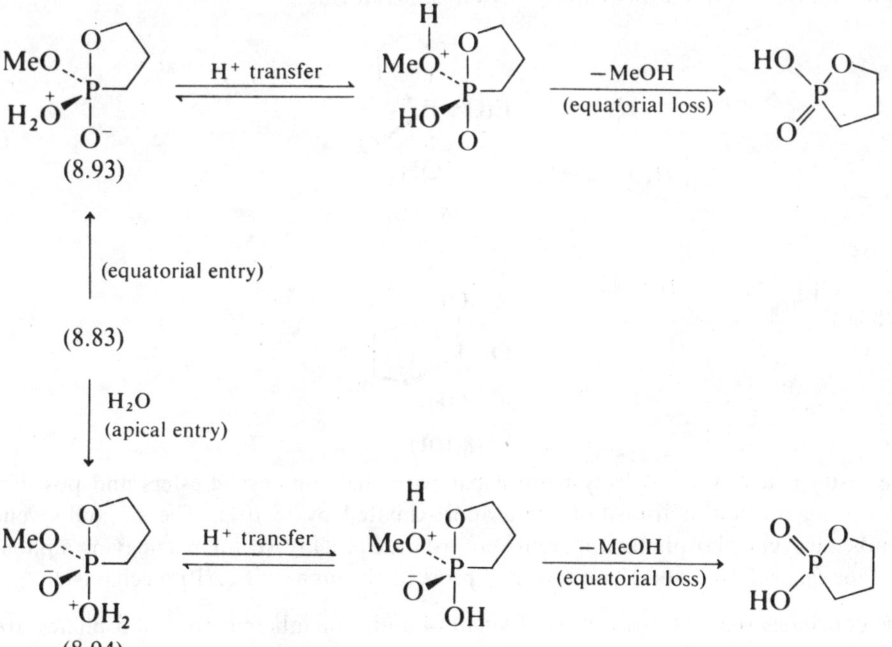

The real test of the Westheimer hypothesis however, comes from an examination of the rates of hydrolysis of cyclic phosphonates and cyclic phosphinates. If one assumes apical entry of the nucleophile, the initial intermediate for phosphonate hydrolysis must be (8.95). The two pseudorotations of (8.95) which would place the MeO group apical, $\psi(O^-)$ or ψ(ring C) are both high energy processes since the first puts a ring carbon apical and the second puts a five-membered ring diequatorial. The only low energy pathway involves proton transfer to (8.98) and cleavage from an apical position to give the ring-opened product (8.99).

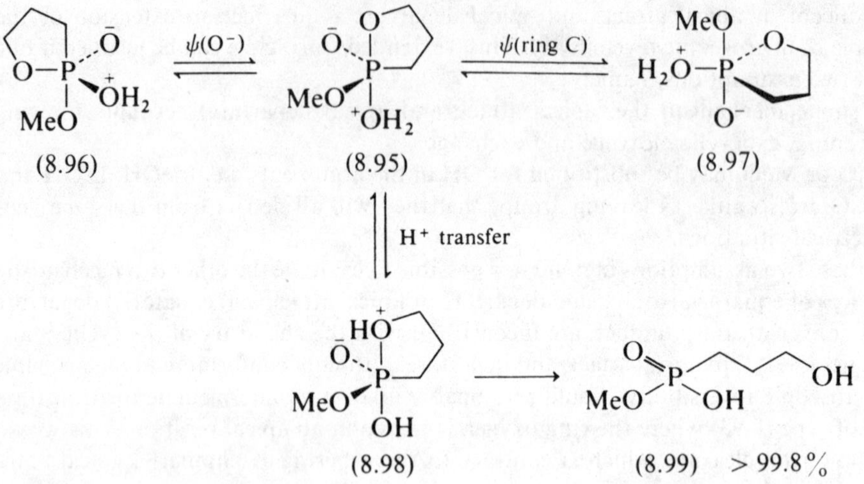

$$\text{(8.96)} \underset{}{\overset{\psi(O^-)}{\rightleftharpoons}} \text{(8.95)} \underset{}{\overset{\psi(\text{ring C})}{\rightleftharpoons}} \text{(8.97)}$$

$\updownarrow$ H$^+$ transfer

$$\text{(8.98)} \longrightarrow \text{(8.99)} \quad > 99.8\%$$

For phosphinate hydrolysis (8.86), again assuming apical entry, there are *no* low energy intermediates since (8.100) has an apical ring carbon and an equatorial ethoxy group and (8.101) has a diequatorial five-membered ring.

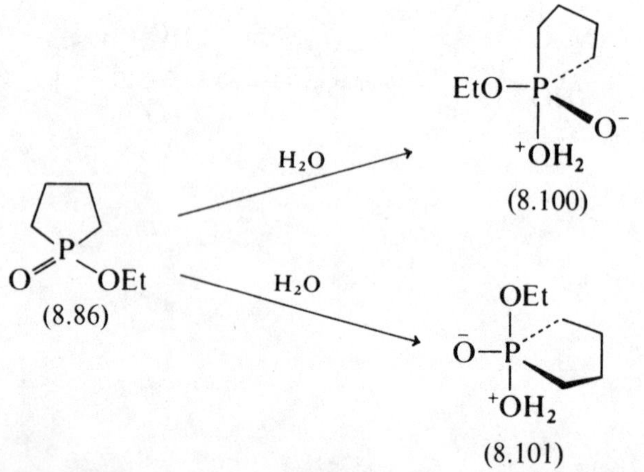

The result is a rate of hydrolysis on a par with that for acyclic esters and possibly proceeding through a transition state approximated by (8.101). The six and seven-membered cyclic phosphates apparently afford little relief of strain in achieving a phosphorane *intermediate* and these too may proceed through an $S_N2(P)$ mechanism.

One concludes that the hydrolysis of small (4 and 5 membered) ring phosphates are somewhat special cases in which the stability of intermediate phosphoranes derived from relief of strain in the small rings in proceeding to the intermediate, allows the addition–elimination mechanism to take precedence over $S_N2(P)$.

In theory, any bimolecular displacement process may be represented by a spectrum of free-energy/reaction coordinate diagrams, A–E of Fig. 8.2. Fig. 8.2C corresponds to the rate profile for an S_N2 displacement and all the others correspond to an addition–elimination involving the addition intermediate, I. Figures 8.2A and 8.2B correspond

8.3 Nucleophilic displacement at phosphorus: cyclic phosphorus esters

Figure 8.2 Free energy/reaction coordinate diagrams for the reaction:

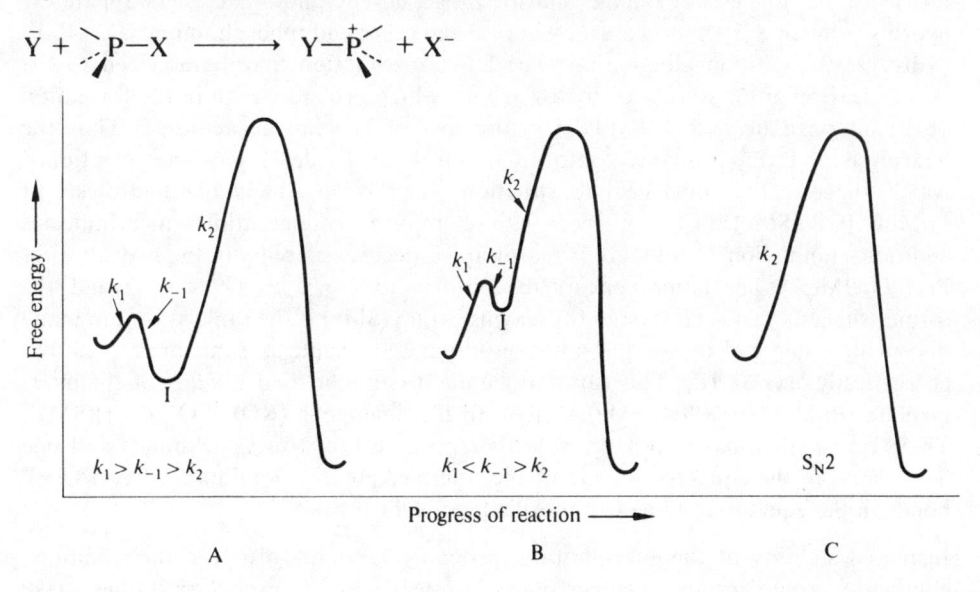

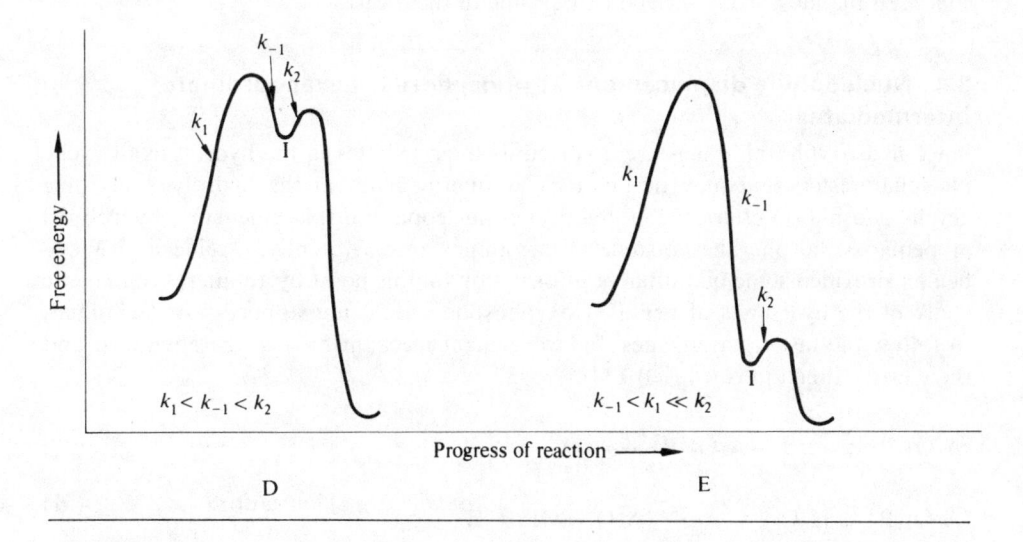

$$\bar{Y} + {>}P{-}X \longrightarrow Y{-}\overset{+}{P}{<} + X^-$$

to situations where intermediate I, might be detected (or possibly, in case A, isolated) because in both cases $k_{-1} > k_2$. Evidence of reversibility and therefore the existence of I, may also be forthcoming in case D provided k_2 is not very much greater than k_{-1}. In case E however, formation of I would not be reversible and its detection would be unlikely even by means of stereochemical mutation via ψ.

Two factors will promote formation of I, viz: (i) an increase in the stability of I relative to the ground state and (ii) a poor leaving group, X, which would raise the barrier to P—X bond cleavage and possibly permit route A or route B to be preferred over the synchronous (S_N2) displacement of Fig. 8.2C.

The study of displacement reactions at tetrahedral phosphorus (Chapters 7 and 8) reveals that at one end of the mechanistic range, acyclic phosphate esters apparently hydrolyze by an $S_N2(P)$ mechanism whilst at the other end, phosphonium salts (R_4P) hydrolyze by addition–elimination. The latter observation may be ascribed to the poorer leaving group ability of hydrocarbons which probably require the formation of an intermediate anion (R_4PO^-) before loss of R^- can be achieved. Thus the hydrolysis of R_4P^+ probably conforms to Fig 8.2B. Evidence has been mentioned which suggests that intermediate situations may occur. Thus the hydrolysis of $PhMeP^+(OR)(SMe) SbCl_6^-$ proceeds with retention of configuration which indicates addition–elimination[65] and there is reason to suspect reversibility in the hydrolysis of $Pr_2^iP(O)OMe$.[66] The latter appears to conform to Fig. 8.2B ($k_2 < k_{-1}$) and the former suggests that even though the leaving group (SMe) is the same as that in many phosphate ester hydrolysis, the addition–elimination route is sometimes preferred energetically over $S_N2(P)$. This can only be due to an increased *stability* of the intermediate, $PhMeP(OH)(OR)(SMe)$ relative to the analogous $(RO)_2P(O^-)(OH)(SMe)$. The extra stability may well arise from the appearance of two C—P bonds and one S—P bond in the equatorial plane of the intermediate *vs* a minimum of two O—P bonds in the equatorial plane for phosphate ester hydrolysis.

Increased stability of the phosphorane probably accounts also for the addition–elimination route for five-membered cyclic esters and the exchange studies make Fig. 8.2B the most likely mechanistic profile in these cases.

8.4 Nucleophilic displacement at phosphorus: hexacoordinate intermediates

Since hydroxyphosphoranes are involved as intermediates in the hydrolysis of cyclic phosphate esters and may be involved as intermediates in the hydrolysis of some acyclic phosphorus esters, the mechanism of nucleophilic displacement (e.g. hydrolysis) at pentaoxyphosphoranes assumes some importance. Recently, Archie and Westheimer provided some quantitative information on this point by an interesting kinetic study of the hydrolysis of pentaaryloxyphosphoranes using stopped-flow techniques to follow the fast reaction rates[93]. Two general mechanisms may be envisaged and these are outlined in eqns. (30)–(34).

$$(ArO)_5P \underset{k_{-1}}{\overset{k_1}{\rightleftarrows}} (ArO)_4P^+ + ArO^- \qquad (30)$$

$$(ArO)_4P^+ + H_2O \xrightarrow{k_2} (ArO)_4POH + H^+ \qquad \text{ionization} \qquad (31)$$

$$(ArO)_4POH \xrightarrow{k_3} (ArO)P_3=O + ArOH \qquad (32)$$

$$(ArO)_5P + H_2O \xrightarrow{k_1} (ArO)_5\overset{-}{P}OH + H^+ \qquad (33)$$

$$(ArO)_5\overset{-}{P}OH \xrightarrow{k_2} (ArO)_3P=O + ArOH + Ar\overset{-}{O} \qquad (34)$$

displacement via hexacoordinate intermediate

The first involves ionization to a tetraaryloxyphosphonium ion (eqn. 30) followed by addition of water (eqn. 31) and breakdown of the intermediate hydroxyphosphorane to products (eqn. 32). The second involves addition of the nucleophile (in this case H_2O) to form a hexacoordinated intermediate (eqn. 33) plus a proton, followed by

8.4 Nucleophilic displacements at phosphorus: hexacoordinate intermediates

breakdown of the anion to products (eqn. 34). Nucleophilic displacement by water on $(ArO)_5P$ via an octahedral *transition state* (eqn. 35) provides an alternative to eqn. (33). However, nothing in the reported data allows a distinction between this and the addition–elimination pathway.

$$H_2O + (ArO)_5P \longrightarrow \begin{bmatrix} ArO & OAr \\ & \overset{\delta+}{\diagdown} \overset{\delta-}{\diagup} \\ H_2O \cdots & P \cdots OAr \\ & \diagup \diagdown \\ ArO & OAr \end{bmatrix} \longrightarrow HOP(OAr)_4 + ArOH \qquad (35)$$

octahedral TS

At high pH (>14) the reaction is first-order in substrate and ^-OH and at low pH (<3) first-order dependence on substrate and H^+ is observed. At intermediate pH values, the rate is constant and is due entirely to the rate of solvolysis by water (k_{H_2O}). The overall reaction rate is in fact represented by eqn. (36).

$$k_{obs} = 0.202 + 16.4[H^+] + 12.0[^-OH] \qquad (36)$$

In neutral or alkaline media, the reaction is accelerated strongly by electron-withdrawing substituents in the phenyl rings and is retarded by electron-donating substituents. More significantly, however, substituents in the *ortho* position reveal a powerful steric retardation for the reaction (factor of ca 10^{-3}). Furthermore, the solvolysis of pentaphenoxyphosphorane is not sensitive to the presence of neutral salt and shows the same rate in 25 % aqueous dimethoxyethane as in solutions of the same solvent containing 0.4 M KCl. Addition of phenol also fails to alter the rate of solvolysis.

Steric retardation is consistent with the formation of a hexacoordinate intermediate or transition state since a change from *tbp* to octahedral geometry would increase steric crowding about phosphorus whereas ionization in or before the rate-determining step would relieve steric hindrance at phosphorus and should lead to higher rates for the ortho-substituted compounds. Acceleration of the rate by electron withdrawal also suggests a hexacoordinate intermediate or transition state and the absence of a salt effect or any influence on the rate by excess phenol, argues strongly against ionization as a critical step in the reaction.

The first-order character in ^-OH at high pH can only be accommodated in an ionization mechanism by assuming rapid, reversible ionization followed by rate-determining attack of ^-OH on the phosphonium ion. This implies that a low concentration of phenoxide ion can compete successfully with a high concentration of ^-OH for the phosphonium ion, which is plainly unreasonable. From these and supplementary arguments, Archie and Westheimer came to the conclusion that a hexacoordinate intermediate or transition state is involved in these hydrolyses.

One must therefore recognize the possibility that hydrolysis of phosphate esters which proceed by addition–elimination mechanisms may also involve such intermediates. Indeed, the hexacoordinate intermediate or transition state was postulated by Ramirez[94] before the Westheimer work appeared. Furthermore, at high base concentrations, the hydrolysis of methyl ethylene phosphate is second-order in hydroxide ion and Gillespie *et al.* have ascribed the second-order behaviour to reaction through a hexacoordinate intermediate (eqn. 37)[95].*

$$\text{(cyclic phosphate)} \underset{}{\overset{H_2O}{\rightleftarrows}} \text{(open intermediate)} \underset{}{\overset{HO^-}{\rightleftarrows}} [\text{hexacoordinate}]^- \overset{-MeO^-}{\longrightarrow}$$

$$[\text{intermediate}] \overset{-H_2O}{\longrightarrow} \text{(cyclic phosphate)} + H_2O \qquad (37)$$

Further investigation will doubtless map out the area in which the mechanism via hexacoordinate species may be operative. In the interim one cannot afford to ignore it as a mechanistic possibility.

8.5 Phosphorylation and the design of phosphorylating agents[3,96]

The importance and wide variety of phosphorus compounds in living systems has provoked a great deal of interest in the synthesis of such compounds[97,98]. The most common, naturally occurring types are mono- and di-esters of phosphoric, diphosphoric and polyphosphoric acid; there are also some important N-substituted phosphoramidates of biological origin.

Most of the synthetic methods involve nucleophilic displacement at a phosphoryl centre by alcohols (to give phosphorus esters) by amines (to give phosphoramidates) and by phosphate anions (to give di- and polyphosphates). This process of nucleophilic displacement at phosphorus (eqn. 38) is termed "phosphorylation" and a large number of phosphorylating agents have been devised and classified on an empirical basis, as a guide to further synthesis.

$$Y^- + (HO)_2P(O)X \longrightarrow (HO)_2P(O)Y + X^- \qquad (38)$$

The simplest phosphorylating agents are phosphoryl chlorides like $(RO)_2P(O)Cl$ and phosphorylation of alcohols, amines or phosphate anions is easily achieved using such molecules (e.g. eqn. 39).

$$EtNH_2 + (RO)_2P(O)Cl \longrightarrow (RO)_2P(O)NHEt \qquad (39)$$

However, this only works well for protected agents such as $(PhO)_2P(O)Cl$ since if one or more free OH groups are present in the phosphorus compound, e.g. $(HO)_2P(O)Cl$,

---

* Reference 95 is a critical review of a wide range of displacement reactions at phosphorus and the section on phosphate ester hydrolysis points out the extreme complexity of these reactions and the difficulty facing a kinetic analysis even when simplifying assumptions are made. In fact the review states "that a mechanistic interpretation of the hydrolysis of trimethyl phosphate and methyl ethylene phosphate from the available kinetic data must be based on a series of simplifications and approximations which have not yet been subject to test". Once again, it seems, that as yet, we have barely scratched the surface!

8.5 Phosphorylation and the design of phosphorylating agents

the non-esterified OH groups compete, by S_N1 or S_N2 mechanisms, with the external nucleophiles and hence give polymeric by-products. Most naturally occurring phosphorus compounds contain at least one free OH group, i.e. are mono or di-esters of phosphoric acids such as the sugar phosphates or nucleotides and in consequence, the objective of much recent work has been to devise reagents which may transfer an *unprotected* phosphoryl group directly.

One useful approach to this problem has been to develop reagents of the type (8.102) where X, Y and Z are commonly C, H, N, O, S and halogen (eqn. 40).

$$HA + \begin{array}{c} HO \\ \diagdown \\ HO \end{array} P \begin{array}{c} O \\ \diagup \\ \diagdown \\ X-Y-Z \end{array} \longrightarrow \begin{array}{c} HO \\ \diagdown \\ HO \end{array} P \begin{array}{c} O \\ \diagup \\ \diagdown \\ A \end{array} + X=Y + HZ \qquad (40)$$

(8.102)

The molecule is usually a phosphorylating agent if the electrons of the P—X bond may be accommodated on Z. To achieve this end, Z must be highly electronegative or must become so by attack of an electrophile or an oxidizing agent. The prerequisites for (8.102) to be a good phosphorylating agent are therefore that (a) the P—X bond should be weak and (b) Z should be strongly electron attracting. Normally bonds between phosphorus and electronegative elements (oxygen, halogens etc.) are strong due in part to $p\pi \to d\pi$ "back donation" of lone pairs. Such back donation may either be eliminated by making X a tetragonal carbon atom or greatly reduced by making Y a trigonal atom so that lone pairs on X are drawn towards Y rather than to P (8.103). Either case will weaken the P—X bond.

$$\begin{array}{c} O \\ \| \\ >P-\ddot{X}-Y=Z \end{array} \longleftrightarrow \begin{array}{c} O \\ \| \\ >P-\overset{+}{X}=Y-\bar{Z} \end{array}$$

(8.103a)   (8.103b)

In some instances the phosphorylation reaction may be bimolecular, $S_N2(P)$, whereas in others the unprotected phosphorylating agent may undergo unimolecular dissociation, $S_N1(P)$, to metaphosphate (eqn. 41) which phosphorylates the substrate (eqn. 42).

$$\begin{array}{c} ^-O \\ \diagdown \\ ^-O \end{array} P \begin{array}{c} O \\ \diagup \\ \diagdown \\ X-Y-Z \end{array} \longrightarrow [PO_3^-] + X=Y + Z^- \qquad (41)$$

$$[PO_3^-] + ROH \longrightarrow \begin{array}{c} O \\ \| \\ RO-P-OH \\ | \\ O^- \end{array} \qquad (42)$$

In any event the end result is transfer of a phosphoryl entity to the substrate. It is now appropriate to discuss the classification of phosphorylating agents (P—X—Y—Z) according to the identity of X, Y and Z and to illustrate their use by means of a few pertinent examples.

1. *No $p\pi$–$d\pi$ contribution to the P—X bond.* This is typified by compounds of the following kinds:

(a) $A_2P(O)C-C-$halogen (X, Y = C; Z = halogen);
(b) $A_2P(O)C-C=C$ (X, Y, Z = C);
(c) $A_2P(O)C-C=O$ (X, Y = C; Z = O)

For example β-chloroalkylphosphonates (8.104) are stable in acid but dissociate in alkali to metaphosphate which then becomes the phosphorylating agent (eqn. 43)[21,22].

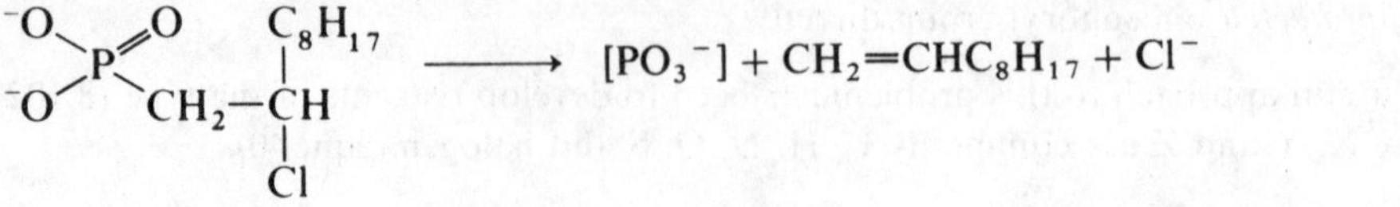

$$\begin{array}{c}\text{}^-O\diagdown\quad\diagup O\quad C_8H_{17}\\ P\quad\quad\;|\\ \text{}^-O\diagup\;\diagdown CH_2-CH\\ \quad\quad\quad\quad\quad|\\ \quad\quad\quad\quad\quad Cl\end{array} \longrightarrow [PO_3^-] + CH_2=CHC_8H_{17} + Cl^- \qquad (43)$$

(8.104)

An example of class 1(b) is afforded by the acid-catalyzed reaction of allylphosphonates (8.105, eqn. 44)[96].

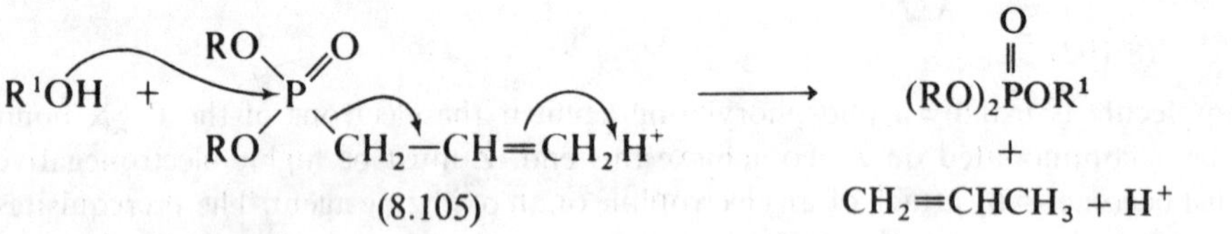

$$R^1OH + \begin{array}{c}RO\diagdown\quad\diagup O\\ P\\ RO\diagup\;\diagdown CH_2-CH=CH_2\,H^+\end{array} \longrightarrow \begin{array}{c}O\\ \|\\ (RO)_2POR^1\end{array} \qquad (44)$$
$$+$$
$$CH_2=CHCH_3 + H^+$$

(8.105)

Finally 1(c) is illustrated by phosphorylation of R^1O^- by the β-ketophosphonate (8.106) through displacement of an enolate ion (eqn. 45)[99].

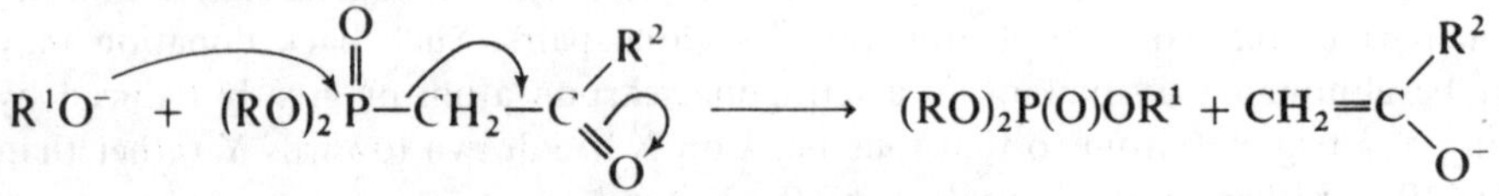

$$R^1O^- + (RO)_2\overset{O}{\overset{\|}{P}}-CH_2-\overset{R^2}{\underset{O}{C}} \longrightarrow (RO)_2P(O)OR^1 + CH_2=C\overset{R^2}{\underset{O^-}{\diagdown}} \qquad (45)$$

(8.106)

2. *Attenuation of $p\pi(X) \to d\pi(P)$ back donation by $p\pi(X) \to p\pi(Y)$ bonding.* In this category phosphorylation occurs by
(a) ground state attenuation;
(b) attenuation by electrophilic attack on Z;
(c) attenuation by oxidative attack on Z; or
(d) by loss of Z as the anion, Z^-.

Examples of 2(a) are provided by the phosphorylation of alcohols by N-acyl phosphoramides (8.107, eqn. 46)[100] and the phosphorylation of acids by mixed anhydrides like acyl phosphates (8.108, eqn. 47)[101].

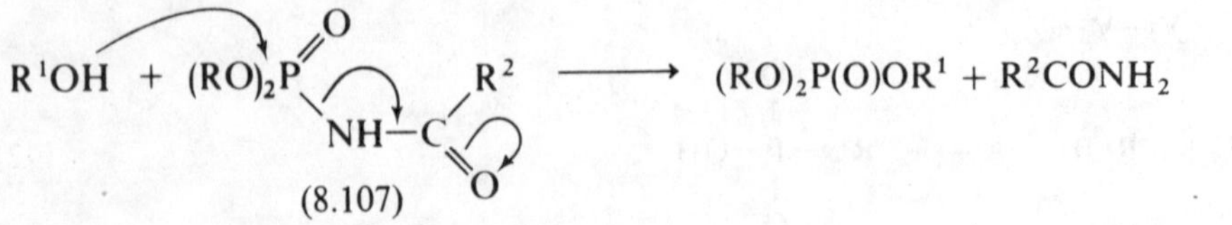

$$R^1OH + (RO)_2P\overset{\diagup O}{\underset{NH-C\diagdown O}{\diagdown R^2}} \longrightarrow (RO)_2P(O)OR^1 + R^2CONH_2 \qquad (46)$$

(8.107)
X = N; Y = C; Z = O

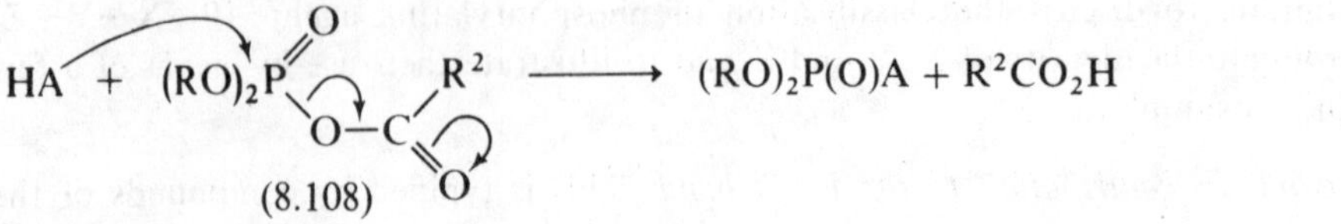

$$HA + (RO)_2P\overset{\diagup O}{\underset{O-C\diagdown O}{\diagdown R^2}} \longrightarrow (RO)_2P(O)A + R^2CO_2H \qquad (47)$$

(8.108)
X=O; Y=C; Z=O

8.5 Phosphorylation and the design of phosphorylating agents

Lone pairs on nitrogen in (8.107) and oxygen (in 8.108) are attenuated by the adjacent carbonyl group and it is this which weakens and thereby promotes P—X bond cleavage.

Examples of 2(b) are afforded by the acid-catalyzed displacement of imidazole from N-phosphoroimidazoles (8.109, eqn. 48)[102] and by the acid-catalyzed alcoholysis of glucose-1-phosphate (8.110, eqn. 49)[103].

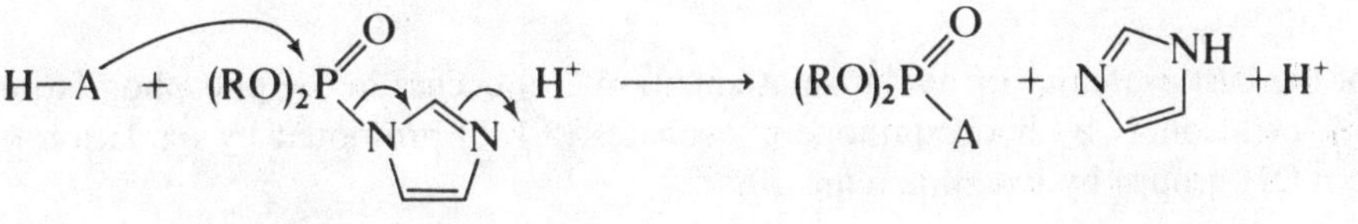

(8.109)

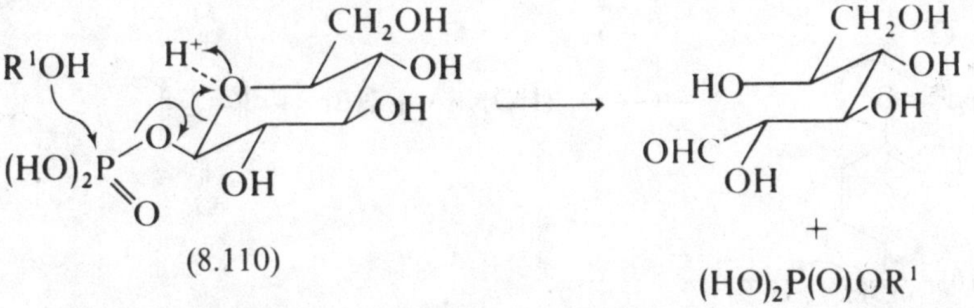

(8.110)

(49)

In both cases the electrophile is the proton but in the first, reaction is via a protected phosphorylating agent (no free OH groups on P) whereas the second is an example of an "unprotected" phosphorylating agent.

Sometimes the phosphorylating agent is generated in situ and an important example of this is the use of carbodiimide (8.111) to generate pyrophosphates from acids. The phosphorylating agent is the addition intermediate (8.112) which has X = O, Y = C and Z = N and reacts with a second molecule of phosphate dianion by an acid-catalyzed phosphorylation[104].

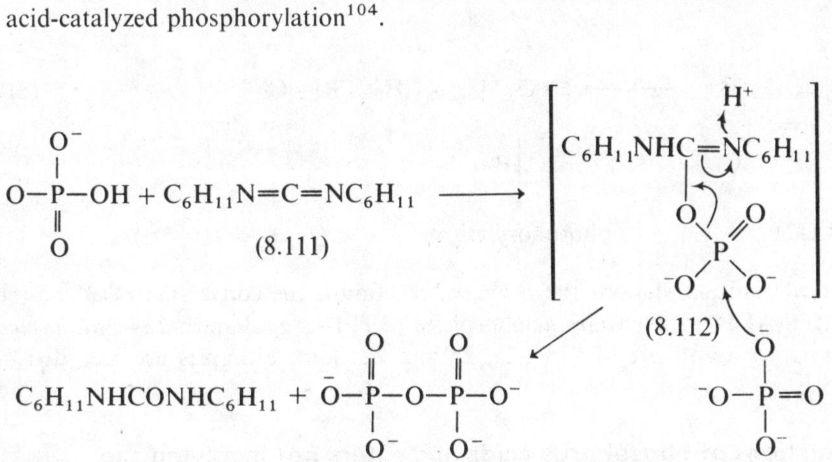

Use of a phosphorothioic acid (8.113) produces the thioamide (8.114) and a mono-thiopyrophosphate (8.115)[105].

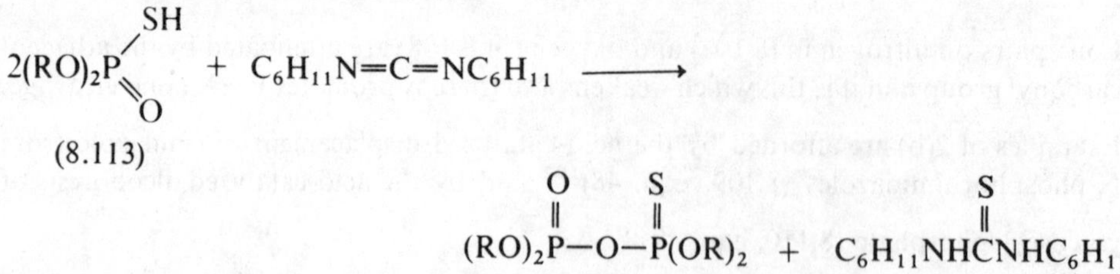

$$2(RO)_2P(SH)(=O) + C_6H_{11}N=C=NC_6H_{11} \longrightarrow$$

(8.113)

$$\underset{(8.115)}{(RO)_2\overset{O}{\underset{\|}{P}}-O-\overset{S}{\underset{\|}{P}}(OR)_2} + \underset{(8.114)}{C_6H_{11}NHC(=S)NHC_6H_{11}}$$

One of the best examples of oxidative attenuation 2(c) occurs in the phosphorylation of acids or alcohols by hydroquinone phosphates (8.116) promoted by oxidation of Z (the *p*-OH group) by bromine (eqn. 50)[106].

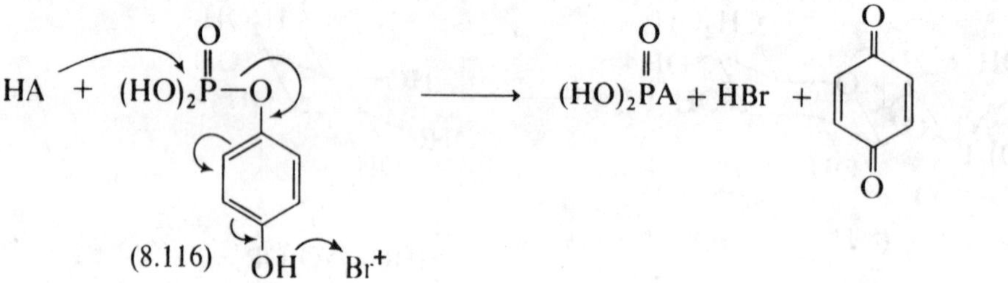

$$HA + (HO)_2P(=O)-O-C_6H_4-OH \xrightarrow{Br^+} (HO)_2PA + HBr + \text{benzoquinone} \qquad (50)$$

(8.116)

A = OR or —OCOR

Process 2(c) occurs *in vivo* where oxidative phosphorylation of ADP to ATP occurs at a number of points in the electron transport chain (Chapter 12).

Finally, attenuation by loss of Z⁻, 2(b), generally requires a highly electronegative Z and is illustrated by the dissociation of β-halogenovinyl phosphates (8.117) under alkaline conditions (eqn. 51)[107].

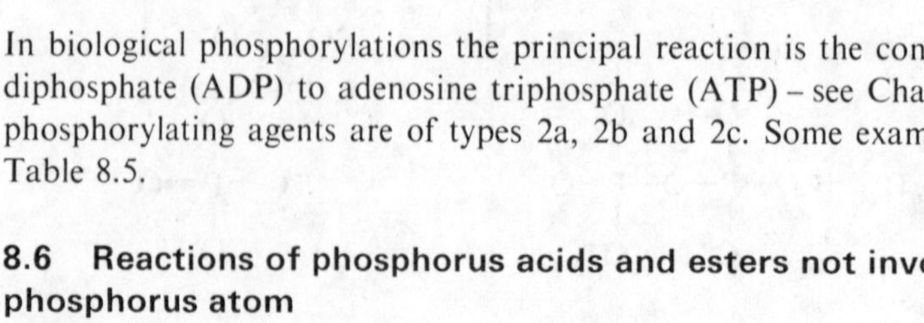

$$\underset{(8.117)}{{}^-O\text{-}P(=O)(O^-)\text{-}CH=C(R)(Cl)} \longrightarrow [PO_3^-] + CH\equiv CR + Cl^- \qquad (51)$$

$$\downarrow HA$$

phosphorylation

In biological phosphorylations the principal reaction is the conversion of adenosine diphosphate (ADP) to adenosine triphosphate (ATP) – see Chapter 12 – and *in vivo* phosphorylating agents are of types 2a, 2b and 2c. Some examples are recorded in Table 8.5.

8.6 Reactions of phosphorus acids and esters not involving the phosphorus atom

Perhaps the most well known and synthetically useful reaction of this kind is the Horner–Emmons modification of the Wittig reaction (see Chapter 7) where the adjacent

8.6 Reactions of phosphorus acids and esters not involving the phosphorus atom

Table 8.5 Some biological phosphorylation reactions

Principal reaction: ADP $\longrightarrow$ ATP

Reaction	Type
ADP + acetyl phosphate $\longrightarrow$ ATP + $CH_3CO_2^-$	2(a)
ADP + 1,3-phosphoglyceric acid $\longrightarrow$ ATP + 3-phosphoglycerate	2(a)
IDP + phosphoenolpyruvate $\longrightarrow$ ITP + CO (oxaloacetate) I = inosine	2(b)
ADP + phosphocreatine $\longrightarrow$ ATP + creatine	2(b)
ADP + aryl phosphate $\longrightarrow$ ATP + quinone + HX	2(c)

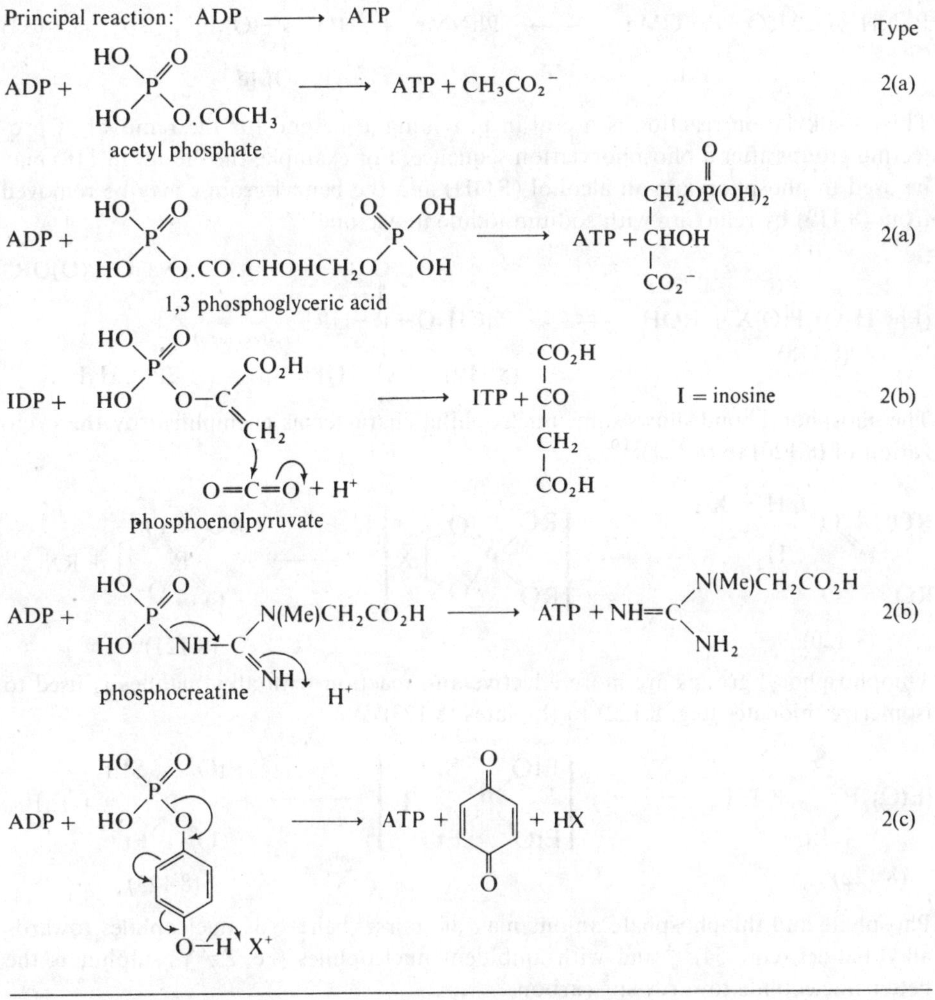

phosphoryl group stabilizes the carbanion resulting from removal of a proton from the α-carbon of a phosphonate (eqn. 52).

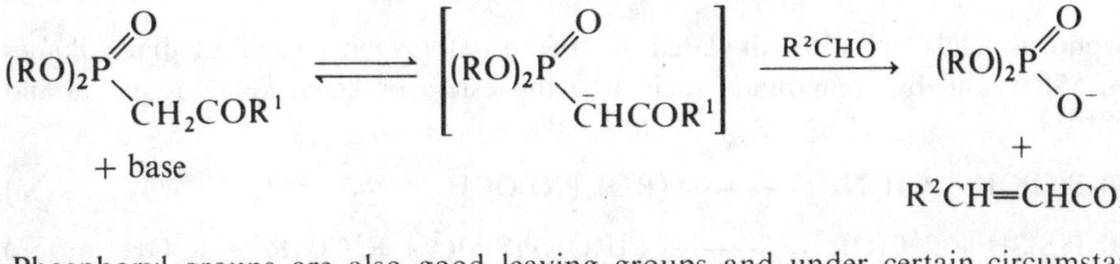

$$(RO)_2P(O)CH_2COR^1 + \text{base} \rightleftharpoons [(RO)_2P(O)\bar{C}HCOR^1] \xrightarrow{R^2CHO} (RO)_2P(O)O^- + R^2CH=CHCOR^1 \quad (52)$$

Phosphoryl groups are also good leaving groups and under certain circumstances will effect alkylation rather than phosphorylation reactions. For instance, aniline reacts with trimethyl phosphate to give N,N-dimethylaniline and dimethyl phosphate (eqn. 53)[108].

$$PhNH_2 + 2MeO-\underset{OMe}{\underset{|}{\overset{O}{\overset{\|}{P}}}}-OMe \longrightarrow PhNMe_2 + 2HO-\underset{OMe}{\underset{|}{\overset{O}{\overset{\|}{P}}}}-OMe \qquad (53)$$

This de-alkylation reaction is useful in providing a method for the removal of protecting groups after a phosphorylation sequence. For example, the diester (8.118) may be used to phosphorylate an alcohol (ROH) and the benzyl groups may be removed from (8.119) by refluxing with sodium iodide in acetone[109].

$$(PhCH_2O)_2P(O)X + ROH \longrightarrow PhCH_2O-\underset{OCH_2Ph}{\underset{|}{\overset{O}{\overset{\|}{P}}}}-OR \xrightarrow[\text{(acetone)}]{NaI} Na_2O_2P(O)OR + 2PhCH_2I$$

(8.118)   (8.119)

The phosphoryl bond shows some nucleophilic character as exemplified by the cyclization of (8.120) to (8.121)[110].

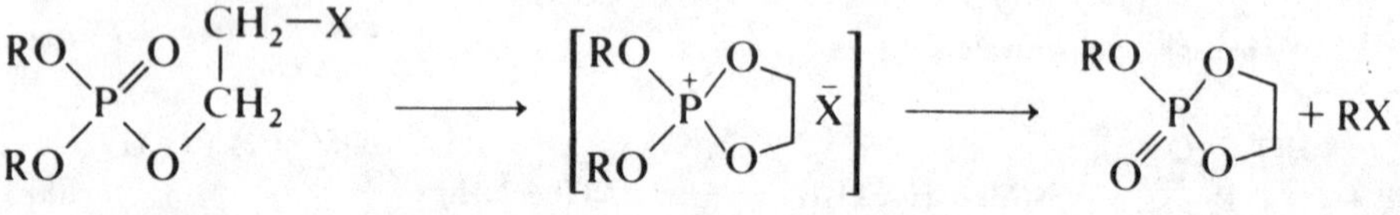

(8.120)   (8.121)

Thiophosphoryl groups are more effective and reaction with alkyl halides is used to isomerize thionates (e.g. 8.122) to thiolates (8.123)[111].

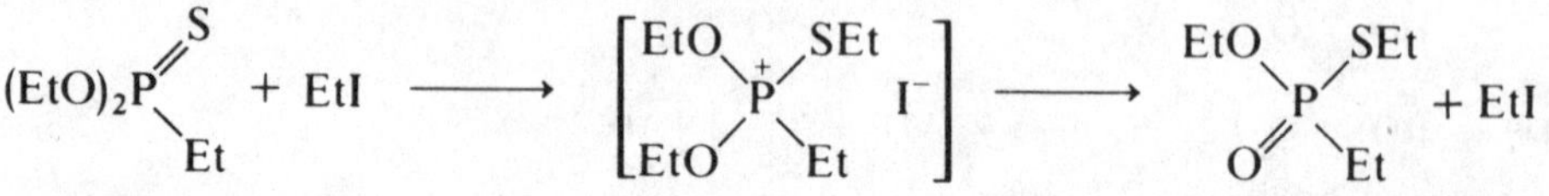

(8.122)   (8.123)

Phosphate and thiophosphate anions may of course, behave as nucleophiles towards alkyl halides (eqn. 54)[3,4] and with ambident nucleophiles (i.e. Z≠Y) sulphur is the better nucleophile towards sp³ carbon.

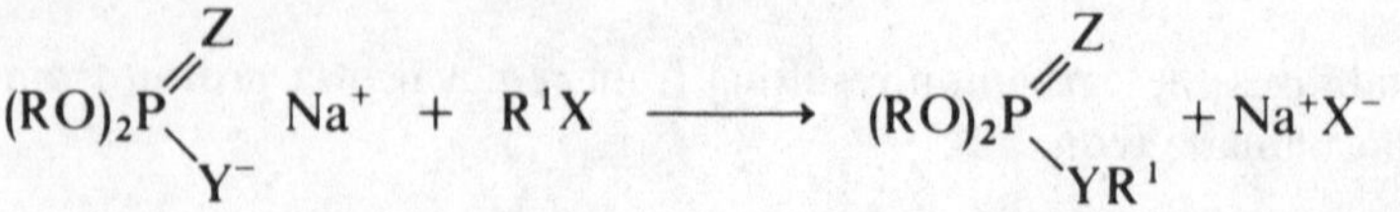

$$(RO)_2P\overset{Z}{\underset{Y^-}{\nwarrow}} Na^+ + R^1X \longrightarrow (RO)_2P\overset{Z}{\underset{YR^1}{\nwarrow}} + Na^+X^- \quad Z = O \text{ or } S \quad (54)$$
$$Y = O \text{ or } S$$

Phosphorus acids are also alkylated by the usual reagents such as diazoalkanes (eqn. 55)[112] and by compounds such as ortho esters or keten ketals (eqn. 56 and 57)[104,113].

$(RO)_2P(O)OH + CH_2N_2 \longrightarrow (RO)_2P(O)OCH_3 + N_2$ (55)

$(RO)_2P(S)SH + R^1C(OR^2)_3 \longrightarrow (RO)_2P(S)SR^2 + R^1CO_2R^2 + R^2OH$ (56)

$(RO)_2P(S)SH + R^1CH=C(OR^2)_2 \longrightarrow (RO)_2P(S)SR^2 + R^1CH_2CO_2R^2$ (57)

Once again, when an ambident (O/S) nucleophile is employed, the thiolate ester is the major product.

The phosphoryl ester saga is now complete but it should be emphasized that more has been recorded about this area of phosphorus chemistry than about any other branch of the subject and the contents of this chapter represent only a small proportion of the available information. As with so many areas of scientific activity, the accumulation of empirical data precedes a comprehensive rationalization of the facts. Recent years have seen the dawn of understanding but much remains to be accomplished.

Problems

1 Salicyl phosphate (1) has a rate maximum at pH 5 corresponding to a dianionic species and it hydrolyzes much faster than the *para*-carboxylate (2) or salicoyl phosphate (3).

(1)

(2)

(3)

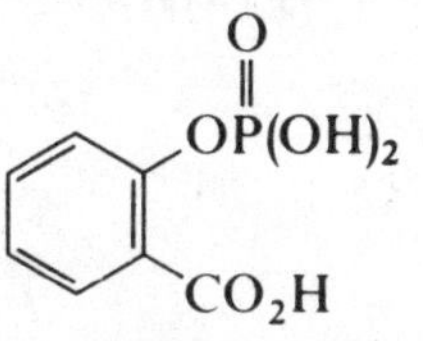

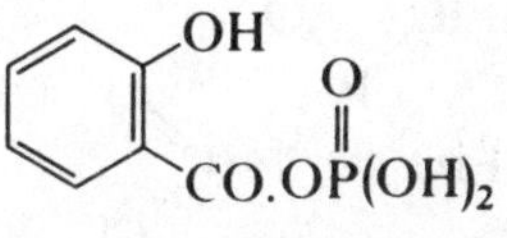

On hydrolysis in $H_2{}^{18}O$, no ^{18}O is incorporated in the carbonyl group. In addition, a negligible solvent isotope effect, $k_{H_2O}/k_{D_2O} = 0.9$ is observed. Rationalize these results.

2 The hydrolysis of the three phosphates (4), (5) and (6) all show maxima at pH 2.5 and the α-isomer (4) shows an enhanced rate relative to (5) and (6). Explain these results.

(4)

(5)

(6)

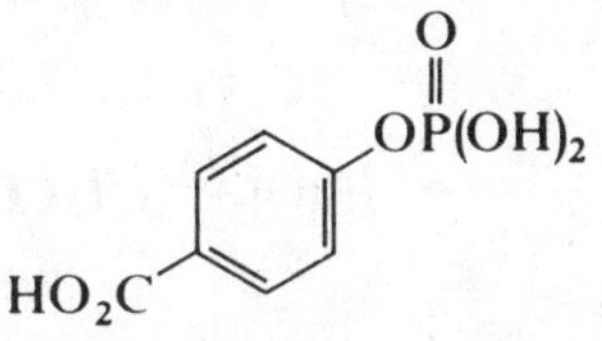

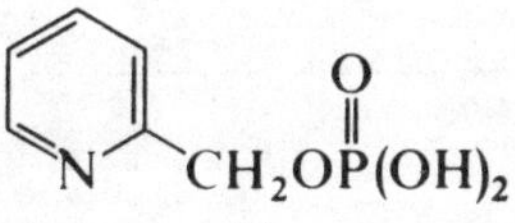

3(a) Work out a scheme for the hydrolysis of $(EtO)_2P(O)F$ using the addition–elimination mechanism and which incorporates two atoms of ^{18}O in the product, $(EtO)_2P(O)OH$.

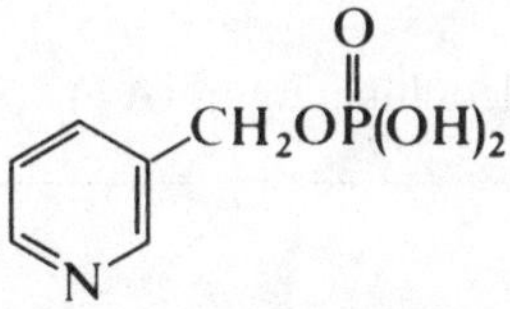

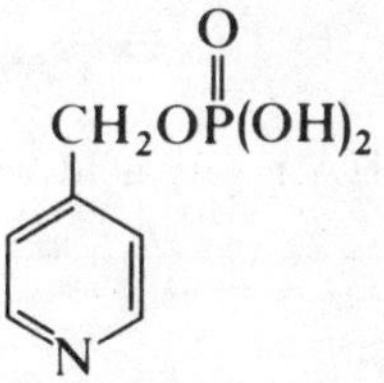

3(b) It might be argued that for the hydrolysis of $(RO)_2P(O)X$, the observation of no exchange of ^{18}O merely reflects a very rapid departure of the leaving group (X) relative to the reversal of nucleophilic addition, i.e.

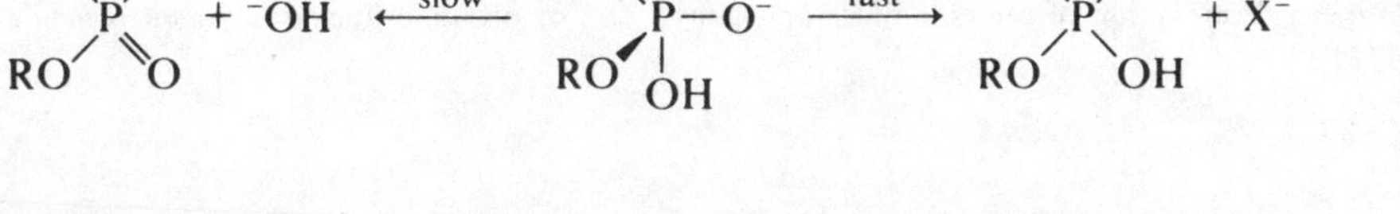

Suggest a reason why this argument is untenable (hint: consider the alkaline hydrolysis of $(EtO)_3P=O$).

4 With reference to the text of Chapter 8, classify the following phosphorylation reactions and devise a mechanism for each:

(i) 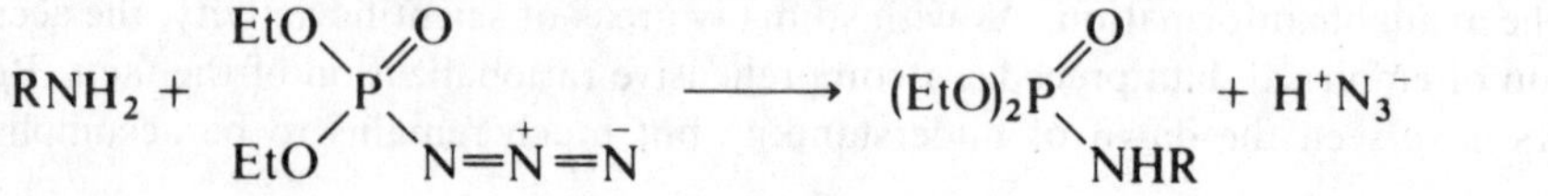 $RNH_2 + (EtO)_2P(=O)N=\overset{+}{N}=\overset{-}{N} \longrightarrow (EtO)_2P(=O)NHR + H^+N_3^-$

(ii) $(n\text{-}PrO)(Me)P(=O)F \xrightarrow{2HOOH/OH^-} (n\text{-}PrO)(Me)P(=O)OH + O_2 + 2H_2O$

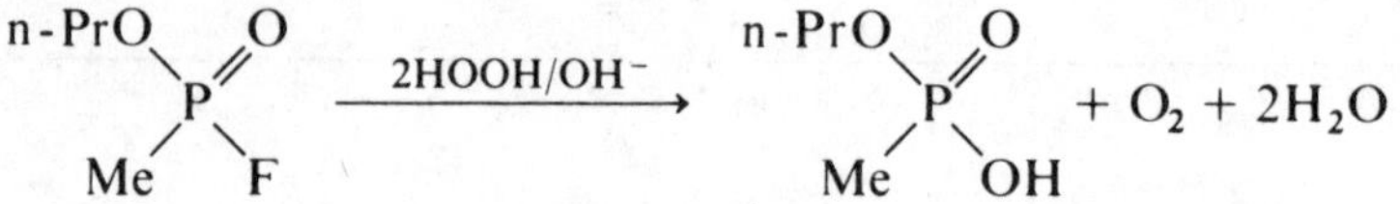

(iii) $(RO)_2POH + R^1OC\equiv CH \longrightarrow [?] \xrightarrow{HA} CH_3CO.OR^1 + (RO)_2PA$

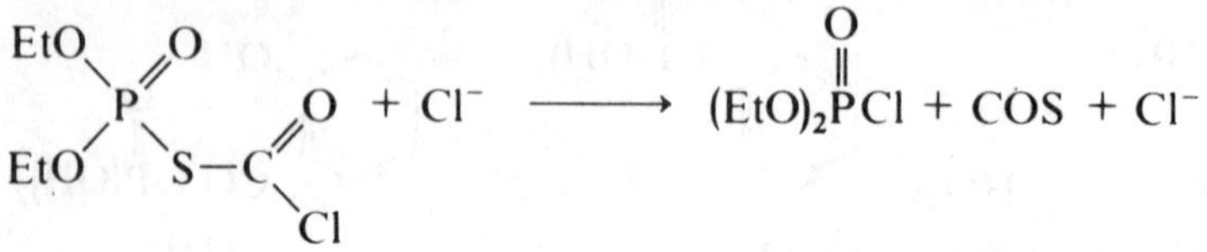

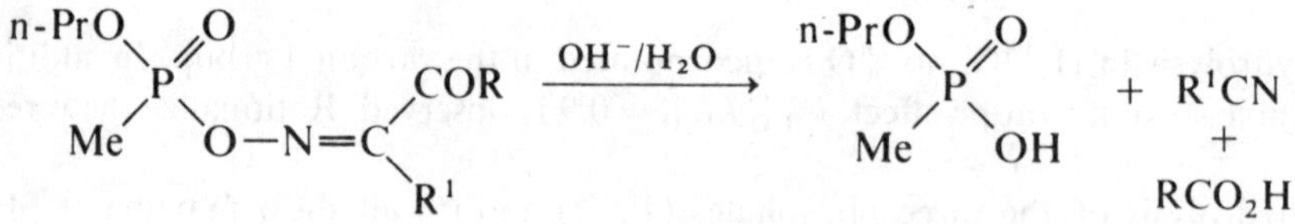

(iv) $(EtO)_2P(=O)\text{—}O\text{—}C(=S)Cl + Cl^- \longrightarrow (EtO)_2PCl(=O) + COS + Cl^-$

(v) $(n\text{-}PrO)(Me)P(=O)\text{—}O\text{—}N=C(R^1)(COR) \xrightarrow{OH^-/H_2O} (n\text{-}PrO)(Me)P(=O)OH + R^1CN + RCO_2H$

(vi) $(RO)_2P(=O)\text{—}O\text{—}C=N\text{-piperidinyl} \xrightarrow{HA} (RO)_2P(=O)A + \text{2-piperidinone}$

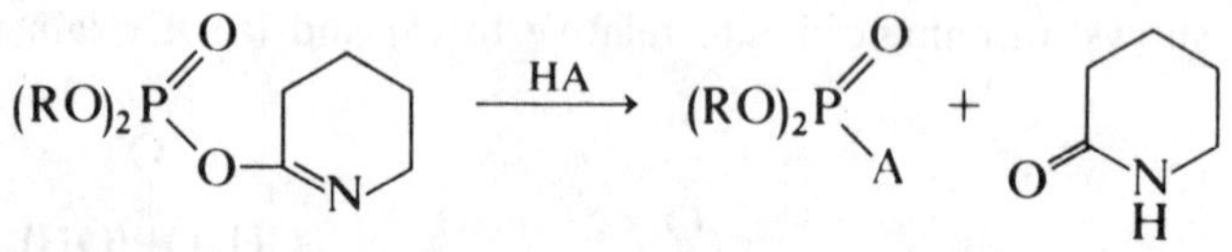

n.b. HA is any acid with a nucleophilic anion (A^-).

References

[1] G. M. Kosolapoff and L. Maier (editors), *Organic Phosphorus Compounds*, Vol. 6, Wiley New York, 1973.

[2] K. Sasse, "Phosphorus Compounds", vol.12 of E. Muller (editor), *Methoden der Organischen Chemie* (Houben-Weyl), Thieme, 1963.

[3] A. J. Kirby and S. G. Warren, *The Organic Chemistry of Phosphorus*, Elsevier, Amsterdam, 1967.

[4] B. J. Walker, *Organophosphorus Chemistry*, Penguin, London, 1972.

[5] R. F. Hudson, *Structure and Mechanism in Organophosphorus Chemistry*, Academic Press, New York, 1965.

[6] S. Trippett (editor), Specialist Reports, Chemical Society, *Organophosphorus Chemistry*, Vols. 1–5, 1969–74.

[7] F. Cramer, "Preparation of esters, amides and anhydrides of phosphoric acid", *Angew. Chem.*, **72**, 236 (1960).

References

[8] F. H. Westheimer, "Pseudorotation in the hydrolysis of phosphate esters", *Accounts Chem. Res.*, **1**, 70 (1968).
[9] J. R. Cox Jr. and O. B. Ramsay, *Chem. Rev.*, **64**, 317 (1964).
[10] S. Trippett, *J. Chem. Soc.*, 4731 (1962).
[11] C. A. Bunton, D. R. Llewellyn, K. G. Oldham and C. A. Vernon, *J. Chem. Soc.*, 3574 (1958).
[12] W. W. Butcher and F. H. Westheimer, *J. Amer. Chem. Soc.*, **77**, 2420 (1955).
[13] J. D. Chanley and E. Feageson, *J. Amer. Chem. Soc.*, **77**, 4002 (1955).
[14] E. Cherbuliez, G. Weber and J. Rabinowitz, *Helv. Chim. Acta*, **46**, 2464 (1963).
[15] S. Akerfeldt, *J. Org. Chem.*, **29**, 493 (1964).
[16] A. Lapidot and D. Samuel, *J. Chem. Soc.*, 1931 (1964).
[17] D. C. Dittmer and O. B. Ramsay, *J. Org. Chem.*, **28**, 1268 (1963).
[18] J. D. Chanley and E. M. Gindler, *J. Amer. Chem. Soc.*, **75**, 4035 (1953).
[19] J. D. Chanley and E. Feageson, *J. Amer. Chem. Soc.*, **85**, 1181 (1963).
[20] A. J. Kirby and A. G. Varvoglis, *J. Amer. Chem. Soc.*, **89**, 415 (1967).
[21] J. A. Maynard and J. M. Swann, *Proc. Chem. Soc., London*, 1394 (1963).
[22] F. Westheimer and G. Kenyon, *J. Amer. Chem. Soc.*, **88**, 3557, 3561 (1966).
[23] K. A. Holbrook and L. Ouellet, *Can. J. Chem.*, **35**, 1496 (1957).
[24] J. D. McGilvary and J. P. Crowther, *Can. J. Chem.*, **32**, 174 (1954).
[25] P. W. Schneider and H. Brinzinger, *Helv. Chim. Acta*, **47**, 1717 (1964).
[26] M. Halmann and A. Lapidot, *J. Chem. Soc.*, 419 (1960).
[27] M. Halmann, A. Lapidot and D. Samuel, *J. Chem. Soc.*, 3158 (1961).
[28] V. M. Clark and S. G. Warren, *Nature (London)*, **199**, 657 (1963).
[29] D. F. Heath, *J. Chem. Soc.*, 3796 (1956).
[30] E. W. Crunden and R. F. Hudson, *J. Chem. Soc.*, 3591 (1962).
[31] P. S. Traylor and F. H. Westheimer, *J. Amer. Chem. Soc.*, **87**, 553 (1965).
[32] A. F. Gerrard and N. K. Hamer, *J. Chem. Soc. B*, 539 (1968).
[33] A. F. Gerrard and N. K. Hamer, *J. Chem. Soc. B*, 1122 (1967).
[34] A. F. Gerrard and N. K. Hamer, *J. Chem. Soc. B*, 369 (1969).
[35] R. F. Hudson and G. Moss, *J. Chem. Soc.*, 3599 (1962).
[36] H. Grunze and E. Thilo, *Angew. Chem.*, **70**, 73 (1958).
[37] W. Lange and R. Livingstone, *J. Amer. Chem. Soc.*, **72**, 1280 (1950).
[38] D. Samuel and B. L. Silver, in V. Gold (editor), *Advances in Physical Organic Chemistry*, Vol. 3, p. 123, Academic Press, New York, 1965.
[39] I. Dostrovsky and M. Halmann, *J. Chem. Soc.*, 1004 (1956).
[40] M. Halmann, *J. Chem. Soc.*, 305 (1959).
[41] L. Keay, *J. Org. Chem.*, **28**, 329 (1963).
[42] I. Dostrovsky and M. Halmann, *J. Chem. Soc.*, 511 (1953).
[43] D. F. Heath, *J. Chem. Soc.*, 3804 (1956).
[44] R. F. Hudson and L. Keay, *J. Chem. Soc.*, 1859 (1960).
[45] L. N. Devonshire and H. H. Rowley, *Inorg. Chem.*, **1**, 680 (1962).
[46] R. F. Hudson and G. Moss, *J. Chem. Soc.*, 1040 (1964).
[47] J. Epstein et al., *J. Amer. Chem. Soc.*, **86**, 3075 (1964).
[48] R. F. Hudson and L. Keay, *J. Chem. Soc.*, 2463 (1956).
[49] D. B. Coult and M. Green, *J. Chem. Soc.*, 5478 (1964).

[50] R. F. Hudson, *Ann. Chim.* (*Rome*), **53**, 47 (1963).
[51] J. A. Ketelaar, H. R. Gersmann and K. Koopmans, *Rec. Trav. Chim. Pays-Bas*, **71**, 1253 (1952).
[52] R. F. Hudson and L. Keay, *J. Chem. Soc.*, 3269 (1956).
[53] L. Larsson, *Sven. Kem. Tidskr.*, **70**, 405 (1958); in English.
[54] G. Asknes and J. Songstad, *Acta Chem. Scand.*, **19**, 893 (1965).
[55] E. Blumenthal and J. B. M. Herbert, *Trans. Faraday Soc.*, **41**, 611 (1945)
[56] K. B. Wiberg, *Chem. Rev.*, **55**, 713 (1955).
[57] C. G. Swain, R. F. W. Bader and E. R. Thornton, *Tetrahedron*, **10**, 182, 200 (1960).
[58] C. Zioudrou, *Tetrahedron*, **18**, 197 (1962).
[59] M. Halmann, A. Lapidot and D. Samuel, *J. Chem. Soc.*, 1299 (1963).
[60] L. Tenud, S. Farooq, J. Seibl and A. Eschenmoser, *Helv. Chim. Acta*, **53**, 2059 (1970).
[61] M. Green and R. F. Hudson, *Proc. Chem. Soc., London*, 307 (1962).
[62] J. Michalsky, M. Mikolajczyk and J. Omelanczuk, *Tetrahedron Lett.*, 1779 (1965).
[63] W. B. Farnham, K. Mislow, N. Mandel and J. Donohue, *Chem. Commun.*, 120 (1972).
[64] J. Donohue, N. Mandel, W. B. Farnham, R. K. Murray jr., K. Mislow and H. P. Benschop, *J. Amer. Chem. Soc.*, **93**, 3792 (1971).
[65] N. J. De'Ath, K. Ellis, D. J. H. Smith and S. Trippett, *Chem. Commun.*, 714 (1971).
[66] R. D. Cook, P. C. Turley, C. E. Diebart, A. H. Fierman and P. Haake, *J. Amer. Chem. Soc.*, **94**, 9260 (1972).
[67] G. O. Dudek and F. H. Westheimer, *J. Amer. Chem. Soc.*, **81**, 2641 (1959).
[68] A. L. Green and B. Saville, *J. Chem. Soc.*, 3887 (1965).
[69] F. Covitz and F. H. Westheimer, *J. Amer. Chem. Soc.*, **85**, 1773 (1963).
[70] J. Epstein, D. H. Rosenblatt and M. M. Demek, *J. Amer. Chem. Soc.*, **78**, 341 (1956).
[71] M. Gordon, V. A. Notaro and C. E. Griffin, *J. Amer. Chem. Soc.*, **86**, 1898 (1964).
[72] R. L. Gustavson, S. Chaberek jr. and A. E. Martell, *J. Amer. Chem. Soc.*, **85**, 598 (1963).
[73] W. W. Butcher and F. H. Westheimer, *J. Amer. Chem. Soc.*, **77**, 2420 (1955).
[74] K. Dimroth, A. Witzel, W. Hülsen and H. Mirback, *Annalen*, **620**, 94 (1959).
[75] F. McC.Blewett and P. Watts, *J. Chem. Soc. B*, 881 (1971).
[76] J. O. Edwards and R. G. Pearson, *J. Amer. Chem. Soc.*, **84**, 16 (1962).
[77] I. Dostrovsky and M. Halmann, *J. Chem. Soc.*, 502 (1953).
[78] Cz. Borecki, J. Michalski and St. Musierowicz, *J. Chem. Soc.*, 4081 (1958).
[79] G. Klopman, K. Tsuda, J. B. Louis and R. E. Davis, *Tetrahedron*, **26**, 4549 (1970).
[80] A. L. Green, G. L. Sainsbury and M. Stansfield, *J. Chem. Soc.*, 1583 (1958).
[81] H. S. Mason and A. R. Todd, *J. Chem. Soc.*, 2267 (1951).
[82] J. Kumamoto, J. R. Cox jr. and F. H. Westheimer, *J. Amer. Chem. Soc.*, **78**, 4858 (1956).
[83] F. Covitz and F. H. Westheimer, *J. Amer. Chem. Soc.*, **85**, 1773 (1963).
[84] H. G. Khorana, G. M. Tener, R. S. Wright and J. G. Moffatt, *J. Amer. Chem. Soc.*, **79**, 430 (1957).
[85] E. Cherbuliez, H. Probst and J. Rabinowitz, *Helv. Chim. Acta*, **42**, 1377 (1959).
[86] P. C. Haake and F. H. Westheimer, *J. Amer. Chem. Soc.*, **83**, 1102 (1961).
[87] F. Ramirez, O. P. Madan, N. B. Desai, S. Mayerson and E. M. Banas, *J. Amer. Chem. Soc.*, **85**, 2681 (1963).
[88] E. A. Dennis and F. H. Westheimer, *J. Amer. Chem. Soc.*, **88**, 3431, 3432 (1966).
[89] G. Asknes and K. Bergesen, *Acta Chem. Scand.*, **20**, 2508 (1966).
[90] D. A. Usher, E. A. Dennis and F. H. Westheimer, *J. Amer. Chem. Soc.*, **87**, 2320 (1965).
[91] T. A. Steitz and W. N. Lipscomb, *J. Amer. Chem. Soc.*, **87**, 2488 (1965).

[92] E. T. Kaiser, M. Panar and F. H. Westheimer, *J. Amer. Chem. Soc.*, **85**, 602 (1963).

[93] W. C. Archie jr. and F. H. Westheimer, *J. Amer. Chem. Soc.*, **95**, 5955 (1973).

[94] F. Ramirez, K. Tasaka and R. Hershberg, *Phosphorus*, **2**, 41 (1972).

[95] P. Gillespie, F. Ramirez, I. Ugi and D. Marquarding, *Angew. Chem., Int. Ed. Eng.*, **12**, 109 (1973).

[96] V. M. Clark, D. W. Hutchinson, A. J. Kirby and S. G. Warren, *Angew. Chem., Int. Ed. Eng.*, **3**, 678 (1964).

[97] D. M. Brown, in R. A. Raphael, E. C. Taylor and H. Wynberg (editors), *Advances in Organic Chemistry*, Vol. 3, p. 75, Wiley Interscience, New York, 1963.

[98] H. G. Khorana, *Some Recent Developments in the Chemistry of Phosphate Esters of Biological Interest*, Wiley, New York, 1961.

[99] N. Kreutzkamp and H. Kayser, *Chem. Ber.*, **89**, 1614 (1956).

[100] C. Zioudrou, *Tetrahedron*, **18**, 197 (1962).

[101] G. Di Sabato and W. P. Jencks, *J. Amer. Chem. Soc.*, **81**, 4660 (1961).

[102] L. Goldman, J. W. Marsico and G. W. Anderson, *J. Amer. Chem. Soc.*, **82**, 2969 (1960).

[103] C. A. Bunton, D. R. Llewellyn, K. G. Oldham and C. A. Vernon, *J. Chem. Soc.*, 3588 (1958).

[104] H. G. Khorana and A. R. Todd, *J. Chem. Soc.*, 2257 (1953).

[105] C. D. Hall, *J. Chem. Soc. B*, 708 (1968).

[106] V. M. Clark, D. W. Hutchinson, G. W. Kirby and A. R. Todd, *J. Chem. Soc.*, 715 (1961).

[107] E. Bergmann and A. Bondi, *Chem. Ber.*, **66**, 278 (1933).

[108] D. G. Thomas, J. H. Bellmann and C. F. Davies, *J. Amer. Chem. Soc.*, **68**, 895 (1946).

[109] V. M. Clark and A. R. Todd, *J. Chem. Soc.*, 2030 (1950).

[110] B. A. Arbuzov and D. K. Jarmukhametova, *Izv. Akad. Nauk SSSR*, 1061 (1958); *Chem. Abs.*, **53**, 3046 (1959).

[111] F. W. Hoffmann and T. R. Moore, *J. Amer. Chem. Soc.*, **80**, 1150 (1958).

[112] F. R. Atherton, H. T. Howard and A. R. Todd, *J. Chem. Soc.*, 1106 (1948).

[113] P. G. Le Grâs, R. L. Dyer, P. J. Clifford and C. D. Hall, *J. Chem. Soc., Perkin II*, 2064 (1973).

Chapter 9
Phosphorus radicals

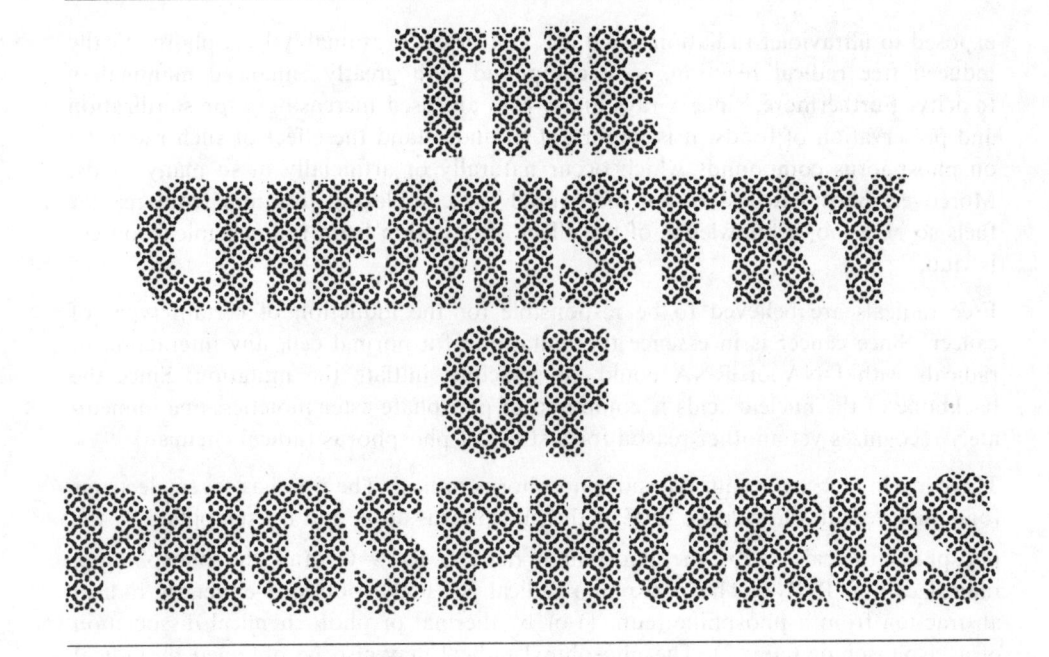

9.1 Introduction

Phosphorus radicals[1-8] are steeped in the history of phosphorus chemistry. They were almost certainly involved in Robert Boyle's observation in 1681 that the glow accompanying the oxidation of phosphorus in air was not evident with solutions of phosphorus in turpentine[9]. A similar quenching effect was reported by Graham[10] and later studies in this area provided an important base for the theory of branching chain reactions developed by Semenov[11]. The first proposal of a chain reaction involving radicals with an unpaired electron centred on phosphorus came, however, from Kharasch through his studies of the reaction of phosphorus trichloride with 1-octene in the presence of acetyl peroxide[12]. This was followed by a great deal of pioneering work in which phosphorus radicals were frequently proposed as transient (but unobservable) intermediates in many reactions[8] until, almost twenty-five years after Kharasch's proposals, Kochi actually "saw" a phosphorus radical by recording the electron spin resonance spectrum of $Me_3\dot{P}OBu^t$.[13]

There is still much to be learned in this fascinating but complex area of phosphorus chemistry. An example of the need for such knowledge can be seen from the fact that Parathion (9.1), a widely used insecticide, is relatively harmless to mammals until

$$(EtO)_2\overset{\overset{S}{\|}}{P}-O-\underset{(9.1)}{\underset{}{\bigcirc}}-NO_2$$

exposed to ultraviolet radiation when it is converted, presumably by a photolytically induced free radical reaction, to a compound with greatly enhanced mammalian toxicity. Furthermore, since x-rays and γ-rays are used increasingly for sterilization and preservation of foods, it is important to understand the effect of such radiation on phosphorus compounds which occur naturally or artificially in so many foods. Moreover, phosphate esters are often used for complexing uranium from reactor fuels so here too, a knowledge of the effect of radiation from radiochemical sources is vital.

Free radicals are believed to be responsible for the induction of certain types of cancer. Since cancer is in essence the mutation of a normal cell, any interaction of radicals with DNA or RNA could, in principle, initiate the mutation. Since the backbone of the nucleic acids is composed of phosphate ester moieties, one immediately recognizes yet another reason for a study of phosphorus radical chemistry.

There are in fact, two main types of phosphorus radical. The first has seven electrons (one unpaired) in the valence shell of the phosphorus atom and is exemplified by the phosphino radical, $R_2\ddot{\dot{P}}\cdot$, the phosphinyl radical, $R_2\dot{P}=O$, and the phosphinium radical cation, $R_3\dot{P}^+_\bullet$. The phosphino radical is usually obtained either by radical abstraction from a phosphine (eqn. 1) or by thermal or photochemical dissociation of a bisphosphine (eqn. 2). The phosphinyl radical may also be obtained by radical abstraction (eqn. 3) but the phosphinium radical cation is normally derived from a one-electron oxidation of a phosphine (eqn. 4).

$$R_2PH + R^1\cdot \longrightarrow R_2P\cdot + R^1H \qquad (1)$$

9.1 Introduction

$$R_2P-PR_2 \xrightarrow{\text{heat or } h\nu} 2R_2P\cdot \qquad (2)$$

$$\underset{\|}{\overset{O}{R_2PH}} + R\cdot \longrightarrow \underset{\|}{\overset{O}{R_2P\cdot}} + RH \qquad (3)$$

$$R_3P \xrightarrow{-e} R_3\overset{+}{P}\cdot \qquad (4)$$

Phosphino radicals are relatively stable, – $Ph_2P\cdot$ has a half-life of 20 mins at 173 K – and undergo one of two types of reaction, addition (eqn. 5) or radical abstraction (eqn. 6).

$$R_2\ddot{P}\cdot + X=Y \longrightarrow R_2P-X-\dot{Y} \qquad (5)$$

$$R_2\ddot{P}\cdot + X-Y \longrightarrow R_2PX + Y\cdot \qquad (6)$$

Since there is a certain degree of double bond character in the phosphorus–oxygen bond, phosphinyl radicals could be regarded as nine-electron radicals of the phosphoranyl type. It is now common practice however, to classify them with the phosphino radical since they show a similar type of reactivity. To be precise, they should be denoted as resonance hybrids, e.g. (9.2a,b,c) but are more conveniently represented by (9.2b).

$$\underset{(9.2a)}{R_2\overset{+\,\cdot}{P}-\overset{-}{O}} \longleftrightarrow \underset{(9.2b)}{R_2\dot{P}=O} \longleftrightarrow \underset{(9.2c)}{R_2\ddot{P}-\dot{O}}$$

The second radical class has nine electrons (one unpaired) around the phosphorus atom and is illustrated by the phosphoranyl radical, $R_4P\cdot$, or $(RO)_4P\cdot$ and by the phosphonium radical anion, $R_3P\cdot^-$. Phosphoranyl radicals may be prepared by radical addition to P(III) compounds (eqn. 7) or phosphoryl compounds (eqn. 8) or by the one-electron reduction of phosphonium salts (eqn. 9). In some special cases they may also be generated by photolysis or radical abstraction of phosphoranes containing the P—H bond (eqn. 10).

$$R_3P + R^1\cdot \longrightarrow R_3\dot{P}R^1 \qquad (7)$$

$$R_3P=O + R^1\cdot \longrightarrow R_3\dot{P}-OR^1 \qquad (8)$$

$$R_4\overset{+}{P}\,\overset{-}{X} \xrightarrow{+e} R_4P\cdot + \overset{-}{X} \qquad (9)$$

$$Ar_4P-H \xrightarrow{h\nu} Ar_4P\cdot \xleftarrow{R\cdot} Ar_4PH \qquad (10)$$

$$R_3P \xrightarrow{+e} R_3\overset{-}{P}\cdot \qquad (11)$$

The radical anions are generated by one-electron reduction of phosphines (eqn. 11) and are a rare species, even for phosphorus radicals.

The reactions of phosphoranyl radicals are more complex than the more stable phosphino or phosphinyl radicals but are generally of two types, α-scission (eqn. 12) which is the reverse of one preparative method (eqn. 7) or β-scission which involves the formation of either P=O or P=S (eqns. 13 and 14).

$R_3\dot{P}R^1 \longrightarrow R_3P + R^1\cdot$ (12)

$(RO)_3\dot{P}OR^1 \longrightarrow (RO)_3P=O + R^1\cdot$ (13)

$(RO)_3\dot{P}SR^1 \longrightarrow (RO)_3P=S + R^1\cdot$ (14)

The scene has so far been set in very general terms but from here we shall turn to specific examples and a discussion of the preparation and chemical reactions of phosphorus radicals. We shall also consider the physico-chemical and spectroscopic evidence for their existence and their structure.

Prior to this however, mention should be made of two other species which have attracted attention. The irradiation or thermolysis of the cyclic pentaphosphine (9.3) gives what is probably phenyl phosphinidene (9.4) a transient six-electron species which behaves somewhat like a carbene and will insert into a disulphide link to form a thiophosphonite (9.5) or react with benzil to form an oxyphosphorane (9.6)[14,15]. The phosphinidene is also produced from the reaction of phenylphosphonous dichloride with zinc.

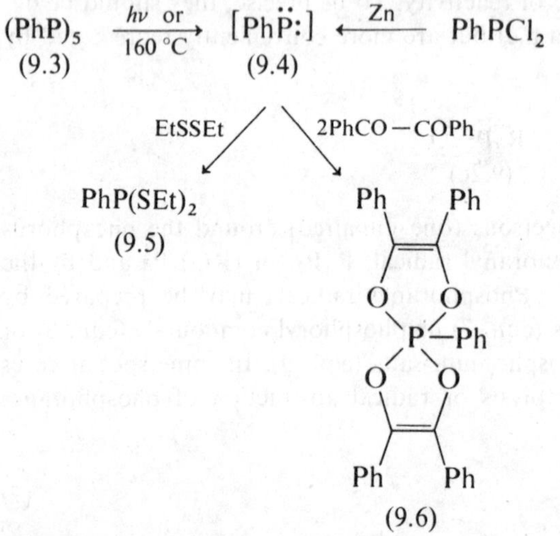

$(PhP)_5 \xrightarrow[160\,°C]{h\nu \text{ or }} [Ph\ddot{P}:] \xleftarrow{Zn} PhPCl_2$
(9.3)   (9.4)

EtSSEt ↙   ↘ 2PhCO—COPh

PhP(SEt)₂
(9.5)

(9.6) — oxyphosphorane structure with Ph groups

The species is commonly observed as its radical cation (RP⁺·) in the mass spectra of phosphines, e.g. PhP⁺· observed in the mass spectrum of $PhPH_2$ and Ph_2PH.[16]

Radicals are also known (e.g. 9.7, 9.8 and 9.9) in which the unpaired electron is not situated on phosphorus. In (9.7) and (9.8) the odd electron is largely confined to the oxygen or sulphur atoms with little spin density on phosphorus[17] and likewise there appears to be no delocalization of the odd electron from carbon into the triphenylphosphonium group of (9.9)[18]. Strictly, therefore, these species are not phosphorus radicals and they will not be discussed further.

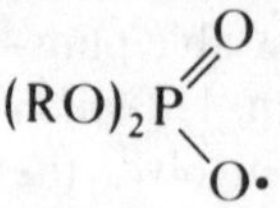

$(RO)_2P(=O)O\cdot$ $(RO)_2P(=S)S\cdot$ $Ph_3\overset{+}{P}-\dot{C}Ph_2$
(9.7)   (9.8)   (9.9)

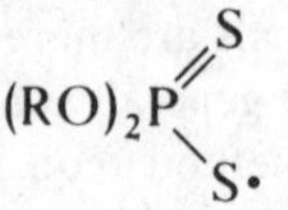

9.2 Seven-electron phosphorus radicals

9.2.1 The phosphino radical

Phosphino radicals are produced by photolysis of phosphines (PH_3, RPH_2 or R_2PH) or bisphosphines (R_2PPR_2, e.g. eqn. 15) or by radical initiated hydrogen atom abstraction from PH_3, RPH_2 or R_2PH (e.g. eqn. 16).

$$Ph_2P-PPh_2 \xrightarrow{h\nu} 2Ph_2P\cdot \qquad (15)$$

$$Ph_2PH \xrightarrow{R\cdot} Ph_2P\cdot + RH \qquad (16)$$

The abstracting radical of eqn. (16) is normally obtained from thermal decomposition of an initiator such as dibenzoyl peroxide (eqn. 17) or azo-bis-isobutyronitrile, (AIBN, eqn. 18).

$$PhCO.OO.COPh \xrightarrow{heat} 2PhCO_2\cdot \longrightarrow 2Ph\cdot + 2CO_2 \qquad (17)$$

$$\underset{\underset{Me}{|}}{NC-C}-N=N-\underset{\underset{Me}{|}}{\overset{\overset{Me}{|}}{C}}-CN \xrightarrow{heat} 2Me_2\dot{C}-CN + N_2 \qquad (18)$$

Once formed, the phosphino radicals either abstract hydrogen (the reverse of eqn.16) or add rapidly to an unsaturated system and the latter is the basis of the free radical addition of phosphines to olefins. The reaction is generalized by eqns. (19)–(21) and occurs with PH_3[19,20] and primary or secondary phosphines[19,20]. Compounds with more than one P—H bond may react with more than one mole of olefin and so PH_3 is capable of forming tertiary phosphines (eqn. 22)[19,20].

$$R_2PH \xrightarrow{h\nu \text{ or } R\cdot} R_2P\cdot \qquad \text{initiation} \qquad (19)$$

$$R_2P\cdot + CH_2=CHR^1 \longrightarrow R_2PCH_2-\dot{C}HR^1 \qquad (20)$$

$$R_2P\dot{C}H_2-\dot{C}HR^1 + R_2PH \longrightarrow R_2PCH_2CH_2R^1 + R_2P\cdot \qquad \Big\} \text{propagation} \qquad (21)$$

$$PH_3 + 3RCH=CH_2 \xrightarrow{AIBN} (RCH_2CH_2)_3P \qquad (22)$$

Notice that, as expected for a radical reaction, the phosphorus radical adds to the least substituted carbon to give the most highly substituted carbon radical (eqn. 20). Furthermore, the addition step is apparently reversible since during the radical-catalyzed addition to cis-butene, trans-butene is produced[21,22] (eqn. 23).

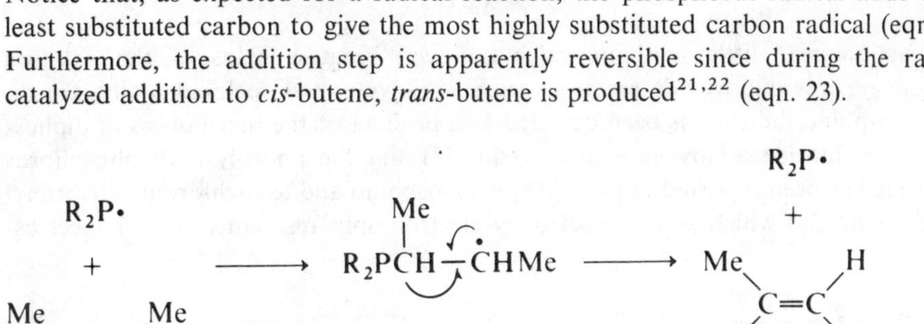

$$R_2P\cdot + \underset{H}{\overset{Me}{\underset{|}{C}}}=\underset{H}{\overset{Me}{\underset{|}{C}}} \longrightarrow R_2PCH-\dot{C}HMe \longrightarrow \underset{H}{\overset{Me}{C}}=\underset{Me}{\overset{H}{C}} + R_2P\cdot \qquad (23)$$

The reaction of dienes with tetramethyl bisphosphine affords 1,4-diphosphines (9.10 and 9.11) and the lack of stereospecificity in this reaction suggests that the dimethylphosphino radical, Me$_2$P•, is indeed, an intermediate[23].

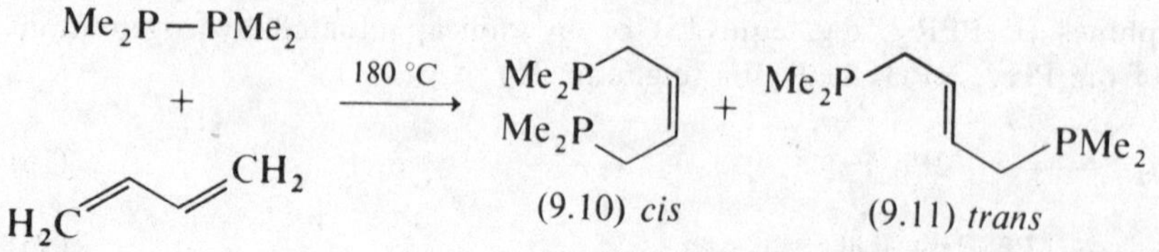

$$Me_2P-PMe_2 + H_2C=CH-CH=CH_2 \xrightarrow{180\,°C} \underset{(9.10)\ cis}{Me_2P-CH_2-CH=CH-CH_2-PMe_2\ (cyclic\ cis)} + \underset{(9.11)\ trans}{Me_2P-CH_2-CH=CH-CH_2-PMe_2}$$

Reaction of phosphino radicals with isonitriles (9.12) however, takes more than one course. Path A yields the iminophosphine (9.14) by the expected hydrogen abstraction but path B yields diethylcyanophosphine (9.15) by β-scission of the intermediate radical (9.13)[24].

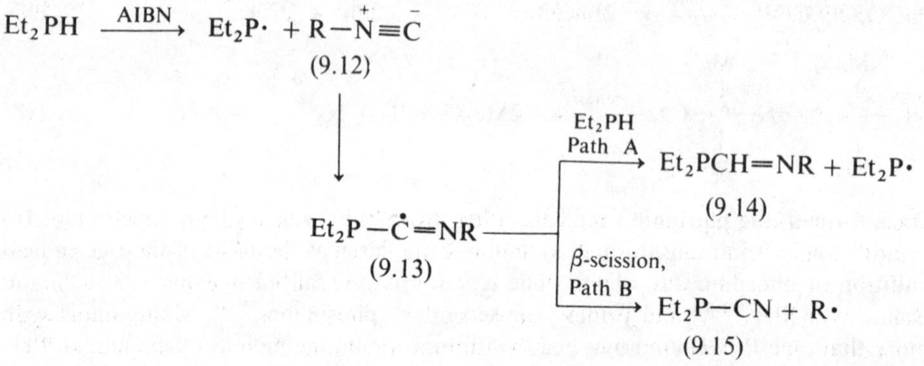

$$Et_2PH \xrightarrow{AIBN} Et_2P• + R-\overset{+}{N}\equiv\overset{-}{C}$$
(9.12)

$$\downarrow$$

$$Et_2P-\overset{•}{C}=NR$$
(9.13)

Path A (Et$_2$PH): $\longrightarrow Et_2PCH=NR + Et_2P•$ (9.14)

β-scission, Path B: $\longrightarrow Et_2P-CN + R•$ (9.15)

In path A, Et$_2$P• is the chain carrier; in path B, the radical R• serves in this capacity.

Analogous reactions occur with phosphorus halides to yield phosphonous chlorides (9.16) probably by the route shown in eqns. (24)–(26).

$$PCl_3 + R• \longrightarrow •PCl_2 + RCl \qquad\qquad \text{initiation} \qquad (24)$$

$$Cl_2P• + CH_2=CHR^1 \longrightarrow Cl_2P-CH_2-\overset{•}{C}HR^1 \qquad\qquad (25)$$

$$Cl_2PCH_2-\overset{•}{C}HR^1 + PCl_3 \longrightarrow Cl_2PCH_2CHClR^1 + •PCl_2 \qquad (26)$$

$\Big\}$ propagation

(9.16)

The reaction is of little synthetic value however, because yields of the addition products are low due to side reactions such as polymerization. Incidentally, the difluorophosphino radical has been detected as a product of the thermolysis of diphosphorus tetrafluoride at low pressure[25] (eqn. 27) and the photolysis of phosphorus trichloride has been reported to give dichlorophosphino and tetrachlorophosphoranyl radicals (eqn. 28) which were detected by electron spin resonance (e.s.r.) spectroscopy[26].

$$F_2P-PF_2 \xrightarrow[\text{low pressure}]{\text{heat}} 2F_2P• \qquad\qquad (27)$$

$$2PCl_3 \xrightarrow{h\nu} Cl_2P• + Cl_4P• \qquad\qquad (28)$$

9.2 Seven-electron phosphorus radicals

A synthetically more useful reaction is that known as oxidative chlorophosphonation in which phosphorus trichloride reacts with hydrocarbons in the presence of oxygen to give phosphonyl chlorides (eqn. 29)[27-29].

$$RH + 2PCl_3 + O_2 \longrightarrow RP(O)Cl_2 + POCl_3 + HCl \qquad (29)$$

In the presence of excess phosphorus compound up to 60 % conversions of alkane are possible and the reaction has been extended to olefins (eqn. 30)[30].

$$MeCH=CHMe + PCl_3 \xrightarrow{O_2} MeCHCl.CHMeP(O)Cl_2 \qquad (30)$$

The mechanisms are complex and not well understood but they undoubtedly involve radical chain reactions probably with chlorine atoms as the chain carriers in the reactions with saturated hydrocarbons (eqn. 29).

It is all very well to propose reactive intermediates such as the phosphino radical as a means of rationalizing reaction products but unless the transient species can be "seen" or trapped, most chemists remain sceptical. Fortunately the sophistication of modern spectroscopic techniques allows us to eliminate such doubts in the case of the elusive phosphino radical. Photolysis of triphenylphosphine, diphenylphosphine or tetraphenylbisphosphine gives the diphenylphosphino radical which has been trapped by t-nitrosobutane (9.17) – a well known radical scavenger. The resultant, relatively stable nitroxyl radical (9.18) was identified by e.s.r. spectroscopy[31,32].

$$\begin{array}{c} Ph_3P, Ph_2PH \\ \text{or} \quad Ph_2P-PPh_2 \end{array} \longrightarrow [Ph_2P\cdot] + t\text{-}BuN=O \longrightarrow Ph_2PN\overset{Bu^t}{\underset{O\cdot}{\diagdown}}$$
$$\qquad\qquad\qquad\qquad (9.17) \qquad\qquad\qquad\qquad (9.18)$$

The phosphino radical has also been trapped in benzene at 77 K and since low temperature extends the lifetime of the radical enormously, it is possible to observe the e.s.r. spectrum of the species directly[33]. Phosphino radicals are therefore well established in the library of transient species.

9.2.2 The phosphinyl radical

Phosphinyl radicals are generated by photolysis of, or radical abstraction from, dialkyl (or aryl)-phosphine oxides (9.20, eqn. 31) or O-alkyl alkylphosphinates (9.21, eqn. 32) or dialkylphosphonates (9.22, eqn. 33); thioanalogues of (9.21) or (9.22) behave in the same way.

$$\underset{(9.19a)}{\overset{+\cdot\;\;-}{R_2P-O}} \longleftrightarrow \underset{(9.19b)}{\overset{\cdot}{R_2P=O}} \longleftrightarrow \underset{(9.19c)}{\overset{\cdot\cdot}{R_2P}-O\cdot}$$

$$\underset{(9.20)}{\overset{O}{\underset{\|}{R_2PH}}} \xrightarrow{h\nu \text{ or } R\cdot} \overset{O}{\underset{\|}{R_2P\cdot}} \qquad (31)$$

$$\underset{(9.21)}{\overset{R^1O}{\underset{R^2}{\diagup}}\!\!P\!\!\overset{O}{\underset{H}{\diagdown}}} \xrightarrow{h\nu \text{ or } R\cdot} \overset{R^1O}{\underset{R^2}{\diagup}}\!\!P\!\!\overset{O}{\underset{\cdot}{\diagdown}} \qquad (32)$$

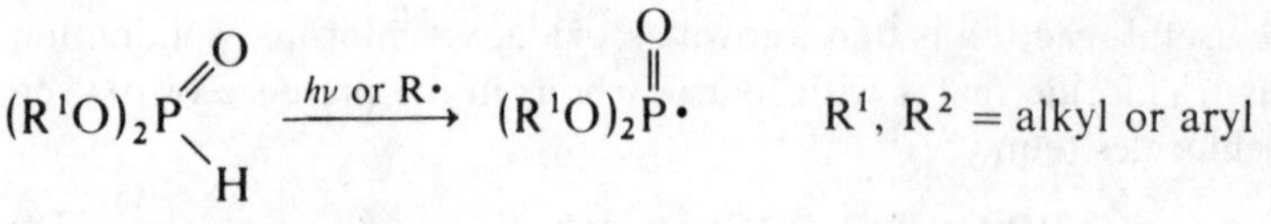

$$(R^1O)_2P(=O)H \xrightarrow{h\nu \text{ or } R\cdot} (R^1O)_2\overset{\cdot}{P}=O \qquad R^1, R^2 = \text{alkyl or aryl} \qquad (33)$$

(9.22)

To be formally correct the phosphinyl radical should be represented as a resonance hybrid of three canonical forms (9.19a,b,c) in one of which (9.19c) the unpaired electron is localized on oxygen. In fact, it behaves as a phosphorus radical of the seven-electron type and displays many of the characteristic reactions of the phosphino radical. For example the radical initiated reaction of dimethyl phosphite (9.23) with vinyl propionate (9.24) gives a 70 % yield of the phosphonate (9.25)[34].

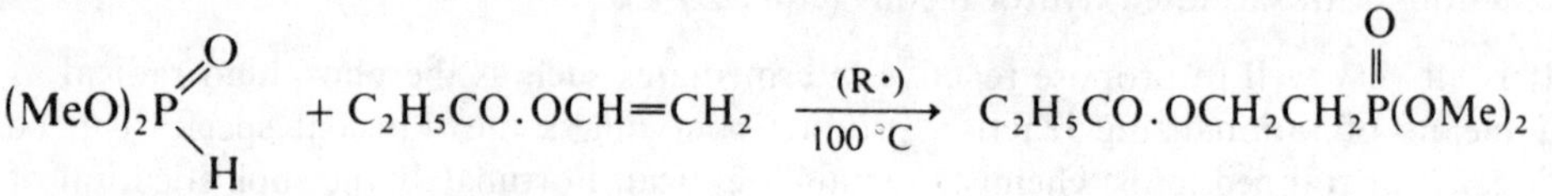

$$(MeO)_2P(=O)H + C_2H_5CO\cdot OCH=CH_2 \xrightarrow[100\,°C]{(R\cdot)} C_2H_5CO\cdot OCH_2CH_2P(=O)(OMe)_2$$

(9.23)   (9.24)   (9.25)

Similar additions to carbonyl compounds take place to give α-hydroxyphosphonates (9.26).

$$(RO)_2P(O)H + Me_2CO \xrightarrow{AIBN} (RO)_2\overset{O}{\underset{\|}{P}}-\overset{OH}{\underset{|}{C}}Me_2$$

(9.26)

The additions to olefins have been shown to occur with retention of configuration at phosphorus (eqns. 34 and 35)[35-38].

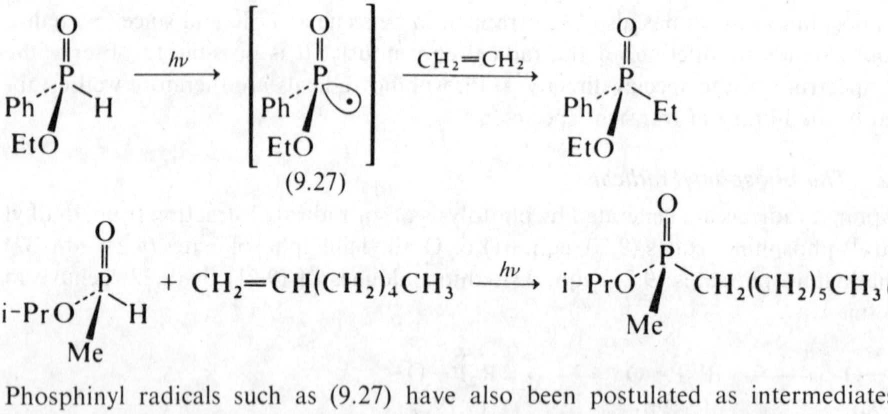

<chemical scheme: Ph, EtO, H, P=O → [phosphinyl radical (9.27)] →CH₂=CH₂→ Ph, EtO, Et, P=O>   (34)

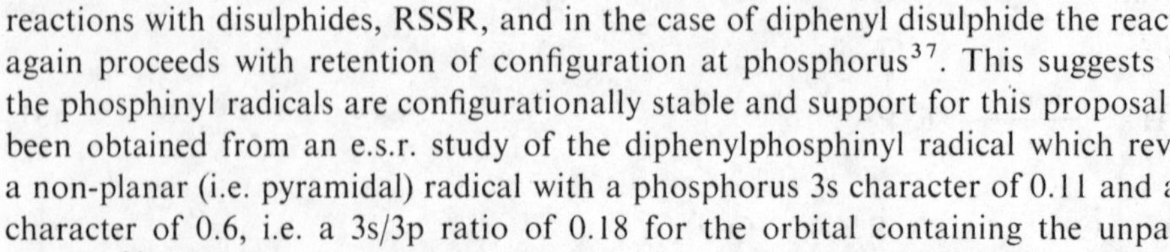

<chemical scheme: i-PrO, Me, H, P=O + CH₂=CH(CH₂)₄CH₃ →hν→ i-PrO, Me, CH₂(CH₂)₅CH₃, P=O>   (35)

Phosphinyl radicals such as (9.27) have also been postulated as intermediates in reactions with disulphides, RSSR, and in the case of diphenyl disulphide the reaction again proceeds with retention of configuration at phosphorus[37]. This suggests that the phosphinyl radicals are configurationally stable and support for this proposal has been obtained from an e.s.r. study of the diphenylphosphinyl radical which reveals a non-planar (i.e. pyramidal) radical with a phosphorus 3s character of 0.11 and a 3p character of 0.6, i.e. a 3s/3p ratio of 0.18 for the orbital containing the unpaired electron[39]. For a C_{3v} system, AB_3, an s/p ratio of 0.18 implies a BAB angle of approximately 113°. This should be compared with an s/p ratio of 0.31 for the $\overset{\cdot}{P}O_3^{2-}$ species[40] (close to tetrahedral) and the analogous diphenyl nitroxide radical, $Ph_2\overset{\cdot}{N}=O$, which appears to be planar[41].

9.2 Seven-electron phosphorus radicals

Like the phosphino radical, the phosphinyl radical has also been trapped by radical scavengers such as t-nitrosobutane and the resultant, relatively stable nitroso radicals (9.28), studied by e.s.r. (eqn. 36)[31].

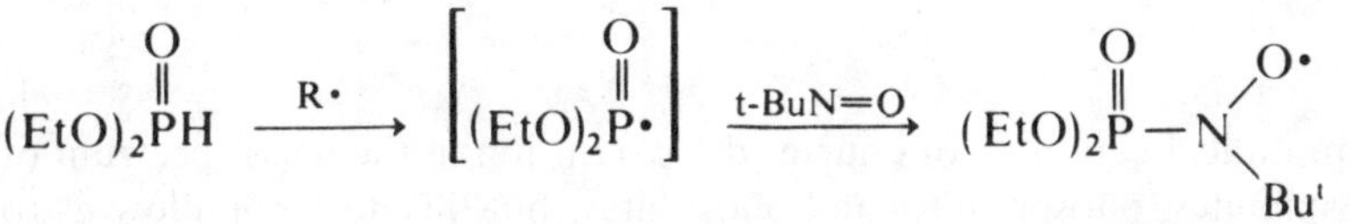

$$(EtO)_2\overset{O}{\overset{\|}{P}}H \xrightarrow{R\cdot} \left[(EtO)_2\overset{O}{\overset{\|}{P}}\cdot\right] \xrightarrow{\text{t-BuN=O}} (EtO)_2\overset{O}{\overset{\|}{P}}-N\overset{O\cdot}{\underset{Bu^t}{\diagdown}} \quad (36)$$

(9.28)

9.2.3 The phosphinium radical cation

The reactions of phosphines with *p*-benzoquinones are characterized by brilliant red colours which appear in the early stages of the reactions. This, coupled with the knowledge that radicals were also involved, led to the proposal of phosphinium radical cations as the coloured intermediates[42]. In view of the well known stability of radical cations (9.29) – the Würster salts, derived from the one-electron oxidation of *p*-phenylenediamines[43] – this seemed an eminently reasonable proposal.

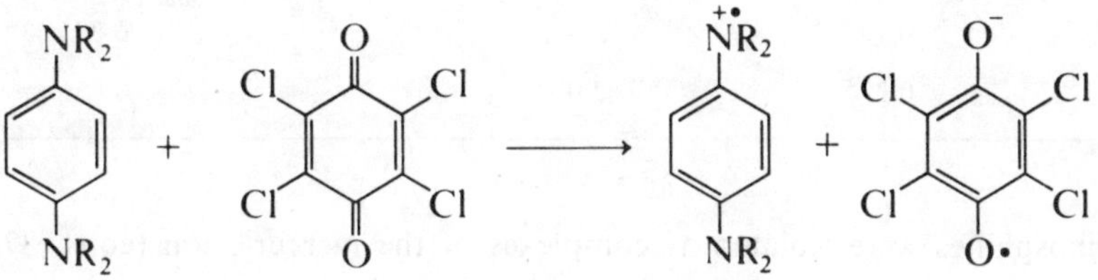

(9.29)

Subsequently however, Lucken showed that the betaine (9.30) obtained from triphenylphosphine and *p*-benzoquinone could be oxidized to a radical cation (9.31) and that the e.s.r. spectrum of this radical cation was similar to that obtained from the *initial* mixture of triphenylphosphine and *p*-benzoquinone[44]. This led to a reinvestigation

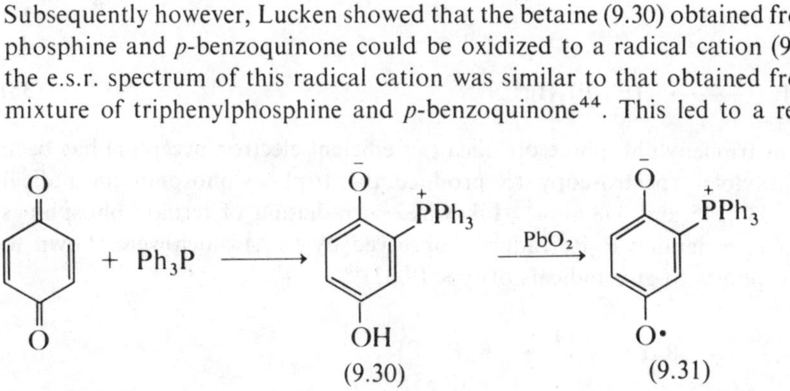

(9.30) (9.31)

of the triphenylphosphine/chloranil reaction and in 1966, Ramirez and Lucken announced in a joint paper that the radical observed by e.s.r. spectroscopy was indeed a phosphobetaine radical but of type (9.32)[45]. The critical piece of evidence was that the *g*-factors and hyperfine splitting constants of radicals derived from triphenylphosphine and chloranil, bromanil (X = Br) or iodanil (X = I) were *all different*; a common phosphinium radical cation, $Ph_3P^{+\cdot}$ would have given the same spectrum in each case.

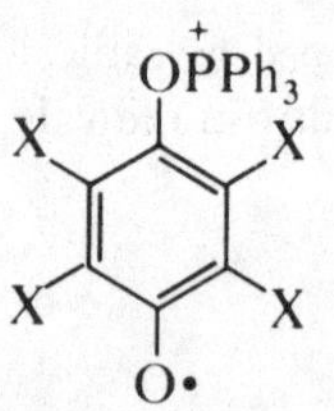

OPPh₃ (on benzene ring with 4 X substituents and O•)   X=Cl, Br or I   (9.32)

The phosphinium radical cation is, of course, the parent ion in the mass spectrum of phosphines, phosphinites, phosphonites and phosphites, but this does not allow us to characterize the species in a chemical sense. It has also been observed by oxidation of phosphines at a mercury anode (with CH_3CN as solvent) and half-wave and oxidation potentials have been reported as a result of this work (Table 9.1)[46].

Table 9.1 Half-wave, $E_{\frac{1}{2}}$ (V), and anodic oxidation, E_A (V) potentials of phosphines

	$E_{\frac{1}{2}}$ (volts)	Current density (amps)	E_A (volts)
Ph_3P	0.12	3.3	0.25
$(p\text{-}MeC_6H_4)_3P$	0.10	2.4	0.30
$(p\text{-}MeOC_6H_4)_3P$	0.29	2.0	0.35

The oxidized phosphines were isolated as complexes of the mercuric ion (eqns. 37 and 38).

$$Ph_3P \xrightarrow{-e} Ph_3\overset{+}{P}\cdot \qquad (37)$$

$$2Ph_3\overset{+}{P}\cdot + Hg \longrightarrow (Ph_3P)_2Hg^{2+} \qquad (38)$$

γ-Irradiation of triphenylphosphine on silica (an efficient electron acceptor) has been shown by ultraviolet spectroscopy to produce the triphenylphosphinium radical cation, λ_{max} = 330, 345 and 348 mm.[47] Likewise, γ-irradiation of tertiary phosphines in sulphuric acid generated $R_3P^{+\cdot}$ radicals (observed by e.s.r.) which were shown to react with phosphines to give radicals of type (9.33)[48].

$$R_3P \xrightarrow{\gamma\text{-rays in } H_2SO_4} R_3\overset{+}{P}\cdot \xrightarrow{R_3P} R_3\overset{\cdot}{P}-\overset{+}{P}R_3$$
(9.33)

Chemical evidence for the existence of the phosphinium radical cation has been obtained from the reactions of optically active phosphines with electron acceptor molecules such as tetracyanoethylene, TCNE (9.34) and tetracyanoquinodimethane, TCNQ (9.36). Tertiary phosphines react with TCNE or TCNQ in acetonitrile/water solvent and in the presence of acid to give virtually quantitative yields of phosphine oxide and reduced acceptor (9.35 or 9.37).

9.2 Seven-electron phosphorus radicals

$$Ar_3P + (CN)_2C=C(CN)_2 + H_2O \xrightarrow[\text{in } H_2O/CH_3CN]{H^+} Ar_3PO + (CN)_2CH.CH(CN)_2$$
(9.34) (9.35)

$$Ar_3P + (CN)_2C=\underset{(9.36)}{\underline{\bigcirc}}=C(CN)_2 + H_2O \xrightarrow[\text{in } H_2O/CH_3CN]{H^+}$$

$$Ar_3PO + (CN)_2CH-\bigcirc-CH(CN)_2$$
(9.37)

With an optically active phosphine (e.g. biphenyl-α-naphthylphenylphosphine) the product oxide is racemic and this suggests the intermediacy of a planar or rapidly inverting (i.e. configurationally unstable) phosphinium radical cation as an essential intermediate[49]. Support for this proposal was obtained from a kinetic study with TCNQ which showed: (i) first-order behaviour in TCNQ, H$^+$ and Ar$_3$P for a wide range of phosphines; (ii) that the rate was independent of [H$_2$O] at low concentrations of water in CH$_3$CN; (iii) low values of E_a (± 4 kJ mol^{-1}, ± 1 kcal mole^{-1}) but high negative values of $\Delta S^{\ddagger}$ (ca -170 J K^{-1} mol^{-1}, -40 cal K^{-1} mole^{-1}) indicating a high degree of orientation in or before the rate-determining step; and (iv) k_H/k_D values (H$^+$ in H$_2$O vs D$^+$ in D$_2$O) of ca 2.0, suggesting proton transfer was involved in the rate-limiting step.

The data led to the proposal of the mechanism outlined in eqns. (39)–(41) with eqn. (40) as the rate-determining step followed by a rapid hydrolysis of the radical cation (eqn. 41), which can be conveniently summarized as the three steps of eqn. (42)[50].

$$Ar_3P + TCNQ \rightleftharpoons \underset{\text{molecular complex}}{Ar_3\overset{\delta+}{P} \longrightarrow \overset{\delta-}{TCNQ}} \quad (39)$$

$$\overset{\delta+}{Ar_3P} \longrightarrow \overset{\delta-}{TCNQ} + H^+ \longrightarrow Ar_3\overset{+}{P}\cdot + TCNQH\cdot \quad (40)$$

$$Ar_3\overset{+}{P}\cdot + TCNQH\cdot + H_2O \longrightarrow Ar_3P=O + TCNQH_2 \quad (41)$$

$$\begin{array}{l}Ar_3\overset{+}{P}\cdot + OH_2 \longrightarrow [Ar_3\overset{+}{\dot{P}}-OH_2] \longrightarrow [Ar_3\dot{P}OH] + H^+ \\ (+TCNQH\cdot) \quad\quad\quad\quad (+TCNQH\cdot) \quad\quad\quad (+TCNQH\cdot)\end{array}$$

$$\longrightarrow Ar_3PO + TCNQH_2 \quad (42)$$

With a variety of substituted phenylphosphines, XC$_6$H$_4$PPh$_2$, a Hammett plot revealed a correlation with σ (and not σ^+) with a ρ-value of -3.2. Thus the reaction is accelerated by electron-donating substituents (negative ρ) and the correlation with σ rather than σ^+ denotes a *pyramidal* rather than a planar radical cation, since the latter would presumably have allowed conjugative delocalization of the positive charge on phosphorus (e.g. 9.38 ↔ 9.39).

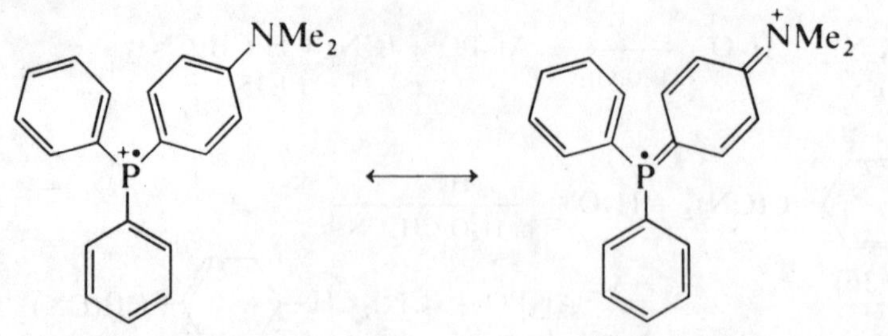

Chapter 9 / Phosphorus radicals

(9.38)   ⟷   (9.39)

The ρ-value of -3.2 is reminiscent of that observed for the quaternization of aromatic amines (eqn. 43) for which $\rho = -2.8$.[51]

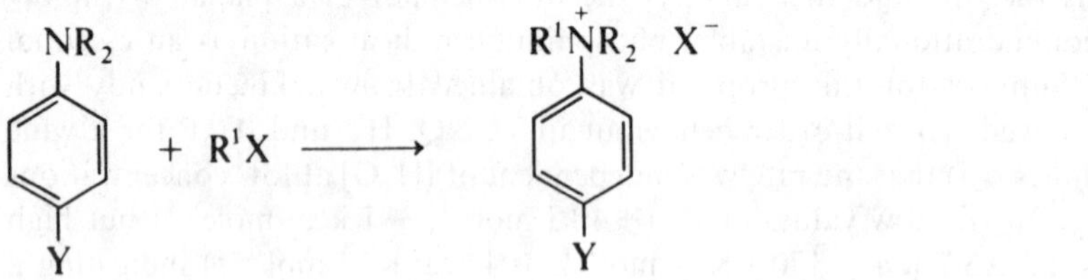

(43)

This also suggests that the radical has a pyramidal configuration similar to that of tertiary amines, i.e. bond angles of ca 108° and like tertiary amines, the barrier to inversion is low and of the order of 21 kJ mol^{-1} (5 kcal mole^{-1})[52].

A recent theoretical study of the ground state geometries of PH_3 and PF_3 and their ground ionic states, $\overset{+}{\cdot}PH_3$ and $\overset{+}{\cdot}PF_3$, lends substantial support to this suggestion[53]. The calculations, using *ab initio* SCFMO wavefunctions, finds that all four species are pyramidal and that for $\overset{+}{\cdot}PH_3$, provided d-functions are included in the basis set, the angle between a P—H bond and the singly occupied orbital is ca 106°. Furthermore, the maximum barrier to inversion in $\overset{+}{\cdot}PH_3$ is estimated to be about 21 kJ mol^{-1} (5 kcal mole^{-1}).

Finally, mention should be made of the reaction of phosphabenzoles (e.g. 9.40) with mercuric acetate which gives the 1,1-diacetoxyphosphabenzoles (9.42); e.s.r. studies indicate that these reactions occur via cation radicals such as (9.41)[54].

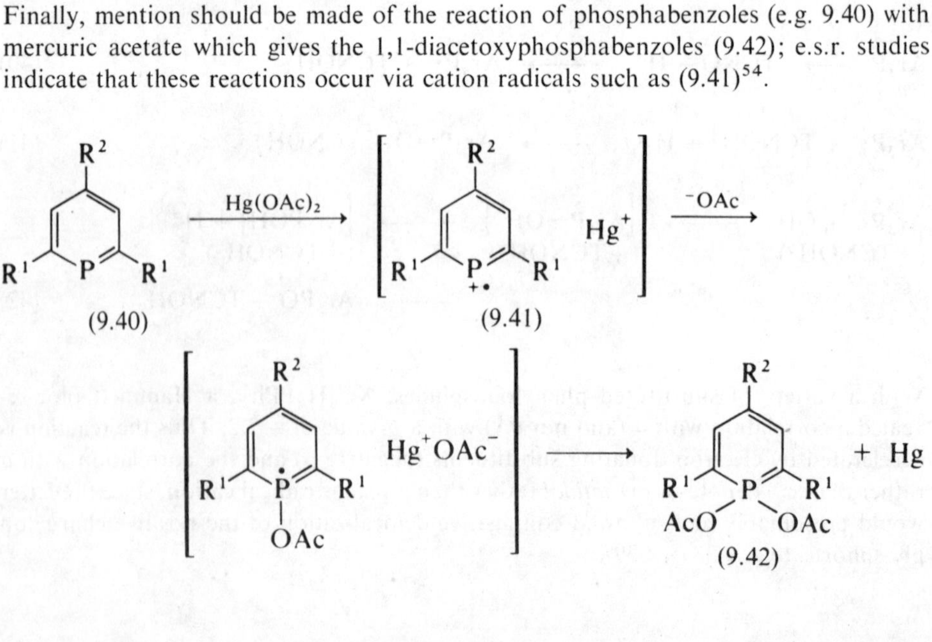

(9.40) → (9.41) → (9.42) + Hg

9.3 Nine-electron phosphorus radicals

Likewise, the phosphabenzole (9.43) has been shown to react with phenoxy or diphenylamino radicals to give (9.44) and (9.45) respectively, each of which may be oxidized electrolytically to radical cations (9.46 and 9.47) whose e.s.r. spectra indicate

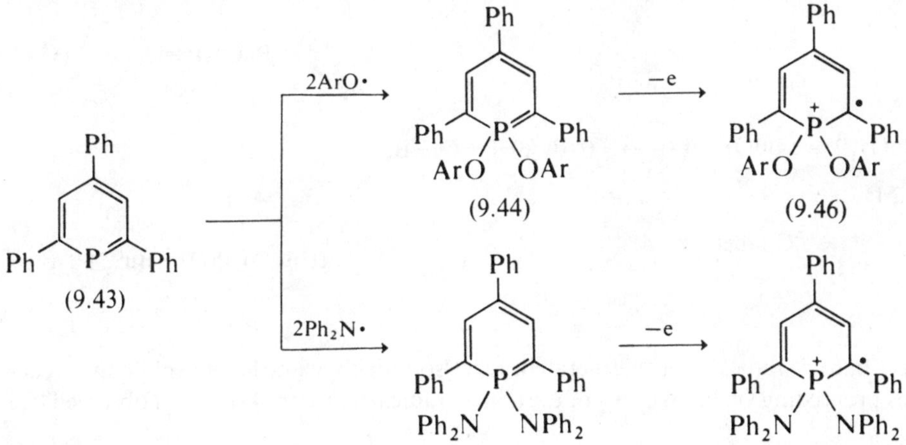

that the unpaired electron is limited to the aromatic system[55]. These are not phosphinium radical cations, however, since in all cases there are a minimum of four formal bonds to phosphorus.

9.3 Nine-electron phosphorus radicals

9.3.1 The phosphoranyl radical[56,57] ($R_4P\cdot$)

Nucleophilic and biphilic displacement reactions between trivalent phosphorus compounds and peroxides have been discussed in Chapter 4. Both phosphines and phosphites will also react with peroxides by a radical mechanism. For instance, triethyl phosphite and di-t-butyl peroxide react at 100 °C to give high yields of triethyl phosphate (9.48) and some 2,2,3,3-tetramethylbutane (9.49)[58] plus other hydrocarbon products of disproportionation and breakdown of the intermediate t-butyl radicals.

$$(EtO)_3P + Bu^tO\text{—}OBu^t \xrightarrow{100\,°C} (EtO)_3PO + Me_3C\text{—}CMe_3$$
$$\qquad\qquad\qquad\qquad\qquad\qquad (9.48)\qquad (9.49)$$

The mechanism of this reaction involves phosphoranyl radicals (9.50) which undergo β-scission to phosphate (eqns. 44–47).

$$Bu^tO\text{—}OBu^t \xrightarrow{heat} 2\,t\text{-}BuO\cdot \qquad (44)$$

$$t\text{-}BuO\cdot + P(OEt)_3 \longrightarrow t\text{-}Bu\dot{O}P(OEt)_3 \qquad (45)$$
$$\qquad\qquad\qquad\qquad\qquad (9.50)$$

$$t\text{-}Bu\dot{O}P(OEt)_3 \xrightarrow{\beta\text{-scission}} t\text{-}Bu\cdot + O{=}P(OEt)_3 \qquad (46)$$

$$2\,t\text{-}Bu\cdot \longrightarrow Me_3C\text{—}CMe_3 \qquad (47)$$

The combination of t-butoxy radicals with the phosphite has been shown to be irreversible since reaction of ^{14}C labelled t-butoxy radicals with tri-t-butyl phosphite (9.51) gives phosphate containing 75 % of the original label[59],† as expected from a statistical cleavage of the intermediate radical.

$$(t\text{-BuO})_3P + t\text{-}\overset{*}{B}uO\cdot \longrightarrow (t\text{-BuO})_3\overset{\cdot}{P}\text{-O-}\overset{*}{B}u^t \begin{array}{c} \nearrow (t\text{-BuO})_3P=O + t\text{-}\overset{*}{B}u\cdot \\ 25\% \\ \searrow (t\text{-BuO})_2P(O)\text{-}\overset{*}{B}u^t + t\text{-Bu}\cdot \\ 75\% \end{array}$$

(9.51)

$* = {}^{14}$C label

The t-butoxy radicals, once generated, react so quickly with the phosphite that reactions proceeding via breakdown of the alkoxy radical (e.g. eqn. 48) are not observed[60].§

$$\underset{\underset{Me}{|}}{\overset{\overset{Me}{|}}{Me\text{-}C\text{-}O\cdot}} \longrightarrow Me_2C=O + Me\cdot \qquad (48)$$

This has been demonstrated more directly by Walling and Pearson[61] who showed that the t-butoxy radical combines with triethyl phosphite much more rapidly than it abstracts hydrogen from an alkane solvent even though the latter reaction is known to have a low activation energy[62]. Recently, Davies et al. were able to provide more quantitative data by an e.s.r. study of the reaction of t-butoxy radicals with triethyl phosphite in cyclopentane[63]. Two reactions were observed (eqns. 49 and 50) and by comparison of the rates, k_2 for the phosphite reaction (eqn. 49) was determined as 1.6×10^8 l mol^{-1} s^{-1} at 30 °C.

$$t\text{-BuO}\cdot + (EtO)_3P \xrightarrow{k_2} \left[(EtO)_3\overset{\cdot}{P}O\text{-Bu}^t\right] \qquad (49)$$

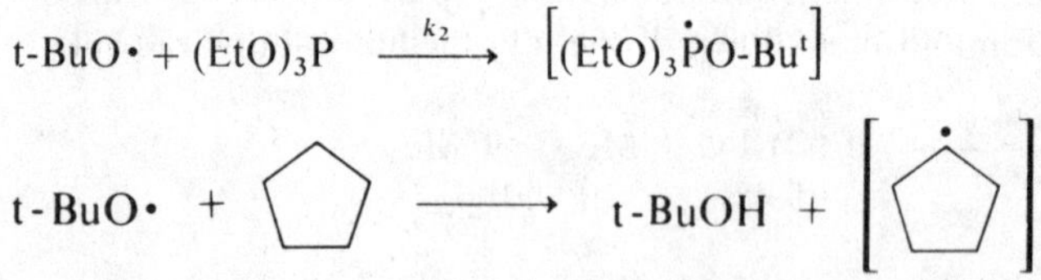

$$t\text{-BuO}\cdot + \bigcirc \longrightarrow t\text{-BuOH} + \left[\overset{\cdot}{\bigcirc}\right] \qquad (50)$$

Oxidations of this type are also stereospecific as evidenced by the reactions of (9.52) and (9.53) with t-butoxy radicals which proceed with retention of configuration at phosphorus[64].

---

† Incidentally this implies *either* that C—O cleavage in the intermediate phosphoranyl radical occurs with equal facility from apical or equatorial positions of the trigonal bipyramidal radical *or*, pseudorotation within the phosphoranyl radical is much faster than C—O bond cleavage.
§ On the other hand, the reactions of alkyl radicals (R·) with phosphite must be slow since no phosphonate, (R^{1}O)$_2$P(O)R is formed in the reaction of (R^{1}O)$_3$P with (R·).

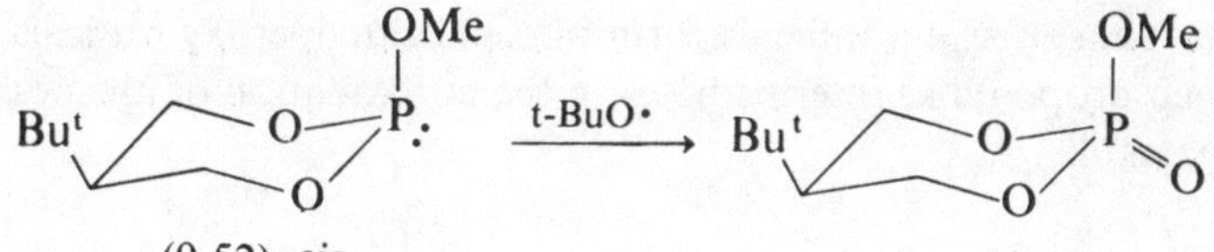

9.3 Nine-electron phosphorus radicals

(9.52) cis →(t-BuO·) cis [cyclohexane with Bu^t, O-P(OMe): → O-P(=O)(OMe)]

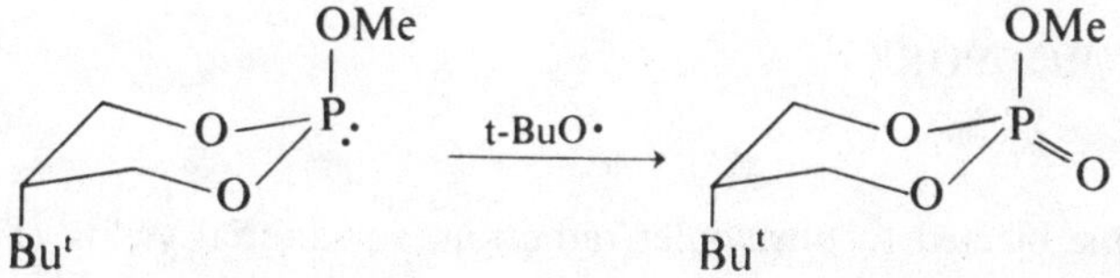

(9.53) trans →(t-BuO·) trans

Phosphines react with alkoxy radicals in a similar way but the reactions are complicated by the occurrence of α-scission as well as β-scission of the intermediate radicals. The gain in enthalpy on forming a phosphoryl (P=O) group is of the order of 630 kJ mol^{-1} (150 kcal mole^{-1}) so β-scission should be favoured. In spite of this, examples are known where α-scission predominates. For instance, tri-n-butylphosphine reacts with t-butyl peroxide to give an 80 % yield of t-butyl di-n-butylphosphinite (9.54) and only 20 % of tri-n-butylphosphine oxide (9.55)[65].

n-Bu$_3$P + t-BuO· ⟶ [n-Bu$_3$ṖOBu^t]
 ├─ α-scission ⟶ n-Bu· + n-Bu$_2$P—OBu^t (9.54) 80%
 └─ β-scission ⟶ n-Bu$_3$P=O + t-Bu·   (9.55) 20%

Displacement (α-scission) as well as β-scission has also been observed in the reactions of alkoxy (RO·) or thioalkyl (RS·) radicals with diethyl alkylphosphonites (9.56)[66].

RX· + R^{1}P(OEt)$_2$ ⟶ [RX—Ṗ(OEt)$_2$ | R^1]
 ├─ α-scission ⟶ RXP(OEt)$_2$ + R^1·
 └─ β-scission ⟶ R· + R^{1}P(OEt)$_2$(=X)

(9.56)

(X = O or S)

Phosphoranyl radicals can also be generated by an alternative route involving the reaction of alkyl radicals with the phosphoryl group. For example, t-butyl peroxy radicals react with triethyl phosphite to give t-butoxy radicals and triethyl phosphate (eqn. 51). The t-butoxy radicals may then decompose to acetone and methyl radicals (eqn. 52) and the latter attack triethyl phosphate at phosphoryl oxygen to produce a phosphoranyl radical and ultimately diethylmethyl phosphate (eqn. 53)[67].

t-BuOO· + (EtO)$_3$P ⟶ (EtO)$_3$P=O + t-BuO·   (51)

t-BuO· ⟶ Me$_2$C=O + Me·   (52)

Me· + (EtO)$_3$P=O ⟶ (EtO)$_3$Ṗ—OMe ⟶ (EtO)$_2$P(=O)OMe + Et·   (53)
                              (9.57)

In the presence of oxygen, phosphoranyl radicals form phosphoranylperoxy radicals (9.58)[68] and these have been proposed as intermediates in the autoxidation of alkenes in the presence of phosphates[69].

$$(RO)_3\dot{P}OR^1 + O_2 \longrightarrow (RO)_3\underset{\underset{O-O\cdot}{|}}{P}OR^1 \qquad (9.58)$$

When trialkyl phosphites are subjected to ultraviolet radiation, substantial yields of the isomeric phosphonate are obtained (eqn. 54)[70].

$$(RO)_3P \xrightarrow{h\nu} (RO)_2P\!\!\begin{array}{c}\nearrow O\\ \searrow R\end{array} \qquad (54)$$

This is the "photochemical Arbusov rearrangement" and is thought to proceed through dissociation of an excited phosphite to a phosphinyl radical and an alkyl radical (eqn. 55) followed by a comparatively slow attack of the alkyl radical on more trialkyl phosphite to produce a phosphoranyl radical (eqn. 56) which undergoes β-scission to a phosphonate (eqn. 57).

$$(RO)_3P \xrightarrow{h\nu} (RO)_2\ddot{P}-O\cdot + R\cdot \qquad (55)$$

$$R\cdot + (RO)_3P \xrightarrow{slow} (RO)_3\dot{P}R \qquad (56)$$

$$(RO)_3\dot{P}R \longrightarrow [R\cdot] + (RO)_2P\!\!\begin{array}{c}\nearrow O\\ \searrow R\end{array} \qquad (57)$$

The reaction of thiols with phosphites may be induced by heating or light and gives high yields of thiophosphates (9.59, eqn. 58)[71].

$$(RO)_3P + R^1SH \xrightarrow{heat\ or\ h\nu} (RO)_3P\!=\!S + R^1H \qquad (58)$$
$$(9.59)$$

Walling and Rabinowitz[58] showed that the reaction is also catalyzed by free radical initiators and that very long chains (5000) are involved. The mechanism involves a thiophosphoranyl radical (9.60) which suffers β-scission of the C—S bond to form a thiophosphoryl group.

$$R^1SH + R\cdot (initiator) \longrightarrow R^1S\cdot$$

$$R^1S\cdot + (RO)_3P \longrightarrow (RO)_3\dot{P}-SR^1$$
$$(9.60)$$

$$(RO)_3\dot{P}-SR^1 \longrightarrow (RO)_3P\!=\!S + R^1\cdot$$

9.3 Nine-electron phosphorus radicals

It should be noted that the C—S bond breaks in preference to the C—O bond; even benzyldiethyl phosphite (9.61) reacts with butane thiol (9.62) to form the thiophosphate (9.63) with only a trace of toluene (3%) originating from the C—O cleavage to give the relatively stable benzyl radical (9.64)[60].

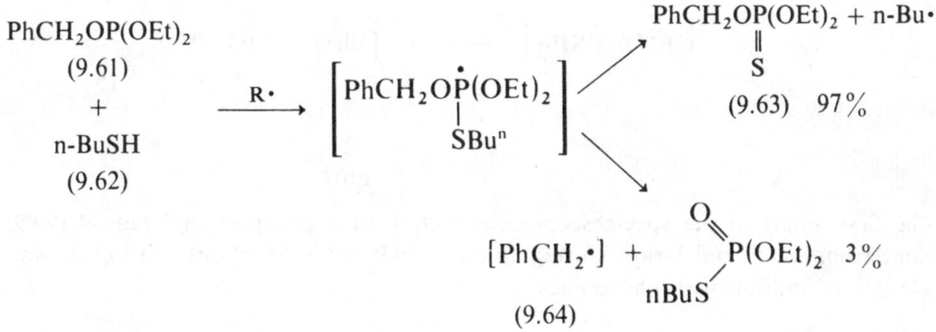

PhCH$_2$OP(OEt)$_2$
(9.61)
+
n-BuSH
(9.62)

$\xrightarrow{R\cdot}$ [PhCH$_2$O$\overset{\cdot}{P}$(OEt)$_2$ | SBu^n]

→ PhCH$_2$OP(OEt)$_2$ ‖ S + n-Bu·
(9.63) 97%

→ [PhCH$_2$·] + nBuS—P(OEt)$_2$=O   3%
(9.64)

The selectivity in favour of the thiophosphoryl product is a reflection of the strength of the C—S bond (290 kJ mol^{-1}, 65 kcal mole^{-1}) vs that of the C—O bond (360 kJ mol^{-1}, 85 kcal mole^{-1}) and it contrasts sharply with nucleophilic displacement on a thioalkoxyphosphonium salt (9.65) where de-alkylation gives the phosphoryl–oxygen bond.

(RS)$_3$$\overset{+}{P}$OR^1 $\overset{-}{X}$ $\longrightarrow$ (RS)$_3$P=O + R^{1}X
(9.65)

An analogous reaction, initiated by ultraviolet radiation, occurs between phosphites and dialkyl disulphides[54,56] probably by the following mechanism:

R'S—SR' $\xrightarrow{h\nu}$ 2R'S·

R'S· + (RO)$_3$P $\longrightarrow$ (RO)$_3$$\overset{\cdot}{P}$SR'

(RO)$_3$$\overset{\cdot}{P}$SR' $\longrightarrow$ (RO)$_3$P=S + R'·

R'· + R'S—SR' $\longrightarrow$ R'SR' + ·SR'

The usual radical initiators (e.g. AIBN) are often ineffective due to the initiating radical being too stable to break the disulphide link and if the chain carrying radical, R'· is relatively stable (e.g. PhCH$_2$·) the length of the radical chain is low and a significant quantity of dimer, R'—R', is formed. Furthermore, reactions with thiols and disulphides are also prone to take place via ionic pathways and care must be exercised in interpretation of the mechanisms. For example, triethyl phosphite and thiophenol undergo simultaneous reactions to give ethylphenyl thioether (9.66) and diethyl phosphite (9.67) in 85% yield by an ionic pathway together with benzene and 15% triethyl phosphorothioate (9.68) by a radical route.

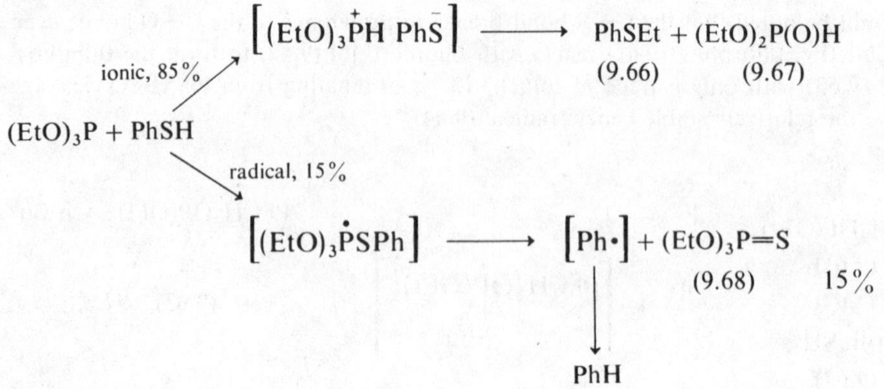

$$(EtO)_3P + PhSH \xrightarrow{\text{ionic, 85\%}} \left[(EtO)_3\overset{+}{P}H\ Ph\overset{-}{S}\right] \longrightarrow PhSEt + (EtO)_2P(O)H \quad 85\%$$
$$(9.66) \quad (9.67)$$

$$\xrightarrow{\text{radical, 15\%}} \left[(EtO)_3\dot{P}SPh\right] \longrightarrow [Ph\cdot] + (EtO)_3P=S \quad 15\%$$
$$(9.68)$$
$$\downarrow$$
$$PhH$$

The first report of the spectroscopic observation of a phosphoranyl radical (9.69) came from Kochi and Krusic during an e.s.r. study of the reactions of t-butoxy and t-butylthiyl radicals with phosphines[72].

$$Me_3P + t\text{-BuO}\cdot \longrightarrow Me_3\dot{P}\text{—OBu}^t$$
$$(9.69)$$

In the same study tri-isobutyl, tri-isopropyl and tricyclohexyl phosphines were shown to give t-butyl radicals by β-scission of the intermediate phosphoranyl radical (eqn. 59) but with triphenyl phosphite a displacement reaction occurred (α-scission) to give a phenoxy radical (eqn. 60).

$$R_3P + t\text{-BuX}\cdot \longrightarrow R_3\dot{P}X\text{—Bu}^t \longrightarrow R_3P=X + [t\text{-Bu}\cdot] \quad (59)$$
$$R = i\text{-Bu, i-Pr, }C_6H_{11}; X = O, S$$

$$t\text{-BuO}\cdot + (PhO)_3P \longrightarrow (PhO)_3\dot{P}OBu^t \longrightarrow PhO\cdot + (PhO)_2POBu^t \quad (60)$$

This work was followed closely by a study of the disproportionation processes arising from ultraviolet irradiation of a series of phosphorus halides[26]. For example, $MePCl_2$ produced $Me\dot{P}Cl$ and $Me\dot{P}Cl_3$ (eqn. 61) and the e.s.r. of the latter was consistent with a near trigonal bipyramidal geometry with the unpaired electron in an equatorial sp^2 hybrid orbital as in (9.70).*

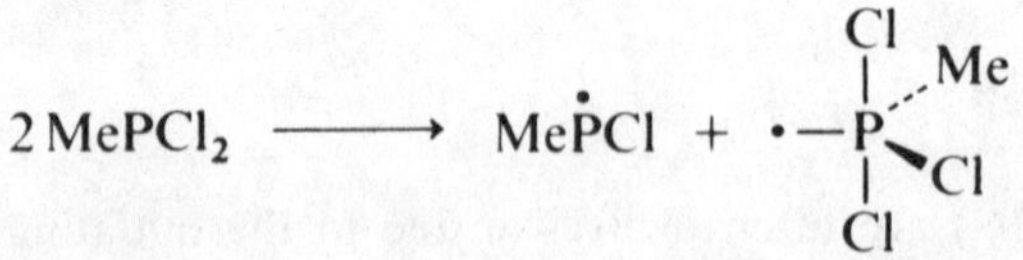

$$2\ MePCl_2 \longrightarrow Me\dot{P}Cl + \cdot\text{—}\underset{\underset{Cl}{|}}{\overset{\overset{Cl}{|}}{P}}\overset{Me}{\underset{Cl}{\diagdown}} \quad (61)$$
$$(9.70)$$

---

* On the basis of e.s.r. studies it has been suggested that in any phosphoranyl radical,

$$\cdot\text{—}\underset{\underset{R}{|}}{\overset{\overset{R}{|}}{P}}\overset{R}{\underset{R}{\diagdown}}$$

9.3 Nine-electron phosphorus radicals

These results provoked further effort in this area and an interesting example was provided by the reactions of a series of phosphoramidites (9.71) with t-butoxy radicals which gave dialkylaminyl radicals (9.73) by α-scission of an intermediate phosphoranyl radical[73].

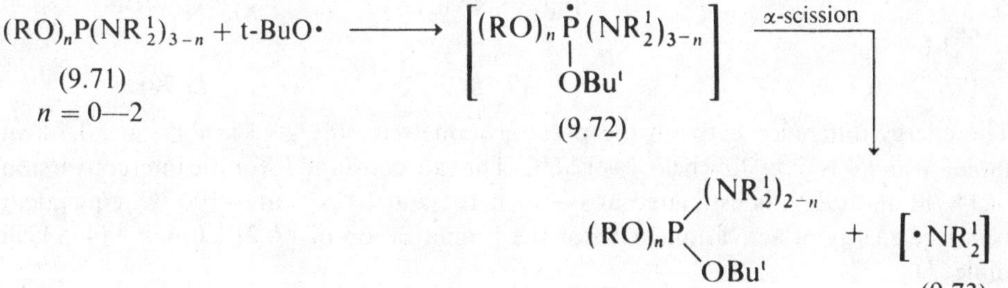

$$(RO)_nP(NR^1_2)_{3-n} + \text{t-BuO}\cdot \longrightarrow \left[(RO)_n\overset{\cdot}{\underset{OBu^t}{P}}(NR^1_2)_{3-n}\right] \xrightarrow{\alpha\text{-scission}}$$

(9.71)   (9.72)
n = 0—2

$$(RO)_nP\underset{OBu^t}{\overset{(NR^1_2)_{2-n}}{\diagup}} + [\cdot NR^1_2]$$

(9.73)

For a given phosphoramidite (e.g. 9.74) assuming apical entry of the t-butoxy radical one obtains (9.75) as the initial configuration of the phosphoranyl radical (detected by e.s.r. at −120 °C) and it is possible to envisage a ψ with the unpaired electron as pivot to give (9.76).

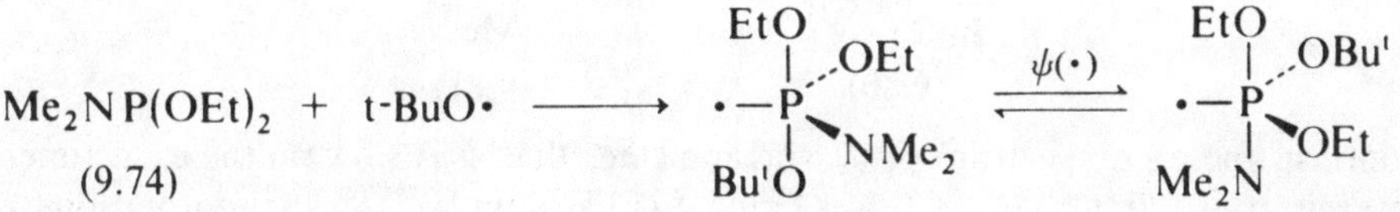

Me₂NP(OEt)₂ + t-BuO· ⟶ (9.75) ⇌^{ψ(·)} (9.76)

(9.74)

Unfortunately although (9.76) was detected at −150 °C, it was not possible to examine the energetics of the interconversion by dynamic e.s.r. due to the ease of the α-scission reaction.

However, in the corresponding reaction of t-butoxy radicals with the 2-dimethylamino-1,3,2-dioxaphospholan (9.77), e.s.r. spectra indicate that the intermediate phosphoranyl radical undergoes a fast isomerization, probably by pseudorotation (9.78) ⇌ (9.79)[74].

---

the *tbp* structure distorts more towards a *spy* structure as R becomes more electropositive, i.e. in the order F (*tbp*), Cl, OR, Me (*spy*) [A. G. Davies *et al.*, *J. Chem. Soc., Perkin II*, 993 (1972)]. Very recent evidence from G. Bockestein *et al., Chem. Commun.*, 118 (1974), reveals that some phosphoranyl radicals e.g. PhṖ(OMe)₂(Ot-Bu), exist in a tetrahedral configuration; in this case however, the unpaired electron is located to a large extent in the phenyl ring, i.e. the structure should be represented by (A) rather than (B)

(A)   (B)

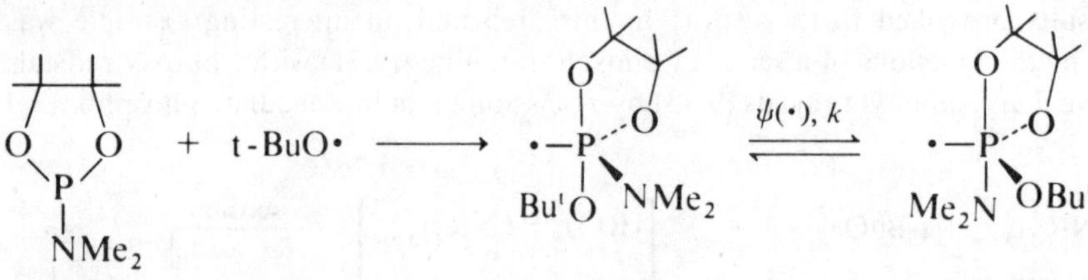

(9.77) + t-BuO• ⟶ (9.78) ⇌ψ(•), k (9.79)

The energy difference between the pseudorotamers is only 2.9 kJ mol^{-1} (ca 0.7 kcal mole^{-1}) and ψ is only slow below $-120\,°C$. The rate constant, k, for the interconversion of (9.78) and (9.79) is estimated at between 10^7 and 10^8 s^{-1} at $-100\,°C$, equivalent to a free energy of activation $\Delta G^{\ddagger}$, for the isomerization of 17–21 kJ mol^{-1} (4–5 kcal mole^{-1}).

A similar energy barrier of 22 kJ mol^{-1} (ca 5.2 kcal mole^{-1}) has been reported for pseudorotation of radical (9.80) to (9.81)[75].

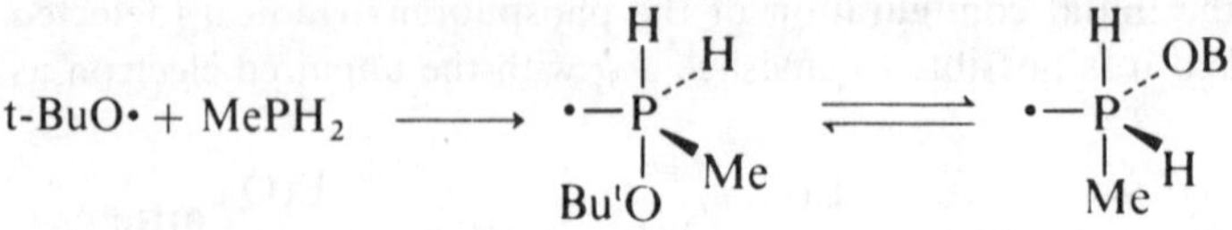

t-BuO• + MePH$_2$ ⟶ (9.80) ⇌ (9.81)

In contrast, the e.s.r. spectrum of (9.82a) indicates that ψ is slow on the e.s.r. timescale even at $+120\,°C$, i.e. $\Delta G^{\ddagger}$ 54 kJ mol^{-1} (13 kcal mole^{-1})[76]. Pseudorotation of (9.82a) with the unpaired electron as pivot gives the identical (9.82b) and this is an unusually high energy barrier between topomers (conformers of identical energy).

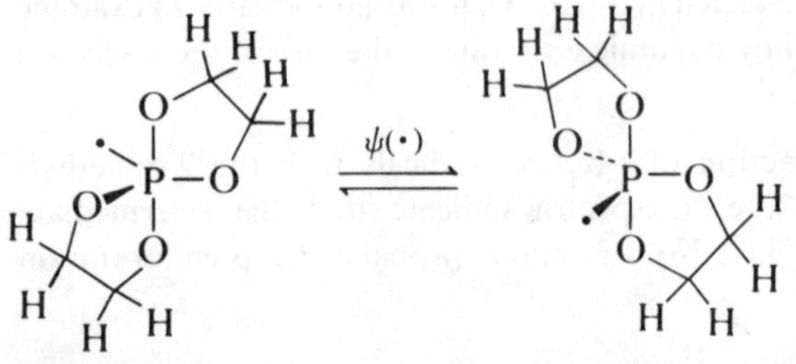

(9.82a) ⇌ψ(•) (9.82b)

The anomaly may be explained by assuming that the ground state configuration of the phosphoranyl radical is in fact a square pyramid (9.83), a suggestion which is consistent with the most recent structural information on spirophosphoranes (see Chapter 6).

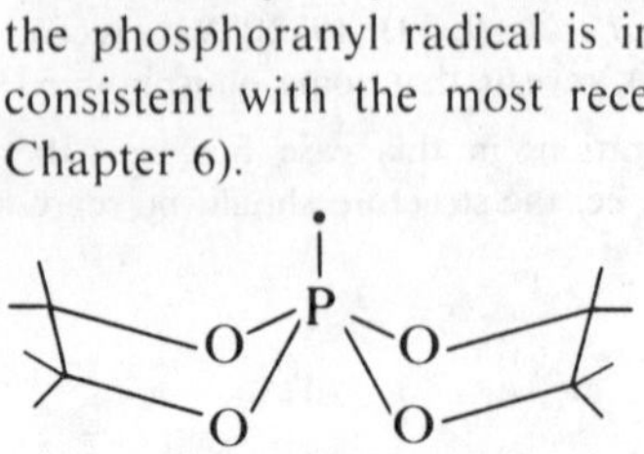

(9.83)

Studies of e.s.r. spectra have also thrown light on the relative stability of intermediate phosphoranyl radicals such as (9.84), (9.85) and (9.86) derived from the reactions of t-butoxy radicals with the appropriate P(III) compound[77,78]. Knowledge of the rela-

9.3 Nine-electron phosphorus radicals

tive stabilities has led in turn to the development of the theory that α-cleavage takes place preferentially from an apical position. The order of stability for the three radicals is: (9.84) > (9.85) > (9.86).

```
   EtO                    EtO                    Et
    |  ,Et                 |  ,Et                 |  ,Et
 •—P                    •—P                    •—P
    |  `Et                 |  `OEt                |  `Et
   Bu'O                   Bu'O                   Bu'O
   (9.84)                 (9.85)                 (9.86)
```

$$\updownarrow \qquad\qquad \updownarrow$$

```
    Et                     Et
    |  ,OBu'                |  ,OBu'
 •—P                    •—P
    |  `OEt                 |  `OEt
    Et                     OEt
   (9.87)                 (9.88)
```

In order to effect α-scission of an ethyl radical from an apical position (9.84) must pseudorotate to (9.87) which is energetically unfavourable because it places *two* electronegative alkoxy groups in equatorial positions. A similar pseudorotation of (9.85) to (9.88) is more facile since in (9.88) only one alkyl group becomes apical at the expense of an alkoxy group. In (9.88) no ψ is necessary to afford apical departure and this radical decomposes most rapidly to an ethyl radical and t-butyl diethylphosphinite.

To complement this theory it has also been proposed that β-scission occurs most readily from an equatorial position[78,79].* On this basis, examination of (9.84)–(9.86) suggests (9.84) would be least susceptible to β-scission since it would require an unfavourable ψ(•) to place the alkoxy groups equatorial. Hence, the observation that only ethyl radicals (α-scission) and no t-butyl radicals (β-scission) are produced, lends support to the proposal[78].

As mentioned earlier, the reactions of triaminophosphines, $(Me_2N)_3P$, or trialkylphosphines (e.g. Et_3P) with t-butoxy radicals give α-scission, displacement of $Me_2N•$ or Et•, in preference to the thermodynamically more favourable β-scission to form the strong phosphoryl (P=O) group[73,80]. This at first, seems a strange situation, but recent calculations on a number of α- and β-scission reactions reveal that to a first approximation, the relative rate of α to β scission for reaction (62) is determined by the relative strengths of the R—A and P—X bonds[81,82], rather than the stability of the P=A bond.

---

* The ease of β-scission certainly increases as the radical produced changes from primary to secondary, to tertiary to benzylic[78]; the β-cleavage would therefore be *most selective* in producing a primary alkyl radical and *least* selective in producing the relatively stable benzyl radical.

$$XP(OEt)_2 + RA\cdot \longrightarrow RA\overset{\cdot}{P}X(OEt)_2 \quad \begin{matrix} \overset{\beta\text{-scission}}{\nearrow} R\cdot + A=PX(OEt)_2 \\ \underset{\alpha\text{-scission}}{\searrow} RAP(OEt)_2 + X\cdot \end{matrix} \quad (62)$$

$X = Cl, R, Ph, PhO, R_2N \qquad A = O, S$

There are also a number of phosphoranyl radicals known which involve only carbon–phosphorus bonds. The first of these arose as transient intermediates in Horner's study of the preparation of optically active phosphines by electrolytic reduction of phosphonium salts (eqn. 63)[83] and since then the method has been thoroughly studied by Horner and Haufe in order to determine the yield of alkane compared to that of benzene in the reduction of salts of the type $RP^+Ph_3\ X^-$ (X = halogen)[84]. It appears that formation of the alkane is favoured (at the expense of PhH) when the alkyl group is large enough to cause serious steric congestion in the intermediate phosphoranyl radical.

$$R_4P^+ \xrightarrow{+e} R_4P\cdot \longrightarrow R_3P\colon + R\cdot \quad (63)$$

Phosphoranyl radicals with four phosphorus–carbon bonds can also be generated from phosphoranes containing a P—H bond. For instance, photolysis of (9.89) produces (9.90) which subsequently gives (9.92) and (9.93)[85]. The radical can also be generated by treatment of the metallated phosphine (9.91) with iodine[86] and in both cases the phosphoranyl radical was detected by e.s.r. and shown to have a g-value of 2.0025 ± 0.0001 and a phosphorus hyperfine coupling constant, $|a(P)|$, of 17.9 Gauss.

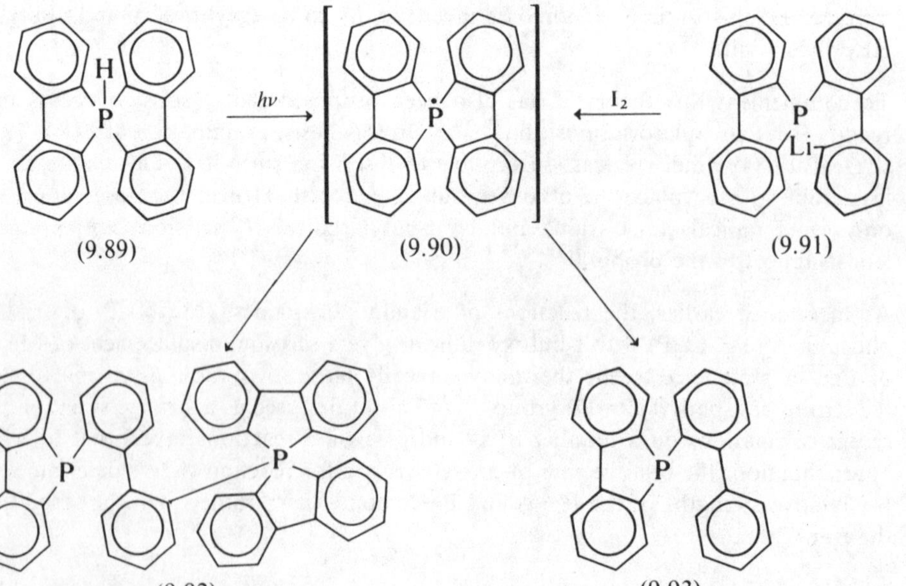

A subsequent e.s.r. study[87] has shown that the unpaired electron of (9.90) is delocalized to a large extent over the two biphenyl ring systems which is consistent with the low value of the coupling constant, $|a(P)|$, reported originally.

9.3 Nine-electron phosphorus radicals

Phosphoranyl radicals have also been proposed as transient intermediates in the high temperature reaction of the phosphabenzole (9.94) with diphenyl mercury which ultimately gives the 1,1-diphenylphosphabenzole (9.95)[88].

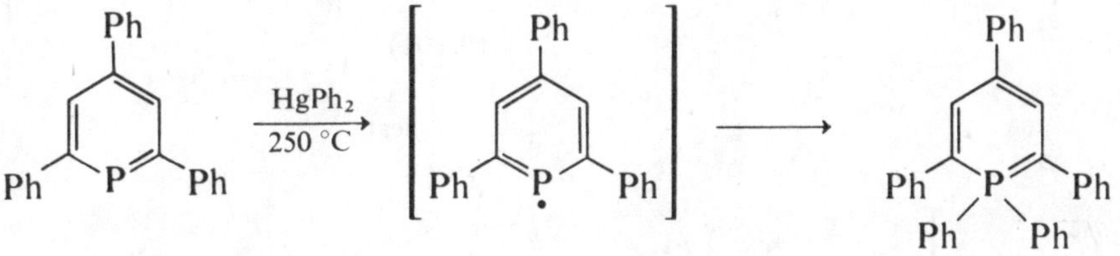

$$\underset{(9.94)}{\text{phosphabenzole}} \xrightarrow[250\,°C]{HgPh_2} [\text{radical intermediate}] \longrightarrow \underset{(9.95)}{\text{1,1-diphenylphosphabenzole}}$$

The above discussion leaves no doubt that the subject of phosphoranyl radicals – particularly their structure and the factors determining their various modes of reaction – is still in its early stages and offers an area of research which is bound to attract investigators for years to come.

9.3.2 The phosphonium radical anion ($R_3P^{-}\cdot$)

Species of this kind are the least well characterized of the catalogue of phosphorus radicals. Electrolytic reduction of triphenylphosphine at a dropping mercury cathode with acetonitrile as solvent gives biphenyl and diphenylphosphinic acid (9.97). The latter product must arise from reaction with water and the radical anion (9.96) is presumably an intermediate[89].

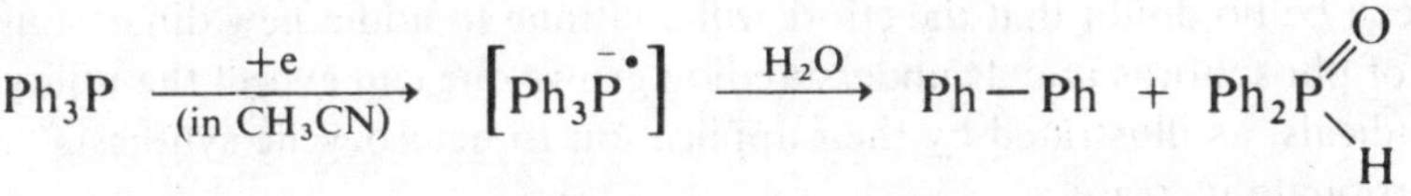

$$Ph_3P \xrightarrow[(\text{in } CH_3CN)]{+e} [Ph_3P^{-\cdot}] \xrightarrow{H_2O} Ph-Ph + Ph_2P(O)H$$
$$\qquad\qquad\qquad (9.96) \qquad\qquad\qquad (9.97)$$

Studies of the e.s.r. spectrum of the radical (9.99) generated by electrolytic reduction of dimethylphenylphosphine (9.98), reveal that the coupling to phosphorus, $|a(P)| \approx 8.5$ Gauss at $-50\,°C$, is strongly temperature dependent which is ascribed to a variation in the geometry of the lone pair on phosphorus[90]. It appears however, that the unpaired electron is very largely delocalized over the phenyl ring rather than being centred on phosphorus.

$$Me_2PPh \xrightarrow{+e} Me_2\overset{-\cdot}{P}Ph$$
$$(9.98) \qquad\quad (9.99)$$

The best established example of such radical anions is provided by the reduction of phospholes (e.g. 9.100) with alkali metals[91,92]. At low temperature the radical (9.101) may be detected by its e.s.r. spectrum which shows a moderately large coupling to phosphorus[91], $|a(P)| = 23.5$ Gauss for $R = Me$. The radical receives its stability from the well known aromaticity of the cyclopentadienylide anion and the aromaticity, conveniently represented by (1.103), is also evident in the e.s.r. spectrum.

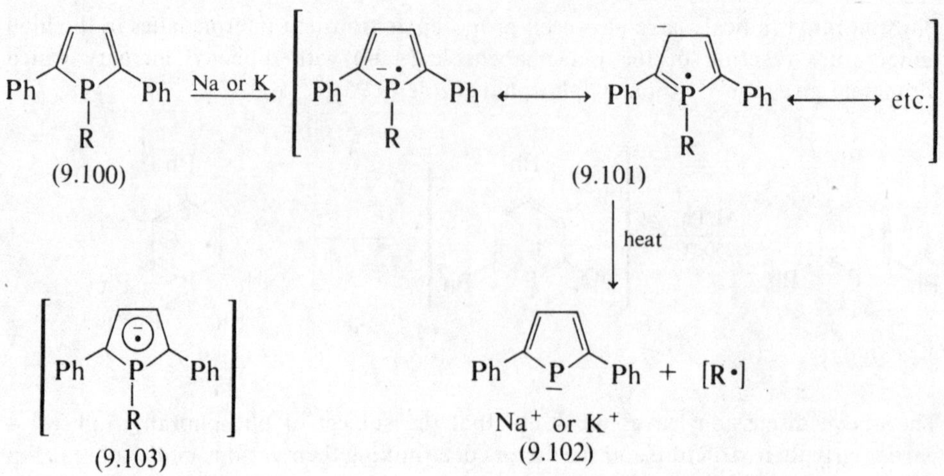

374   Chapter 9 / Phosphorus radicals

As the temperature is raised, phosphorus–carbon bond cleavage occurs to form carbon radicals (R·) and metal phosphacyclopentadienides (9.102).

9.4 Conclusion

In retrospect it is easy to see that the "take-off point" for research in phosphorus radical chemistry occurred about the time of Kochi's observation of the phosphoranyl radical by e.s.r. (1968–69). Much had been accomplished before then to provide convincing evidence for phosphorus radical intermediates but the advent of highly specialized e.s.r. techniques such as flow-through e.s.r. cells and photolysis or electrolysis in an e.s.r. cavity provided the impetus for a deeper, more fundamental enquiry in this area. There can be no doubt that the effort will continue to add a new dimension to the chemistry of phosphorus and as understanding grows one can expect the utility of phosphorus radicals, as illustrated by their application to heterocyclic synthesis[93], to show a proportionate increase.

Problems

1 Suggest mechanisms for the following reactions:

(a) $R-\overset{O}{\underset{\|}{C}}CH_2OP(OMe)_2 \xrightarrow{h\nu} CH_2=\overset{R}{\underset{|}{C}}O-\overset{O}{\underset{\|}{P}}(OMe)_2$

(b) $(PhP)_5 + PhC\equiv CPh \xrightarrow{heat}$ [four-membered P–P ring with Ph groups] + [phospholene with Ph groups]

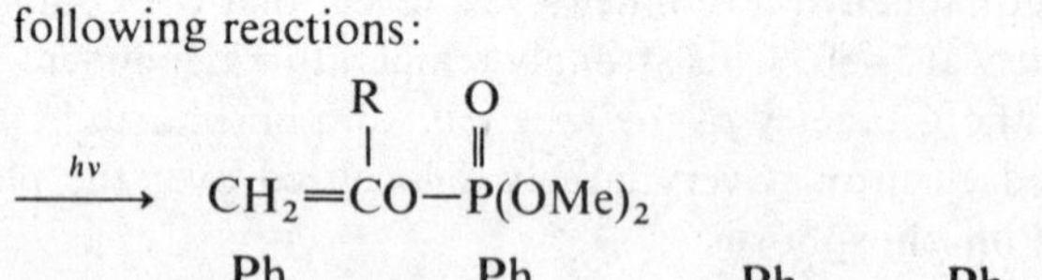

(c) [bridged polycyclic structure with P=O, Ph groups] + MeOH $\xrightarrow{heat}$ [tetraphenylnaphthalene] + $Ph-\overset{O}{\underset{H}{\overset{\|}{P}}}-OMe$

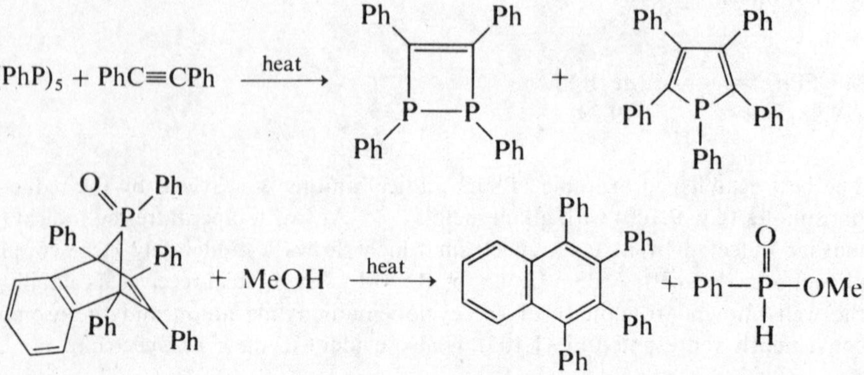

Problems

(d) 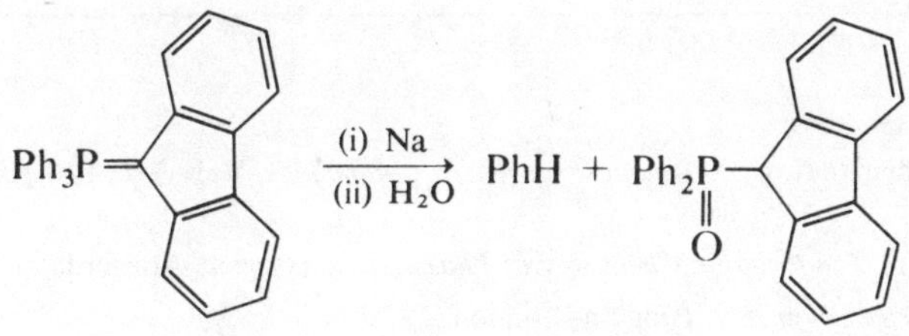 Ph₃P=fluorenylidene $\xrightarrow[\text{(ii) H}_2\text{O}]{\text{(i) Na}}$ PhH + Ph₂P(=O)-fluorenyl

(e) Ph₂PPPh₂ + PhCH₂CO₂H $\xrightarrow{100\ °C}$ PhMe + Ph₂P(=O)H + CO

(f) 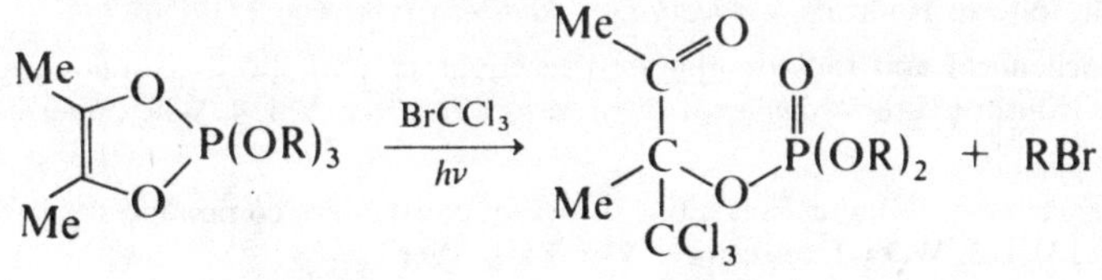 [Me-C=C-Me dioxaphosphole with P(OR)₃] $\xrightarrow[h\nu]{\text{BrCCl}_3}$ Me-C(=O)-O-P(OR)₂(=O) with Me-C(CCl₃)- + RBr

(g) 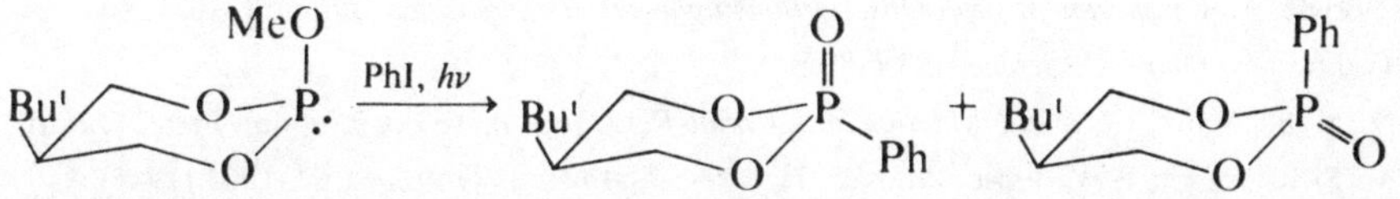 [Bu^t dioxaphosphorinane with MeO, P:] $\xrightarrow{\text{PhI, }h\nu}$ [Bu^t dioxaphosphorinane with O=P-Ph] + [Bu^t dioxaphosphorinane with Ph, P=O]   95%

2 (a) When alkyl phosphites, (RO)₃P, are reacted with dialkyl disulphides, R¹SSR¹, in the presence of an initiator and carbon monoxide under high pressure, substantial yields of the thiol ester, R¹—C(O)—SR¹ are obtained. Explain this result.

(b) When a trialkyl phosphite reacts with the alkene thiol (A) in the presence of a radical initiator, a 43 % yield of methylcyclopentane (B) is obtained. Explain the formation of (B).

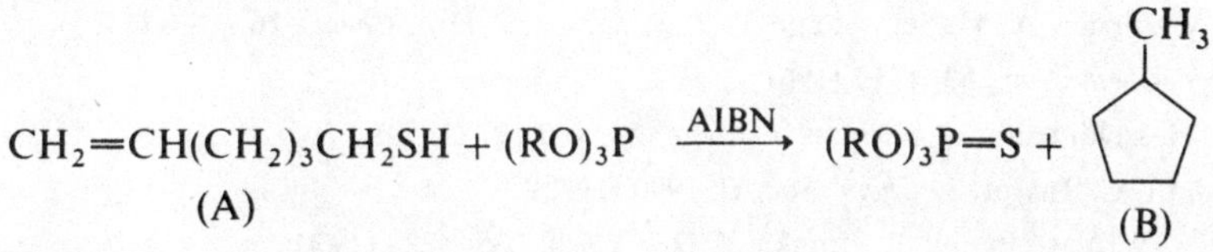

$$\text{CH}_2=\text{CH}(\text{CH}_2)_3\text{CH}_2\text{SH} + (\text{RO})_3\text{P} \xrightarrow{\text{AIBN}} (\text{RO})_3\text{P}=\text{S} + \text{methylcyclopentane}$$
(A) → (B)

3 When triethyl phosphite reacts with carbon tetrachloride in the presence of 1-butane thiol and a radical scavenger and in the absence of a radical initiator, S-n-butyldiethylphosphorothiolate (C) is the main product. The same reactants, in the absence of a scavenger but in the presence of AIBN give a substantial yield of triethyl phosphorothionate (D).

$$(\text{EtO})_3\text{P} + \text{CCl}_4 + n\text{-BuSH} \xrightarrow{\text{scavenger}} (\text{EtO})_2\text{P}(=\text{O})\text{S-Bu}^n + \text{CHCl}_3 + \text{EtCl}$$
(C)

$$(\text{EtO})_3\text{P} + \text{CCl}_4 + n\text{-BuSH} \xrightarrow{\text{AIBN}} (\text{EtO})_3\text{P}=\text{S} + n\text{BuH}$$
(D)

Suggest mechanistic pathways for both reactions.

References

[1] G. M. Kosolapoff and L. Maier (editors), *Organic Phosphorus Compounds*, Vols. 1–6, Wiley Interscience, New York, 1972–73.

[2] A. J. Kirby and S. G. Warren, *The Organic Chemistry of Phosphorus*, Elsevier, Amsterdam, 1967.

[3] B. J. Walker, *Organophosphorus Chemistry*, Penguin, London, 1972.

[4] R. F. Hudson, *Structure and Mechanism in Organophosphorus Chemistry*, Academic Press, New York, 1965.

[5] S. Trippett (editor), Specialist Reports, Chemical Society, *Organophosphorus Chemistry*, Vols. 1–5, 1969–74.

[6] J. I. G. Cadogan, "Phosphorus Radicals", *Advan. Free-Radical Chemistry*, 2 (1968).

[7] M. Halmann, "Photochemical and radiation-induced reactions of phosphorus compounds", in M. Grayson and E. J. Griffiths (editors), *Topics in Phosphorus Chemistry*, Vol. 4, Wiley Interscience, New York, 1967.

[8] C. Walling and M. C. Pearson, "Radical reactions of organophosphorus compounds", in *Topics in Phosphorus Chemistry*, Vol. 3, Wiley Interscience, New York, 1966.

[9] R. Boyle, *New Experiments and Observations upon the Icy Noctiluca*, London, 1681–2.

[10] T. Graham, *Quart. J. Science*, 83 (1829).

[11] N. N. Semenov, *Chemical Kinetics and Chain Reactions*, p. 163, Clarendon Press, Oxford, 1935.

[12] M. S. Kharasch, E. V. Jenson and W. H. Urry, *J. Amer. Chem. Soc.*, **67**, 1864 (1945).

[13] J. K. Kochi and P. J. Krusic, *J. Amer. Chem. Soc.*, **91**, 3944 (1969).

[14] U. Schmidt, I. Boie, C. Osterroht, R. Shröer and H. F. Grützmacher, *Chem. Ber.*, **101**, 1381 (1968).

[15] B. Block and Y. Gounelle, *C. R. Hebd. Seances Acad. Sci.*, **266** C, 220 (1968).

[16] B. Zeek and J. B. Thomson, *Tetrahedron Lett.*, 111 (1969).

[17] M. Sato, M. Yanagita, Y. Fujita and T. Kwan, *Bull. Chem. Soc. Japan*, **44**, 1423 (1971).

[18] H. M. Buck, A. H. Huizer, S. J. Oldenburg and P. Schipper, *Phosphorus*, **1**, 97 (1971).

[19] A. R. Stiles, F. F. Rust and W. E. Vaughan, *J. Amer. Chem. Soc.*, **74**, 3282 (1952).

[20] M. Rauhut, H. A. Currier, A. M. Semsel and V. P. Wystrack, *J. Org. Chem.*, **26**, 5138 (1961).

[21] J. Pellon, *J. Amer. Chem. Soc.*, **83**, 1915 (1961).

[22] R. Fields, R. N. Haszeldine and J. Kirman, *J. Chem. Soc. C*, 197 (1970).

[23] W. Hewerston and I. C. Taylor, *J. Chem. Soc. C*, 1990 (1970).

[24] T. Saegusa, Y. Ito, M. Yasudo and T. Hotaka, *J. Org. Chem.*, **35**, 4238 (1970).

[25] D. Solan and P. L. Timms, *Chem. Commun.*, 1540 (1968).

[26] G. F. Kokooszka and F. E. Brinckmann, *J. Amer. Chem. Soc.*, **92**, 1199 (1970).

[27] R. Graf, *Chem. Ber.*, **85**, 9 (1952).

[28] J. O. Clayton and W. L. Jensen, *J. Amer. Chem. Soc.*, **70**, 3880 (1948).

[29] L. Z. Soborovskii, Y. M. Zinovier and M. A. Englin, *Dokl. Akad. Nauk SSSR*, **67**, 293 (1949); *Chem. Abs.*, **44**, 1401 (1950).

[30] Y. M. Zinovier and L. Z. Soborovskii, *Zh. Obsch. Khim.*, **29**, 615 (1959); *Chem. Abs.*, **54**, 340 (1960).

[31] H. Karlsson and C. Lagercrantz, *Acta Chem. Scand.*, **24**, 3411 (1970).

[32] S. K. Wang, W. Sytryk and J. K. S. Wan, *Can. J. Chem.*, **49**, 994 (1971).

[33] S. K. Wang and J. K. S. Wan, *Spectrosc. Lett.*, **3**, 135 (1970).

[34] R. Sasin, W. F. Olszewski, J. R. Russel and D. Swern, *J. Amer. Chem. Soc.*, **81**, 6275 (1959).

[35] G. R. Van den Berg, D. H. J. M. Platenburg and H. P. Benschopp, *Chem. Commun.*, 606 (1971).

[36] H. P. Benschopp and D. H. J. M. Platenburg, *Chem. Commun.*, 1098 (1970).
[37] L. P. Reiff and H. S. Aaron, *J. Amer. Chem. Soc.*, **92**, 5275 (1970).
[38] W. B. Farnham, R. K. Murray and K. Mislow, *Chem. Commun.*, 146 and 605 (1971).
[39] M. Geoffrey and E. A. C. Lucken, *Mol. Phys.*, **22**, 257 (1971).
[40] A. Horsfield, J. R. Morton and D. H. Whiffen, *Mol. Phys.*, **4**, 475 (1961).
[41] A. Deguchi, *Bull. Chem. Soc., Japan*, **34**, 910 (1961).
[42] F. Ramirez and S. Dershowitz, *J. Amer. Chem. Soc.*, **78**, 5614 (1956).
[43] E. Weitz, *Angew. Chem.*, **66**, 658 (1954).
[44] E. A. C. Lucken, *J. Chem. Soc.*, 5123 (1963).
[45] E. A. C. Lucken, F. Ramirez, V. P. Catto, D. Rhum and S. Dershowitz, *Tetrahedron*, **22**, 637 (1966).
[46] L. Horner and J. Haufe, *Chem. Ber.*, **101**, 2921 (1968).
[47] R. K. Wong and A. O. Allen, *J. Phys. Chem.*, **74**, 774 (1970).
[48] A. R. Lyons, G. W. Neilson and M. C. R. Symons, *Chem. Commun.*, 507 (1972).
[49] R. L. Powell and C. D. Hall, *J. Amer. Chem. Soc.*, **91**, 5403 (1969).
[50] M. P. Naan, R. L. Powell and C. D. Hall, *J. Chem. Soc. B*, 1683 (1971).
[51] W. C. Davies and W. P. G. Lewis, *J. Chem. Soc.*, 1599 (1934).
[52] G. Binsch, in E. L. Eliel and N. L. Allinger (editors), *Topics in Stereochemistry*, Vol. 3, p. 97, Interscience, New York, 1968.
[53] L. J. Aarons, M. F. Guest, M. B. Hall and I. H. Hillier, *J. Chem. Soc., Faraday II*, 643 (1973).
[54] K. Dimroth and W. Stüde, *Angew. Chem., Int. Ed. Eng.*, **7**, 881 (1968).
[55] K. Dimroth, A. Hettche, W. Städe and F. W. Steuber, *Angew. Chem. Int. Ed. Eng.*, **8**, 770 (1969).
[56] A. G. Davies and B. P. Roberts, *Accounts Chem. Res.*, **5**, 387 (1972).
[57] D. G. Pobedimski, N. A. Mukmeneva and P. A. Kirpichnikov, *Russ. Chem. Rev. (Eng. Trans.)*, **41**, 555 (1972).
[58] C. Walling and R. Rabinowitz, *J. Amer. Chem. Soc.*, **81**, 1243 (1959).
[59] W. G. Bentrude and R. A. Wielesck, *J. Amer. Chem. Soc.*, **91**, 2406 (1969).
[60] C. Walling, O. H. Basedow and E. Savas, *J. Amer. Chem. Soc.*, **82**, 2181 (1960).
[61] C. Walling and M. S. Pearson, *J. Amer. Chem. Soc.*, **86**, 2262 (1964).
[62] P. Gray and A. Williams, *Chem. Rev.*, **59**, 239 (1959).
[63] A. G. Davies, D. Griller and B. P. Roberts, *Angew. Chem., Int. Ed. Eng.*, **10**, 738 (1971).
[64] W. G. Bentrude, J. H. Hargis and P. E. Rusek, *Chem. Commun.*, 296 (1969).
[65] S. A. Buckler, *J. Amer. Chem. Soc.*, **84**, 3093 (1962).
[66] W. G. Bentrude, E. R. Hansen, W. A. Khan and P. E. Rogers, *J. Amer. Chem. Soc.*, **94**, 2867 (1972).
[67] Y. A. Levin, E. K. Trutneva, I. P. Gozman, A. G. Abulkhanov and B. E. Ivanov, *Bull. Acad. Sci. USSR, Div. Chem. Sci.*, 2687 (1970).
[68] A. G. Davies, D. Griller and B. P. Roberts, *J. Chem. Soc., Perkin II*, 993 (1972).
[69] G. B. Watts and K. U. Ingold, *J. Amer. Chem. Soc.*, **94**, 2528 (1972).
[70] R. B. La Count and C. E. Griffin, *Tetrahedron Lett.*, 3071 (1965).
[71] F. W. Hoffmann, R. J. Ess, T. C. Simmons and R. S. Hanzel, *J. Amer. Chem. Soc.*, **78**, 6414 (1956).
[72] J. K. Kochi and P. J. Krusic, *J. Amer. Chem. Soc.*, **91**, 3944 (1969).
[73] R. W. Dennis and B. P. Roberts, *J. Organomet. Chem.*, **43**, C2 (1972).
[74] R. W. Dennis and B. P. Roberts, *J. Organomet. Chem.*, **47**, C8 (1973).
[75] P. J. Krusic, W. Mahler and J. K. Kochi, *J. Amer. Chem. Soc.*, **94**, 6033 (1972).
[76] A. G. Davies, R. W. Dennis, D. Griller and B. P. Roberts, *J. Organomet. Chem.*, **42**, C47 (1972).
[77] A. G. Davies, R. W. Dennis, D. Griller and B. P. Roberts, *J. Organomet. Chem.*, **40**, C33 (1972).

[78] A. G. Davies, D. Griller and B. P. Roberts, *J. Chem. Soc., Perkin II*, 2224 (1972).
[79] G. B. Watts, D. Griller and K. U. Ingold, *J. Amer. Chem. Soc.*, **94**, 8784 (1972).
[80] J. R. Roberts and K. U. Ingold, *J. Amer. Chem. Soc.*, **95**, 3228 (1973).
[81] W. G. Bentrude, E. R. Hansen, W. A. Khan, T. B. Min and P. E. Rogers, *J. Amer. Chem. Soc.*, **95**, 2286 (1973).
[82] W. G. Bentrude, J-J. L. Fu and P. E. Rogers, *J. Amer. Chem. Soc.*, **95**, 3625 (1973).
[83] L. Horner, H. Fuchs, H. Winkler and A. Rapp, *Tetrahedron Lett.*, 965 (1963).
[84] L. Horner and J. Haufe, *Chem. Ber.*, **101**, 2903 (1968).
[85] D. Hellwinkel, *Chem. Ber.*, **102**, 528 (1969).
[86] D. Hellwinkel, *Chem. Ber.*, **102**, 548 (1969).
[87] R. Rothius, T. K. J. Luderer and H. M. Buck, *Rec. Trav. Chim. Pays-Bas.*, **91**, 836 (1972).
[88] G. Märkl and A. Merz, *Tetrahedron Lett.*, 1231 (1969).
[89] K. S. V. Santhanam and A. J. Bard, *J. Amer. Chem. Soc.*, **90**, 1118 (1968).
[90] F. Gerson, G. Plattner and H. Bock, *Helv. Chim. Acta*, **53**, 1629 (1970).
[91] D. Kilcast and C. Thomson, *Tetrahedron*, **27**, 5705 (1971).
[92] E. H. Braye, I. Caplier and R. Saussez, *Tetrahedron*, **27**, 5523 (1971).
[93] D. Redmore, *Chem. Rev.*, **71**, 315 (1971).

Chapter 10
Phosphorus–nitrogen compounds

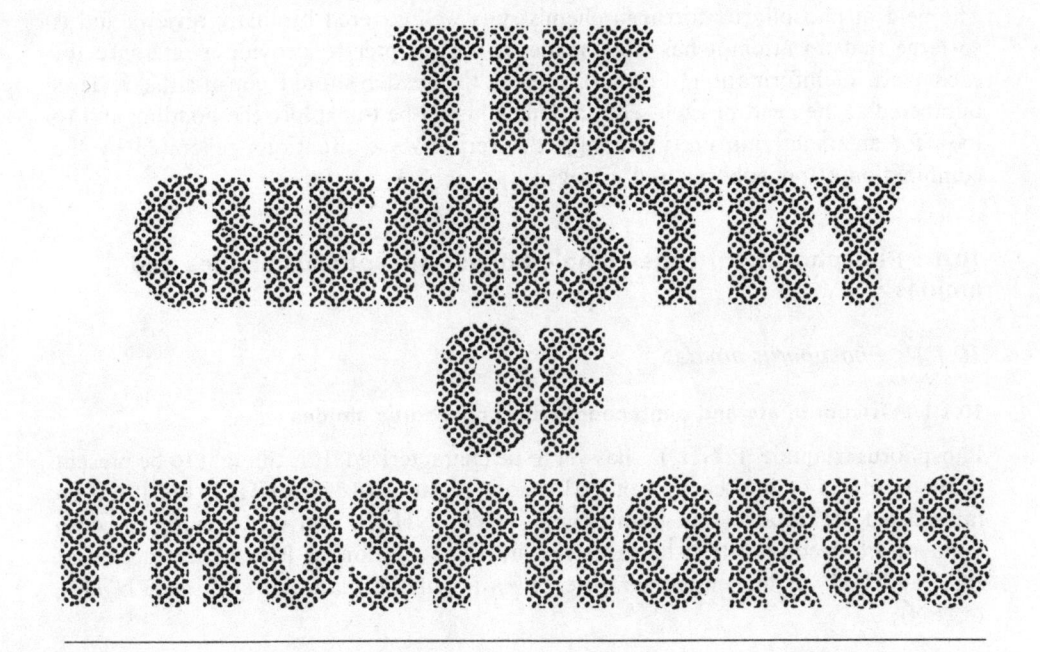

Phosphorus and nitrogen are members of the same group, M5, of the Periodic Table, but they have little else in common as most of this book shows. When these two elements come together in bond formation they produce one of the most intriguing bonding systems in chemistry, especially in the cyclophosphazenes where something akin to inorganic aromaticity (based on a 3d–2p π system) is found. Half of this chapter will be devoted to these compounds.

It is convenient when discussing phosphorus–nitrogen chemistry to distinguish five types of compound, reflecting as they do, five different bonding situations. *Firstly*, the amides and amine derivatives of tri-, tetra- and pentacoordinate phosphorus, in which there is formally a single σ P—N bond, although even this simple bond can have an element of extra bonding involving a donor 2p(N)→3d(P) π system. *Secondly*, there are the monophosphazenes having the formula $X_3P=NR$ with a localized double bond. These compounds are nitrogen analogues of the phosphorus ylids. *Thirdly*, come the cyclodiphosphazanes which are four-membered rings, $(X_3PNR)_2$ or $(X(O,S)PNR)_2$. This group of compounds can be seen as the cyclic dimers of the monophosphazenes. *Fourthly*, there are the diphosphazenes with delocalized π-bonding, $[X_3P \cdots N \cdots PX_3]^+$ or $X_3P=N-P(O)X_2$, which are the lower members of a series of polyphosphazenes ranging from only a few $\diagdown\!\!P=N\!\!-\diagup$ repeating units up to high polymers. And *fifthly*, the cyclopolyphosphazenes with the general formula $(NPX_2)_n$ where n is 3,4,5 etc. Delocalization of the π-bonding becomes the key issue in these compounds.

The field of phosphorus–nitrogen chemistry is well covered by many reviews and is so large that no attempt has been made in this chapter to provide a reference for each piece of information. For more detail the reader should consult the reviews numbered at the head of each section. Our aim will be to explore the bonding and to look for an underlying unity amongst the very diverse situations generated by the combination of phosphorus and nitrogen.

10.1 Phosphorus–nitrogen single bonds–phosphorus amides and amines[1,2]

10.1.1 Phosphorus amides

10.1.1.1 Tricoordinate and pentacoordinate phosphorus amides

Phosphorus triamide, $P(NH_2)_3$, has yet to be characterized. It is thought to be present in the products of the reaction of PCl_3 and NH_3 at -78 °C in $CHCl_3$ but it has so far resisted separation from by-products such as NH_4Cl, and on warming to room temperature decomposition takes place with the evolution of PH_3. The reaction of PCl_3 and NH_3 at higher temperatures gives polymeric materials such as $(PN_3H_2)_n$ or $(PN)_n$.

Neither has phosphorus pentamide, $P(NH_2)_5$, been isolated, although the reaction of PCl_5 and NH_3 has been more thoroughly studied[3]. The product from the reaction in liquid ammonia at -70 °C is $P(NH_2)_4Cl$ (eqn. 1) which can be separated from NH_4Cl by sublimation of the latter *in vacuo*. Tetraamidophosphonium chloride (10.1)

10.1 Phosphorus–nitrogen single bonds – phosphorus amides and amines

$$PCl_5 + 8NH_3 \longrightarrow \underset{(10.1)}{P(NH_2)_4Cl} + 4NH_4Cl \qquad (1)$$

can be recrystallized from methanol and decomposes at 200 °C. In the presence of bases such as diethylamine it is converted to the diphosphazene (10.2) (eqn. 2).

$$P(NH_2)_4Cl + Et_2NH \longrightarrow \underset{(10.2)}{[(NH_2)_3P\!=\!\!=\!N\!=\!\!=\!P(NH_2)_3]Cl} + NH_3 + Et_2NH_2Cl \qquad (2)$$

Reaction of (10.1) with more PCl_5 results in the remarkable cation (10.3) (eqn. 3)

$$P(NH_2)_4Cl + 4PCl_5 \longrightarrow \underset{(10.3)}{\begin{bmatrix} & N\!=\!PCl_3 & \\ & | & \\ Cl_3P\!=\!N\!-\!P\!-\!N\!=\!PCl_3 & \\ & | & \\ & N\!=\!PCl_3 & \end{bmatrix}^{+}} Cl^- + 8HCl \qquad (3)$$

which is also a phosphazene. The products of the amidolysis of PCl_5 by NH_4Cl at higher temperatures yields other polyphosphazenes as we shall see in Section 10.5.

Although the simple amides $P(NH_2)_3$ and $P(NH_2)_5$ have eluded separation it is possible to get a partially substituted derivative of the former as the bis(tri-fluoromethyl) compound (eqn. 4) in 96% yield. If more $(CF_3)_2PCl$ is added to the product $[(CF_3)_2P]_2NH$ is formed. The methyl analogue of this derivative is formed in reaction (5) and this is also the product of attempts to make R_2PNH_2 compounds by amine–amide exchange (eqn. 6). Thus it would seem that highly electronegative groups on phosphorus are required to stabilize the $P(III)-NH_2$ grouping and make possible separation of a compound with this component in it.

$$(CF_3)_2PCl + NH_3 \longrightarrow (CF_3)_2PNH_2 \quad [\text{m.p.}, -88\,°C; \text{b.p.}\ 67\,°C] \qquad (4)$$

$$2Me_2PH + NaNH_2 \xrightarrow[-78\,°C]{NH_4Cl} (Me_2P)_2NH \quad [\text{m.p.}, 39\,°C] \qquad (5)$$

$$2Me_2PNMe_2 + NH_3 \longrightarrow (Me_2P)_2NH + 2Me_2NH \qquad (6)$$

As we have seen $P(V)-NH_2$ is a little less demanding. When PF_5 and NH_3 gas react $PF_3(NH_2)_2$ is formed (eqn. 7)[4]. This compound is the first amide derivative of PF_5 and it has been the object of ^{19}F n.m.r. studies which showed it to have restricted rotation about the apparently single P—N bonds[5]. We shall return to this problem at the end of this section.

$$PF_5 + 2NH_3 \longrightarrow PF_3(NH_2)_2(41\%) + NH_4PF_6 \qquad (7)$$

10.1.1.2 Tetracoordinate phosphorus amides

We have already seen an example of this class in compound (10.1) and this is by far the largest group of P—NH_2 compounds – the amido derivatives of phosphinic[6], phosphonic[7] and phosphoric[8] acids and their thio counterparts. Successive replacements of the hydroxyl groups of phosphoric acid give monoamidophosphoric acid, $P(O)(NH_2)(OH)_2$, diamidophosphoric acid, $P(O)(NH_2)_2(OH)$, and phosphoryl triamide, $P(O)(NH_2)_3$. The preparations of these are shown in Scheme 10.1 together with their acid strengths compared to the equivalent ones of H_3PO_4. It can be seen that one NH_2 group has little effect on acid strength.

Scheme 10.1 Amidophosphoric acids, etc.

$POCl_3$ + liquid $NH_3 \longrightarrow P(O)(NH_2)_3$

$POCl_3 + PhOH \longrightarrow P(O)Cl_2(OPh) \xrightarrow{NH_3} P(O)(NH_2)_2(OPh) \xrightarrow[\text{(ii) } H_2S]{\text{(i) NaOH}} P(O)(NH_2)_2(OH)$
$K_a = 1.2 \times 10^{-5}$

$POCl_3 + 2PhOH \longrightarrow P(O)Cl(OPh)_2 \xrightarrow{NH_3} P(O)(NH_2)(OPh)_2 \xrightarrow[\text{(ii) } H_2S]{\text{(i) NaOH}} P(O)(NH_2)(OH)_2$
$K_{a1} = 1.2 \times 10^{-3}$
$K_{a2} = 2.1 \times 10^{-8}$

[Cf. $P(O)(OH)_3$: $K_{a1} = 8.3 \times 10^{-3}$ and $K_{a2} = 5.5 \times 10^{-8}$]

Monoamidophosphoric acid reverts to H_3PO_4 on treatment with perchloric acid or to ammonium polyphosphates on heating at 100 °C (eqn. 8).

$$P(O)(NH_2)(OH)_2 \xrightarrow{100\,°C} \left[\begin{array}{c} O \\ \parallel \\ -P-O- \\ | \\ ONH_4 \end{array} \right]_n \qquad (8)$$

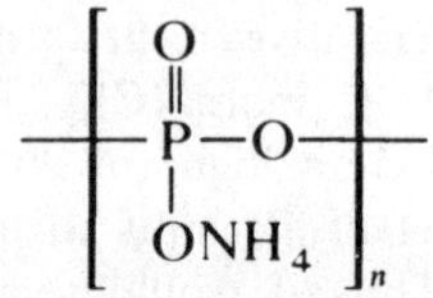

Esters of the acid are best prepared from $POCl_3$ prior to the treatment with NH_3 as shown in Scheme 10.1. The monoamidophosphate anion exists in the betaine form, i.e. $P(O)(NH_3^+)(O^-)_2$, and this has a special place in P—N chemistry since it is one of the few examples of this bond free of π complications, the nitrogen lone pair being occupied with a proton. Bond energy[9] and bond length[10] data for this bond are often quoted as authentic single bond values (10.4).

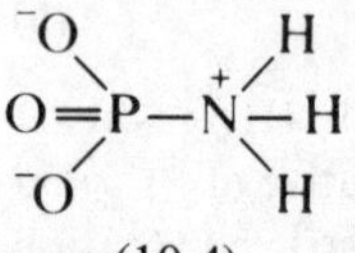

$$\begin{array}{c} ^-O \quad\quad H \\ \diagdown \quad + \diagup \\ O=P-N-H \\ \diagup \quad\quad \diagdown \\ ^-O \quad\quad H \end{array} \qquad r(P-N) = 177\text{ pm}; \; E(P-N) = 288 \text{ kJ mol}^{-1} \; (69 \text{ kcal mole}^{-1})$$

(10.4)

Both the diamidophosphoric acid and phosphoryl triamide are hydrolytically unstable, the former being converted to $P(O)(NH_2)(ONH_4)(OH)$ and the latter to the mono- or di-amido acids or H_3PO_4 depending on conditions. It would appear that the main disadvantage of P—NH_2 bond is the hydrolytic instability relative to P—OH. However in eqn. (9) it is the P(S)OR moiety which reacts and the P—NH_2 remains intact.

$P(S)(NH_2)(OR)_2 + NaOH \longrightarrow P(O)(NH_2)(OR)(SNa) + ROH \qquad (9)$

Stability of the P—N single bond can be achieved by incorporating organic groups on the nitrogen. However before we turn to these it is necessary to ask why P—NH_2 bonds in X_2PNH_2 compounds do not rearrange, like X_2POH, to give tetracoordinate derivatives (eqn. 10). If it were simply a question of the proton attaching itself to the more basic atom then by a judicious choice of electron-donating groups on phosphorus

10.1 Phosphorus–nitrogen single bonds – phosphorus amides and amines

$$\text{>P-OH} \longrightarrow \text{-P(H)=O} \quad \text{but} \quad \text{>P-NH}_2 \xrightarrow{\times} \text{-P(H)=NH} \quad (10)$$

(such as Me_2N) and electron-withdrawing groups on nitrogen (such as p-toluenesulphonyl) it might be possible to tilt the balance in favour of P being more basic than N. Indeed the product of reaction (11) is such a compound (10.5) in which the proton prefers P to N. The product (10.5) is a colourless, stable compound which shows $vP-H$ at 2460 cm^{-1}, $vP=N$ at 1140 cm^{-1} and ^1H n.m.r. confirms that H is attached to P.[11]

$$(Me_2N)_2PCl + NaNH.SO_2.C_6H_4\text{-}p\text{-}Me \longrightarrow$$

$$(Me_2N)_2P-NH.SO_2.C_6H_4\text{-}p\text{-}Me \rightleftharpoons (Me_2N)_2P(H)=N-SO_2.C_6H_4\text{-}p\text{-}Me \quad (11)$$

$$(10.5)$$

10.1.2 Phosphorus amino derivatives

10.1.2.1 Tricoordinate phosphorus amino derivatives

The dearth of information on >P-NH_2 is in stark contrast to the wealth on >P-NR_2, typified by $P(NMe_2)_3$. This was first prepared in high yield by Michaelis in 1895 by the method demonstrated in eqn. (12)[12]. It is a stable liquid, b.p. 52 °C at 14 mmHg, and its many reactions are laid out in Scheme 10.2, which demonstrates its versatility. Most of the reactions leave the P—N bond intact, although hydrogen

$$PCl_3 + 6Me_2NH \longrightarrow P(NMe_2)_3 + 3Me_2NH_2Cl \quad (12)$$

halides and alcohols cleave it, and this is a general feature of this bond. The Lewis basicity of $P(NMe_2)_3$ is demonstrated by the stability of its borane adduct. As a ligand it presumably works via a strong σ-bond since its π-acceptor properties are likely to be quite poor (see Chapter 5). While not sufficient to achieve complete substitution about a metal several partially substituted metal carbonyl complexes are known[13]. On the other hand $PF_2(NMe_2)$ is capable of replacing up to all four CO ligands in $Ni(CO)_4$ (eqn. 13)[14]. Dimethylaminophosphorus difluoride is a particular-

$$Ni(CO)_4 + 4PF_2(NMe_2) \longrightarrow Ni\{PF_2(NMe_2)_3\}_4 \quad [\text{m.p. 111 °C}] \quad (13)$$

ly useful compound[15] and its preparation is shown in Scheme 10.2. From a theoretical point of view it is also very interesting; the structure of $PF_2(NMe_2)$[16] has a planar C_2N-P framework which shows that the lone pair on nitrogen is not exerting its usual stereochemical influence in making the arrangement at nitrogen pyramidal. Moreover the P—N bond length is 163 pm compared to the single P—N bond length of 177 pm just discussed. Together these facts suggest a donor $2p(N) \rightarrow 3d(P)$ π-bond in addition to the σ-bond.

Scheme 10.2 Reactions of P(NMe₂)₃ and its derivatives, compiled mainly from ref. 1

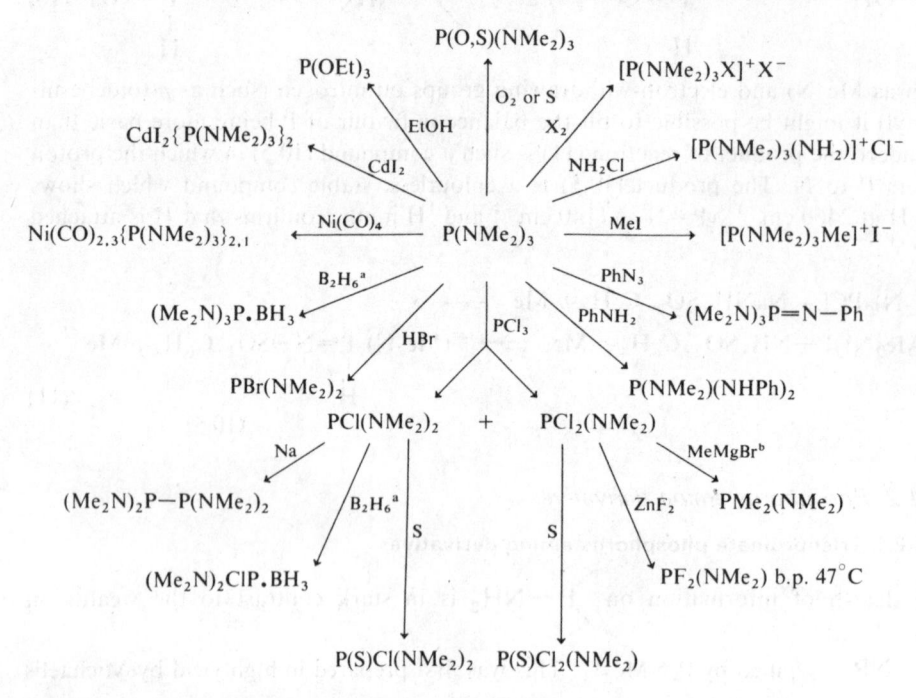

Central compound: P(NMe₂)₃

Reactions radiating from P(NMe₂)₃:
- + EtOH → P(OEt)₃
- + O₂ or S → P(O,S)(NMe₂)₃
- + X₂ → [P(NMe₂)₃X]⁺X⁻
- + NH₂Cl → [P(NMe₂)₃(NH₂)]⁺Cl⁻
- + MeI → [P(NMe₂)₃Me]⁺I⁻
- + CdI₂ → CdI₂{P(NMe₂)₃}₂
- + Ni(CO)₄ → Ni(CO)₂,₃{P(NMe₂)₃}₂,₁
- + B₂H₆[a] → (Me₂N)₃P·BH₃
- + PhN₃ → (Me₂N)₃P=N—Ph
- + PhNH₂ → P(NMe₂)(NHPh)₂
- + HBr → PBr(NMe₂)₂
- + PCl₃ → PCl(NMe₂)₂ + PCl₂(NMe₂)

From PCl(NMe₂)₂:
- + Na → (Me₂N)₂P—P(NMe₂)₂
- + B₂H₆[a] → (Me₂N)₂ClP·BH₃
- + S → P(S)Cl(NMe₂)₂

From PCl₂(NMe₂):
- + S → P(S)Cl₂(NMe₂)
- + ZnF₂ → PF₂(NMe₂) b.p. 47 °C
- + MeMgBr[b] → PMe₂(NMe₂)

[a] Alternatively using NaBH₄/HCl; [b] excess Grignard gives PMe₃.

Organo-substituted derivatives of the type R₂PNMe₂ can be made either by starting with the corresponding organophosphorus halide e.g. eqn. (14) or by preforming the

$$Ph_2PCl + Ph_2NNa \longrightarrow Ph_2P-NPh_2 \quad [m.p.\ 132\ °C] \qquad (14)$$

P—N bond and subsequently introducing the organic groups as for Me₂P—NMe₂ in Scheme 10.2. This compound which is a liquid (m.p. −97 °C, b.p. 99 °C) has been taken as representative of its class and its reactions are shown in Scheme 10.3;

Scheme 10.3 Reactions of Me₂P—NMe₂ compiled mainly from ref. 1

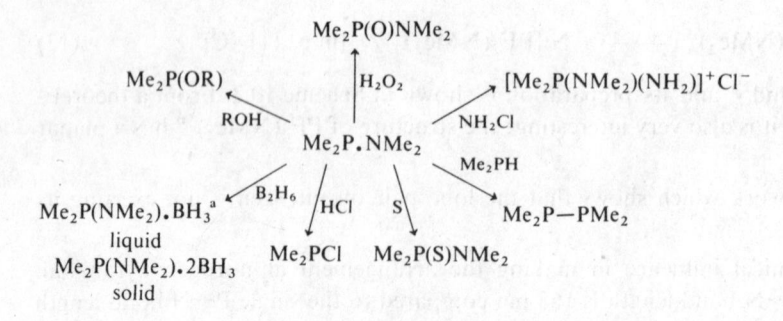

Central compound: Me₂P·NMe₂

Reactions:
- + ROH → Me₂P(OR)
- + H₂O₂ → Me₂P(O)NMe₂
- + NH₂Cl → [Me₂P(NMe₂)(NH₂)]⁺Cl⁻
- + Me₂PH → Me₂P—PMe₂
- + B₂H₆ → Me₂P(NMe₂)·BH₃[a] (liquid); Me₂P(NMe₂)·2BH₃ (solid)
- + HCl → Me₂PCl
- + S → Me₂P(S)NMe₂

[a] BH₃ attached to P – see Chapter 11.

10.1 Phosphorus–nitrogen single bonds – phosphorus amides and amines

similarities to Scheme 10.2 are obvious, and these compounds add little to our knowledge of P(III)—NR$_2$ bonds.

10.1.2.2 Tetracoordinate phosphorus amino derivatives

Most interest in amino-phosphorus chemistry has centred around phosphoryl derivatives and the number of compounds with the P(O)NR$_2$ grouping is truly enormous[2]. The most useful of them is P(O)(NMe$_2$)$_3$ which can be prepared from POCl$_3$ and excess dimethylamine (eqn. 15); it is a liquid and finds use as a good polar aprotic non-aqueous solvent. Alkali metals will dissolve in it giving blue solutions which are paramagnetic like their Na/liquid NH$_3$ counterparts. It is a donor solvent and because of its ligand behaviour it is capable of dissolving transition metal salts. Complexes like FeCl$_3${P(O)(NMe$_2$)$_3$} and Zn{P(O)(NMe$_2$)$_3$}$_4$ 2ClO$_4$ have been isolated from such solutions.

$$POCl_3 + 6Me_2NH \longrightarrow P(O)(NMe_2)_3 + 3Me_2NH_2Cl \tag{15}$$

$$P(O,S)X_3 + 2R_2NH \longrightarrow P(O,S)X_2(NR_2) + R_2NH_2X \tag{16}$$

$$P(O,S)X_3 + 4R_2NH \longrightarrow P(O,S)X(NR_2)_2 + 2R_2NH_2X \tag{17}$$

$$P(O,S)X_3 + 6R_2NH \longrightarrow P(O,S)(NR_2)_3 + 3R_2NH_2X \tag{18}$$

Reaction (15) is only one of a series of reactions (16)–(18) which can be performed with any phosphoryl or thiophosphoryl amide and almost any amine. The products from some of the reactions, especially (16) and (17) can be used as starting materials for the introduction of other groups by replacement of the residual halide e.g. eqns. (19) and (20).

$$P(O)Cl_2(NR_2) + ROH \text{ or } RSH \longrightarrow P(O)(OR,SR)_2(NR_2) \tag{19}$$

$$P(O)Cl(NR_2)_2 + ROH \text{ or } RSH \longrightarrow P(O)(OR,SR)(NR_2)_2 \tag{20}$$

Not all are amenable to further reaction, however, and some are particularly immune to nucleophilic attack especially the thio derivatives such as P(S)Cl$_2$(NR$_2$), which can be steam-distilled, and are indifferent to ROH or PhOH. The tris(amino)thiophosphoryl compounds react with alkyl halides to form quaternary phosphonium salts (eqn. 21).

$$P(S)(NR_2)_3 + R'X \longrightarrow [P(SR')(NR_2)_3]^+ X^- \tag{21}$$

Organometallic reagents will react with the P—X bonds in the mono- and di-amino derivatives eqns. (22) and (23) and this offers a route to phosphinic and phosphonic derivatives because the amino groups are resistant towards these organometallic reagents. Usually though these types of compound are prepared by the introduction of the amino group as the final step in the synthesis.

$$P(O)X_2(NR_2) + R'MgBr \text{ or } R'Li \longrightarrow P(O)R'_2(NR_2) \tag{22}$$

$$P(O)X(NR_2)_2 + R'MgBr \text{ or } R'Li \longrightarrow P(O)R'(NR_2)_2 \tag{23}$$

Thio derivatives can sometimes be made by treating the corresponding phosphoryl compound with P$_4$S$_{10}$ (24). The thio acids can be obtained directly from P$_4$S$_{10}$ by treating this with the requisite amount of amine (eqns. 25), and the tris(dialkylamino) derivative by using excess amine (eqn. 26).

$$P(O)Cl_{1,2}(NR_2)_{2,1} + P_4S_{10} \xrightarrow{150°C} P(S)Cl_{1,2}(NR_2)_{2,1} \qquad (24)$$

$$P_4S_{10} + 4R_2NH \longrightarrow 4P(S)(SH)_2(NR_2) \text{ low yield}$$

$$P_4S_{10} + 8R_2NH \longrightarrow 4P(S)(SH)(NR_2)_2 \qquad\qquad\qquad (25)$$

$$P_4S_{10} + 12R_2NH \longrightarrow 4P(S)(NR_2)_3 \qquad (26)$$

The P—N bond in all the above compounds is split by hydrogen halides. In general the hydrolytic stability is low and occurs in a stepwise process, e.g. eqn. (27), although in eqn. (28) it is one of the ester groups which is replaced. It has been shown that

$$P(O)(NPh_2)_3 \xrightarrow{KOH} P(O)(OK)(NPh_2)_2 \longrightarrow P(O)(OK)_2(NPh_2)$$
$$\longrightarrow K_3PO_4 \qquad (27)$$

$$P(O)(OR')_2(NR_2) + Ba(OH)_2 \longrightarrow P(O)(OR)(OBa_{\frac{1}{2}})(NR_2) + R'OH \qquad (28)$$

phosphinic amides $R_2P(O)NH_2$, $R_2P(O)NHMe$ and $R_2P(O)NMe_2$ are hydrolyzed by acids much faster than carboxylic amides $RCONH_2$, $RCONHMe$, $RCONMe_2$[17]. The reason for this is that the former are protonated at the N atom, prior to hydrolysis, which makes the amine a good leaving group, whereas the carboxylic amides are protonated on oxygen which is less advantageous.

The reason why the carboxylic amide nitrogen is less basic is due to the involvement of the lone pair in a delocalized π-bond to carbon $R—C\overset{O}{\underset{NMe_2}{\cdots}}$ Confirmation of this comes from ^{1}H n.m.r. spectroscopy which shows a doublet for the methyl protons because of restricted rotation about the C—N bond. In the phosphinic amides the nitrogen is more basic because its lone pair is more localized on the P atom and is less involved in π-bonding to phosphorus. However such π-bonding is not entirely absent since in some instances $R_2P(O)NMe_2$ compounds also show restricted rotation about the P—N bond[18].

Phosphinic amides of primary amines can be methylated with diazomethane (eqn. 29) without the P—N bond being affected. Primary amine derivatives can undergo transamination in which aryl groups can be replaced by more strongly bonded alkyl groups, or in which a more volatile amine can be boiled off and the equilibrium shifted to the desired side (eqn. 30).

$$R_2P(O)NHR' + CH_2N_2 \longrightarrow R_2P(O)NR'CH_3 + N_2 \qquad (29)$$

$$P(S)(NHR)_3 + 3R'NH_2 \rightleftharpoons P(S)(NHR')_3 + 3RNH_2 \qquad (30)$$

$$P(S)(NR_2)_3 \xrightarrow{heat} \begin{array}{c} R \\ | \\ S\diagdown \quad N \diagup NR_2 \\ P \quad\quad P \\ R_2N \diagup \quad N \diagdown S \\ | \\ R \end{array} \qquad (31)$$

10.1 Phosphorus–nitrogen single bonds – phosphorus amides and amines

Simple heating of phosphoryl and thiophosphoryl amides will cause elimination of an amine molecule and produce a cyclodiphosphazane (eqn. 31), which we shall discuss in detail later in this chapter.

10.1.2.3 Pentacoordinate phosphorus amino derivatives

The reaction of PCl_5 and primary amines tends to produce monophosphazenes, $Cl_3P=NR$, or cyclodiphosphazanes, $(Cl_3P \cdot NR)_2$, presumably by eliminating HCl from $Cl_4P \cdot NHR$ intermediates. Pentacoordinate, secondary amine derivatives, $P(NR_2)_5$, are unknown also. The reaction between PF_5 and Me_2NH gives firstly a Lewis adduct, $PF_5 \cdot Me_2NH$, which on heating eliminates HF and forms $PF_4 \cdot NMe_2$[19]. Further heating with Me_2NH at 100 °C gives $PF_3(NMe_2)_2$ in which the dimethylamino groups are equatorially placed in a *tbp* structure as expected. Other secondary amines do not go beyond the monosubstituted stage although they can be obtained by other methods.

Schmutzler[20] has championed the cause of trimethylsilylamines, Me_3SiNR_2, as useful starting compounds in many syntheses and with them he has been able to produce diamino P(V) derivatives of bulkier secondary amines (eqn. 32).

$$PF_4(NEt_2) + Me_3SiNEt_2 \longrightarrow PF_3(NEt_2)_2 + Me_3SiF \qquad (32)$$

More recently he has used Me_3SiNR_2 (R = Me, Et etc.) in reactions with PF_5, $R'PF_4$ (R' = Me, Et, Ph) and Ph_2PF_3 to form $PF_4(NR_2)$, $R'PF_3(NR_2)$ and $Ph_2PF_2(NR_2)$.[21] The ease of reaction decreases as the phosphorus decreases as an acceptor, i.e. $PF_5 >$ $PhPF_4 > RPF_4 > R_2PF_3 > R_3PF_2$. The structure of the products are as expected, with the fluorine atoms apical. Some of the products rearrange to ionic forms (eqn. 33)[22]. Several P(V) amino derivatives were mentioned in Chapter 6 and the reader is referred back to this for other information on the preparation and reactions of aminophosphoranes.

$$PhPF_4 + Me_2NH \xrightarrow{0°C} PhPF_3(NMe_2) \xrightarrow[\text{temp.}]{\text{room}} [PhPF(NMe_2)_2]^+[PhPF_5]^- \qquad (33)$$

In addition to amide and amino groups attached to phosphorus by formal single bonds there are also many hydrazine and substituted organohydrazine derivatives. These add very little that is new to what we have said. Another class of P—N compounds are the isocyanate, P—NCO, and isothiocyanate, P—NCS, derivatives. The chemistry of these resembles the phosphorus halides, in keeping with the pseudohalide properties of these groups. Again their chemistry adds little to that already discussed.

10.1.3 The bonding in phosphorus amides and amines; evidence of π contributions

The bonding between phosphorus and nitrogen in all these compounds has been described as a formal single bond, but we have seen that in some molecules things are not so simple. It would appear that there can be a π component to the bonding. This manifests itself in PF_2NMe_2 as a planar arrangement about nitrogen and a shorter P—N bond than expected. The strongest evidence of a π system is the effect this has on rotation about the bond and in Chapter 2 we mentioned compounds in which restricted rotation about P—N bonds had been observed. Proton and fluorine nuclear magnetic resonance has been most revealing in demonstrating such restriction.

Two clearly defined doublet signals for the NMe_2 group can be seen in the spectra of $PX_2.NMe_2$ compounds below $-60\ °C$, and this implies restricted P—N rotation. The reason for this restriction has been attributed to interaction between the nitrogen lone pair and that on phosphorus[23] whereas others support an explanation based on $2p(N) \rightarrow 3d(P)$ donor π-bonding[24]. In view of the fact that tetracoordinate and even pentacoordinate P—N compounds display preferred configurations about this bond it would seem that lone pair–lone pair interaction cannot be the main reason for restricted rotation.

Compound (2.9(b), page 55) showed two fluorine environments at $-70\ °C$ which arise from the α-methyl of the pyridyl ring. This is forced into the axial plane as required by the N—P π-bonding which prefers the equatorial plane[25]. Work on $F_3P(NH_2)_2$ has shown an energy barrier to free rotation about the P—N bonds of ca 46 kJ mol^{-1} (11.2 kcal mole^{-1}).[5] The measurement of coalescence temperatures, T_c, has been used to calculate the energy barrier to rotation, E_r, in $X_2P.NMe_2$ compounds[26]: $Cl_2P.NMe_2$, $T_c = -120\ °C$, $E_r = 35$ kJ mol^{-1} (8.4 kcal mole^{-1}); $PhClP.NMe_2$, $-50\ °C$, 50 kJ mol^{-1} (12 kcal mole^{-1}); $(CF_3)_2P.NMe_2$, $-120\ °C$, 38 kJ mol^{-1} (9.0 kcal mole^{-1}).

All this and more[27-29], goes to show that very few P—N bonds in phosphorus amides and amines are formal single σ-bonds. This may also serve to explain the lack of correlation between the bond and the infrared spectra of these compounds. The expected frequency of the bond is 743 cm^{-1} and Corbridge[30] who calculated this value correlated bands in the 650–850 cm^{-1} range with this bond. However, as Thomas[31] has pointed out, many organophosphorus compounds with a P—N bond have no band in this region that could be ascribed to P—N vibrations. He plumped for an overall range of 789–1102 cm^{-1} and the only region in all the P—N compounds he investigated which consistently showed a band was ca 960 cm^{-1}, and this he took to be the most reliable diagnosis for the P—N bond.

Thomas notes that in P—N—C compounds there are often two bands at 650–850 cm^{-1} (v_1) and 870–1150 cm^{-1} (v_2) which are not clearly divisible into v_1P—N and v_2N—C because these bonds are strongly coupled. That this is so was demonstrated by the deuteration of $Cl_2(S)P—NH—CH_3$ which affected both v_1 and v_2.[32] Laffite[33], who was less reticent about assigning the v_1 bands to vP—N in the compounds he investigated, concluded that this was very sensitive both to changes at nitrogen and phosphorus.

10.2 Monophosphazenes; the phosphorus–nitrogen double bond

A review in 1966 by Johnson[34] draws a parallel between the monophosphazenes, $R_3P=NR'$, and the ylids, $R_3P=CR'_2$. Since then, however, these areas have grown apart from each other and both have expanded considerably. Thus Berman[35] in 1972 in a review of $Cl_3P=N-$ compounds stated that over 300 papers had appeared in this one field in eight years. A great deal of effort has gone into the synthesis of the monophosphazenes almost all of which involves direct formation of the P=N bond by means of $-NH_2$, $-N_3$, N—Cl or $-N=N-$ compounds, although there is a method which uses the deprotonation of $\overset{\backslash}{\underset{/}{\overset{+}{P}}}-NH-$ compounds to $\overset{\backslash}{\underset{/}{P}}=N-$, see eqn. (42).

10.2 Monophosphazenes; the phosphorus–nitrogen double bond

10.2.1 Synthesis of monophosphazenes

One of the few named reactions in inorganic phosphorus chemistry is the Kirsanov reaction, named after the Russian chemist, who has developed an original discovery. Equation (34) gives the general reaction which is thought to go via $X_3PCl.NHR$ intermediates, since in the reaction between PCl_5 and $ArNH_2$ the solutions showed infrared evidence of such an intermediate, and the effect of formic acid hydrolysis on the reacting solution was the formation of $Cl_2(O)P.NHAr$.[36] The P(V) compounds used in the Kirsanov reaction have included PCl_5, Ph_3PCl_2, PF_3Cl_2 and $(PhO)_3PCl_2$, and the amido reagents have been of the type NH_3, $PhNH_2$, $RCONH_2$, $ArSO_2NH_2$ and $SO_2(NH_2)_2$ in which both groups react[37]. Although HCl is usually the eliminated by-product this is not always the case and in reaction (35) a pseudo Kirsanov reaction occurs with elimination of phenol.

$$X_3PCl_2 + H_2NR \longrightarrow X_3P{=}NR + 2HCl \qquad (34)$$

$$\text{e.g. } Ph_3PCl_2 + PhNH_2 \longrightarrow Ph_3P{=}NPh \quad [\text{m.p. } 132\,°C]$$

$$P(OAr)_5 + RSO_2NH_2 \longrightarrow (ArO)_3P{=}N{-}SO_2R + 2ArOH \qquad (35)$$

The combination of PCl_5 and NH_3 is a most complex reaction, as we have seen and will see, but it may go via the elusive monophosphazene, $Cl_3P{=}NH$, which has often been postulated although never separated or identified as an intermediate. The nearest that anyone has come to this is in reaction (36)[38] which gives (10.7) a product sufficiently stable to be distilled, – b.p. 69 °C at 0.03 mmHg.

$$(Cl_3C)_2PCl_3 + NH_3 \xrightarrow{-10\,°C} (Cl_3C)_2ClP{=}NH \qquad (36)$$
$$\qquad\qquad\qquad\qquad\qquad (10.7)$$

Sometimes the product from the Kirsanov reaction is not the monophosphazene but its dimer, a cyclodiphosphazane. A reaction which follows this path is that between PCl_5 and aromatic amines (eqn. 37).

$$PCl_5 + ArNH_2 \longrightarrow [Cl_3P.NAr]_2 + 2HCl \qquad (37)$$

The second method of wide application was used by Staudinger[39,40] in the earlier syntheses of monophosphazenes. It consists of the reaction of a P(III) derivative such as PCl_3, $P(OR)_3$, $P(NR_2)_3$, PPh_3, PPh_2Cl or PR_2Cl with an organic azide (eqn. 38). The azides which have been used for this purpose are PhN_3, RSO_2N_3, $PhCON_3$, Ph_3SiN_3 and Me_3SiN_3. The last one forms $R_3P{=}N{-}SiMe_3$ derivatives which can be converted by $MeOH/H_2SO_4$ to the corresponding $R_3P{=}NH$ compound.

$$PX_3 + R'N_3 \xrightarrow{\text{ether}} X_3P{=}N{-}N{=}N{-}R' \longrightarrow X_3P{=}N{-}R' + N_2 \qquad (38)$$

$$\text{e.g. } PPh_3 + PhN_3 \longrightarrow Ph_3P{=}NPh \quad (\text{ref. } 39)$$

In a few cases the triazo intermediate of reaction (38) has itself been isolated as in reaction (39)[41]. Incredibly the explosive looking intermediate (10.8) does not decompose to the monophosphazene until heated to 131 °C.

$$\text{Ph}_3\text{P} + \text{N}_3\text{-C}_6\text{H}_2(\text{NO}_2)_3 \longrightarrow$$

$$\underset{(10.8)}{\text{Ph}_3\text{P}=\text{N}-\text{N}=\text{N}-\text{C}_6\text{H}_2(\text{NO}_2)_3} \xrightarrow[-\text{N}_2]{131\,°\text{C}} \text{Ph}_3\text{P}=\text{N}-\text{C}_6\text{H}_2(\text{NO}_2)_3 \qquad (39)$$

Again this reaction has been used in an attempt to prepare $\text{Cl}_3\text{P}=\text{NH}$, this time from PCl_3 and hydrazoic acid, HN_3, irradiated at 5 °C.[42] The product was a PNCl containing polymer which suggests that $\text{Cl}_3\text{P}=\text{NH}$ undergoes rapid dehydrochlorination if it forms. The use of PPh_2Cl in place of PCl_3 produced the intermediate $\text{Ph}_2\text{Cl-P}=\text{N}-\text{N}=\text{N}-\text{H}$, m.p. 126 °C, which decomposed at 130 °C to give $(\text{NPPh}_2)_4$. Some $\text{Ph}_2\text{ClP}=\text{NH}$ may also have been formed but it was unstable with respect to cyclotetraphosphazene.

The Kirsanov and azide methods are the ways in which most monophosphazenes have been prepared but several other routes to them exist. N-chloro reagents react with P(III) compounds, e.g. eqns. (40A) and (40B)

$$\underset{\text{(chloramine T)}}{\text{PR}_3 + \text{NaClN.SO}_2.\text{C}_6\text{H}_4.\text{Me}} \longrightarrow \text{R}_3\text{P}=\text{N}-\text{SO}_2.\text{C}_6\text{H}_4.\text{Me} + \text{NaCl} \qquad (40\text{A})$$

$$\text{P(OAr)}_3 + \text{Et}_2\text{NCl} \xrightarrow[\text{vac.}]{100\,°\text{C}} (\text{ArO})_3\text{P}=\text{N}-\text{Et} + \text{EtCl} \qquad (40\text{B})$$

Organic diazo compounds have been used e.g. eqn. (41). The formation and deprotonation of amido phosphonium compounds can give excellent yields as in eqn. (42) which gives 92% of $\text{Ph}_3\text{P}=\text{NH}$, m.p. 126 °C. In addition to these reactions which are examples of general routes there are many others which produce monophosphazenes quite unexpectedly, e.g. eqn. (43).[43]

$$\text{P(OR)}_3 + \text{CH}_2\text{N}_2 \longrightarrow (\text{RO})_3\text{P}=\text{N}-\text{N}=\text{CH}_2 \qquad (41)$$

$$\text{PPh}_3 + \text{H}_2\text{N.SO}_3\text{H} \longrightarrow \text{Ph}_3\overset{+}{\text{P}}\text{NH}_2\,\text{HSO}_4^- \xrightarrow[\text{liq. NH}_3]{\text{NaNH}_2} \text{Ph}_3\text{P}=\text{NH} \qquad (42)$$

$$\text{P(S)(NCS)}_3 + \text{Cl}_2 \longrightarrow \text{Cl}_3\text{P}=\text{N}-\text{CCl}_3 + \text{SCl}_2 + \text{S}_2\text{Cl}_2 + \text{ClSCCl}_3 \qquad (43)$$

These then are some of the reactions in which the $\text{P}=\text{N}$ link forms. The products are chemically reactive as one might expect and we shall consider their reactions as they affect the phosphorus and nitrogen sides of the bond.

10.2.2 Chemical attack at phosphorus

The P,P,P-trichloromonophosphazenes are particularly useful in that many derivatives can be made from them by nucleophilic attack.

10.2 Monophosphazenes; the phosphorus–nitrogen double bond

The extent of alcoholysis depends upon whether the alcohol or phenol itself is used, in which case partial replacement occurs (eqn. 44), or the corresponding alkoxide or phenoxide, in which case complete substitution is possible (eqn. 45).

$$Cl_3P{=}N{-}SO_2R + R'OH \longrightarrow Cl_2(R'O)P{=}N{-}SO_2R + HCl \qquad (44)$$

$$Cl_3P{=}N{-}Ar + 3PhONa \longrightarrow (PhO)_3P{=}N{-}Ar + 3NaCl \qquad (45)$$

In reaction (44) steps must be taken to ensure rapid removal of the HCl formed, e.g. by bubbling CO_2 through the reaction mixture, otherwise the reaction takes a different course with the formation of a single P—N bond (eqn. 46).

$$Cl_3P{=}N{-}SO_2R + R'OH \xrightarrow{-R'Cl} [Cl_2(HO)P{=}N{-}SO_2R]$$
$$\longrightarrow Cl_2(O)P{-}NH{-}SO_2R \qquad (46)$$

Aminolysis e.g. eqn. (47) and the reaction with Grignard reagents e.g. eqn. (48) occur without affecting the P=N bond.

$$Cl_3P{=}N{-}CO.Ph + PhNH_2 \xrightarrow[\text{benzene}]{\text{boiling}} (PhNH)_3P{=}N{-}CO.Ph \qquad (47)$$

$$Cl_3P{=}N{-}CO{-}Ph + PhMgBr \longrightarrow Ph_3P{=}N{-}CO.Ph \qquad (48)$$

Complete hydrolysis of the $Cl_3P{=}NR$ derivatives results in complete breakdown of the molecule into phosphoric acid, HCl, and amine hydrochloride. Partial hydrolysis can be achieved by the use of formic acid (eqn. 49) but this leads to single P—N bond formation, c.f. eqn. (46). The hydrolytic stability of the other monophosphazenes depends very much on the nature of the groups attached to phosphorus. Acid hydrolysis results in the cleaving of the P=N bond probably via a phosphonium step (eqn. 50), but this need not always be the outcome of the reaction.

$$Cl_3P{=}N{-}X + HCO_2H \longrightarrow Cl_2(O)P{-}NHX + CO + HCl$$
$$[X = R, COR, SO_2R \text{ etc.}] \qquad (49)$$

$$Ph_3P{=}N{-}R + H_2O \longrightarrow Ph_3\overset{+}{P}{-}NHR\ OH^- \longrightarrow Ph_3PO + RNH_2 \qquad (50)$$

For instance in eqn. (51) the P—N bond survives, and it is an S—N bond which is the victim of attack.

$$(PhO)_3P{=}N{-}SO_2.OPh + 3H_2O \longrightarrow (PhO)_2P(O)NH_2 + 2PhOH + H_2SO_4 \qquad (51)$$

Alkaline hydrolysis, while effective for $Cl_3P{=}NR$ is without effect on $(RO)_3\text{-}P{=}N.COR$ derivatives and $(ArNH)_3P{=}N{-}SO_2R$ compounds are even able to resist the onslaught of boiling alcoholic alkali solutions.

10.2.3 Chemical attack at nitrogen

Substitution at nitrogen in the compound $Ph_3P{=}NH$ has been possible[44] with a variety of reagents, eqns. (52), (53) and (54).

$$Ph_3P{=}NH + EtI \longrightarrow Ph_3P{=}N{-}Et + HI \qquad (52)$$

$$Ph_3P{=}NH + ClSO_2NH_2 \longrightarrow Ph_3P{=}N.SO_2NH_2 + HCl \qquad (53)$$

$$Ph_3P{=}NH + X_2 \longrightarrow Ph_3P{=}N{-}X + Ph_3\overset{+}{P}NH_2X^- \quad X = Cl, Br, I \qquad (54)$$

In the last reaction one of the products, Ph₃P=N—Br, m.p. 172 °C, serves as a very useful starting point for other phosphazenes, see page 403.

Alkylation by EtI of the product of eqn. (52) to give Ph₃PNEt₂⁺I⁻ requires forcing conditions and this compound easily reverts to the monophosphazene on heating. However alkylation is generally easier than this and the products undergo hydrolysis to R₃PO and the corresponding amine.

Protonation and deprotonation of monophosphazenes is similar to that of the ylids. But whereas ylids are readily protonated the monophosphazenes are less avid for a proton, and whereas the ylids normally require a very strong base to deprotonate them* the monophosphazene will give up the proton under milder conditions. By way of illustration compare the bases required to effect reactions (55) and (56).

$$Ph_3\overset{+}{P}-NH_2\ Br^- \xrightarrow{Et_3N} Ph_3P=N-H + Et_3NHBr \tag{55}$$

$$Ph_3\overset{+}{P}-CH_3\ Br^- \xrightarrow{LiMe} Ph_3P=CH_2 + CH_4 + LiBr \tag{56}$$

The Ph₃PNH₂⁺ cation is hydrolytically stable and can be prepared directly from Ph₃P and NH₂Cl or Ph₃PCl₂ and NH₃ as well as by the method of eqn. (42). It can be converted into other salts Ph₃PNH₂⁺ X⁻ (X = PF₆⁻, ClO₄⁻, IO₄⁻, I⁻ etc.) by metathetical reaction with the appropriate sodium salt. The other protonated monophosphazenes R₃PNHR⁺X⁻ are hydrolyzed to R₃PO and amine.

Some of the reactions of the monophosphazenes involve both the phosphorus and nitrogen centres. Compounds of the type R₂(R'O)P=NX might have been expected to undergo migration of the R' group from the oxygen to the nitrogen (eqn. 57), the driving force behind the rearrangement being the exchange of a P=N bond for a P=O bond. Since this does not occur in these compounds it is tempting to use this fact as a basis of comparison between the P=O and P=N bonds. However rearrangement of (EtO)₃P=NPh takes place at 75 °C in the presence of EtI (eqn. 58).

$$R_2(R'O)P=N-X \longrightarrow R_2(O)P-NR'X \tag{57}$$

$$(EtO)_3P=N-Ph \xrightarrow[75\ °C]{EtI} (EtO)_2(O)P-NEtPh \tag{58}$$

Reactions analogous to the Wittig reaction between ylids and other double bonds (eqn. 59) also occur between monophosphazene and such double bond compounds as CO₂, RNCO, Ph₂CO and SO₂ e.g. eqn. (60).

$$R_3P=CR^1R^2 + R^3R^4C=O \longrightarrow R^1R^2C=CR^3R^4 + R_3PO \tag{59}$$

$$R_3P=NR^1 + R^2R^3CO \longrightarrow R^1N=CR^2R^3 + R_3PO \tag{60}$$

Both reactions are thought to go via four-centre intermediates e.g. eqn. (61)⁴⁵; the isocyanic acid produced in this case then adds to excess Ph₃P=NH to form the

---

* Note however that Ph₃P⁺CH(CN)₂ can be deprotonated by dilute NaOH and cyclopentadienyl-PPh₃⁺ only requires aqueous ammonia to form cyclopentadienylidene=PPh₃.

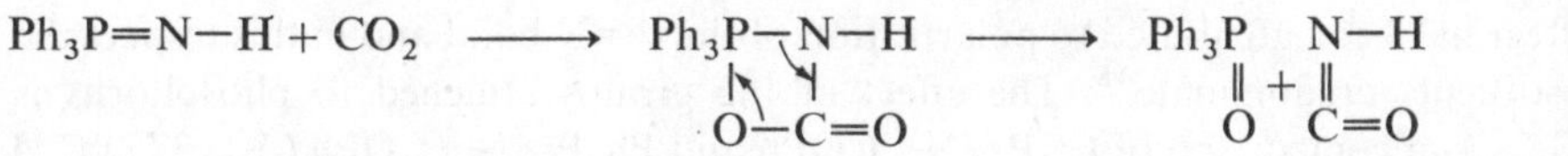

10.2 Monophosphazenes; the phosphorus–nitrogen double bond

$$Ph_3P=N-H + CO_2 \longrightarrow Ph_3P-N-H \longrightarrow Ph_3P \quad N-H \qquad (61)$$
$$\qquad\qquad\qquad\qquad\quad | \quad\; | \qquad\qquad\quad \| + \|$$
$$\qquad\qquad\qquad\qquad\; O-C=O \qquad\qquad\; O \quad C=O$$

$$Ph_3P=N-H + CS_2 \longrightarrow Ph_3P=S + HNCS \qquad\qquad\qquad (62)$$

salt $Ph_3PNH_2^+ NCO^-$. The same reaction occurs with CS_2 in place of CO_2 (eqn. 62), which somewhat undermines the idea that the driving force behind the above reactions is the formation of the very strong P=O bond.

10.2.4 Physical properties of the monophosphazenes and the nature of the bond

Berman[46] has produced an exhaustive compilation of the melting points, boiling points, ^{31}P chemical shifts, etc. for the phosphazo trihalides, $X_3P=N-$ compounds. Unfortunately little of this is of help in understanding the bonding. In fact some information such as ^{31}P values become even more mystifying than usual, in that consistent molecular environmental changes do not produce comparable changes in shift: e.g. $Cl_3P=N-COCCl_3$, $\delta = 25.1$ ppm (85 % H_3PO_4); $PhCl_2P=N-COCCl_3$, $\delta = 31.6$ ppm; $Ph_2ClP=N-COCCl_3$, $\delta = 40.0$ ppm; and $Ph_3P=N-COCCl_3$, $\delta = 23.8$ ppm.

Infrared spectroscopy has been more enlightening in that the $\nu P=N$ vibration gives rise to a peak in the 1147–1500 cm^{-1} region[47]. Although this is a wide spread of values there is not the doubt about correlating this stretching mode that there was with $\nu P-N$. Table 10.1 lists $\nu P=N$ values for a wide range of groups attached to phosphorus and nitrogen, arranged where possible to demonstrate the variations caused by changes at nitrogen. The groups attached to nitrogen cause the maximum changes, up to 300 cm^{-1} in the value of $\nu P=N$ as the $Cl_3P=N-X$ compounds show.

Table 10.1 $\nu P=N$ assignments for the monophosphazenes[48-51]

Monophosphazene	$\nu P=N/(cm^{-1})$	Monophosphazene	$\nu P=N/(cm^{-1})$
$Cl_3P=N-C(CF_3)_3$	1500	$Ph_3P=N-Ph$	1344
$Cl_3P=N-CCl_3$	1450 (1360)	$Ph_3P=N-NHCOPh$	1333
$Cl_3P=N-COCF_3$	1390	$Ph_3P=N-COPh$	1332
$Cl_3P=N-Bu^t$	1370	$Ph_3P=N-SiMe_3$	1315 (1302)
$Cl_3P=N-COCCl_3$	1370 (1320)	$Ph_3P=N-Me$	1230
$Cl_3P=N-CO-C_6H_4\text{-}p\text{-}NO_2$	1295	$Ph_3=N-SO_2-C_6H_4\text{-}p\text{-}Me$	1147
$Cl_3P=N-COPh$	1295	$Bu^n_3P=N-Ph$	1339
$Cl_3P=N-SO_2-C_6H_4\text{-}p\text{-}Me$	1199	$Me_3P=N-SiMe_3$	1286
$F_3P=N-SO_2F$	1357	$PhCl_2P=N-Ph$	1357
$Br_3P=N-SO_2F$	1190	$MeCl_2P=N-COPh$	1307
		$(Cl_3C)_2ClP=N-H$	1208

The effect has been attributed to polarization of the P=N bond and in this respect the N-substituents predominate[48]. The effect of the groups attached to phosphorus is much less as a comparison of Cl$_3$P=N—COPh and Ph$_3$P=N—COPh (Δ = 37 cm^{-1}) shows. The largest difference, between F$_3$P=N—SO$_2$F and Br$_3$P=N—SO$_2$F, is 167 cm^{-1}. Why are the frequencies so sensitive to changes of one substituent at nitrogen and so much less sensitive to changes in *three* substituents at phosphorus?

Frequencies are related to force constants, which in turn reflect bond order. Can the bond order vary very much in the monophosphazenes? It would appear so. The simplest way to imagine the bonding in these compounds is to use the analogy of X$_3$P=O in which there is a donor σ-bond and a donor π back-bond to phosphorus, see Chapter 2. A purely σ-bond would give an ionic form to the bonding, $-\overset{+}{P}-\overset{-}{N}$, with bond lengths in keeping with formal single bonds, i.e. *ca* 178 pm. In fact in the few derivatives whose dimensions are known it is considerably less than this, (10.9)[52] and (10.10)[53]. Moreover the bond angles at nitrogen indicate an sp^2 (or tr) σ framework which is in keeping with a pair of electrons on nitrogen occupying the 2p$_z$ orbital and overlapping with the empty 3d$_{xz}$ on phosphorus to form a donor π-bond as shown in (10.11). The second lone pair remains on nitrogen, but there are good

Ph_156.7 pm
Ph—P—N
Ph 124°
    Br (p-C$_6$H$_4$Br)

(10.9)

F_164.1 pm
Ph—P—N
Ph   119° CH$_3$

(10.10)

(10.11)

grounds for thinking that this is also involved in the bonding. It could either donate into another empty 3d orbital on phosphorus, one which is lying in the xy plane that encompasses P, N and R, or it could be drawn into the bonding between N and group R. For example, in X$_3$P=N—SO$_2$—C$_6$H$_4$-*p*-Me it could well be the latter, in which case the N—S bond would be strengthened at the expense of the N—P bond. Compounds of this type have much lower vP=N values than the rest of the monophosphazenes, which is in line with this argument. Whatever the explanation, however it can be seen that the group attached to nitrogen has the greatest potential for affecting its bonding to phosphorus, as the infrared frequencies show.

Infrared spectroscopy has also been used to study a tautomeric relationship (63) in which it is not obvious at first glance whether the amino (I) or phosphazene form (II) would be preferred. In fact all of fifteen esters of this type which were studied showed structure (II) with vP=N at 1325–1385 cm^{-1}.[54]

(RO)$_2$P—NHPh $\rightleftharpoons$ (RO)$_2$HP=N—Ph   (63)
    (I)   (II)

Bond enthalpy values can also give strong supporting evidence for a particular type of bonding, but in those monophosphazenes whose thermochemistry has been investigated the results are of little help. For the compound $Ph_3P=N-Et$ $E(P=N)$ was found to be 412 kJ mol^{-1} (98.4 kcal mole^{-1}) nicely in keeping with the double bond concept; in $Me_3P=N-Et$, on the other hand $E(P=N)$ turned out to be 291 kJ mol^{-1} (69.6 kcal mole^{-1})[55]. Bearing in mind that there is no change at nitrogen in these compounds it is difficult to produce a convincing explanation for the difference in view of what has been said about the infrared data. If the phenyl groups attached to phosphorus can greatly strengthen the P—N bond in this situation why does this not show up as dramatically in the infrared? Cf $Ph_3P=N-SiMe_3$ $\nu P=N$ 1315 (1302) cm^{-1} and $Me_3P=N-SiMe_3$ 1286 cm^{-1}.

The next class of phosphorus–nitrogen derivatives are the cyclodiphosphazanes, $(R_3P.NR)_2$, which have already been referred to as the dimers of the monophosphazene. Unfortunately compounds come down strongly on one side or the other and as yet little work has been done on studying the monomer $\rightleftharpoons$ dimer relationship since few derivatives display an equilibrium. Kirsanov[56] noted that the K_b of the amine used in reaction (64) determined the fate of the product. Very weakly basic amines, $K_b < 10^{-13}$, gave monophosphazenes, while amines with $K_b > 10^{-10}$ invariably gave the cyclodiphosphazanes, and in benzene solution these showed no sign of dissociation even on heating. Bases of intermediate K_b i.e. 10^{-13}–10^{-10} also gave cyclodiphosphazanes but in benzene solution they did show some indication of dissociation as shown by their mol. wts.

$$PCl_5 + RNH_2 \longrightarrow Cl_3P=NR \text{ or } (Cl_3P.NR)_2 \qquad (64)$$

If the formation of dimers involves a donor N→P σ-bond then electron-withdrawing groups on phosphorus should make it a better acceptor, and this would be an explanation of why $Cl_3P=NR$ compounds tend to be dimeric and $Ph_3P=NR$ tend to be monomeric. However the groups attached to phosphorus play a minor role as shown by the fact that $(Ph_2ClP.NMe)_2$ is dimeric[57]. By the right combinations of groups on phosphorus and nitrogen it might be possible to produce a compound in which the monomer and dimer co-exist in solution. Alternatively the choice of solvent should affect an equilibrium if one is present and this would appear to be what is happening with $(Cl_3P.NAr)_2$ compounds, which are dimeric in benzene but appear to be monomeric in the more polar solvent dioxan.

Interesting as this work promises to be in breaking down the barrier between monophosphazenes and cyclodiphosphazanes, at the present time they appear as distinct types of phosphorus compound. This is not unexpected when one considers the bonding situation at phosphorus which is pentacoordinate in the latter group, with all that this implies.

10.3 The cyclodiphosphazanes

Two kinds of phosphorus–nitrogen compound parade under this banner, those of formula $(X_3P.NR)_2$ which we have been relating to the monophosphazenes, and those in which the phosphorus is tetracoordinate $(X(O,S)P.NR)_2$. Although from a valency point of view these are similar, from a bonding position they are very different.

10.3.1 Synthesis of cyclodiphosphazanes

Most aromatic amines react with PCl$_5$* to form (Cl$_3$P.NAr)$_2$ rather than monomers. The reaction of PCl$_5$ and MeNH$_2$, or its hydrochloride[58], gives not only cyclic dimer but a derivative containing three linked diphosphazane rings (10.12) (eqn. 64).

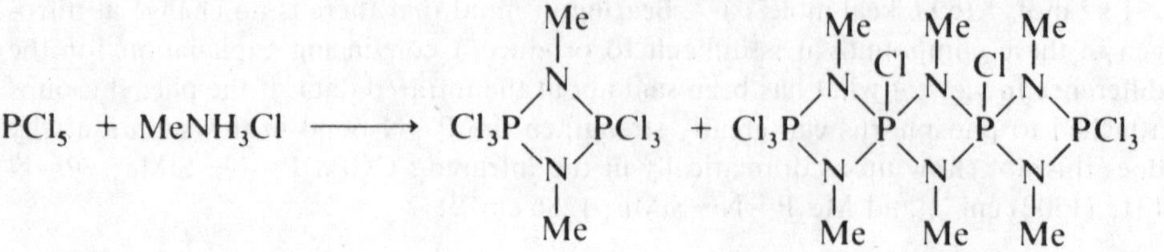

$$PCl_5 + MeNH_3Cl \longrightarrow \underset{(10.12)}{Cl_3P\text{-ring-}PCl_3 + Cl_3P\text{-three linked rings-}PCl_3} \quad (64)$$

In the presence of a trace of water a cage compound (10.13) is formed this being consistent with the formula P$_4$N$_6$Cl$_8$Me$_6$ yet having a singlet resonance in its ^{31}P n.m.r. spectrum showing all phosphorus atoms to have the same environment[59].

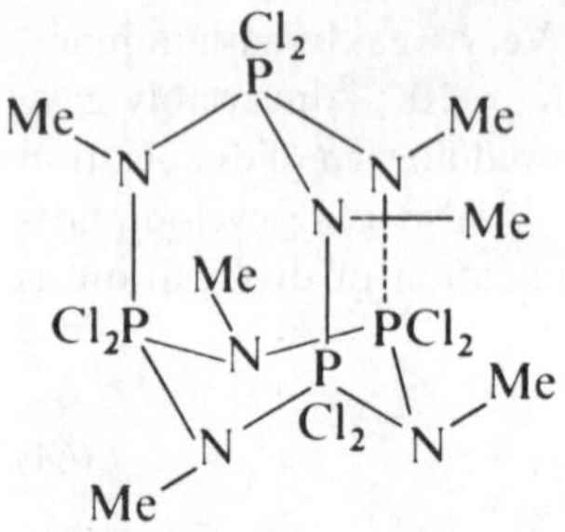

(10.13)

Not only PCl$_5$ but also PF$_5$ can be used to prepare trifluoromonophosphazenes although this requires a more active nitrogen derivative such as N-methyl or N-phenyl hexamethyldisilazane, e.g. eqn. (65)[60].

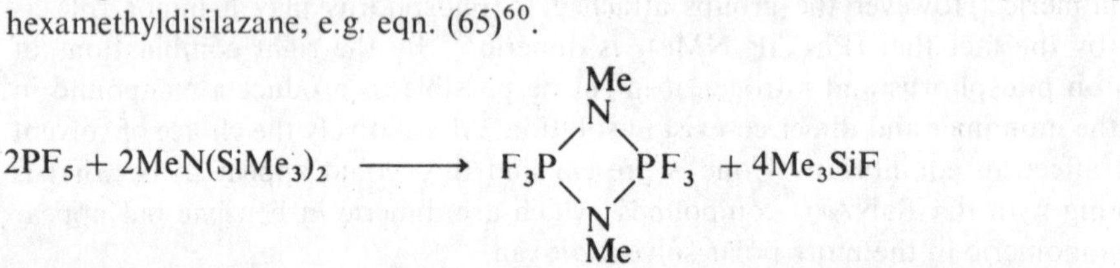

$$2PF_5 + 2MeN(SiMe_3)_2 \longrightarrow F_3P\text{-ring-}PF_3 + 4Me_3SiF \quad (65)$$

With PhPF$_4$ the reaction goes the same way but with Ph$_2$PF$_3$ the product is the monophosphazene Ph$_2$FP=NMe.[61] This is surprising because we have just seen how the corresponding chloro derivative is dimeric (Ph$_2$ClP.NMe)$_2$. The only conclusion to be drawn from this is that neither electronegativity nor steric factors at phosphorus

---

* The reaction of PCl$_3$ with aniline also gives a dimer which can react further to replace the existing P-Cl bonds. The compounds are also cyclic dimers but of phosph(III)azanes.

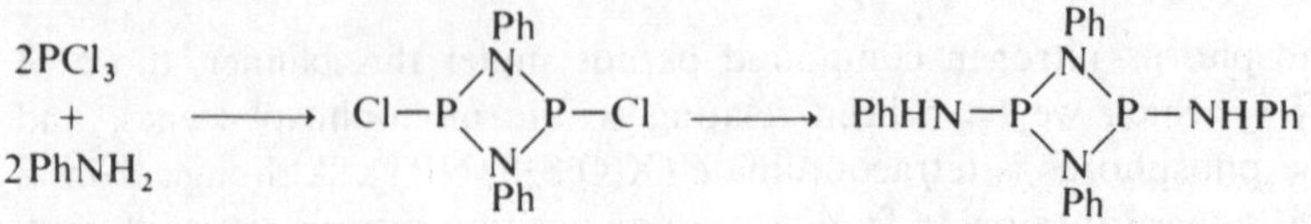

$$2PCl_3 + 2PhNH_2 \longrightarrow Cl-P\text{-ring-}P-Cl \longrightarrow PhHN-P\text{-ring-}P-NHPh$$

10.3 The cyclodiphosphazanes

are instrumental in determining whether a compound is a monophosphazene or a cyclodiphosphazane.

The second type of cyclodiphosphazanes, in which there are phosphoryl or thiophosphoryl groups, can be prepared by treating POCl$_3$ or PSCl$_3$ with the appropriate amine or amine hydrochloride (eqn. 66), although reaction proceeds beyond the intermediate chloro stage to (10.14). To obtain the chloro derivatives it is best to treat the trichloro cyclodiphosphazanes with SO$_2$ or H$_2$S, e.g. eqn. (67).

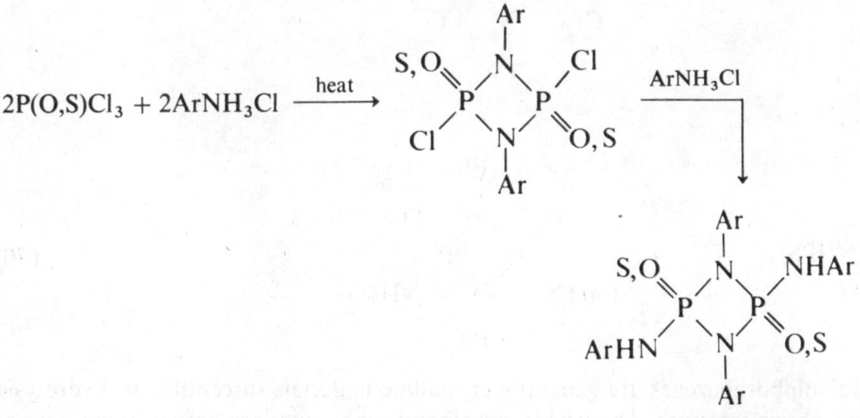

$$2P(O,S)Cl_3 + 2ArNH_3Cl \xrightarrow{heat} \text{[cyclic intermediate]} \xrightarrow{ArNH_3Cl} \text{(10.14)} \tag{66}$$

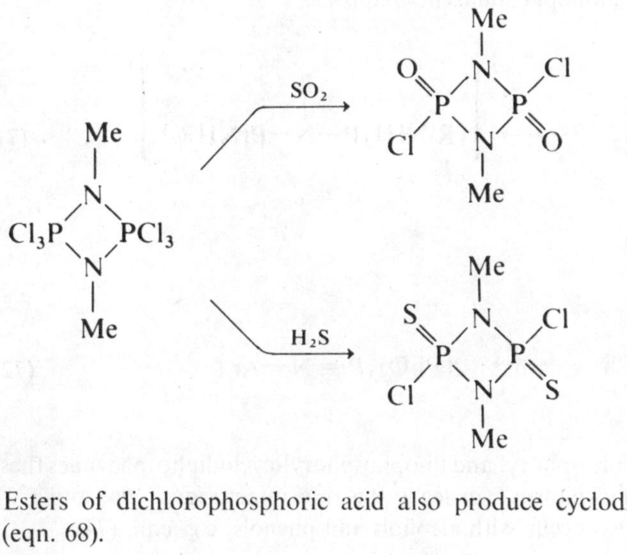

$$\text{Cl}_3\text{P-N(Me)-PCl}_3 \text{ (cyclic)} \xrightarrow{SO_2 / H_2S} \text{products} \tag{67}$$

Esters of dichlorophosphoric acid also produce cyclodiphosphazanes with amines (eqn. 68).

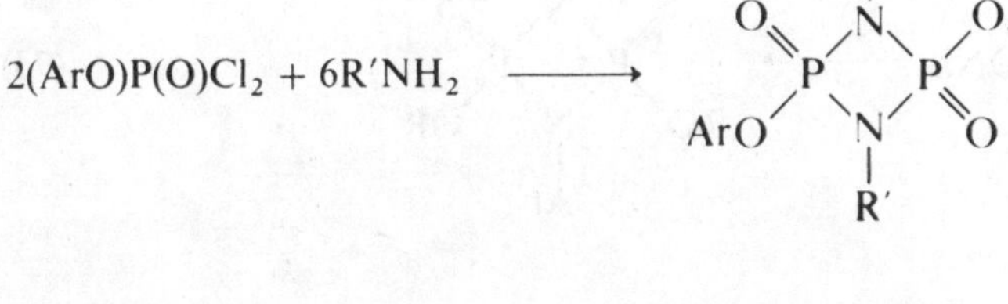

$$2(ArO)P(O)Cl_2 + 6R'NH_2 \longrightarrow \text{[cyclic product]} \tag{68}$$

One of the oldest methods of producing these compounds is by the pyrolysis of aminophosphorus derivatives. Stokes[62] in 1893 carried out reaction (69) and modifications of this reaction using $P(O)Cl(NHPh)_2$, and eliminating $PhNH_3Cl$, or $P(O)(NHPh)_3$, and eliminating $PhNH_2$, are also possible, e.g. eqn. (70)[63].

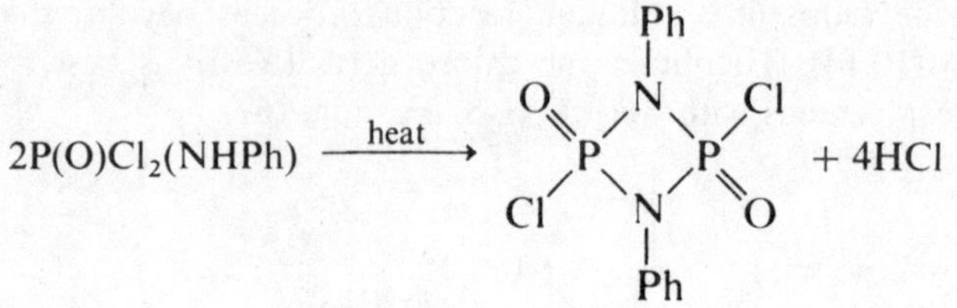

$$2P(O)Cl_2(NHPh) \xrightarrow{heat} \underset{Ph}{\underset{|}{\underset{N}{\overset{O}{\underset{\|}{P}}}}} + 4HCl \quad (69)$$

$$2P(O)(NHPh)_3 \xrightarrow{heat} \quad (70)$$

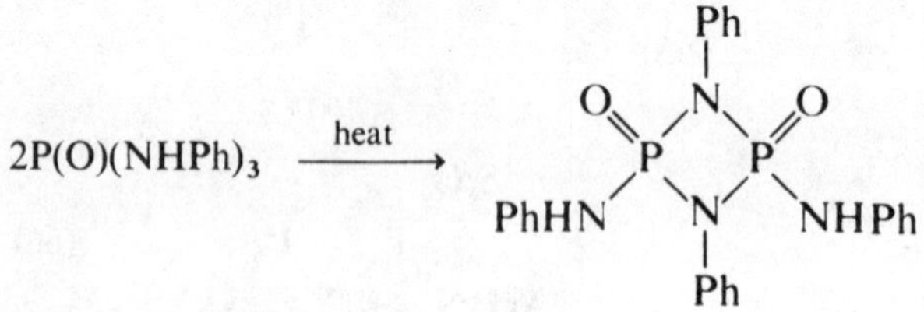

The cyclodiphosphazenes are generally crystalline materials susceptible to hydrolysis by atmospheric moisture. The trichlorocyclodiphosphazenes are naturally more reactive. For example with amines ring cleavage occurs (eqn. 71) and with alcohols or alkoxides the products are monophosphazenes (eqn. 72).

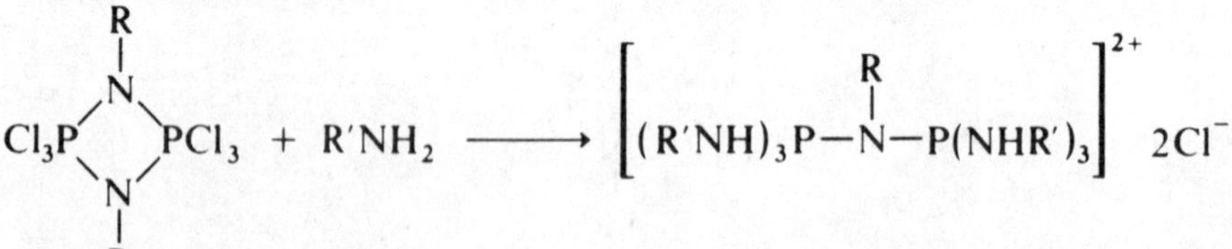

$$Cl_3PPCl_3 + R'NH_2 \longrightarrow \left[(R'NH)_3P-\underset{R}{\underset{|}{N}}-P(NHR')_3\right]^{2+} 2Cl^- \quad (71)$$

$$Cl_3PPCl_3 + NaOPh \longrightarrow 2(PhO)_3P=N-Ar \quad (72)$$

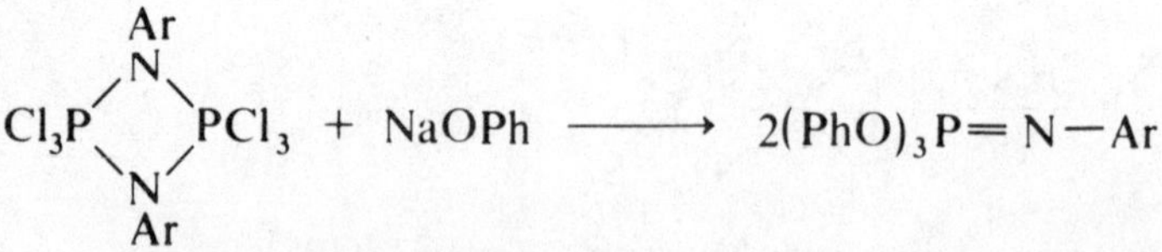

On the other hand with the phosphoryl and thiophosphoryl-cyclodiphosphazanes these substitution reactions at phosphorus can leave the ring intact, eqns. (66) and (73), but ring breakdown may also occur with alcohols and phenols, e.g. eqn. (74).

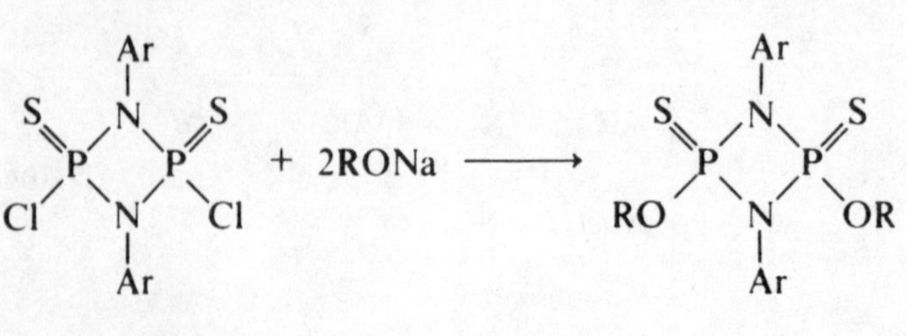

$$ + 2RONa \longrightarrow (73)$$

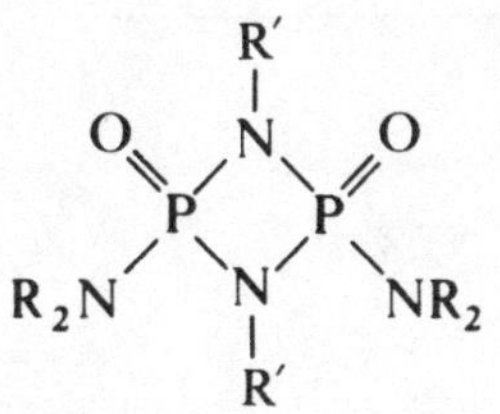

10.3 The cyclodiphosphazanes

$$\text{(structure shown)} + \text{ROH} \longrightarrow \text{P(O)(NR}_2\text{)(NHR')(OR)} \quad (74)$$

Work on the chemical reactivity of these ring systems has been somewhat spasmodic but there are hints of what might be possible. For instance P-trichloro-N-methyl cyclodiphosphazane can displace two carbonyl ligands from $Cr(CO)_6$ to form a light red complex with the structure $(Cl_3P.NMe)_2Cr(CO)_4$, in which the nitrogen atoms are probably coordinated to the metal[64].

10.3.2 Physical properties of the cyclodiphosphazenes and the nature of the bonding

The structures and bonding highlight the difference between the two types of cyclodiphosphazenes. In the compound $(X_3P.NR)_2$ the bonding does seem to approximate to a double (equatorial) bond and a single (apical) bond, the latter formed by the nitrogen lone pairs donating to phosphorus: Fig. 10.1 shows the structure of

---

Figure 10.1 The structure of $(Cl_3P.NMe)_2$ following ref. 65.

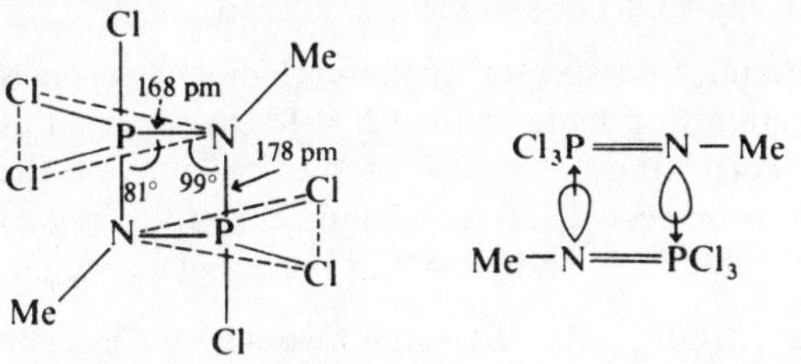

$(Cl_3P.NMe)_2$[65] and the structure of $(PhF_2P.NMe)_2$[66] is the same in terms of P—N bond lengths and ring angles. In both compounds the structure at nitrogen is planar and the arrangement at phosphorus is *tbp*. The axial P—N bond distance is exactly that of a single bond while the equatorial distance is much shorter revealing its double bond character. These differences however may have resulted as a necessary adjunct of the *tbp* configuration and so do not prove anything about the bonding. The planar arrangement at nitrogen is the key. This is best explained as an sp^2 (or tr) σ framework with the lone pair in the $2p_z$ orbital which is oriented in the equatorial plane, just right for $2p(N) \to 3d(P)$ donor π-bonding as in the monophosphazenes. The structures reveal these cyclodiphosphazenes as the legitimate dimers of the monophosphazenes. Even bond enthalpy data supports this for $(Cl_3P.NPh)_2$ in which $E(P-N, axial)$ is 289 kJ mol^{-1} (69 kcal mole^{-1}) and $E(P-N, equatorial)$ is 326 kJ mol^{-1} (77.5 kcal mole^{-1})[67].

The cyclodiphosphazenes of type $(X(O)P.NR)_2$ are basically unrelated to the monophosphazenes. These are really the first members of a series of cyclic polymers, sometimes called metaphosphinates, having the general formula $(X(O)P.NR)_n$ which are

Figure 10.2 The structure of (Cl(S)P.NMe)$_2$ following ref. 68.

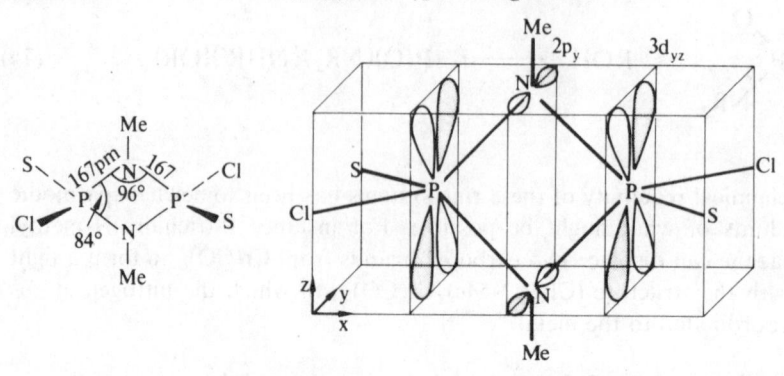

best known as the trimer and tetramer. We shall see more of these later in this chapter. What makes these compounds so different from the other cyclodiphosphazanes is the equality of ring bond lengths as revealed for example in Fig. 10.2 for (Cl(S)P.NMe)$_2$.[68]

The bond enthalpy E(P—N) calculated for the compound (PhHN(O)P.NPh)$_2$ is 330 kJ mol^{-1} (79 kcal mole^{-1})[67] which is higher than for either of the bonds in the other type of cyclodiphosphazane. These facts suggest a delocalized π system around the ring especially as the ring is planar, which would serve to maximize overlap, e.g. between 2pπ orbitals on N and 3d$_{yz}$ orbitals on P, as shown.

The researchers who have studied the infrared spectra of the cyclodiphosphazanes have usually assigned the νP—N vibration to a band in the 820–910 cm^{-1} region of the spectrum. Table 10.2 lists these correlations. In view of the structure of the (X$_3$P.NR)$_2$ compounds one might have expected infrared bands corresponding to νP—N, at about the values quoted in Table 10.2, and νP=N at much higher values.

Moreover the symmetry of the ring in (X(O)P.NR)$_2$ derivatives would have led one to expect a ring vibration not far below the latter range. The only compound to which two νP—N assignments have been made is ((PhHN)(O)P.NPh)$_2$ and these are at 750 and 1225 cm^{-1} [67], and the investigators assigned the former to the ring P—N bond vibration and the latter to the exocyclic P—N bonds. Thomas[47] dismissed this second assignment as unlikely since he could see no reason for suspecting that these were vibrating with a frequency characteristic of double bonds. More likely it is νP=N.

A reappraisal of the infrared spectra of (X(O)P.NR)$_2$ compounds leads one to believe that there might be a *ring* stretching vibration at about 1200 cm^{-1}. For example the published spectra of (Cl(O)P.NR)$_2$ and (Me$_2$N(O)P.NR)$_2$ [69] all show νP=O as a very strong band at 1282–1290 cm^{-1} in the former (calculated value 1294 cm^{-1}) and 1235–1239 cm^{-1} in the latter (calculated value 1242 cm^{-1}). Below this frequency in all the spectra is a strong band at 1149–1190 cm^{-1} which would seem to correlate with the ring system and not the exocyclic groups.

Part of the uniqueness of the (X$_3$P.NR)$_2$ compounds is that they possess two *tbp* centres joined together. Polytopal rearrangement at one phosphorus atom cannot be independent of the other. In the derivatives (RF$_2$P.NMe)$_2$ only one fluorine atom at each phosphorus can be apical, the other must be equatorial. Yet the ^{19}F n.m.r.

10.4 The linear polyphosphazenes

Table 10.2 νP—N assignment in cyclodiphosphazanes[61, 69-71]

	νP—N/(cm^{-1})		νP—N/(cm^{-1})
(Cl$_3$P.NMe)$_2$	847	(Cl(O)P.NPh)$_2$	893
(F$_3$P.NMe)$_2$	847	((Me$_2$N)(O)P.NMe)$_2$	871
(Cl$_3$P.NCH$_2$Cl)$_2$	821	((Me$_2$N)(O)P.NEt)$_2$	910
(Ph$_2$FP.NH)$_2$	864	(MeS(O)P.NMe)$_2$	852
		((PhNH)(O)P.NPh)$_2$	750 (see text)
(Cl(O)P.NMe)$_2$	852		
(Cl(O)P.NEt)$_2$	886		
(Cl(O)P.NPr^i)$_2$	844		

spectrum shows them to be equivalent except at low temperatures[66,72]. Pseudorotation is taking place and this must be in a concerted manner with the two alkyl-phosphorus bonds always acting as the pivots (see Chapter 2).

10.4 The linear polyphosphazenes

These compounds, like the cyclopolyphosphazenes, have the structural repeating unit

$$-\overset{|}{\underset{|}{P}}=N-$$

. Their relationship to the monophosphazenes is obvious but in terms of bonding they have the additional possibility of delocalized π-bonding covering the whole of the phosphorus–nitrogen backbone. However this feature will be covered in the cyclopolyphosphazene section, 10.5. In this section we shall concentrate on their chemistry and in this respect we shall focus mainly on the diphosphazenes, which fall into two groups: (i) those which are cations, such as the easily prepared Cl$_3$P=N—PCl$_3^+$ PCl$_6^-$ (10.15) and (ii) those which have a terminal phosphoryl or thiophosphoryl group, such as the model compound Cl$_3$P=N—P(O)Cl$_2$.

10.4.1 Synthesis of the linear polyphosphazenes

In these diphosphazenes there are no organic groups attached to the nitrogen so that only inorganic nitrogen compounds such as NH$_3$, NH$_4$Cl, NH$_2$Cl or S$_4$N$_4$ can be used in their synthesis. The reaction can be a two-step process involving the prior formation of a P—N bond and its subsequent reaction with another phosphorus centre, or it can be a simple one-step process in which the intermediate monophosphazene is not separated, nor in some cases has even its fleeting existence been proved e.g. with Cl$_3$P=NH.

The reaction of PCl$_5$ and NH$_4$Cl, in which the elusive Cl$_3$P=NH is postulated as an intermediate, gives high yields of (10.15) (eqn. 75)[73,74].

$$3PCl_5 + NH_4Cl \xrightarrow{\text{solvent}} Cl_3P=N-\overset{+}{P}Cl_3 \, PCl_6^- + 4HCl\uparrow \quad (75)$$
$$(10.15) \text{ m.p. } 310\text{--}315 \, °C$$

With more NH_4Cl (10.15) will react to form the triphosphazene (10.16) (eqn. 76)[75].

$$Cl_3P=N-\overset{+}{P}Cl_3\,PCl_6^- + NH_4Cl \longrightarrow Cl_3P=N-PCl_2=N-\overset{+}{P}Cl_3\,Cl^- + 4HCl\uparrow \quad (76)$$
$$(10.16)$$

On the other hand if PCl_5 is treated with liquid NH_3 the product can be the amido substituted derivative of (10.15) as shown in eqn. (77)[76] [compare eqn. (1)].

$$2PCl_5 + 16NH_3 \text{ (liquid)} \longrightarrow (NH_2)_3P=N-\overset{+}{P}(NH_2)_3\,Cl^- + 9NH_4Cl \quad (77)$$

Organophosphorus derivatives can be made by incorporating the organic group before reacting it with NH_3 e.g. eqn. (78)[76] and the product from this particular reaction (10.17) can be used in the formation of other phosphazenes, by making use of the amide groups in Kirsanov reactions as we shall see.

$$Ph_2PCl_3 + NH_3 \xrightarrow[20\,°C]{CHCl_3} Ph_2(NH_2)P=N-\overset{+}{P}(NH_2)Ph_2\,Cl^- \quad (78)$$
$$(10.17;\ m.p.\ 246\ °C,\ yield\ 82\%).$$

One of the more interesting diphosphazenes is $Cl_3P=N-P(O)Cl_2$* which can be obtained in a variety of reactions such as eqns. (79)[77], (80)[78] and (81)[79] and several more. A mixture of PCl_5 and $POCl_3$ can be used with NH_4Cl to give good yields[80]. Another way of preparing $Cl_3P=N-P(O)Cl_2$ or its thio analogue is to treat $Cl_3P=N-PCl_3^+\,PCl_6^-$ with SO_2 or H_2S (eqns. 82).

$$4PCl_5 + (NH_4)_2SO_4 \longrightarrow 2Cl_3P=N-P(O)Cl_2 + 8HCl + SO_2 + Cl_2 \quad (79)$$

$$2PCl_5 + NH_2OH \longrightarrow Cl_3P=N-P(O)Cl_2 + 3HCl + Cl_2 \quad (80)$$

$$4PCl_3 + 2N_2O_4 \longrightarrow Cl_3P=N-P(O)Cl_2 + 2POCl_3 + NOCl \quad (81)$$

$$Cl_3P=N-\overset{+}{P}Cl_3 \begin{array}{c} \xrightarrow{SO_2} Cl_3P=N-P(O)Cl_2 + SOCl_2 \\ \\ \xrightarrow{H_2S} Cl_3P=N-P(S)Cl_2 + HCl \end{array} \quad (82)$$

$Cl_3P=N-P(O)Cl_2$ can also be obtained by the reaction of a P—N compound with PCl_5, e.g. reaction (83). Diamidophosphoric acid and phosphoryl triamide give similar reactions. Other diphosphazene compounds can be prepared by what are essentially Kirsanov type reactions (eqns. (84)[81] and (85)[82]), and reaction (86) can be thought of in the same way.

$$P(O)(NH_2)(OH)_2 + 3PCl_5 \longrightarrow Cl_3P=N-P(O)Cl_2 + 2POCl_3 + 4HCl \quad (83)$$

$$RP(O)(NH_2)(OR) + PCl_5 \longrightarrow Cl_3P=N-P(O)(OR)(R) + 2HCl \quad (84)$$

$$P(O)(NH_2)(F)_2 + PF_3Cl_2 \longrightarrow F_3P=N-P(O)F_2 + 2HCl \quad (85)$$

$$P(OAr)_5 + P(O)(NH_2)(OR)_2 \longrightarrow (ArO)_3P=N-P(O)(OR)_2 + 2ArOH \quad (86)$$

---

* The vapour of this compound is a potent insecticide killing houseflies within 10 seconds. Fortunately for the flies it is too readily hydrolyzed to be used efficiently.

10.4 The linear polyphosphazenes

Kirsanov type reactions are not the only way of preparing diphosphazenes from P—N containing compounds; monophosphazenes, especially $Ph_3P=NBr$ (eqn. 87)[83] and phosphonyl azides (e.g. eqn. 88)[84] have also been used. In the case of the phosphinyl azides reaction usually proceeds beyond the diphosphazene stage to longer polymers.

$$Ph_3P=NBr + PX_3 \quad (X = Ph, Cl, Br) \longrightarrow Ph_3P=N-\overset{+}{P}X_3 \; Br^- \quad (87)$$

$$Ph_2P(O)N_3 + Ph_2PCl \longrightarrow Ph_2ClP=N-P(O)Ph_2 \longrightarrow$$
$$Ph_2ClP=N-(PPh_2=N)-P(O)Ph_2 \quad (88)$$

There are many reactions which produce diphosphazenes and the above selection are taken as representative of the different routes to these compounds. Fuller details are available in reviews (refs. 1, 2, and 37 pp 109–117).

10.4.2 Reactions and properties of the linear polyphosphazenes

Probably the most intensively investigated of the diphosphazenes has been (10.17) especially since it offers the possibility of ring closure given the right conditions. With another phosphorus compound it should, in theory, close to form a cyclo*tri*phosphazene but sometimes it unaccountably produces a cyclo*tetra*phosphazene. Its reactions are summarized later in Scheme 10.4 on page 408 – most of the products being cyclopolyphosphazenes.

Those diphosphazenes with a reactive $Cl_3P=N-$ or $=N-P(O)Cl_2$ end group, or both, can undergo nucleophilic substitution to form various other derivatives, e.g. reactions (89)–(91).

$$Cl_3P=N-P(S)Cl_2 + 5NaOPh \longrightarrow (PhO)_3P=N-P(S)(OPh)_2 + 5NaCl \quad (89)$$

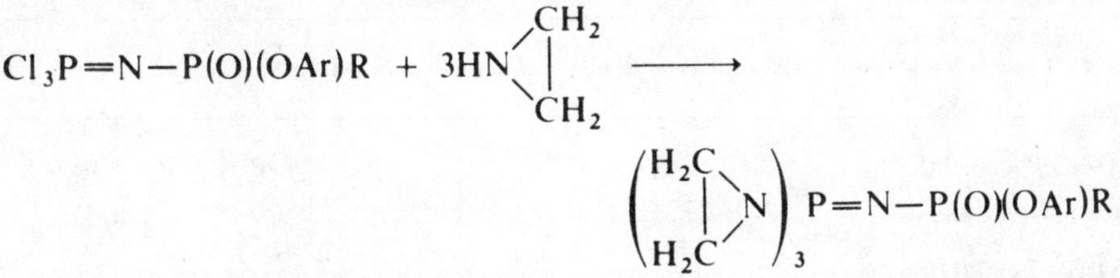

$$Cl_3P=N-P(O)(OAr)R + 3HN\begin{matrix}CH_2\\|\\CH_2\end{matrix} \longrightarrow$$

$$\left(\begin{matrix}H_2C\\|N\\H_2C\end{matrix}\right)_3 P=N-P(O)(OAr)R \quad (90)$$

$$Cl_3P=N-P(O)Cl_2 + 5Me_2NH \longrightarrow (Me_2N)_3P=N-P(O)(NMe_2)_2 \quad (91)$$

Conversion to phosphazanes occurs in any reaction with water, or formic acid if only partial hydrolysis is sought (eqn. 92).

$$Cl_3P=N-P(O)(OAr)_2 + HCO_2H \longrightarrow Cl_2(O)P-NH-P(O)(OAr)_2 \quad (92)$$

Conversion to phosphazanes might have been expected with some of the alkoxy derivatives by migration of an alkyl group from oxygen to nitrogen, as occurs with some cyclophosphazenes. Instead alkyl migration from oxygen to oxygen takes place (eqn. 93)[85].

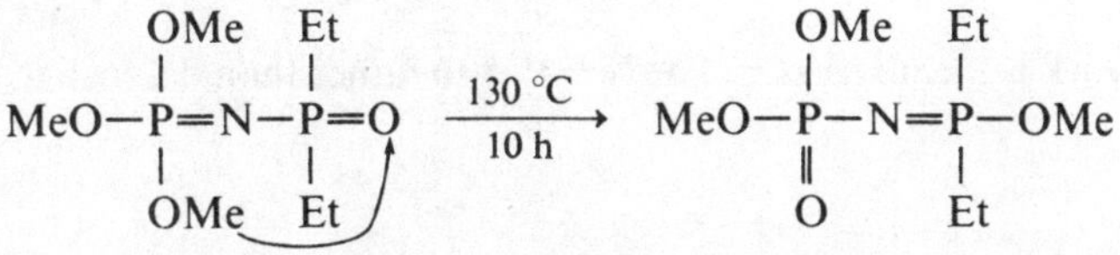

$$\begin{matrix}OMe & Et\\|&|\\MeO-P=N-P=O\\|&|\\OMe & Et\end{matrix} \xrightarrow[10\,h]{130\,°C} \begin{matrix}OMe & Et\\|&|\\MeO-P-N=P-OMe\\\|&|\\O & Et\end{matrix} \quad (93)$$

Removal of the proton from the phosphazane $Ph_2P(O)-NH-P(O)Ph_2$ produces an anion which behaves very like the acetylacetonate anion in that it will form chelate complexes. It will replace the acetylacetonate ligands (acac) from the beryllium complex in two stages and eqn. (94)[86] shows not only this, but in the partially substituted complex (10.18) the similarity of the two anions and this behaviour supports the concept of a delocalized π system over the phosphazene framework.

$$Be(acac)_2 + Ph_2(O)P-NH-P(O)Ph_2 \xrightarrow[12\,h]{103\,°C}$$

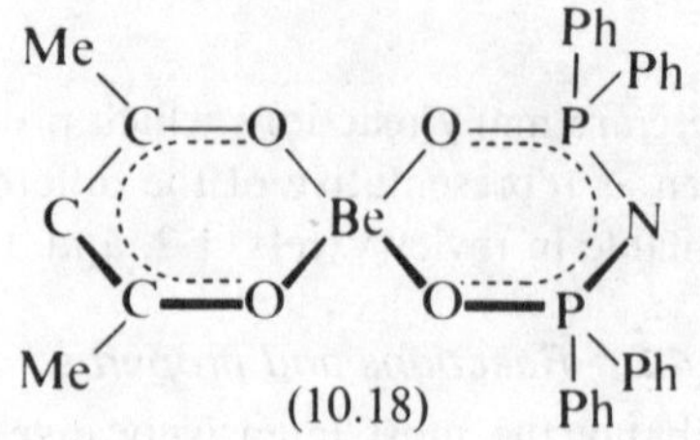

(10.18)

$\downarrow 175\,°C$

$$Be(Ph_2(O)PNP(O)Ph_2)_2 \quad (94)$$

The physical properties of the diphosphazenes have received less attention than those of other phosphorus–nitrogen compounds. They exhibit $\nu P-N=P$ vibrational modes at *ca* 1200 cm^{-1} which is the same spectral region as their cyclic counterparts. Table 10.3 lists a few of the $\nu P-N$ values. The infrared spectrum of $Cl_3P=N-P(O)Cl_2$

Table 10.3 $\nu P-N=P/(cm^{-1})$ vibrational modes[37,47] of some diphosphazenes

$Me_2(MeS)P=N-P(S)Ph_2$	$\nu P-N=P = 1208$ cm^{-1}
$Ph_3P=N-P(O)Ph_2$	1611
$(MeO)_3P=N-P(O)Et_2$	1320
$Ph_3P=N-P(S)F_2$	1306
$(NH_2)_3\overset{+}{P}=N-P(NH_2)_3\,Cl^-$	1265

consists of three strong bands above 650 cm^{-1}, which is beyond the region of P—Cl vibrations, so that these must be PNPO framework modes. The bands are centred at 770 cm^{-1} a region typical of $\nu P-N$, 1260 cm^{-1} which is almost certainly related to $\nu P=O$ since the calculated value for $-N=P(Cl)_2=O$ is 1286 cm^{-1}, and at 1330 cm^{-1} which is the expected region for $\nu P=N$. This simple analysis seems to suggest that the bonding along this framework is localized as $>P=N-P=O$ rather than delocalized as $>P\equiv N\equiv P\equiv O$.

10.5 The cyclopolyphosphazenes

As the chains of the polyphosphazenes get longer there comes a point at which they represent the linear counterpart of the corresponding cyclophosphazene. Cyclization with elimination of PCl_4^+ or $POCl_3$ will convert $Cl_3P=N-(PCl_2=N)_n-PCl_3^+$ or $Cl_3P=N-(PCl_2=N)_n-POCl_2$ into the cyclomer $(NPCl_2)_{n+1}$. Neither process has been directly observed although they have been inferred in the preparation of the cyclophosphazenes. The reverse reactions, i.e. heating the cyclic derivatives with PCl_5 or $POCl_3$ to obtain linear polyphosphazenes, are known. This brings us to the fifth class of phosphorus–nitrogen derivatives, and by far the largest, the cyclopolyphosphazenes.

10.5 The cyclopolyphosphazenes[37,87-90]

10.5.1 The synthesis of the cyclopolyphosphazenes

As we have seen with the diphosphazenes there are several ways in which the P—N—P system can be built up. So it is with the cyclopolyphosphazenes, but one method is paramount and this is the reaction between PCl_5 and NH_4Cl (eqn. 95). We have already met this reaction in the preparation of $Cl_3P=N-PCl_3^+ PCl_6^-$ and indeed this is an intermediate in the formation of the cyclomers. Equation (95) hides an unusually complex set of reactions which have still to be completely resolved.

$$PCl_5 + NH_4Cl \xrightarrow[\text{solvent}]{\text{refluxing}} \frac{1}{n}(NPCl_2)_n + 4HCl \qquad (95)$$

It was first reported by Liebig[91] in 1834 that when PCl_5 and NH_3 were heated together the product was mainly phospham, $(NPNH)_n$, but some white crystals were also formed which were stable to steam distillation. These were analyzed incorrectly at the time but were later shown to be $(NPCl_2)_3$. The use of NH_4Cl in place of NH_3 was found to improve the yield and by heating a solid mixture of PCl_5 and NH_4Cl in sealed tubes an American chemist, Stokes, was able to prepare, separate and characterize the cyclomers $(NPCl_2)_3$, $(NPCl_2)_4$, $(NPCl_2)_5$, $(NPCl_2)_6$ and $(NPCl_2)_7$.[92] He rightly guessed that they had cyclic structures and he made several derivatives of them including the rubbery polymer.

In 1924 Schenck and Romer[93] suggested the use of a refluxing, inert solvent, such as 1,1,2,2-tetrachloroethane, b.p. 146 °C, as the best medium for the reaction and this gave much better yields, ca 95%. This has been the method used most since then. The products consist of about two thirds cyclomers, $(NPCl_2)_{3,4,...}$, and one third oligomers which have the general formula $Cl_3P=N-(PCl_2=N)_n-PCl_3^+ Cl^-$ where n is an average of about 10. The cyclomers can be separated by virtue of their solubility in petrol, the linear products being insoluble. The cyclomers can be separated by fractional crystallization, distillation or by the preferential extraction of the trimer from petrol solution into pure sulphuric acid and its release on dilution of the acid solution. Other methods can be used as well.

Improvements in the synthesis have been designed to increase the proportion of the cyclomers[94,95]. Metal halides are claimed to be catalysts for the reaction but these often end up incorporated into the linear products. The typical reaction of PCl_5 and NH_4Cl used nowadays consists of heating together equimolar amounts of PCl_5 and NH_4Cl of small particle size in refluxing 1,1,2,2-$C_2H_2Cl_4$ (1000 cm³ per 0.5 mol PCl_5)

Table 10.4 Physical properties of the cyclopolychlorophosphazenes, $(NPCl_2)_n$

	Melting point/(°C)	Boiling point/(°C)	$\nu P{=}N/(cm^{-1})$	$\delta\ ^{31}P$ n.m.r./ (ppm 85% H_3PO_4)
$(NPCl_2)_3$	113	256	1218	−19.7
$(NPCl_2)_4$	123.5	328	1315	+7.4
$(NPCl_2)_5$	41.3	224 at 13 mm	1298, 1354	+17
$(NPCl_2)_6$	92.3	282 at 13 mm	1325	+16
$(NPCl_2)_7$	8–12	289–294 at 13 mm	1310	+18
$(NPCl_2)_8$	57–58	–	1305	+18

for 3 hours. The products from such a reaction can be over 95% cyclomers with a makeup of ca 60% $(NPCl_2)_3$, 20% $(NPCl_2)_4$ and 20% $(NPCl_2)_{5,6,7}$..[95]. The physical properties of these compounds are listed in Table 10.4.

The mechanism of the reaction has been intensively investigated. The reaction goes in two clearly defined steps (eqns. 96 and 97) which do not greatly overlap in time because the product of the first reaction (96) is virtually insoluble in the reaction medium, which temporarily removes it from further attack. Reaction (96) is more than 80% complete after one hour.

$$3PCl_5 + NH_4Cl \longrightarrow Cl_3P{=}N{-}\overset{+}{P}Cl_3\,PCl_6^-(\downarrow) + 4HCl(\uparrow) \qquad (96)$$

$$Cl_3P{=}N{-}\overset{+}{P}Cl_3\,PCl_6^- + 2NH_4Cl \longrightarrow (NPCl_2)_3 + 8HCl(\uparrow) \qquad (97)$$

It is believed that $Cl_3PNH_2^+$ and $Cl_3P{=}NH$ may be intermediates in this step. Eventually reaction (97) takes over from (96), but this too is a complicated process. The first part is the production of linear triphosphazene (10.19), which is a soluble cation, and this produces the rapid increase in the conductance of the solution which is observed (eqn. 98). The use of ^{31}P n.m.r. spectroscopy to follow the complete synthesis led to the profile shown in Fig. 10.3.[96] The appearance in the reaction mixture of $(NPCl_2)_3$ is delayed by about 20 minutes after the production of linear trimer and it would seem that cyclization of the latter is not responsible for the former.

$$Cl_3P{=}N{-}\overset{+}{P}Cl_3\,PCl_6^- + NH_4Cl \longrightarrow Cl_3P{=}N{-}PCl_2{=}N{-}\overset{+}{P}Cl_3\ Cl^- \qquad (98)$$
$$(10.19)$$

The mechanism for cyclization is more in keeping with eqn. (99) in which the cyclic derivative forms from a linear polymer which is at least one phosphorus longer.

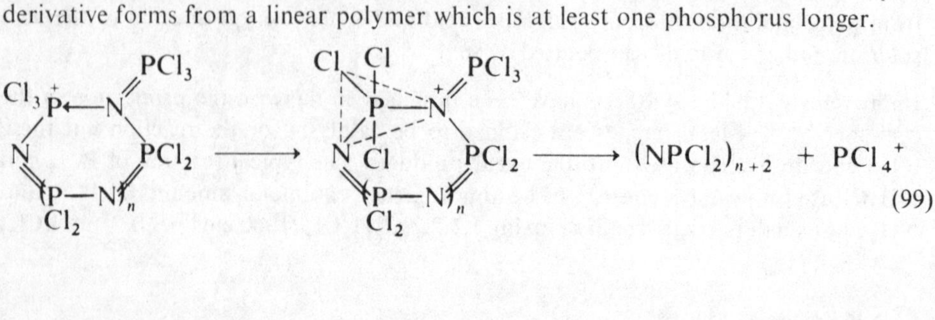

$$\longrightarrow (NPCl_2)_{n+2} + PCl_4^+ \qquad (99)$$

10.5 The cyclopolyphosphazenes

Figure 10.3 Composition of the reaction of PCl$_5$ and NH$_4$Cl throughout the course of the reaction (95).

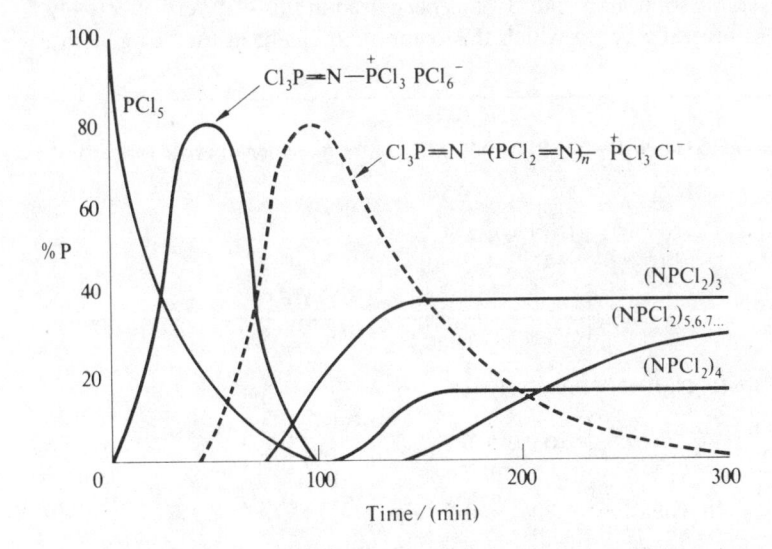

[Graph showing %P vs Time/(min) from 0 to 300. Curves labeled: PCl$_5$, Cl$_3$P=N—$\overset{+}{P}$Cl$_3$ PCl$_6^-$, Cl$_3$P=N—(PCl$_2$=N)$_n$—$\overset{+}{P}$Cl$_3$ Cl$^-$, (NPCl$_2$)$_3$, (NPCl$_2$)$_{5,6,7...}$, (NPCl$_2$)$_4$]

This mechanism is supported by the observation that cyclic products can be formed from linear material without the need for NH$_4$Cl,[97] and by the depolymerization of the high rubber polymer to the small rings by 'curling-off' or 'looping-off' mechanisms.

The reaction described above is the principal, but not the only method for preparing cyclopolyphosphazenes. Some reactions have been designed as alternative routes and others appear to have turned up by accident. The tailor-made reactions have generally been modifications of the above reaction incorporating other groups on to the phosphorus atom, e.g. reactions (100) and (101). Thus Ph$_2$PCl$_3$ gives (NPPh$_2$)$_{3,4...}$, Me$_2$PCl$_3$ gives (NPMe$_2$)$_{3,4...}$ and (C$_3$F$_7''$)$_2$PCl$_3$ gives (NP(C$_3$F$_7''$)$_2$)$_{3,4...}$

$$RPCl_4 + NH_4Cl \longrightarrow \frac{1}{n}(NPClR)_n + 4HCl \qquad (100)$$

$$R_2PCl_3 + NH_4Cl \longrightarrow \frac{1}{n}(NPR_2)_n + 4HCl \qquad (101)$$

These derivatives are prepared this way because the alternative method of reacting (NPCl$_2$)$_{3,4...}$ with a Grignard reaction is either very slow, as in the case of the phenyl compounds, or does not occur, as with methyl Grignards. Partially substituted derivatives produced by (100) are the non-*geminal* isomers, i.e. each phosphorus atom of the product carries an organic group. The bromocyclophosphazenes are obtained by reaction (102) using PBr$_3$ + Br$_2$ rather than PBr$_5$ which disproportionates into these above 35 °C. The iodophosphazenes are unknown.

$$PBr_3 + Br_2 + NH_4Br \longrightarrow (NPBr_2)_{3,4,5...} + 4HBr\uparrow \qquad (102)$$

We have already seen in the section on diphosphazenes that linear cations can be made separately and these have been used in the formation of cyclomers. One diphosphazenium compound in particular, $(NH_2)Ph_2P=N-PPh_2(NH_2)^+$ Cl^- which is a white, crystalline, stable solid, m.p. 246 °C, has been used in this way very effectively. Scheme 10.4 outlines several ways in which this compound has been used to generate

---

Scheme 10.4 Reactions of $(NH_2)Ph_2P=N-\overset{+}{P}Ph_2(NH_2)$ Cl^- in the formation of cyclic derivatives

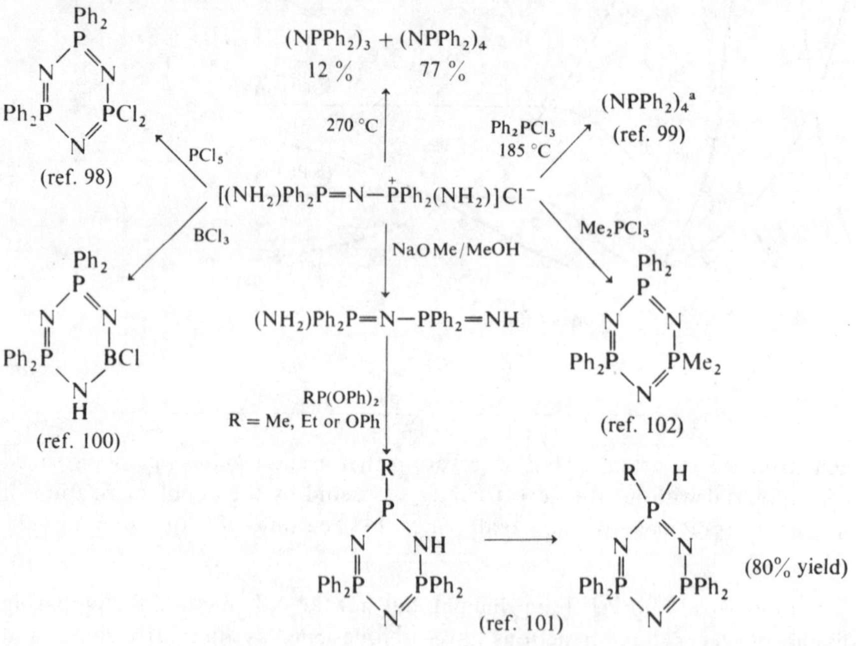

[Scheme showing reactions of the diphosphazenium compound with various reagents:
- With PCl$_5$: gives cyclic product with Ph$_2$P, PCl$_2$ groups (ref. 98)
- At 270 °C: gives (NPPh$_2$)$_3$ + (NPPh$_2$)$_4$, 12% and 77%
- With Ph$_2$PCl$_3$ at 185 °C: gives (NPPh$_2$)$_4$^{a} (ref. 99)
- With BCl$_3$: gives cyclic product with BCl and H (ref. 100)
- With Me$_2$PCl$_3$: gives cyclic product with PMe$_2$ (ref. 102)
- With NaOMe/MeOH: gives $(NH_2)Ph_2P=N-PPh_2=NH$
- With RP(OPh)$_2$, R = Me, Et or OPh: gives cyclic product (ref. 101), which rearranges to product (10.20) with R and H on P (80% yield)]

^{a} Mainly (NPPh$_2$)$_3$ at 125 °C

---

cyclomers, in some cases producing tetrameric compounds where trimers might have been anticipated, e.g. in the reaction with Ph_2PCl_3. Interestingly, one of the products (10.20) is stable in the phosphazene form, in which there is a P—H bond, rather than the phosphazane form, with the proton on nitrogen. This was proved conclusively by its ^{31}P n.m.r. spectrum which showed $^1J_{PH}$ of 509 Hz (R = Me).[101] The preference for structure (10.20) demonstrates the low basicity of the ring nitrogens in these compounds which should have their lone-pair electrons intact. In fact there is strong evidence that these are involved in donor π-bonding as we shall see.

The second class of preparative reactions are those in which the production of cyclophosphazenes seems to have been achieved more by accident than design. A large number of reactions come into this category, but most are of specific rather than general applicability, and we shall leave these out of our discussion.

The decomposition of phosphorus azides has been adapted to making several polymers. In a way this method is an extension of the azide synthesis of monophosphazenes,

10.5 The cyclopolyphosphazenes

except that instead of using RN_3, and ending up with $X_3P=N-R$ products, N_3^- is used and the product is $(X_2PN)_n$. The general reaction is (103) but the products are often linear polymers, e.g. eqn. (104)[103], although cyclic derivatives have been produced, e.g. eqn. (105)[104]. In some cases trimethylsilyl azide, Me_3SiN_3, has been used in lieu of metal azides.

$$R_2PCl + NaN_3 \xrightarrow{-NaCl} R_2PN_3 \xrightarrow{heat} (NPR_2)_n + N_2\uparrow \quad (103)$$

$$(CF_3)_2PCl + LiN_3 \xrightarrow[24h]{0\,°C} (CF_3)_2PN_3 \xrightarrow[37\,mm]{60\,°C} (NP(CF_3)_2)_n + N_2\uparrow \quad (104)$$
$$\text{waxy polymer}$$

$$Ph_2PCl + NaN_3 \longrightarrow Ph_2PN_3 \xrightarrow{165\,°C} (NPPh_2)_4 + N_2\uparrow \quad (105)$$

Phosphorus amides, essentially those with the $PCl_2(NH_2)$ grouping, can eliminate two molecules of HCl per molecule of compound to produce polyphosphazenes, and the general reaction is summed up by eqn. (106).

$$\left.\begin{array}{l} R_2PCl + NH_2Cl \\ R_2PNH_2 + Cl_2 \\ R_2P(O)NH_2 + PCl_5 \end{array}\right\} \xrightarrow{} R_2PCl_2(NH_2) \xrightarrow{Et_3N} \frac{1}{n}(NPR_2)_n + Et_3NHCl \quad (106)$$

(last arrow labeled $POCl_3$)

The reaction proceeds by a stepwise elimination of HCl and in the reaction of $Ph_2P(O)NH_2$ and PCl_5 the intermediate $Ph_2ClP(NH_2)-NH-PCl_2Ph_2$ has been isolated[105]. The products from these reactions are cyclic, e.g. reaction (107)[106].

$$(C_6F_5)_2PNH_2 + Cl_2 \longrightarrow (C_6F_5)_2PCl_2(NH_2) \longrightarrow (NP(C_6F_5)_2)_3 \quad (107)$$

A similar sort of reaction occurs by heating the ionic diamido derivatives formed from reaction of a phosphorus amide and chloramine (eqn. 108), but in this case ammonium chloride is eliminated.

$$Ph_2PNH_2 + NH_2Cl \longrightarrow Ph_2\overset{+}{P}(NH_2)_2\,Cl^- \xrightarrow{200\,°C} (NPPh_2)_{3,4} + NH_4Cl \quad (108)$$

Cyclodiphosphazanes are, in theory, the saturated analogues of cyclodiphosphazenes, except that four-membered P—N rings of the latter have never been made. In only one instance has a cyclodiphosphazane been converted directly to a cyclophosphazene, (eqn. 109)[107], the driving force behind this reaction being the very high H—F bond energy and the very strong HF_2^- hydrogen-bond.

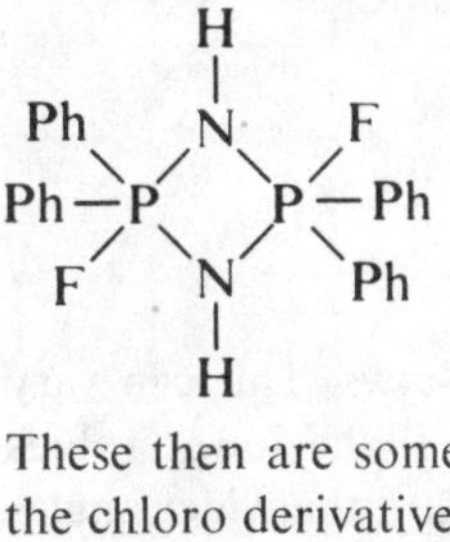

$$\underset{\substack{}}{\overset{\displaystyle H}{\underset{\displaystyle}{\,}}}\quad \text{Ph}_2\text{P}_2\text{N}_2\text{F}_2\text{H}_2(\text{Ph})_2 \xrightarrow[C_6H_6]{CsF} (NPPh_2)_3 + CsHF_2 \quad (109)$$

These then are some of the ways of preparing the cyclopolyphosphazenes, of which the chloro derivatives $(NPCl_2)_3$ and $(NPCl_2)_4$ have been investigated most. However, before we look at this work, and as a necessary preliminary to its understanding, we shall consider the bonding in these ring systems. These are the example *par excellence* of a delocalized $p\pi$–$d\pi$ bonding system.

10.5.2 The bonding in the cyclophosphazenes

That there is something special about the bonding of the cyclopolyphosphazenes can be seen from the structure and properties of $(NPF_2)_3$. This very volatile solid, m.p. 28 °C and b.p. 50 °C, is surprisingly stable and the ring part of the molecule is especially so. The nitrogen atoms are very weakly basic even though they are bonded to only two other atoms. The structure, Fig. 10.4, shows a planar P—N ring system with

Figure 10.4 The structure of $(NPF_2)_3$ taken from refs. 108 and 109.

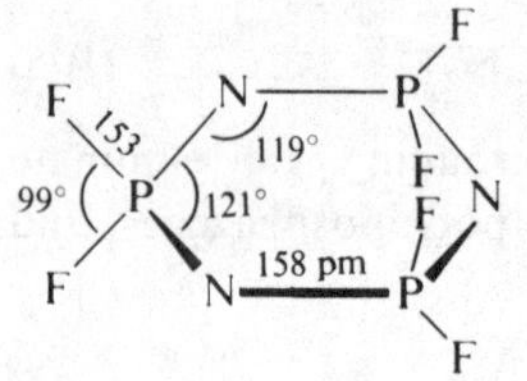

equal bond distances which are 20 pm shorter than the accepted single bond length. The ring is very nearly a perfect hexagon like benzene.

Hexafluorocyclotriphosphazene is unique in that it possesses all those qualities characteristic of an 'aromatic' ring system. Other cyclophosphazenes fall short in one way or another from the near perfect hexagon ring of $(NPF_2)_3$ by having a non-planar ring or different bond angles at phosphorus and nitrogen. Nevertheless almost all the cyclophosphazenes retain the essential quality or combination of qualities which set them apart as a special branch of phosphorus–nitrogen chemistry. These qualities derive basically from the bonding and especially the ring bonding which incorporates a π system. But first we must consider the σ framework.

In all the phosphazenes phosphorus has an approximately tetrahedral arrangement of bonds about itself. In the cyclo compounds however the NPN angle is almost always 120° which is larger than the tetrahedral angle of 109°28'. The opening of one angle has a closing effect on the opposite angle, i.e. the angle between the exocyclic bonds, reducing this below the tetrahedral angle by roughly the same amount as the ring angle is above, (10.21). However to a first approximation the situation at P is met by 4(sp³) or 4(te) σ bonding pairs, two to nitrogen atoms of the ring, and two to exocyclic groups.

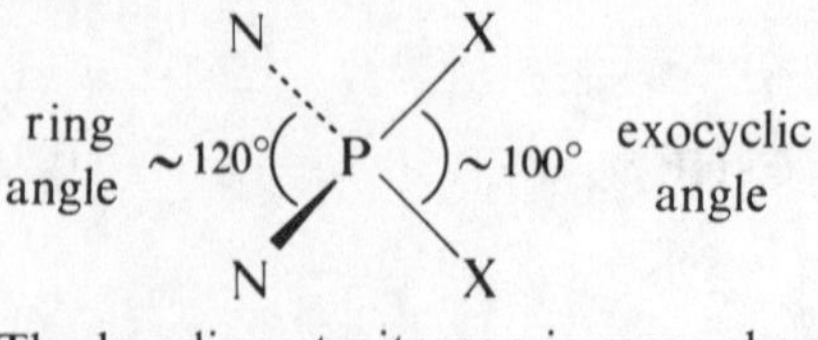

The bonding at nitrogen is more changeable as the PNP angle shows. This can vary from 118.4° in $(NPCl_2)_3$ to 148.6 in $(NPCl_2)_5$, a span of 30°. If the lower values correspond to a 3(sp²) or 3(tr) situation, which ideally requires 120°, the wider angles indicate more s-character in the bonding. Alternatively if the non-bonding pair occupies one of the 3(tr) orbitals then this should constrict the PNP angle to less than 120°, according to VSEPR theory. However were it to be drawn away from the nitro-

10.5 The cyclopolyphosphazenes

gen atom, in donor bonding, this would allow the angle to open. Hence to a first approximation we can use the ring bond angle at nitrogen as a guide to lone-pair involvement. The σ framework is shown in (10.22), with the nitrogen lone pair occupying an in-plane exocyclic sp² orbital.

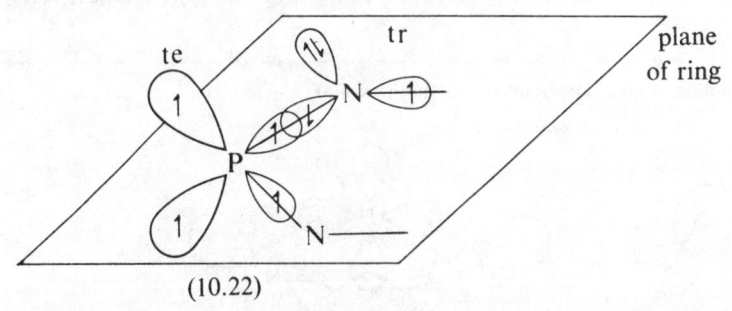

(10.22)

Having taken care of the basic σ framework there still remains one electron on phosphorus and one on nitrogen to be accounted for. These are the electrons which produce the π-bonding, unless we postulate transfer of the P electron into the N $2p_z$ orbital as in the ionic model of the bonding, i.e. $-\overset{|}{\underset{|}{P}}{}^+-\overset{-}{N}-$. This simple model is at variance with the facts and its rejection then leaves the way open to a π-bonding model involving overlap between 3d(P) and $2p_z$(N). The 3d orbitals most suited for this purpose are the $3d_{xz}$ and $3d_{x^2-y^2}$ which do not interfere with the σ framework in any way as Fig. 2.4 shows. The $3d_{xy}$ and $3d_{yz}$ orbitals have lobes which overlap the σ frameworks, to the ring and exocyclic atoms respectively, and $3d_{z^2}$ is of the wrong symmetry, although it might conceivably provide supplementary π-bonding to the exocyclic groups[89].

Only the $3d_{xz}$ orbital is suited to a π system perpendicular to the plane of the ring, Fig. 10.5(a). It overlaps in phase with $2p_z$ and is termed the heteromorphic π system

Figure 10.5 π-Bonding at phosphorus in the phosphazenes (not to scale).

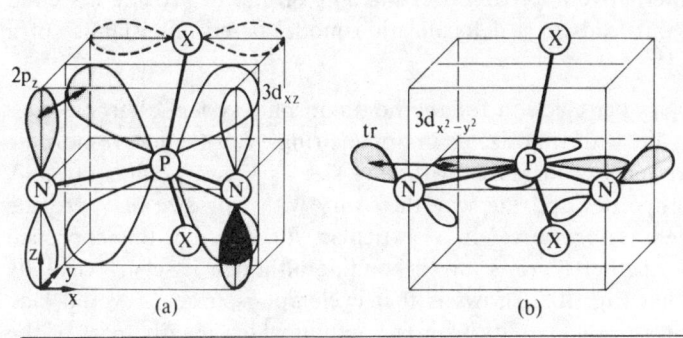

or π_a (a being short for asymmetric, referring to the asymmetry of the wave function with respect to reflection in the plane of the ring)[110]. This can be thought of as a normal π system involving the unpaired electrons on nitrogen and phosphorus. Figure 10.5(a) also shows that $3d_{xz}$ is ideally situated to form π bonds to the exocyclic groups.

There is also scope for a second π system based on the in-plane $3d_{x^2-y^2}$ orbital and utilizing the lone pair on nitrogen (Fig. 10.5(b)). This is called the homomorphic system or π_s (symmetric with respect to the plane).

When considering the complete trimeric ring it can be seen, Fig. 10.6(a), that the hetero-

Figure 10.6 Ring π-bonding in the cyclophosphazenes (not to scale).

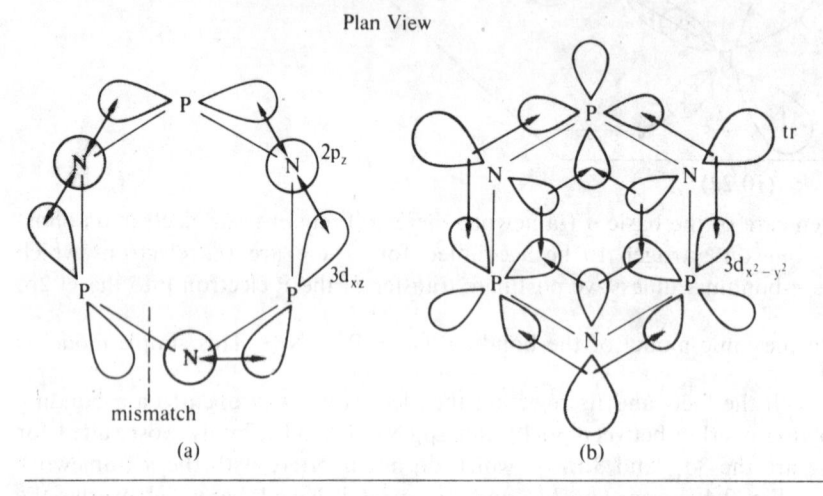

Plan View

morphic π system produces a mis-match in that the wave functions are out of phase at some point around the ring whereas for the homomorphic π system no such mismatching occurs. With the tetrameric, hexameric, octameric etc. rings the heteromorphic π system also produces no mis-matching which occurs only with the trimer, pentamer, heptamer etc.

The above theory of π-bonding is attributable to Craig and Paddock[89,111] but it is not the only one. An alternative approach uses the $3d_{yz}$ orbital to provide a second π-bonding system, and this leads to a delocalization model based on 'islands' of π electrons about the ring[112].

The delocalization energies per electron for the homomorphic system alternate with the number of electrons, i.e. with ring size as in organic rings. For the heteromorphic system the delocalization energy increases with ring size as Fig. 10.7 illustrates. A study of the physical properties and the way they vary with ring size may provide clues as to which π system is 'operative' in a particular situation, but this approach is begging the question whether there is any π-bonding although it seems virtually certain that there is. What Fig. 10.7 shows is that cyclo*di*phosphazenes would lack a significant contribution to ring stability from π-bonding which may be part of the reason why these remain unknown.

10.5.3 Physical properties of the cyclopolyphosphazenes

What support is there from physical measurement for π systems of the types described above, or any π-bonding for that matter? In carbocyclic compounds such as benzene

10.5 The cyclopolyphosphazenes

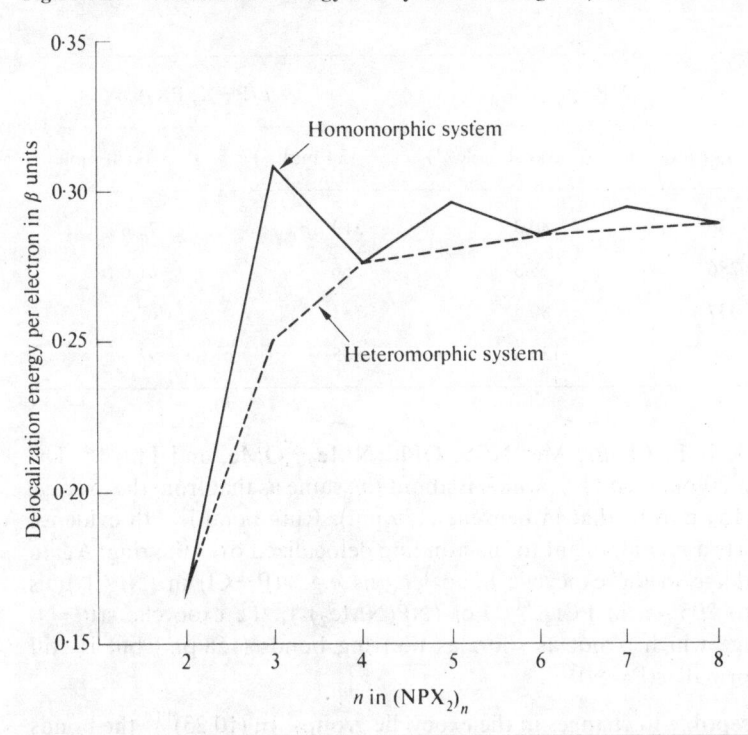

Figure 10.7 Delocalization energy of π systems and ring size, taken from ref. 113.

there is plenty of evidence of aromaticity from bond enthalpy data, bond lengths, n.m.r. spectroscopy, ultraviolet spectroscopy, etc. In cyclophosphazene chemistry the equivalent data is less dramatic, as one might expect of a π system which is much more diffuse.

We have already seen that bond enthalpy data for P—N and P=N bonds is sparse and contradictory, but what little there is seems to indicate an $E(P-N)$ value of ca 290 kJ mol^{-1} (69 kcal mole^{-1}) and an $E(P=N)$ value of ca 400 kJ mol^{-1} (95 kcal mole^{-1}). The cyclophosphazenes have $E(PN)$ in the range 303 ± 3 kJ mol^{-1} (72 ± 1 kcal mole^{-1}) measured for the compounds $(NPCl_2)_{3,4}$, $(NPPh_2)_4$, $(NP(OC_6H_{11})_2)_3$ and $(NPMe_2)_3$.[114,115] The difference between $(NPCl_2)_3$ and $(NPCl_2)_4$ was insignificant (<1 kJ mol^{-1}) which seems to suggest that at least the homomorphic π system is not the dominant π system. These values for $E(PN)$ do little more than hint at a π contribution to the ring bonds. If anything, phosphorus is using its 3d orbitals to strengthen the exocyclic bonding as the values in Table 10.5 show. These values together with the $E(PN)$ values suggest a tetrahedral π system in operation at phosphorus, like that of PO_4^{3-}, rather than one which is limited only to the ring system[89].

Another generally accepted criteria of π-bonding is bond foreshortening and bond equalization around the ring. Here the evidence is more conclusive. The P—N bond length free of all possible π contributions is 177 pm ($NaPO_3NH_3$), while the ring bond lengths in the cyclophosphazenes fall in the range 157–160 pm in $(NPX_2)_{3,4}$

Table 10.5 Exocyclic bond enthalpies in (NPX$_2$)$_n$ compounds compared to PX$_3$ bond enthalpies[114,115]

Compound	E(P—X) /(kJ mol^{-1})	/(kcal mole^{-1})	E(P—X; PX$_3$) /(kJ mol^{-1})	/(kcal mole^{-1})
(NPCl$_2$)$_3$	336	80.3	319	72.6
(NPMe$_2$)$_3$	286	68.3	256	61.6
(NPPh$_2$)$_4$	337	80.5	282	67.5
(NP(OC$_6$H$_{11}$)$_2$)$_3$	566	135.3	385	92.0

compounds where X is F, Cl, Br, Me, NCS, OPh, NMe$_2$, OMe, and Ph[116].* The contraction of about 20 pm is *ca* 11 % which is about the same as that from the average C—C single bond (154 pm) to that in benzene (139 pm). Ring bond length evidence then strongly supports a π-component to the bonding delocalized over the ring. Again there is some shrinkage in the exocyclic bond lengths e.g. r(P—Cl) in (NPCl$_2$)$_3$ is 199 pm compared to 203 pm in PCl$_3$.[117] For (NP(NMe$_2$)$_2$)$_4$ the exocyclic r(P—N) is 168 pm on average which is not as short as the ring bonds (158 pm) but is still shorter than the 'normal' r(P—N)[118].

Ring bonds are susceptible to changes in the exocyclic groups. In (10.23)[119] the bonds to the phosphorus carrying the pair of phenyl groups are longer, which suggests that relative to Cl the phenyl group is electron withdrawing or, more likely, that the π-bonding at phosphorus is localized in the P—Ph bonds. The alternation in ring bond lengths in 1,1-N$_4$P$_4$Me$_2$F$_6$, so conclusive of π-bonding, was mentioned in Chapter 2, compound (2.8).

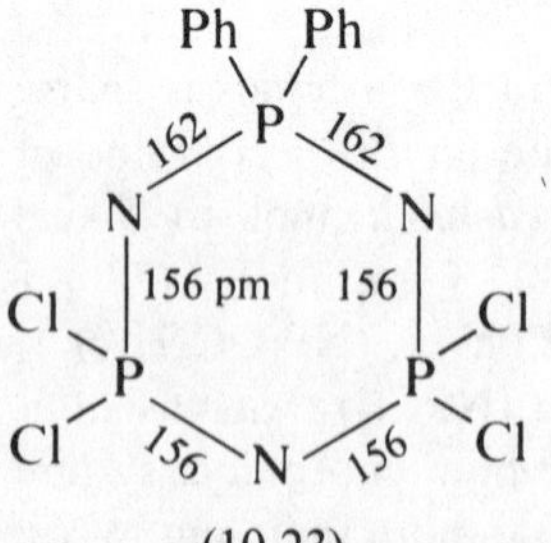

(10.23)

The phosphazane rings of the acid (NHP(O)(OH))$_3$ and esters (NRP(O)(OR))$_3$ have bond lengths r(P—N) of 166–168 pm which is the same as in the cyclodiphosphazanes. These compounds, formed by rearrangement of phosphazene derivatives, still appear to retain a trace of their former π-bonding.

---

* There are a few rings with even shorter bonds than this e.g. r(P—N) in (NPF$_2$)$_4$ is 151 pm and in (NPCl$_2$)$_5$ is 152 pm.

10.5 The cyclopolyphosphazenes

The ring angle at phosphorus changes very little in the phosphazenes; whatever the ring size it is $120 \pm 2°$.* It is the angle at nitrogen which shows the greatest change as this is only dicoordinate and the flexibility of the molecule hinges on the nitrogen. In the trimers the angle at nitrogen is 120–122°, rising to 131–133° in the tetramers, although in $(NPF_2)_4$ it is 147° indicating a planar ring for this compound. In $(NPCl_2)_5$ it is even wider, 149°, and again the ring is almost planar. The lone pair should act to constrain angle widening at N unless this has been drained away into a donor π-bond such as in the homomorphic π-system. However too much should not be read into bond angles and the delocalization of the lone pair since ring geometry may play an important role. However the effect on ring bonds and angles in a series of compounds in which F atoms of $(NPF_2)_4$ are replaced by methyl groups is consistent with a weakening of the ring bonds due to a weakening of the donor π-system and increased localization of the lone pair on nitrogen (Table 10.6).

Table 10.6 Ring bond lengths and angles at nitrogen in $N_4P_4F_xMe_{8-x}$

Compound	r(P—N)/(pm)	$\angle$PNP	Reference
$(NPF_2)_4$	151	147°	120
$1,1-N_4P_4F_6Me_2$	152 (av)	145°	121
$1,1,5,5-N_4P_4F_4Me_4$	156 (av)	135°	121
$(NPMe_2)_4$	160	132°	122

The infrared spectra of the cyclopolyphosphazenes are dominated by a strong broad band in the region 1200–1400 cm^{-1} which is an asymmetric ring stretching vibration often labelled vP=N in the literature. This high frequency is comparable to the stretching frequency range of the monophosphazenes which is centred on 1324 cm^{-1} (see page 393), and much higher than that assigned to the cyclodiphosphazanes, although as we have seen there are reasons for doubting the P—N correlations of these. The infrared data is very much in favour of a π-system but more interesting still is the variation in vP=N with variation in exocyclic substituents and ring size.

The series of compounds $N_3P_3F_xCl_{6-x}$ ($x = 0$–6), $N_4P_4F_xCl_{8-x}$ ($x = 0$–8), and $N_4P_4F_xMe_{8-x}$ ($x = 4$–8) demonstrate the effect on vP=N of replacing one atom or group at a time with another (Table 10.7)[123,124]. It can be seen that increasing electron withdrawal raises the frequency of the vibration, and in a fairly regular manner. The effect has been attributed directly to the electronegativity of the groups and the 3d orbital contraction which this property is supposed to promote, the contracted orbitals becoming better π-bonders which in turn strengthens the ring bonding and increases vP=N.

Ring size itself has a marked influence on vP=N particularly between trimers and tetramers as Table 10.8 shows for a few systems. The increase from tetramers to

---

* In the cyclotriphosphazanes it drops to *ca* 105°.

Table 10.7 Changes in ring substituents and changes in $\nu P=N/(cm^{-1})$

	\multicolumn{9}{c}{x}	
	0	1	2	3	4	5	6	7	8	Ref.
$N_3P_3F_xCl_{6-x}$[a]	1215	1230	1244	1255	1268	1280	1305			123
$N_4P_4F_xCl_{8-x}$[a]	1315	1330	1346	1362	1337	1390	1400	1408	1419, 1438	124
$N_4P_4F_xMe_{8-x}$	1222	—	—	—	1355[b]	1375[c] 1372[d]	1390[e] 1386[f]	1410	1419, 1438	125

[a] *Geminal* isomers; [b] 1,1,5,5 isomer; [c] 1,1,3 isomer; [d] 1,1,5 isomer; [e] 1,1 isomer; [f] 1,5 isomer.

pentamers is less marked and the effect of ring size disappears with the hexamers. This trend varies as the delocalization energy of a π_a system (Fig. 10.6) rather than a π_s system. Too much cannot be read into the trimer to tetramer increases because these vary according to the substituents as Table 10.9 indicates, although only the methyl and phenyl compounds show an increase much out of line with the rest, suggestive of a weaker π system in these products.* Table 10.9 is arranged in order of

Table 10.8 Effect of ring size on $\nu P=N/(cm^{-1})$ in $(NPX_2)_n$ compounds taken from ref. 126

	\multicolumn{4}{c}{n}
Compound	3	4	5	6
$(NPCl_2)_n$	1215	1315	1298, 1354[a1]	1325
$(NP(OMe)_2)_n$	1235, 1275[a2]	1337	1340	1335
$(NPMe_2)_n$	1185	1222	1255	

[a] Split by Fermi resonance probably; averages [1] 1328 and [2] 1255 represent reasonable values of undisturbed peak.

decreasing $\nu P=N$ values, an order which can be seen as related to the 'electronegativity' of the exocyclic groups. The values of Table 10.9 cannot be correlated to so-called π-values, used to predict the locations of $\nu P=O$ frequencies nor has a simple corresponding equation been devised to meet the cyclophosphazene situation. All in

---

* The difference between Et and Me seems somewhat anomalous but the relative behaviour of these two groups towards phosphorus is basically different as the bond moments of P—Et and P—Me showed in Chapter 2, Table 2.6, and as will be seen in Chapter 11 where we compare the relative Lewis basicities of phosphorus compounds.

10.5 The cyclopolyphosphazenes

Table 10.9 $\nu P=N/(cm^{-1})$ values for $(NPX_2)_{3,4}$ derivatives taken from ref. 126

	F	CF$_3$	OPh	OMe	Et	NCS	Cl	NMe$_2$	Ph	Me	Br	NH$_2$	SEt
$(NPX_2)_3$	1305	1300	1250–80	1235	1225	1218	1215	1195	1190	1185	1175	1170	1150
$(NPX_2)_4$	1429[a]	1412	1330–50	1337	1320	—	1315	1265	1213	1222	1253–80	1240	—
Difference[b]	124	112	80	102	95	—	100	70	23	37	90	80	—

[a] Average of 1419 and 1438; [b] difference between trimer and tetramer values.

all infrared studies have been less satisfactory in explaining the bonding in these compounds than might have been anticipated.

Ultraviolet spectroscopy, which has been of great use in studying organic aromaticity, has failed to provide convincing evidence of π activity in the cyclophosphazenes. These compounds do not generally absorb in the near ultraviolet, and if they do this can be attributed to exocyclic substituents. Most cyclophosphazenes have an absorption maximum at wavelengths of *ca* 200 mμ which is virtually independent of ring size. This rules out its being due to a π system whether it be the regular π_a or the donor π_s. The maxima vary with the nature of the exocyclic groups which means that these are the likely origin of the transitions, Table 10.10.

Table 10.10 Ultraviolet absorption maxima, $\lambda_{max}/(m\mu)$, of $(NPX_2)_{3,4}$ where X = F,Cl,Br taken from ref. 127

$(NPF_2)_3$	$\lambda_{max} = 149.4$	$(NPF_2)_4$	$\lambda_{max} = 147.5$
$(NPCl_2)_3$	175	$(NPCl_2)_4$	175
$(NPBr_2)_3$	200	$(NPBr_2)_4$	200

In Chapter 5 we saw that ionization potentials derived from photoelectron spectroscopy were instrumental in proving the π-bonding between phosphorus and nickel in the complex $Ni(PF_3)_4$, and so it is with the cyclopolyphosphazenes. The values for the first ionization potentials of several trimeric derivatives are listed in Table 10.11

Table 10.11 First ionization potentials of some cyclotriphosphazenes/(eV) taken from ref. 128

$(NP(NMe_2))_3$	7.85	$N_3P_3Br_3Cl_3$	9.72
$(NPMe_2)_3$	8.35	$N_3P_3Br_2Cl_4$	9.80
$(NP(OPh)_2)_3$	8.83	$N_3P_3BrCl_5$	9.83
$(NP(OMe)_2)_3$	9.29	$(NPCl_2)_3$	10.26
$(NPBr_2)_3$	9.56	$(NP(OCH_2CF_3)_2)_3$	10.43
$N_3P_3Br_5Cl$	9.47	$(NPF_2)_3$	11.4
$N_3P_3Br_4Cl_2$	9.60		

including those of the series $N_3P_3Br_xCl_{6-x}$ ($x = 0$–6) which show that exocyclic groups exert an irregular effect e.g. $IP(N_3P_3Br_5Cl) < IP(N_3P_3Br_6)$.[129] These first ionization potentials refer to an electron being removed from a π system, or from a lone pair on nitrogen. However if the latter were the case then ring size should have relatively little effect. The importance of ring size and the key to the π_a vs π_s contributions was revealed by Paddock's work on the fluoride cyclomers. The lowest IP's for these vary as shown in Table 10.12, and he concluded that IP's less than 14 eV were due to removal of an electron from a π system. His interpretation of the

10.5 The cyclopolyphosphazenes

Table 10.12 Lowest ionization potentials (in eV) of $(NPF_2)_n$ where $n = 3-8$ taken from ref. 128

$(NPF_2)_3$	$(NPF_2)_4$	$(NPF_2)_5$	$(NPF_2)_6$	$(NPF_2)_7$	$(NPF_2)_8$
	10.7		10.9		10.9 π_s
11.4	11.5	11.4	11.7	11.3	(–) π_a

results is that the even-numbered cyclomers show both the homomorphic π_s-system with the lower IP (ca 10.8 eV) and the heteromorphic π_{as}-system underlying this at ca 11.5 eV. In the odd-numbered cyclomers the two systems have equal energies. A variation of this kind in the homomorphic system is to be expected since the delocalization energy per electron in odd-numbered cyclomers should be larger than in even-numbered ones.

Although a great deal of ^{31}P n.m.r. spectroscopy has been carried out on the cyclophosphazenes it has been of very limited value in terms of clarifying the bonding. For correlation tables the reader is referred to the compilations in Allcock's book[37]. One n.m.r. technique, dynamic nuclear polarization, has been used by Paddock to show that in contrast to F in C_6H_5F, which is conjugated to the ring π orbitals and can transmit appreciable spin density, the F in $(NPF_2)_{3-7}$ shows no evidence for spin coupling via either the π_a or π_s systems[130]. On the other hand Cl in $(NPCl_2)_{3-7}$ is conjugated.

In another piece of work by Paddock the $N_3P_3F_5$ group was itself studied as a group attached to fluorobenzene rings[131]. It was shown to be strongly electron withdrawing, reducing electron density at all points in the benzene ring. Inductive withdrawal was of the same order as $-NO_2$, $-F$ and $-CN$, while conjugative withdrawal was also large, comparable to that of $-NO_2$, $-CN$ and $-PF_2$. There was, moreover, evidence that conjugation extended beyond the first phosphorus atom, and that as far as the phosphazene ring is concerned the fluorophenyl group conjugates with the homomorphic π-system.

It has been suggested that P—P bonding across the ring is significant, as revealed by theoretical calculations, and that this is important to the stability of these compounds[132]. Even in a molecule such as $(NPF_2)_4$ there is the possibility of trans-anular P—P bonding as well as between adjacent P atoms. It can be seen that these suggestions add a new dimension to the homomorphic π_s system of Fig. 10.6(b) and imply an external π system using the (tr) lone pair on nitrogen and an internal system based on 3d–3d overlap, which becomes the framework for P—P bonding.

To summarize the bonding: it would appear that we are dealing with compounds which are unusual in that they can have two delocalized π systems. Firstly there is the 'normal' π_a system based on $2p_z(N)-3d_{xz}(P)$ overlap, above and below the plane of the ring, secondly a 'donor' $tr(N) \rightarrow 3d_{x^2-y^2}(P)$ π_s system concentrating electron density around the outside of the ring. The $3d_{x^2-y^2}$, and possibly the $3d_{yz}$, may also produce some trans-anular P—P bonding. The bonding is complex and it is thus hardly surprising that the chemical behaviour of these compounds fails to fit any simple theory.

10.5.4 Nucleophilic substitution at phosphorus in the cyclopolyphosphazenes

Hexachlorocyclotriphosphazene has six replaceable exocyclic chlorine atoms and these can be replaced to give *geminal* or non-*geminal* product isomers. *Gem* isomers are obtained by pairwise substitution i.e. at the same phosphorus, non-*gem* isomers at different phosphorus atoms. Figure 10.8 shows what this means for compounds of

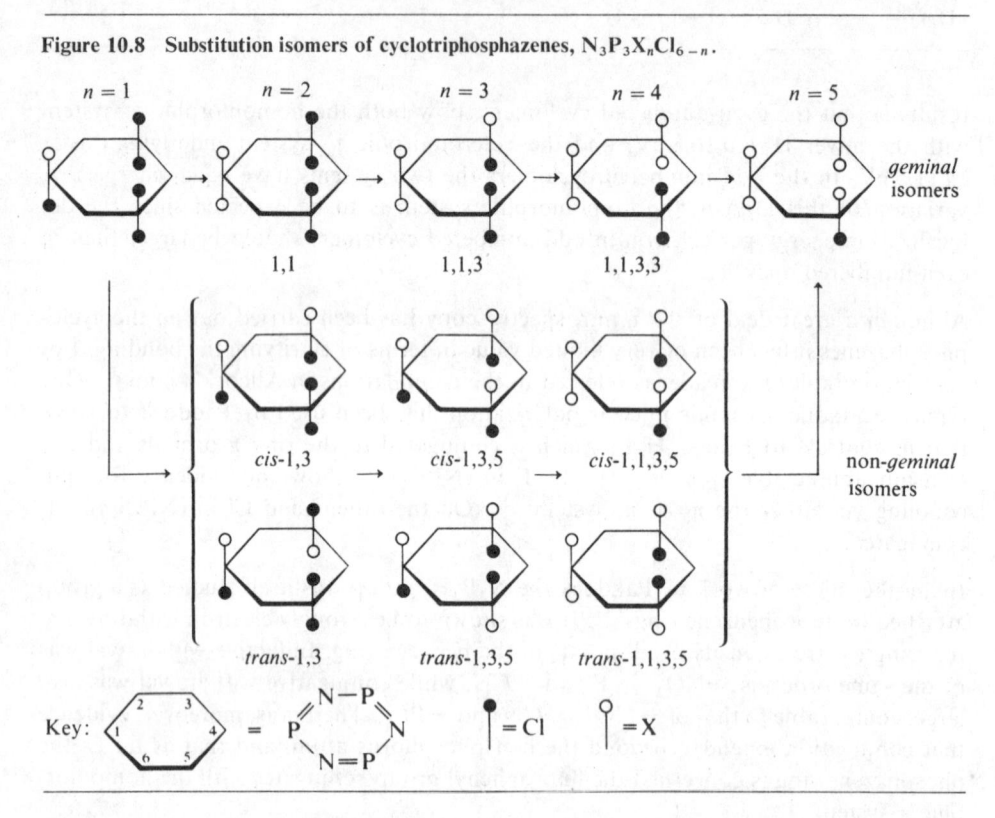

Figure 10.8 Substitution isomers of cyclotriphosphazenes, $N_3P_3X_nCl_{6-n}$.

general formula $N_3P_3X_nCl_{6-n}$. For the tetracyclophosphazene system there is the added complication of the extra phosphorus centre which means that $N_4P_4X_2Cl_6$, for example, can be one of five isomers: 1,1,- i.e. *gem*; *cis*-1,3-; *trans*-1,3-; *cis*-1,5-; *trans*-1,5-. To date most work has been done on the trimer especially on the nucleophilic substitution of the readily prepared $(NPCl_2)_3$. The chlorine atoms in this compound are replaceable by a large variety of groups such as F, NCS, OR, OPh, SR, SPh, NR_2 etc. Some of these reactions are relatively straightforward giving only one type of substitution product, others are more complex. However before we go on to consider the nucleophilic substitution in detail let us try to predict the likely isomer pattern from the nature of the cyclotriphosphazene and the attacking nucleophile.

As we saw in Chapter 2, p. 70, the mechanism could be either one involving a tricoordinate phosphorus intermediate, i.e. an ionic intermediate, or one involving a pentacoordinate phosphorus intermediate (eqns. 109 and 110). The mechanism will depend to a large extent on the nature of X in the first attack, but what happens after that will also depend on the effect of the first group Y.

10.5 The cyclopolyphosphazenes

$$\text{X}_2\text{P=N-N=} \longrightarrow \text{XP}^+\text{=N-N=} \quad \text{X}^- \xrightarrow{Y^-} \text{XYP=N-N=} \qquad S_N1(P) \text{ type mechanism} \qquad (109)$$

$$\text{X}_2\text{P=N-N=} \xrightarrow{Y^-} \underset{tbp}{\text{X}-\text{P}(Y)(-N^-)-\text{N=}} \longrightarrow \text{X}^- \quad \text{XYP=N-N=} \qquad S_N2(P) \text{ type mechanism} \qquad (110)$$

Just as in the electrophilic substitution of benzene there are two parts that Y plays — first to modify the rate of further attack and second as a director of the site of further attack. Not enough work has been done on kinetics to make discussion of the first of these points profitable, although it appears that fluoride and alkoxy substituents activate the molecule to further attack whereas amino and organic groups deactivate it. The second point, the orientation of attack has been thoroughly studied.

The question about further attack can be framed thus: will the second nucleophilic attack of Y^- take place at the PXY centre or will it go for another, untouched, PX_2 centre on the ring? If it prefers the former it will produce the *gem* isomer, if the latter it will give non-*gem* isomers. Superficially the answer seems obvious: if the presence of Y in >PXY reduces the electron density of the phosphorus more than X then it will encourage further nucleophilic attack at this centre if the mechanism is of the $S_N2(P)$ type (eqn. 110). Conversely if Y is a net electron donor relative to X it will discourage attack at the same phosphorus if the mechanism is $S_N2(P)$. If the mechanism is ionic (eqn. 109), however, electron withdrawal by Y will hinder ionization at >PXY relative to >PX_2, but if electron releasing it will stimulate ionization. The results of nucleophilic attack on these assumptions will be to give the isomers shown in Table 10.13, which refers specifically to $(NPCl_2)_3$. No account has been taken of steric factors which may hinder $S_N2(P)$ reactions or of the type of electronic effect, whether it be inductive or conjugative, or the *tbp* intermediate and the relative apicophilicity of Y and X. All these second order effects may be important and it is little wonder that the substitution patterns we find only rarely conform to the prediction of Table 10.13.

Before we turn to look at individual cases there is a further factor to consider and that is the relative reactivity of the various ring systems. Substitution of Cl by labelled Cl* has been used to study the relative rate of exchange of these (eqn. 111) and the

$$(NPCl_2)_n + Me_4NCl^* \xrightarrow{MeCN} (NPCl^*_2)_n + Me_4NCl \qquad (111)$$

order, compared to the exchange with $POCl_3$, was found to be [133]: $POCl_3 \gg (NPCl_2)_4 > (NPCl_2)_5 > (NPCl_2)_6 > (NPCl_2)_3$. If, as some have postulated, ring flexibility is the reason for this order then we would have expected reactivity to increase with ring

Table 10.13 Nucleophilic substitution of $(NPCl_2)_3$

	Electronic effect of incoming group Y relative to Cl	
Mechanism	Y electron releasing	Y electron withdrawing
$S_N1(P)$ type (3-coordinate intermediate)	*gem* products	non-*gem* products
$S_N2(P)$ type (5-coordinate intermediate)	non-*gem* products	*gem* products

N.B. The presence of Y in the molecule is likely to affect the mechanism, e.g. the initial attack on $(NPCl_2)_3$ by Y^- may be $S_N1(P)$ but if Y is electron withdrawing it will discourage further $S_N1(P)$ and the reaction of Y^- on $N_3P_3Cl_5Y$ may then be $S_N2(P)$.

size from the tetramer to the pentamer and hexamer. 'Flexibility' is not a valid reason since it can have little bearing on the situation at phosphorus; in all these compounds the ring bond at phosphorus remains the same. The only significant change with ring size is in the bond angle at nitrogen which may be related to the donor π system as we have seen. As $\angle$ PNP increases so will the s character of the P—N bond and so will the electronegativity of the nitrogen atom as the Hinze-Jaffé values show (Table 2.7 page 74). This may serve to reduce electron density about phosphorus and expose it more to nucleophilic attack, but again this would suggest an order $(NPCl_2)_5 > (NPCl_2)_4 > (NPCl_2)_3$. At present there is no obvious explanation for the relative reactivities of various ring sizes.

10.5.4.1 Nucleophilic substitution by fluoride and pseudohalides

The complete substitution of chlorine by fluorine can be effected by heating the chloride with KF in liquid SO_2[134] or with KSO_2F[135] in an autoclave (eqn. 112). By this method, and using the mixture of chlorocyclopolyphosphazenes obtained from the synthetic reaction (95), it was shown[134] that rings $(NPF_2)_{3-17}$ were formed. Partial fluorination of the trimer gives solely *gem* products, which can be obtained by means of NaF in refluxing dipolar aprotic solvents such as $MeNO_2$ and $PhNO_2$.[136] The same reaction with $(NPCl_2)_4$ gives only $(NPF_2)_4$ without partially substituted intermediates being formed in significant amounts, illustrating both the extra reactivity of the larger ring and the activating effect of substituted fluorine atoms.

$$(NPCl_2)_n + 2nKSO_2F \longrightarrow (NPF_2)_n + 2nKCl + 2nSO_2 \qquad (112)$$

The partially substituted tetrameric products can be obtained by a solid state reaction of $(NPCl_2)_4$ and KSO_2F under reduced pressure[136] (eqn. 113). Again only *gem* isomers are produced and the $1,1,3-N_4P_4F_3Cl_5$ isomer predominates over the $1,1,5-N_4P_4F_3Cl_5$. The explanation of *gem* products from fluorination follows from an S_N2 mechanism involving an incoming electron-withdrawing group – Table 10.13.

$$(NPCl_2)_4 + xKSO_2F \xrightarrow[\text{press.}]{\text{low}} N_4P_4F_xCl_{8-x} + KCl \qquad x = 1-8 \qquad (113)$$

The non-*gem* fluoro derivatives can be obtained indirectly by blocking some of the sites round the ring with amino groups which are substituted non-*geminally*. Even so

10.5 The cyclopolyphosphazenes

this is not always effective as eqn. (114a) shows. It is necessary to use the right fluorinating agent, SbF$_3$, which is thought to act by coordinating to the most basic nitrogen, N*, situated between the amino-substituted phosphorus atoms[137].

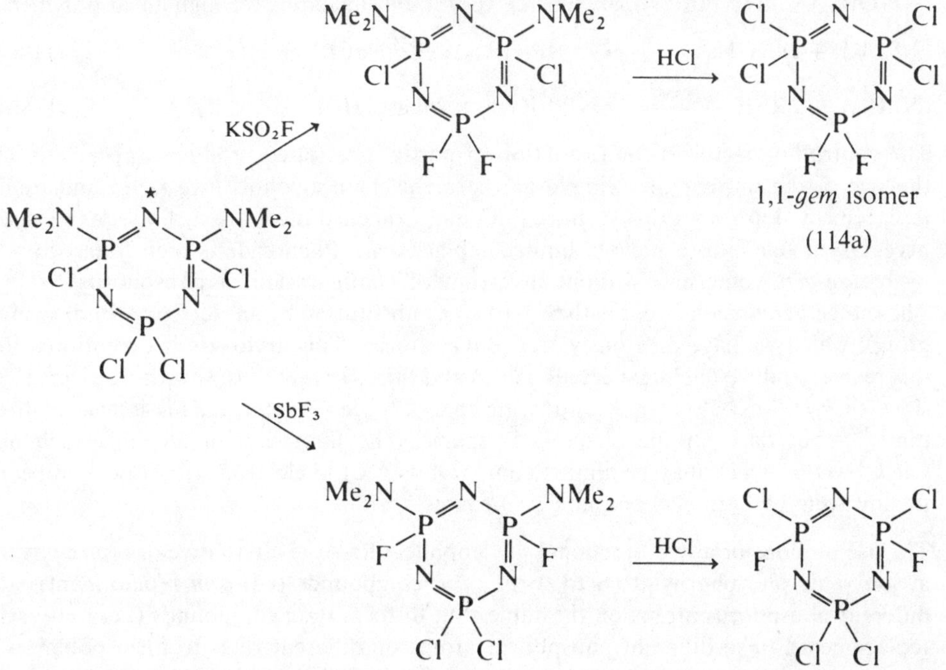

1,1-*gem* isomer

(114a)

1,3-non-*gem* isomer

(114b)

Of the pseudohalide groups replacement by cyanide has not been reported but displacement by azido and isothiocyanato is known. The product from reaction (115) is an explosive oil[138], as one might expect from a compound of formula P$_3$N$_{21}$! However the azido group does not necessarily mean instability, e.g. (10.24) from reaction (116) is stable up to 200 °C and thereabove evolves N$_2$.[139]

$$(NPCl_2)_3 + 6NaN_3 \xrightarrow{acetone} (NP(N_3)_2)_3 + 6NaCl \qquad (115)$$

$$1,3\text{-}N_4P_4Ph_6Cl_2 + LiN_3 \xrightarrow{MeCN} 1,3\text{-}N_4P_4Ph_6(N_3)_2 \qquad (116)$$
$$(10.24)$$

Another interesting fact about (10.24) is that whether this be the *cis* or *trans* isomer an equilibrium mixture of 28 % *cis* : 72 % *trans* is formed on heating.

When acetone solutions of the chlorides and KSCN are mixed the fully substituted products form rapidly (eqn. 117) and can be crystallized out at low temperatures[140].

$$(NPCl_2)_{3,4} + 2nKSCN \longrightarrow (NP(NCS)_2)_{3,4} + 2nKCl\downarrow \qquad (117)$$

Partially substituted derivatives such as N$_3$P$_3$(NCS)$_2$Cl$_4$ have been prepared but their substitution pattern is unknown.

10.5.4.2 Nucleophilic substitution by alkoxides, aryloxides and thio analogues

Although fluorination is fairly straight forward, and can be explained by electron withdrawal, the other substitutions are less so. Alcohols, for example, are slow to react and often produce by-products which are not cyclophosphazenes. To effect alcoholysis it is necessary to use basic reagents (eqns. 118). A large variety of aliphatic and aromatic alcohols and thiols have been employed as well as diols. The products $(NP(OR)_2)_n$ and $(NP(SR)_2)_n$ are usually very stable, including the high linear polymers

$$(NPCl_2)_n + 2nRONa \longrightarrow (NP(OR)_2)_n + 2nNaCl \qquad (118a)$$

$$(NPCl_2)_n + 2RSH \xrightarrow{base} (NP(SR)_2)_n + 2n\text{base}\cdot HCl \qquad (118b)$$

The controlling factor in the formation of partly substituted products appears to be the size of the nucleophile. Simple and straight chain alcohols give rapid and total replacement. Larger groups such as PhO and branched alcohols such as Pr^iO- are necessary if one is to achieve a limited replacement. Phenoxide is then found to go by a non-*gem* route and without inversion of configuration at phosphorus[141,142]. These together are consistent with $S_N1(P)$ type substitution by an electron-withdrawing group, which we have previously seen PhO— to be. This aryloxide is exceptional in this respect and we might expect alkoxides and thio derivatives to substitute *geminally* since they are less electron withdrawing than Cl, see Chapter 2. This is true of the thiols[143] but data on the alcohols is sparse. The fluorocarbon alcohols such as CF_3CH_2O-, which may be almost comparable to Cl in electron withdrawal, appear to substitute by both *gem* and non-*gem* routes[144].

The use of diols, or any bifunctional nucleophile, offers the alternative of replacement at the same phosphorus atom to form *spiro* compounds (i.e. *gem* replacement), at different phosphorus atoms on the same ring to form *ansa* compounds (i.e. non-*gem* replacement), or at different phosphorus atoms on different rings to form polymers. The choice will depend to a large extent on the diol; thus *para*-substituted aromatic diols must give polymers whereas *meta*-substituted ones must give *spiro* compounds.

Aliphatic diols such as propane-1,3-diol and butane-1,4-diol at $-10\,°C$ give *spiro* derivatives[145] (eqn. 119), as does $HOCH_2CF_2CF_2CH_2OH$.[146] This supports the *gem*

$$(NPCl_2)_3 + HO(CH_2)_{3,4}OH \longrightarrow \left(NP\begin{matrix}O-H_2C\\ \\O-H_2C\end{matrix}(CH_2)_{1,2}\right)_3 + 6HCl \qquad (119)$$

tendency we would expect of alcoholysis although we might have expected the last diol to produce some *ansa* derivatives, on the basis that CF_3CH_2OH gave some non-*gem* products.

The aryl diols can be made as five-, six- and seven-membered ring compounds depending upon the diol, (10.25), (10.26) and (10.27)[37,147].

(10.25)   (10.26)   (10.27)

10.5 The cyclopolyphosphazenes

We shall talk more about these in the context of hydrolysis. The derivative (10.25) is a crystalline compound capable of acting as the host molecule in the formation of clathrates with organic molecules[148]. These guest molecules are held in channels of 450 pm diameter which penetrate the host lattice and are effectively held despite the absence of a physical barrier to their escaping, as is found with other calthrates where the guest molecules are held captive in cavities in the lattice. Guest molecules which have formed clathrates with (10.25) include hydrocarbons, halogenocarbons, esters, ketones, ethers, nitriles etc. and generally temperatures of 170 °C are required to cause the clathrate to break down.

10.5.4.3 Nucleophilic substitution by amines and ammonia

Less reactive than alcohols, amines rarely substitute all the chlorine atoms of $(NPCl_2)_{3,4}$ except under very forcing conditions. They are therefore useful reagents for studying the mechanism of substitution. Reactions can be done under anhydrous conditions with liquid amine or in a two-phase system using aqueous amine solutions and $(NPCl_2)_3$ in an organic solvent. A wide range of aliphatic, aromatic, heterocyclic amines and diamines, amino acid and hydrazino derivatives have been produced.

Diamines such as $NH_2(CH_2)_{2,3,4}NH_2$ do not form *spiro* compounds, like the diols, but give *ansa* derivatives (10.28)[149] (eqn. 120). This demonstrates the 'basic' difference between nucleophilic attack by alkoxides which are *gem* directing and amino groups which are usually non-*gem* directing. This, however, is only a first approximation as we shall see.

$$(NPCl_2)_3 + NH_2(CH_2)_{2,3,4}NH_2 \xrightarrow{\text{base}} \text{(10.28)} \quad (120)$$

The reaction of $(NPCl_2)_{3,4}$ and liquid ammonia produces the fully ammonolyzed derivatives $(NP(NH_2)_2)_{3,4}$. In ether solution (eqn. 121) $(NPCl_2)_3$ gives the 1,1-diamido derivative as shown by ^{31}P n.m.r. spectroscopy[150,151].

$$(NPCl_2)_3 + 4NH_3 \xrightarrow{\text{ether}} 1,1\text{-}N_3P_3(NH_2)_2Cl_4 + 2NH_4Cl \quad (121)$$

On the other hand the tetramer gives the non-*gem* 1,5-diamido derivative[152] (eqn. 122) in which the two NH_2 groups are as far apart as possible.

$$(NPCl_2)_4 + 4NH_3 \xrightarrow{\text{ether}} 1,5\text{-}N_4P_4(NH_2)_2Cl_6 + 2NH_4Cl \quad (122)$$

Aromatic primary amines such as aniline etc. behave in the same manner giving *gem* products with the trimer[153] and 1,5-non-*gem* products with the tetramer[154]. This behaviour is consistent only with a change of mechanism from the trimer ($S_N1(P)$) to the tetramer ($S_N2(P)$) – Table 10.12. This may also be the reason why there is a difference between the reactivity of $(NPCl_2)_3$, which may prefer the slower ionic ($S_N1(P)$ type) mechanism, and $(NPCl_2)_4$ which prefers the $S_N2(P)$ type mechanism. Higher electron density about the phosphorus in $(NPCl_2)_3$ would encourage ionization and $S_N1(P)$ while lower electron density at P in $(NPCl_2)_4$, due to bond changes between P—N bonds, would discourage ionization whilst at the same time promote nucleophilic attack in general.

Primary, unbranched, aliphatic amines react rapidly and completely, but under mild conditions partially substituted products such as the mono(methylamino) derivative[155], can be isolated e.g. eqn. (123a).

$$(NPCl_2)_3 + 2MeNH_2 \xrightarrow{\text{ether/water}} N_3P_3(NHMe)Cl_5 + MeNH_3Cl \qquad (123a)$$

Further reaction with methylamine gives a mixture of all three bis(methylamino) isomers: 1,1-, cis-1,3- and trans-1,3-$N_3P_3(NHMe)_2Cl_4$. These have been separated and characterized[156]. Clearly both *gem* and non-*gem* substitution has occurred.

Primary, branched, aliphatic amines are less reactive and reaction goes only to the tetra-substituted derivative, $N_3P_3(NHR)_4Cl_2$, unless forcing conditions are used. Isopropyl amine goes by both *gem* and non-*gem* routes[157] but t-butylamine by the *gem* pathway only[158].

Of the secondary amines dimethylamine and piperidine have been most investigated. Higher aliphatic amines are much slower to react. Partial substitution products of the dimethylamine reaction show that mainly a non-*gem* route is followed but significant amounts of *gem* isomers are formed[159], at least in the anhydrous but not in the aqueous reaction[160]. This sensitivity to reaction conditions introduces yet another variable which has to be taken into account in the substitution mechanism.

The heterocyclic amine piperidine behaves like Me_2NH with $(NPCl_2)_3$ although reaction is a thousand times faster[161]. This is despite the similar basicity of these amines in water which suggests that steric effects have an important role in determining the reaction pathway. Other heterocyclic amines such as pyrrolidine and morpholine seem, like piperidine, to prefer the non-*gem* route, but aziridine, $(CH_2)_2NH$, goes *geminally* despite being a weaker base[162].

It is difficult to rationalize the amine results and to explain why some amines behave one way and very similar ones behave another way. All we can deduce about the substitution pattern of aminolysis is that at least three factors are responsible. Firstly, the size of the ring which we have already mentioned. Secondly, steric effects, i.e. if the amine is primary it allows *gem* but if it is secondary it hinders this and non-*gem* substitution occurs. Thirdly, the electron-releasing ability of the amine, of which its basicity in water is a measure. This shows that weak bases such as aniline and aziridine substitute *geminally*, while strong bases tend to go non-*geminally*.

Even allowing for these factors there is yet another effect which is operating in non-*gem* substitution since the products of such a mechanism are predominantly *trans* isomers. This is not due to steric effects. What is happening is that the amino group in some way is promoting or directing the nucleophilic attack at a neighbouring phosphorus. In the case of the mechanism being $S_N2(P)$ the amino group is affecting the *cis* chlorine atom and this has become known as the 'cis effect'[163].

Another phenomenon, which may also have some bearing on the distribution of *cis*- and *trans*-non-*gem* isomers, is *cis–trans* isomerization, which the aminophosphazenes display. We have already mentioned the *cis–trans* isomerization of 1,3-$N_4P_4Ph_6(N_3)_2$

10.5 The cyclopolyphosphazenes

which results in a 28 % *cis* + 72 % *trans* isomer mix regardless of which isomer is the starting point. Refluxing *trans*-1,3,5-N$_3$P$_3$Ph$_3$Br$_3$ in acetonitrile for 1½ h converts it to the *cis* isomer[164]. Reconversion to the *trans* isomer requires 50 h at 180 °C. The *trans* → *cis* conversion does not occur in n-hexane as solvent which suggests that a polar solvent is necessary because the mechanism is ionic (eqn. 123b).

$$N_3P_3Br_3Ph_3 \rightleftharpoons N_3P_3Br_2Ph_3{}^+ + Br^- \qquad (123b)$$

Cis-trans isomerization of aminophosphazenes likewise does not occur in apolar solvents, such as benzene, but in acetone, pyridine and chloroform it is feasible. 'Catalysts' are generally required such as substituted ammonium salts although there is no exchange between these and the phosphazene amino groups[165]. This suggests that in isomerizations such as (124) the mechanism involves only the making and breaking of P—Cl bonds, and, in view of the necessity for polar solvents, that it is primarily ionic.

$$1,3,5\text{-}N_3P_3Cl_3(NMe_2)_3 \xrightarrow[\text{MeCN}]{\text{Me}_2\text{NH}_2\text{Cl}} 1,3,5\text{-}N_3P_3Cl_3(NMe_2)_3 \qquad (124)$$
$$\textit{cis} \text{ or } \textit{trans-} \qquad\qquad 1:2 \text{ ratio } \textit{cis}:\textit{trans-}$$

All this may have no bearing on the substitution pattern of non-*gem* isomers formed from (NPCl$_2$)$_3$ and amines but it cannot be ruled out as a possible contributing factor.

10.5.4.4 Nucleophilic substitution at phosphorus by organic groups

From the still unsolved mysteries of aminolysis we turn to the somewhat simpler field of organic substitution by organolithium and Grignard reagents. We shall also consider the Friedel–Crafts type acylations.

Organolithium alkyl or aryls react with (NPCl$_2$)$_3$ but cause breakdown of the phosphazene ring. The fluorides, however, are less reactive and a variety of reagents have been used with these to effect partial and complete substitution. Dimethylation of (NPF$_2$)$_3$ with MeLi takes place *geminally* to give 1,1-N$_3$P$_3$F$_4$Me$_2$ which is inconsistent with an S$_N$2(P) mechanism in view of the fact that a methyl group should be very much electron releasing compared to fluorine, and so would make the phosphorus to which it is attached less attractive as a site for further nucleophilic attack. On the other hand it would encourage loss of fluoride in a S$_N$1(P) type mechanism. Trimethylation to form N$_3$P$_3$F$_3$Me$_3$ derivatives does not occur but instead the reaction takes a different course[166].

The reaction of MeLi with (NPF$_2$)$_4$ has been more thoroughly investigated but is even more puzzling[167]. It has been possible to introduce up to four methyl groups. Dimethylation, for example, gives all five possible isomers but the *gem* isomer predominates (eqn. 125). Trimethylation gives only two isomers, instead of the theoretical five, and these are the two *gem* isomers 1,1,3- and 1,1,5-N$_4$P$_4$F$_5$Me$_3$. Further methylation produced only the 1,1,5,5-N$_4$P$_4$F$_4$Me$_4$ derivative.

$$(NPF_2)_4 + 2MeLi \longrightarrow 1,1\text{-}N_4P_4F_6Me_2(41\%) + \textit{cis}\text{-}1,3\text{-} + \textit{trans}\text{-}1,3\text{-}$$
$$+ \textit{cis}\text{-}1,5\text{-} + \textit{trans}\text{-}1,5\text{-}N_4P_4F_6Me_2 \qquad (125)$$

These results seem to show a struggle between the S$_N$1(P) mechanism and the S$_N$2(P), which moves in favour of the former as more methyl groups are bound to the ring.

The reaction of PhLi with $(NPF_2)_3$ may be influenced by steric factors as the preponderance of the least hindered, *cis*-1,3 isomer in reaction (126) demonstrates[168].

$(NPF_2)_3 + 2PhLi \longrightarrow$ 1,1-$N_3P_3F_4Ph_2$ + *cis*-1,3- + *trans*-1,3-$N_3P_3F_4Ph_2$ (126)
                                    (5 %)   (70 %)   (25 %)

Grignard reagents, and especially alkyl Grignards, react but tend to give non-cyclic by-products. Reaction (127) gives only a 5 % yield of the desired product, the main products being linear polyphosphazenes, and what little hexaphenylcyclotriphosphazene is produced is thought to be due to recyclization of linear species[169]. The same thing must be happening in the reaction of $(NPCl_2)_4$ and PhMgBr (eqn. 128) in which the yield of the trimeric product (10.29) can be as high as 93 %.[170]

$(NPCl_2)_3 + PhMgBr \longrightarrow (NPPh_2)_3 + Ph_3P=N-PPh_2=N-PPh_2=N-MgBr$ (127)

$(NPCl_2)_4 + PhMgBr \longrightarrow$

1,1,5,5-$N_4P_4Cl_4Ph_4$ + [structure 10.29] + traces of others (128)

(10.29)

The Grignard reaction of $(NPF_2)_3$ and PhMgBr gives the *gem*-1,1-$N_3P_3Ph_2F_4$ product consistent with an $S_N1(P)$ mechanism and electron-releasing groups[171]. Reaction of $(NPBr_2)_3$ gives only decomposition products.

Friedel–Crafts type arylation gives *geminally* substituted products (eqn. 129), but the reaction is slow, taking days to achieve a yield of 40 %.

$(NPCl_2)_3 + C_6H_6 \xrightarrow[\text{reflux}]{Al_2Cl_6}$ 1,1-$N_3P_3Cl_4Ph_2$ (129)

Further substitution can be achieved by refluxing the mixture for 6 *weeks* and this produces a 40 % yield of 1,1,3,3-$N_3P_3Cl_2Ph_4$ and a mere 6 % yield of the fully phenylated product, $(NPPh_2)_3$.[172] The increasing deactivation of the ring by successive phenyl groups may be due to steric factors. The reaction has been extended to a few other aromatic hydrocarbons but many do not react. The tetramer, $(NPCl_2)_4$, is less reactive than the trimer which suggests that it is being forced to react by an ionic mechansim contrary to its natural tendency. The reaction is more complex as shown by the formation of some (10.29)[173].

The mechanism of the Friedel–Crafts reaction is thought to go via eqn. (130) which produces a cation which is a strong enough electrophile to stimulate benzene to behave as a nucleophile (eqn. 131).

$(NPCl_2)_3 + AlCl_3 \longrightarrow N_3P_3Cl_5^+ + AlCl_4^-$ (130)

$N_3P_3Cl_5^+ + C_6H_6 \longrightarrow N_3P_3Cl_5Ph + H^+$ (131)

The monophenyl product of this reaction then goes on to form a *gem*-1,1-diphenyl derivative because the second ionization is easier from ⧹PClPh, with its electron-

10.5 The cyclopolyphosphazenes

releasing phenyl group, than from another $\text{\textbackslash}PCl_2$ centre. Amino groups should also aid ionization from $\text{\textbackslash}PCl(NR_2)$ and this is demonstrated in eqn. (132) when the only acylation is at these centres[174].

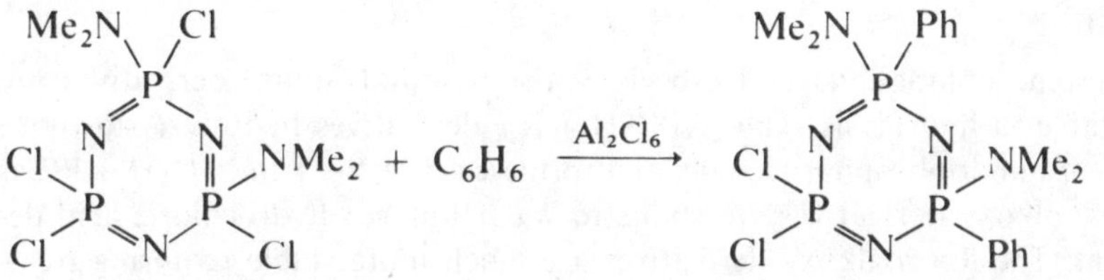

$$\text{[Me}_2\text{N/Cl-substituted cyclotriphosphazene]} + C_6H_6 \xrightarrow{Al_2Cl_6} \text{[Me}_2\text{N/Ph-substituted cyclotriphosphazene]} \quad (132)$$

10.5.4.5 Hydrolysis of cyclopolyphosphazene compounds

Finally we turn to hydrolysis, which embraces not only that of the halides but of many of the products of these which we have discussed. However let us begin with the halides. The trimeric fluoro- and chloro-phosphazenes react more slowly with water than the tetrameric counterparts [$(NPCl_2)_3$ can be steam distilled], but the acid produced from the tetramer is more stable to further hydrolysis than that from the trimer. In both cases the acid undergoes proton migration from oxygen to ring nitrogen, e.g. eqn. (133), so that the product is no longer a cyclic phosphazene but a cyclic phosphazane.

$$(NPCl_2)_3 + 6H_2O \longrightarrow \text{``}(NP(OH)_2)_3\text{''} \longrightarrow \text{[cyclic phosphazane product]} \quad (133)$$

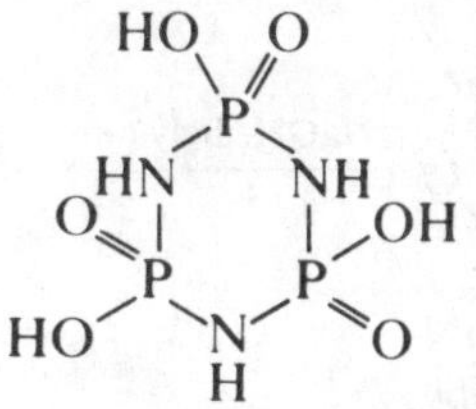

This proton migration does not always occur and in the hydrolysis of $N_3P_3Ph_5Cl$, (10.30), the product is (10.31) – eqn. (134).

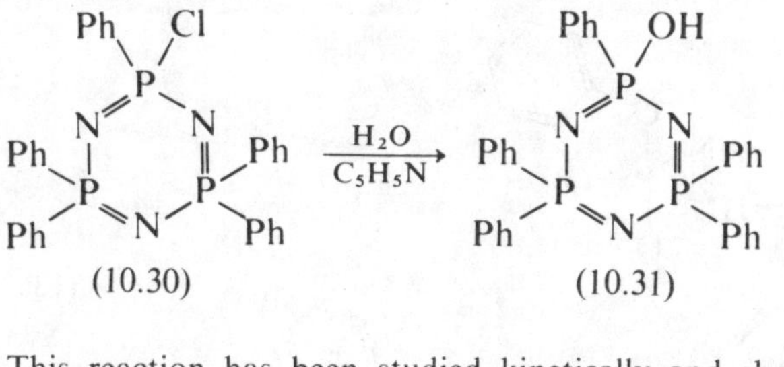

$$\text{(10.30)} \xrightarrow{\frac{H_2O}{C_5H_5N}} \text{(10.31)} \quad (134)$$

This reaction has been studied kinetically and shown to go via the intermediate $[N_3P_3Ph_5\cdot NC_5H_5]^+$.[175] The reason for non-migration of the proton in (10.31) must be that the oxygen is more basic than the ring nitrogen, a feature which is attributable to the donor electronic effect of the phenyls. On the other hand $N_3P_3(OPh)_5(OH)$ undergoes migration to give (10.32).

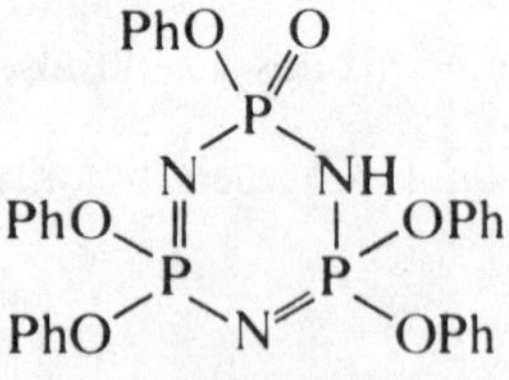

(10.32)

Other phosphazene compounds and especially the oxy and amino derivatives are much more stable to hydrolysis. The $(NP(NH_2)_2)_{3,4}$ derivatives hydrolyze slowly to give a variety of amidophosphorus acids and in the case of the trimer $(NH_2)_3PO$ is formed[176]. The alkoxy derivatives are stable to water but hot hydrochloric acid decomposes them. The fluoroalkoxy derivatives are much more stable requiring treatment with NaOH in methanol at 75 °C for a week to break them down.

Aryloxy trimer compounds are more resistant to hydrolysis than tetrameric ones. Although for $(NP(OPh)_2)_3$ it is easy to replace one phenoxy group by a hydroxyl group further replacement is difficult. Studies on substituted phenol derivatives show that electron withdrawal enhances the rate of hydrolysis and this is consistent with an $S_N2(P)$ type rate-determining step.

Hydrolysis of the spiro arylenedioxycyclophosphazenes show a marked difference between the five- and the six- and seven-membered rings. The last two are completely stable to hydrolysis whereas the first is very rapidly hydrolyzed with break-down of the phosphazene ring (eqn. 135)[37,177].

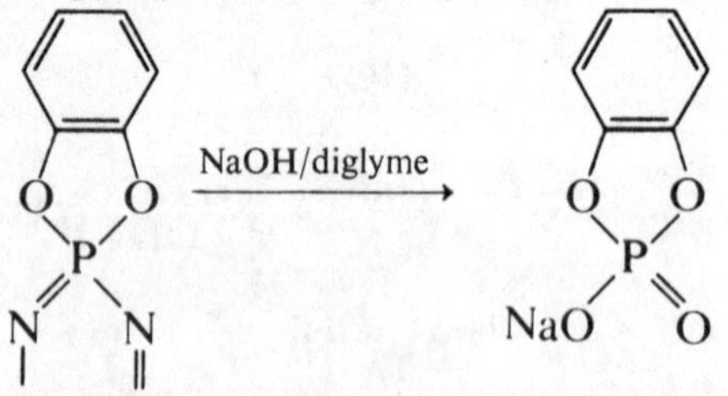

(135)

This poses some rather interesting questions of mechanism. Proton migration, followed by ψ of a *tbp* intermediate to manoeuvre the protonated nitrogen into a leaving apical position, will explain *how* this could come about (eqn. 136) but not *why* it should occur.

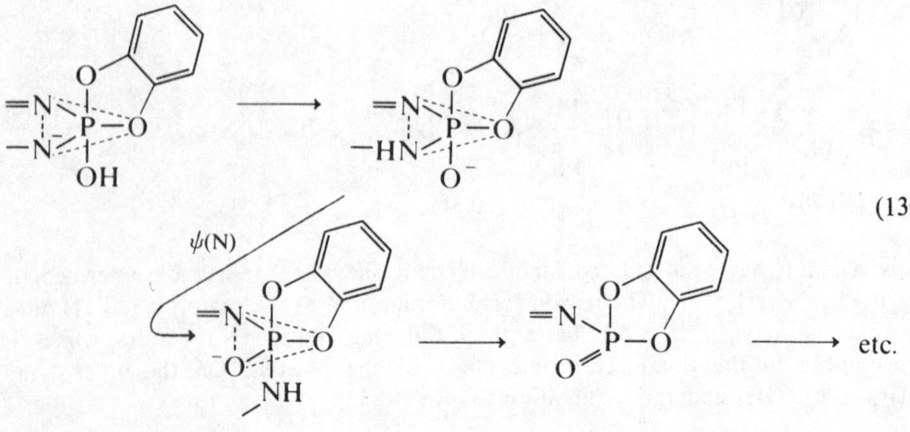

(136)

10.5 The cyclopolyphosphazenes

To summarize: nucleophilic substitution is still only partly understood in the cyclophosphazenes. Electron changes at phosphorus can act through σ, π_a and π_s systems but so far we are a long way from being able to use information about these to explain everything that happens. The expected mechanism and the nature of the group can be used to bring some order to the situation, but some reactions show a healthy disrespect for these rationalizations, which is what makes the chemistry of the phosphazenes so fascinating.

10.5.5 Reactions involving nitrogen in the cyclopolyphosphazenes

Although most work has naturally been on reactions at the phosphorus atom, a significant amount has been achieved on studying the behaviour at nitrogen. Here attention has been focussed on the availability of the lone pair since this may be involved in the ring π_s bonding. One measure of the availability is the ability to form Lewis adducts and complexes to transition metals.

10.5.5.1 Lewis basicity and complex formation

For the formation of a Lewis acid–base adduct with the cyclophosphazenes it is usually necessary to employ a strong Lewis acid. Even so $(NPCl_2)_3$ forms no adduct with $SnCl_4$ or $SbCl_5$ but with SO_3 it gives an adduct in which all three ring atoms are involved as bases i.e. $(NPCl_2)_3 \cdot 3SO_3$. With $(NPBr_2)_3$ the groups on phosphorus are less electron withdrawing which should enhance the electron density on the ring and at nitrogen. Again there is no adduct formation with such strong Lewis acids as BBr_3, $SnCl_4$ or $TiCl_4$ but $AlBr_3$ gives 1:1 and 1:2 adducts, $(NPBr_2)_3 \cdot AlBr_3$ and $(NPBr_2)_3 \cdot 2AlBr_3$.

Electron-releasing groups such as amino and methyl improve the basicity of the ring nitrogens and adducts such as $(NP(NHR)_2)_3 \cdot BF_3$, $(NPMe_2)_3 \cdot SnCl_4$ and $(NPMe_2)_3 \cdot TiCl_4$ are formed. In the last example infrared spectroscopy confirmed the N—Ti bond[178].

Several transition metal complexes are known for amino and methyl derivatives. The complex $(NP(NMe_2)_2)_4 \cdot W(CO)_4$ has been shown to coordinate to the metal through one ring and one exocyclic nitrogen atom[179]. However the complex $(NP(NMe_2)_2)_6 \cdot CoCl^+$ has four of the ring nitrogens coordinated to the cobalt[180]. Paddock and co-workers have also studied complexes of the methylcyclophosphazenes such as $(NPMe_2)_4 \cdot Mo(CO)_3$ and $(NPMe_2)_4 \cdot W(CO)_3$, where it might have been hoped that something akin to the π complexes of delocalized organic ligands would be formed[181]. Unfortunately for the theory, coordination is probably through the ring nitrogens.

10.5.5.2 Alkylation and protonation of ring nitrogens

Methylation and ethylation of ring nitrogen atoms occurs with $(NPMe_2)_3$ and $(NPMe_2)_4$ when treated with the corresponding methyl or ethyl iodide e.g. eqn. (137)

$$(NPMe_2)_3 + MeI \longrightarrow N_3P_3Me_7^+ \; I^- \tag{137}$$
$$(10.33)$$

The product (10.33) is soluble in water (as is $(NPMe_2)_3$ itself) but it can be obtained as its insoluble HgI_3^- salt[182]. With amino derivatives there is the possibility of alkylation of either ring or exocyclic nitrogens. Hexakis(dimethylamino)cyclotriphosphazene and $Me_3O^+BF_4^-$ give the latter type (eqn. 138) whereas with $N_3P_3Cl_2(NHPr^i)_4$ ring methylation occurs (eqn. 139)[183].

$$(NP(NMe_2)_2)_3 + 2Me_3O^+BF_4^- \longrightarrow [N_3P_3(NMe_2)_4(NMe_3)_2]^{2+} 2BF_4^- \quad (138)$$

$$N_3P_3Cl_2(NHPr^i)_4 + Me_3O^+BF_4^- \longrightarrow [N_3P_3Cl_2Me(NHPr^i)_4]^+ BF_4^- \quad (139)$$

Protonation of the cyclophosphazenes bring us to the basicity of these compounds in general. Those displaying the strongest basicity are the amino derivatives, which sometimes show a greater basicity than the parent amine; in some cases they are capable of forming isolateable salts, e.g. $(NP(NHPr^n)_2)_3 \cdot HCl$ which can be recrystallized from hot heptane[184]. The introduction of one proton lowers the base strength considerably but in certain instances dibasic behaviour is observed, e.g. $(NP(NHPr^n)_2)_3 \cdot 2HCl$ [37] and $(NPCl_2)_4 \cdot 2HClO_4$[185], and there is one rare example of tribasicity in $(NP(NHC_5H_{11})_2)_3 \cdot 3HCl$[186].

Infrared and p.m.r. studies show that protonation occurs at a ring nitrogen, and the x-ray crystallographic investigation of $N_3P_3Cl_2(NHPr)_4 \cdot HCl$ confirmed these observations[187]. Again it was also demonstrated that with $(NP(NMe_2)_2)_6$ the proton attaches itself to a ring and not to an exocyclic nitrogen atom[188].

The comparative basicity of the cyclophosphazenes towards a reference acid cannot be measured in water as the solvent since so few are soluble. Instead Feakins and Shaw[189,190] have used nitrobenzene solutions and developed a method of potentiometric titration using a glass electrode for these bases and $HClO_4$. They published their results as pK_a' values* and a list of these are given in Table 10.14. As can be seen, only the very basic derivatives exhibit a pK_{a_2}' for a second proton. The lower measurable limit for this method is -6.0 so that bases weaker than this such as $(NPCl_2)_3$ cannot be distinguished. The order of basicity in Table 10.14 reflects closely the electron-releasing abilities of the groups attached to phosphorus and this is in line with the data discussed in previous chapters.

By extending their method to cyclophosphazenes with mixed substituents, Feakins and Shaw have devised a sophisticated theory of the contributions to the basicity of a particular ring nitrogen from groups attached to the α phosphorus and the γ phosphorus atoms[191–193] of the following structure:

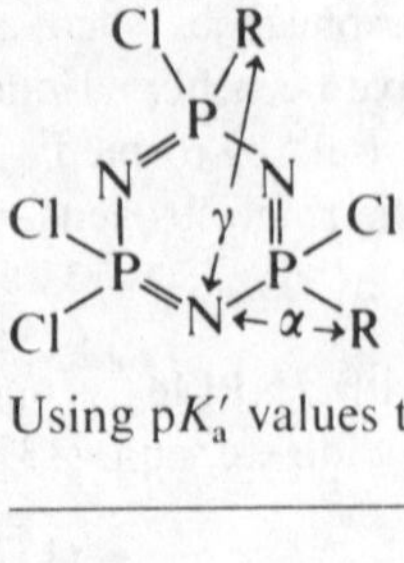

Using pK_a' values they have devised a set of α and γ substituent contributions making

---

* These refer to the equilibrium, $(NPX_2)_3 \cdot H^+ \overset{K_a'}{\rightleftharpoons} (NPX_2)_3 + H^+$, the ' denoting that the solutions are in nitrobenzene. The dissociation constant, K_a' is $(a_H^+ \cdot a_{(NPX_2)_3})/(a_{(NPX_2)_3H^+})$ and hence $pK_a' = pH' + \log(a_{(NPX_2)_3H^+}/a_{(NPX_2)_3})$. By choosing and defining an arbitrary value of pH' a pK_a' can be calculated. pH' was defined for the half-neutralized 0.100 M solution of $N_3P_3(NH \cdot C_6H_4\text{-}p\text{-}Me)_6$ as 3.0.

10.5 The cyclopolyphosphazenes

Table 10.14 pK_a' values as a measure of the basicity of cyclotriphosphazenes[189,190]

	pK_{a_1}'	pK_{a_2}'		pK_a'
(NP(NHMe)$_2$)$_3$	8.8	−2.0	(NP(SEt)$_2$)$_3$	−2.8
(NP(NMe$_2$)$_2$)$_3$	7.6	−3.3	(NP(SPh)$_2$)$_3$	−4.8
(NPEt$_2$)$_3$	6.4		(NP(OPh)$_2$)$_3$	−5.8
(NPPh$_2$)$_3$	1.5		(NPCl$_2$)$_3$	−6.0
(NP(C$_6$H$_4$.p-Cl)$_2$)$_3$	−1.4		(NP(CF$_3$)$_2$)$_3$	−6.0
(NP(OMe)$_2$)$_3$	−1.9			

up pK_a' and these they have called substituent constants. They have demonstrated that these are reasonably constant for a particular group and some are listed in Table 10.15. It is very tempting to relate α values to inductive (σ) effects and γ values to mesomeric (π) effects but at this stage it is necessary to tread carefully since we have nothing to back up the figures of Table 10.15 in support of this theory. However the substituent constants do enable us to work out the most basic of the ring nitrogen atoms in a mixed derivative and by calculating the expected pK_a this has been useful in providing supporting evidence in determinations of the structures of isomers.

Reactions which involve both the phosphorus and nitrogen centres in the cyclophosphazenes are the rearrangements of the alkoxy derivatives. When hexamethoxycyclotriphosphazene is distilled it rearranges completely to the cyclotriphosphazane (eqn. 140)[194]. The tetramer behaves similarly but the pentamer and hexamer decompose. The ethoxy derivative is less ready to rearrange, the trimer giving 29 % of the rearranged form after 1 hour at 200 °C. The tetramer gives only 6 % after 4 hours at this temperature. Other alkoxyl compounds which rearrange are (NP(OPr^i)$_2$)$_3$, (NP(OPr^n)$_2$)$_3$ and (NP(OCH$_2$Ph)$_2$)$_3$. On the other hand (NP(OCH$_2$Ph)$_2$)$_4$ does not rearrange but decomposes above 200 °C. The derivatives (NP(OPh)$_2$)$_{3,4}$ and (NP(OCH$_2$CF$_3$)$_2$)$_{3,4}$ are stable up to 300 °C.

Table 10.15 Substituent constants of cyclophosphazene ligands in pK_a' units taken from refs. 192 and 193

pK_a'	NHMe	NMe$_2$	Ph	OMe	SEt	OPh	SPh	OCH$_2$CF$_3$
α_R	5.8	5.6	4.2	3.6	3.6	3.1	3.0	—
γ_R	3.1	2.8	2.3	1.8	1.8	1.3	1.5	0.3

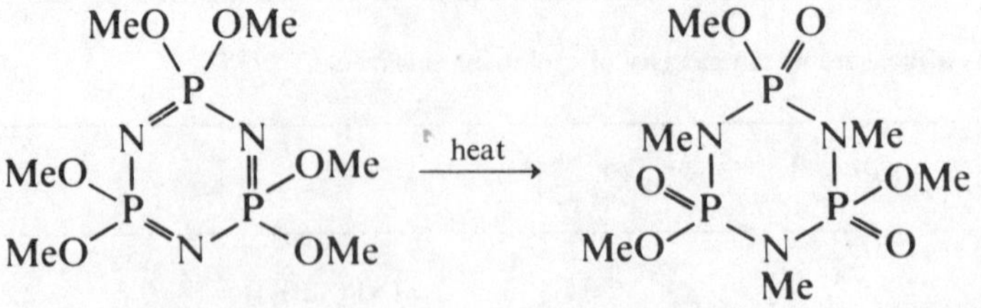

[Equation 140: Rearrangement of hexamethoxy cyclotriphosphazene to N-methylated phosphazane on heating]

(140)

Where rearrangement can occur it is facilitated by the presence of excess of the corresponding alkyl iodide[195]. The use of a different alkyl iodide gives mixed products. This last observation does not seem to fit with the suggested mechanism for rearrangement which involves *intra*molecular attack of the adjacent ring nitrogen atom on the α-carbon of the alkoxy group (eqn. 141).

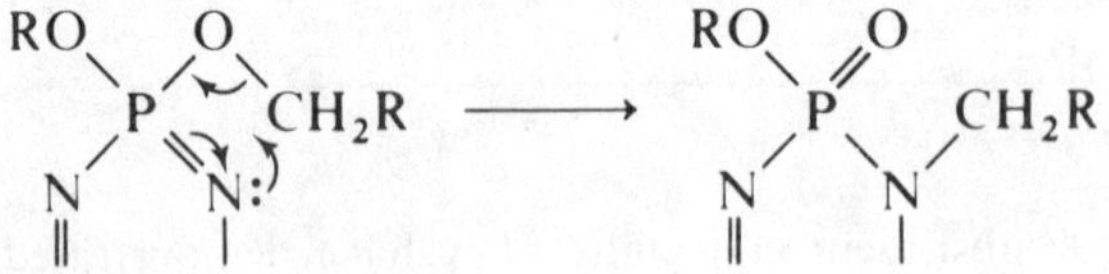

(141)

The α-carbons in phosphorus esters are susceptible to nucleophilic attack and so to a first approximation we might expect the basicity of the nitrogen ring atom to determine whether rearrangement takes place. As a general rule this is true but basicity of the ring nitrogens is only a part of the answer since $(NP(OCH_2Ph)_2)_3$, $pK_a' = -2.1$, rearranges whereas the more basic $(NP(OCH_2Ph)_2)_4$, $pK_a' = -1.6$, does not. In those reactions involving alkyl iodide as catalyst the first step is probably alkylation of a ring nitrogen (eqn. 142).

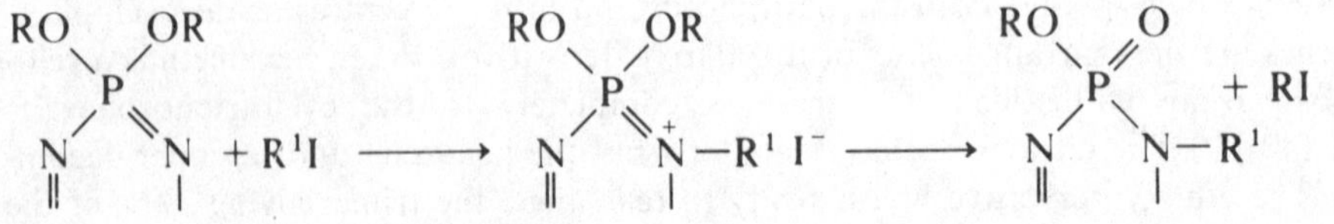

(142)

10.5.6 Polyphosphazene high polymers

Above 250 °C the cyclic chlorophosphazenes polymerize to form long chain $(NPCl_2)_n$ polymers which are thermally stable up to about 350 °C, when they break down again to form cyclic derivatives (eqn. 143). The cross-linked polymer is a benzene-insoluble rubbery material which can be separated easily from the benzene soluble linear polymer. Even at the higher temperatures at which cross-linking takes place not all the cyclic trimer has polymerized and this suggests that we are probably dealing with a system in partial equilibrium. That this is so was demonstrated in 1939 by Schmitz-

$$(NPCl_2)_3 \xrightarrow[2\ days]{250\ °C} \underset{linear}{(NPCl_2)_n} \xrightarrow{270\ °C} \underset{cross\text{-}linked}{(NPCl_2)_n} \xrightarrow{350\ °C} \underset{cyclic}{(NPCl_2)_{3,4,5..}}$$

(143)

DuMont[196] who showed that when $(NPCl_2)_3$ and $(NPCl_2)_4$ were heated at 600 °C the mixture of cyclic products was the same in each case. The trimer polymerized more readily than the tetramer, and it is the latter which predominates in the depoly-

10.5 The cyclopolyphosphazenes

merized mixture of cyclic compounds, unless the more volatile trimer is continually removed by being allowed to distill off when it becomes the main product.

A great deal of effort has been put into studying the polymerization and the interested reader is referred to Allcock's book for a detailed survey[37]. Polymerization can be effected in solution and in the presence of catalysts at 210 °C, although the rate is slow, taking about three times as long as bulk polymerization. Ethers, carboxylic acids, ketones, alcohols, nitrobenzene and powdered metals such as tin and zinc act as catalysts. Water has an adverse effect, and t-butyl peroxide has no effect at lower temperatures, which suggests that the mechanism is ionic rather than free radical. This is supported by the rise in conductance (for $(NPCl_2)_3$ this increases from ca 1.7×10^{-11} ohm^{-1} cm^{-1} at 163 °C to 6250×10^{-11} ohm^{-1} cm^{-1} at 253 °C which is the temperature of polymerization) and the fact that e.s.r. spectroscopy failed to detect free radicals even at 250 °C.[197]

The initial step of the polymerization of $(NPCl_2)_3$ is believed to be the formation of the $N_3P_3Cl_5^+$ cation (eqn. 144) which then attacks another $(NPCl_2)_3$ molecule, probably with ring opening, to give a short linear cation (10.34) (eqn. 145) which would then grow by attacking further cyclic units. Chain growth would be terminated by reaction with a chloride ion (eqn. 146).

$$(NPCl_2)_3 \longrightarrow N_3P_3Cl_5^+ + Cl^- \quad (144)$$

$$(NPCl_2)_3 + N_3P_3Cl_5^+ \longrightarrow \underset{(10.34)}{N_6P_6Cl_{11}^+} \quad (145)$$

$$\sim\!\!\sim\!\!\sim\!N=PCl_2^+ + Cl^- \longrightarrow \sim\!\!\sim\!\!\sim\!N=PCl_3 \quad (146)$$

Cross-linking would come about via loss of a chloride ion from a phosphorus atom situated along the chain (eqn. 147) thus providing a site at which another nitrogen could attack.

$$\sim\!\!\sim\!N=PCl_2-N=PCl_2\!\sim\!\!\sim \longrightarrow$$

$$\sim\!\!\sim\!\!\sim\!N=\overset{+}{P}Cl-N=PCl_2\!\sim\!\!\sim\!\!\sim \quad (147)$$
$$Cl^-$$

The sensitivity of polymerization to the acidity of the glass walls of the polymerization vessel suggests an alternative ionic mechanism for the formation of the initial cation (10.34) (eqn. 148)[198].

$$(NPCl_2)_3 + H^+ \longrightarrow N_3P_3Cl_6H^+ \xrightarrow{(NPCl_2)_3} N_6P_6Cl_{11}^+ + HCl \quad (148)$$

Depolymerization has been studied less than polymerization but probably proceeds by a 'curling-off' or 'tailing-off' mechanism (eqn. 149), which is reminiscent of the

$$\sim\!\!\sim\!PCl_2=\overset{\frown}{N}-PCl_2=N-PCl_2=N-\overset{+}{P}Cl_3 \longrightarrow$$

$$\sim\!\!\sim\!\overset{+}{P}Cl_3 + (NPCl_2)_3 \quad (149)$$

cyclization step in the formation of $(NPCl_2)_n$ cyclomers in the preparative reaction (eqn. 99). The higher temperatures necessary for depolymerization suggest an alternative mechanism may be more important than ionic depolymerizations like eqn. (149) which should occur at lower temperatures. 'Looping-off' may be the alternative depolymerization method (eqn. 150).

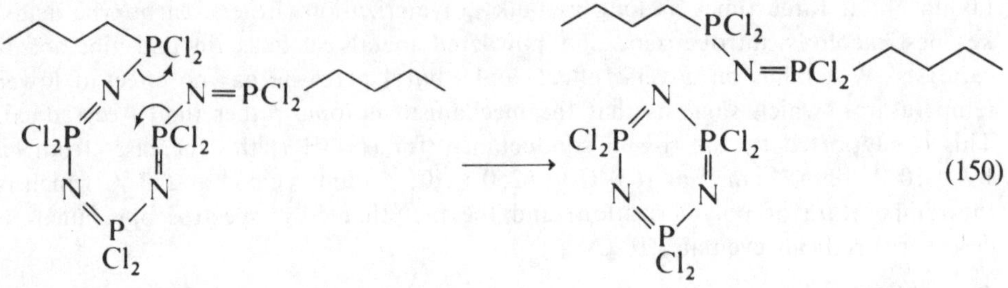

$$\text{(150)}$$

Both the 'curling-off' and 'looping-off' mechanisms are thought to operate in the depolymerization of the polyphosphates.

These high polyphosphazene polymers are unstable with respect to hydrolysis. This instability probably stems from the reactive end-of-chain groups, since if certain metal halides such as $SbCl_5$, $ZnCl_2$, Al_2Cl_6 etc. are added prior to polymerization then the products have both enhanced thermal and hydrolytic stability[199]. Incorporation of a metal chloride grouping at the end of the chain would prevent curling-off depolymerization and resist hydrolysis.

The structure of $(NPCl_2)_n$ has been shown by x-ray analysis to have a *cis-trans* arrangement (10.35)[200]. In a similar study $(NPF_2)_n$ also showed a *cis-trans* planar conformation[201]. This particular polymer is remarkable for its elasticity which it retains down to *ca* $-70\,°C$. The structure of the cross-linked $(NPCl_2)_n$ polymer shows one cross-linking unit per 700 phosphorus atoms.

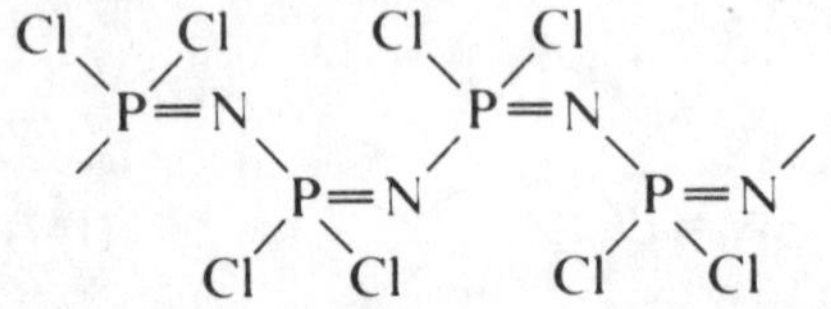

(10.35)

Apart from the chloride and fluoride cyclomers only the bromide and isothiocyanate rings will form high polymers on heating. When other cyclomers are heated they decompose into non-phosphazene compounds and do not polymerize. Part of the reason for this may be due to the fact that no ionic mechanism exists to initiate the polymerization because there is no good anionic leaving group as in the case of the halides and pseudohalide NCS^-. The real reason however is that derivatives of these linear polymers are too thermally unstable with respect to their cyclic counterparts at elevated temperatures. Although they cannot be obtained directly by polymerization of the corresponding cyclomers it is still possible to obtain a variety of high polymers by replacing the chlorine atoms of $(NPCl_2)_n$ by other groups. Allcock and co-workers have made a large number of alkoxy, aryloxy and fluoroalkoxy derivatives of these, plus primary and secondary amino derivatives by this method[37]. The methyl and fluoromethyl polymers, $(NPMe_2)_n$[202] and $(NP(CF_3)_2)_n$,[203] have also been prepared. However all these polymers tend to depolymerize at low temperatures (*ca* 110 °C) except the last two; $(NPMe_2)_n$ depolymerizes at >300 °C. Steric interactions are the chief factor contributing to cyclomer stability relative to the linemer. The order of increasing ease of depolymerization is: Ph < NHMe ≃ NMe₂ < OMe < Et < OEt ≃ OPr^n < NHPh < OCH₂CF₃ < OPh < Bu^t. This is roughly the order of increasing steric hindrance – it is certainly not an order which can be related to electronic effects. The small groups F, Cl, NH₂, Me and CF₃ give much more stable linemers.

The alkoxy and aryloxy, $(NP(OR)_2)_n$, polymers have been made from $(NPCl_2)_n$ and the appropriate sodium salt, NaOR.[204] The products are hydrolytically stable but not thermally stable, breakdown beginning at about 100 °C and becoming rapid about 200 °C. The amino high polymers can be produced from $(NPCl_2)_n$ and the appropriate amine but with bulky amines reaction is slow[205]. Primary amines and NH₂ groups lead to cross-linking by direct bond formation or via the weaker —N—H···N— hydrogen-bonding. The alkyl and aryl polymers $(NPR_2)_n$, R = Me, Et, Ph etc. are obtained from the chloride and the corresponding Grignard reagents.

So far the polyphosphazenes have had the disadvantage of being too expensive to produce. However, they display several desirable features such as low temperature flexibility and can be hydrolytically and thermally very stable. They are not inflammable and in some instances show flame retarding properties. Specialized uses of some of these polymers, where this overrules economic consideration is probable in the near future.

Problems

1 How may the compounds PCl(NMe₂)₂, PF₂(NMe₂), (Me₂N)₂HP=N.SO₂.C₆H₄.*p*-Me and Me₂P(O)NMe₂ be synthesized from PCl₃?

2 What is the overall increase in bond enthalpy at phosphorus in the cyclophosphazenes $(NPCl_2)_3$, $(NPMe_2)_3$, $(NPPh_2)_4$ and $(NP(OC_6H_{11})_2)_3$ compared to the normal trivalent bond enthalpies at phosphorus? Assume $E(P\!\!=\!\!N) = 303$ kJ mol^{-1} and $E(P-N) = 290$ kJ mol^{-1} and use the values in Table 10.5 for the exocyclic bonds. What do the results show?

3 Which of the nitrogen atoms (A) or (B) is the more basic in (10.36) and how is this relevant to the fluorination of (10.36) by SbF₃? What is the product from the reaction with KSO₂F?

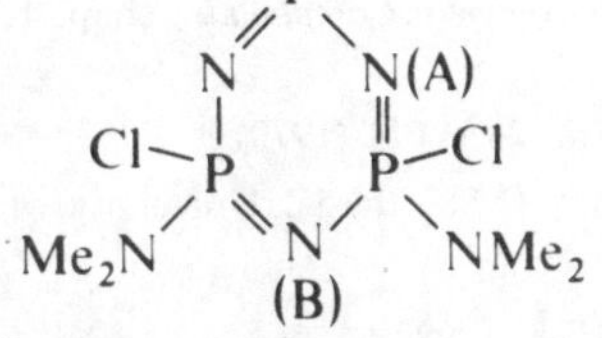

(10.36)

T.C.O.P.—P

References

[1] E. Fluck, *Topics in Phosphorus Chem.*, **4**, 291 (1967).

[2] E. Fluck and W. Hanbold, in G. M. Kosolapoff and L. Maier (editors), *Organic Phosphorus Compounds*, Chap. 16, **6**, 579 (1973).

[3] A. Schmidtpeter and C. Weingand, *Angew. Chem., Int. Ed. Eng.*, **8**, 615 (1969).

[4] M. Lustig and H. W. Roesky, *Inorg. Chem.*, **5**, 1289 (1970).

[5] E. L. Muetterties, P. Meakin and R. Hoffman, *J. Amer. Chem. Soc.*, **94**, 5674 (1972).

[6] P. C. Crofts, in G. M. Kosolapoff and L. Maier (editors), *Organic Phosphorus Compounds*, Chap. 14, **6**, 1 (1973).

[7] W. Gerrard and H. R. Hudson, in G. M. Kosolopoff and L. Maier (editors), *Organic Phosphorus Compounds*, Chap. 13, **4**, 88 (1973).

[8] D. A. Palgrave, *Mellor's Inorg. & Theoret. Chem.*, Vol VIII, Supp. III, P. Section XXVIII, Longman, London, 1971.

[9] S. B. Hartley, W. S. Holmes, J. K. Jacques, M. F. Mole and J. C. McCoubrey, *Quart. Rev. Chem. Soc.*, **17**, 204 (1963).

[10] D. W. J. Cruickshank, *Acta Crystallogr.*, **17**, 671 (1964).

[11] A. Schmidtpeter and H. Rossknecht, *Angew. Chem., Int. Ed. Eng.*, **8**, 614 (1969).

[12] A. Michaelis and K. Luxembourg, *Ber. Deut. Chem. Ges.*, **28**, 2205 (1895).

[13] R. B. King, *Inorg. Chem.*, **2**, 936 (1963).

[14] J. F. Nixon, *Advan. Inorg. Chem. Radiochem.*, **13**, 363 (1970).

[15] R. Schmutzler, *Inorg. Chem.*, 3, 415 (1964).

[16] E. D. Morris and C. E. Nordman, *Inorg. Chem.*, **8**, 1673 (1969).

[17] T. Koizumi and P. Haake, *J. Amer. Chem. Soc.*, **95**, 8073 (1973).

[18] J. Emsley and J. K. Williams, *J. Chem. Soc.*, 1576 (1973).

[19] N. H. Brown, G. W. Fraser and N. W. A. Sharp, *Chem. & Ind. (London)*, 367 (1964).

[20] R. Schmutzler, *Angew. Chem.*, **76**, 893 (1964); *ibid*, **77**, 530 (1965). *Inorg. Chem.*, **3**, 410 and 415 (1964).

[21] R. Schmutzler, *J. Chem. Soc., Dalton*, 2687 (1973).

[22] R. Schmutzler, *J. Chem. Soc.*, 5630 (1965); *Angew. Chem.*, **76**, 570 (1964).

[23] A. H. Cowley, M. J. S. Dewar and W. R. Jackson, *J. Amer. Chem. Soc.*, **90**, 4185 (1968).

[24] H. Goldwhite and D. G. Rowsell, *Chem. Commun.*, 713 (1969).

[25] M. J. C. Hewson, S. C. Peake and R. Schmutzler, *Chem. Commun.*, 1454 (1971).

[26] A. H. Cowley, M. J. S. Dewar, W. R. Jackson and W. B. Jennings, *J. Amer. Chem. Soc.*, **92**, 1085 (1970).

[27] A. Hung and J. W. Gilje, *Chem. Commun.*, 662 (1972).

[28] S. Trippett and P. J. Whittle, *J. Chem. Soc., Perkin I*, 2302 (1973).

[29] J. S. Harman and D. W. A. Sharp, *Inorg. Chem.*, **10**, 1538 (1971).

[30] D. E. C. Corbridge, *Topics in Phosphorus Chem.*, **6**, 235 (1969).

[31] L. C. Thomas, *Interpretation of the Infrared Spectra of Organophosphorus Compounds*, chap. 1, Heyden, London, 1974.

[32] R. A. Nyquist, M. N. Wass and W. W. Muelder, *Spectrochim. Acta*, **26A**, 611 (1970).

[33] C. Laffite, "Infrared spectra of organophosphorus compounds", D.Sc. thesis, University of Montpellier, 1965. Extensively quoted in ref. 31.

[34] A. W. Johnson, *Ylid Chemistry*, Chap. 6, Academic Press, New York, 1966.

[35] M. Berman, *Prog. Inorg. Chem. Radiochem.*, **14**, 1 (1972).
[36] I. N. Zhmurova and A. V. Kirsanov, *Zh. Obshch. Khim.*, **32**, 2576 (1962); *Chem. Abs.*, **58**, 7848f (1963)
[37] H. R. Allcock, *Phosphorus–Nitrogen Compounds*, p. 113, Academic Press, New York, 1972.
[38] E. S. Kozlov, S. N. Gaidamaka and A. V. Kirsanov, *J. Gen. Chem. USSR*, **39**, 1616 (1969).
[39] H. Staudinger and J. Meyer, *Helv. Chim. Acta*, **2**, 635 (1919).
[40] H. Staudinger and E. Hauser, *Helv. Chim. Acta*, **4**, 861 (1921).
[41] G. Wittig and K. Schwarzenback, *Ann. Chem.*, **650**, 1 (1961).
[42] R. K. Bunting and C. D. Schmulbach, *Inorg. Chem.*, **5**, 533 (1966).
[43] E. Fluck and F. L. Golaman, *Z. Anorg. Allgem. Chem.*, **356**, 307 (1968).
[44] R. Appel and A. Haus, *Z. Anorg. Allgem. Chem.*, **311**, 290 (1961).
[45] R. Appel and A. Haus, *Chem. Ber.*, **93**, 405 (1960).
[46] M. Berman, *Topics in Phosphorus Chem.*, **7**, 311 (1972).
[47] Ref. 31, Chapter 10.
[48] W. Wiegrabe and H. Bock, *Chem. Ber.*, **101**, 1414 (1968).
[49] H. Schmidbaur and W. Wolfsberger, *Chem. Ber.*, **100**, 1000 (1967).
[50] E. S. Kozlov, A. A. Kisilenko, A. I. Sedlov and A. V. Kirsanov, *Zh. Obshch. Khim.*, **37**, 1611 (1967).
[51] V. A. Shokol, A. A. Kisilenko and G. I. Derkach, *Zh. Obshch. Khim.*, **39**, 874 (1969).
[52] M. J. E. Hewlins, *J. Chem. Soc. B*, 942 (1971).
[53] G. W. Adamson and J. C. J. Bart, *J. Chem. Soc. A*, 1452 (1970).
[54] E. M. Popov, M. I. Kabachnik and V. A. Gilyanov, *J. Gen. Chem. USSR*, **32**, 1581 (1962).
[55] P. A. Fowell and C. T. Mortimer, *J. Chem. Soc.*, 2913 (1969).
[56] I. N. Zhmurova and A. V. Kirsanov, *Zh. Obshch. Khim.*, **30**, 3044 (1960) [*Chem. Abs.*, **55**, 1755c (1961)].
[57] P. B. Hormuth and H. P. Latscha, *Z. Anorg. Allgem. Chem.*, **365**, 26 (1969).
[58] A. C. Chapman, W. S. Holmes, N. L. Paddock and H. T. Searle, *J. Chem. Soc.*, 1825 (1961).
[59] M. Becke-Goehring and L. Leichner, *Angew. Chem., Int. Ed. Eng.*, **3**, 590 (1964).
[60] R. Schmutzler, *J. Chem. Soc., Dalton*, 2687 (1973).
[61] R. Schmutzler, *Z. Naturforsch.*, **19B**, 1101 (1964).
[62] H. M. Stokes, *Amer. Chem. J.*, **15**, 198 (1893).
[63] H. Bock and W. Wiegrabe, *Chem. Ber.*, **99**, 1068 (1966).
[64] H-G. Horn and M. Becke-Goehring, *Z. Anorg. Allgem. Chem.*, **367**, 165 (1969).
[65] L. G. Howard and R. A. Jacobson, *J. Chem. Soc. A*, 1203 (1966).
[66] J. W. Cox and E. R. Corey, *Chem. Commun.*, 123 (1967).
[67] H. Fleig and M. Becke-Goehring, *Z. Anorg. Allgem. Chem.*, **376**, 215 (1970).
[68] J. Weiss and G. Hartman, *Z. Naturforsch.*, **21B**, 891 (1967).
[69] M. Green, R. N. Hazeldine and G. S. A. Hopkins, *J. Chem. Soc. A*, 1766 (1966).
[70] M. P. Yagupsky, *Inorg. Chem.*, **6**, 1770 (1967).
[71] T. Moeller and A. W. Westlake, *J. Inorg. Nucl. Chem.*, **29**, 957 (1967).
[72] O. Schlak, R. Schmutzler, R. K. Harris and M. Murray, *Chem. Commun.*, 23 (1973).
[73] M. Becke-Goehring and E. Fluck, *Angew. Chem.*, **74**, 382 (1962).
[74] J. Emsley and P. B. Udy, *J. Chem. Soc. A*, 3025 (1970); *ibid.*, 768 (1971).
[75] M. Becke-Goehring and W. Lehr, *Z. Anorg. Allgem. Chem.*, **325**, 287 (1963).
[76] I. I. Bezman and J. H. Smalley, *Chem. & Ind.* (*London*), 839 (1960).

[77] J. Emsley, J. Moore and P. B. Udy, *J. Chem. Soc. A*, 2863 (1971).

[78] M. Becke-Goehring, W. Gehrmann and W. Goetze, *Z. Anorg. Allgem. Chem.*, **326**, 127 (1963).

[79] M. Becke-Goehring, A. Debo, E. Fluck and W. Goetze, *Chem. Ber.*, **94**, 1383 (1961).

[80] L. Seglin, M. R. Lietz and H. Strange, U.S. Pat. 3,123,327/1966 [*Chem. Abs.*, **64**, 10810C (1966)].

[81] V. A. Shokol, G. A. Golik and G. I. Derkach, *Zh. Obshch. Khim.*, **38**, 871 (1968).

[82] M. Lustig, *Inorg. Chem.*, **8**, 443 (1969).

[83] R. Appel and G. Buchler, *Z. Anorg. Allgem. Chem.*, **320**, 3 (1963).

[84] K. L. Paciorek, *Inorg. Chem.*, **3**, 96 (1964).

[85] I. M. Filatova, E. L. Zaitseva and A. Ya. Yakubovich, *J. Gen. Chem. USSR*, **36**, 1854 (1966); and A. P. Simanov, *ibid*, **38**, 1256 (1968).

[86] K. L. Paciorek and R. H. Kratzev, *Inorg. Chem.*, **5**, 538 (1966).

[87] H. R. Allcock, *Chem. Rev.*, **72**, 315 (1972).

[88] R. A. Shaw, R. Keat and C. Hewlett, *Prep. Inorg. React.*, **2**, 1 (1965).

[89] N. L. Paddock, *Quart. Rev., Chem. Soc.*, **18**, 168 (1964).

[90] I. Haiduc, *The Chemistry of Inorganic Ring Systems*, Vol 2, Wiley–Interscience, London, 1970.

[91] J. Liebig, *Liebigs Ann. Chem.*, **11**, 139 (1834).

[92] H. M. Stokes, *Amer. Chem. J.*, **17**, 275 (1895); *ibid*, **18**, 629 and 780 (1896); *ibid*, **19**, 782 (1897).

[93] R. Schenck and G. Romer, *Chem. Ber.*, **57B**, 1343 (1924).

[94] L. G. Lund, N. L. Paddock, J. E. Proctor and H. I. Searle, *J. Chem. Soc.*, 2542 (1960).

[95] J. Emsley and P. B. Udy, *J. Chem. Soc. A*, 768 (1971).

[96] J. Emsley and P. B. Udy, *J. Chem. Soc. A*, 3025 (1970).

[97] E. Kobayashi, *J. Chem. Soc. Japan*, **87**, 135 (1966).

[98] C. D. Schmulbach and C. Derderian, *J. Inorg. Nucl. Chem.*, **25**, 1395 (1963).

[99] D. L. Herring and C. M. Douglas, *Inorg. Chem.*, **3**, 428 (1964).

[100] F. G. Sherif and C. D. Schmulbach, *Inorg. Chem.*, **5**, 322 (1966).

[101] A. Schmidtpeter and J. Ebeling, *Angew. Chem.*, **6**, 565 (1967); *ibid*, **7**, 209 (1968).

[102] M. Berman and K. Utvary, *J. Inorg. Nucl. Chem.*, **31**, 271 (1969).

[103] G. Tesi, C. P. Haber and C. M. Douglas, *Proc. Chem. Soc., London.*, 219 (1960).

[104] W. H. Kratzer and K. L. Paciorek, *Inorg. Chem.*, **4**, 1767 (1965).

[105] M. Becke-Goehring, *U.S. Govt. Res. Rep.*, AD 642–394 (1965).

[106] D. D. Magnelli, G. Tesi, J. U. Lowe and W. E. McQuiston, *Inorg. Chem.*, **5**, 457 (1966).

[107] R. Schmutzler, *Z. Naturforsch.*, **19B**, 1101 (1964).

[108] M. W. Dougill, *J. Chem. Soc.*, 3211 (1963).

[109] M. I. Davis and J. W. Paul, *Bull. Amer. Phys. Soc.*, (*II*), **13**, 832 (1968).

[110] D. P. Craig, *J. Chem. Soc.*, 997 (1959).

[111] D. P. Craig and N. L. Paddock, *Nature*, **181**, 1052 (1958); *J. Chem. Soc.*, 4118 (1962);

[112] M. J. S. Dewar, E. A. C. Lucken and M. A. Whitehead, *J. Chem. Soc.*, 2423 (1960).

[113] K. A. R. Mitchell, *Chem. Revs.*, **69**, 157 (1969).

[114] S. B. Hartley, N. L. Paddock and H. T. Searle, *J. Chem. Soc.*, 430 (1961).

[115] A. F. Bedford and C. T. Mortimer, *J. Chem. Soc.*, 4649 (1960).

[116] Ref. 37 appendix I summarizes all the structural information on cyclophosphazenes and cyclophosphazanes.

[117] G. J. Bullen, *J. Chem. Soc.*, 1450 (1971).

[118] G. J. Bullen, *J. Chem. Soc.*, 3193 (1962).

References

[119] N. V. Mani, F. R. Ahmed and W. H. Barnes, *Acta Crystallogr.*, **19**, 693 (1965).

[120] H. McD. McGeachin and F. R. Tromans, *J. Chem. Soc.*, 4777 (1961).

[121] W. C. Marsh and J. Trotter, *J. Chem. Soc. A*, 569, 573 (1971).

[122] W. C. Marsh, T. N. Ranganathan, J. Trotter and N. L. Paddock, *Chem. Commun.*, 815 (1970).

[123] J. Emsley, *J. Chem. Soc. A*, 109 (1970).

[124] G. Allen, M. Barnard, J. Emsley, N. L. Paddock and R. F. M. White, *Chem. & Ind. (London)*, 952 (1963).

[125] H. T. Searle, J. Dyson, T. N. Ranganathan and N. L. Paddock, *J. Chem. Soc., Dalton*, 203 (1975).

[126] Ref. 37, Chap. 3 and refs therein.

[127] B. Lakatos, A. Hesz, Z. Vetessy and G. Horvath, *Acta Chim. (Budapest)*, **60**, 309 (1969).

[128] G. R. Branton, E. C. Brion, D. C. Frost, K. A. R. Mitchell and N. L. Paddock *J. Chem. Soc. A*, 151 (1970).

[129] G. E. Coxon, T. F. Palmer and D. B. Sowerby, *J. Chem. Soc. A*, 358 (1969).

[130] E. H. Poindexter, R. D. Bates Jr., N. L. Paddock and J. A. Potenza, *J. Amer. Chem. Soc.*, **95**, 1714 (1973).

[131] T. Chivers and N. L. Paddock, *Inorg. Chem.*, **11**, 848 (1972).

[132] D. R. Armstrong, G. H. Longmuir and P. G. Perkins, *Chem. Commun.*, 464 (1972).

[133] D. B. Sowerby, *J. Chem. Soc.*, 1396 (1965).

[134] A. C. Chapman, D. H. Paine, H. T. Searle, D. R. Smith and R. F. M. White, *J. Chem. Soc.*, 1768 (1961).

[135] F. Seel and J. Langer, *Angew. Chem.*, **68**, 461 (1956); *Z. Anorg. Allgem. Chem.*, **295**, 317 (1958).

[136] J. Emsley and N. L. Paddock, *J. Chem. Soc. A*, 2590 (1968).

[137] B. Green and D. B. Sowerby, *Chem. Commun.*, 628 (1969), *J. Chem. Soc. A*, 987 (1970).

[138] C. Grundmann and R. Ratz, *Z. Naturforsch.*, **10B**, 116 (1955).

[139] C. M. Sharts, A. J. Bilbo and D. R. Gentry, *Inorg. Chem.*, **5**, 2140 (1966).

[140] G. Tesi, R. J. A. Otto, F. G. Sherif and L. F. Audrieth, *J. Amer. Chem. Soc.*, **82**, 528 (1960).

[141] E. T. McBee, K. Okuhara and C. J. Morton, *Inorg. Chem.*, **5**, 450 (1966).

[142] D. Dell, B. W. Fitzsimmons and R. A. Shaw, *J. Chem. Soc. A*, 4070 (1965); *ibid*, 1680 (1966).

[143] A. P. Carroll and R. A. Shaw, *J. Chem. Soc. A*, 914 (1966).

[144] E. T. McBee, L. Brinkman and H. P. Braendlin, *U.S. Govt. Res. Rep.*, AD 254,982 (1960).

[145] R. Pornin, *Bull. Soc. Chim. Fr.*, **258**, 2861 (1966).

[146] R. Ratz, H. Schroeder, H. Ulrich, E. Kober and C. Grundmann, *J. Amer. Chem. Soc.*, **84**, 551 (1962).

[147] H. R. Allcock, *J. Amer. Chem. Soc.*, **85**, 4050 (1963); *ibid*, **86**, 2591 (1964).

[148] H. R. Allcock and L. A. Siegel, *J. Amer. Chem. Soc.*, **86**, 5140 (1964).

[149] M. Becke-Goehring and B. Boppel, *Z. Anorg. Allgem. Chem.*, **322**, 239 (1963).

[150] G. R. Feistel and T. Moeller, *J. Inorg. Nucl. Chem.*, **20**, 2731 (1967).

[151] W. Lehr, *Z. Anorg. Allgem. Chem.*, **350**, 18 (1967).

[152] W. Lehr and J. Pietschmann, *Chem. Ztg.*, **94**, 362 (1970).

[153] H. Lederle, G. Ottmann and E. Kober, *Inorg. Chem.*, **5**, 1818 (1966).

[154] K. John, T. Moeller and L. F. Audrieth, *J. Amer. Chem. Soc.*, **82**, 5616 (1960).

[155] T. Moeller and S. Lanoux, *Inorg. Chem.*, **2**, 1061 (1963).

[156] W. Lehr, *Z. Anorg. Allgem. Chem.*, **352**, 27 (1967).

[157] S. K. Das, R. Keat, R. A. Shaw and B. C. Smith, *J. Chem. Soc. A*, 1677 (1966).

[158] S. K. Das, R. Keat, R. A. Shaw and B. C. Smith, *J. Chem. Soc. A*, 5032 (1965).

[159] R. Keat and R. A. Shaw, *J. Chem. Soc.*, 2215 (1965).
[160] H. Koopman, F. J. Spruit, F. van Deursen and A. J. Bakker, *Rec. Trav. Chim. Pays-Bas*, **84**, 341 (1965).
[161] R. Keat and R. A. Shaw, *J. Chem. Soc. A*, 908 (1966).
[162] A. A. Kropacheva and L. E. Mukhina, *J. Gen. Chem. USSR*, **32**, 512 (1962); *ibid*, **33**, 699 (1963); *ibid*, **38**, 314 (1966).
[163] R. Keat and R. A. Shaw, *J. Chem. Soc. A*, 908 (1966).
[164] B. S. Manhas, S. K. Chu and T. Moeller, *J. Inorg. Nucl. Chem.*, **30**, 322 (1968).
[165] R. Keat and R. A. Shaw, *J. Chem. Soc. A*, 4067 (1965).
[166] N. L. Paddock, T. N. Ranganathan and S. M. Todd, *Can. J. Chem.*, **49**, 164 (1971).
[167] T. N. Ranganathan, S. M. Todd and N. L. Paddock, *Inorg. Chem.*, **12**, 316 (1963).
[168] C. W. Allen and T. Moeller, *Inorg. Chem.*, **7**, 2178 (1968); *Inorg. Synth.*, **12**, 293 (1970).
[169] M. Biddlestone and R. A. Shaw, *J. Chem. Soc. A*, 178 (1968).
[170] R. A. Shaw and M. Biddlestone, *J. Chem. Soc. A*, 1750 (1970).
[171] C. W. Allen, *Chem. Commun.*, 152 (1970).
[172] K. G. Acock, R. A. Shaw and F. B. G. Wells, *J. Chem. Soc.*, 121 (1964).
[173] V. B. Desai, R. A. Shaw and B. C. Smith, *Angew. Chem., Int. Ed. Eng.*, **7**, 887 (1968).
[174] I. I. Bezman and C. T. Ford, *Chem. & Ind. (London)*, 163 (1963).
[175] C. D. Schmulbach and V. R. Miller, *Inorg. Chem.*, **7**, 2191 (1968).
[176] M. Kouřil, personal communication.
[177] H. R. Allcock and E. J. Walsh, *J. Amer. Chem. Soc.*, **91**, 3102 (1969); *Chem. Commun.*, 580 (1970).
[178] M. F. Lappert and G. Svivastava, *J. Chem. Soc. A*, 210 (1966).
[179] H. P. Calhoun, N. L. Paddock, J. Trotter and J. N. Wingfield, *Chem. Commun.*, 875 (1972).
[180] W. Harrison and J. Trotter, *J. Chem. Soc., Dalton*, 61 (1973).
[181] N. L. Paddock, T. N. Ranganathan and J. N. Wingfield, *J. Chem. Soc., Dalton*, 1578 (1972).
[182] G. Allen, J. Dyson and N. L. Paddock, *Chem. & Ind. (London)*, 1832 (1964).
[183] J. N. Rapko and G. R. Feistel, *Chem. Commun.*, 474 (1968); *Inorg. Chem.*, **9**, 1401 (1970).
[184] T. Moeller and S. G. Kokalis, *J. Inorg. Nucl. Chem.*, **25**, 875 (1963).
[185] H. Bode, K. Butow and G. Lienau, *Chem. Ber.*, **81**, 547 (1948).
[186] K. Denny and S. Lanoux, *J. Inorg. Nucl. Chem.*, **31**, 1531 (1969).
[187] N. V. Mani and A. J. Wagner, *Chem. Commun.*, 658 (1968).
[188] H. R. Allcock, E. C. Bissell and E. T. Shawl, *J. Amer. Chem. Soc.*, **94**, 8603 (1972).
[189] D. Feakins, W. A. Last and R. A. Shaw, *J. Chem. Soc.*, 4464 (1964).
[190] D. Feakins, W. A. Last, N. Neemuchwala and R. A. Shaw, *J. Chem. Soc.*, 2804 (1965).
[191] D. Feakins, W. A. Last, S. N. Nabi and R. A. Shaw, *J. Chem. Soc. A*, 1831 (1966).
[192] D. Feakins, S. N. Nabi, R. A. Shaw and P. Watson, *J. Chem. Soc. A*, 10 (1968); *ibid*, 2468 (1969).
[193] D. Feakins, W. A. Last, S. N. Nabi, R. A. Shaw and P. Watson, *J. Chem. Soc. A*, 196 (1969).
[194] B. W. Fitzsimmons, C. Hewlett and R. A. Shaw, *J. Chem. Soc.*, 4459 (1964).
[195] B. W. Fitzsimmons, C. Hewlett and R. A. Shaw, *J. Chem. Soc.*, 7432 (1965).
[196] O. Schmitz-DuMont, *Z. Elektrochem.*, **45**, 651 (1939); *Angew. Chem.*, **52**, 498 (1939).
[197] H. R. Allcock and R. J. Best, *Can. J. Chem.*, **42**, 447 (1964).
[198] J. Emsley and P. B. Udy, *Polymer*, **13**, 593 (1972).
[199] N. L. Paddock, German Patent 1,064,037 (1959); [*Chem. Abs.*, **55**, 16927d (1961)].
[200] K. H. Meyer, W. Lotmar and G. W. Pankow, *Helv. Chim. Acta*, **19**, 930 (1936).

[201] H. R. Allcock, R. L. Kugel and E. G. Stroh, *Inorg. Chem.*, **11**, 1120 (1972).

[202] H. H. Sisler, S. E. Frazier, R. G. Rice and M. G. Sanchez, *Inorg. Chem.*, **5**, 326 (1966).

[203] G. Tesi, C. P. Haber and C. M. Douglas, *Proc. Chem. Soc., London*, 219 (1960); US Patent 3,087,937 (1963).

[204] H. R. Allcock and R. L. Kugel, *J. Amer. Chem. Soc.*, **87**, 4216 (1965); and K. J. Valan, *Inorg. Chem.*, **5**, 1709 (1966).

[205] H. R. Allcock and R. L. Kugel, *Inorg. Chem.*, **5**, 1716 (1966).

Chapter 11
Less-common phosphorus bonds: boron–phosphorus and phosphorus–phosphorus

Most of this book has been concerned with the bonds in which phosphorus is linked to carbon, nitrogen and oxygen. In among, but indirectly, we have dealt with the bonds to hydrogen, fluorine and sulphur. The remaining covalent bonds to phosphorus have been somewhat neglected, not because they are unknown but because they add little that is original to our discussions. However there are two bonds which deserve more than a cursory glance – boron–phosphorus and phosphorus–phosphorus. The compounds in which they are to be found are called boraphosphanes and polyphosphines. The reason we have been able to come this far in the story of phosphorus without mentioning them is that they are relatively rare. Nevertheless each type of compound has its own distinctive contribution to make to the chemistry of the element. Although so different in their chemistry they do have one thing in common, a propensity for small ring formation.

11.1 The boron–phosphorus bond[1,2]. The boraphosphanes

There are two kinds of boron–phosphorus derivatives – those having only one B—P bond and those which are polymeric and these are generally cyclic trimers. Boron compounds are sufficiently strong Lewis acids to bring out Lewis base behaviour in tervalent phosphorus compounds. Diborane will react with a large variety of substituted phosphines and even phosphine itself to form Lewis adducts (eqn. 1). This produces one type of P—B bond – a σ-donor bond.

$$B_2H_6 + 2PR_3 \longrightarrow 2R_3P.BH_3 \tag{1}$$

A second type is the normal σ-bond which is found in $R_2P.BR_2$ compounds.* However, in these compounds there is a phosphorus atom with a lone pair of electrons adjacent to a boron atom with a vacant orbital, a situation which should lead to the formation of a supplementary donor π-bond, $R_2P \underset{\sigma}{\overset{\pi}{\rightleftharpoons}} BR_2$. Does this in fact occur? A more likely alternative to this arrangement is the formation of a donor σ-bond between phosphorus and boron atoms in different molecules. The result is the production of polymers which may be linemers (11.1) or cyclomers (11.2), usually the latter. If this occurs then the compounds of normal valency, i.e. R_2P-BR_2, become a special case of the Lewis adduct type. Such polymerization tendencies can be weakened by reducing the Lewis acidity of the boron centre and/or the Lewis basicity of the phosphorus centre by a suitable choice of groups attached to boron and phosphorus. This is illustrated by comparing $H_2P.BMe_2$ which can exist as a monomer with $Me_2P.BH_2$ which can only be got as the cyclic trimer, $(Me_2P.BH_2)_3$. In these compounds the methyl groups release electron density thereby weakening the Lewis acidity of boron in $H_2P.BMe_2$ sufficiently to prevent adduct formation, but stengthening the basicity of phosphorus in $Me_2P.BH_2$, thereby encouraging intermolecular adduct formation of type (11.2).

---

* By analogy with boron–nitrogen chemistry it might have been expected that RB=PR compounds would exist, but knowing what we do about phosphorus chemistry this bonding situation is highly improbable.

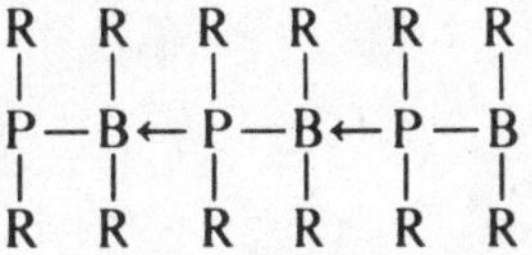

11.1 The boron–phosphorus bond. The boraphosphanes

$$\begin{array}{cccccc} R & R & R & R & R & R \\ | & | & | & | & | & | \\ P-B & \leftarrow & P-B & \leftarrow & P-B \\ | & | & | & | & | & | \\ R & R & R & R & R & R \end{array}$$

(11.1)

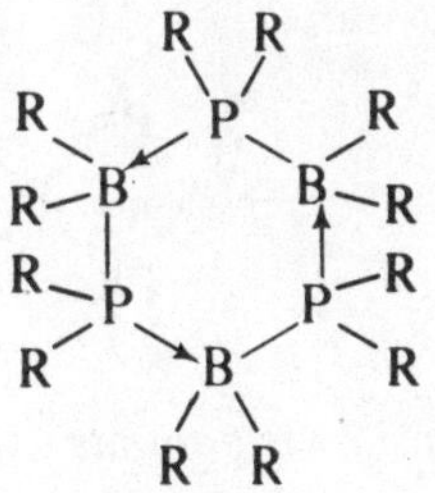

(11.2)

Compounds with organic groups on phosphorus require organic or amino groups on boron as a prerequisite for monomer stability. The determining factor appears to be the boron half of the molecule since changes here have most effect. For example, phenyl groups reduce the acidity of boron more than they enhance the basicity of phosphorus. The reason for this is that the B—Ph system may have a 2p–2p π component, i.e. $\overset{\diagdown}{\underset{\diagup}{\bar{B}}}=\!\!\bigcirc\!\!+$, which would considerably reduce the Lewis acidity.

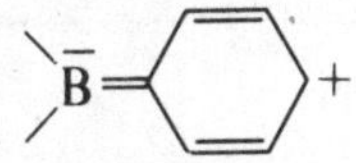

Even with organic groups on phosphorus to increase its basicity, polymer formation may be prevented by the presence of amino groups on the boron atom, e.g. $Et_2P.B(NMe_2)_2$ produced by reaction (2) is a monomer[3]. In this particular molecule steric effects may also conspire to prevent polymerization.

$$(Me_2N)_2BCl + LiPEt_2 \longrightarrow Et_2P.B(NMe_2)_2 + LiCl \qquad (2)$$

Coates and Livingstone[4] studied the products from reaction (3) by measuring their dipole moments. These showed that only weak donor π-bonding existed in these monomers. In other words these compounds, when denied Lewis adduct polymerization because of the attached groups, did not compensate by internal π-bonding. The aryl groups on boron prevent the former and those on phosphorus prevent the latter.

$$Ar_2BCl + Ar_2PH + Et_3N \longrightarrow Ar_2P.BAr_2 + Et_3N.HCl \qquad (3)$$

With compounds of the type $R_2P.BX_2$, where X is either hydrogen or halogen, the result is almost always cyclomers. These compounds are not made via the monomers, however, but by elimination of a simple molecule RX from Lewis adducts of the type $R_3P.BX_3$ (eqn. 4). Loss of a molecule such as HX may be encouraged by the use of a strong base (eqn. 5).

$$R_3P.BX_3 \xrightarrow{-RX} \frac{1}{n}(R_2P.BX_2)_n \qquad n = 2,3,4 \qquad (4)$$

The product from reaction (5) is a cyclodiboraphosphane[5] and this particular compound is stable in air and resists attack by nucleophilic reagents such as EtOH, Et_2NH, KCNS and PhMgBr, demonstrating that the B—I bond has been deactivated in some way and showing moreover that there is nothing intrinsically weak about the P—B system.

$$2\text{Ph}_2\text{HP}\cdot\text{BI}_3 + 2\text{Et}_3\text{N} \longrightarrow \underset{\substack{\text{Ph} \\ \\ \text{Ph}}}{\overset{\substack{\text{Ph} \\ \\ \text{Ph}}}{\begin{array}{c} \text{I} \quad \text{I} \\ \text{B} \\ \text{P} \quad \text{P} \\ \text{B} \\ \text{I} \quad \text{I} \end{array}}} + 2\text{Et}_3\text{NHI} \qquad (5)$$

11.1.1 Phosphorus–boron Lewis adducts

It is difficult to produce an internally consistent order of Lewis basicity and acidity from phosphorus–boron adduct data. Empirically derived orders, based on those stable adducts $R_3P\cdot BR'_3$ which do form, are shown in Table 11.1. The order of basicity

Table 11.1 Potential for adduct formation, $R_3P\cdot BR'_3$

Lewis basicity of R_3P compounds:

$(RO)_3P$ (R = alkyl) > $Me_3P \simeq Ph_3P$ > Et_3P^a > $(Me_2N)_3P$ > PH_3 > PF_3 > PCl_3 > $(PhO)_3P$

Lewis acidity of BR'_3 compounds:

BH_3 > BI_3 > BBr_3 > BCl_3 > BF_3 > BMe_3 > BPh_3

[a] see text

of the phosphorus compounds is almost the same as that of the bond dipole moments, with the trialkyl phosphites the strongest bases and triphenyl phosphite the weakest (see page 40, Chapter 2). The 'basicity' as revealed by their ability to act as ligands for transition metals is significantly different to that with respect to boron compounds when only a donor σ-bond is formed: compare the order of Table 11.1 to that of page 204, Chapter 5.

A molecule like BCl_3 will form the adduct, $Me_3P\cdot BCl_3$, which is stable in air up to its m.p. of 242 °C; $H_3P\cdot BCl_3$ will form but is air sensitive. On the other hand PF_3, PCl_3 and $(PhO)_3P$ will not form adducts with BCl_3 [$(PhO)_3P$ does not even form adducts at −10 °C]. Although BCl_3 is not a strong enough acid to couple with PCl_3, BBr_3 is just that bit stronger and $Cl_3P\cdot BBr_3$ will form. And whereas BCl_3 will form an adduct with PH_3 a weaker acid such as BMe_3 will not.

Steric factors can influence adduct formation, thus Me_3P is a stronger base than Et_3P towards BMe_3, but Et_2MeP is stronger than either. The explanation is that ethyl groups are better than methyl groups at releasing electron density to phosphorus and thereby enhancing its basicity, but three ethyl groups are just too much of a good thing in that it leads to overcrowding around phosphorus and physical repulsion of the boron moiety. This strain is relieved by the exchange of an ethyl for a methyl.

At the head of the list of acidity is BH_3. At first sight this seems odd, but how else is one to explain the formation of the gaseous adduct $F_3P\cdot BH_3$? This is not the only unusual aspect of BH_3 with phosphorus bases. Surprisingly, BH_3 shows a greater

11.1 The boron–phosphorus bond. The boraphosphanes

affinity for these than for nitrogen bases which with their tighter lone pairs are more suitable for donor σ-bond formation. The relative strengths of the adducts can be judged by the ability of the stronger of the phosphorus bases to displace Et_3N from its borane adduct[6] (eqn. 6).

$$\left.\begin{array}{l} R_3P \\ (RO)_3P \\ (Me_2N)_3P \end{array}\right\} + Et_3N \cdot BH_3 \longrightarrow \left.\begin{array}{l} R_3P \\ (RO)_3P \\ (Me_2N)_3P \end{array}\right\} \cdot BH_3 + Et_3N \quad (6)$$

The temptation to treat borane adducts as a special case is nigh on irresistable. But are they so exceptional? The P—B distance in $F_3P \cdot BH_3$ is 183.6 pm which is the shortest of the adducts. The P—B bond length in $H_3P \cdot BH_3$, $R_3P \cdot BH_3$ and $Ph_3P \cdot BH_3$ is 193 pm which is not unduly short but it is the same order as found in the boraphosphanes. These are another group of compounds with a very stable bond system. They are obtained from the Lewis adduct by the elimination of simple molecules with large bond enthalpies such as H_2, HX (X = halogen), Me_3SiX and even Me_3SiH. In some instances monomers are formed but more usually the products are the cyclopolyboraphosphanes.

11.1.2 The cyclopolyboraphosphanes

11.1.2.1 Synthesis

About 60 of these compounds are known. The heating of the adduct formed between dimethylphosphine and borane results in the loss of H_2 and the formation of a cyclic trimer (eqn. 7). This type of reaction has been used to prepare various $(R_2P \cdot BH_2)_3$ derivatives, with R as Me[7], Ph[8], CF_3[9] etc., as well as mixed substituent derivatives.

If a small amount of Me_3N is added to the reaction this hinders the cyclization step and linemers are produced. In these reactions some complex polymer is always produced even in the absence of Me_3N, but this can in some cases be broken down by further heating to give the cyclic trimer.

$$3 Me_2HP \cdot BH_3 \xrightarrow{heat} (11.3) + 3H_2 \quad (7)$$

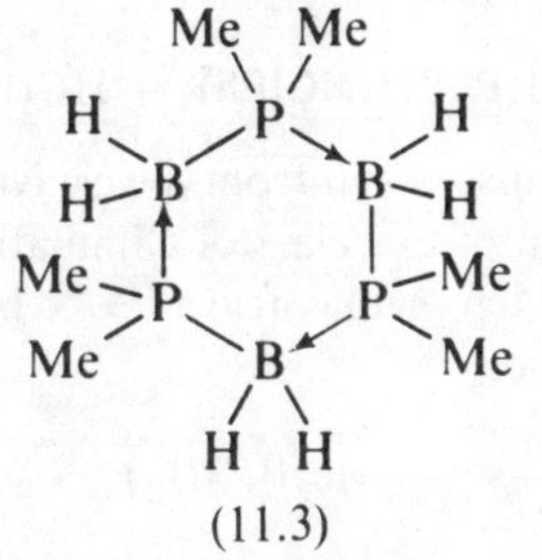

The use of primary phosphine adducts, $RH_2P \cdot BH_3$, or $H_3P \cdot BH_3$ itself in reaction (7) gives only poorly characterized products which are probably non-cyclic. The co-heating of a mixture of primary and secondary phosphine adducts can lead to condensed rings (eqn. 8), amongst other products; the bicyclic derivative (11.4) is a solid m.p. 99 °C.[10]

$$Me_2HP \cdot BH_3 + MeH_2P \cdot BH_3 \xrightarrow[-H_2]{heat}$$

[Structure (11.4): cyclic boraphosphane oligomer with Me, H substituents on P and B atoms] etc. (8)

The use of diborane in the formation of $R_2HP \cdot BH_3$ adducts, and then the pyrolysis of these, is not the only method of producing them. It is much more convenient to use $NaBH_4$ and R_2PCl in diglyme solution (eqn. 9) and not to bother isolating the intermediate adduct. The use of a mixture of Et_2PCl and $EtPCl_2$ did not give mixed products – only $(Et_2P \cdot BH_2)_3$ was isolated[11]. A modification of this method is the combination of BCl_3 or BF_3 and Me_2PH with $LiAlH_4$ which gives $(Me_2P \cdot BH_2)_3$.[13] Method (9) can be extended to phosphinyl (eqn. 10) and diphosphine disulphide derivatives (eqn. 11).

$$R_2PCl \; (R = Et\;^{[11]}, Ph\;^{[12]} \text{ etc.}) + NaBH_4 \longrightarrow \tfrac{1}{3}(R_2P \cdot BH_2)_3 + NaCl + H_2 \quad (9)$$

$$R_2P(O)Cl + NaBH_4 \longrightarrow \tfrac{1}{3}(R_2P \cdot BH_2)_3 + NaCl + H_2O + H_2$$
$$R = Me, Ph \text{ etc.}[11] \quad (10)$$

$$R_2P(S)P(S)R_2 + LiBH_4 \longrightarrow \tfrac{1}{3}(R_2P \cdot BH_2)_3 + Li_2S + H_2 \quad R = Me, Et\;[14,15] \quad (11)$$

All the above methods can be classified as involving the elimination of H_2 at the boraphosphane formation stage. Dehydrohalogenation, encouraged by the use of Et_3N offers an alternative route (eqn. 12), in this case with the production of cyclic dimer.* The use of a base to pick up the eliminated HX is not essential since the HCl may be removed by heating (eqn. 13)[17].

$$Ph_2HP + BX_3 \longrightarrow Ph_2HP \cdot BX_3 \xrightarrow{Et_3N} \tfrac{1}{2}(Ph_2P \cdot BX_2)_2 + Et_3NHCl \quad (12)$$
X = Br, I but not Cl (ref. 16)

$$PhPH_2 + PhBCl_2 \xrightarrow{reflux} \tfrac{1}{3}(PhPH \cdot BClPh)_3 + HCl\uparrow \quad (13)$$

Formation of boraphosphanes may result from the cleavage of P—Si bonds in adducts with boron compounds. Typical of this class of elimination reactions are eqns. (14)[18], (15)[19] and (16)[18]. Adducts in which there is also a P—N bond can give boraphosphanes on heating by eliminating amine.

$$Me_2(Me_3Si)P \cdot BH_3 \xrightarrow{150-300\,°C} \tfrac{1}{3}(Me_2P \cdot BH_2)_3 + Me_3SiH \quad (14)$$

$$Et_2(Me_3Si)P \cdot BF_3 \xrightarrow{100\,°C} \tfrac{1}{2}(Et_2P \cdot BF_2)_2 + Me_3SiF \quad (15)$$

$$Me_2(Me_3Si)P \cdot BPr^n{}_2Cl \xrightarrow{140\,°C} \tfrac{1}{3}(Me_2P \cdot BPr^n{}_2)_3 + Me_3SiCl \quad (16)$$

---

* Cyclic dimer is also obtained by the elimination of a metal halide [H. Noth and W. Schragle, *Angew. Chem.*, **74**, 587 (1962)] thus $Et_2PLi + BCl_3 \to \tfrac{1}{2}(Et_2P \cdot BCl_2)_2 + LiCl$

11.1 The boron–phosphorus bond. The boraphosphanes

With the borane adduct of Me_2NPMe_2 it is not clear whether phosphorus or nitrogen is acting as the base centre. However, considering what we have already said about respective basicities, plus the fact that boraphosphanes rather than borazanes result from heating the adduct (eqn. 17)[11], it is likely that the BH_3 attaches itself to the phosphorus. Neither reason is proof of a P—B link in the adduct since thermodynamics may favour the boraphosphane ring; but if this is so then it serves to demonstrate the relative stabilities of B—P and B—N ring systems.

$$Me_2NPMe_2 \cdot BH_3 \xrightarrow{160\,°C} \tfrac{1}{3}(Me_2P \cdot BH_2)_3 + Me_2NH \quad (17)$$

In the bisborane adduct (11.5), where it can be assumed that both P and N are forming donor σ-bonds to boron, it is the P—N link which breaks[11] (eqn. 18). Adduct (11.5) has its diphosphine counterpart $H_3B \cdot PMe_2 \cdot PMe_2 \cdot BH_3$, which we shall meet again in the second half of this chapter (page 460), and this on pyrolysis produces boraphosphanes[20] (eqn. 19). The adduct formed between diphosphine itself and diborane,

$$\underset{(11.5)}{H_3B \cdot PMe_2 \cdot NMe_2 \cdot BH_3} \xrightarrow{200\,°C} \tfrac{1}{3}(Me_2P \cdot BH_2)_3 + Me_2NH \cdot BH_3 \quad (18)$$

$$H_3B \cdot PMe_2 \cdot PMe_2 \cdot BH_3 \xrightarrow{130\,°C} Me_2HP \cdot BH_3 \xrightarrow{290\,°C} \tfrac{1}{n}(Me_2P \cdot BH_2)_{3,4} \quad (19)$$

$P_2H_4 \cdot B_2H_6$, decomposes on heating but not to give the elusive $(H_2P \cdot BH_2)_n$ compounds but a polymeric material of approximate composition $(PBH_{3.5})_n$ is obtained[21].

11.1.2.2 Physical and chemical behaviour of the cycloboraphosphanes

What is most striking about the cyclopolyboraphosphanes is their thermal stability. For example $(Ph_2P \cdot BH_2)_3$ begins to eliminate H_2 at 350 °C, probably forming condensed rings, but real breakdown of the system only occurs above 510 °C. Thermal stability has been shown to be related to the groups attached to the boron atoms and the order of decreasing stability is $H \simeq Me > Cl > Br \simeq I$.

Resistance to hydrolysis is another noteworthy feature of the cycloboraphosphanes; $(Me_2P \cdot BH_2)_{3,4}$ hydrolyze only on prolonged heating at 300 °C with hydrochloric acid in sealed tubes, to give $B(OH)_3$ and Me_2PO_2H.[22] The corresponding P-phenyl compound $(Ph_2P \cdot BH_2)_3$ resists hydrolysis even under these vigorous conditions, as does $(Me_2P \cdot BMe_2)_3$. Derivatives with trifluoromethyl groups are less stable, reaction taking place at 85 °C with methanolic HCl solutions[9]. Hydrolytic stability depends on the substituents on both phosphorus and boron. Alkyl and aryl groups attached to phosphorus are more stable than CF_3 or F. For boron substituents the order of hydrolytic stability is $H > Me > Cl > Br > I \simeq F$.

Similar orders of oxidative stability have been deduced for groups attached to phosphorus and boron[23]. For phosphorus substituents the order is cyclo-$C_6H_{11} > Ph > Me \simeq Et > H$; for boron substituents it is $Br \simeq I > Cl > Me > H > F$.

Why are the cyclopolyboraphosphanes so very stable? The $\underset{H}{\overset{H}{B}}$ unit is isoelectronic with $\underset{..}{\overset{..}{N}}$ so that some resemblance between $(R_2P \cdot BH_2)_3$ and $(R_2PN)_3$, the corresponding cyclotriphosphazenes, might be expected. But that the former should so outshine the latter is truly remarkable. Isoelectronic though they be, this only applies to the $\diagup\!\!\!\diagdown$BH$_2$ derivatives and in terms of structure and formal bonding there is no resemblance except in terms of stability.

Instead of asking why they are so stable perhaps we ought to re-examine our reasons for expecting them to be unstable. Firstly the B—H bond is generally considered to have a hydridic proton, $B^{\delta+}$—$H^{\delta-}$, and in many compounds this bond is very susceptible to hydrolysis. The fact that it is not in the boraphosphane suggests that the proton is not hydridic and that the bond moment is not as predicted. It would appear that the electron density of this bond has been tapped by the ring bonding system, with the result that both ring and exocyclic bonds are mutually stabilized. Phosphorus 3d orbitals may play a role in this but it is not clear how, since there are no electron pairs surplus to the requirements of the σ framework of the molecule for them to use. The bonding is explained simply in terms of a Lewis adduct polymer (11.3). The ring is not even planar, which would at least have encouraged belief in a π system[24]. However the ring angles at both phosphorus and boron are larger than the tetrahedral angle (11.6).

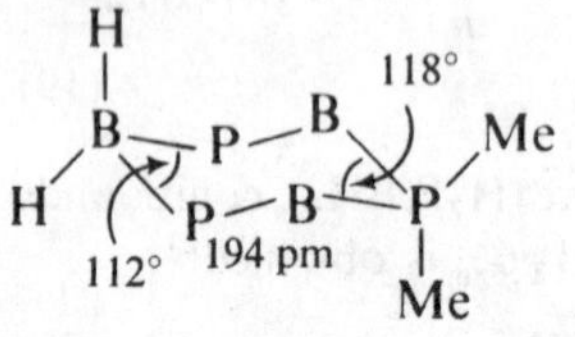

(11.6)   (groups on other P and B not shown)

Steric crowding may explain some of the stability towards nucleophilic attack by hindering chemical attack at phosphorus or boron. The vagaries of the *tbp* intermediate may also retard substitution at phosphorus since this would presumably be of the form (11.7) which is in violation of polarity rules both as it stands and even more so after pseudorotation which would result in at least one boron atom being apical. Boron is one of the least electronegative elements and so presumably will have low apicophilicity.

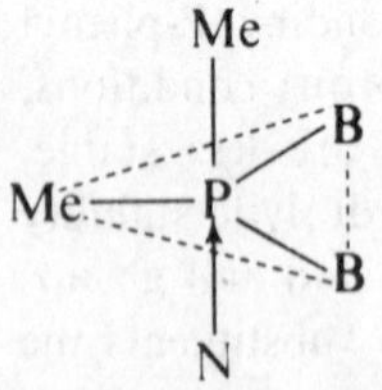

(11.7)

The cyclopolyboraphosphanes will participate in substitution reactions. Replacement of the protons of the $\diagup\!\!\!\diagdown$BH$_2$ groups can be effected by excess halogen or N-halosuccinimide e.g. eqn. (20)[8,23], but in the case of iodine partial replacement to give (Ph$_2$P.BHI)$_3$ occurs under the same circumstances although (Et$_2$P.BI$_2$)$_3$ can be obtained after prolonged treatment[14]. Halogenation can also be achieved by means of Al$_2$X$_6$ (eqn. 21)[14].

11.1 The boron–phosphorus bond. The boraphosphanes

$$(Ph_2P.BH_2)_3 + 6X_2 \xrightarrow{\text{reflux } CCl_4} (Ph_2P.BX_2)_3 + 6HX \quad (X = Cl, Br) \quad (20)$$

$$(R_2P.BH_2)_3 + Al_2X_6 \longrightarrow (R_2P.BX_2)_3 \quad (R = Me, Et)\ (X = Cl, Br, I) \quad (21)$$

The boron-halogeno derivatives are remarkably stable when compared to their BX_3 counterparts. Thus they are unaffected by H_2O, NH_3, Et_2NH, KSCN, NaOEt and Grignard reagents, all of which react readily with BX_3. Nor will $LiAlH_4$ reduce $\diagdown\!\!\!\diagup BX_2$ back to $\diagdown\!\!\!\diagup BH_2$. In the dimeric derivatives, $(Ph_2P.BX_2)_2$, where X is Br or I, the same low reactivity is noted towards most of these nucleophilic reagents but the bromide is decomposed by moist air and with Et_2NH it forms $B(NEt_2)_3$.[16B] The infrared and Raman spectra of dimers, trimers and tetramers all have a characteristic ring stretching band in the 620–720 cm^{-1} region – Table 11.2. This low range of values is more in keeping with single-bonding systems (Chapter 3, Table 3.10) than one with a delocalized π contribution.

Table 11.2 Vibrational frequencies of boraphosphane rings

	m.p./(°C)	νP—B/(cm^{-1})		m.p./(°C)	νP—B/(cm^{-1})
Dimers[a]					
$(Et_2P.BPr^n_2)_2$	72–4	708	$(Me_2P.BCl_2)_3$	393–4	689
$(Et_2P.BCl_2)_2$	136–8	718	$(Me_2P.BBr_2)_3$	411 (dec)	699, 708
$(Et_2P.BBr_2)_2$	172–5	702	$(Me_2P.BMe_2)_3$	334	702
Trimers[b]			Tetramer[b]		
$(Me_2P.BH_2)_3$	87–8	665	$(Me_2P.BH_2)_4$	161	627, 682
$(Me_2P.BF_2)_3$	127–8	671, 678			

[a] H. Noth and W. Schragle, *Chem. Ber.*, **98**, 352 (1965); [b] A. C. Chapman, *Trans. Faraday Soc.*, **59**, 806 (1963).

In conclusion it would seem an unnecessary luxury to invoke π bonding in the P—B bond in any of the compounds having this type of link. The unusual stability associated with the cyclomers of this system probably arises from a combination of steric hindrance of the molecule as a whole protecting the boron centres, plus the fact that the borons are already tetracoordinated, and the energetically unfavourable intermediates which would be required at phosphorus, prevent successful attack at these. But even when this has been said there still remains the nagging doubt that we are missing something and that perhaps both the $R_3P.BH_3$ adducts and the cycloboraphosphane system have some form of extrabonding.*

---

* Hyperconjugation?

454   Chapter 11 / Less-common phosphorus bonds: boron–phosphorus and phosphorus–phosphorus

Before we leave boron–phosphorus compounds there is the combination of boron and phosphorus(V) to consider. Tervalent boron compounds and pentavalent phosphorus compounds do form 1:1 adducts, e.g. $PCl_5 \cdot BCl_3$, but these are ionic and have the structure $PCl_4^+ BCl_4^-$. Phosphoryl compounds also form adducts such as $POCl_3 \cdot BCl_3$ but here there is a P—O—B bond and not a direct P—B link.

11.2 The phosphorus–phosphorus bond[25-28]: The polyphosphines

Broadly speaking the compounds containing a P—P link can be divided into three groups: the element itself and its polymorphs; the linear polyphosphines; and the cyclic polyphosphines. The linear species are essentially the diphosphines, the tri- and higher polyphosphines are known but chain growth adds little in the way of novelty and much in the way of instability so will only be briefly mentioned. The cyclopolyphosphines present something of an enigma – not only is their chemistry basically different but a lot of the results published by workers in the field are contradictory.

The manufacture of elemental phosphorus was described in Chapter 1. The phosphorus produced by this method is α-white phosphorus which consists of P_4 tetrahedral molecules. This is only one of the polymorphs of phosphorus, the others being β-white phosphorus (also consisting of P_4 molecules but packed differently), red or violet phosphorus (prepared by heating white phosphorus – its structure varies according to the method of preparation, but it is probably polymeric with chains of linked polyhedra) and black phosphorus (obtained by heating white phosphorus at 200 °C at 12 000 atmos – it resembles graphite and consists of buckled layers of linked hexagonal rings). The phase diagram for phosphorus is shown in Fig. 11.1 which includes the structures[29].

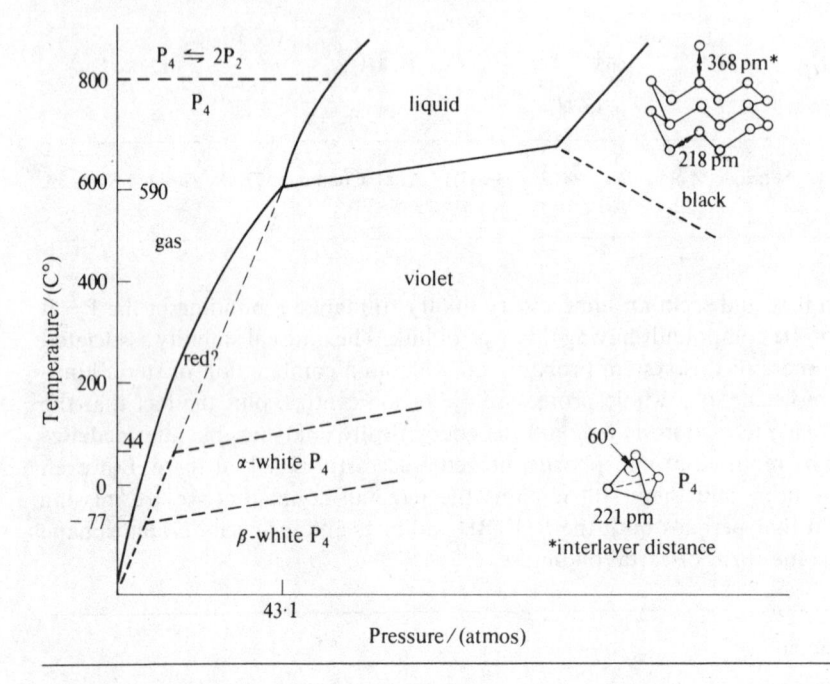

Figure 11.1 Phase diagram for elemental phosphorus.

11.2 The phosphorus–phosphorus bond: The polyphosphines

In the phosphorus sulphide P_4S_3 the tetrahedral arrangement is partly preserved (11.8) and a few other compounds like this contain P—P links.

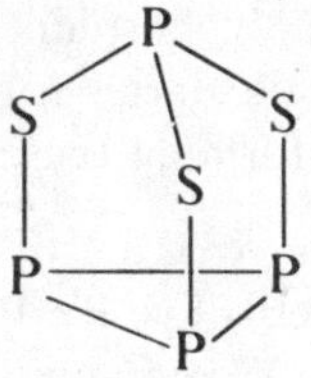

(11.8)

In the context of P—P bonding elemental phosphorus would seem a useful starting material for preparative work, but in reality it is not often used. When treated with KOH solution P_4 yields small amounts of diphosphine, P_2H_4, along with PH_3. Another preparation from P_4 is diphosphorus tetraiodide, P_2I_4 (eqn. 22) but the best known reaction in which a P—P compound is synthesized from P_4 is the preparation of hypophosphoric acid (11.9) from red phosphorus and sodium chlorite (eqn. 23) or H_2O_2 as the oxidizing agent. By means of hypochlorite or hypobromite oxidation of red phosphorus in an alkaline medium the cyclic polyphosphorus derivative (11.10) is produced[30].

$$P_4 + 4I_2 \xrightarrow{CS_2} 2P_2I_4 \tag{22}$$

$$2P(\text{red}) + 2NaClO_2 + 2H_2O \longrightarrow \underset{(11.9)}{\text{NaO}-\overset{\overset{O}{\|}}{\underset{H}{P}}-\overset{\overset{O}{\|}}{\underset{H}{P}}-\text{ONa}} + 2HCl \tag{23}$$

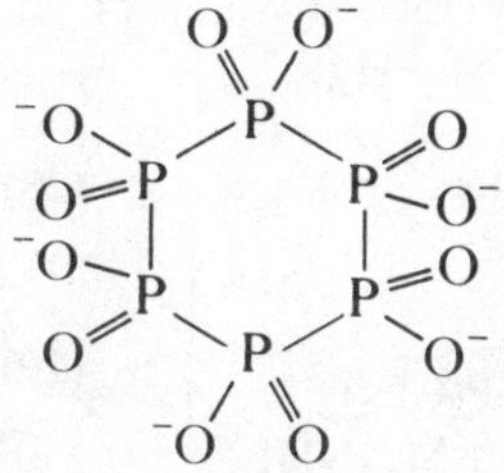

(11.10)

This unexpected phosphorus acid has been obtained as its alkali metal (Na,K,Cs) and ammonium salts. The x-ray analysis of $Cs_6P_6O_{12}$ showed it to have a chair configuration of ring atoms. The ring is cleaved by acids and alkalis to give fragments such as

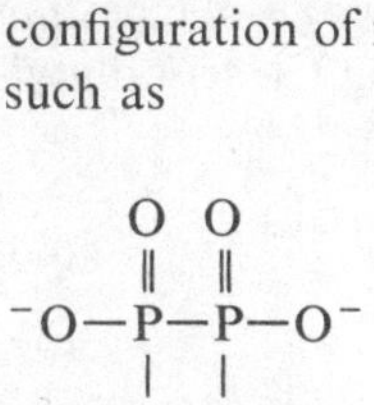

$$^-\text{O}-\overset{\overset{O}{\|}}{\underset{O^-}{P}}-\overset{\overset{O}{\|}}{\underset{O^-}{P}}-\text{O}^- \quad \text{and} \quad ^-\text{O}-\overset{\overset{O}{\|}}{\underset{O^-}{P}}-\overset{\overset{O}{\|}}{\underset{O^-}{P}}-\overset{\overset{O}{\|}}{\underset{O^-}{P}}-\text{O}^-$$

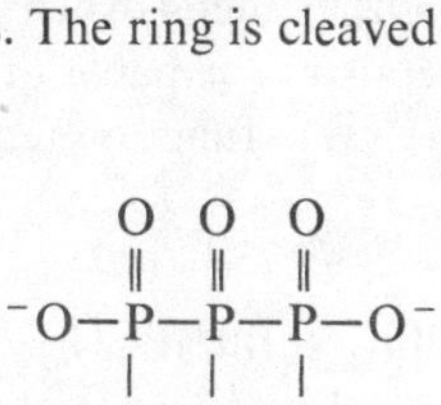

among others[31].

11.2.1 Diphosphine and linear polyphosphine derivatives

11.2.1.1 The synthesis of diphosphines

Although elemental P_4 has been used as a starting point for some polyphosphine compounds, such as cyclic $(PBu^n)_4$,[32] it has not found use as a suitable starter for linear derivatives. However there are several satisfactory methods of forming these which are classified under the following headings:

Dehydrohalogenation: The first method used in the deliberate synthesis of a P—P bond was based on hydrogen halide elimination between a P—H and P—X compound (eqn. 24). For example reaction (eqn. 25) produces tetraphenyldiphosphine, which is a stable solid that melts at 121 °C and boils at 260 °C/1 mm without decomposition.

$$R_2PH + R'_2PX \longrightarrow R_2P-PR'_2 + HX \tag{24}$$

$$Ph_2PH + Ph_2PCl \xrightarrow{100\,°C} Ph_2P-PPh_2 + HCl \tag{25}$$

This method has been widely used for a variety of alkyl-, perfluoroalkyl-, aryl- and amino-diphosphines, and it is one of the methods by which asymmetric diphosphine can be prepared (eqn. 26)[33].

$$Me_2PCl + (CF_3)_2PH \xrightarrow[Me_3N]{-78\,°C} Me_2P-P(CF_3)_2 + Me_3NHCl \tag{26}$$

Disulphide derivatives can be obtained also by dehydrohalogenation (eqn. 27)[34] and this reaction can be used in the formation of cyclophosphines.

$$Ph_2P(S)Cl + Ph_2P(S)H \xrightarrow{100\,°C} \overset{S}{\overset{\|}{Ph_2P}}-\overset{S}{\overset{\|}{PPh_2}}(79\,\%) + HCl \tag{27}$$

The method can be extended to produce longer chains (eqn. 28)[35] a reaction which has been shown to proceed in two steps (eqns. 29)[36] by the isolation of an $(CF_3)_2PP(Me)H$ intermediate in the reaction of $(CF_3)_2PI$ and $MePH_2$.

$$2Ph_2PH + Br_2PPh \xrightarrow{Et_3N} Ph_2P-P(Ph)-PPh_2 + 2Et_3NHBr \tag{28}$$

$$(CF_3)_2PI + MePH_2 \longrightarrow (CF_3)_2P-P(Me)H + MePH_3I$$

$$(CF_3)_2P-P(Me)H + (CF_3)_2PI \xrightarrow{Me_3N} (CF_3)_2P-P(Me)-P(CF_3)_2 + Me_3NHI \tag{29}$$

Coupling reactions: The treatment of phosphorus halides with active metals such as the alkali metals can result in a P—P bond according to eqn. (30) where R may be alkyl[37], phenyl[37] or amino[38] groups. Other metals such as mercury and magnesium have also been used. But not only metals are capable of abstracting Cl, tri-n-butyl-phosphine is capable of performing the same function (eqn. 31)[32].

$$2R_2PCl + 2Na \xrightarrow{dioxan} R_2P-PR_2 + 2NaCl \tag{30}$$

$$2Ph_2PCl + Bu^n_3P \longrightarrow Ph_2P-PPh_2 + Bu^n_3PCl_2 \tag{31}$$

A modification of this reaction, and a way of preparing unsymmetrical derivatives, is to preform R_2PM, which is presumably an intermediate in (30) anyway. The reaction of this with R'_2PCl then yields the desired product, e.g. reaction (32)[39].

11.2 The phosphorus–phosphorus bond: The polyphosphines

A similar method begins with a lithium phosphide and reacts it with 1,2-dibromo-ethane (eqn. 33)[40]. This reaction probably goes via a partial exchange to form R_2PBr and $LiCH_2CH_2Br$, the former then going on to couple with more phosphide to give the diphosphine, while the latter eliminates LiBr to produce ethylene.

$$Ph_2PNa + ClP(NMe_2)_2 \longrightarrow Ph_2P-P(NMe_2)_2 + NaCl \qquad (32)$$

$$2R_2PLi\ (R = Et, C_6H_{11}\ \text{etc.}) + BrCH_2CH_2Br \longrightarrow R_2P-PR_2 + 2LiBr + CH_2=CH_2 \qquad (33)$$

The 'anomalous' Grignard reaction: When $PSCl_3$ and aryl Grignards react the products are the expected $P(S)Ar_3$ derivatives. However when simple alkyl Grignards are used the products are diphosphine disulphides (eqn. 34)[33,41,42]; this is the 'anomalous' Grignard reaction. It will also work with $R'P(S)Cl_2$, where R' is alkyl or aryl, and the result is diphosphine disulphides with two asymmetric phosphorus centres (eqn. 35).

$$2\ PSCl_3 + 6RMgX \longrightarrow R_2P(S)-P(S)R_2 + 6MgXCl + R-R \qquad (34)$$

$$2R'P(S)Cl_2 + 6RMgBr \longrightarrow RR'P(S)-P(S)RR' + 6MgBrCl + R-R \qquad (35)$$

Maier[43], who did a lot of this work, was able in several instances to separate the products into two isomeric forms, one high melting and the other low melting. In the case of 1,2-dimethyl-1,2-diphenyldiphosphine disulphide these were shown by x-ray analysis to be the meso form (11.11) and a racemic mixture (11.12).

```
      S                S                S
      ||               ||               ||
  Me—P—Ph          Me—P—Ph          Ph—P—Me
      |                |                |
  Me—P—Ph          Ph—P—Me          Me—P—Ph
      ||               ||               ||
      S                S                S
  ─────────        ─────────────────────
  (11.11) m.p. 206–208 °C    (11.12) m.p. 145–146 °C
```

The 'anomalous' Grignard reaction will give yields of over 80 % if the conditions are right, such as using magnesium bromides rather than chlorides or iodides, and keeping the temperature below 20 °C. With branched or cyclic alkyls the reaction takes a different course and to obtain diphosphine disulphides of these groups one must resort to other methods, such as preforming the $\diagdown\!\!P\!-\!P\!\diagup$ link by dehydrogenation or coupling methods followed by sulphurization of the product to form $\diagdown\!\!P(S)\!-\!P(S)\!\diagup$.

This last step is rather like swimming against the current since the disulphides are most used in the formation of the diphosphines by desulphurization (eqn. 36), rather than vice versa.

$$R_2P(S)-P(S)R_2 + M\ \text{or}\ Bu^n_3P \longrightarrow R_2P-PR_2 + MS\ \text{or}\ Bu^n_3PS \qquad (36)$$
$$(M = Na, Zn, Fe\ \text{etc.})$$

Diphosphines from aminophosphines: Reaction (37) gives 85 % yields of tetramethyl-diphosphine within minutes[20,44] and in the presence of HCl goes almost quantitatively. Reaction (38) gave 50 % yields of the triphosphine[45].

$$Me_2P—NMe_2 + Me_2PH \longrightarrow Me_2P—PMe_2 + Me_2NH \qquad (37)$$

$$MeP(NMe_2)_2 + 2Ph_2PH \longrightarrow Ph_2P—P(Me)—PPh_2 + 2Me_2NH \qquad (38)$$

Reaction (37) was once offered as evidence for the formation of a P≡P triple bond, this being the driving force behind the reaction. Not enough thermochemical evidence is available to show why reaction (37) goes. The single bond energy difference, $E(P—P) - E(P—N)$ could be as much as $209 - 290$ i.e. -82 kJ mol^{-1}, using normal values for bond enthalpies, and this is almost compensated for by $E(N—H) - E(P—H)$ which is about $390 - 318$ i.e. 72 kJ mol^{-1}. It is not clear from bond enthalpies what makes reaction (37) go as it does, and one can see that the formation of a triple bond would be an attractive explanation, especially as the bond dissociation energy of the gaseous P_2 molecule is 490 kJ mol^{-1} (117 kcal mole^{-1}) which shows this to be more than a simple σ-bond, although this particular molecule only exists in phosphorus vapour above 800 °C. There may be something different about the P—P bond in P_2Me_4 but more evidence is required before it can be treated as a special case.

Miscellaneous ways of forming P—P bonds: Other methods are listed in Table 11.3, and this list is by no means complete. Altogether a large variety of methods are known for producing a single P—P link and a few are extendable to longer linear chains.

Table 11.3 Other methods of forming the P—P bond

$2PI_3$ (or $2PhPI_2$) $\xrightarrow{ether}$ P_2I_4 (or PhIP—PIPh)+I_2		(38a)[46]
$2Ph_2P—CONHPh \xrightarrow{180\,°C} Ph_2P—PPh_2 + CO + (PhNH)_2CO$		(39)[47]
$3Me_2PF \xrightarrow{heat} Me_2P—PMe_2 + Me_2PF_3$		(40)[48]
$Ph_2PCl + P(OR)_3 \longrightarrow Ph_2P—P(O)(OR)_2 + RCl$		(41)[49]
$PhPCl_2 + 2P(OR)_3 \longrightarrow (RO)_2(O)P—P(Ph)—P(O)(OR)_2 + 2RCl$		(42)[49]

11.2.1.2 The properties of the diphosphines

The nature of a covalent bond can often be deduced from its bond length and bond enthalpy. The $r(P—P)$ distance, however, is remarkably constant at 220–225 pm and seems insensitive to the bonding situation at phosphorus as Table 11.4 shows for several different types of molecules. Bond enthalpy data, what little there is of it, is not much help either. We have already seen that in the P_2 molecule a strong bond is present. In elemental P_4 molecules the value of $E(P—P)$ has been estimated to be in the range 185–200 kJ mol^{-1} (44–48 kcal mole^{-1}). For other molecules it is $E(P—P; P_2H_4)$, 196 kJ mol^{-1} (47 kcal mole^{-1})[50] and $E(P—P; P_2Cl_4)$, 243 kJ mol^{-1} (58 kcal mole^{-1})[51].

11.2 The phosphorus–phosphorus bond: The polyphosphines

Table 11.4 P—P Bond Lengths, $r(P-P)/(pm)$

Molecule	$r(P-P)/(pm)$	Molecule	$r(P-P)/(pm)$
P_4	221	$Et_2(S)P-P(S)Et_2$	222[a]
P_4S_3	225	$(CF_3P)_4$	221
P_2I_4	221	$(CF_3P)_5$	222
$(PO_2^-)_6$	220		

[a] Several other compounds of this type, $R_2P(S)P(S)R_2$ also have $r(P-P) = 221$ pm irrespective of R [J. D. Lee, *J. Inorg. Nucl. Chem.*, **32**, 3209 (1970)].

These low bond enthalpies would seem the likely explanation for the thermal instability of P_2H_4 and P_2Cl_4, but clearly this is not the real reason since $CF_3(H)P-P(H)CF_3$ is stable up to 200 °C[52] and R_2P-PR_2 compounds can be heated to 300 °C before disproportionation occurs (eqn. 43)[52].

$$R_2P-PR_2 \xrightarrow{300\ °C} R_3P + (RP)_n + P + \text{polymers} \qquad (43)$$

Although most diphosphines are thermally stable they are not stable in other respects as will be seen. Before their chemical reactivity is discussed let us consider the nature of the P—P bond.

Although $r(P-P)$ and $E(P-P)$ offer no support for π-bonding in the P—P bond, the ultraviolet spectra of diphosphines apparently does so. These compounds have intense ultraviolet spectra (typical of compounds with π systems) compared to the corresponding monophosphines, e.g. $(CF_3)_2P-P(CF_3)_2$ has an intense band (ε 7800) at 216 nm. The methyl derivative has the band maximum shifted to longer wavelengths and it becomes less intense, suggesting that there is less delocalization in this[33]. Since each phosphorus atom carries a lone pair, as well as empty 3d orbitals, the opportunities for π-bonding are present. The best evidence for π-bonding would be a high rotational barrier but this has yet to be satisfactorily proved.* We shall return to the bonding when we come to discuss the cyclopolyphosphines.

Chemical attack of the diphosphines usually results in the break down of the P—P bond. Alkali metals form organophosphides by homolytic cleavage (eqn. 44)[34]. As one might expect this process occurs more readily with aryl than alkyl diphosphines due in part to the electron-withdrawing effect of the former, which weakens the bond in the first place, and in part to the relative stabilities of the anions formed, $-Ph_2P^-$ being stabilized by delocalization.

$$R_2P-PR_2 + 2M \longrightarrow 2R_2PM \qquad (44)$$

---

* The inversion barriers of diphosphines of the type $RCH_3P \cdot PCH_3R$ (R = Ph etc.) are considerably lower than those of analogous monophosphines. This has been attributed to (p–d)π effects [J. B. Lambert, G. F. Jackson and D. C. Mueller, *J. Amer. Chem. Soc.*, **92**, 3093 (1970)].

Halogens also cleave the P—P bond and give almost quantitative yields of products (eqn. 45). With CF$_3$ groups on P these also tend to be lost[53].

$$R_2P-PR_2 + X_2 \longrightarrow 2R_2PX \xrightarrow{\text{excess } X_2} R_2PX_3 \qquad (45)$$

Hydrolysis of diphosphines by H$_2$O is slow at room temperature but is rapid on heating with alkalis. The diphosphine (CF$_3$)$_2$P—P(CF$_3$)$_2$ has quite good hydrolytic stability. It has been closely studied and the mechanism involves an initial cleavage into (CF$_3$)$_2$PH and (CF$_3$)$_2$POH.[54]

Other reactions which split the P—P bond are (46)[20], (47)[55] and (48)[42]. The last one is not typical of its kind since with most compounds the P—P bond remains intact and quaternization of one, and only one, of the phosphorus atoms occurs, e.g. (eqn. 49)[56].

$$R_2P-PR_2 + CH_2=CH_2 \longrightarrow R_2PCH_2CH_2PR_2 \qquad (46)$$

$$R_2P-PR_2 + LiR' \longrightarrow R_2PLi + R_2PR' \qquad (47)$$

$$R_2P-PR_2 + MeI \longrightarrow R_2PI + R_2PMe \quad (R = Ph, CF_3, C_6H_{11}) \qquad (48)$$

$$Me_2P-PMe_2 + MeI \longrightarrow Me_3\overset{+}{P}-PMe_2 \; I^- \qquad (49)$$

Lewis base behaviour depends upon the groups attached to phosphorus. Diphosphine itself will form diadducts, P$_2$H$_4$·2BF$_3$, and P$_2$H$_4$·2BH$_3$. Tetrakis(trifluoromethyl) diphosphine is unresponsive to these Lewis acids[33,54], but P$_2$Me$_4$ will form a diadduct with B$_2$H$_6$[20] as previously mentioned (page 451), but only a mono-adduct with BMe$_3$.[57] The adduct P$_2$(NMe$_2$)$_4$·2BH$_3$ has been shown to involve P → BH$_3$, not N → BH$_3$, bonds[38] thus demonstrating the relative basicities of the phosphorus and nitrogen centres in this compound (cf. page 451). Protonation of the P atoms of diphosphines is probably a precursor to the inevitable cleavage which occurs when they are treated with protic acids e.g. eqn. (50), but P$_2$(CF$_3$)$_4$ is resistant to HCl up to 300 °C.[33,54]

$$R_2P-PR_2 + 2HCl \longrightarrow R_2PH_2Cl + R_2PCl \qquad (50)$$

Another way in which the diphosphines display their donor properties is in the formation of complexes with transition metals[58]. In some complexes the P—P bond remains intact and the diphosphine behaves as a bidentate ligand e.g. (11.13); in others it breaks down and behaves as bridging units e.g. (11.14)[59].

```
     OC   Me Me   CO              Me   Me
      \   |  |   /                 \   P   /
   ON—Fe←P—P→Fe—NO              ON     \    NO
      /   |  |   \                \   / \  /
     ON  Me Me    NO               Fe———Fe
                                  /   \ / \
                                 ON    P   NO
                                      / \
                                    Me   Me
       (11.13)                       (11.14)
```

So far we have spoken of the properties of the diphosphines but most of what has been said would apply to the triphosphines, e.g. eqn. (42)[49]. These compounds have the overriding tendency to disproportionate (eqn. 51) and consequently little of their chemistry has been studied. Nevertheless, H(PCF$_3$)$_{3,4}$H linear compounds

11.2 The phosphorus–phosphorus bond: The polyphosphines

have been made[52] by the ring-opening hydrolysis of $(CF_3P)_{4,5}$ with methanol. Recently it has proved possible to separate the triphosphine, P_3H_5, and tetraphosphine P_4H_6, by gas–liquid chromatography between -60 to $+60\ °C$.[60] This work also uncovered a triphosphine, P_3H_3, which may be a cyclic trimer.

$$R_2P-P(R)-PR_2 \longrightarrow \frac{1}{n}(RP)_n + R_2P-PR_2 \qquad (51)$$

In the diphosphines we have discussed, the phosphorus atoms have been tri- or tetracoordinate. In theory, pentacoordinate compounds R_4P-PR_4 should be possible but the nearest to this so far obtained is R_3P-PX_5 with tetra- and hexacoordinate phosphorus atoms. Phosphorus pentafluoride, which is a Lewis acid, reacts with PMe_3, a Lewis base, to form the adduct $Me_3P \rightarrow PF_5$ which is a white solid stable *in vacuo* at room temperature[61]. The P—P bond in this compound cannot be other than a pure σ-bond since neither atom has a lone pair of electrons available for multiple bonding. Competition studies between PF_5 and BF_3 for PMe_3 revealed that the former is the stronger Lewis acid towards this base; and competition between PMe_3 and NMe_3 for PF_5 reveal the former to be the stronger Lewis base towards this acid. Steric factors are thought to be responsible for this state of affairs[61].

P—P coupling: Diphosphines offer a rare opportunity to study homonuclear magnetic coupling between nuclei of $I = \frac{1}{2}$. However in a lot of the work reporting $^1J_{PP}$ values the sign of the coupling was not determined and the large spread of values seemed inexplicable, and still does to some extent, ranging as they do from an unbelievably small 19 Hz for $Me_2(S)P-P(S)Me_2$[62] to over 715 Hz for the adduct $Me_3P \rightarrow PF_5$.[61]

Finer and Harris[63] were the first to recognize that the large range in magnitude of $^1J_{PP}$ was due in part to sign variation. It is now clear that with a P—P bond in which one atom has a lone pair the coupling constant is negative and numerically greater than 100 Hz whereas, with no lone pairs on either phosphorus atom it is positive. Table 11.5 lists a selection of $^1J_{PP}$ values for P—P derivatives.

11.2.2 Cyclopolyphosphines[27,64]

11.2.2.1 The synthesis of the cyclopolyphosphines

The methods available for preparing cyclophosphines are basically the same as those for preparing the diphosphines.

Dehydrohalogenation: As with the diphosphines this was the method which first brought to light these compounds, when Kohler and Michaelis[65] in 1877 produced 'phosphobenzene' (eqn. 52), to which they mistakenly, but understandably, gave the the formula PhP=PPh. This reaction has been used in preparing many cyclophosphines such as alkyl, substituted aryl and perfluoroorganic derivatives.

$$PhPH_2 + PhPCl_2 \longrightarrow (PhP)_n + HCl \qquad (52)$$

In some instances the primary phosphine can be generated *in situ* by reducing half of the phosphonous dichloride with LiH or $LiAlH_4$, and then proceeding with the dehydrohalogenation.

Table 11.5 P—P coupling constants, $^1J_{PP}/(Hz)$ taken from ref. 61, Table III

Coordination at P—P	Compound	$^1J_{PP}/(Hz)$
3–3	H_2P-PH_2	−108
3–3	H_2P-PF_2	−211
3–3	F_2P-PF_2	−230
3–3	Me_2P-PMe_2	−180
3–3	$Me_2P-P(CF_3)_2$	−256
3–4	$H_2P-PF_2 \cdot BH_3$	−255
3–4	$Me_2P-P(S)Me_2$	−220
4–4	$[H(O)_2P-PO_3]^{3-}$	+466
4–4	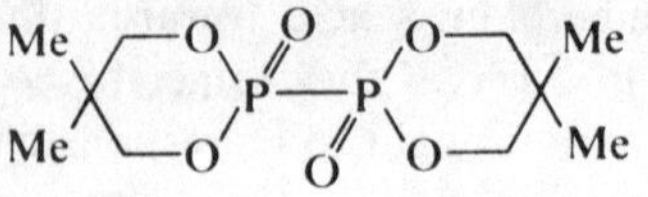 (cyclic diester structure with Me groups)	+475
4–6	$Me_3P \rightarrow PF_5$	+715

Coupling reactions: The general reaction is eqn. (53) which has been carried out with mercury (eqn. 54), alkali metals (eqn. 55) and even magnesium (eqn. 56)[52,66].

$$RPX_2 + 2M \longrightarrow (RP)_n + 2MX \qquad (53)$$

$$CF_3PI_2 + Hg \longrightarrow (CF_3P)_4(60\%) + (CF_3P)_5(40\%) + HgI_2 \qquad (54)$$

$$MePCl_2 + Li \xrightarrow{THF} (MeP)_5 + LiCl \qquad (55)$$

$$PhPCl_2 + Mg \xrightarrow{THF} (PhP)_n(\text{form A}) + MgCl_2 \qquad (56)$$

Yields from these reactions are often over 70%. Just as with the diphosphines (eqn. 31) tri-n-butylphosphine appears an efficient chlorine abstractor (eqn. 57)[67], and another reaction which has its diphosphine counterpart, eqn. (33), is eqn. (58)[68].

$$RPCl_2 \text{ (R = Ph, Et etc.)} + Bu^n_3P \longrightarrow (RP)_n + Bu^n_3PCl_2 \qquad (57)$$

$$KHPC_6H_{11} + BrCH_2CH_2Br \longrightarrow (C_6H_{11}P)_4 + CH_2=CH_2 + KBr \qquad (58)$$

From aminophosphines: Reaction (59) is one in which a P—N bond gives way to a P—P bond by the elimination of an amine (eqn. 59)[69]. In certain cases thermal decomposition of an aminophosphine will produce a polyphosphine (eqn. 60)[70].

$$PhPH_2 + (Me_2N)_2PPh \xrightarrow{170\,°C} (PhP)_n(\text{form A})(83\%) + 2Me_2NH \qquad (59)$$

$$PhP(NHEt)_2 \xrightarrow{heat} (PhP)_n(\text{form A}) + EtN=PPh(NHEt)_2 + EtNH_2 \qquad (60)$$

Miscellaneous methods of synthesis are outlined in Table 11.6. In some of these preparative reactions there is evidence of the monomer as intermediate. For example, when $(PhP)_n$ is heated to 160 °C its dissociation into free radicals is implied from the

11.2 The phosphorus–phosphorus bond: The polyphosphines

Table 11.6 Other syntheses of cyclopolyphosphines

$$4C_6H_{11}P(O)H_2 \xrightarrow[1 \text{ mm Hg}]{6 \text{ h } 60 \,°C} (C_6H_{11}P)_4(22\%) + 4H_2O \quad (61)^a$$

$$10\text{MePF}_2 \xrightarrow[40\,°C]{\text{sealed tube}} (\text{MeP})_5(96\%) + 5\text{MePF}_4 \quad (62)^b$$

$$P_4 + Bu^nMgBr + Bu^nBr \longrightarrow (Bu^nP)_4(42\%) + MgBr_2 \quad (63)^c$$

$$PhPH_2 + (PhCH_2)_2Hg \longrightarrow (PhP)_n(\text{form A}) + PhCH_3 + 4Hg \quad (64)^d$$

$$PhP(CONHPh)_2 \xrightarrow{180\,°C\ 2\,h} (PhP)_n(\text{form A})(40\%) + CO + (PhNH)_2CO \quad (65)^e$$

$$PhHPCH_2CH_2PHPh \xrightarrow{150-160\,°C} (PhP)_n(\text{form D})(35\%) + PhPH_2 + EtPhPH \quad (66)^f$$

[a] W. A. Henderson Jr., M. Epstein and F. S. Seichter, *J. Amer. Chem. Soc.*, **85**, 2462 (1963); [b] V. N. Kulakova, Yu. M. Zinov'ev and L. Z. Soborovskii, *Zh. Obshch. Khim.*, **29**, 3957 (1959); [c] M. M. Rauhut and A. M. Semsel, *J. Org. Chem.*, **28**, 473 (1963); [d] G. B. Postnikova and I. F. Lutsenko, *Zh. Obshch. Khim.*, **29**, 3957 (1959); [e] H. Fritzsche, U. Hasserodt and F. Korte, *Angew. Chem.*, **75**, 1205 (1963); [f] K. Issleib and K. Standtke, *Chem. Ber.*, **96**, 279 (1963).

products formed by the reaction with diethyldisulphide or benzil (eqn. 67)[71] (eqn. 68)[72]; see Chapter 9 for fuller details.

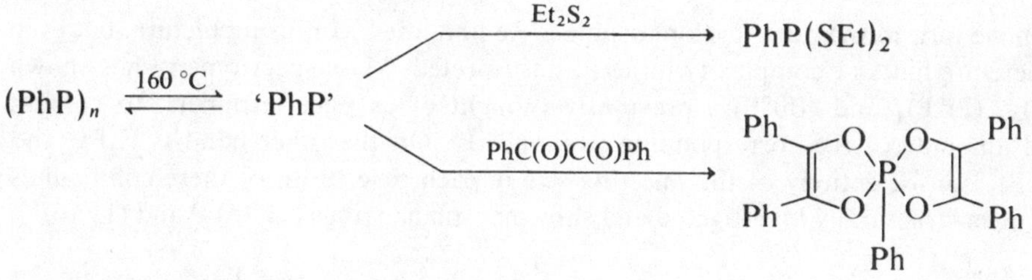

$$(PhP)_n \underset{160\,°C}{\rightleftharpoons} \text{'PhP'} \xrightarrow{Et_2S_2} PhP(SEt)_2 \quad (67)$$

$$\text{'PhP'} \xrightarrow{PhC(O)C(O)Ph} \text{(benzil adduct)} \quad (68)$$

11.2.2.2 The properties of the cyclopolyphosphines

Before going on to discuss the properties and other features of the cyclophosphines it will be necessary to consider the enigma of 'phosphobenzene'. Depending upon its mode of preparation this can have one of four recognized forms designated: A, m.p. 150 °C; B, m.p. 193 °C; C, m.p. 252–256 °C; or D, m.p. 260–305 °C. When first prepared it was assumed to be PhP=PPh and so it remained until cryoscopic determinations showed it to be tetrameric in solution, $(PhP)_4$. However x-ray crystallography on A revealed it to be cyclic pentamer, $(PhP)_5$, with an almost planar ring of phosphorus atoms[73], while analysis of B proved it to be a cyclic hexamer[73] with a chair-shaped ring of phosphorus atoms, and phenyl groups in equatorial positions their planes perpendicular to that of the P_6 ring.

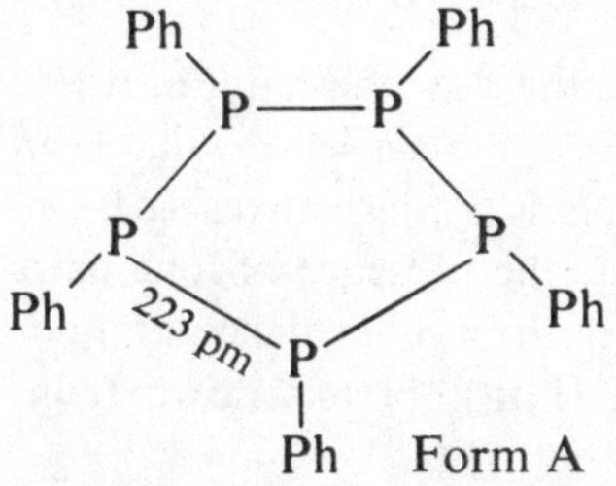

Form A 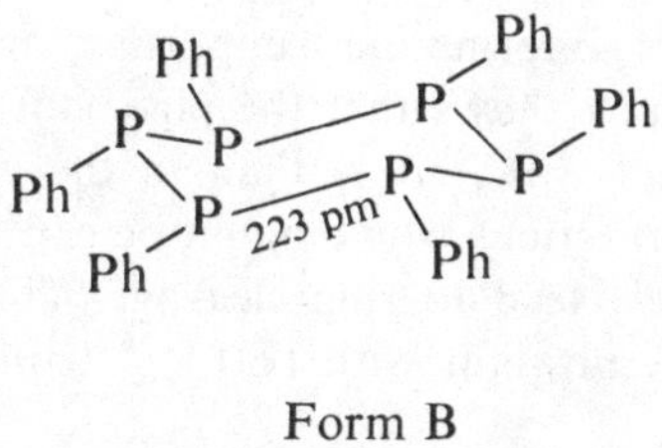   Form B

Why should forms A and B give solutions of tetramers? Another curious feature of these solutions is that they do not represent an equilibrium system – A only recrystallizes from solutions of A, even when seeded with crystals of B, and B only crystallizes from solutions of B, even when seeded with A. The various forms can be interconverted according to eqn. (68a).

$$B \xrightarrow{\text{heat or HCl/C}_6\text{H}_6} A \underset{\text{heat}}{\overset{\text{C}_5\text{H}_5\text{N/0 °C}}{\rightleftharpoons}} C \qquad (68a)$$

The mass spectrum of A showed the presence of $(PhP)_{3,4 \text{ and } 5}$,[74] and this mixture was not a product of the spectrometer since the mass spectrum of $(MeP)_5$ showed a single species[74]. One explanation of A and B in solution is that their apparent molecular weight represents an average of several cyclomers which just happens to give an average value corresponding to the tetramer. Whether equilibrium can exist between various ring sizes seems doubtful in view of the lack of pentamer ⇌ hexamer conversion in solution. The equilibria (69) and (70) should make interconversion between A and B possible since they contain a common entity, $(PhP)_3$.

$$2(PhP)_5 \rightleftharpoons (PhP)_4 + 2(PhP)_3 \qquad (69)$$

$$(PhP)_6 \rightleftharpoons 2(PhP)_3 \qquad (70)$$

When we turn to the other cyclophosphines we find a less confusing picture, but even so there are hints of complexity hitherto unsuspected. Mass spectrometry has shown $(EtP)_4$, $(Pr^nP)_4$ and $(Bu^nP)_4$, previously thought of as pure tetramers, to contain small amounts of the corresponding pentamers[74]. On the other hand $(CF_3P)_4$ and $(CF_3P)_5$ consist entirely of the one ring size in each case. Both of these compounds have been structurally investigated and show non-planar rings (11.15)[75] and (11.16)[76].

(11.15)   (11.16)

In all the structures so far determined the ring bond lengths are all equal and about the same value as that of other P—P bonds, i.e. *ca* 222 pm. The unequivocal inference from bond length data is that either all P—P compounds have a π contribution to the bond, or none of them has. With the diphosphines the ultraviolet spectra contained intense bands characteristic of π systems, and similar, well resolved bands are found in the ultraviolet spectra of the cyclophosphines, Table 11.7.

Another piece of information supporting π-bonding is the low basicity of these compounds, which shows that the phosphorus lone pairs are even less available in $(RP)_4$ compounds (pK_a *ca* 1) than in the corresponding R_3P derivatives (pK_a *ca* 2–10)[77]. Other reactions which might be expected to show the presence of lone pairs can unfortunately lead to ring cleavage. Thus with MeI there is evidence of non-disruptive quaternization with $(EtP)_4$[79] (eqn. 71a), but $(PhP)_n$ breaks down (eqn. 71b).[80]

11.2 The phosphorus–phosphorus bond: The polyphosphines

Table 11.7 Ultraviolet spectra (λ_{max}) of the alkyl cyclophosphines[52,77,78]

$(CF_3P)_4$	$(CF_3P)_5$	$(CH_3P)_5$	$(RP)_4$[a]
259	260	296	290
239	240		280
221			275

[a] $R = Et_2CH$, Bu^i, $C_8H_{17}{}^n$, C_6H_{11} etc.

$(EtP)_4 + MeI \longrightarrow (EtP)_4 \cdot MeI$ (71a)

$(PhP)_n + MeI \longrightarrow (PhPI)_2 + PhPMe_3I + PhPMe_2I_2$ (71b)

Lewis acids of the boron type form adducts but break up of the ring occurs e.g. $(PhP)_5$ gives $(PhP)_3 \cdot BF_3$ and the tetracycloalkylphosphines, $(RP)_4$, give $(RP)_2 \cdot BX_3$ ($R = Et$, Pr^n, Bu^n; $X = F$, Cl, Br)[66]. The trifluoromethyl derivatives, $(CF_3P)_{4,5}$, however, do not form adducts with either BF_3 or B_2H_6[52] but appear to exhibit Lewis *acid* behaviour themselves. With Me_3P and Me_3N they form monomer adducts in equilibrium with the reactants, e.g. eqn. (72)[81], although in these the bonding is likely to be far more sophisticated than a simple donor σ-bond between P and P or between P and N.

$(CF_3P)_4 + 4Me_3P \rightleftharpoons 4CF_3P \cdot PMe_3$ (72)

From infrared studies, bands in the range 390–410 cm^{-1} have been assigned to a symmetric ring stretching vibration, while those at 465–490 cm^{-1} correspond to an asymmetric mode[82]. Forms A and B of the phenylcyclophosphines have very similar infrared spectra although their p.m.r. spectra distinguishes the two forms. ^{31}P n.m.r. spectroscopy[77] shows a large range of chemical shifts: (referenced to 85% H_3PO_4)

$(MeP)_5 < (EtP)_4 < (Bu^nP)_4 < (PhP)_5 < (Bu^tP)_4 < (Pr^iP)_4 < (C_6F_5P)_4 < (C_6H_{11}P)_4$
$\quad -21 \quad\quad -17 \quad\quad -14 \quad\quad +9 \quad\quad +58 \quad\quad +66 \quad\quad +67 \quad\quad +70$ ppm

No firm conclusions regarding electronic or bonding effects can be drawn from these results. For some compounds more than one signal was found, e.g. $(Pr^nP)_n$ had peaks at -53, -16 and -12 ppm and these may reflect the different ring sizes that this compound was shown to have, by mass spectrometry.

Chemically the cyclophosphines are very reactive. Tetrakis(trifluoromethyl)cyclotetraphosphine, $(CF_3P)_4$, inflames in air but the phenyl derivatives are air stable although oxidized in solution. Thermally the compounds are very stable, many have high m.p.'s and some can be distilled under reduced pressure. At temperatures exceeding 200 °C changes in ring size and disproportionation occur. For example $(CF_3P)_4$ is stable for at least 24 h at 280 °C but at 314 °C breakdown occurs (eqn. 73a). $(PhP)_5$, when heated to 300 °C disproportionates (eqn. 73b) in a similar manner. The order of increasing thermal stability is $(PhP)_5 < (CF_3P)_4 < (C_6H_{11}P)_4 < (EtP)_4$. Thermal

$$(CF_3P)_4 \xrightarrow{314 °C} P_4 + (CF_3)_3P \tag{73a}$$

$$(PhP)_5 \xrightarrow{300 °C} P_4 + Ph_3P + Ph_2P-PPh_2 \tag{73b}$$

rearrangement amongst side groups may also occur, and this happens when a mixture of $(MeP)_5$ and $(EtP)_5$ is heated. This gives products containing both methyl and ethyl groups attached to the same ring[83].

Most reagents which attack the cyclophosphines go for the ring. Yet despite its apparent vulnerability, in certain derivatives it is much more resistant to attack than expected. The hydrolysis of $(CF_3P)_4$ is a case in point, and this reaction has been studied in detail[52]. The amount of fluoroform, CF_3H, produced depends upon the reaction conditions, eqn. (74) being one way in which hydrolysis can completely disrupt the ring. Hydrolysis in diglyme solvent at 50 °C, however, produces di- and tri-phosphines (eqn. 75) as well.

$$(CF_3P)_4 + H_2O \xrightarrow{OH^-} 2CF_3P(O)(H)OH + 2CF_3PH_2 \tag{74}$$

$$\downarrow$$

$$2CF_3H + HP(O)(OH)_2$$

$$(CF_3P)_4 + H_2O \xrightarrow[50 °C]{diglyme} H(CF_3P)_2H + H(CF_3P)_3H \tag{75}$$

With many reagents the phosphine rings break into diphosphine fragments. Thus in the reaction of $(PhP)_n$ with iodine, a diphosphine can be isolated (eqn. 76), although excess iodine will break this up, but with chlorine and bromide only monophosphorus products, $PhPCl_4$ and $PhPBr_4$, are obtained[84].

$$(PhP)_n + I_2 \longrightarrow Ph(I)P-P(I)Ph \xrightarrow{I_2} 2PhPI_2 \tag{76}$$

A useful diphosphine intermediate is obtained when $(EtP)_4$ is treated with alkali metals (especially potassium) and this has been used to synthesize other diphosphines as shown in eqns. (77) and (78)[85].

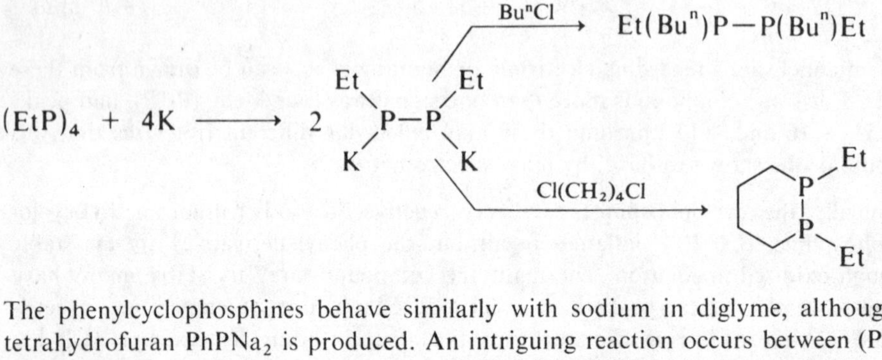

$$(EtP)_4 + 4K \longrightarrow 2 \begin{array}{c} Et \quad Et \\ \diagdown \quad \diagup \\ P-P \\ \diagup \quad \diagdown \\ K \quad K \end{array} \begin{array}{c} \xrightarrow{Bu^nCl} Et(Bu^n)P-P(Bu^n)Et \quad (77) \\ \\ \xrightarrow{Cl(CH_2)_4Cl} \end{array} \begin{array}{c} Et \\ P \\ P \\ Et \end{array} \quad (78)$$

The phenylcyclophosphines behave similarly with sodium in diglyme, although in tetrahydrofuran $PhPNa_2$ is produced. An intriguing reaction occurs between $(PhP)_5$ and potassium which results in the production in 90 % yield of a relatively stable

11.2 The phosphorus–phosphorus bond: The polyphosphines

cyclotriphosphine dianion (eqn. 79)[86]. A delocalized π system may be helping to stabilize this ring.

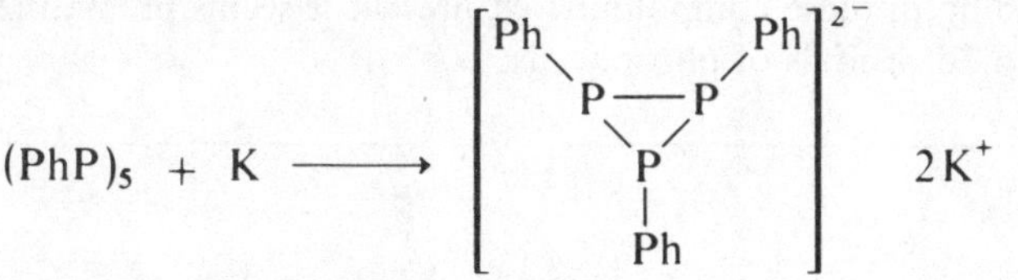

$$(PhP)_5 + K \longrightarrow \begin{bmatrix} Ph & Ph \\ P-P \\ | \\ P \\ | \\ Ph \end{bmatrix}^{2-} 2K^+ \quad (79)$$

The cyclophosphines can act as ligands, forming transition metal complexes. In some instances the result is a ring expansion[87] (eqn. 80), and this has been proved for (11.17) by x-ray analysis[88].

$$(EtP)_4 + Mo(CO)_6 \longrightarrow (EtP)_5Mo(CO)_4 \quad (80)$$
$$(11.17)$$

However ring size in most cases remains unaffected, eqn. (81)[89] and (82)[90]. Ring contraction can occur, eqns. (83)[90] and (84)[87], although with the phenylcyclophosphines one must suspend judgement over ring size changes bearing in mind what has already been said about this system.

$$(MeP)_5 + CuBr \longrightarrow (MeP)_5CuBr \quad (81)$$

$$(C_6H_{11}P)_4 + Fe(CO)_5 \longrightarrow (C_6H_{11}P)_4Fe_2(CO)_8 \quad (82)$$

$$(PhP)_5 + M(CO)_6 \xrightarrow{Et_2O} (PhP)_4M(CO)_4 \quad M = Cr, Mo, W \quad (83)$$
form A

$$(MeP)_5 + CuCl \longrightarrow (MeP)_4CuCl \quad (84)$$

Finally the cyclophosphines react with double bonds to form organophosphorus ring systems (eqn. 85)[91] which can be explained as being formed via diamers and monomers (cf. Chapter 6). Even larger fragments of the original phosphine ring can be incorporated in the formation of some organophosphorus rings (eqn. 86)[92].

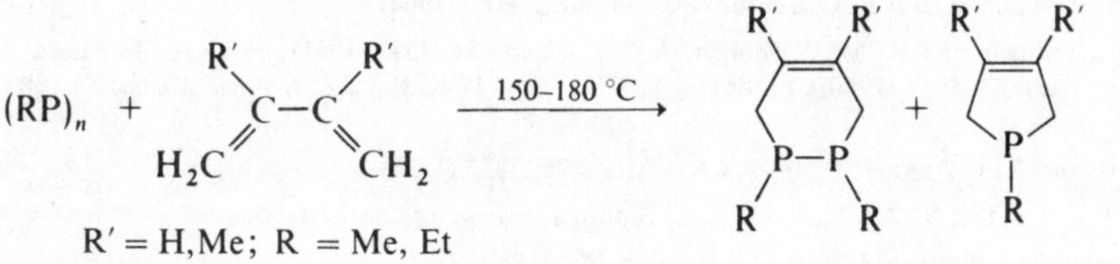

$$(RP)_n + \begin{array}{c} R' \quad R' \\ C=C \\ H_2C \quad CH_2 \end{array} \xrightarrow{150-180\,°C} \begin{array}{c} R' \quad R' \\ \diagup \\ P-P \\ R \quad R \end{array} + \begin{array}{c} R' \quad R' \\ \diagup \\ P \\ R \end{array} \quad (85)$$

$R' = H, Me;\ R = Me, Et$

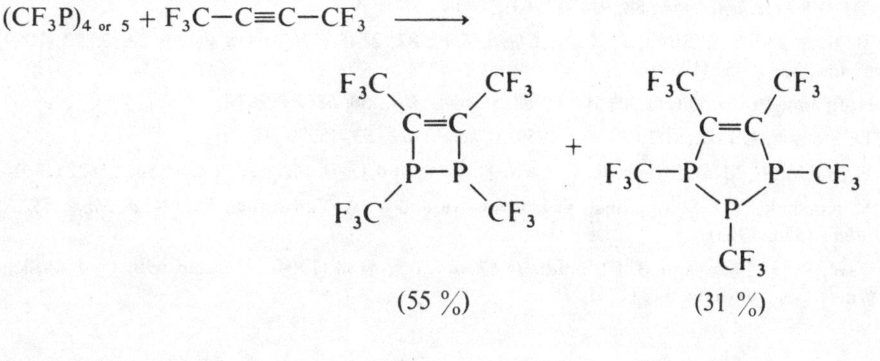

$$(CF_3P)_{4\text{ or }5} + F_3C-C\equiv C-CF_3 \longrightarrow$$

$$\begin{array}{c} F_3C \quad CF_3 \\ C=C \\ | \quad | \\ P-P \\ F_3C \quad CF_3 \end{array} + \begin{array}{c} F_3C \quad CF_3 \\ C=C \\ F_3C-P \quad P-CF_3 \\ P \\ | \\ CF_3 \end{array} \quad (86)$$

$$(55\%) \quad\quad\quad (31\%)$$

Much remains to be discovered about these tantalizing ring systems. Both the cyclophosphines and cycloboraphosphanes pose many unanswered questions, not least about the nature of the bonding in these compounds. At present it seems premature to talk of π systems based on 3d orbitals of phosphorus.

Problems

1 Contrary to expectations based on single-bond enthalpy data Me_2PH will displace Me_2NH from $Me_2P{-}NMe_2$ (see eqn. 37). Assuming that only bond enthalpies need be taken into account, calculate the energy released if the P—P bond being formed had the bond enthalpy of the gaseous P_2 molecule [*E* values on page 458.]

2 Why should PF_5 form a stronger adduct with PMe_3 than NMe_3?

3 Assuming that $3p(P) \rightarrow 2p(B)$ donor π-bonding is operative in $X_2P{-}BX_2$ compounds what effect should this have on the structure of these compounds? What mitigates against such π-bonding.

4 Assuming *maximum* $3p(P) \rightarrow 3d(P)$ donor π-bonding is operative in $X_2P{-}PX_2$ compounds, what effect should this have on the structure of these molecules?

References

[1] G. W. Parshall, in E. L. Muetterties (editor), *The Chemistry of Boron and its Compounds*, Chapt. 9, Wiley & Sons, New York, 1967.

[2] H. Steinberg and R. J. Brotherton, *Organoboron Chem.*, **2**, 479 (1966).

[3] H. Noth and W. Schragle, *Chem. Ber.*, **97**, 2218 (1964).

[4] G. E. Coates and J. G. Livingstone, *J. Chem. Soc.*, 1000 (1961).

[5] W. Green, R. A. Shaw and B. C. Smith, *J. Chem. Soc.*, 4180 (1964).

[6] PR_3, R. A. Baldwin and R. M. Washburn, *J. Org. Chem.*, **26**, 3549 (1961); $P(OR)_3$, T. Reetz, *J. Amer. Chem. Soc.*, **82**, 5039 (1960); $P(NR_2)_3$, T. Reetz and B. Katlafsky, *J. Amer. Chem. Soc.*, **82**, 5036 (1960).

[7] A. B. Burg and R. I. Wagner, *J. Amer. Chem. Soc.*, **75**, 3872 (1953).

[8] W. Gee, J. B. Holden, R. A. Shaw and B. C. Smith, *J. Chem. Soc.*, 3171 (1965).

[9] A. B. Burg and G. Brendel, *J. Amer. Chem. Soc.*, **80**, 3198 (1958).

[10] A. B. Burg and R. I. Wagner, British Patent 852,970 (1960) [*Chem. Abs.*, **56**, 505 (1962)]; US Patent 3,065,271 (1962) [*Chem. Abs.*, **58**, 9140 (1963)].

[11] A. B. Burg and P. J. Slota, *J. Amer. Chem. Soc.*, **82**, 2145 (1960); US Patent 2,877,272 (1959) [*Chem. Abs.*, **53**, 16062 (1959)].

[12] E. Hofmann, British Patent 908,106 (1962) [*Chem. Abs.*, **58**, 6862 (1963)].

[13] R. D. Stewart, US patent 2,879,301 (1959) [*Chem. Abs.*, **53**, 15979 (1959)].

[14] R. H. Biddulph, M. P. Brown, R. C. Cass, R. Long and H. B. Silver, *J. Chem. Soc.*, 1822 (1961).

[15] V. V. Korshak, A. I. Solomatina, N. I. Bekasova and V. A. Zamyatina, *Izv. Akad. Nauk SSSR, Ser. Khim.*, 1856 (1963).

[16] W. Gee, R. A. Shaw and B. C. Smith, *J. Chem. Soc.*, 4180 (1964); 16B and with G. J. Bullen, *Proc. Chem. Soc., London*, 432 (1961).

References

[17] A. D. Tevebaugh, *Inorg. Chem.*, **3**, 302 (1964).
[18] H. Noth and W. Schragle, *Z. Naturforsch.*, **16B**, 473 (1961).
[19] H. Noth and W. Schragel, *Chem. Ber.*, **98**, 352 (1965).
[20] A. B. Burg, *J. Amer. Chem. Soc.*, **83**, 2226 (1961).*
[21] G. J. Beichl and C. E. Evers, *J. Amer. Chem. Soc.*, **80**, 5344 (1958).
[22] I. B. Johns, E. A. McElhill and J. O. Smith, *J. Chem. Eng. Data*, **7**, 277 (1962).
[23] M. H. Goodrow, R. I. Wagner and R. D. Stewart, *Inorg. Chem.*, **3**, 1212 (1964).
[24] W. C. Hamilton, *Acta Crystallogr.*, **8**, 199 (1955).
[25] J. E. Huheey, *J. Chem. Educ.*, **40**, 159 (1963).
[26] A. H. Cowley, *Chem. Revs.*, **65**, 617 (1965).
[27] A. H. Cowley and R. P. Pinnell, *Topics in Phosphorus Chem.*, **4**, 1 (1967).
[28] E. Fluck, *Prep. Inorg. React.*, **5**, 103 (1968).
[29] D. R. Peck, *Mellor Vol. VIII Supp. III (P)*, Section IV, p 149, Longman, London, 1971.
[30] B. Blaser and K.-H. Worms, *Z. Anorg. Allgem. Chem.*, **300**, 237 (1959)*
[31] J. Weiss, *Z. Anorg. Allgem. Chem.*, **306**, 30 (1960).
[32] M. M. Raubut and A. M. Semsel, *J. Org. Chem.*, **28**, 473 (1963).*
[33] L. R. Grant and A. B. Burg, *J. Amer. Chem. Soc.*, **84**, 1834 (1962).
[34] H. Niebergall and B. Langenfeld, *Chem. Ber.*, **95**, 64 (1962).
[35] E. Wiberg, M. Van Ghemen and H. Muller-Schiedmayer, *Angew. Chem.*, **75**, 814 (1963).
[36] A. B. Burg and K. K. Joshi, *J. Amer. Chem. Soc.*, **86**, 353 (1964).
[37] W. Hewertson and H. R. Watson, *J. Chem. Soc.*, 1490 (1962).
[38] H. Noth and H-J. Vetter, *Chem. Ber.*, **94**, 1505 (1961).
[39] H. Noth and H-J. Vetter, *Chem. Ber.*, **96**, 1816 (1963).
[40] K. Issleib and D-W. Muller, *Chem. Ber.*, **92**, 3175 (1959).
[41] M. I. Kabachnik and E. S. Shepeleva, *Izv. Akad. Nauk SSSR, Ser. Khim.*, 56 (1949); *Chem. Abs.*, **43**, 5739 (1949).*
[42] K. Issleib and A. Tzschach, *Chem. Ber.*, **92**, 704 and 1397 (1959).
[43] L. Maier, *Chem. Ber.*, **94**, 3043 (1961).
[44] A. B. Burg, *J. Inorg. Nucl. Chem.*, **11**, 258 (1959).
[45] L. Maier, *Helv. Chim. Acta*, **49**, 1119 (1966).
[46] N. G. Feshchenko and A. V. Kirsanov, *Zh. Obshch. Khim.*, **30**, 3041 (1960) [*Chem. Abs.*, **55**, 14145 (1961)]; *ibid*, **31**, 1399 (1961) [*Chem. Abs.*, **55**, 27169 (1961)].
[47] H. Fritzsche, U. Hasserodt and F. Korte, *Angew. Chem.*, **75**, 1205 (1963).
[48] F. Seel, K. Rudolph and W. Gambler, *Angew. Chem.*, **79**, 686 (1967).
[49] E. Fluck and H. Binder, *Inorg. Nucl. Chem. Lett.*, **3**, 307 (1967).*
[50] S. R. Gunn and L. G. Green, *J. Phys. Chem.*, **65**, 779 (1961).
[51] A. A. Sandoval, H. C. Moser and R. W. Kiser, *J. Phys. Chem.*, **67**, 124 (1963).
[52] (a) W. Mahler and A. B. Burg, *J. Amer. Chem. Soc.*, **80**, 6161 (1958).*
(b) A. B. Burg and L. K. Peterson, *Inorg. Chem.*, **5**, 943 (1966).*
[53] A. B. Burg and J. E. Griffiths, *J. Amer. Chem. Soc.*, **82**, 3514 (1960).
[54] F. W. Bennett, H. J. Emelens and R. N. Hazeldine, *J. Chem. Soc.*, 1565 (1953)*; *ibid*, 3896 (1954).
[55] K. Issleib and F. Krech, *Z. Anorg. Algem. Chem.*, **328**, 21 (1964).
[56] K. Issleib and W. Seidel, *Chem. Ber.*, **92**, 2681 (1959).
[57] A. G. Garrett and G. Urry, *Inorg. Chem.*, **2**, 400 (1963).
[58] J. Chatt and D. T. Thornton, *J. Chem. Soc.*, 1005* and 2713 (1964).

[59] R. G. Hayter, *Inorg. Chem.*, **3**, 711 (1964)*; and L. F. Williams, *ibid*, **3**, 717 (1964).
[60] H. Bandler, H. Standeke and M. Kemper, *Z. Anorg. Allgem. Chem.*, **388**, 125 (1972).*
[61] C. W. Schultz and R. W. Rudolf, *J. Amer. Chem. Soc.*, **93**, 1898 (1971).*
[62] R. K. Harris and R. G. Hayter, *Can. J. Chem.*, **42**, 2282 (1964).
[63] E. G. Finer and R. K. Harris, *Chem. Commun.*, 110 (1968).
[64] I. Haiduc, *The Chemistry of Inorganic Ring Systems*, Vol I, pp 82–102, Wiley–Interscience, London, 1970.
[65] H. Kohler and A. Michaelis, *Chem. Ber.*, **10**, 807 (1877).*
[66] W. Mahler and A. B. Burg, *J. Amer. Chem. Soc.*, **79**, 251 (1957).
[67] A. H. Cowley and R. P. Pinnell, *Inorg. Chem.*, **5**, 1459 and 1463 (1966).
[68] K. Issleib and G. Doll, *Chem. Ber.*, **94**, 2664 (1961).
[69] L. Maier, *Prog. Inorg. Chem.*, **5**, 27 (1963).
[70] A. P. Lane and D. S. Payne, *Proc. Chem. Soc.*, 403 (1964).
[71] U. Schmidt, I. Boie, C. Osterroht and H. F. Grutzmacher, *Chem. Ber.*, **101**, 1381 (1968).*
[72] B. Block and Y. Gounelle, *C. R. Hebd. Seances Acad. Sci.*, **266C**, 220 (1968).
[73] J. J. Daly, *J. Chem. Soc.*, 6147 (1964),* (PhP)$_5$; *ibid.*, 4789 (1965) and *ibid*, 428 (1966),* (PhP)$_6$.
[74] M. Baudler, K. Kipker and H. W. Valpertz, *Naturwissenschaften*, **53**, 612 (1966).
[75] G. J. Palenik and J. Donohue, *Acta Crystallogr.*, **15**, 564 (1962).*
[76] C. J. Spencer and W. N. Lipscomb, *Acta Crystallogr.*, **14**, 250 (1961)*; and with P. G. Simpson, *ibid*, **15**, 509 (1962).
[77] W. A. Henderson Jr., M. Epstein and F. S. Seichter, *J. Amer. Chem. Soc.*, **85**, 2462 (1963).
[78] J. W. B. Reesor and G. F. Wright, *J. Org. Chem.*, **22**, 385 (1957).
[79] K. Issleib and B. Mitscherling, *Z. Naturforsch.*, **15B**, 267 (1960).
[80] W. Hoffmann and W. Grunewald, *Chem. Ber.*, **94**, 186 (1961).
[81] A. B. Burg and W. Mahler, *J. Amer. Chem. Soc.*, **83**, 2388 (1961).*
[82] R. L. Amster, N. B. Colthup and W. A. Henderson, *Spectrochim. Acta*, **19**, 1841 (1963); and *Can. J. Chem.*, **42**, 2577 (1964).
[83] U. Schmidt, R. Schroer and H. Ackenbach, *Angew. Chem., Int. Ed. Eng.*, **5**, 316 (1966).
[84] W. Kuchen and H. Buckwald, *Chem. Ber.*, **91**, 2296 (1958).
[85] K. Issleib and K. Krech, *Chem. Ber.*, **98**, 2545 (1965).
[86] K. Issleib and E. Fluck, *Angew. Chem.*, **78**, 597 (1966).
[87] C. S. Cundy, M. Green, F. G. A. Stone and A. Taunton-Rigby, *J. Chem. Soc. A*, 1776 (1968).*
[88] M. A. Bush and P. Woodward, *J. Chem. Soc. A*, 1221 (1968).*
[89] K. Issleib and M. Keil, *Z. Anorg. Allgem. Chem.*, **333**, 10 (1964).
[90] G. W. A. Fowles and D. K. Jenkins, *Chem. Commun.*, 61 (1965).
[91] U. Schmidt and I. Boie, *Angew. Chem., Int. Ed. Eng.*, **5**, 1038 (1966).
[92] W. Mahler, *J. Amer. Chem. Soc.*, **86**, 2306 (1964).*

---

* These papers are reproduced, and in the case of German and Russian ones are translated into English, in a compilation of 33 historical papers which traces the development of the chemistry of P—P compounds: *Compounds Containing Phosphorus–Phosphorus Bonds* (Benchmark Papers in Inorganic Chemistry) Ed. A. H. Cowley, Dowder, Hutchinson & Ross Inc., Stroudsburg, Pennsylvania 1973.

Chapter 12
Biophosphorus chemistry

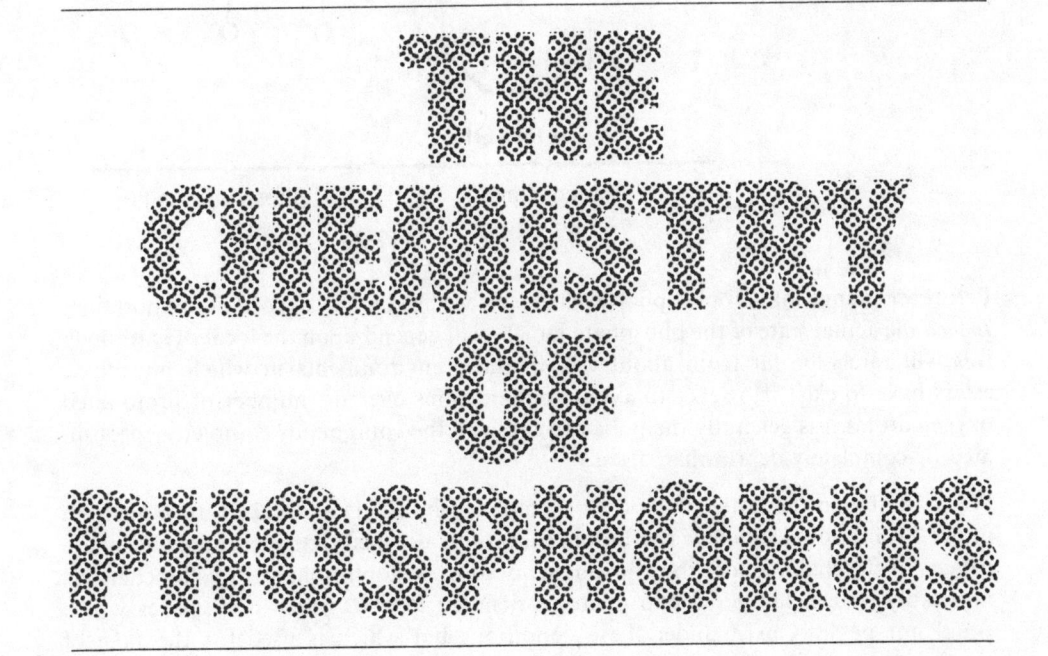

THE CHEMISTRY OF PHOSPHORUS

12.1 Phosphorus in life

Phosphorus plays a vital role in all life forms, and phosphate esters and diesters are the principal mode in which it performs its essential functions. There are a few derivatives with phosphorus–carbon and phosphorus–nitrogen bonds and these are noteworthy for their rarity. The first phosphorus–carbon compound identified of this type was 2-aminoethylphosphonic acid, $NH_2CH_2CH_2P(O)(OH)_2$, which was isolated from certain protozoa[1] and this remains the predominant example of a naturally occurring P—C compound[2]. Phosphocreatine is an essential phosphorus–nitrogen compound and this bond may well be the reason for its usefulness as a stored form of phosphate. This derivative (12.1) is held in reserve as a reagent for the 'emergency' regeneration of the one biophosphorus compound on which all life depends, adenosine triphosphate (ATP) which has the structure (12.2)[3].

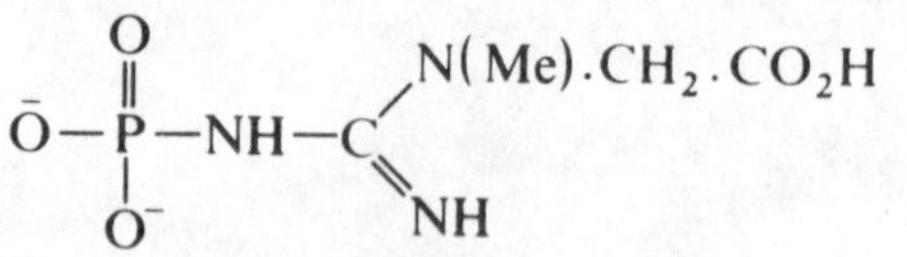

(12.1)

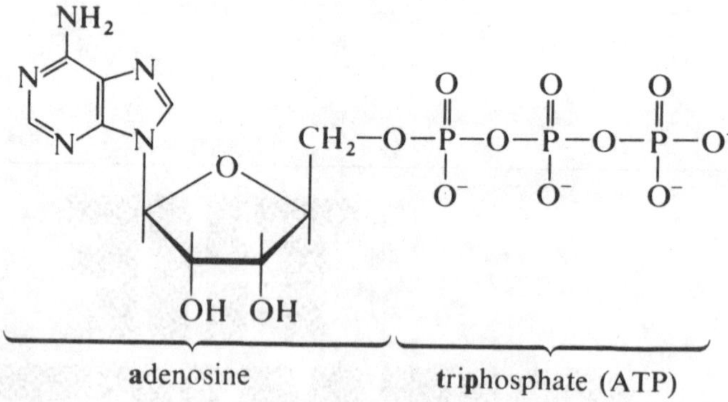

(12.2)

Before we plunge into biophosphorus chemistry we need to say a word about notation. *In vivo* the actual state of the phosphate moiety will depend upon the local pH although this will rarely be far from about 7.3, in those environments in which phosphate esters have to exist. However to avoid complications over the number of protonated oxygen atoms it is generally the policy to show all the compounds completely protonated or completely deprotonated.

We have chosen the latter method as exemplified by (12.1) and (12.2). In a medium of pH 7 ATP would still have one unionized proton attached to the terminal phosphate oxygen of the chain. Under the phosphate notation orthophosphate, or as biochemists invariably term it inorganic phosphate, is denoted PO_4^{3-}, which is a species which could not possibly exist under these conditions but which is used for the sake of consistency.

Biochemists also employ a shorthand notation for the phosphate group, showing it as P, and using the symbol P_i for inorganic phosphate. However there is a great deal of laxity in the use of this symbol since it is sometimes employed in place of the

12.1 Phosphorus in life

$-\overset{\overset{O}{\|}}{\underset{\underset{O^-}{|}}{P}}-$ grouping, sometimes in place of $-\overset{\overset{O}{\|}}{\underset{\underset{O^-}{|}}{P}}-O^-$, and sometimes used for both. Thus ATP is written as adenosine—(P)—(P)—(P) (12.3) or adenosine—(P)—O—(P)—O—(P) (12.4), and the diphosphate entity, $^{2-}O_2P(O)OP(O)O_2^{2-}$ becomes either (PP) or (P)—O—(P). Since the phosphorus centre is our focus of interest we shall refrain from this truncated notation but the reader should be prepared for it in many biochemistry texts, e.g. refs. 4 and 5. One biochemical symbol we shall make use of, however, is the coupled reaction symbol, known as Baldwin's notation: which signifies

$$A \xrightarrow{\quad X \searrow \nearrow Y \quad} B$$

that two processes are linked but not in the sense that they are reacting directly as in eqn. (1). The mechanism of most biochemical reactions is extremely complex, and where there is no definite knowledge as yet about the steps involved, the Baldwin symbol is ideal for the purpose of describing overall changes.

$$A + X \longrightarrow B + Y \qquad (1)$$

As we shall see, ATP crops up in many metabolic pathways where its role is not entirely clear in terms of chemical reaction steps and the coupled reaction symbol will often read:

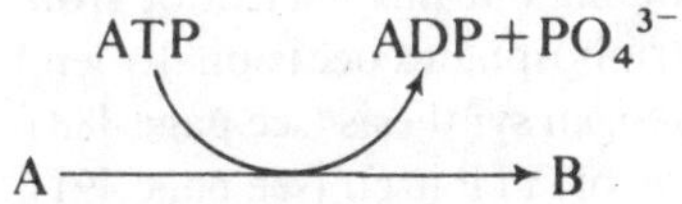

$$A \xrightarrow{\quad ATP \searrow \nearrow ADP + PO_4^{3-} \quad} B$$

which signifies that associated with the change of one molecule of A to B a molecule of ATP loses its terminal phosphate group being converted to ADP, adenosine diphosphate.

Life forms at the bottom of the food chain absorb phosphorus as inorganic phosphate from the soil or water surrounding them (Chapter 1). Eventually this has to be converted to esters, a reaction which cannot be performed by direct combination with organic hydroxyl groups. The phosphate group needs activating and this seems to be achieved best by the formation of an anhydride link of the sort in ATP. Phosphorylation can then be accomplished by the agency of such intermediates. The biosynthesis of DNA is performed in this manner and will be discussed later. The structure of DNA is that of a polymeric diester of phosphoric acid (12.5).

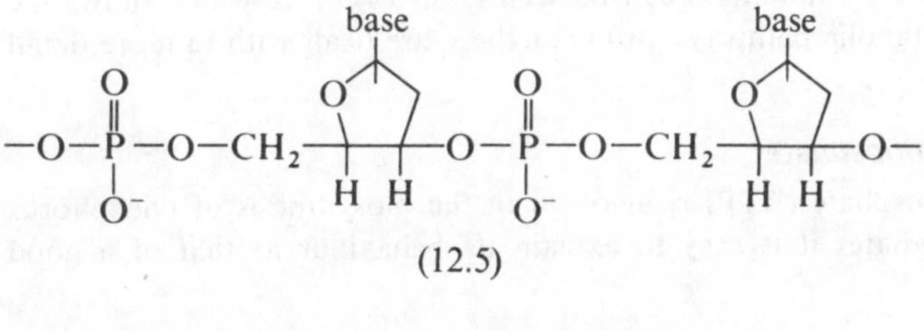

(12.5)

Chapter 12 / Biophosphorus chemistry

The fundamental unit of nucleic acids is the nucleotide which consists of an organic base attached to a sugar unit which is itself attached to a phosphate group. The base- + - sugar part is known as the nucleoside and takes its name from the base. The more common bases are adenosine and guanine, which are purine derivatives, and cytosine, uracil and thymine, which are pyrimidine derivatives – Fig. 12.1. The sugar units are of two kinds: ribose, which gives its name to ribonucleic acid (RNA), and deoxyribose which gives its name to the all important deoxyribonucleic acid (DNA).

The molecule ATP can now be seen in the context of nucleotide chemistry, and although adenosine monophosphate (AMP) (12.6) is capable of attaching phosphate units to form ATP it is not the only nucleotide to do this. As a precursor to their involvement in the biosynthesis of RNA and DNA it is necessary for nucleotides to be converted to triphosphates by reaction with ATP. Nature has selected this particular

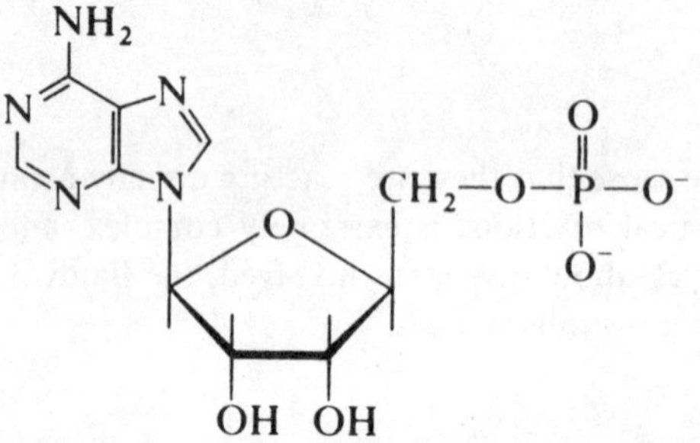

adenosine monophosphate (AMP)
(12.6)

molecule as the work horse although the reasons for singling out this nucleotide from the rest are not clearly understood. Some of the other triphosphates occasionally lend a hand such as uridine triphosphate (UTP) (12.7), in glycogen synthesis (see page 488), and guanosine triphosphate (GTP) (12.8), in the synthesis of ATP itself (see page 491), for example. However ATP bears the brunt of the work of the biosynthesis of phosphate esters, and of the activation by phosphorylation of intermediates, which is necessary to drive a reaction in the desired direction.

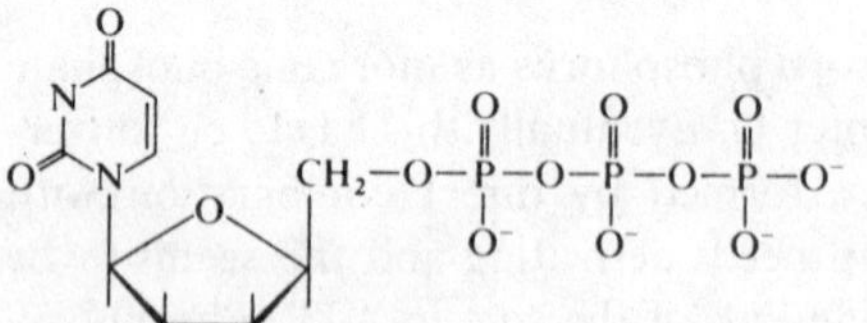

uridine triphosphate (UTP)
(12.7)

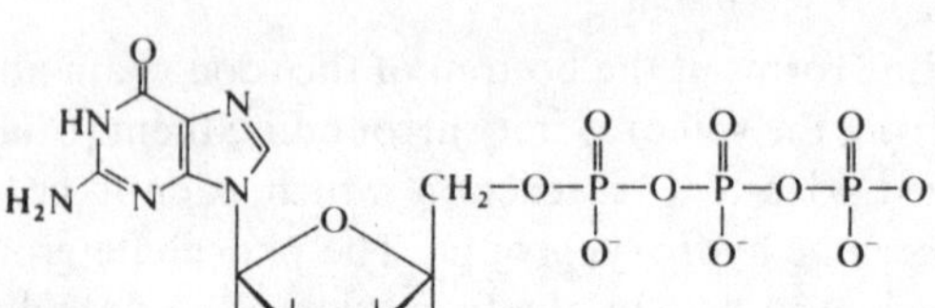

guanosine triphosphate (GTP)
(12.8)

Just how ubiquitous this compound is can be seen from Fig. 12.2 which shows the major intermediary metabolic pathways. Some of these are dealt with in more detail in this chapter.

12.1.1 Adenosine triphosphate

Where adenosine triphosphate (ATP) is involved in the biosynthesis of phosphorus compounds or intermediates it is easy to explain its behaviour as that of a good

12.1 Phosphorus in life

Figure 12.1 Nucleotides.

Organic base part:

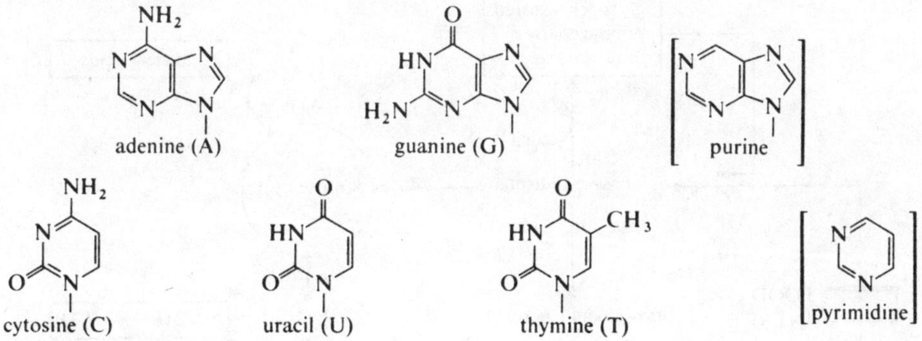

adenine (A)   guanine (G)   [purine]

cytosine (C)   uracil (U)   thymine (T)   [pyrimidine]

Sugar units:

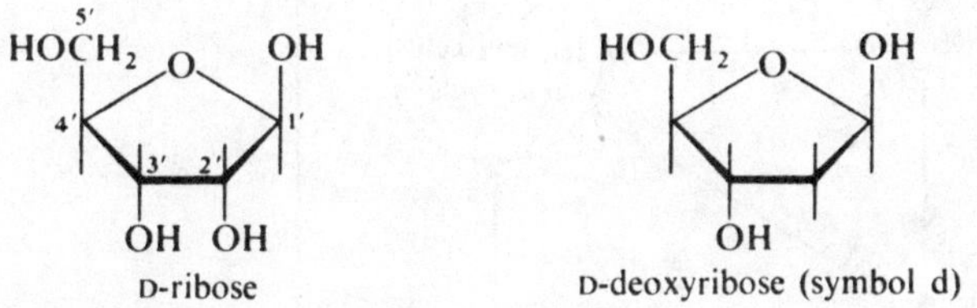

D-ribose   D-deoxyribose (symbol d)

Nucleotide: base + sugar-5'-phosphate e.g.

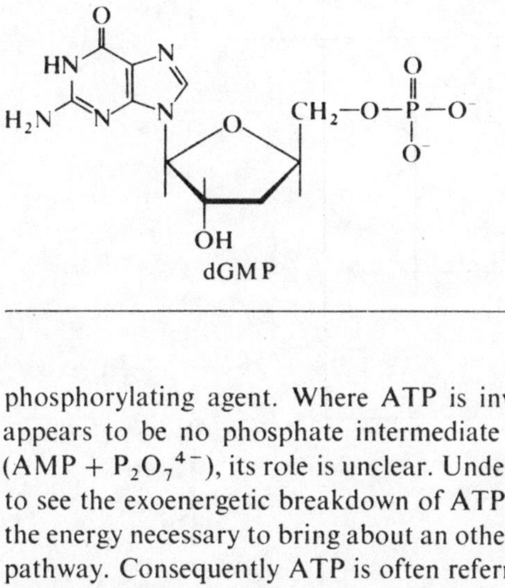

dGMP

---

phosphorylating agent. Where ATP is involved in a metabolic pathway, but there appears to be no phosphate intermediate and yet it ends up as (ADP + PO_4^{3-}) or (AMP + $P_2O_7^{4-}$), its role is unclear. Under these circumstances it has been customary to see the exoenergetic breakdown of ATP to the di- or mono-phosphate as providing the energy necessary to bring about an otherwise unfavourable chemical reaction in the pathway. Consequently ATP is often referred to as 'energy rich', a rather misleading phrase, although of course the hydrolysis of ATP to ADP or AMP will release energy. Not that this direct reaction ever occurs in biological systems, although reaction sequences amounting to hydrolysis do take place and as a result this energy is transferred to the system. For example eqns. (2) and (3) are one way in which this could occur since together these amount to hydrolysis step (4).

476   Chapter 12 / Biophosphorus chemistry

Figure 12.2 Intermediary metabolic pathways.

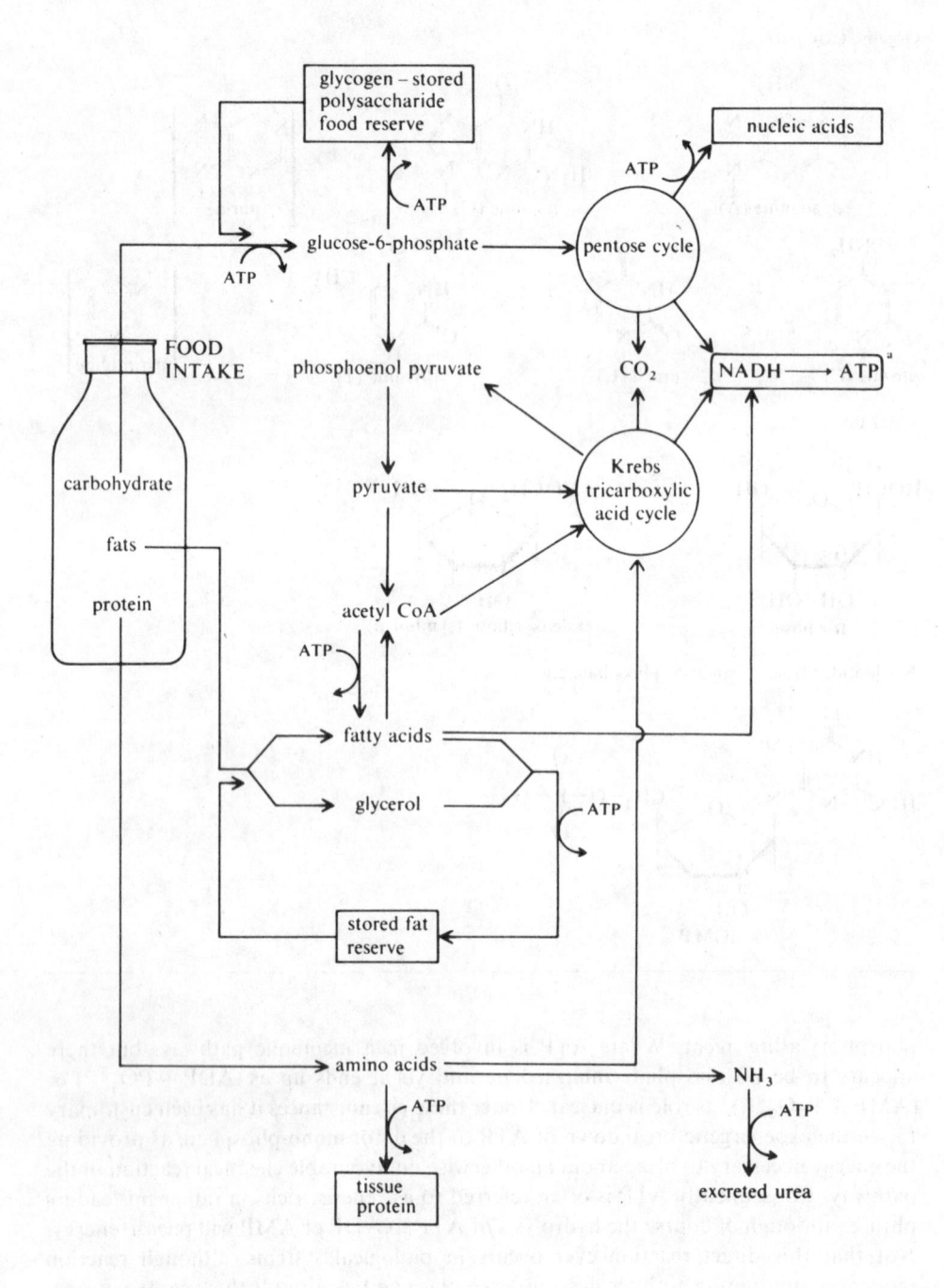

[a] Electron transport chain.

12.1 Phosphorus in life

$$\text{ROH} + \text{ATP}^{4-} \longrightarrow \text{R}-\text{O}-\overset{\overset{\text{O}}{\|}}{\underset{\underset{\text{O}^-}{|}}{\text{P}}}-\text{O}^- + \text{ADP}^{3-} + \text{H}^+ \qquad (2)$$

$$\text{R}-\text{O}-\overset{\overset{\text{O}}{\|}}{\underset{\underset{\text{O}^-}{|}}{\text{P}}}-\text{O}^- + \text{H}_2\text{O} \longrightarrow \text{R}-\text{OH} + \text{PO}_4^{3-} + \text{H}^+ \qquad (3)$$

$$\text{ATP}^{4-} + \text{H}_2\text{O} \longrightarrow \text{ADP}^{3-} + \text{PO}_4^{3-} + 2\text{H}^+ \qquad (4)$$

The standard free energy of hydrolysis, $\Delta G^{\circ\prime}$,* is a convenient measure of the energy stored in a molecule and Table 12.1 lists this quantity for several types of biochemical compounds. It can be seen that ATP is not the best molecule from this point of view. The conditions under which the values of $\Delta G^{\circ\prime}$ were obtained were pH 7.0, temperature 37 °C and in the presence of excess Mg^{2+} ions[3], these being typical of the natural environment. (The values vary with environmental changes[6], but not relative to one another.) The ability of Mg^{2+} to complex with the oxygen atoms of the triphosphate chain of ATP is essential to its role although the exact nature of the interaction has yet to be identified. It is not difficult to imagine the chain behaving as a polydentate ligand even to the extent of completely enfolding the cation. In this way the triphosphate moiety would be protected and perhaps prevented from cyclizing to the stable tricyclophosphate ring $(PO_3^-)_3$.

To speak of the molecules of Table 12.1 as high energy molecules has been vigorously attacked [7,8], and equally vigorously defended [9-11]. Nevertheless energy and its transfer and conversion from one form to another is very important to living things. Mammals use most energy in environment control, i.e. keeping warm and moving about, whereas bacteria use most energy in chemical synthesis since their very existence as a life form depends upon rapid reproduction.

There are two original sources of energy: light and oxygen, O_2. Photosynthesis and oxidative metabolism convert these forms via photosynthetic phosphorylation and oxidative phosphorylation respectively into ATP which is in effect the molecular repository of this original energy. The close link between the reactions of energy metabolism and phosphate metabolism was recognized by the early workers in biochemistry.

The 'high energy bond' concept was originally put forward by Lipman in 1941 following the recognition that ATP was produced when foodstuffs were broken down *in vivo* and that this was then used in synthetic reactions[12]. He went even further than this and suggested the energy was stored in the bond between the terminal phosphate group and the rest of the chain. Although this is obviously an untenable theory it is still common for authors to offer explanations in structural or bonding terms as to why and where ATP is energy rich.† ATP, like the other molecules in

---

* The prime refers to pH being 7.0.
† The molecule H_2 is enormously energy rich as its explosive reaction with oxygen demonstrates. Where is its energy stored? Hardly in a strained structural arrangement or unusual electron distribution within the molecule. The question is meaningless.

Table 12.1 Standard free energies of hydrolysis of some important biochemical compounds taken from ref. 3

Compound	$\Delta G^{o\prime}$ /(kJ mol^{-1})	/(kcal mole^{-1})	Structure
Phosphoenolpyruvate	−62.0	−14.8	(12.9)
Phosphocreatine	−43.1	−10.3	(12.1)
Acetyl phosphate	−42.3	−10.1	$CH_3.CO.O-\overset{\overset{O}{\|}}{\underset{\underset{O^-}{\|}}{P}}-O^-$
Diphosphate	−33.5	−8.0	$P_2O_7^{4-}$
Acetyl CoA	−31.4	−7.5	(12.12)
ATP^{4-} ($\to$ ADP^{3-} + PO$_4^{3-}$)	−30.5	−7.3	(12.2)
ATP^{4-} ($\to$ AMP^{2-} + P$_2$O$_7^{4-}$)	−30.5	−7.3	
Sucrose	−30	−7.0	
Glucose-1-phosphate	−21	−5.0	
Glucose-6-phosphate	−14	−3.3	(eqn. 14)

Table 12.1, is rich in energy *with respect to the products of reaction*, e.g. with respect to the products of hydrolysis. However in the case of phosphoenolpyruvate (12.9) it is not unreasonable to associate some of the energy with an enol→keto rearrangement after the hydrolysis (eqn. 5).

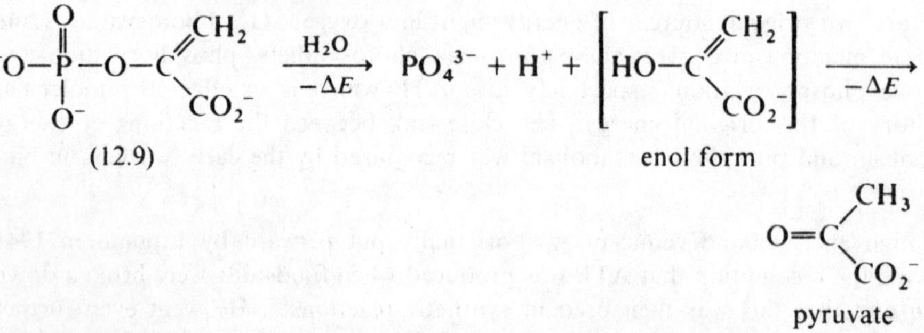

$$^-O-\overset{\overset{O}{\|}}{\underset{\underset{O^-}{\|}}{P}}-O-C\overset{CH_2}{\underset{CO_2^-}{\diagdown}} \xrightarrow[-\Delta E]{H_2O} PO_4^{3-} + H^+ + \left[HO-C\overset{CH_2}{\underset{CO_2^-}{\diagdown}} \right] \xrightarrow{-\Delta E}$$

(12.9)   enol form

$$O=C\overset{CH_3}{\underset{CO_2^-}{\diagdown}} \quad (5)$$

pyruvate

Phosphocreatine has a vital role to play in that it is capable of reacting with ADP to regenerate ATP and because of its chemical stability it is produced and stored for the purpose of meeting an 'emergency' demand for ATP. When resting, the level of ATP in the muscle cell is only about 5×10^{-6} mol per g of muscle, which is enough to last about a second when the muscle is contracted. The two to three second gap between this supply of ATP running out, and before the on-coming supply from accelerated catabolic processes arrives, is bridged by phosphocreatine and reaction (6). This reaction, catalyzed by the enzyme *creatine phosphokinase* is reversible so that when resting the store of phosphocreatine can be regenerated from ATP. (Not all species

12.1 Phosphorus in life

use phosphocreatine as the reagent to regenerate ATP, some use phosphoarginine instead.)

$$\text{phosphocreatine} + \text{ADP} \underset{}{\overset{\text{creatine phosphokinase}}{\rightleftharpoons}} \text{creatine} + \text{ATP} \qquad (6)$$

The ability of phosphocreatine to act as reserve capacity for phosphorylation demonstrates that thermodynamic instability, as revealed by the free energy of hydrolysis, is no guide to reactivity which may be kinetically controlled. Indeed both phosphocreatine and ATP need to be kinetically stable to hydrolysis since they operate in an aqueous medium and yet must not undergo direct hydrolysis which would short-circuit their real functions as phosphorylating agents. Both are stable in water for several days and in this respect phosphocreatine outlasts ATP as we would expect from its behaviour *in vivo* but not from $\Delta G^{\circ\prime}$ values. The hydrolysis of ATP has been studied intensively[13].

Standard free energies of hydrolysis do have important uses in that they can be a signpost to the direction of a reaction and in the case of an equilibrium situation give the equilibrium constant. For instance $\Delta G^{\circ\prime}$ for ribulose-1,5-diphosphate is -13.4 kJ mol^{-1} (-3.2 kcal mole^{-1}) which means that the reaction of ribulose-5-phosphate with ATP (eqn. 7) will have an overall $\Delta G^{\circ\prime}$ of -17.1 kJ mol^{-1} (-30.5 minus -13.4) in the direction of ribulose-1,5-diphosphate. In terms of the equilibrium constant we can use $\Delta G^{\circ\prime} = RT \ln K = -5.71 \log_{10} K$ kJ mol^{-1} to get K which turns out to be *ca* 10^3, i.e. equilibrium (7) lies well to the right hand side.

$$\text{ribulose-5-phosphate} + \text{ATP} \overset{K}{\rightleftharpoons} \text{ribulose-1,5-phosphate} + \text{ADP} \qquad (7)$$

It is now time to turn to details of the chemistry of ATP that we have so far only alluded to: firstly its formation and then its use in biosynthesis and the metabolic pathways to which it is coupled. This last use is one in which ATP still keeps its part in the mechanism secret and the concept of it as an energy supplier seems as realistic as any. How its chemical energy is transformed to mechanical energy in muscle activity, for example, still remains unsolved.

12.1.2 The formation of adenosine triphosphate[14]

The chemical reactions which can produce ATP from ADP are dealt with in Chapter 8 (Table 8.5), where the explanation of the phosphorylating ability of reagents such as phosphoenol pyruvate, acetyl phosphate and phosphocreatine was discussed in terms of the bonding situation existing between the phosphoryl group and the substrate. In this section we shall be looking, not so much at the actual chemical reaction which produces ATP from ADP, but rather at the biochemical set-up which is responsible for the regeneration. In fact in many cases the actual *chemical* reactions are still not known with certainty.

Adenosine triphosphate is produced by two processes which derive their energy originally from non-biological sources* and these are (i) photosynthetic phosphorylation and (ii) oxidative phosphorylation. There are also two secondary processes

---

* Light from the sun is certainly non-biological but the second source, oxygen in the atmosphere, could be regarded as of biological origin.

for forming ATP which use biochemical energy and these are (iii) enzymic formation by glycolysis and (iv) substrate level phosphorylation.

12.1.2.1 Photosynthetic phosphorylation[15-17]

Plant cells contain a solid-state device called a chloroplast which is capable of absorbing quanta of light and using the absorbed energy to promote an electron to an excited state. The electron eventually returns to its lower state, not by emitting a photon, but by a circuitous route in which three important events occur: ATP is formed, nicotinamide adenine dinucleotide-2-phosphate (NADP$^+$) (12.10) is reduced to NADPH and water is oxidized to O_2.

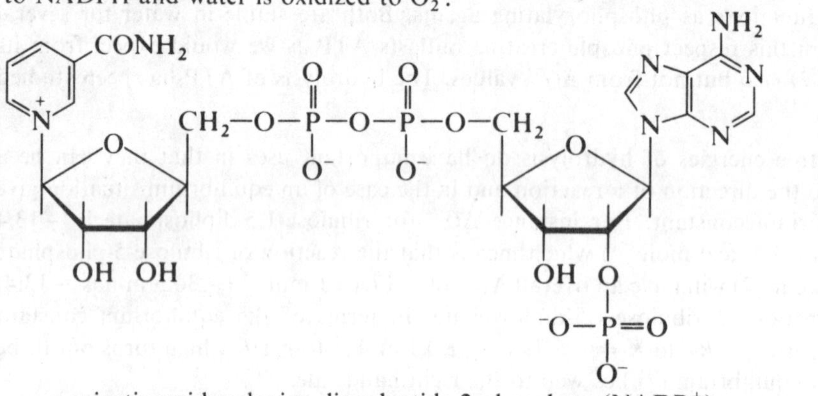

nicotinamide adenine dinucleotide-2-phosphate (NADP$^+$)
(12.10)

The site of photosynthesis in the chloroplast is a pigmented organelle bounded by a double membrane in which the inner membrane is highly convoluted. The light-activated molecules are chlorophylls and there are two interdependent systems one of which produces the ATP and the other the reducing agent NADPH.

Figure 12.3 shows the system in outline only, in fact many of the details are as yet unknown. Light is absorbed by chlorophyll and the excited electron transferred to other molecules which carry out the various reactions although for the sake of simplicity all this has been termed 'chlorophyll'.

The formation of the reducing agent NADPH occurs by the pyridine ring picking up a proton and two electrons to form:

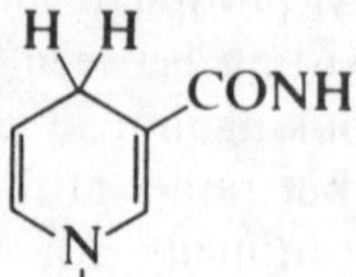

This compound has an important role in the second method by which ATP is produced – oxidative phosphorylation.

12.1.2.2 Oxidative phosphorylation[18-20]

Oxidative metabolism is an important source of energy (eqn. 8), and is the most important source of ATP in aerobic organisms. Cells that use oxidative reactions to produce energy also have a solid-state device, called the mitochondrion, to 'trap'

12.1 Phosphorus in life

Figure 12.3 Coupled photosystems.

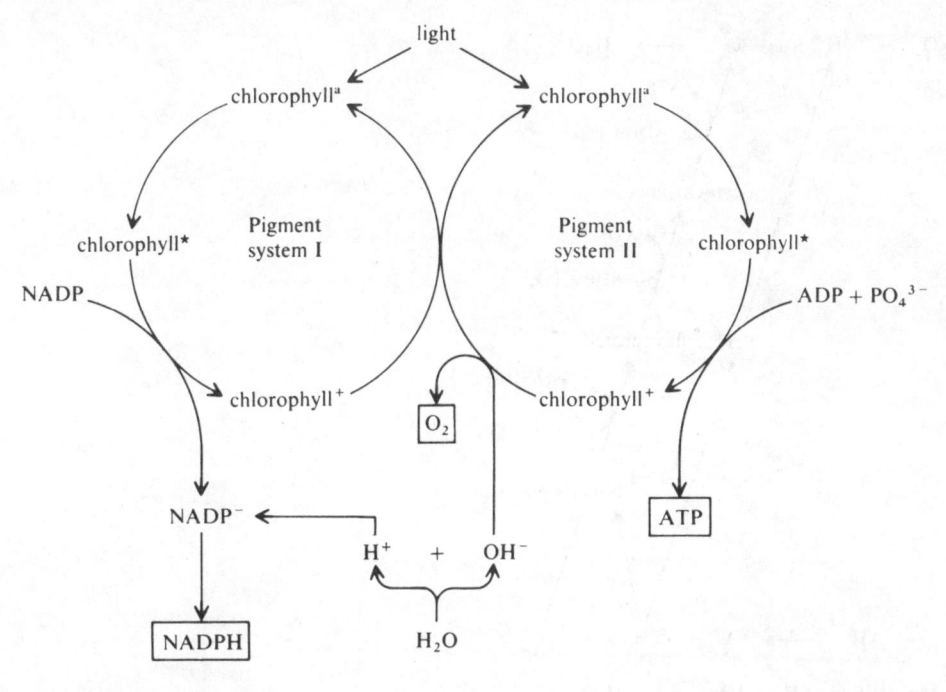

[Diagram: Two coupled cycles labeled "Pigment system I" (left) and "Pigment system II" (right), driven by light. Pigment system I: chlorophyll^a → chlorophyll* (with NADP input) → chlorophyll$^+$ → NADP$^-$ → NADPH. Pigment system II: chlorophyll^a → chlorophyll* (with ADP + PO$_4^{3-}$ input) → chlorophyll$^+$ → ATP. Between the systems: O$_2$ released, H$_2$O splits into H$^+$ + OH$^-$.]

^{a} See text.

---

this energy by the production of ATP. A mitochondrion is a small (1–2 μm) structure surrounded by a double membrane of which the inner one is responsible for the processes associated with oxidative phosphorylation. The number of mitochondria in a cell is a function of the cellular requirement for energy.

$$AH_2 + \tfrac{1}{2}O_2 \longrightarrow A + H_2O \qquad \Delta G° \simeq 125\text{--}210 \text{ kJ mol}^{-1} \qquad (8)$$

Although the overall reaction for oxidation of the substrate is given by equation (8) the process involves many steps. Figure 12.4 illustrates the principal steps involved and how these are coupled to the formation of ATP. The key step in the process whereby electrons (and protons) are transferred to oxygen, in its reduction to water, is the formation of NADH. Nicotinamide adenine dinucleotide (NAD$^+$) is the forerunner of (12.10), without the 2-phosphate substituent; NADH is its reduced form. The key route in Fig. 12.4 is electron transport and not every link in the chain is clearly understood. Proton transport represents no problem since the system is in an aqueous medium.

The electron transport chain can be seen as a series of oxidation–reduction reactions in which there is a large potential drop $\Delta E'_0$ associated with three of the steps and it is these three steps which have the ability to generate a phosphorylating species capable of phosphorylating ADP to give ATP, in other words to give a reaction sequence which amounts to the regeneration of ATP from ADP and PO$_4^{3-}$.

Figure 12.4 Oxidative phosphorylation.

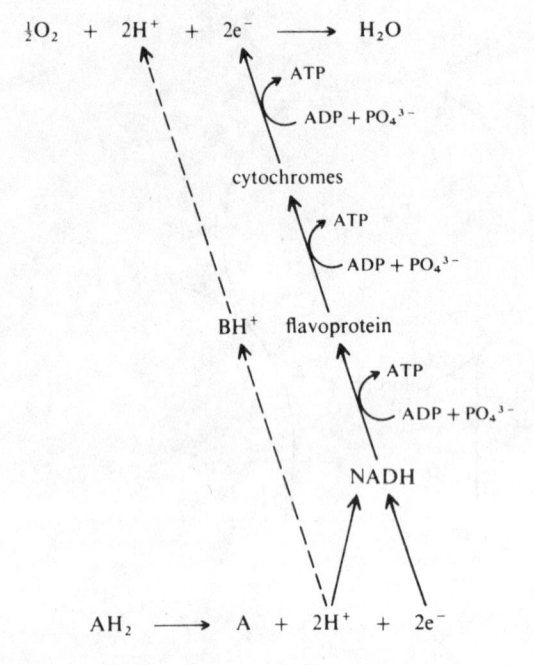

$$\tfrac{1}{2}O_2 + 2H^+ + 2e^- \longrightarrow H_2O$$

```
                    ↑ ATP
                    ↓ ADP + PO4^3-
              cytochromes
                    ↑ ATP
                    ↓ ADP + PO4^3-
   BH+      flavoprotein
                    ↑ ATP
                    ↓ ADP + PO4^3-
                  NADH
                    ↑  ↑

   AH2  ⟶  A + 2H+ + 2e-
```

Net reaction: $AH_2 + \tfrac{1}{2}O_2 \longrightarrow A + H_2O$

These three steps lie between the production of NADH and the final step producing H_2O. Basically the overall process amounts to eqn. (9) which can be split into the half-reactions (10) and (11). The relationship between $\Delta E'_0$ and the free energy for the overall process is given by eqn. (12) which gives a value for $\Delta G^{o\prime}$ of -219 kJ mol^{-1} (-52.4 kcal mole^{-1}).

$$NADH + H^+ + \tfrac{1}{2}O_2 \longrightarrow NAD^+ + H_2O \tag{9}$$

$$NAD^+ + 2H^+ + 2e^- = NADH + H^+ \qquad E'_0 = -0.320 \text{ V} \tag{10}$$

$$\tfrac{1}{2}O_2 + 2H^+ + 2e^- = H_2O \qquad E'_0 = 0.816 \text{ V} \tag{11}$$

$$\Delta G^{o\prime} = n\mathfrak{F}\Delta E'_0 \tag{12}$$

Of this amount of free energy 3×-30.5 or -91.5 kJ mol^{-1} (-21.9 kcal mole^{-1}) is recouped as ATP representing an 'efficiency' of around 40 %.

What oxidative phosphorylation means in chemical terms is that whenever a molecule of NADH is produced this will eventually lead to the production of three molecules of ATP. The oxidation of pyruvate is where most of the NADH comes from, and this brings us to the tricarboxylic, or Krebs cycle[21,22] which is the final pathway of oxidation of carbohydrate, fat and many amino acids. Figure 12.5 outlines the intermediates of the cycle. Pyruvate is fed into the system via its reaction with coenzyme A, symbol CoA—SH (12.11), which becomes acetylated at the —S—H bond to form

12.1 Phosphorus in life

Figure 12.5 The tricarboxylic acid (Krebs) cycle.

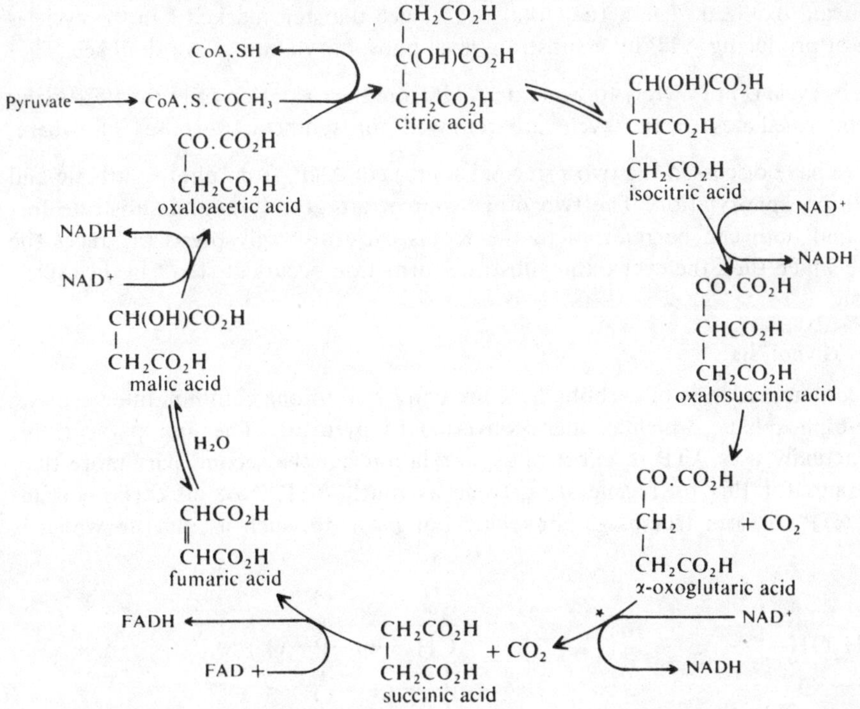

[Diagram of the Krebs cycle showing the following transformations:

Pyruvate → CoA.S.COCH₃ (with CoA.SH)

CoA.S.COCH₃ combines to form citric acid:
CH₂CO₂H
|
C(OH)CO₂H
|
CH₂CO₂H
citric acid

oxaloacetic acid:
CO.CO₂H
|
CH₂CO₂H

⇌ isocitric acid:
CH(OH)CO₂H
|
CHCO₂H
|
CH₂CO₂H

→ (NAD⁺ → NADH) oxalosuccinic acid:
CO.CO₂H
|
CHCO₂H
|
CH₂CO₂H

→ α-oxoglutaric acid + CO₂:
CO.CO₂H
|
CH₂
|
CH₂CO₂H

→ (NAD⁺ → NADH)* succinic acid + CO₂:
CH₂CO₂H
|
CH₂CO₂H

→ (FAD → FADH) fumaric acid:
CHCO₂H
‖
CHCO₂H

→ (H₂O) malic acid:
CH(OH)CO₂H
|
CH₂CO₂H

→ (NAD⁺ → NADH) back to oxaloacetic acid]

* This step is discussed in detail in the text.

CoA—S—COCH₃ (12.12), (eqn. 13). Note that a molecule of NADH is produced in the process.

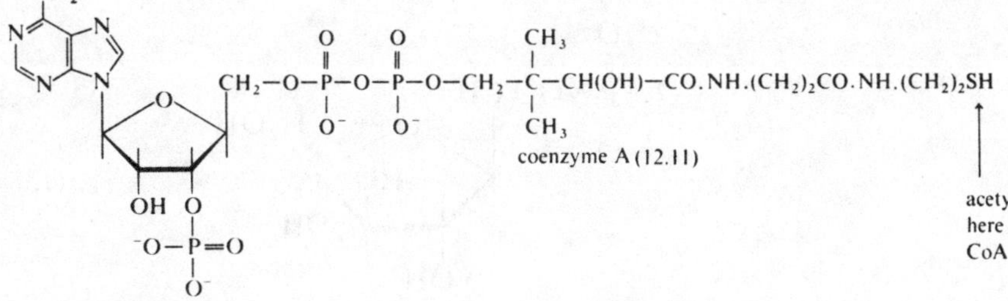

[Structure of coenzyme A (12.11): adenine-ribose-phosphate backbone with CH₂—O—P(O)(O⁻)—O—P(O)(O⁻)—O—CH₂—C(CH₃)₂—CH(OH)—CO.NH.(CH₂)₂CO.NH.(CH₂)₂SH, and an additional O—P(=O)(O⁻)O⁻ group on the ribose]

acetylation here gives CoA.S.COCH₃

$$\text{CH}_3.\text{CO}.\text{CO}_2^- + \text{CoA}.\text{SH} + \text{NAD}^+ \longrightarrow \text{CoA}.\text{S}.\text{COCH}_3 + \text{CO}_2 + \text{NADH}$$
pyruvate (13)

Acetyl coenzyme A is a very reactive acetylating agent and like ATP it appears in many biochemical processes; CoA.S.COCH₃ is to acetylation what ATP is to phosphorylation, and both have been selected by nature as the best reagent for the job.

The cycle converts three molecules of NAD$^+$ to NADH plus a molecule of FAD$^+$ to FADH*. These in turn convert a total of 11 ADP molecules into ATP and including the pyruvate step prior to the cycle (eqn. 13) makes a grand total of 14 ATP molecules per pyruvate oxidized. The actual total is 15 since the step marked * in the cycle is capable of producing ATP by a substrate level phosphorylation as we shall see.

The Krebs cycle is not merely to generate ATP from the oxidation of pyruvate – many of the intermediates of the cycle are removed for synthetic purposes elsewhere.

Above we have discussed the two principal sources of ATP, from photosynthetic and oxidative phosphorylation. The two other sources are glycolysis and substrate formation and both can be referred to the Krebs cycle also: glycolysis produces the pyruvate which fuels the cycle, and substrate formation occurs at step * in the cycle's operation.

12.1.2.3 Glycolysis

All the common sources of carbohydrate are converted to one common intermediate, fructose-6-phosphate, which is then converted to pyruvate. The first part of this change actually uses ATP to effect phosphorylation but the second part more than compensates for this by regenerating twice as much ATP. Not all carbohydrates require ATP to form fructose-6-phosphate but most do, such as glucose which is

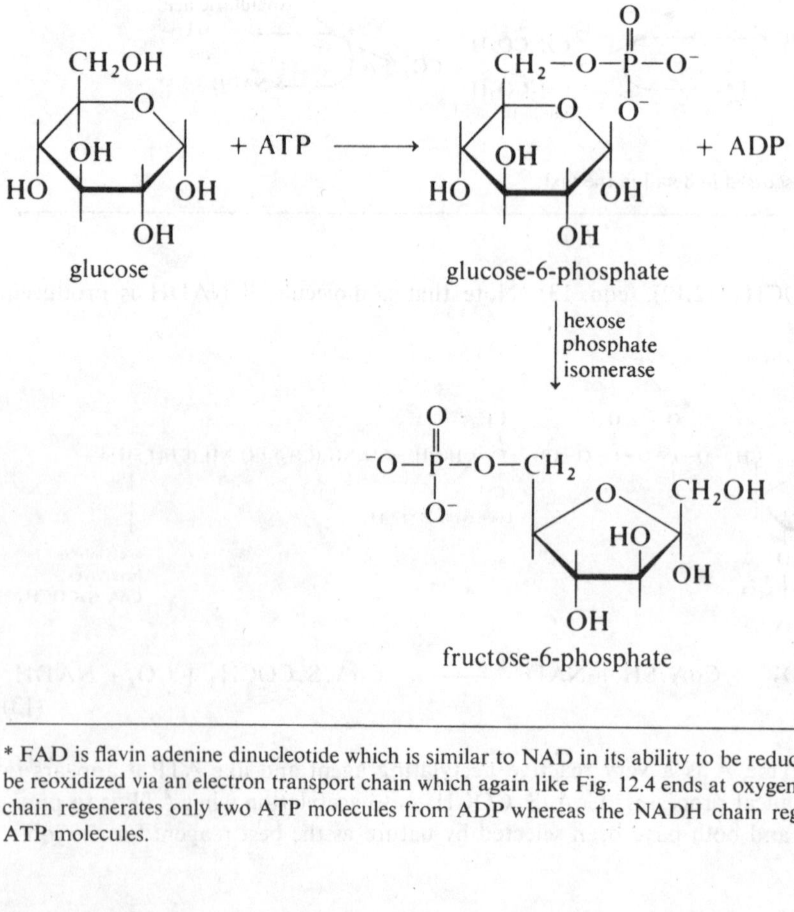

glucose + ATP → glucose-6-phosphate + ADP (14)

↓ hexose phosphate isomerase

fructose-6-phosphate (15)

---

* FAD is flavin adenine dinucleotide which is similar to NAD in its ability to be reduced and then to be reoxidized via an electron transport chain which again like Fig. 12.4 ends at oxygen. However this chain regenerates only two ATP molecules from ADP whereas the NADH chain regenerates three ATP molecules.

12.1 Phosphorus in life

converted first to glucose-6-phosphate and is then isomerized to fructose-6-phosphate by the enzyme *hexose phosphate isomerase*, (eqns. 14 and 15). A second molecule of ATP now phosphorylates fructose-6-phosphate to form fructose-1,6-diphosphate which undergoes cleavage, by an enzyme *aldolase*, to give two products (12.12) and (12.13) which can be interconverted by another enzyme, *triose phosphate isomerase* (eqns. 16 and 17).

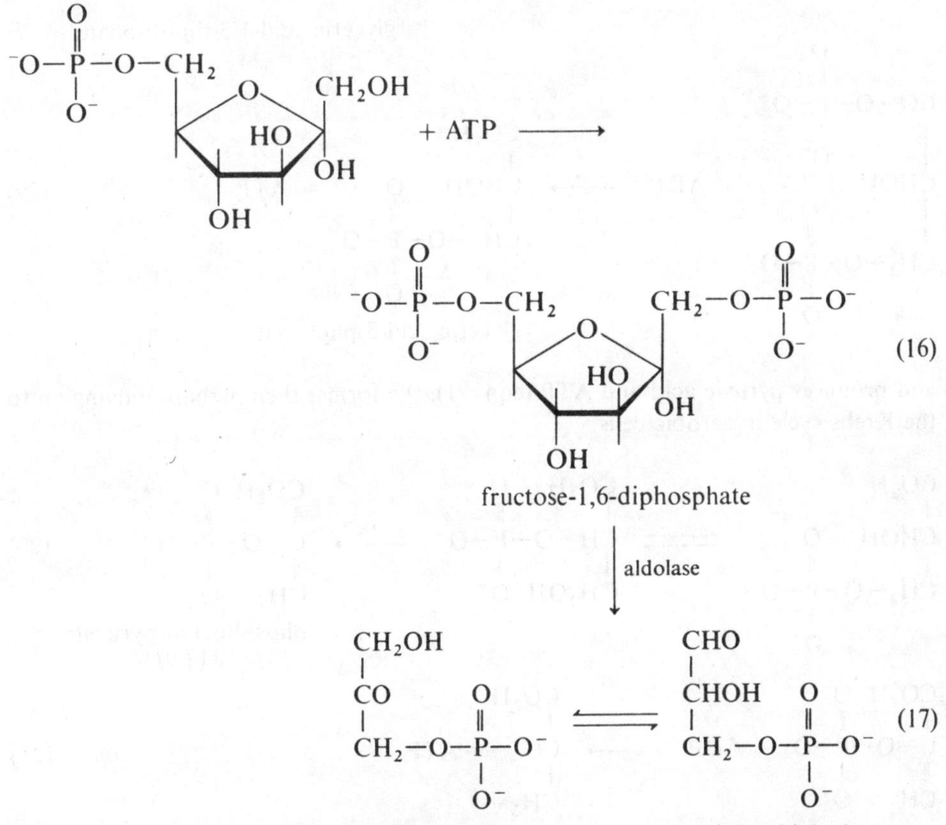

$$\text{[glucose-6-phosphate]} + ATP \longrightarrow$$

$$\text{fructose-1,6-diphosphate} \quad (16)$$

$$\downarrow \text{aldolase}$$

$$\begin{array}{cc} CH_2OH & CHO \\ | & | \\ CO & CHOH \\ | & | \\ CH_2-O-P(=O)(O^-)O^- & CH_2-O-P(=O)(O^-)O^- \end{array} \quad (17)$$

dihydroxyacetone phosphate (12.12)   glyceraldehyde 3-phosphate (12.13)

The glyceraldehyde phosphate (12.13) is oxidized by NAD^+ in the presence of PO_4^{3-} and forms glyceric acid-1,3-diphosphate (eqn. 18) which is capable of phosphorylating ADP to generate ATP (eqn. 19). So far we have put in two molecules of ATP (eqns. 14 and 16) and regenerated two (eqn. 19 × 2). The final steps to pyruvate now generate more ATP via the isomerization of glyceric acid-3-phosphate (eqn. 20) and its dehydration to phosphoenol pyruvate (12.9) the compound which tops the list of high energy phosphorylating agents (Table 12.1). This will transfer its phosphate group to ADP

$$\begin{array}{c} \text{CHO} \\ | \\ \text{CHOH} \quad \text{O} \\ | \quad \quad \| \\ \text{CH}_2-\text{O}-\text{P}-\text{O}^- \\ | \\ \text{O}^- \end{array} + \text{NAD}^+ + \text{PO}_4^{3-} \longrightarrow \begin{array}{c} \text{O} \\ \| \\ \text{CO}-\text{O}-\text{P}-\text{O}^- \\ | \quad \quad | \\ \text{CHOH} \quad \text{O}^- \\ | \quad \quad \text{O} \\ | \quad \quad \| \\ \text{CH}_2-\text{O}-\text{P}-\text{O}^- \\ | \\ \text{O}^- \end{array} + \text{NADH} \quad (18)$$

glyceric acid-1,3-diphosphate

$$\begin{array}{c} \text{O} \\ \| \\ \text{CO}-\text{O}-\text{P}-\text{O}^- \\ | \quad \quad | \\ \quad \quad \text{O}^- \\ \text{CHOH} \\ | \quad \quad \text{O} \\ | \quad \quad \| \\ \text{CH}_2-\text{O}-\text{P}-\text{O}^- \\ | \\ \text{O}^- \end{array} + \text{ADP} \longrightarrow \begin{array}{c} \text{CO}_2\text{H} \\ | \\ \text{CHOH} \quad \text{O} \\ | \quad \quad \| \\ \text{CH}_2-\text{O}-\text{P}-\text{O}^- \\ | \\ \text{O}^- \end{array} + \text{ATP} \quad (19)$$

glyceric acid-3-phosphate

and produces pyruvic acid and ATP (eqn. 21), the former then perhaps moving on to the Krebs cycle in aerobic cells.

$$\begin{array}{c} \text{CO}_2\text{H} \\ | \\ \text{CHOH} \quad \text{O} \\ | \quad \quad \| \\ \text{CH}_2-\text{O}-\text{P}-\text{O}^- \\ | \\ \text{O}^- \end{array} \rightleftharpoons \begin{array}{c} \text{CO}_2\text{H} \quad \text{O} \\ | \quad \quad \| \\ \text{CH}-\text{O}-\text{P}-\text{O}^- \\ | \quad \quad | \\ \text{CH}_2\text{OH} \quad \text{O}^- \end{array} \xrightarrow{-\text{H}_2\text{O}} \begin{array}{c} \text{CO}_2\text{H} \quad \text{O} \\ | \quad \quad \| \\ \text{C}-\text{O}-\text{P}-\text{O}^- \\ \| \quad \quad | \\ \text{CH}_2 \quad \text{O}^- \end{array} \quad (20)$$

phosphoenol pyruvate
(12.9)

$$\begin{array}{c} \text{CO}_2\text{H} \quad \text{O} \\ | \quad \quad \| \\ \text{C}-\text{O}-\text{P}-\text{O}^- \\ \| \quad \quad | \\ \text{CH}_2 \quad \text{O}^- \end{array} + \text{ADP} \longrightarrow \begin{array}{c} \text{CO}_2\text{H} \\ | \\ \text{CO} \\ | \\ \text{CH}_3 \end{array} + \text{ATP} \quad (21)$$

pyruvic acid

Under anaerobic conditions the pyruvic acid or pyruvate from eqn. (21) reacts with NADH (from eqn. 18) to form lactic acid and regenerate NAD$^+$ (eqn. 22). Whatever

$$\begin{array}{c} \text{CO}_2\text{H} \\ | \\ \text{CO} \\ | \\ \text{CH}_3 \end{array} + \text{NADH} \longrightarrow \begin{array}{c} \text{CO}_2\text{H} \\ | \\ \text{CHOH} \\ | \\ \text{CH}_3 \end{array} + \text{NAD}^+ \quad (22)$$

lactic acid

the final destination of the pyruvic acid its formation from glucose (or hexoses) results in the net overall production of 2ATP per hexose unit.

12.1 Phosphorus in life

12.1.2.4 Substrate formation of ATP

In the Krebs cycle the step marked ★ (Fig. 12.5) in which α-oxoglutaric acid is converted to succinic acid is much more complex than it appears. As well as involving $NAD^+ \rightarrow NADH$ this sector involves the intermediate formation of the succinyl derivative of coenzyme A (eqn. 23, Fig. 12.6). This reacts with an enzyme to form succinic acid and the enzyme ·CoA which picks up a phosphate, PO_4^{3-}, prior to phosphorylating $GDP \rightarrow GTP$, which in turn phosphorylates $ADP \rightarrow ATP$. The result of all this is the production of an extra ATP molecule per turn of the Krebs cycle.

Figure 12.6 Substrate formation of adenosine triphosphate.

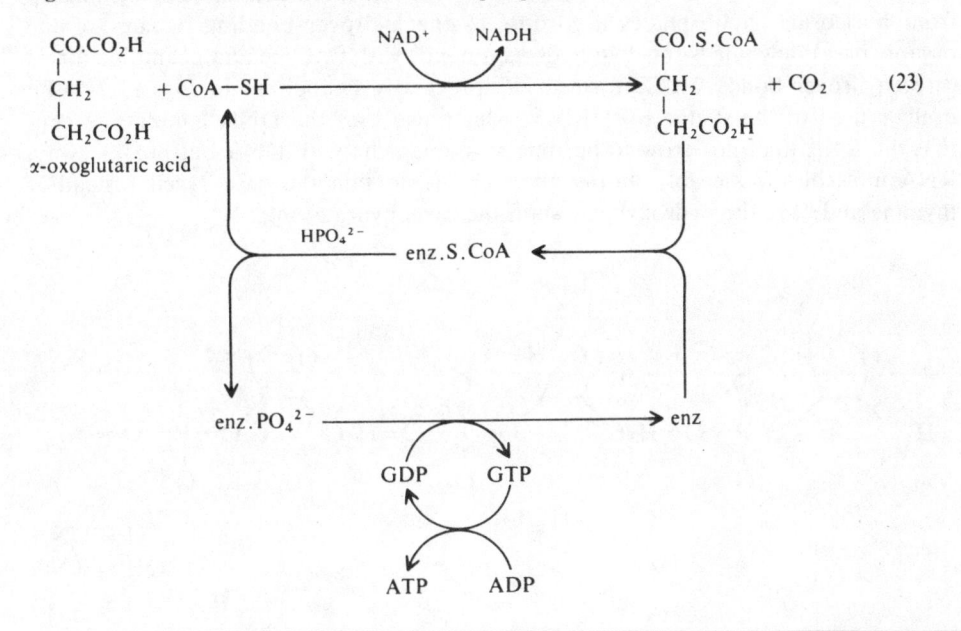

$$\begin{array}{c} CO.CO_2H \\ | \\ CH_2 \\ | \\ CH_2CO_2H \end{array} + CoA-SH \xrightarrow{NAD^+ \quad NADH} \begin{array}{c} CO.S.CoA \\ | \\ CH_2 \\ | \\ CH_2CO_2H \end{array} + CO_2 \quad (23)$$

with intermediates: HPO_4^{2-}, enz.S.CoA, enz.PO_4^{2-} → enz, coupled cycles GDP ⇌ GTP and ATP ⇌ ADP.

These then are the ways in which ATP is generated, or rather regenerated, from ADP. We now turn to the ways in which the cell uses ATP. To clarify the discussion we shall distinguish two kinds of process in which ATP plays an important part: biosynthesis and metabolic activity. In biosynthesis ATP may itself become part of the macromolecule being synthesized (as in RNA) or, as is more likely, it acts as a phosphorylating agent in the production of phosphate intermediates. The mechanism of phosphorylation was discussed in Chapter 8 where it was proposed that ATP, like other phosphate monoesters and pyrophosphates, might react via an $S_N1(P)$ mechanism which involves the very reactive metaphosphate intermediate $[PO_3^-]$. Evidence for this species is indirect but is well established, and it seems reasonable to suppose that ATP may owe its special ability as a phosphorylating agent to its ability to react through this super-active fragment. The reaction can be seen as an $S_N2(P)$ process promoted by Mg^{2+}, i.e. electrophilic catalysis.

The participation of ATP in many metabolic pathways may be as a phosphorylating agent, but its mode of operation is often unclear except that it serves in some way to 'drive' the process in the right direction. In this role ATP 'moves in a mysterious way its wonders to perform' but there can be no doubt that some of the reactions which it promotes seem little short of miraculous, e.g. nitrogen fixation.

12.1.3 Biosynthesis

All the nucleoside triphosphates are involved in the construction of DNA or RNA[24]. The backbone of both type of macromolecule are very similar consisting of carbohydrate moieties joined between the 5' and 3' position by phosphate units, (12.14). DNA as the double helix is found only in the cell nucleus and it is synthesized by the separation of the two strands which are held together by hydrogen-bonds. Each strand then acts as the template for the synthesis of the matching strand, building this up from nucleoside triphosphates according to the hydrogen-bonding dictates of the organic bases: adenine forms two hydrogen-bonds with thymine and guanidine forms three hydrogen-bonds with cytosine. The process is symbolized in Fig. 12.7. ATP itself is used in the synthesis of RNA, which also uses the DNA template, except that the RNA does not grow to become a double helix with DNA but breaks away. RNA molecules, noticeably shorter than DNA, contain the base uracil instead of thymine and have the hydroxyl group on the carbohydrate ring.

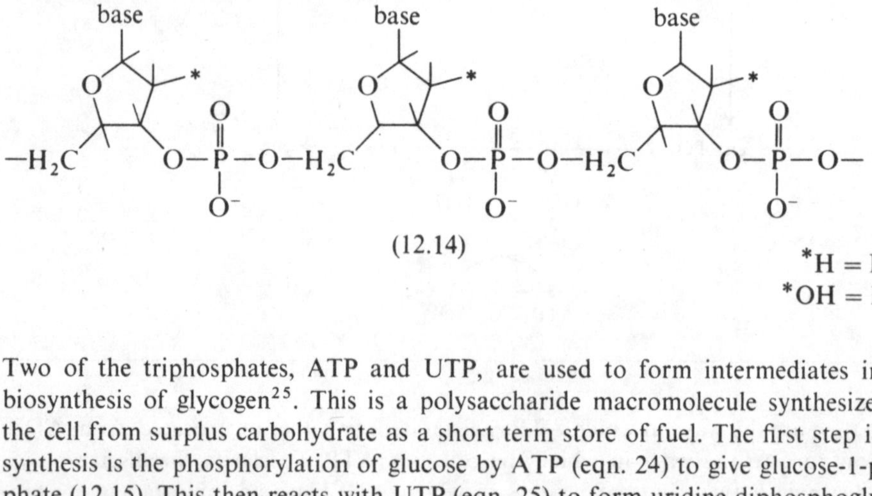

(12.14)

*H = DNA
*OH = RNA

Two of the triphosphates, ATP and UTP, are used to form intermediates in the biosynthesis of glycogen[25]. This is a polysaccharide macromolecule synthesized by the cell from surplus carbohydrate as a short term store of fuel. The first step in the synthesis is the phosphorylation of glucose by ATP (eqn. 24) to give glucose-1-phosphate (12.15). This then reacts with UTP (eqn. 25) to form uridine diphosphoglucose

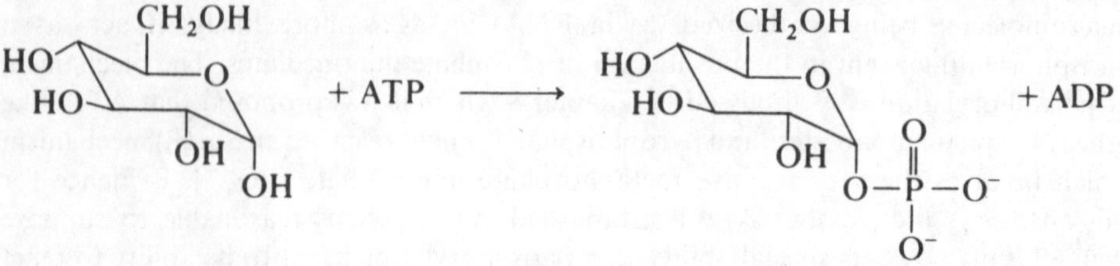

(12.15)   (24)

(12.16), which with another glucose unit eliminates UDP to form the disaccharide (12.17), eqn. (26). The process can be repeated at the marked (←) hydroxyl group to form the polysaccharide which in glycogen can have 50 000 sugar units. ATP's contri-

12.1 Phosphorus in life

Figure 12.7 Symbolic representation of DNA synthesis.

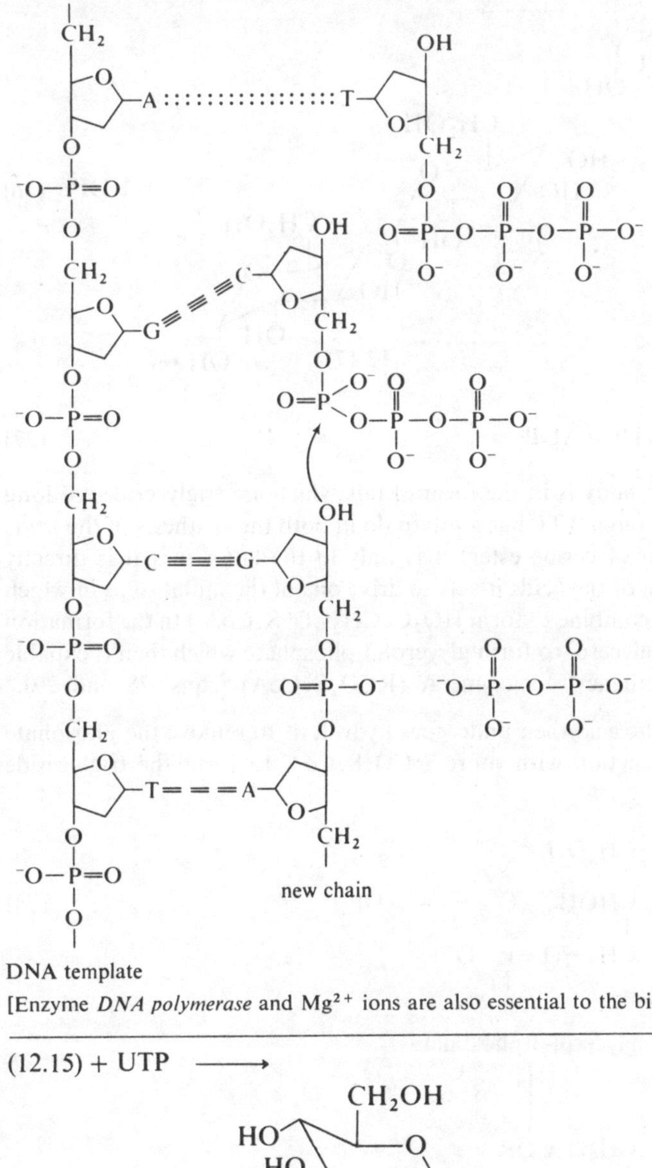

DNA template
[Enzyme *DNA polymerase* and Mg^{2+} ions are also essential to the biosynthesis.]

(12.15) + UTP ⟶

$P_4O_7^{4-}$ + (12.16)   (25)

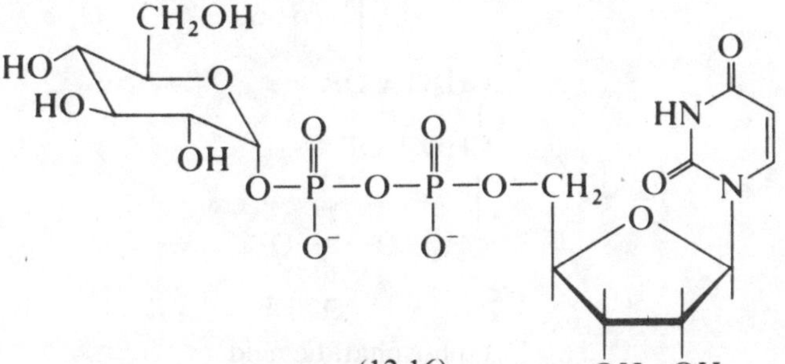

bution to this process is not only in reaction (24) but it also serves to regenerate the UTP for reaction (25) from the UDP by-product of reaction (26), eqn. (27).

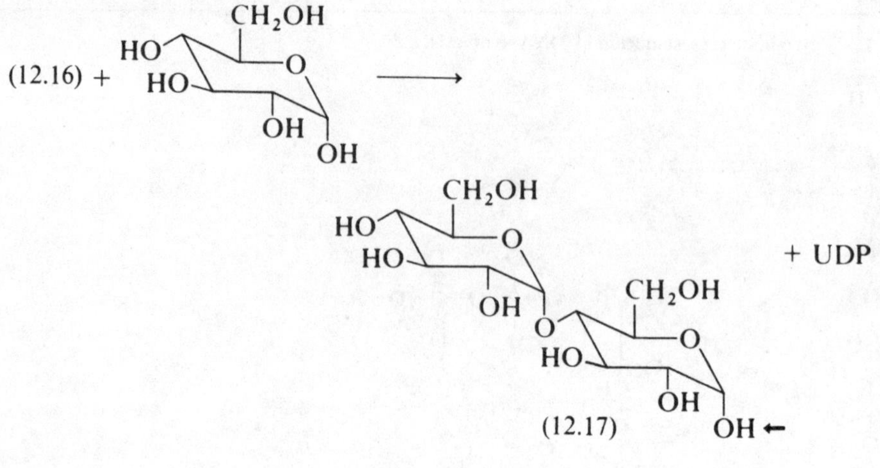

(12.16) + [glucose structure] ⟶ [disaccharide structure] + UDP (26)

(12.17)

$$\text{UDP} + \text{ATP} \longrightarrow \text{UTP} + \text{ADP} \qquad (27)$$

The long term store of the body is in the form of fats which are triglycerides of long chain fatty acids[26,27]. Although ATP has a job to do in both the synthesis of the acids themselves as well as their glyceride esters it is only in the latter that it is directly involved. (In the formation of the acids it acts to drive one of the initial steps in which $CH_3CO.S.CoA$ and CO_2 combine to form $HO_2C.CH_2CO.S.CoA$.) In the formation of the esters it reacts with glycerol to form glycerol-3-phosphate which then is capable of forming the diester with acyl coenzyme A (RCO.S.CoA) (eqns. 28 and 29).*

The product, L-phosphatidic acid then undergoes hydrolysis to remove the phosphate group prior to the final reaction with more RCO.S.CoA to form the triglyceride (eqn. 30).

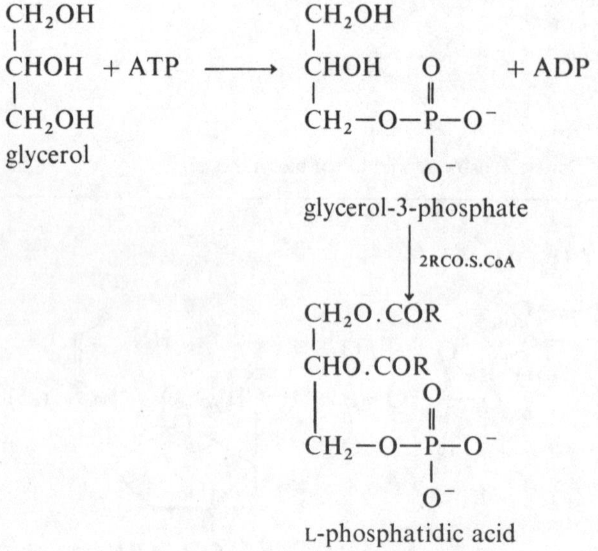

$$\begin{array}{l}\text{CH}_2\text{OH}\\|\\\text{CHOH}\\|\\\text{CH}_2\text{OH}\\\text{glycerol}\end{array} + \text{ATP} \longrightarrow \begin{array}{l}\text{CH}_2\text{OH}\\|\\\text{CHOH}\\|\\\text{CH}_2-\text{O}-\overset{\text{O}}{\underset{|}{\overset{\|}{\text{P}}}}-\text{O}^-\\\phantom{\text{CH}_2-\text{O}-}\text{O}^-\\\text{glycerol-3-phosphate}\end{array} + \text{ADP} \qquad (28)$$

$$\downarrow 2\text{RCO.S.CoA}$$

$$\begin{array}{l}\text{CH}_2\text{O.COR}\\|\\\text{CHO.COR}\\|\\\text{CH}_2-\text{O}-\overset{\text{O}}{\underset{|}{\overset{\|}{\text{P}}}}-\text{O}^-\\\phantom{\text{CH}_2-\text{O}-}\text{O}^-\\\text{L-phosphatidic acid}\end{array} \qquad (29)$$

---

* Glycerol-3-phosphate can also be generated from dihydroxyacetone phosphate and NADH.

12.1 Phosphorus in life

$$\begin{array}{c} CH_2O.COR \\ | \\ CHO.COR \\ | \\ CH_2-O-\underset{|}{\overset{O}{\overset{\|}{P}}}-O^- \\ O^- \end{array} + H_2O \longrightarrow \begin{array}{c} CH_2O.COR \\ | \\ CHO.COR \\ | \\ CH_2OH \end{array} \xrightarrow{RCO.S.CoA} \begin{array}{c} CH_2O.COR \\ | \\ CHO.COR \\ | \\ CH_2O.COR \end{array} \quad (30)$$

Not all syntheses are of large molecules. The synthesis of urea, NH_2CONH_2, also requires ATP[28]. Urea is synthesized by the liver in order to prevent the build-up of NH_3 which is a by-product of amino acid processes. The urea is subsequently excreted. The first step in the synthesis is reaction (31) in which carbamyl phosphate is prepared.

$$NH_3 + CO_2 + ATP \xrightarrow{ATP \quad ADP + PO_4^{3-}} H_2NCO-O-\underset{|}{\overset{O}{\overset{\|}{P}}}-O^- + ADP \quad (31)$$

This reaction is extremely complex and not clearly understood. The carbamyl phosphate then reacts with ornithine (12.18) to eliminate phosphate and form citrulline (12.19), eqn. (32). Although citrulline is obviously related to urea there are several steps between the two, including one with aspartic acid which also involves ATP; in this case the ATP is converted to AMP and $P_2O_7^{4-}$.

$$NH_2CH_2CH_2CH(NH_2)CO_2H + NH_2CO-O-\underset{|}{\overset{O}{\overset{\|}{P}}}-O^-$$
$$(12.18) \qquad\qquad\qquad\qquad O^-$$

$$\longrightarrow NH_2CONHCH_2CH_2CH(NH_2)CO_2H + PO_4^{3-} \quad (32)$$
$$(12.19)$$

The above syntheses involve ATP as a phosphorylating agent as well as making use of it as an energy supply. Many other biosyntheses employ it only in the latter capacity. One such synthesis is that of the organic bases pyrimidine[29] and purine[30]. Indeed since adenine is a purine derivative we have the paradoxical situation of needing ATP in order to synthesize itself – a variant of which came first the chicken or the egg? The major steps of the synthesis of AMP and GMP are outlined in Fig. 12.8. It is also rather curious that GTP is needed to make AMP and ATP is required to complete GMP. Several of the steps in Fig. 12.8 involve the elimination of water between two organic compounds and the role of ATP and GTP in these reactions can be seen as that of dehydrating agents.

12.1.4 Metabolic activation by adenosine triphosphate

We have already seen how the dissociation of ATP to ADP and PO_4^{3-} or to AMP and $P_2O_7^{4-}$ is used to 'drive' energetically unfavourable metabolic changes. One such case is its coupling to the initial step in the oxidation of fatty acids, eqn. (33). The

$$RCH_2CH_2CO_2H + CoA.SH \xrightarrow{ATP \quad AMP + P_2O_7^{4-}} RCH_2CH_2CO.S.CoA \quad (33)$$
$$\text{fatty acids} \qquad\qquad\qquad\qquad (12.20)$$

Figure 12.8 Biosynthesis of AMP and GMP.

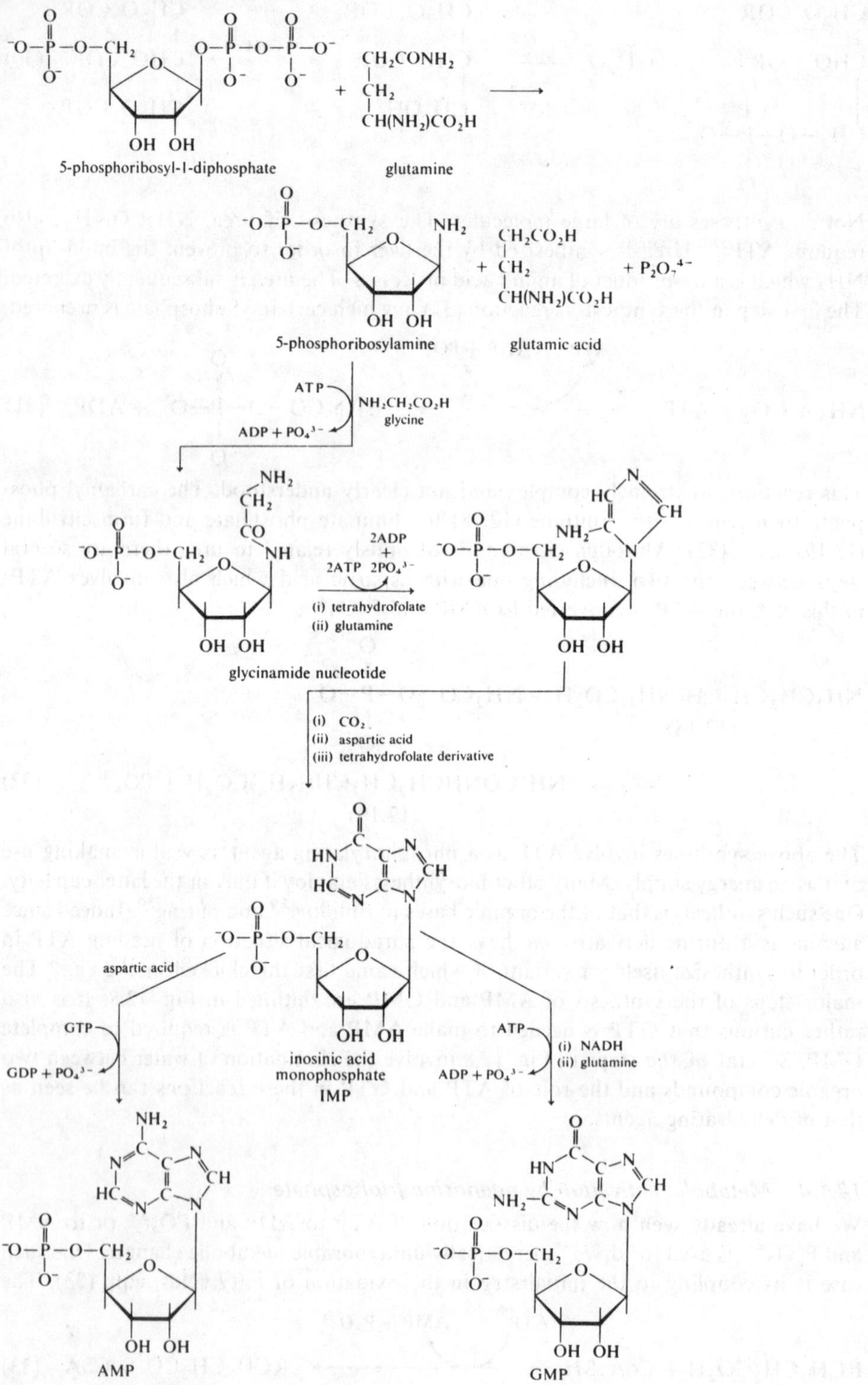

12.1 Phosphorus in life

product (12.20) from this reaction is eventually converted to $R.CO.CH_2.CO.S.CoA$ which with more coenzyme A gives $R.CO.S.CoA$ and $CH_3.CO.S.CoA$. The fatty acid is thus two carbon atoms shorter in length and the process is repeated until it is all broken down into acetyl coenzyme A which then is fed into the Kreb's cycle. [The pyrophosphate of eqn. (33) is hydrolyzed by *pyrophosphatases* present in the tissue.]

The most remarkable process from an energy point of view is the fixation of nitrogen[31]. Nitrogen gas is renowned for its chemical inertness and yet ultimately it is the source of all nitrogen in living things, and the process by which nitrogen gas is converted into ammonia is called nitrogen fixation. The nodule-forming bacteria *Rhizobia* which grow in a symbiotic relationship with leguminous plants is the most important agent in fixing nitrogen. Other forms such as blue-green algae can also fix nitrogen. How the process works is unknown but it seems likely that the N_2 will attach itself as a ligand to the iron or molybdenum complexes which are present. The formation of labile N_2 complexes with transition metals is now an active field of research[32] in which the ultimate aim is a low temperature synthesis of ammonia for fertilizers to replace the current high temperature, energetically wasteful, methods. How the bacteria and algae perform this remarkable feat is still not clear except that ATP and ferredoxin[33] are intimately involved. The cycle is given in Fig. 12.9. Cells fixing N_2 do not accumulate NH_3, which is slightly toxic in moderate concentration, but quickly convert it to other compounds.

Figure 12.9 Nitrogen fixation.

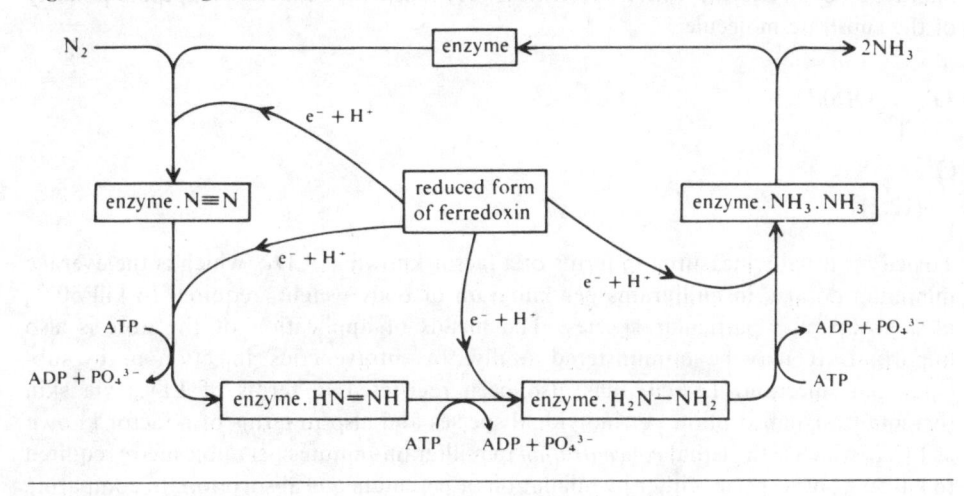

The conversion of the chemical energy of ATP to the mechanical energy of muscular action is yet another use that is made of ATP.

It has been shown how phosphorylation is very important to the proper working of the living cell and how one molecule in particular, ATP, has been singled out by nature to perform this task. We are still far from a complete understanding of why ATP has been so favoured. Indeed the necessity for the triphosphate link rather than the di- or pyrophosphate link is not clear; nor is the role of Mg^{2+} in the process understood.

Clearly the cell treats the phosphate ion as a friend and has no need to protect itself against phosphorus from the environment, since it is only found as phosphate. Never having the need to distinguish a particular type of phosphorylating agent there is no defence against infiltration by the sort which will phosphorylate but then cannot be removed. Minute amounts of such reagents can irreversibly block essential metabolic pathways and the organism quickly ceases to function and dies a painful death. We now turn to such reagents.

12.2 Phosphorus in death: pesticides and chemical warfare agents

Since phosphorus in the form of its phosphate esters, particularly ADP and ATP, plays such a vital role in the chemistry of life, it is perhaps somewhat surprising to learn that some phosphorus compounds are also vehicles of death. As we shall see later, however, it is the ability of particular phosphorus compounds to mimic certain, naturally occurring carboxylic esters and hence inhibit essential enzymes (e.g. acetylcholinesterase) which is the origin of their toxicity.

The general structure of organophosphorus poisons is represented by (12.21) where G^1 and G^2 are groups which are difficult to displace from phosphorus (e.g. alkoxy, dialkylamino or alkyl) whereas X is a fairly good leaving group (e.g. F or p-NO$_2$-C$_6$H$_4$O—). Slight variations in structure can have very dramatic effects on the efficiency of organophosphorus poisons[34-37] and this is undoubtedly due to the fact that the interaction of an enzyme with a substrate is very sensitive to the size, shape and polarity of the substrate molecule.

$$G^1\!\!\underset{G^2}{\diagdown}\!P\!\underset{X}{\overset{O(S)}{\diagup}}$$

(12.21)

Toxicity is usually measured in terms of a factor known as LD$_{50}$ which is the average minimum dosage, in milligrams per kilogram of body weight, required to kill 50 % of a group of a particular species. The means of application of the dose is also important. It may be administered orally, by intravenous injection or by subcutaneous injection. Toxicity has also been registered in terms of LD$_{50}$ via skin (percutaneous) absorption per individual species and also in terms of a factor known as LCt$_{50}$ which is the lethal *concentration* in milligram-minutes per cubic metre required to kill 50 % of a species either by inhalation or percutaneous absorption. In comparing the toxicity of a variety of compounds it is obviously important to use a common lethal dose factor and a single method of exposure. Furthermore, the effectiveness of chemical poisons varies widely from one species to another. This is probably due (a) to structural differences in the affected enzyme between species and (b) to the varying ability of different species to metabolize (i.e. break down) chemical poisons before they reach the critical site of lesion within the living system. Differentiation between species is the factor which has allowed the development of organophosphorus compounds which are lethal to the pest but relatively harmless to man and it is now appropriate to examine this subject in more detail.

12.2.1 Organophosphorus pesticides[36,38-40]

The term 'pesticide' is a very general one and refers to any substance which affords protection to man from the point of view of his personal health or his crops by eliminating the culprit pest (e.g. the use of DDT* against malaria carrying mosquitoes or to combat infestations of locusts). Thus, insecticides, fungicides, herbicides (weed killers), rodenticides, acaricides (spider-mite killers), nematicides (worm killers) and molluscocides (snail killers) all fall in the category of pesticides and those which are used for crop protection are sometimes called "agrochemicals"[41,42].

Although the need for controlling pests had long been recognized as important it was not until the latter half of the 19th century that the first agrochemicals were introduced.

These included materials such as sulphur for controlling powdery mildew in vines, Bordeaux mixture (copper sulphate solution plus lime) for controlling downy mildew in a variety of crops and Paris Green (a mixture of cupric and arsenious oxides) against the Colorado beetle. Further advances came in the early part of this century with the use of nicotine solutions against aphids (greenfly and blackfly) and tar distillate sprays to control aphids and lichens on dormant fruit trees. Ferrous sulphate and sodium chlorate were introduced as general herbicides and the dinitrophenols and cresols were employed to control weeds in cereal crops. All these have since been categorized as "first-generation" pesticides[43].

The second generation of pesticides blossomed during the 1940's when several groups of compounds including the organochlorines, (DDT, benzene hexachloride (BHC)§, aldrin† and dieldrin**), the carbamates (compounds of general formula $R^1O-C(O)-NR_2^2$) and the organophosphates or organophosphonates were recognized as active pesticides. In fact it was just prior to the Second World War whilst

---

* DDT = 1,1-bis(p-chlorophenyl)-2,2,2-trichloroethane, Cl—C₆H₄—CH(CCl₃)—C₆H₄—Cl

§ BHC = (hexachlorocyclohexane structure) a mixture of isomers including the active γ-isomer

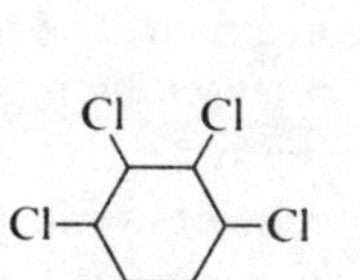

† ALDRIN = (structure) mixture of endo–exo isomers

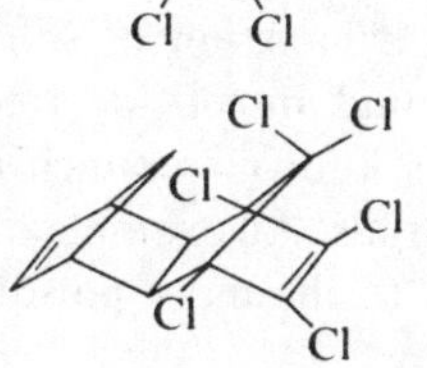

** DIELDRIN = (structure) mixture of endo–exo isomers

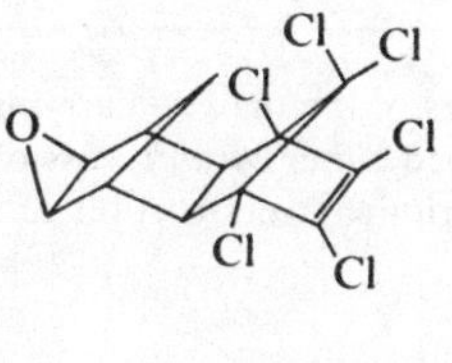

conducting a search for more active pesticides that Dr. Gerhard Schrader of I. G. Farben established the extremely high toxicity of an organophosphorus compound – later to be called Tabun – which became the first of the lethal organophosphorus chemical warfare agents.

The "third-generation" pesticides, currently in the development stage, are the insect hormones and sex-attractants[10]. It is hoped that these latest chemical weapons against the pest will be more specific in their action and also proof against the evolution of resistance.**

Some typical examples of organophosphorus compounds which are effective as pesticides are shown in Table 12.2. Their principal use has been as insecticides and some indication of their toxicity towards mammals and insects is included in the table. All the listed compounds have been produced commercially and used as insecticides but some (especially TEPP, Paraoxon, Dimefox and Schradan) have fallen into disrepute because of their high mammalian toxicity. Others, such as Malathion and Ekatin are still in widespread use since on the basis of the LD_{50} values available from rats, 120 g of Malathion or 20 g of Ekatin would be required to kill 50 % of a group of men weighing 12 stone (80 kg) each and both these represent very high doses.

The mammalian toxicity factor highlights a serious limitation for any pesticide since (a) there is always the danger that pesticide residues in crops would be hazardous to life and (b) there is the risk of contamination of personnel engaged in applying the pesticide. The first is not a serious problem since most organophosphorus compounds are broken down quite rapidly by the plant and legislation has been introduced to control the minimum time between spraying and harvesting so as to ensure a minimum of toxic residue. It is the second problem which to a large extent promoted the intensive research into and development of, safer phosphorus pesticides. In fact very few cases of accidental phosphorus poisoning have been reported recently but in earlier years when precautions were not so stringent, a number of very unpleasant incidents occurred[44].

A second limitation of organophosphorus insecticides is that the insect species may develop resistance to the poison by a process of natural selection – a kind of survival of the fittest – at least as far as a phosphorus diet is concerned. The resistance has been offset to some extent by a phenomenon known as *potentiation* in which the activity of the applied toxin is greatly enhanced within the living system. One way of achieving potentiation is by application of a mixture of two insecticides which sometimes results in enhanced activity for one or both of the poisons. The reasons for this are not fully understood but are probably connected with the exhaustion of some metabolizing enzyme by one insecticidal agent leaving the second insecticide free to pursue its destructive career. Another type of potentiation is known in which compounds are metabolized by the animal to even more toxic materials. An example of this behaviour is provided by Parathion which is oxidized *in vivo* to the more poisonous Paraoxon.

** It is now well established for example, that many species of the malaria carrying Anopheles mosquito are strongly resistant to DDT. Resistance is developed either by a process of natural selection in species which reproduce rapidly and/or by the indiscriminate destruction of natural predators during chemical treatments intended to eradicate the pest[8].

12.2 Phosphorus in death: pesticides and chemical warfare agents

$$(EtO)_2P(=S)-O-C_6H_4-NO_2 \xrightarrow{[O]} (EtO)_2P(=O)-O-C_6H_4-NO_2$$

Parathion → Paraoxon

Both methods of potentiation are useful means of augmenting insecticidal activity but both must be handled with caution since potentiation effects are just as likely to occur in mammalian systems as in the insect world thus creating unexpected dangers to man and the animals he seeks to rear or protect.

The final limitation of organophosphorus insecticides and one which is common to most second-generation pesticides is that of a lack of discrimination between the pest and the rest of the insect world, including the pest's predators. There are approximately 3×10^6 insect species known and of these only 0.1 % constitute agricultural pests or vectors of human disease. The remaining 99.9 % are either innocuous or positively helpful to agriculture (e.g. pollination by bees). It is the difficulty inherent in designing specificity which has led to the current research activity on insect hormones since only compounds such as these can offer hope of effectiveness against a single species and at the same time remain immune to the development of resistant strains.

Some organophosphorus insecticides however, are selective by virtue of the fact that they display *systemic activity* which means that they are absorbed by and translocated within, the tissues of the host. Insects which feed on the host by sucking or chewing are then destroyed by the agent resident in the host's system. Examples of such systemics include Schradan which was outstandingly successful in the protracted control of the cabbage aphis; Parathion and Dimefox (Table 12.2) are also useful systemics.

The two systems of transport within a plant are the xylem tubes (from roots to leaf) and the phloem system which transports the products of photosynthesis from mature leaves to growing tissue. Both involve transport via an aqueous medium so to exhibit systemic activity compounds should be more soluble in water than glyceride oils (which approximate the plant tissue). Alternatively the pesticide may be converted to a water soluble substance after application to the plant. Thus Demeton (Table 12.2) is converted to a systemic by oxidation of the thioether group to a water soluble sulphoxide. Another example is provided by Menazon (12.22) which is unsuitable for application to foliage because of an all round low solubility (0.1 % in H_2O, less in oil).

$(MeO)_2P(=S)-S-CH_2-[\text{triazine with } NH_2, NH_2]$ LD_{50} (rats) ≈ 1500 mg kg^{-1}
(12.22)

But the compound is effective as a systemic insecticide by application to the soil because the large volume of soil water gradually dissolves the insecticide and thus allows absorption into the plant via the root and xylem system and so affords protection over a long period. The compound also possesses the advantage of extremely low mammalian toxicity.

The use of systemic insecticides has, by and large, been restricted to crop protection

Table 12.2 Organophosphorus pesticides[a]

Trade or trivial name	Chemical name	Formula	Use[b]	Mammalian toxicity (rat) LD$_{50}$/(mg kg^{-1})	Insecticical toxicity, typical values LD$_{50}$/(mg kg^{-1})[c]
TEPP (Tetron®)	Tetraethylpyrophosphate	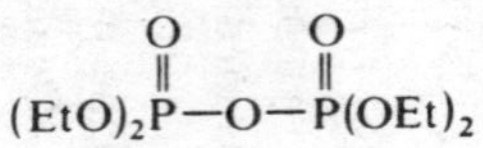 (EtO)$_2$P(O)—O—P(O)(OEt)$_2$	I	2	15
Parathion (Thiophos®)	Diethyl p-nitrophenyl phosphorothionate	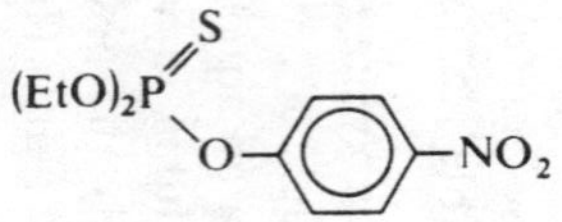 (EtO)$_2$P(S)—O—C$_6$H$_4$—NO$_2$	I, F, A	6	3.5
Paraoxon	Diethyl p-nitrophenyl phosphate	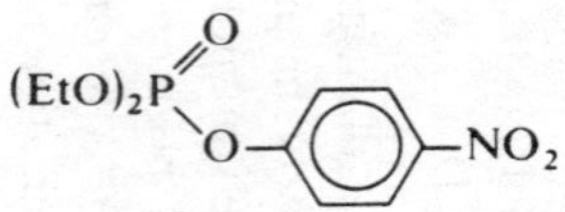 (EtO)$_2$P(O)—O—C$_6$H$_4$—NO$_2$	I	3	0.7
Dimefox	Tetramethylphosphoro-diamidic fluoride	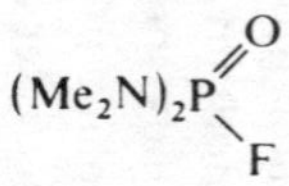 (Me$_2$N)$_2$P(O)F	I	7	n.a.
Schradan	Tetramethylphosphoro-diamidic anhydride	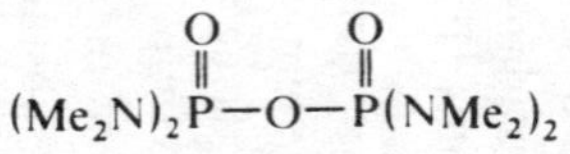 (Me$_2$N)$_2$P(O)—O—P(O)(NMe$_2$)$_2$	I	20	120
Demeton (Systox)	mixture of O,O-diethyl S-(and O)-2-(ethyl-thio)ethyl phosphoro-thioates.	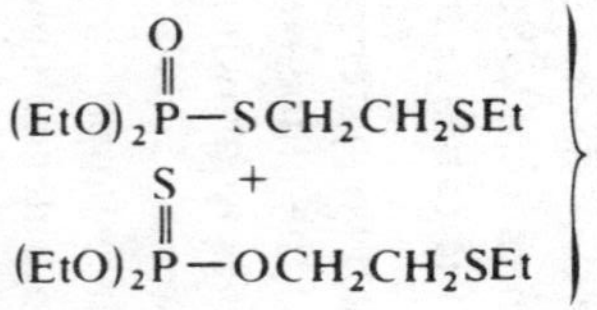 (EtO)$_2$P(S)—SCH$_2$CH$_2$SEt + (EtO)$_2$P(S)—OCH$_2$CH$_2$SEt	I, A	5	n.a.

12.2 Phosphorus in death: pesticides and chemical warfare agents

Name	Structure	Use[b]	LD50 (rat, mg/kg)	LD50 (insect, mg/kg)
Dimethoate (Cygon®, Perfekthion®)	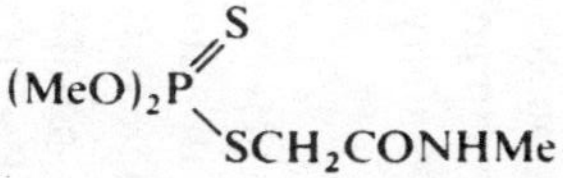 (MeO)₂P(S)—SCH₂CONHMe	I, A	~500	<1.0
Malathion (Cythion®)	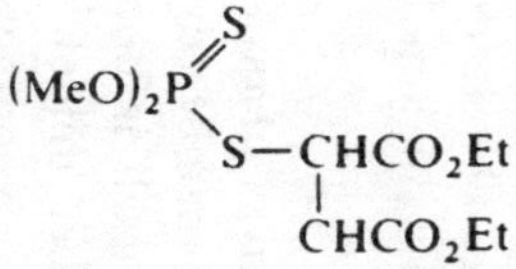 (MeO)₂P(S)—S—CH(CO₂Et)CH₂CO₂Et	I, A	~1000	30
Dichlorvos (DDVP) Vapona®	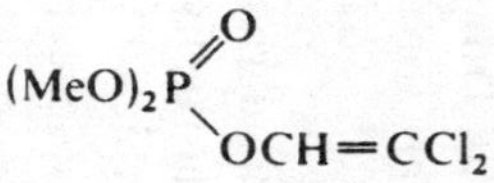 (MeO)₂P(O)—OCH=CCl₂	I	~100	n.a.
Diazion®	(EtO)₂P(S)—O—(2-isopropyl-6-methylpyrimidin-4-yl)	I, A	100	240
Dipterex®	(MeO)₂P(O)—CH(OH)CCl₃	I	600	30
Ekatin®	(MeO)₂P(S)—SCH₂CH₂SEt	I	225	n.a.

O,O-dimethyl S-(N-methyl-carbamoylmethyl)phosphorodithioate
O,O-dimethyl-S-(1,2-di-carbethoxyethyl)phosphorodithioate
2,2-dichlorovinyl dimethyl phosphate
diethyl-2-isopropyl-4-methylpyrimid-6-yl phosphorothioate
dimethyl-1-hydroxy-2,2,2-trichloroethyl phosphonate
O,O-dimethyl-S-2(ethyl-thio)ethyl phosphorodithioate

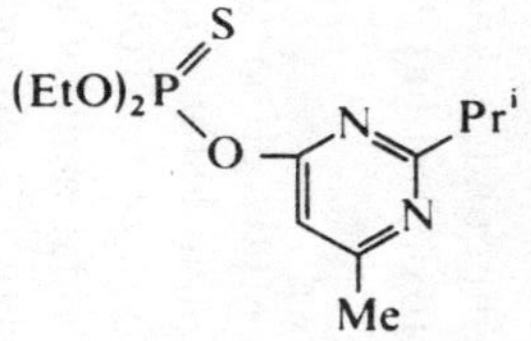

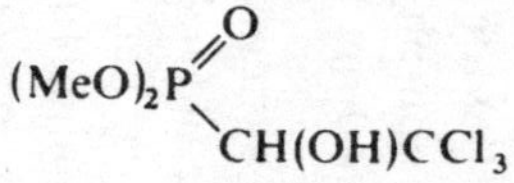

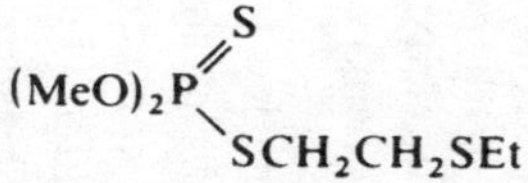

[a] Most of these compounds and many more are reported by E. E. Kenaga and W. E. Allison in *Bull. of the Entomological Society of America*, **15** (2), 85-148 (1969).
[b] I = insecticide; F = fungicide; N = nematicide; A = acaricide.
[c] n.a. indicates not available.

although there are some organophosphorus compounds which are sufficiently selective between insects and higher animals to be effective in the control of warble fly in cattle.

So far the discussion has been limited to the use of organophosphorus compounds as *insecticides*. Applications to other types of pest are on record however, and these are discussed briefly below.

Fungicides[42,45]: Wepsyn 155 (12.23, R = Ph, R' = H) which is effective against powdery mildews was claimed as the first systemic fungicide[46]. The discovery prompted the preparation of many analogues such as the series represented by (12.23) and (12.24). The position of the phosphoryl group on the triazine ring is important since higher activity as acaricides (against mites), as insecticides (against aphids) and as fungicides (against cucumber powdery mildew or apple mildew) was shown in each case by members of the (12.23) group where phosphorus is attached to N(1).

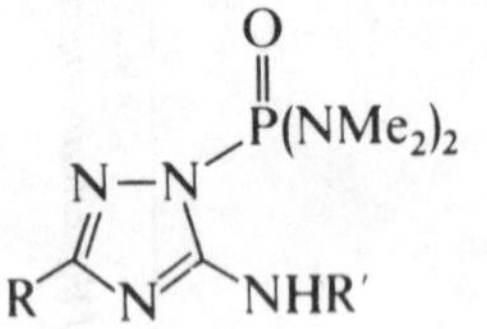

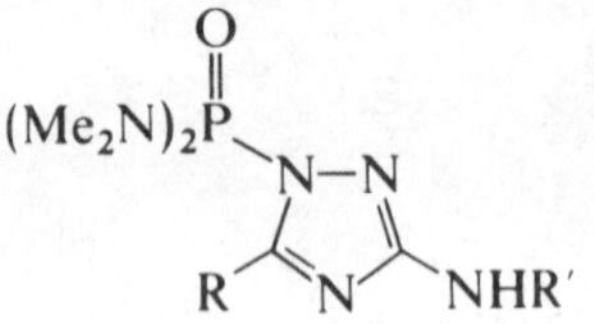

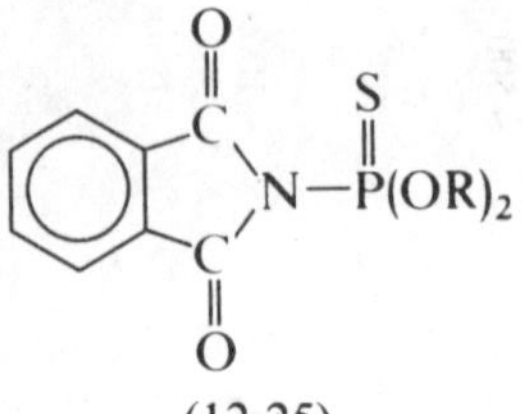

Another group of fungicides are the dialkyl N-phthalimidophosphoramidothionates (12.25)[47]. With R = Et, the compound is known as Dowco 199. It displays high fungicidal toxicity, is non-phytotoxic and has a remarkably low mammalian toxicity with LD_{50} (rats) = 5660 mg kg^{-1}.

This finding led to a systematic study of phosphorus derivatives of other N-heterocyclics with ring shape and size similar to that of the planar, five-membered phthalimido group. In consequence, derivatives of imidazole (which by itself is non-fungicidal) were found to have high fungicidal activity against cucumber mildew and potato blight[48]. The series is represented by (12.26) and the most effective compound (R = H; R' = R'' = CH$_3$; X = Ph) again showed very low mammalian toxicity despite high activity as a fungicide. This, coupled with the knowledge that certain phosphorus-free derivatives of imidazole (e.g. 12.27) are active fungicides, led Tolkmith to propose that fungicidal activity does not necessarily involve phosphorylation of an essential enzyme. If this is true, it represents an interesting departure from the normal mode of action of organophosphorus pesticides.

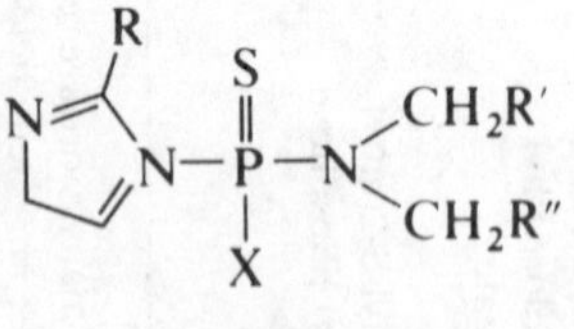

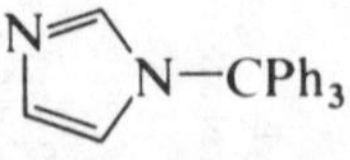

*Nematicides**: Certain organophosphorus compounds have proved to be useful nematicides particularly against eel worms which may infest and damage the roots,

---

* Nematodes are small, unsegmented worms, frequently parasitic on plant roots or stems.

12.2 Phosphorus in death: pesticides and chemical warfare agents

stem or leaves of plants. Parathion has been used against leaf-dwelling eel-worms on chrysanthemums but for stem eelworms, better systemic behaviour is required. This is provided by thionazin (12.28) which is used for the treatment of narcissus bulbs, despite its high mammalian toxicity. One of the most successful compounds for application to the soil (and hence protecting roots) was introduced by Dow and is known as Nellite (12.29). The compound is very hydrophilic and is absorbed to a

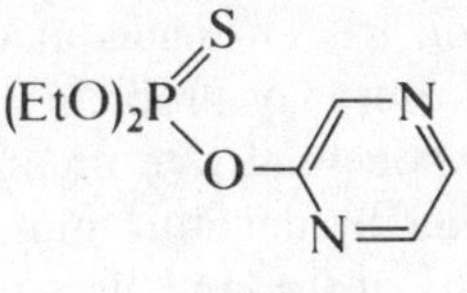

LD_{50} (rats) ≈ 10 mg kg^{-1}

(12.28)

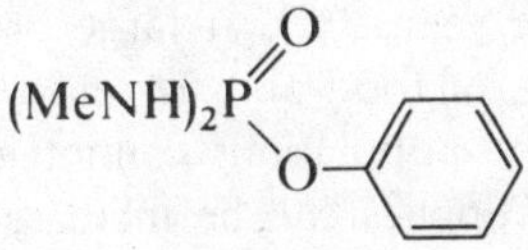

(12.29)

minimal extent by organic matter in the soil. Hence it is readily available to attack the pest infesting the root.

Chemosterilants: Future trends in the pesticide field include the development of chemosterilants. So far organophosphorus compounds have played a minor role in this area but substances such as tepa (12.30), metepa (12.31) and apholate (12.32) have been shown to produce sterility in adult insects at doses below the level required for any other effect.

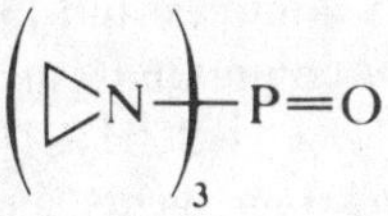

(12.30)
tepa
triaziridylphosphine oxide

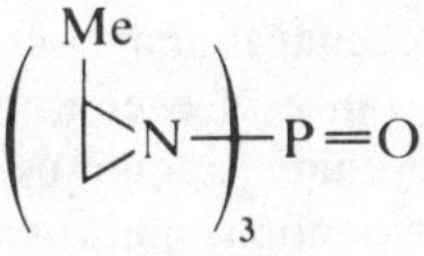

(12.31)
metepa
tri(methylaziridyl)phosphine oxide

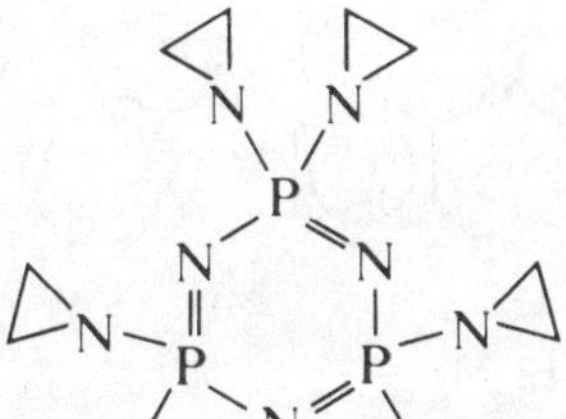

(12.32)
apholate
hexaaziridylcyclotriphosphazene

Pesticides and the environment: Over the past fifteen years many warnings have been sounded about the indiscriminate use of pesticides[39,44]. Although these warnings have frequently exaggerated the hazards, they have nevertheless served to encourage careful control in the use of pesticides. Such control is now exercized in the U.K. by the Pesticides Safety Precautions Scheme (formed to protect the consumer from contaminated produce) and the Agricultural Chemicals Approval Scheme (concerned with the performance of a pesticide in terms of its intended purpose). Both these schemes are voluntary arrangements between government and industry but have sufficient influence to ensure that a reputable manufacturer would not attempt to distribute a new product without at least provisional clearance under the Safety Precautions Scheme.

One of the most formidable problems facing the world today is the population explosion. Current world population is estimated at 3500 million and this is expected to double by the year 2000. A large proportion of the current population has an inadequate supply of food and for many starvation or malnutrition is an everyday problem[49]. Clearly, populations cannot go on increasing at the present exponential rate unless food supplies can be increased. This can only be achieved either by devoting more land to agriculture or by increasing the efficiency and productivity of the land already farmed. The use of chemical fertilizers and pesticides is part of the latter objective and if mass starvation is to be avoided chemicals will, of necessity, remain essential to agricultural productivity for the foreseeable future. It is vitally important therefore, that such chemical weapons be treated with respect and the degree of caution necessary to negate their potential to pollute the environment. For it is only in this way that Rachel Carson's "elixirs of death" will be contained as man's servants rather than his ultimate masters.

12.2.2 Chemical warfare agents[50-53]

The development of chemical and biological weapons is a very complex and highly emotive subject. The range of chemical agents covers (i) *lethal* compounds such as phosgene, hydrogen cyanide, sesquimustard gas ($ClCH_2CH_2SCH_2CH_2SCH_2CH_2Cl$) and the highly toxic organophosphorus G and V agents; (ii) *harassing* agents such as *ortho*-chlorobenzylidene malononitrile (12.33), the arsenicals such as Adamsite (12.34) and the less highly toxic organophosphorus compounds such as Sarin and Tabun and (iii) *incapacitating* agents such as BZ (12.35) which create symptoms of giddiness, disorientation and hallucinations within the victim.

(12.33)   (12.34)   (12.35)

Defoliants such as 2,4-D (12.36), 2,4,5-T (12.37) and cacodylic acid (12.38) are also part of the chemical armoury and are, in fact, very powerful herbicides.

(12.36)   (12.37)   (12.38)

The last has a high mammalian toxicity ($LD_{50} = 184$ mg kg^{-1}) and the phenoxyacetic acids are known to be powerful skin irritants. The subsequent discussion however, is restricted to organophosphorus chemical warfare (cw) agents and since most of the

12.2 Phosphorus in death: pesticides and chemical warfare agents

chemical formulae associated with these agents are now declassified, a range of such compounds is detailed in Table 12.3, together with some indication of their toxicity. To place the figures in perspective, it is estimated that 0.01 g of a V-agent when placed on a man's skin is sufficient to be lethal; this amounts to a drop no larger than a pinhead.

It will be noticed that all the agents conform to the general structure (12.21) of pesticides and that the most potent are derivatives of phosphonic acid (i.e. contain one carbon–phosphorus bond) with X as fluorine or a thioether group.

The symptoms of poisoning by organophosphorus compounds are dramatic and very unpleasant indeed. The harassing agents cause constriction of the pupils, blurred vision and pain behind the eyes. In fact these are frequently the first signs of phosphorus poisoning. Subsequently, tightness of the chest and difficulty in breathing develop. More severe poisoning causes drooling, sweating, nausea, involuntary defaecation or urination followed by convulsions, coma and eventual death through asphyxia. The onset of such symptoms can be anything from 10 minutes to half an hour after exposure with the severity of the effects dependent upon the dosage. For instance in a comparatively inactive man an exposure to Sarin of 15 mg m^{-3} for one minute (15 mg-min m^{-3}) dims the vision and causes pain behind the eyes which may last for a week or more. At 40 mg-min m^{-3}, the chest becomes tight, breathing is difficult and nausea, heartburn and twitching of the muscles develop. At 55 vomiting, cramps, tremors and involuntary defaecation and urination occur. At 70 severe convulsions set in followed by collapse, paralysis and death. All these symptoms are indications of a loss of muscular coordination and glandular control which provides a clue to the mode of action of organophosphorus toxins.

Overwhelming evidence is now available to indicate that organophosphorus compounds act by inhibition of enzymes involved in the function of nerves. In most cases the enzyme affected is acetylcholinesterase (AChE) which under normal circumstances catalyzes the hydrolysis of acetylcholine (12.39) to acetic acid and choline (12.40).

$$\text{X}^- \text{Me}_3\overset{+}{\text{N}}\text{CH}_2\text{CH}_2\text{O}\overset{\text{O}}{\overset{\|}{\text{C}}}\text{Me} \xrightarrow{\text{H}_2\text{O, AChE}} \text{X}^- \text{Me}_3\overset{+}{\text{N}}\text{CH}_2\text{CH}_2\text{OH} + \text{HO}\overset{\text{O}}{\overset{\|}{\text{C}}}\text{Me}$$

(12.39) (12.40)

In order to understand how AChE affects muscular coordination it is necessary to know something about the way in which nerve impulses cause muscles to contract. The following picture is of necessity grossly oversimplified but adequate for the purpose of this discussion.

Messages from the brain are transmitted by a change in electrical potential along a nerve fibre and when the area of reversed potential reaches a nerve ending it stimulates the release of acetylcholine. If we consider the stimulation of muscle fibre, there is a gap of ca 10^{-8} m, known as the synapse, between nerve ending and the muscle (Fig. 12.10). Acetylcholine diffuses across this gap and on contact with the muscle it causes a change in electrical potential which initiates a muscular contraction. In order to induce a muscular relaxation it is necessary to "de-fuse" the acetylcholine and this is where AChE plays a vital role.

The AChE enzyme catalyzes the hydrolysis of acetylcholine and the hydrolysis products no longer activate the muscle. Another muscular contraction requires another,

Table 12.3 Organophosphorus chemical warfare agents

U.S. Army code	Trivial name	Chemical structure	LD$_{50}$[a] Intravenous injection mg kg^{-1}	Subcutaneous injection mg kg^{-1}	Percutaneous absorption mg per man	Type of agent
GA	Tabun	Me$_2$N—P(=O)(—OEt)—CN	0.014	0.34 (mice)	1500	harassing
GB	Sarin	Me—P(=O)(—OPr^i)—F	n.a.	0.33 (mice)	2000	harassing
GD	Soman	Me—P(=O)(—OCH(Me)—Bu^t)—F	n.a.	0.14 (mice)	1250	harassing
GF	CMPF	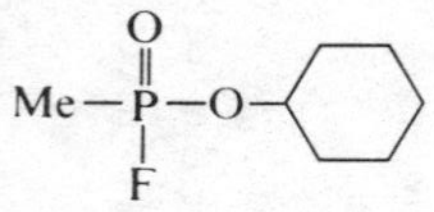 Me—P(=O)(—O-cyclohexyl)—F	n.a.	0.4 (mice) 0.1 (rabbits)	n.a.	lethal
V agents	—	Me—P(=O)(—SCH$_2$CH$_2$NR$_2'$)—OR R, R' = alkyl groups	0.03–0.2 (cats)	n.a.	v. small	lethal

[a] n.a. indicates not available.

12.2 Phosphorus in death: pesticides and chemical warfare agents

Fig. 12.10 The synapse between nerve ending and muscle.

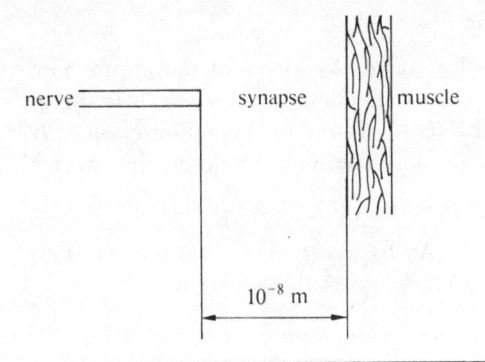

nerve-stimulated release of acetylcholine and the alternating process of stimulation, contraction and relaxation may occur thousands of times per second in a normal nerve–muscle system. The stimulation of glandular function and the transmission of messages between different nerve routes occurs in much the same way. If acetylcholine is not removed however, further stimulation of the nerve releases still more acetylcholine with the result that it accumulates at the muscle or gland and causes a perpetual state of muscular contraction (tetanus) or excessive glandular activity. It is therefore not surprising that inhibition of AChE should cause such symptoms as pupil constriction, cramps and convulsions (all due to muscular contractions) together with excessive salivation and perspiration (glandular functions). The terminal symptoms of phosphorus poisoning are caused by more complex and less well understood blockages of the central nervous system.

Exactly how do phosphorus toxins inhibit the vitally important enzymes such as AChE? In common with most biochemical systems this is an extremely complex subject[54,55], rendered even more difficult by the fact that AChE has not been isolated in a pure form which would facilitate *in vitro* studies and enable important information about the structure of the enzyme to be acquired. Much of what we know has in fact been derived by inference from a study of another esterase – chymotrypsin – which is available in a pure form and whose esterase activity is also inhibited by organophosphorus esters[56].

The mechanism of hydrolysis of acetylcholine involves formation of an initial complex between the enzyme and substrate (eqn. 34) followed by the release of choline and the formation of esterified enzyme (eqn. 35). The final step is the hydrolysis of the acetylated enzyme to give acetic acid and the free enzyme which can then pursue its catalytic function (eqn. 36). All organophosphorus toxins (both CW agents and

$$\text{AChE} + {}^-X\ \text{Me}_3\overset{+}{\text{N}}\text{CH}_2\text{CH}_2\text{O}\overset{\text{O}}{\overset{\|}{\text{C}}}\text{Me} \rightleftharpoons \text{AChE}\cdots\text{Me}_3\overset{+}{\text{N}}\text{CH}_2\text{CH}_2\text{O}\overset{\text{O}}{\overset{\|}{\text{C}}}\text{Me}\ X^- \quad (34)$$
$$\text{complex}$$

$$\text{complex} \longrightarrow \text{AChE} - \overset{\text{O}}{\overset{\|}{\text{C}}} - \text{Me} + \text{HOCH}_2\text{CH}_2\overset{+}{\text{N}}\text{Me}_3\ X^- \quad (35)$$

$$\text{AChE}-\overset{\overset{\text{O}}{\|}}{\text{C}}-\text{Me} \xrightarrow{\text{H}_2\text{O}} \text{AChE} + \text{HO}\overset{\overset{\text{O}}{\|}}{\text{C}}\text{Me} \qquad (36)$$

pesticides) are inherently good phosphorylating agents by virtue of the group X in general structure (12.21). It is believed that when such compounds reach their site of action they form an initial complex with AChE (eqn. 37) and then combine *chemically* to form a *stable*, phosphorylated enzyme (eqn. 38). The nucleophile on the enzyme

$$\text{AChE} + \underset{(12.1)}{\overset{G^1}{\underset{G^2}{\diagup}}\text{P}\overset{\diagup\text{O(S)}}{\diagdown}\text{X}} \rightleftharpoons \underset{\text{complex}}{\text{AChE} \cdots \overset{G^1}{\underset{G^2}{\diagup}}\text{P}\overset{\diagup\text{O(S)}}{\diagdown}\text{X}} \qquad (37)$$

$$\text{complex} \longrightarrow \text{AChE}-\underset{G^2}{\overset{G^1}{|}}\text{P}=\text{O(S)} + \text{HX} \qquad (38)$$

surface which displaces X in the phosphorylation process is probably the —OH group of the amino acid, serine (12.41) which comprises a vital part of the enzyme protein. The phosphorylated enzyme produced in this way is unable to catalyze the hydrolysis of acetylcholine because its active site is blocked. What is crucial however is that it cannot easily be unblocked since the hydrolysis of the phosphorylated enzyme, to produce a phosphorus acid and free enzyme (eqn. 39) is very slow.

$$\underset{\text{NH}_2 \quad (12.41)}{\text{HOCH}_2-\overset{|}{\text{CHCO}_2\text{H}}}$$

$$\text{AChE}-\underset{G^2}{\overset{G^1}{|}}\text{P}=\text{O(S)} \xrightarrow[\text{very slow}]{\text{H}_2\text{O}} \text{AChE} + \text{HO}-\underset{G^2}{\overset{G^1}{|}}\text{P}=\text{O(S)} \qquad (39)$$

Reactivation of the inhibited enzyme requires displacement of the phosphorus group and this can only be achieved by the use of powerful nucleophiles such as oximes. The pyridinium oxime (12.42) has proved effective as a therapeutic agent (eqn. 40) and it is believed that the quaternary nitrogen cation binds to an anionic site of the enzyme thus placing the oxime group in a favourable position for attack on phosphorus.

$$\underset{(12.42)}{\overset{\text{I}^-}{\underset{\text{R}}{\text{N}}}\diagdown\text{CH}=\text{NOH}} + \underset{\text{enzyme}}{\overset{G^1}{\underset{G^2}{\diagup}}\text{P}\overset{\diagup\text{O(S)}}{\diagdown}} \longrightarrow$$

$$\underset{\text{I}^-\underset{\text{R}}{|}}{\overset{+}{\text{N}}}\diagdown\text{CH}=\text{NO}-\underset{G^2}{\overset{G^1}{|}}\text{P}=\text{O(S)} + \text{free enzyme} \qquad (40)$$

12.2 Phosphorus in death: pesticides and chemical warfare agents

Many factors influence the efficiency of organophosphorus compounds as enzyme inhibitors. Size, shape and polarity are very important to the stability of the initial enzyme–substrate complex[54,55]. Moreover the phosphorus compound may be hydrolyzed or metabolized to an acid before it reaches the enzyme site. All phosphorus acids tested so far have proved to be non-toxic so if X is displaced easily by —OH the compound is unlikely to be an active poison. This is presumably why the highly toxic compounds are those containing leaving groups which are not displaced very rapidly (X = F, CN or SR). With X = —SCH$_2$CH$_2$NR$_2$, i.e. a V agent, one might imagine initial protonation of the amino group followed by displacement of a thioepoxide and amine (eqn. 41). Effective antidotes are discussed in Appendix 2.

$$\text{AChE} + \underset{RO}{\overset{Me}{\underset{|}{P}}}\overset{O}{\underset{SCH_2H_2C-\overset{+}{N}HR_2}{=}} \longrightarrow$$

$$\text{AChE} - \underset{OR}{\overset{Me}{\underset{|}{P}}} = O + H_2C\overset{S}{\diagup\diagdown}CH_2 + NHR_2 \quad (41)$$

Metabolic processes may either destroy the phosphorus compound or affect its activity, generally by oxidation processes. An example of the latter has already been cited in terms of the *in vivo* conversion of Parathion to the more active Paraoxon. Another example is the oxidative demethylation of Dimefox.

$$\underset{\text{Dimefox}}{\overset{Me_2N}{\underset{Me_2N}{\diagdown}}\overset{O}{\underset{F}{\diagup}}P} \xrightarrow{[O]} \underset{(12.43)}{\overset{HOCH_2}{\underset{Me_2N}{\diagdown}}\overset{O}{\underset{F}{\diagup}}\underset{Me}{N}P} \xrightarrow{-CH_2O} \underset{(12.44)}{\overset{MeHN}{\underset{Me_2N}{\diagdown}}\overset{O}{\underset{F}{\diagup}}P}$$

$$\downarrow H_2O \qquad\qquad\qquad \downarrow H_2O$$

$$\underset{(12.45)}{\overset{HOCH_2NMe}{\underset{Me_2N}{\diagdown}}\overset{O}{\underset{OH}{\diagup}}P} \qquad \underset{(12.46)}{\overset{MeHN}{\underset{Me_2N}{\diagdown}}\overset{O}{\underset{OH}{\diagup}}P} + HF$$

The intermediate oxidation products (12.43) and (12.44) are considerably more toxic than Dimefox itself but are also much more susceptible to hydrolysis to the inactive acids (12.45) and (12.46).

Far less is known about the nervous system of insects than that of mammals. Although acetylcholine and AChE have been detected in insects it is not clear whether they play the same vital role as in mammals. Some insects remain healthy with all their AChE inhibited; others, killed by pesticide, have been shown to retain much of their AChE intact. It seems likely that enzyme inhibition is responsible for the toxic effects but in many cases, the enzymes inhibited in insects must be different to those affected in mammals.

Finally what about the ethics of research in the area of CW agents? Much has been written about this highly controversial subject[50-53] and this book is not the place to discuss the matter in any detail. No sane individual wants another world war, nor the kind of war which would employ lethal chemical, biological and nuclear weapons; the consequences for world society are too appalling to contemplate. But beyond this there is the concept of "benign war" – war waged with incapacitating agents which neither damage property nor kill people but instead, eliminate the ability or the will to resist. No one argues the case against the development of such weapons more convincingly than Clarke (ref. 50, Chapter 11). But does this mean that we should end all scientific research on toxic or incapacitating substances or organisms?

Such a move is plainly unreasonable since it would mean stopping almost all the important work in public health, molecular biology, pesticides and organic chemistry. Every new compound which is synthesized in a chemical laboratory is potentially lethal. We need to know more, not less, about what classes of compounds are toxic and about their mode of action within living systems. It is not simply a question of developing suitable antidotes since, in several instances, studies of toxic action have led to a greater understanding of physiological processes and the development of drugs to counteract disease. For instance, the CW agents known as the mustard gases (e.g. $ClCH_2CH_2SCH_2CH_2Cl$) have been used as alkylating agents in what might be termed "first-generation" cancer chemotherapeutic agents. And organophosphorus anti-cholinesterases have found some application in the treatment of glaucoma* and muscle fatigue (myastheria gravis). Furthermore, a brief review of the literature over the last twenty years is sufficient to reveal important contributions to scientific knowledge from major defence establishments in Britain (Porton), the U.S.A. (Edgewood, Maryland), Australia, Canada and Europe. Similar examples of benefits derived from research in the field of microbiology are also on record.

Perhaps the greatest contribution which the scientist can make to the alleviation of international suspicion surrounding research in this area is to foster and encourage the open publication of scientific results. For this seems to be the only way to solve what Clarke calls "the scientists dilemma" – the scientific conviction that the work must continue against the moral conviction that it is wrong to direct it towards the development of a super weapon. It must be seen quite clearly that the latter is not the final objective. Then perhaps we shall have removed the element of fear and generated the atmosphere for the negotiation of a lasting peace.

On the other side of the morality argument it has been pointed out[24] that revealing all that is known about chemical and biological warfare would amount to a "proliferation of dangerous knowledge". It has been claimed therefore that there are sound, peacekeeping reasons for maintaining secrecy until it can be assumed that all countries have abandoned offensive intentions.

---

* A disease which causes an increase of fluid pressure inside the eye which may damage the optic nerve and lead to blindness.

It is a vicious circle of classic proportions. In order to generate trust it is necessary to divulge information but before the secrets are disclosed one would like to be certain that every country in the world could be trusted not to abuse the knowledge. The policy to be adopted is no longer the singular province of the scientist. His role is important in an advisory capacity but the decisions belong to government and ultimately to society.

References

[1] M. Horiguchi and M. Kandatsu, *Nature, (London)*, **184**, 901 (1959).

[2] L. D. Quin, *Topics in Phosphorus Chemistry*, **4**, 23 (1967).

[3] A. Todd, *Science*, **127**, 787 (1958).

[4] J. M. Suttie, *Introduction to Biochemistry*, Holt, Rinehart and Winston, New York, 1972.

[5] M. Yudkin and R. Offord, *Comprehensible Biochemistry*, Longman, London, 1973.

[6] M. R. Atkinson and R. K. Morton, *Comparative Biochemistry*, Vol II, p 5 (1960), Academic Press, New York.

[7] B. E. C. Banks, *Chem. Brit.*, **5**, 514 (1969); and with C. A. Vernon, *ibid*, **6**, 541 (1970).

[8] R. A. Ross and C. A. Vernon, *Chem. Brit.*, **6**, 539 (1970).

[9] L. Pauling, *Chem. Brit.*, **6**, 468 (1970).

[10] A. F. Huxley, *Chem. Brit.*, **6**, 477 (1970).

[11] D. Wilkie, *Chem. Brit.*, **6**, 472 (1970).

[12] F. Lipman, *Advan. in Enzymol*, **1**, 99 (1941).

[13] P. M. Schneider and H. Brinzinger, *Helv. Chim. Acta*, **47**, 1717 (1964).

[14] E. Racker, *Advan. in Enzymol*, **23**, 323 (1961).

[15] M. Avron and J. Newmann, *Ann. Rev. Plant Physiol.* **19**, 137 (1968).

[16] A. W. Frenkel and K. Cost, *Comprehensive Biochem.*, **14**, 397 (1966).

[17] R. K. Clayton, *Light and Living Matter*, Vol 2, McGraw-Hill, New York, 1971.

[18] D. E. Green, *Advan. in Enzymol*, **21**, 73 (1959).

[19] T. E. King and M. Klingenberg, *Electron and Coupled Energy Transfer in Biological Systems*, Dekker, New York, 1971.

[20] E. C. Slater, *Comprehensive Biochemistry*, **14**, 327 (1966).

[21] H. A. Krebs, "The History of the Tricarboxylic Acid Cycles", *Perspect. Biol. Med.*, **14**, 154 (1970).

[22] J. M. Lowenstein, "The Tricarboxylic Acid Cycle", *Metab. Pathways, 3rd Ed.*, **1**, 146 (1967).

[23] B. Axelrod, "Glycolysis", *Metab. Pathways, 3rd Ed.*, **1**, 112 (1967).

[24] J. N. Davidson, *The Biochemistry of the Nucleic Acids*, 7th Ed., Chapman and Hall, London, 1972.

[25] V. Ginsberg, *Advan. in Enzymol*, **26**, 35 (1964).

[26] D. E. Green and D. W. Allmann, Biosynthesis of Fatty Acids, *Metab. Pathways, 3rd Ed.*, **2**, 37 (1968).

[27] P. R. Vagelos, *Annu. Rev. Biochem.*, **33**, 139 (1964).

[28] S. Ratner, *Advan. in Enzymol.*, **15**, 319 (1954).

[29] P. Reichard, *Advan. in Enzymol.*, **21**, 263 (1959).

[30] J. M. Buchanan and S. C. Hartman, *Advan. in Enzymol.*, **21**, 199 (1959).

[31] H. R. Mahler and E. H. Cordes, *Biological Chemistry*, 2nd Ed., p 757, Harper and Row, New York, 1971.

[32] E. E. Van Temelen, *Accounts Chem. Res.*, **3**, 361 (1970).

[33] B. B. Buchanan and D. I. Arnon, *Advan. in Enzymol.*, **31**, 119 (1970).

[34] H. G. Khorana, *Some recent developments in the chemistry of phosphate esters of biological interest*, Wiley, New York, 1961.

[35] D. F. Heath, *Organophosphorus poisons*, Pergamon, Oxford, 1961.

[36] G. Schrader, "The modification of biological activity by structural changes in organophosphorus compounds", *World Review Pest Control*, **4**, 140 1965.

[37] B. C. Saunders, *Some aspects of the chemistry and toxic action of organic compounds containing phosphorus and fluorine*, Cambridge University Press, Cambridge, 1957.

[38] G. S. Hartley and T. F. West, *Chemicals for pest control*, Pergamon, Oxford, 1969.

[39] K. Mellanby, *Pesticides and pollution*, Fontana, London, 1969.

[40] G. Schrader, "Insecticidal phosphorus esters", *Angew. Chem.*, **69**, 86 (1957).

[41] J. O. Walker, "Agrochemicals", *Chem. Brit.*, **9**, 224 (1973).

[42] D. Woodcock, "Agriculture and horticultural fungicides", *Chem. Brit.*, **7**, 415 (1971).

[43] C. M. Williams, "Third-generation pesticides", *Sci. Amer.*, **217**, No 1, p 13 (1967).

[44] R. Carson, *The Silent Spring*, Chapter 3, Hamish Hamilton, London, 1963.

[45] A. F. Grapov and N. N. Mel'nikov, *Russ. Chem. Rev. (Eng. Trans.)*, **42** (9) 772 (1973).

[46] B. G. van dan Bos, M. J. Koopmans and H. O. Huisman, *Rec. Trav. Chim. Pays-Bas Belg.*, **79**, 807 (1960).

[47] H. Tolkmith, *Nature (London).*, **211**, 522 (1966).

[48] H. Tolkmith, P. B. Budde, D. R. Mussell and R. A. Nyquist, *J. Med. Chem.*, **10**, 1074 (1967).

[49] D. E. G. Irvine and B. Knights (eds.), *Pollution and the Use of Chemicals in Agriculture*, Butterworths, London, 1974.

[50] R. Clarke, *We all fall down*, Pelican, London, 1969.

[51] S. Rose (editor), *Chemical and biological warfare*", Harrap, London, 1968.

[52] J. Perry Robinson, "Chemical warfare", *Science*, p. 33, April 1967.

[53] M. S. Meselson, "Chemical and biological weapons", *Sci. Amer.*, **222**, No 5, p 15 (1970).

[54] D. R. Davies and A. L. Green, "Mechanism of hydrolysis by cholinesterase and related enzymes", *Advan. in Enzymol.*, **20**, 283 (1958)

[55] M. I. Kabachnik et al., "Hydrophobic regions on the surface of choline-esterases", *Russ. Chem. Rev. (Eng. Trans.)*, **39**(6) 485 (1970).

[56] A. K. Balls and E. F. Jansen, Stoichiometric inhibition of chymotrypsin, *Advan. Enzymol.*, **13**, 321 (1952).

[57] Ref. 18, pp 135–137.

Appendix 1
The nomenclature of phosphorus compounds

512   Appendix 1

This appendix contains an account of the rules of nomenclature as used in each branch of phosphorus chemistry. Ideally the reader should memorize the contents of this section before he begins the book but a more boring approach to phosphorus would be hard to imagine. As with the learning of a new game it is often better to start playing and to pick up the rules as one goes along. Since the chemical name of a compound is often a screen rather than a window to understanding we have used the chemical formula where possible, simply because this communicates the essential information of identity and structure so much better than any name could hope to do. However names are required for some purposes, and this appendix outlines the way in which the language of phosphorus has developed.

There are five dialects spoken in phosphorus chemistry corresponding to five areas into which the subject is divided: (i) the basic inorganic naming; (ii) the phosphorus hydrides and their organic derivatives; (iii) the phosphorus oxoacids and thioacids and their derivatives including the polyphosphates; (iv) phosphorus–nitrogen compounds; and (v) phosphorus ylids. Although all phosphorus chemists speak a common language, or at least pay lip service to one, in actual fact local variations are so marked as to constitute almost a separate tongue.

The inorganic names[1] are reserved now for the phosphorus halides and there are six types of compound dealt with in this way. Table 1 illustrates this by naming the chlorides, including those with more than two phosphorus atoms. The naming of the halides could have been extended to the hydrides and thence to organophosphorus derivatives, but unfortunately the hydrides, and in particular the parent PH_3 (phosphine) were named differently.

Table 1 Names of the phosphorus chlorides and hydrides

PCl_3	phosphorus trichloride	PH_3	phosphine
$POCl_3$	phosphoryl (tri)[a]chloride	PH_3O	phosphine oxide
$PSCl_3$	thiophosphoryl (tri)[a]chloride	PH_3S	phosphine sulphide
PCl_5	phosphorus pentachloride	PH_5	phosphorane
PCl_4^+	tetrachlorophosphonium[b] cation	PH_4^+	phosphonium[b] cation
PCl_6^-	hexachlorophosphate anion	PH_6^-	hexaphosphate anion
P_2Cl_4	diphosphorus tetrachloride	P_2H_4	diphosphine
P_nCl_{n+2}	$n$phosphorus ($n+2$)chloride	P_nH_{n+2}	polyphosphines

[a] Officially a redundant prefix, but many prefer to include it.
[b] phosph + onium, phosphon- has another meaning.

The 'inorganic' phosphorus nomenclature remains somewhat out on a limb although it is still 'officially'* recognized. There is one point to note however: the yl ending,

---

* By the Chemical Society (London), the American Chemical Society, and the International Union of Pure and Applied Chemists.

Appendix 1

which has now been relegated to radical terminology, is still acceptable in the name phosphoryl trichloride.

In 1952 a committee, set up jointly by the American and British Chemical Societies, reported its recommendations on the naming of phosphorus compounds with one P atom[2]. At that time there was obviously a need to bring order to the naming of phosphorus acids and their derivatives and this aspect rather dominated the proposals; phosphorus acids therefore served as the basis of classification. Twenty years later the system is still basically sound when it comes to the naming of the majority of phosphorus compounds and especially the organic derivatives of phosphorus acids.

The 1952 system recognizes the inorganic and hydride names in Table 1 and covers organic derivatives by prefixes to the hydride names (alkoxy, aryloxy and amino groups are not included in this but are named as esters or amido derivatives of phosphorus acids). Examples of names obtained thus are

Me_3P, trimethylphosphine

Et_3PO, triethylphosphine oxide

$Bu^n_4P^+$, tetra-n-butylphosphonium cation

Ph_5P, pentaphenylphosphorane.

If one or more of the hydrogen atoms of the compounds PH_3, PH_3O and PH_3S are replaced by OH or SH groups the result is an acid or thioacid. Replacement of the H's of PH_5 to give a series of $PH_{5-x}(OH)_x$ acids was also dealt with in the 1952 proposals but the names suggested for these can be ignored since the need for them is unlikely to arise.

The acids in which phosphorus has oxidation state (III) have the '-ous' ending and their salts and esters have the '-ite' ending. Those with oxidation state (V) have corresponding '-ic' and '-ate' endings. This is the common inorganic acid terminology.

The possible combination of acids produced by the foregoing proposals are listed in Table 2. It must be pointed out that some of the acids in Table 2 are not known in the form given, e.g. $P(OH)_3$ is called phosphorous acid although it exists in the rearranged form $HP(O)(OH)_2$ which is phosphonic acid. However, both names are required because organic derivatives of both are known. The table also contains anhydride names.

Organophosphorus compounds can be seen as formed by replacement of the hydrogen atoms in the acids of Table 2, either of the H directly bonded to phosphorus or of the acidic H's. In the former case the name of the organic groups are attached as a prefix to the name of the acid; in the latter case the organic part is put as a separate word before the name of the acid.

For example:

$Me_2P(O)OH$, dimethylphosphinic acid
$H_2P(O)OMe$, methyl phosphinic acid
$Me_2P(O)OMe$, methyl dimethylphosphinic acid

Table 2 Notation for oxo- and thio-phosphorus acids

Derivatives of PH$_3$, phosphine

Replacement of one H by OH or SH	Replacement of two H's by OH or SH	Replacement of three H's by OH or SH
(HO)PH$_2$ phosphinous acid (HS)PH$_2$ phosphinothious acid	(HO)$_2$PH phosphonous acid (HO)(HS)PH phosphonothious acid (HS)$_2$PH phosphonodithious acid	(HO)$_3$P phosphorous acid (HO)$_2$(HS)P phosphorothious acid (HO)(HS)$_2$P phosphorodithious acid (HS)$_3$P phosphorotrithious acid

Derivatives of PH$_3$O, phosphine oxide

(HO)PH$_2$O phosphinic acid (HS)PH$_2$O phosphinothioic acid	(HO)$_2$PHO phosphonic acid (HO)(HS)PHO phosphonothioic acid (HS)$_2$PHO phosphonodithioic acid	(HO)$_3$PO phosphoric acid (HO)$_2$(HS)PO phosphorothioic acid (HO)(HS)$_2$PO phosphorodithioic acid (HS)$_3$PO phosphorotrithioic acid

Derivatives of PH$_3$S, phosphine sulphide

(HO)PH$_2$S phosphinothionic acid (HS)PH$_2$S phosphinodithioic acid	(HO)$_2$PHS phosphonothionic acid (HO)(HS)PHS phosphonothiolothionic acid (HS)$_2$PHS phosphonotrithioic acid	(HO)$_3$PS phosphorothionic acid (HO)$_2$(HS)PS phosphorothiolothionic acid (HO)(HS)$_2$PS phosphorodithiolothionic acid (HS)$_3$PS phosphorotetrathioic acid

No-ambiguity rule: to signify S use thio, but if this leads to ambiguity then it is necessary to distinguish the P—SH grouping by the term thiolo-, and the P=S grouping by the term thiono-.

Anhydrides: $\rangle$P(O)—O—P(O)$\langle$ phosphoric anhydride; $\rangle$P(S)—O—P(S)$\langle$ dithionophosphoric anhydride; $\rangle$P(O)—S—P(O)$\langle$ thiolophosphoric anhydride.

Appendix 1

The names of the esters of the oxoacids are given in more detail in the introduction to Chapter 8. With mixed oxo- and thio-acids it may be necessary to distinguish the ester and organo-groups directly, e.g:

EtP(S)(OMe)(SMe), O-methyl S-methyl ethylphosphonodithioate

[Note that in this case the use of O- and S- rules out any ambiguity so that the root ending of the acid, which is phosphonothiolothionic acid, can be condensed to dithio-.] Slavish adherence to the rule that the correct name is the shortest unambiguous one can be carried too far. Often O- and S- prefixes are included even though they may be technically redundant. But if this makes for easier understanding then it is best to include them.

Organophosphorus chemistry is very much the chemistry of compounds of the type $R_2P(O)-$, i.e. phosphinic derivatives, and $RP(O){<}$, i.e. phosphonic derivatives.

This convenient nomenclature has percolated through almost all branches of phosphorus chemistry so much so that whereas $POCl_3$ is phosphoryl chloride, $MePOCl_2$ is methylphosphonic dichloride and Me_2POCl is dimethylphosphinic chloride.*

The phosphorus acids give rise to another series of derivatives in which one or more of the OH groups of the acid are replaced by halide, pseudohalide (CN, NCO, etc), amide or substituted amide groups. In this case, and provided the compound is still an acid or its anion or ester, the name is modified by the insertion of the name of the group after the phosphorus root. Examples are:

HP(O)(F)(OH), phosphonofluoridic acid
P(O)(Cl)(CN)(OH), phosphorchloridocyanidic acid
P(O)(NH$_2$)(OH)$_2$, phosphoramidic acid
P(O)(OMe)(NMe$_2$)$_2$, methyl tetramethylphosphorodiamidate.

Table 3 Names of phosphorus radicals

Univalent:		Divalent:	
H$_2$P—	phosphino	HP=	phosphinidene
H$_2$P(O)—	phosphinyl	HP(O)=	phosphinylidene
H$_2$P(S)—	phosphinothioyl	HP(S)=	phosphinothioylidene

Radicals: The parent names of univalent and divalent radicals are given in Table 3 where it can be seen that divalent radicals are produced by adding -idene to the univalent name. Derivatives of these radicals are named by replacement of the H's by other groups with no distinction being made with respect to the type of group, e.g. OH is hydroxy and does not imply an acid function. Names of radicals are for instance:

MeO.P(Cl)—, chloro(methoxy)phosphino
MeO(HO)P(O)—, hydroxy(methoxy)phosphinyl

---

* The rule changes yet again for Me$_3$PO which is trimethylphosphine oxide.

This more or less completes the relevant sections of the 1952 system. Modifications to these rules have been made by the compilers of *Chemical Abstracts*: in 1962 (vol **56**) they dealt briefly with compounds containing two phosphorus atoms; in 1967 (**66**) phosphorus-nitrogen cations and ring systems were renamed, the latter according to IUPAC recommendations[3] which unfortunately produced such clumsy names that they have rarely been used outside *Chem. Abs.*, and in 1972 (**76**) when it was finally decided that

$$\text{HO}-\overset{\overset{\text{O}}{\|}}{\underset{\underset{\text{OH}}{|}}{\text{P}}}-\text{O}-\overset{\overset{\text{O}}{\|}}{\underset{\underset{\text{OH}}{|}}{\text{P}}}-\text{OH}$$

would be called diphosphoric acid instead of pyrophosphoric acid.

Phosphorus chemistry has moved on a lot in the last 20 years and it has progressed in some areas that were only cursorily dealt with by the 1952 nomenclature committee. Although the enormous development of the sixties did not pose any insurmountable problems in nomenclature, the upsurge in phosphorus–nitrogen chemistry and ylid research eventually led to the emergence of alternative names. Furthermore the incorporation of phosphorus in ring systems necessitated the introduction of new nomenclature for these and the names of the more important rings are given in Table 4.

Phosphorus–nitrogen nomenclature: In so far as a compound can be considered as the amido derivative of one already covered by the 1952 convention then this system is preferred. Singly bonded derivatives are covered quite adequately by this, e.g.

P(O)(OH)$_2$(NH$_2$), phosphoramidic acid

MeP(O)(NHMe)(SMe), S-methyl N,P-dimethylphosphonamidothioate

However when the compound is no longer an acid or an ester then it is named according to the *Chemical Abstracts* order of prime functions[4], e.g.

MeP(O)Cl.NH$_2$, methylphosphonamidic chloride

P(NH$_2$)$_2$.CN, phosphorocyanidous diamide

This is perhaps the least comprehensible part of the 1952 recommendations especially in view of the nature of these compounds and their relationships to a compound such as P(O)(NH$_2$)$_3$ which is named phosphoryl triamide by analogy with POCl$_3$. The derivative of this, P(O)(NMe$_2$)$_3$ is called hexamethylphosphoryl triamide, usually abbreviated to HMPT by those chemists who make use of this versatile polar aprotic solvent.

The most prolific area of inorganic phosphorus–nitrogen chemistry has been the study of derivatives with the backbone unit $-\overset{|}{\underset{|}{\text{P}}}=\text{N}-$, especially the cyclic and linear polymers. The original name for these compounds was phosphonitriles but most workers in this field have gone over to a system of nomenclature which brings the name into line with other heteronuclear inorganic ring systems. Shaw[5] proposed the term phosphazene for the $-\overset{|}{\underset{|}{\text{P}}}=\text{N}-$ unit and phosphazane for the $-\overset{\overset{\|}{}}{\underset{|}{\text{P}}}-\overset{|}{\text{N}}-$ unit. (A full explanation of this system of nomenclature is given by Allcock[6].) The system is illustrated by the name:

Ph$_3$P=N—Et, N-ethyltriphenylmonophosphazene

Appendix 1

Table 4 Ring nomenclature of organic rings and cyclic inorganic polymers

Organic rings

phosphirane 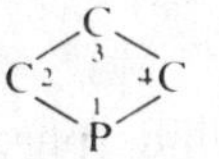   phosphetane

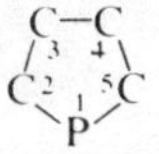

phospholane 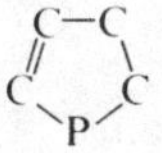   2-phospholene

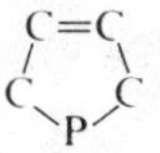

3-phospholene 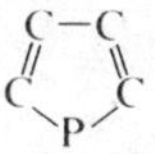   phosphole

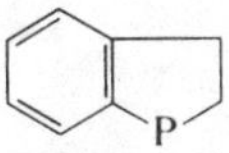

phosphindoline 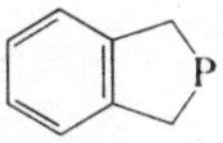   isophosphindoline

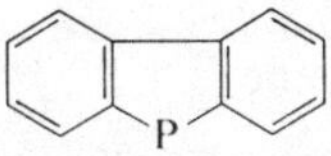

9-phosphafluorene

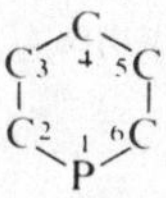

phosphorinane 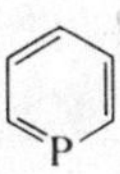   phosphorin

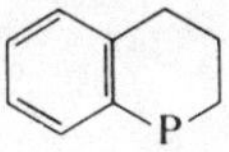

phosphinoline 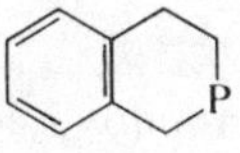   isophosphinoline

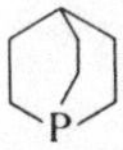

1-phosphabicyclo[2.2.2]octane

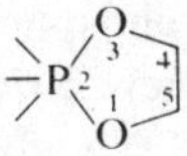

1,3,2-dioxaphospholane 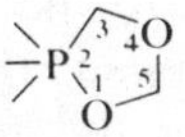   1,4,2-dioxaphospholane

Cyclic inorganic polymers (prefix 'cyclo')

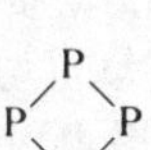

cyclotetraphosphine 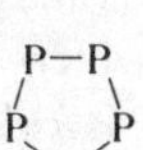   cyclopentaphosphine

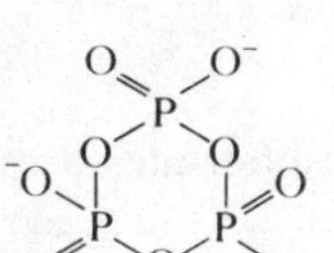

cyclotriphosphate (previously trimetaphosphate) 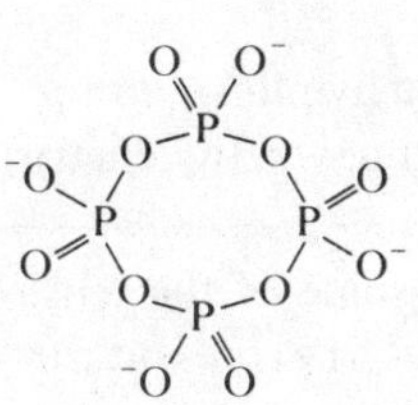   cyclotetraphosphate

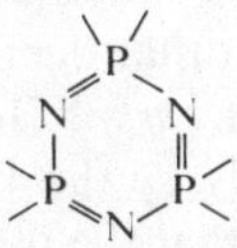

cyclotriphosphazene 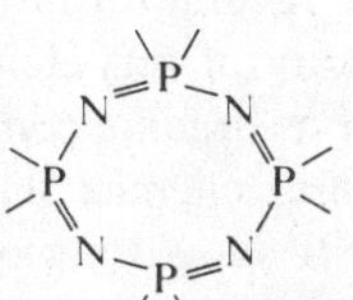   cyclotetraphosphazene

This conflicts with the 1952 name which is based on H₃P=NH variously called phosphine imide or iminophosphorane, so that this derivative can be termed N-ethyl-triphenylphosphine imide or N-ethyltriphenyliminophosphorane.

Related to this group of compounds, and formally their cyclic dimers, are compounds of the type

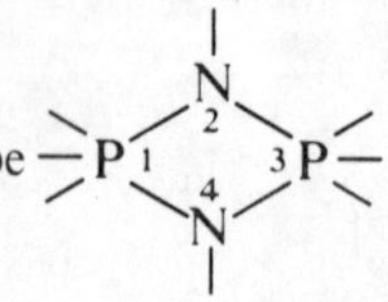

$$\begin{array}{c} | \\ N \\ \diagup^2 \diagdown \\ -P^1 \quad {}^3P- \\ \diagdown_4 \diagup \\ N \\ | \end{array}$$

which are called cyclodiphosphazanes, e.g. (Cl₃PNMe)₂ is 1,1,1,3,3,3-hexachloro-2,4-dimethylcyclodiphosphazane.

The formally unsaturated rings are better known and the naming is revealed by, e.g.

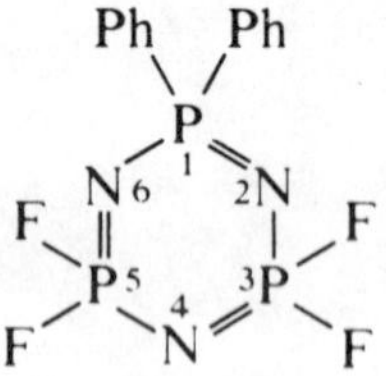

$$\begin{array}{c} Ph \quad Ph \\ \diagdown \diagup \\ P \\ \diagup^1 \diagdown \\ N^6 \quad {}^2N \\ \| \qquad | \\ F-P^5 \quad {}^3P-F \\ \diagup_4 \diagdown \\ F \quad N \quad F \end{array}$$

1,1-diphenyl-3,3,5,5-tetrafluorocyclotriphosphazene

If in phosphorus–nitrogen compounds the phosphorus is in a lower oxidation state, then this can be incorporated into the name by means of the Roman number, for instance:

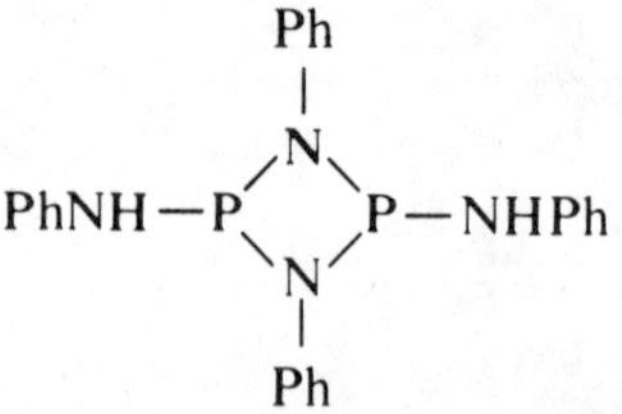

$$\begin{array}{c} Ph \\ | \\ N \\ \diagup \diagdown \\ PhNH-P \quad P-NHPh \\ \diagdown \diagup \\ N \\ | \\ Ph \end{array}$$

1,3-bis(phenylamino)-2,4-diphenyl-cyclodiphosph(III)azane.

Ylids: The carbon derivatives analogous to the phosphazenes are the phosphorus ylids based on the hypothetical compound H₃P=CH₂ which is often written H₃P⁺—⁻CH₂. The names have reflected this duality of the bonding, the above compound being called either phosphine methylene, methylenephosphonium ylid or phosphonium methylide. The name which has come into common usage however is none of these, but is methylenephosphorane, which emphasizes the pentavalency of the phosphorus and derives from PH₅ which is called phosphorane. Ylid naming is illustrated by:

Ph₃P=CH₂, methylenetriphenylphosphorane
(CF₃)₃P=CHCO₂Et, carbethoxymethylenetris(trifluoromethyl)phosphorane.

This appendix has revealed some of the tensions arising from inconsistencies within phosphorus nomenclature at the present time. Undoubtedly there will eventually be a big shake-up perhaps even a revolution in the naming of all phosphorus compounds, recognizing the individuality of this element to such an extent that the root *phospho* disappears altogether just as organic compounds omit the root *carbo*. However until such a time the phosphorus chemist should be guided equally by the 1952 proposals and by common sense. It should always be borne in mind that the object of any name in any language is primarily the communication of information. Thus it is better to err on the side of being too explicit than too economical in the use of space if this latter approach eventually leaves the reader bewildered. For this very reason it is often better to use chemical formulae than chemical names.

Problems

1. Name the following compounds:
   (a) $PH(CH_3)(CH_2Cl)$
   (b) $P(O)MeEt(SPr^i)$
   (c) 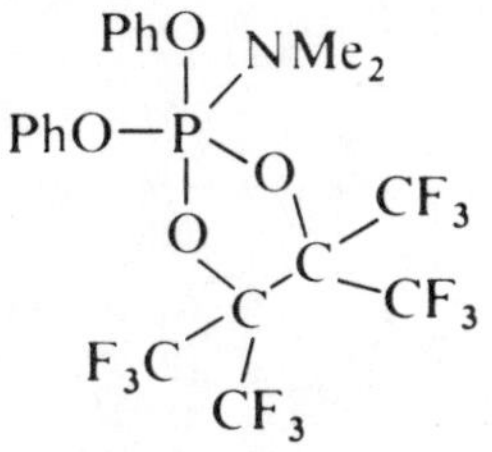

   (d) 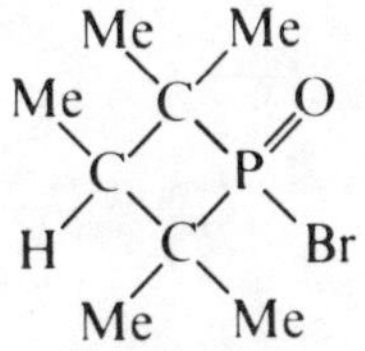

   (e) $Cl_3P=N-Et$
   (f) $P(O)(OH)(F)(Cl)$

2. Draw the chemical formulae of the following compounds:
   (a) S-isopropyl N,N-diethylmethylphosphonoamidothioate
   (b) trimethylsilylmethylenetrimethylphosphorane
   (c) 1,3-bis(dimethylamino)tetrachlorocyclotriphosphazene
   (d) dimethyltrifluorophosphorane
   (e) hexaphenoxyphosphate anion
   (f) 2,4,6-triphenylphosphorin

References

[1] International Union of Pure and Applied Chemistry, *Nomenclature of Inorganic Chemistry*, Butterworths, London, 1966.

[2] Published in *J. Chem. Soc.*, 5122 (1952) and *Chem. Eng. News*, **30**, 4515 (1952).

[3] International Union of Pure and Applied Chemistry, convention for naming ring compounds, *J. Amer. Chem. Soc.*, **82**, 5566 (1960).

[4] The naming and indexing of chemical compound by Chemical Abstracts, para. 68, *Chem. Abs.*, **39**, 5876 (1945).

[5] R. A. Shaw, B. W. Fitzsimmons and B. C. Smith, *Chem. Rev.*, **62**, 247 (1962).

[6] H. R. Allcock, *Phosphorus–nitrogen Compounds*, Chap 1, Academic Press, New York, 1972.

Appendix 2
Antidotes for toxic organophosphorus chemicals—pesticides and nerve gases

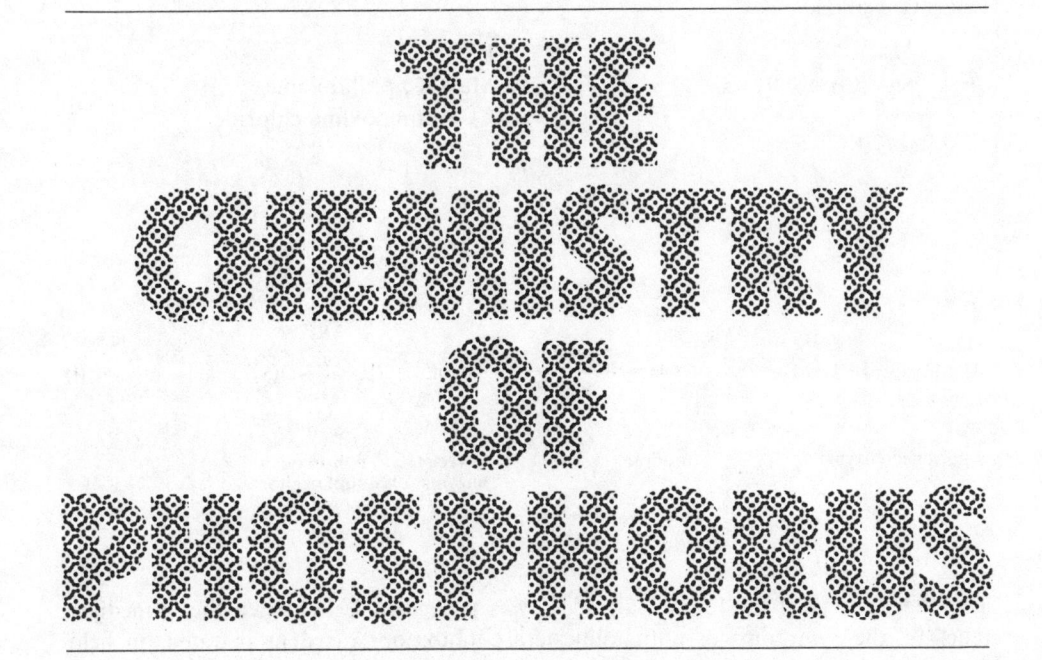

Appendix 2

The effective antidotes for poisoning by organophosphorus pesticides and nerve gases are N-methypyridinium-2-aldoxime salts, of which the methylsulphonate salt is particularly recommended. Doses of 1–5 g of the antidote should be administered by injection or by mouth as soon as possible after exposure to the poison. The antidote is available as the drug **pralidoxime** which is held in readiness at designated Poison Centres throughout the United Kingdom, and is supplied by the Supplies Division, Department of Health and Social Security, 14 Russell Square, London WC1B 5EP and produced by Glaxo limited, Greenford, Middlesex, England.

The accolade for the discovery of an effective type of antidote to organophosphorus poisoning must go to I. B. Wilson and his co-workers[1] who were the first to recognise the value of the —NOH group as a nucleophile strong enough to displace the blocking phosphorus group from poisoned anticholinesterase enzyme. The nature of the toxic organophosphorus compounds (pesticides and nerve gases) and their mode of operation within the body were discussed in Chapter 12. Many instances of accidental poisoning occurring as a result of the use of pesticides and the escape of nerve gases have been recorded. The incidence of poisoning among those employed in the manufacture, distribution and application of pesticides has been much reduced as the inherent dangers have been realized. Nevertheless accidents will always occur especially among the vulnerable sections of society – the ignorant, the illiterate, the very young and animals.

In 1955 it was reported[2] that N-methylpyridinium-2-aldoxime iodide* (I) was a specific reactivator of acetylcholinesterase that had been inhibited by alkylphosphonate reagents, eqn. (1).

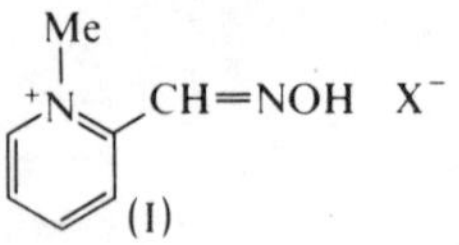

$X^- = I^-$, PAM
$X^- = MeSO_3^-$, pralidoxime
$X^- = Cl^-$, pralidoxime chloride

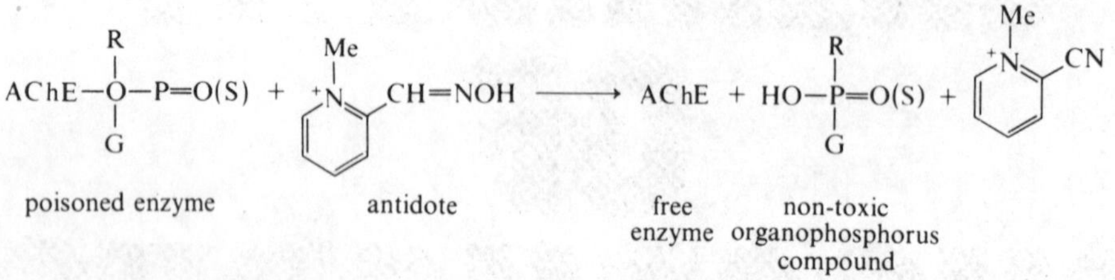

| poisoned enzyme | antidote | | free enzyme | non-toxic organophosphorus compound | |

The effectiveness of the iodide salt of (I) was proved in 1958[3] when the immediate relief of the symptoms of anticholinesterase-inhibition was demonstrated on field

---

* Alternately called 2-hydroxyiminomethyl-N-methylpyridinium iodide.

workers poisoned by Parathion. The iodide salt is not very soluble, however, and the form now available is the methylsulphonate salt which is much more soluble[4], has a long shelf life even in solution[5], and has the slight advantage that the active component, the oxime cation, comprises 57% by weight compared to 52% in the iodide[6]. In this last respect the chloride salt is even better being 80% oxime, and this is the form of the antidote manufactured for use in the USA. In continental Europe the perferred form is bis(4-hydroxyiminomethylpyridinium-N-methyl)ether dichloride, which is called toxigonin. This has the disadvangate of adverse side effects is some cases and moreover is poorly absorbed when taken orally.

Studies on the relative effectiveness of intravenous, intramuscular and oral administration of pralidoxime have been undertaken and these show injection to be most effective[7]. However, because of the delayed action of organophosphorus compounds, including nerve gases when disseminated as aerosols or smokes, oral administration is also possible. Studies on pralidoxime tablets revealed that in the human body the biological half-life of this is $1\frac{1}{2}$ hours reaching a maximum concentration about 2 hours after taking the dose[8]. Most of the dose of the antidote was excreted in the urine[9].

Much of the research into nerve gases, into antidotes for them, and into modified nerve gases capable of resisting the known antidotes, is of course secret. However the more toxic organophosphorus chemicals are hardly likely to find everyday use, and one can assume that pralidoxime will remain as the effective antidote for all foreseeable kinds of organophosphorus chemical.

References

[1] I. B. Wilson and E. K. Meislich, *J. Amer. Chem. Soc.*, 75, 6428, (1953).
[2] A. F. Childs, D. R. Davies, A. L. Green and J. P. Rutland, *Brit. J. Pharmacol.* **10**. 462, (1955).
[3] T. Namba and K. Hiraki, *J. Amer. Med. Assn.*, 1834, (1958).
[4] N. H. Creasy and A. L. Green, *J. Pharm. Pharmacol.*, **11**, 485, (1959).
[5] B. R. Cole and L. Leadbeater, *J. Pharm. Pharmacol.*, **18**, 101, (1966).
[6] R. I. Ellin and J. H. Wills, *J. Pharm, Sci.*, **53**, 995 and 1143, (1964).
[7] A. Sundwall, *Biochem. Pharmac.*, **5**, 225, (1960).
[8] A. Sundwall, *Biochem. Pharmac.*, **8**, 413, (1961).
[9] F. R. Sidell, W. A. Groff and A. Kaminskis, *J. Pharm. Sci.*, **61**, 1137 and 1765 (1972).

Answers to problems

Chapter 1

1

	% P	% Ca	% F
U.S.A.	15.4	35.2	3.8
Morocco	15.3	37.9	4.1
U.S.S.R.	17.6	37.4	3.7
Pacific	17.1	38.4	3.8
$Ca_5(PO_4)_3F$	18.4	39.7	3.8

2 $[Ca^{2+}]^5[PO_4^{3-}]^3[OH^-] = [10^{-2}]^5[2.7 \times 10^{-6}]^3[10^{-6}] = 2 \times 10^{-33}$

Sea is supersaturated with respect to hydroxyapatite assuming *all* phosphorus is inorganic phosphate. However it is likely to be present in other forms and assuming sea is saturated with respect to hydroxyapatite then $[PO_4^{3-}]$ calculated on this basis is 10^{-12} M considerably less than the total phosphorus present.

Chapter 2

1 $\cos \beta = \sqrt{1 - \tfrac{4}{3} \sin^2 (\alpha/2)}$; $\gamma = 109°28'$, i.e. the tetrahedral angle.

2 (i) six,   (ii) two.

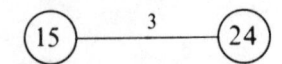

3 (15) —3— (24)

Chapter 3

1 References will be found in *Topics in Phosphorus Chemistry*, Vol. 5; ^{31}P n.m.r.

(a) $(CH_3O)_3P$
(b) $(CF_3)_2PBr$
(c) $(CH_3S)PBr_2$
(d) Cl_2PCH_2Cl
(e) $(CH_3)_2NPF_2$
(f) $(CF_3)_3P=O$

(g) $C_6H_5OPF_2$ (with P=O)

(h) $(CH_3O)_2PH$ (with P=O)

(i) $(C_2H_5O)_2PH$ (with P=S)

(j) $(CH_3)_2P-P(CH_3)_2$ (with P=S on each)

(k) $(C_6H_5)_3\overset{+}{P}CH_3\ Br^-$

(l) $(CH_3)_3\overset{+}{P}H\ Cl^-$

(m) cyclotriphosphazene with C_6H_5O, OC_6H_5, C_6H_5O-P, $P-OC_6H_5$, C_6H_5O, OC_6H_5 substituents

(n) CF_3PF_4

(o) $(CH_3)_2NP(C_6H_5)F_3$

(p) 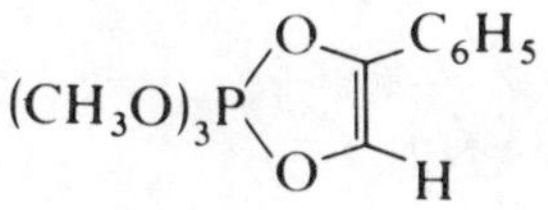

(q) $ClC_6H_4\overset{+}{N_2}\,\overset{-}{PF_6}$

2 In (1) F is equatorial (two apical–equatorial rings); in (2) F is apical because the phospholane ring may adopt a diequatorial situation [see G. O. Doak and R. Schmutzler, *Chem. Commun.*, 476 (1970)].

3 Phosphine (3) isomerizes to phosphine sulphide:

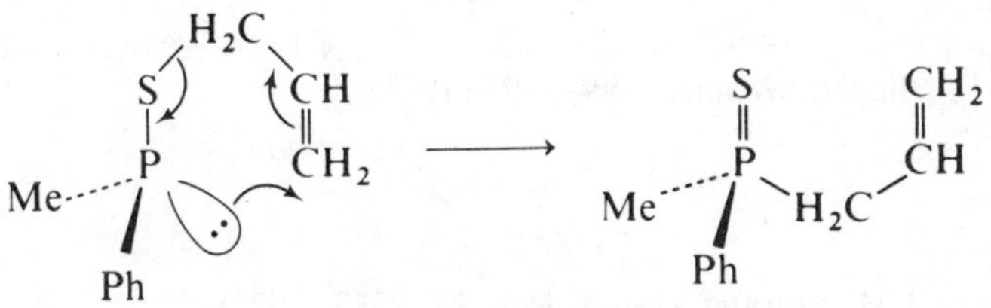

[W. B. Farnham, A. W. Herriott and K. Mislow, *J. Amer. Chem. Soc.*, **91**, 6878 (1969)].

4 Greater degree of pπ–dπ bonding as electron-donating power of group on α-carbon increases causes increased shielding of P atom from left to right [R. Koster, D. Simic and M. A. Grassberger, *Annalen*, **739**, 211 (1970)].

5 πEtO = 2.9, [νP=O, $(EtO)_2P(O)NMe_2)$] = 1258 cm^{-1} (observed, 1250 cm^{-1});
[νP=O, $(Me_2N)_2P(O)Cl$] = 1258 cm^{-1} (observed, 1241 cm^{-1});
[νP=O, $(EtO)POCl_2$] = 1298 cm^{-1} (observed, 1295 cm^{-1}).

6 (a) MeP(O)(OH)(OMe): note hydrogen-bonding broad bands above 1500 cm^{-1} and 995 cm^{-1}, sharp νP—Me at 1300, νP=O at 1200, and νP—O—Me at 1055 cm^{-1}.

(b) $MeP(S)Cl_2$: νP—Me at 1290 cm^{-1}, ($\delta_{as}CH_3$ at 1390 cm^{-1}, ν_s and ν_{as} CH_3 at *ca* 3000 cm^{-1}). No evidence of P—O links but νP=S at 660 + 780 cm^{-1}, typical doublet often found with this vibration, and ν_sPCl_2 and $\nu_{as}PCl_2$ at 445 and 500 cm^{-1}.

Chapter 4

1 Attributed to stabilization of the planar transition state for the inversion process by (2p—3p)π delocalization of the lone pair on phosphorus into the phosphole ring [W. Egan, R. Tang, G. Zon and K. Mislow, *J. Amer. Chem. Soc.*, **92**, 1442 (1970)].

2 See text, sections 4.1.1–4.1.4 and 4.3. The Arbusov and Perkow reactions are the most important examples.

3 (a)

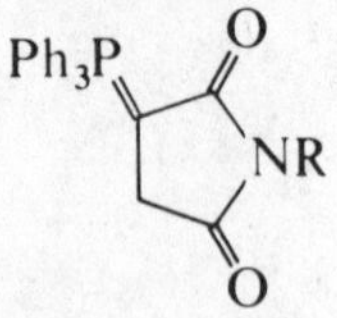

[Structure: triphenylphosphonium ylide of N-substituted succinimide — Ph₃P= at α-position of succinimide with NR]

[H. Hedaya and S. Theodoropulos, *Tetrahedron*, **24**, 2241 (1968)].

(b)

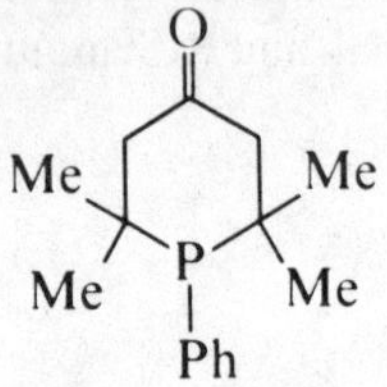

[Structure: 2,2,6,6-tetramethyl-1-phenyl-phosphinan-4-one]

[F. Asinger, A. Saus and E. Michel, *Monatsch*, **99**, 1695 (1968)].

(c) [3-pyridyl]−N=PPh₃

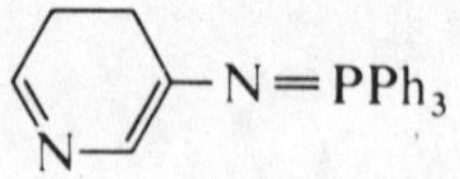

[T. Sasaki, K. Kanematsu and M. Murata, *Tetrahedron*, **28**, 2383 (1972)].

(d)

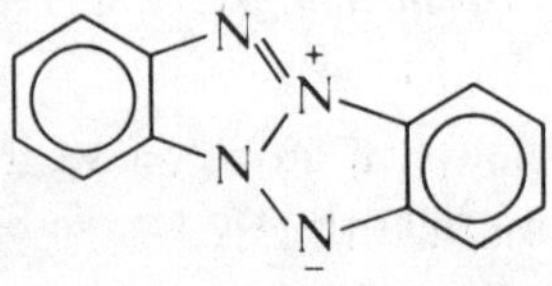

[Structure: dibenzo-fused tetrazapentalene with N⁺ and N⁻]

[J. C. Kauer and R. A. Carboni, *J. Amer. Chem. Soc.*, **89**, 2633 (1967)].

(e)

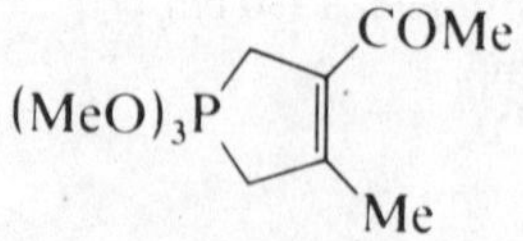

(MeO)₃P−[cyclopentene with COMe and Me substituents]

[F. Ramirez, R. V. Patwardhan and S. R. Heller, *J. Amer. Chem. Soc.*, **86**, 514 (1964)].

(f)

(EtO)₂P(O)CH₂CH=CHMe

[A. N. Pudovik and I. M. Aladzhyera, *Zh. Obsch. Khim.*, **33**, 3096 (1963)].

(g)

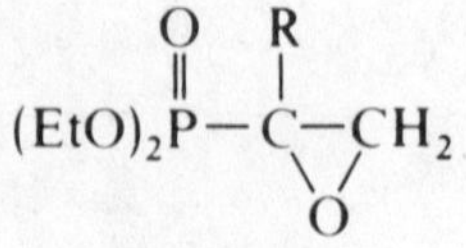

(EtO)₂P(O)−C(R)−CH₂
                  _O_/

[G. W. Kenner, *Proc. Chem. Soc.*, 136 (1957)].

(h)

R¹₂N−P(O)(OR)₂

[K. A. Petrov and G. A. Skolakii, *Zh. Obsch. Khim.*, **26**, 3378 (1956) [*Chem. Abs.*, **51**, 8029 (1957)].

(i) $Cl_2C=CH-CH=CPh-OP(OEt)_2 + EtCl$

  [J. F. Allen and O. H. Johnson, *J. Amer. Chem. Soc.*, **77**, 2871 (1955)].

(j) Dibenzo[b,d][1,3,2]dioxaphosphole with P–Cl substituent.

  [L. D. Freedman, G. O. Doak and J. R. Edmiston, *J. Org. Chem.*, **26**, 284 (1961)].

4 (a) Nucleophilic attack on Br; via Ph–C=C–Ph episulfone (thiirene dioxide, SO$_2$).

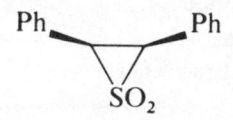

  [F. Bordwell, R. B. Jarvis and P. W. R. Corfield, *J. Amer. Chem. Soc.*, **90**, 5298 (1968)].

  (b) Nucleophilic attack by the thiophosphoryl group to form $R_3P^+-SSiCl_3\ SiCl_3^-$ followed by nucleophilic attack on S to displace R_3P [K. Mislow *et al.*, *J. Amer. Chem. Soc.*, **91**, 7023 (1969)].

  (c) Nucleophilic displacement on the peroxidic oxygens; attack on one oxygen leads to $PhRC=CH_2$; attack on the second leads to $CH_2=C=O$ [W. Adam, R. J. Ramirez and S-C. Tsai, *J. Amer. Chem. Soc.*, **91**, 1254 (1969)].

  (d) Nucleophilic attack on C of the epoxide ring, followed by methylation of phosphorus and collapse of the betaine to an olefin and $Ph_2P(O)Me$ [E. Vedige and P. L. Fuchs, *J. Amer. Chem. Soc.*, **93**, 4070 (1971)].

  (e) Nucleophilic addition to the α,β-unsaturated aldehyde followed by an intermolecular de-alkylation [G. Kamai and V. A. Kukhtin, *Zh. Obsch. Khim.*, **27**, 2376 (1957) [*Chem. Abs.*, **52**, 7127 (1958)]].

  (f) Ring opening to zwitter ion followed by nucleophilic attack on bromine to form $CH_3C(Br)(COCH_3)-O-P^+(OR)_3\ Br^-$ which de-alkylates to products [F. Ramirez and M. B. Desai, *J. Amer. Chem. Soc.*, **85**, 3252 (1963)].

  (g) Deoxygenation via a nitrene intermediate [P. J. Bunyan and J. L. G. Cadogan, *J. Chem. Soc.*, 42 (1963)].

  (h) Nucleophilic attack on S (rather than C) indicated by the formation of *cis*-olefin from the *meso* episulphide [N. P. Neureiter and F. G. Bordwell, *J. Amer. Chem. Soc.*, **81**, 578 (1959)].

  (i) Nucleophilic displacement on halogen (bromine) by Ph_3P [I. J. Borowitz and R. Virkhaus, *J. Amer. Chem. Soc.*, **85**, 2183 (1963)].

(j) Nucleophilic attack on P(III) by the oxygen of the epoxide [M. I. Kabachnik *et al.*, *Chem. Abs.*, **55**, 14288 (1961)].

(k) Deprotonation of the starting material followed by intramolecular rearrangement and expulsion of Cl⁻ [I. S. Bengelsdorf, *J. Org. Chem.*, **21**, 475 (1956)].

Chapter 5

1

Complex	A_1 νCO calcd/(cm^{-1})	A_1 νCO observed/(cm^{-1})
Ni(CO)$_3$(PH$_2$Ph)	2076.0	2077.0
Ni(CO)$_3$(PHPh$_2$)	2073.0	2073.3
Ni(CO)$_3${PMe$_2$(CF$_3$)}	2080.9	2080.7
Ni(CO)$_3${PMe(CF$_3$)$_2$}	2097.9	not recorded

χOPr^i will be slightly less than χOEt, 6.8, which in turn is 0.9 less than χOMe at 7.7. A reasonable guess would be about 6.0 ($\pm$0.5) which leads to A_1 νCO for Ni(CO)$_3${P(OPr^i)$_3$} of 2074.1 cm^{-1}. The observed value is 2075.9 cm^{-1} and in fact χOPr^i is 6.3.

2 Neither nd$_{x^2-y^2}$ nor nd$_{z^2}$ is suitable for π bonding; this leaves nd$_{xy}$, nd$_{xz}$ and nd$_{yz}$ to be considered. For the ligands A and B attached to metal M in which the A—M bond lies along the x axis only the nd$_{xy}$ and nd$_{xz}$ are relevant.

cis-[MAB]
one nd orbital in common

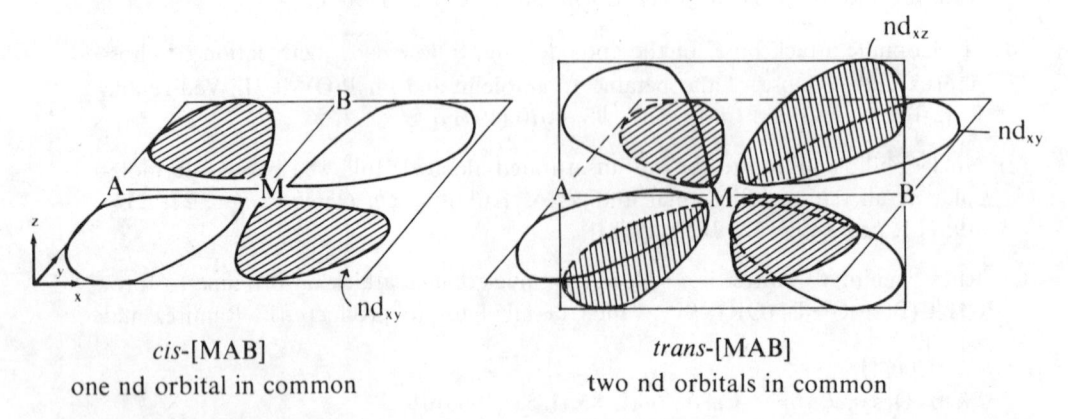

trans-[MAB]
two nd orbitals in common

3 Orthophenylation is favoured for geometric and chelation reasons and not for any electronic effects associated with the benzene ring:

Chapter 6

1 Initial hydrolysis step:

$$(C_2H_5O)_2P(OC_2D_5)_3 \xrightarrow{H_2O} C_2H_5OH + C_2D_5OH + (C_2H_5O)_2P(OC_2D_5)_2\!\!-\!\!OH$$
$$\phantom{(C_2H_5O)_2P(OC_2D_5)_3 \xrightarrow{H_2O}\,} 40\% \quad\;\; 60\% \qquad 60\%$$

$$+ (C_2H_5O)P(OC_2D_5)_3\!\!-\!\!OH$$
$$ 40\%$$

Then,

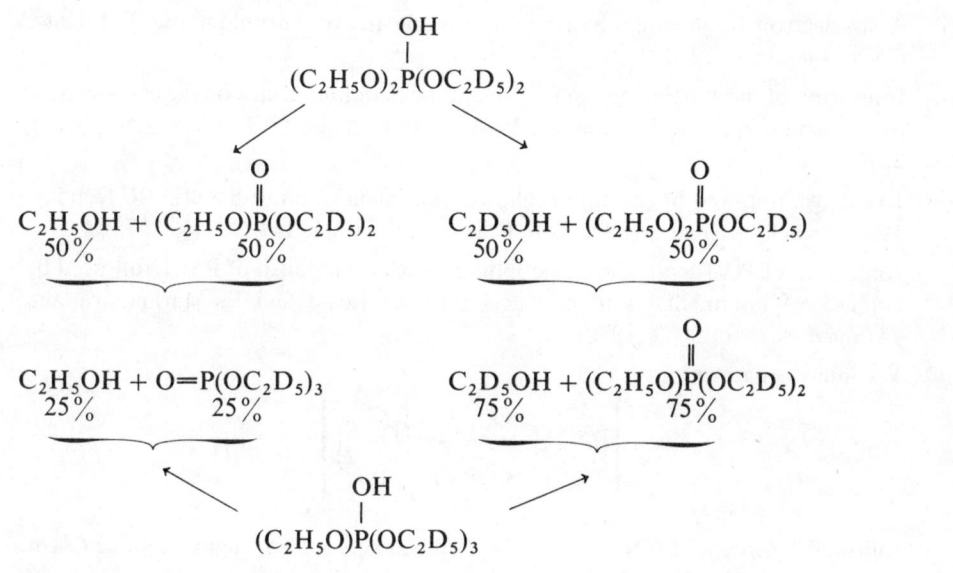

$$(C_2H_5O)_2P(OC_2D_5)_2\!\!-\!\!OH$$

branches to:

- C_2H_5OH (50%) + $(C_2H_5O)P(OC_2D_5)_2\!=\!O$ (50%)
- C_2D_5OH (50%) + $(C_2H_5O)_2P(OC_2D_5)\!=\!O$ (50%)

and $(C_2H_5O)P(OC_2D_5)_3\!\!-\!\!OH$ gives:

- C_2H_5OH (25%) + $O=P(OC_2D_5)_3$ (25%)
- C_2D_5OH (75%) + $(C_2H_5O)P(OC_2D_5)_2\!=\!O$ (75%)

C_2H_5OH	C_2D_5OH	
$0.5 \times 0.6 = 0.3$	$0.5 \times 0.6 = 0.3$	i.e. in ethanol
$0.25 \times 0.4 = 0.1$	$0.75 \times 0.4 = 0.3$	40 % C_2H_5OH and
$0.4 = 0.4$	$0.6 = 0.6$	60 % C_2D_5OH
0.8	1.2	

In phosphate esters:

$(C_2H_5O)_2P(OC_2D_5)\!=\!O \qquad (C_2H_5O)P(OC_2D_5)_2\!=\!O \qquad (C_2D_5O)_3P\!=\!O$

$0.5 \times 0.6 = 0.3 \qquad\quad 0.5 \times 0.6 = 0.3 \qquad\quad 0.25 \times 0.4 = 0.1$

$ 0.75 \times 0.4 = 0.3$

% D $= 0.3 \times \tfrac{1}{3} = 0.1 \qquad\quad \overline{0.6}$

 % D $= 0.6 \times \tfrac{2}{3} = 0.4 \qquad$ % D $= 0.1 \times 1 = 0.1$

deuterium in phosphate ester $= (0.6/1) \times 100 = 60\%$

532   Answers to problems

2   (i) Each apex represents one pseudorotamer with numbers corresponding to apical ligands; 21 is the enantiomer of 12.

(ii) Each line represents a pseudorotation connecting two pseudorotamers; e.g. the line connecting 12 and 34 is a pseudorotation using 5 *as the pivot*. Racemization is impossible because all pathways interconverting enantiomers (e.g. 12 and 21) lead through pseudorotamers in which the ring either becomes diequatorial (23, 32, 24, 42, 34 or 43) or would be impossibly strained by having the ring span a diapical position (15 and 51). The allowed ψ pathways are denoted by the two "chair" hexagons in heavy type and these two hexagons are effectively "isolated" from each other.

3   (i) A six-electron fragmentation of a phosphorane [E. W. Turnblom and T. J. Katz, *Chem. Commun.*, 1270 (1972)].

(ii) Ionization of the starting phosphorane and nucleophilic attack on the carbon atom of the isothiocyanate [F. Ramirez, V. A. V. Prasad and H. J. Bauer, *Phosphorus*, **2**, 185 (1973)].

(iii) Exchange, followed by intramolecular displacement of phosphate [D. B. Denney, R. L. Powell, A. Taft and D. Twitchell, *Phosphorus*, **1**, 151 (1971)].

(iv) Ionization of P(V) again and nucleophilic attack on halogen of RSCl, followed by collapse of intermediates to products [D. N. Harpp and P. Mathiaparanam, *Tetrahedron Lett.*, 2089 (1970)].

(v) Via intermediate anion,

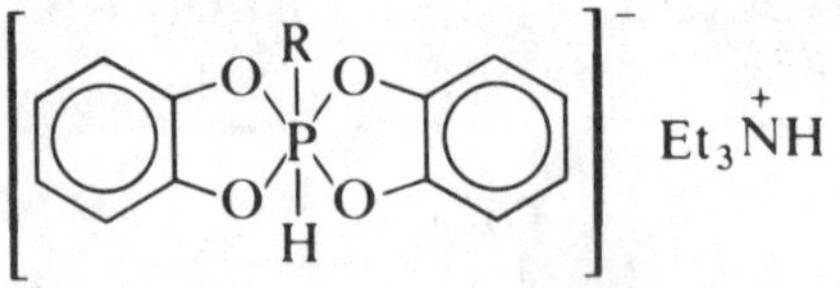

followed by loss of Et_3N and H_2 [M. Wieber and K. Foroughi, *Angew. Chem. Int. Ed. Eng.*, **12**, 419 (1973)].

4

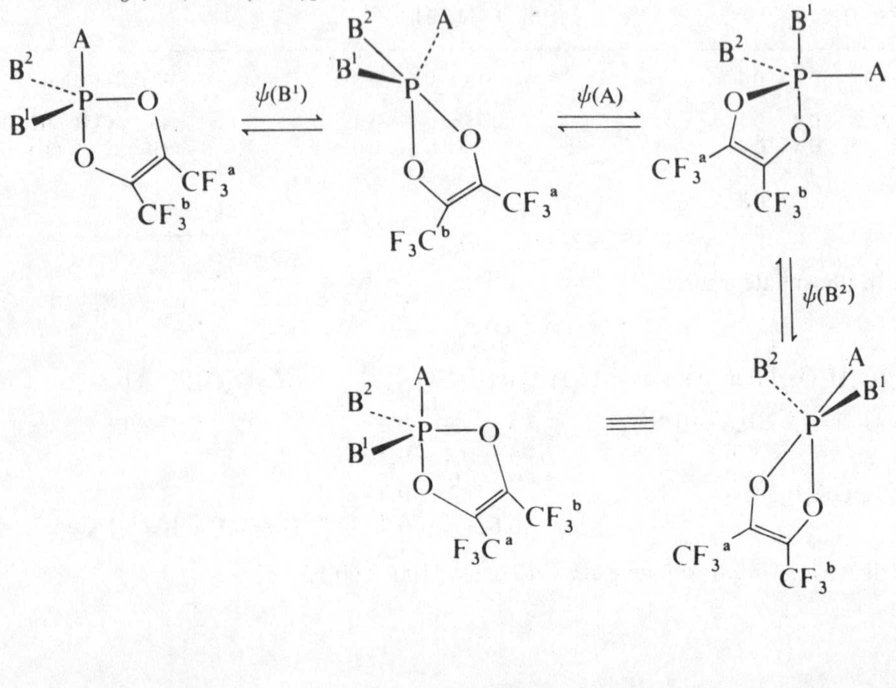

CF$_3$[b] is now in the same situation as CF$_3$[a] in the starting pseudorotamer.

This method has also been used to estimate the relative apicophilicities of A and B [J. I. Dickstein and S. Trippett, *Tetrahedron Lett.*, 2203 (1973)].

Chapter 7

1 (i) *trans*-(7.60) → [Apical entry intermediate] → *cis*-oxide

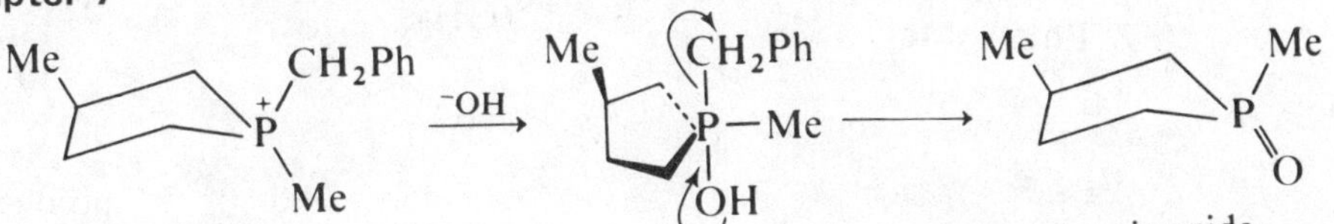

i.e. Apical entry by $^-$OH to give diequatorial ring; leads to inversion.

(ii) *trans*-(7.60) → ... → *cis*-oxide, inversion

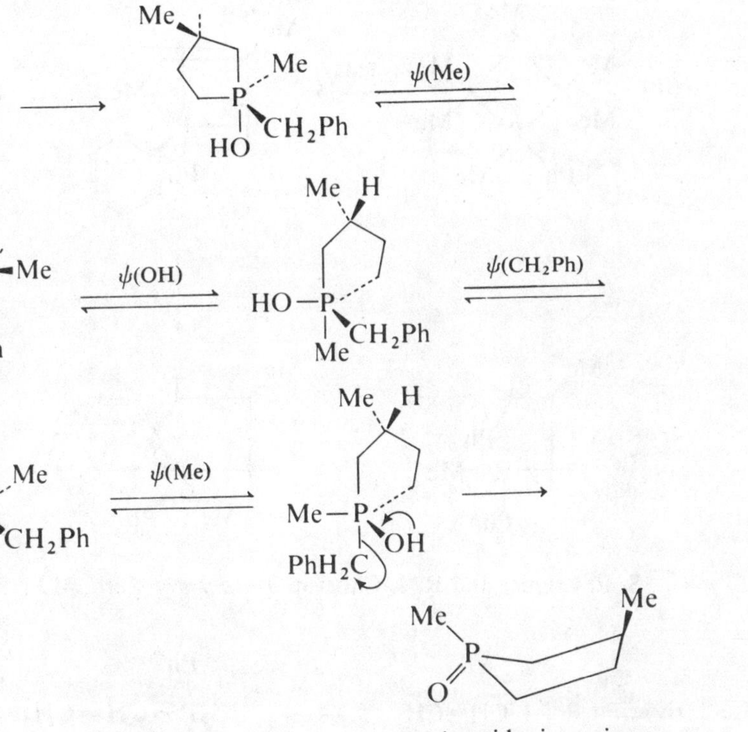

Four pseudorotations lead to *cis*-oxide (i.e. inversion) and therefore full equilibration would lead to racemization at phosphorus.

2 (a) (i)

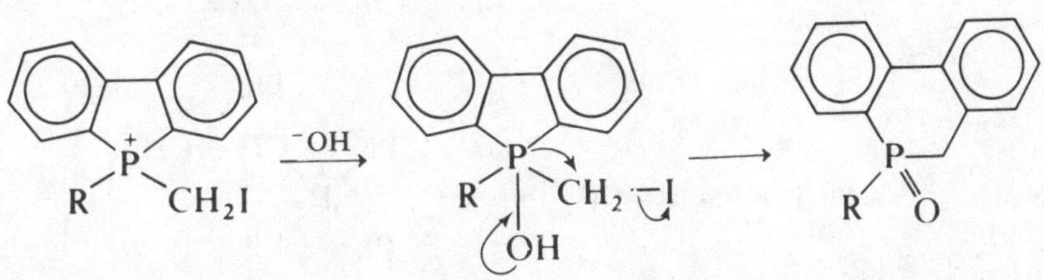

D. W. Allen and J. T. Millar, *J. Chem. Soc. C*, 252 (1969).

(ii) 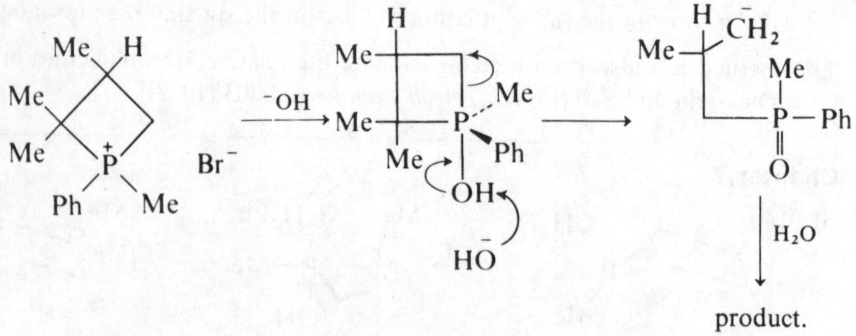 [reaction scheme]

S. E. Fishwick and J. Flint, *Chem. Commun.*, 182 (1969).

(iii) 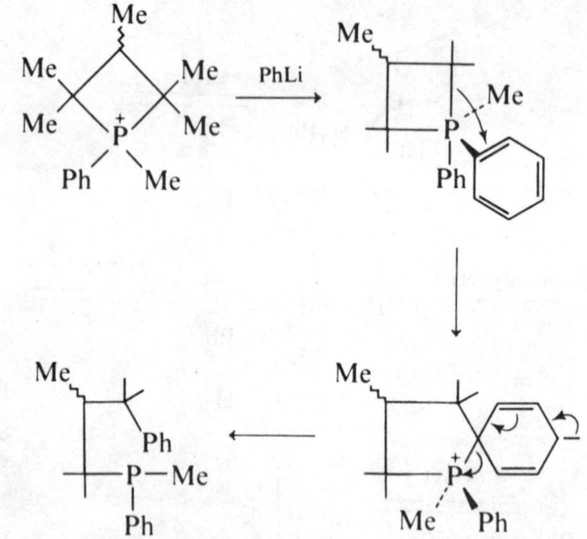 [reaction scheme]

S. E. Cremer and R. J. Chorvat, *Tetrahedron Lett.*, 413 (1968).

(iv) 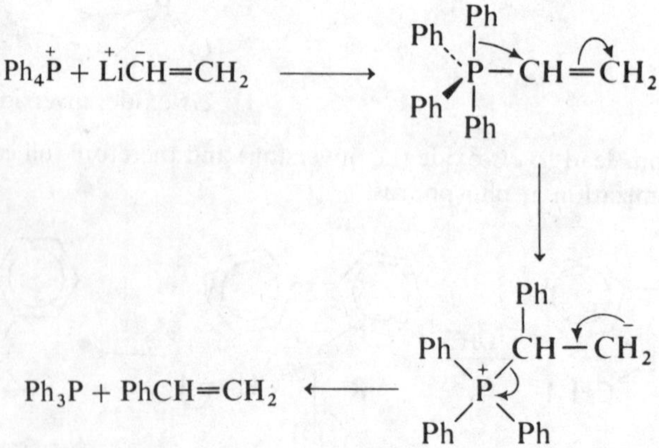 $Ph_4\overset{+}{P} + \overset{-}{Li}\overset{-}{C}H=CH_2 \longrightarrow$ [reaction scheme] $\longrightarrow Ph_3P + PhCH=CH_2$

D. Seyforth, J. S. Fogel and J. K. Heeren, *J. Amer. Chem. Soc.*, **86**, 307 (1964).

(b) *Answer*:

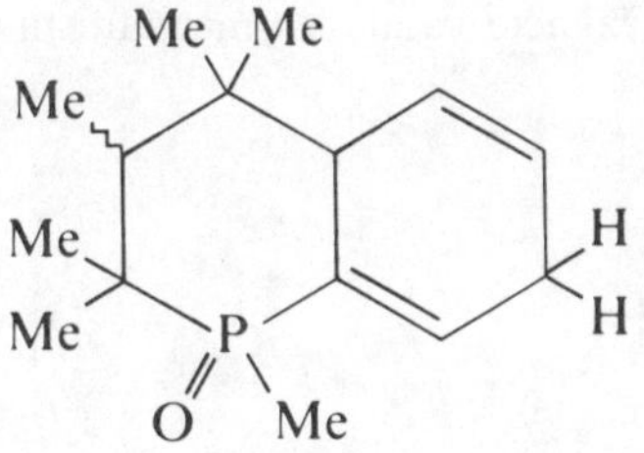

$$\text{[structure: bicyclic phosphine oxide with Me groups, P(=O)Me]}$$

By attack on the *ortho*-carbon of the phenyl ring.
Negated by deuterium labelling experiments.
S. E. Fishwick, J. Flint, W. Hawes and S. Trippett, *Chem. Commun.*, 1113 (1967); and S. E. Cremer, *Chem. Commun.*, 1132 (1968).

3 0.152 g = 23.3/1000 × 0.02 moles

mol. wt. = $\dfrac{0.152 \times 1000}{23.3 \times 0.02}$ = 326.2

Fifteen protons at δ 7.5 suggests three phenyl groups, i.e. 3 × C_6H_5. Since Hg(OAc)$_2$ not required for titration, compound is an ylid and not a phosphonium salt; ^{31}P n.m.r. data is consistent with this. Monobasic suggests, *one* phosphorus atom. i.e. P(C_6H_5) unit = mol. wt of 262. Leaves fraction of mol. wt. = 64; i.e. C_5H_4. Thus the structure is

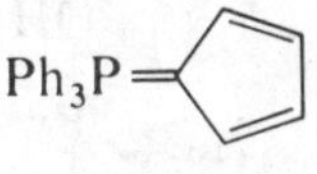

$$Ph_3P=\text{cyclopentadienylidene}$$

4 See the following references:

   (i)   T. Eicher, E. V. Angerer and A-M. Hansen, *Annalen*, **746**, 102 (1971).
   (ii) G. Büchi and H. Wuest, *Helv. Chim. Acta*, **54**, 1767 (1971).
   (iii) W. Flitsch and U. Neumann, *Chem. Ber.*, **104**, 2170 (1971).
   (iv) D. T. Connor and M. von Strandtmann, *J. Org. Chem.*, **38**, 1047 (1973).
   (v)   P. Ykman, G. L. Abbé and G. Smets, *Tetrahedron*, **27**, 845 (1971).

Chapter 8

1 Intramolecular catalysis of type:

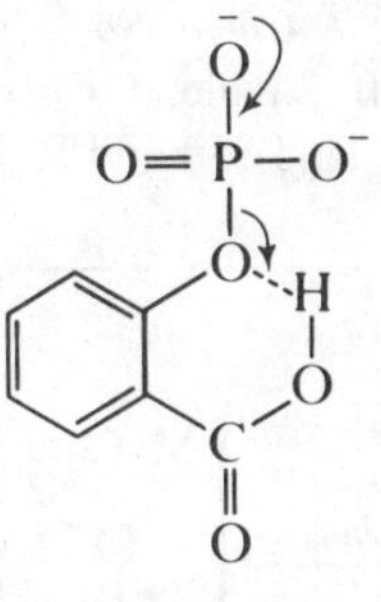

$$\text{[structure showing salicylate phosphate with O=P-O}^-\text{, intramolecular H-bond to carbonyl]}$$

The lack of a rate enhancement with (3) is evidence against nucleophilic catalysis. See J. D. Chanley *et al.*, *J. Amer. Chem. Soc.*, **74**, 4347 (1952); M. L. Bender and J. M. Lawlor, *J. Amer. Chem. Soc.*, **85**, 3010 (1963).

Answers to problems

2 Hydrolysis is alleged to proceed by a general acid-catalyzed mechanism via zwitterionic structures e.g.,

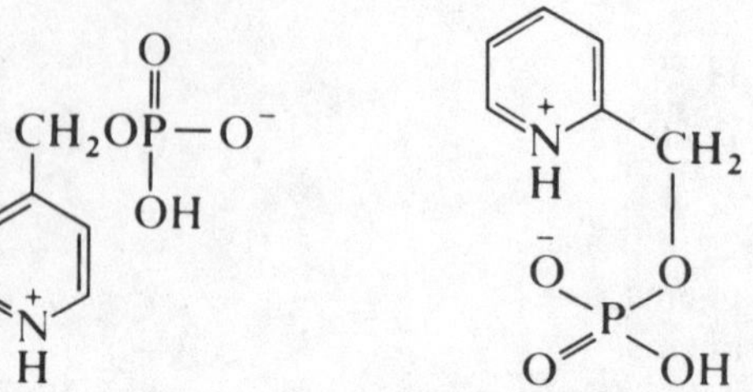

[Structures of zwitterionic pyridinium phosphate intermediates]

The enhanced rate for the α-isomer is due to intramolecular proton transfer [M. Murakami and M. Tayaki, *J. Amer. Chem. Soc.*, **91**, 5130 (1969)].

3 (a)

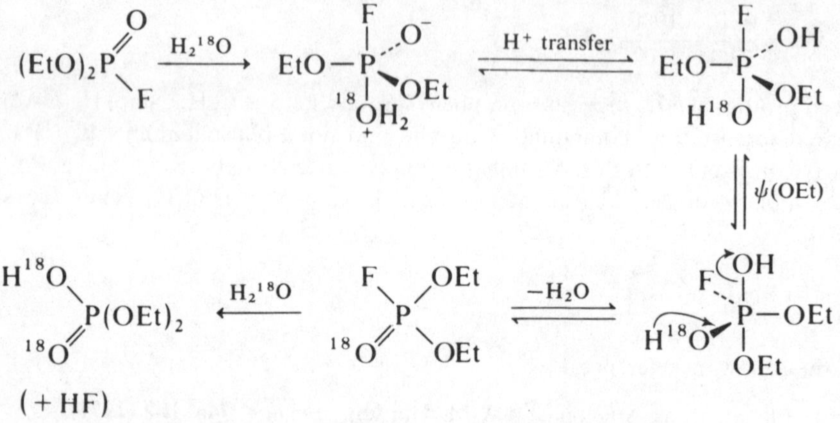

[Mechanism scheme showing $(EtO)_2P(=O)F$ reacting with $H_2^{18}O$ through pentacoordinate intermediates, pseudorotation $\psi(OEt)$, loss of H_2O, and $-HF$ to give $^{18}O=P(OEt)_2(^{18}OH)$ (+ HF)]

(b) If X is OEt, leaving group would be $^-$OEt which is a stronger base, and a poorer leaving group than $^-$OH. Thus the lack of ^{18}O exchange cannot be attributed to the leaving group ability of X, at least in the case of phosphate esters.

4 (i) Class 2a: F. L. Scott, R. Riorden and P. D. Morton, *J. Org. Chem.*, **27**, 4255 (1962).
(ii) Class 2c: L. Larsson, *Svensk kem Tidskr.* **70**, 405 (1958).
(iii) Class 2b: H. H. Wasserman and D. Cohen, *J. Amer. Chem. Soc.*, **82**, 4435 (1960).
(iv) Class 2d: J. I. G. Cadogan, *J. Chem. Soc.*, 3067 (1961).
(v) Class 2a: A. L. Green and B. Saville, *J. Chem. Soc.*, 3887 (1956).
(vi) Class 2b: G. W. Kenner, A. R. Todd and R. F. Webb, *J. Chem. Soc.*, 1231 (1956).

Chapter 9

1 (a)

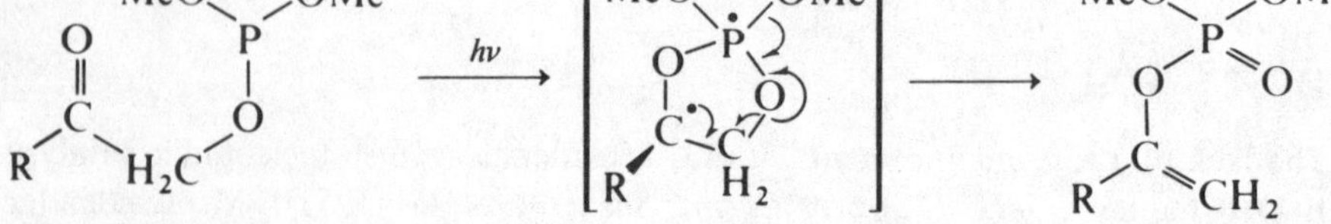

[Photochemical rearrangement scheme: acyl phosphite $R-C(=O)-CH_2-O-P(OMe)_2$ under hv through biradical intermediate to give enol phosphate $R-C(=CH_2)-O-P(=O)(OMe)_2$]

C. E. Griffin, W. G. Bentrude and C. M. Johnson, *Tetrahedron Lett.*, 969 (1969).

(b) $(PhP)_5 \xrightarrow{heat} 5PhP\colon$

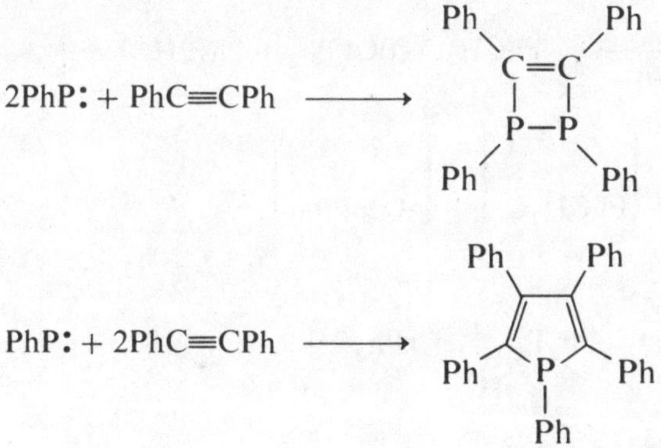

$2PhP\colon + PhC{\equiv}CPh \longrightarrow$ [cyclic product with C=C double bond, two Ph on carbons, P–P bond, two Ph on phosphorus atoms]

$PhP\colon + 2PhC{\equiv}CPh \longrightarrow$ [phosphole ring with four Ph on carbons and one Ph on phosphorus]

A. Ecker and U. Schmidt, *Chem. Ber.*, **106**, 1453 (1973).

(c) [bicyclic phosphine oxide with Ph on P] $\xrightarrow{heat}$ [dimethylnaphthalene] + $\left[\begin{array}{c} O \\ \| \\ PhP\colon \end{array} \right]$ $\xrightarrow{MeOH}$ $Ph-\underset{MeO}{\overset{\overset{\displaystyle O}{\|}}{P}}H$

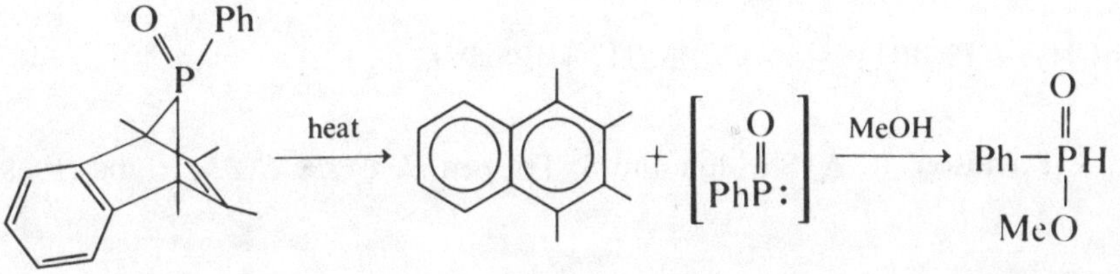

J. K. Stille, J. L. Eichelberger, J. Higgins and M. E. Freeburger, *J. Amer. Chem. Soc.*, **94**, 4761 (1972).

(d) $Ph_3P{=}$[tetramethylcyclopentadiene] $\xrightarrow{Na}$ $\left[Ph_3\dot{P}-\text{[cp]}^- \; Na^+ \right]$ $\longrightarrow$ $\left[Ph_2\overset{+\bullet}{P}-\text{[cp]}^- \right]$

$+$

$PhNa$

$\Big\downarrow H_2O$

$Ph_2\overset{H}{\underset{O}{P}}$[cp]$ + \tfrac{1}{2}H_2 \longleftarrow \left[Ph_2\dot{P}\underset{OH}{\overset{H}{-}}\text{[cp]} \right] \xleftarrow{H_2O} \left[Ph_2\overset{+\bullet}{P}-\text{[cp]}^- \right]$

$+$

$PhH + NaOH$

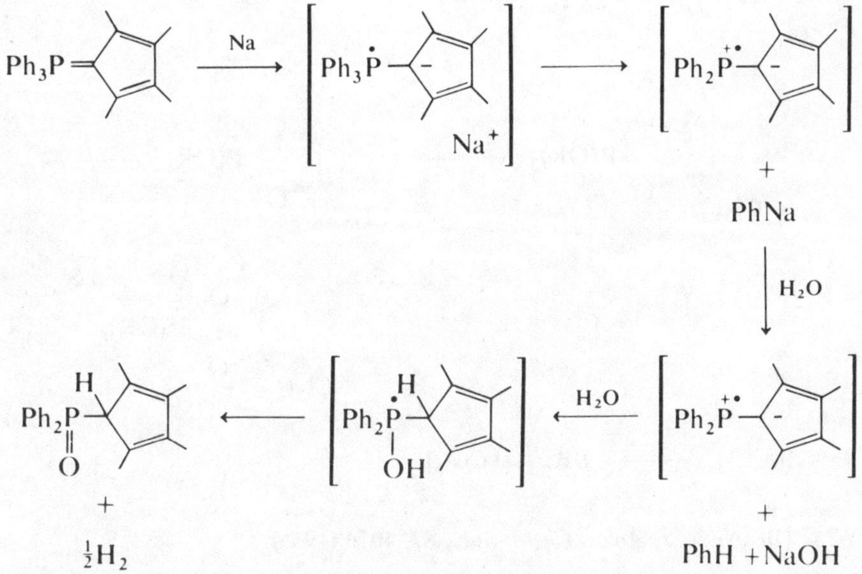

J. Kauffmann, G. Ruckelhauss and D. Glindemann, *Chem. Ber.*, **106**, 1453 (1973).

(e) $Ph_2P.PPh_2 \xrightarrow{heat} 2Ph_2P\cdot$

$2Ph_2P\cdot + PhCH_2C(O)OH \longrightarrow PhCH_2C(O)OPPh_2 + Ph_2PH$

$PhCH_2C(O)OPPh_2 \xrightarrow{heat} \left[PhCH_2\overset{\overset{O}{\|}}{C}\cdot \right] + \left[\cdot O-PPh_2 \right]$

$Ph_2PO\cdot + Ph_2PH \longrightarrow Ph_2\overset{\overset{O}{\nearrow}}{\underset{H}{P}} + [Ph_2P\cdot]$

$PhCH_2\overset{\overset{O}{\|}}{C}\cdot \longrightarrow CO\uparrow + [PhCH_2\cdot]$

$PhCH_2\cdot + Ph_2PH \longrightarrow PhCH_3 + [Ph_2P\cdot]$

R. S. Davidson, R. A. Sheldon and S. Trippett, *J. Chem. Soc. C*, 1700 (1968).

(f) $BrCCl_3 \xrightarrow{h\bar{\nu}} Br\cdot + \cdot CCl_3$

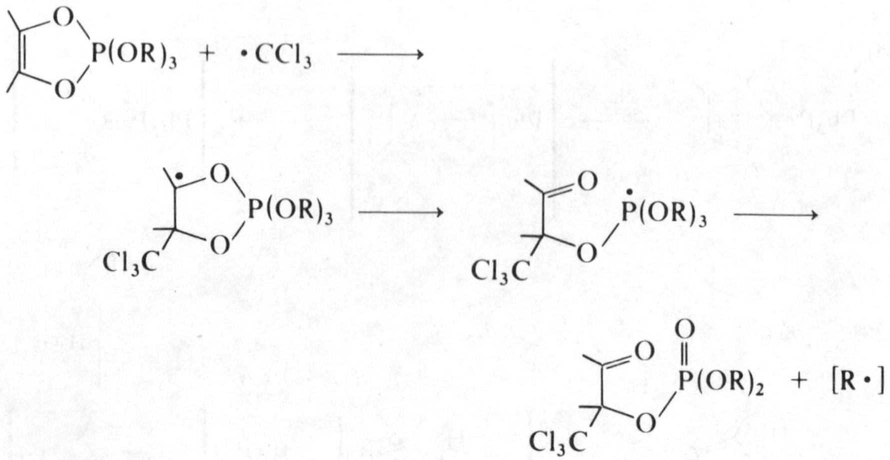

[structure: dioxaphospholene + $\cdot CCl_3 \longrightarrow$]

[structure with Cl_3C and $P(OR)_3 \longrightarrow$ structure with Cl_3C and $\dot{P}(OR)_3 \longrightarrow$]

[final structure with Cl_3C and $P(OR)_2$] $+ [R\cdot]$

$R\cdot + BrCCl_3 \longrightarrow RBr + [\cdot CCl_3]$

W. G. Bentrude, *J. Amer. Chem. Soc.*, **87**, 4026 (1965).

(g)

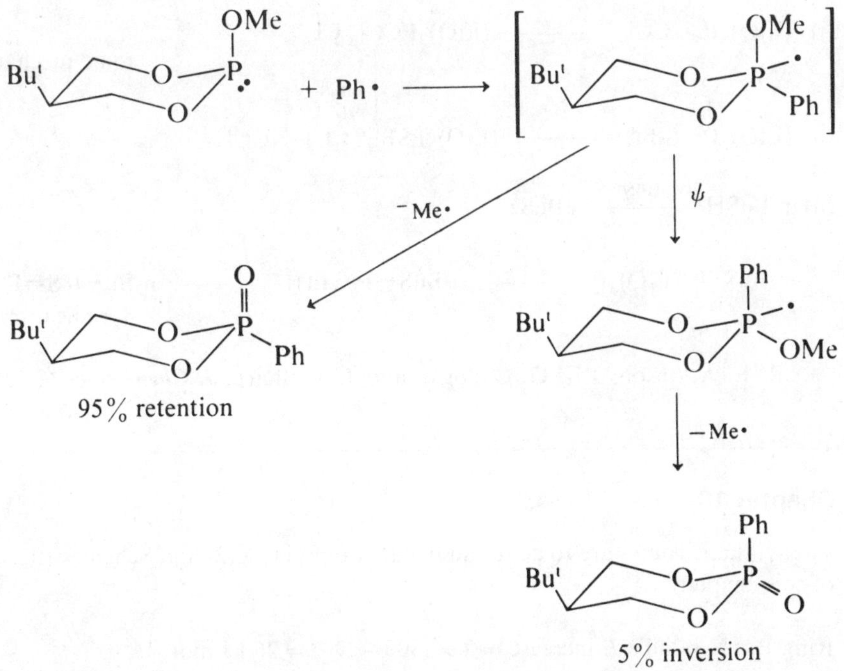

[Reaction scheme: cyclic phosphite with OMe and Bu^t substituent reacting with Ph• to form a phosphoranyl radical intermediate, which undergoes ψ (pseudorotation) and loss of Me• to give products with 95% retention and 5% inversion]

W. G. Bentrude and K. C. Yee, *Tetrahedron Lett.*, 3999 (1970).

2 (a) $R^1SSR^1 \xrightarrow{AIBN} 2R^1S\cdot$

$R^1S\cdot + (RO)_3P \longrightarrow (RO)_3\dot{P}SR^1$

$(RO)_3\dot{P}SR^1 \longrightarrow (RO)_3P=S + R^1\cdot$

$R^1\cdot + CO \longrightarrow R^1-\overset{O}{\underset{\cdot}{C}}$

$R^1\overset{O}{\underset{\cdot}{C}} + R^1S-SR^1 \longrightarrow R^1\overset{O}{\underset{\|}{C}}-SR^1 + R^1S\cdot$

C. Walling, O. H. Basedow and E. Savas, *J. Amer. Chem. Soc.*, **82**, 2181 (1960).

(b) $CH_2=CH(CH_2)_3CH_2SH \xrightarrow{AIBN} CH_2=CH(CH_2)_3CH_2S\cdot$

$CH_2=CH(CH_2)_3CH_2S\cdot + (RO)_3P \longrightarrow CH_2=CH(CH_2)_3CH_2\dot{S}P(OR)_3$

$CH_2=CH(CH_2)_3CH_2\dot{S}P(OR)_3 \longrightarrow CH_2=CH(CH_2)_3CH_2\cdot + S=P(OR)_3$

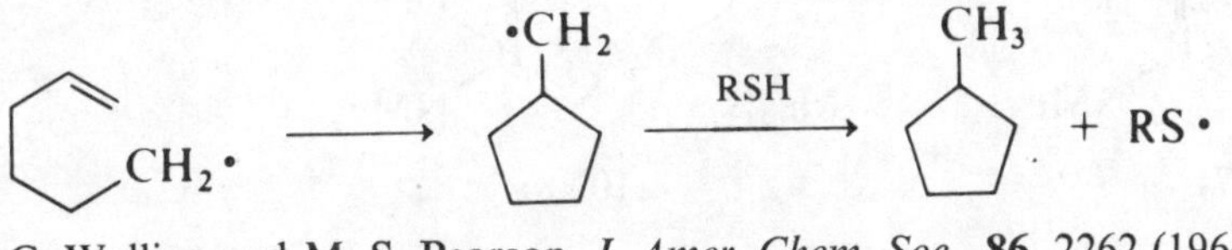

[cyclization: hexenyl radical → cyclopentylmethyl radical ·CH₂ → (with RSH) → methylcyclopentane CH₃ + RS·]

C. Walling and M. S. Pearson, *J. Amer. Chem. Soc.*, **86**, 2262 (1964).

(i) $(EtO)_3P + CCl_4 \longrightarrow (EtO)_3\overset{+}{P}CCl_3\ \overset{-}{Cl}$ — ionic mechanism

$(EtO)_3\overset{+}{P}CCl_3\ \overset{-}{Cl} \xrightarrow{\text{n-BuSH}} (EtO)_3\overset{+}{P}SBu^n\ \overset{-}{Cl} + HCCl_3$

$(EtO)_3\overset{+}{P}SBu^n\ \overset{-}{Cl} \longrightarrow (EtO)_2P(=O){-}SBu^n$

(ii) $\text{n-BuSH} \xrightarrow{\text{AIBN}} \text{nBuS}\cdot$

$\text{n-BuS}\cdot + (EtO)_3P \longrightarrow \text{n-BuS}{-}\dot{P}(OEt)_3 \longrightarrow \text{n-Bu}\cdot + S{=}P(OEt)_3$

radical mechanism

R. E. Atkinson, J. I. G. Cadogan and J. T. Sharp, *J. Chem. Soc. B*, 138 (1969).

Chapter 10

1 The synthetic routes are to be found in equations (11), (12) and Schemes 10.2 and 10.3 of this chapter.

2 Ring P$\overset{...}{=}$N bond's E increase $= 2 \times (303-290) = 26$ kJ mol^{-1}.

Compound	Ring bond's increase	Exocyclic increase	Overall increase
$(NPCl_2)_3$	26	$2 \times (336-319)$	60 kJ mol^{-1}
$(NPMe_2)_3$	26	$2 \times (286-256)$	86 kJ mol^{-1}
$(NPPh_2)_3$	26	$2 \times (337-282)$	136 kJ mol^{-1}
$(NP(OC_6H_{11})_2)_3$	26	$2 \times (566-385)$	388 kJ mol^{-1}

From these figures it would appear that exocyclic groups which can π bond to the phosphorus do so, and form stronger π bonds than those of the ring.

3 N(B) is more basic (Table 10.14) and this leads to adduct formation between this ring nitrogen atom and SbF_3 as a primary step to non-*gem* fluorination of the adjacent chlorines to give (10.37). On the other hand KSO_2F behaves as a nucleophile and directly attacks the more electrophilic phosphorus to give the *gem*-difluoro product (10.38).

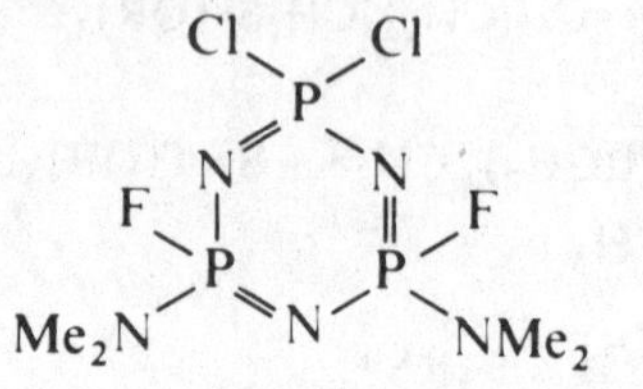

(10.37)

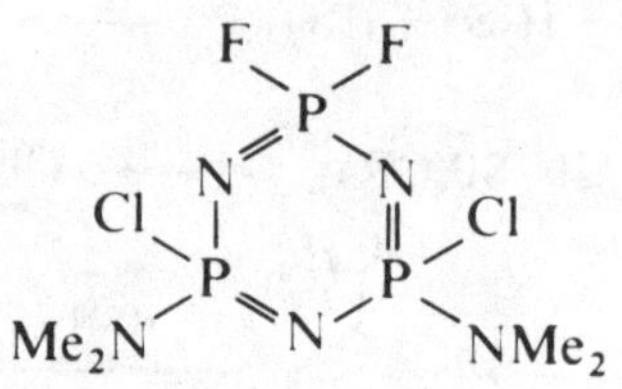

(10.38)

Appendix

Chapter 11

1   188 kJ mol^{-1}.

2   The shorter P—N bond (*ca* 170 pm) compared to the P—P bond (*ca* 220 pm) will result in greater steric repulsion of the methyl groups and so weaken the P—N bond, see ref. 60.

3   Maximum π overlap is possible only when the molecule is planar:

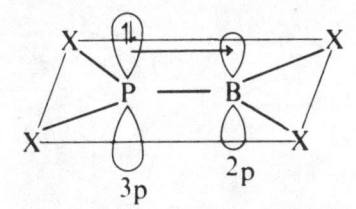

Orbital sizes are different and this prevents effective overlap.

4   Both lone pairs can donate if they are perpendicular to each other which will give a staggered structure of planar phosphorus groups.

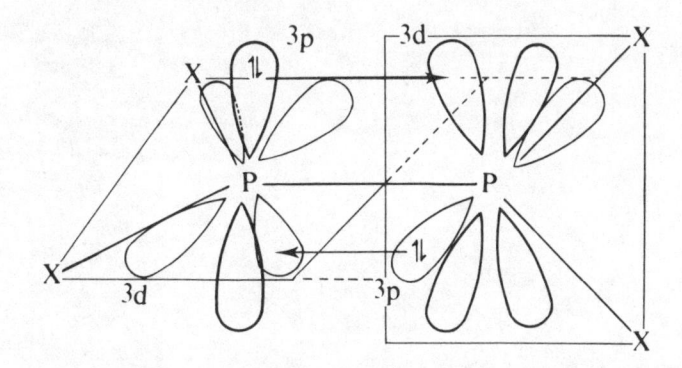

Appendix

1   (a) (chloromethyl)methylphosphine
    (b) S-isopropyl ethylmethylphosphinothioate
    (c) 2-dimethylamino-2,2-diphenoxy-4,4,5,5-tetrakis(trifluoromethyl)-1,3,2-dioxaphospholan [The ring takes precedence over the fact that phosphorus is penta-coordinate]
    (d) 1-bromo-2,2,3,4,4-pentamethylphosphetan 1-oxide
    (e) N-ethyltrichloro(mono)phosphazene [(mono) can be omitted]
    (f) phosphorchloridofluoridic acid

2   (a) P(O)Me(NEt$_2$)(SPr^i)
    (b) Me$_3$P=CH.SiMe$_3$

(c)

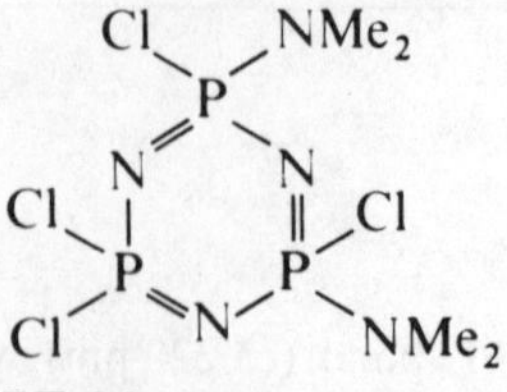

[The positions of the chlorine atoms are implied by the positions of the amino groups and so may be excluded from the name.]

(d) PF_3Me_2

(e) $[P(OPh)_6]^-$

(f)

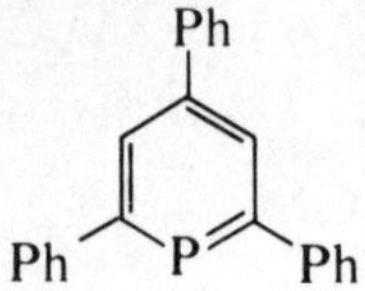

Index

Acaricides, 495

Acetyl choline, as initiator of muscular contraction, 503

Acetyl cholinesterase
  inhibition by organophosphorus compounds, 503–508
  in hydrolysis of acetyl choline, 503–506
  in insects, 507

Acetyl coenzyme A, acetyl CoA, 476, 478, 482–484

Acetylene dicarboxylate esters, 1 : 1 and 2 : 1 adducts with phosphines, 129

Acetyl phosphate, in conversion of ADP to ATP, 343, 478

Acidity, Lewis, 448

2-Acyloxyvinyl phosphate, from 1,3,2-dioxaphosphorane and acyl chloride, 236

Acyl phosphates, as phosphorylating agents, 340

N-Acyl phosphoramides, as phosphorylating agents, 340

Adamsite, as a harassing agent, 502

Adenosine, 474

Adenosine diphosphate, ADP, 437, 478, 842–843

Adenosine monophosphate, AMP, 478, 491–492

Adenosine triphosphate, ATP, 472, 473, 474–494
  in biosynthesis, 488–491, 492
  as energy store, 476–479
  formation of, 479–487
  hydrolysis, nucleophilic catalysis by picolines, 326
  in metabolic activation, 491–493
  metal catalyzed hydrolysis of, 315
  in phosphorylation of amino acids, 310
  by oxidative phosphorylation of ADP, 342

Ag{P(OMe)$_3$}$_4$ClO$_4$, 185

Agrochemicals, 495

Aldrin, as an insecticide, 495

Algae, 18–21
  algal bloom, 18, 23–24

Alkoxylcyclophosphazenes
  (NP(OMe)$_2$)$_3$, 417, 418, 433, 434
  (NP(OR)$_2$)$_{3,4}$, [R = Et, Pr, CH$_2$Ph, OCH$_2$CF$_3$, OC$_6$H$_{11}$] 413, 417, 418, 433, 437

Alkoxyphosphoranes
  in alkylation of carboxylic acids, 229
  in alkylation of phenols, 229
  in alkylation of activated methylene groups, 230

Alkylation of cyclopolyphosphazenes, 431

Alkyl migration
  in cyclopolyphosphazenes, 433–434
  in diphosphazenes, 403

Alkyldifluorodihalogenophosphoranes, disproportionation reactions, 216

Alkylfluorophosphoranes
  reactions with ArOSiMe$_3$, 223
  reactions with NaHF$_2$, 243
  reactions with Me$_3$SiNR$_2$, 223
  reactions with RNH$_2$ and R$_2$NH, 223, 243
  from thiophosphoryl and thiophosphinyl halides and SbF$_3$, 227

Alkylidene trialkylphosphoranes
  complexes with Li$^+$ salts, 277
  by desilylation of silylmethylenetrialkylphosphoranes, 279
  from phosphines and carbenes, 278

Allcock, H. R., 419, 435, 436

Allyl phosphite, thermal rearrangement to phosphonate, 129

Allylphosphonates, as phosphorylating agents, 340

α-effect, in nucleophilic displacement at phosphorus, 328

Ambident nucleophiles
  enolate anions, 125
  phosphorous acid anions, 328

Amidocyclophosphazenes
  (NP(NH$_2$)$_2$)$_3$, 417, 425, 430
  (NP(NH$_2$)$_2$)$_4$, 417, 425, 430

Amino-dialkoxydifluorophosphoranes, 229

Amino-di-(alkylthio)-difluorophosphoranes, 229

5-Aminopentan-1-ol, reaction with Ph$_3$P(OEt)$_2$, 234

Aminophosphines
  preparation, 148–149
  reactions of, 458, 462

Aminophosphoryl difluorides, preparation, 229

Aminotriphenylphosphonium cation, Ph$_3$PNH$_2^+$, 392

Ammonium polyphosphates, 382

2-Aminoethylphosphoric acid, 472

Angelici, R. J., 192, 197

Aniline, N-alkylation, 344

Animal feed, 9, 13

Anodic oxidation potentials of arylphosphines, 360

Ansa derivatives of cyclopolyphosphazenes, 424, 425

Answers to problems
  Chapter 1, 526
  Chapter 2, 526
  Chapter 3, 526
  Chapter 4, 527
  Chapter 5, 530
  Chapter 6, 531
  Chapter 7, 533
  Chapter 8, 535
  Chapter 9, 536
  Chapter 10, 540
  Chapter 11, 541

Anticholinesterase, 522

Antidote for organophosphorus poisons, appendix 2, 522

Antimony trifluoride
  reaction with bis-phosphine sulphides, 227

Index

reaction with thiophosphoryl and thiophosphinyl halides, 227
Apatite, 5
Apholate, as a chemosterilant, 501
Apicophilicity, 57, 65–68
  determination of, 66–67, 169, 212–213
  orders of, 66–67
Arbusov reaction, 117
  with dialkyl phosphites (Michaelis-Becker reaction), 119
  mechanism and stereochemistry, 118
Arbusov rearrangement (photochemical), of phosphites to phosphonates, 366
Aryl isocyanates, reaction with 1,3,2-dioxaphosphoranes, 236
Aryloxycyclophosphazenes, 417, 418, 422, 430, 433
Arylphosphonous dihalides, synthesis from PX_3 and aromatic compounds, 150–151
Asknes, G.
  kinetic study of Arbusov reaction, 118
  kinetic study of hydrolysis of phosphites, 146
  rate data from hydrolysis of cyclic phosphonium salts, 264
Atomic orbitals of phosphorus, 37
Attenuation, $p\pi$-$d\pi$ bonding in phosphorylating agents, 340
Azides in synthesis, 389, 408–409
Azidocyclophosphazenes, 423, 426
Aziridine, from ethanolamine and $Ph_3P(OEt)_2$, 234
Aziridyl derivatives, of phosphazenes, 403, 426
Azo-bis-isobutyronitrile (AIBN), as initiator for radical reactions, 355

Bergelson, L. D., ρ values for Wittig reaction with reactive ylids, 290
Berman, M., 388, 393
Berry, S., 58
Bestmann, H. J., synthetic applications of the Wittig reaction, 279
Biacetyl
  reaction with a bicyclic phosphite, 218
  reaction with ethylenephosphonothioites, 235
  reaction with $(MeO)_3P$, 166, 167
Bigorgne, M., 191–192
Bimolecular displacement processes, free energy/reaction coordinate diagrams for, 335
Biological phosphorylation reactions, 343
Biphenyl-α-naphthylphenylphosphine (optically active), reaction with tetracyanoquinodimethane (TCNQ), 301
Biophosphorus chemistry, 472–510
  notation, 472–473
Biosynthesis, 488–491
Biphilic reactivity of P(III) compounds

with cyclic disulphides, 156
with diethyl peroxide, 153–154
with dioxetanes, 156
with ozonides, 156
with sulphenate esters, 157
1,2-Bis-(diphenylphosphino)ethane, 190
Black phosphorus, 454
Bond angles
  in cyclopolyphosphazenes, 410, 415
  in boraphosphanes, 452
Bond enthalpies, 35, 51, 395, 399, 400, 412, 458–459
Bonding, Chapter 2, 30–75
  P-B, 452
  P-metals, 191–204
  P-N, 387–388
Bond lengths, 34, 36, 72
  P-B, 452
  P-N, 382, 394, 399, 400, 412–413
  P-P, 458–459, 464
Bond moments, (μ), 38–40, 50
Bone, 7
Bone Valley phosphate deposit, 7
Borane, BH_3/B_2H_6, 446, 448, 465
  adducts with trivalent phosphorus compounds, 384, 448, 449–451, 460, 462, 465
Boraphosphanes, 446–454
  $(Me_2P \cdot BH_2)_3$, 446, 449, 450, 451, 453
  $(Me_2P \cdot BH_2)_4$, 451, 453
  $(Me_2P \cdot BX_2)_3$, 450, 451, 453 [X = F, Cl, Br, Pr]
  $(Et_2P \cdot BX_2)_2$, 450, 453 [X = Cl, F, Br, Pr]
  $(Et_2P \cdot BX_2)_3$, 450, 452 [X = H, I]
  $(Ph_2P \cdot BX_2)_2$, 488–489, 450, 453 [X = I, Cl]
  $(Ph_2P \cdot BX_2)_3$, 450, 451, 452, 453
Bordeaux mixture, as a fungicide, 495
Boron-phosphorus bond, 446–454
Brandt, Hennig, 10
Borowitz, I. J., kinetic studies of the Perkow reaction, 136
Boyle, R., the glow of phosphorus in air, 352
Bromocyclophosphazenes
  $(NPBr_2)_3$, 407, 417, 418, 431, 436
  $(NPBr_2)_4$, 417, 418
Brushite, 6
Butadiene, coupling of, 186–187
  reaction with P(III) compounds, 157–163
Butyl derivatives
  $(Bu^nP)_4$, 456, 463, 464, 465
Bacteria, 21
  Rhizobia, 493
Basicity
  cyclopolyphosphazenes, 432–433
  cyclopolyphosphines, 464
  phosphines, 37–38, 203 footnote
Basicity, Lewis, cyclopolyphosphazenes, 431, 460, 461

Be(acac)(Ph$_2$(O)PNP(O)Ph$_2$), 404
Benzenehexachloride (BHC), as an insecticide, 495
Benzil
  reaction with P(III) compounds, 218
  reaction with (RO)$_3$P, 163, 166
3,4-Benzoheptatriene, by Wittig reaction, 280
o-Benzoquinone, reaction with (RO)$_3$P, 163, 218
p-Benzoquinone
  reaction with phosphines, 128, 359
  reaction with phosphites, 131
(−) Benzyl-t-butylmethylphenylphosphonium iodide, stereochemistry of hydrolysis, 272
Benzyl diethyl phosphite, reaction with butane thiol, 367
Benzylidene malononitriles
  addition of P(III) compounds, 127
  reaction with cyclopentadienylidenetriphenylphosphorane, 295
3-Benzylidene-2,4-pentanedione, reaction with P(III) compounds, 170, 218–219
Benzylidenetriphenylphosphorane, reaction with cyclohexene oxide, 297
1-Benzyl-4-methyl-1-phenylphosphepanium bromides (cis and trans), hydrolysis, 269
1-Benzyl-1-phenyl-2,2,3,4,4-pentamethylphosphetanium bromides, hydrolysis, 269
1-Benzyl-1-phenyl-2,2,3,3-tetramethylphosphetanium bromide, hydrolysis, 270
Benzyl phosphates, removal of benzyl groups by NaI, 344
Benzyne, reaction with phosphines, 129, 256
Be(Ph$_2$(O)PNP(O)Ph$_2$)$_2$, 404
1,4-Butane diol, reaction with (EtO)$_5$P, 234
(+) Butan-2-ol, in SN1(P) displacement reactions at phosphoryl P, 312
Butene-2-oxides (cis and trans), by thermal decomposition of 1,3,2-dioxaphosphoranes, 233
t-Butyl di-n-butylphosphinite, from tri-n-butylphosphine and di-t-butylperoxide, 365
BZ, as an incapacitating agent, 502

Cacodylic acid, as a defoliant, 502
Cage phosphites, 184, 196, 202, 203
Cage phosphorus amide, 396
Calcium phosphates, 6, 13–15
D(+) camphorsulphonate, for separation of enantiomeric phosphonium salts, 254
Carbamyl phosphate, 491
Carbazole, from o-nitrobiphenyl and triethyl phosphite, 139
Carbethoxymethylenecyclopropene, 2,3-diphenyl, by Wittig reaction, 280
Carbethoxymethylenetriphenylphosphorane
  reaction with methyl benzoylacrylate, 294
  reaction with p-substituted benzaldehydes, 285
  reaction with styrene oxide, 297
Carbodiimides, as phosphorylating agents, 341
Carbon disulphide, crystalline complexes with phosphines, 133
Carbon tetrabromide, reaction with P(III) compounds, 123
Carbon tetrachloride, reaction with P(III) compounds, 123
8-Carboxy-1-naphthyl phosphate, hydrolysis by SN1(P) mechanism, 314
o, m and p-Carboxyphenylphosphonate esters, intramolecular acid-catalyzed hydrolysis, 327
β-Carotene, by Wittig reaction, 281
Carson, R., 'elixirs of death', 502
Catechol
  reaction with cyclic phosphonites, 220
  reaction with cyclic phosphoramidites, 221
  reaction with halogenocyclophosphazenes, 243
  reaction with (PhO)$_5$P, 222
CdI$_2${P(NMe$_2$)$_3$}$_2$, 384
Chatt, J., 178, 180
Cheletropic reactions, 238
Chemosterilants, 501
  apholate, 501
  metepa, 501
  tepa, 501
Chloral, in Perkow reaction, 133, 137
β-Chloroalkylphosphonates
  decomposition to metaphosphate, 314
  as a phosphorylating agent, 340
Chloranil
  reaction with Ph$_3$P, 131
  reaction with phosphites, 132
o-Chlorobenzylidene malononitrile, as a harassing agent, 502
Chlorocyclophosphazenes
  (NPCl$_2$)$_3$, 405, 406, 410, 413, 416, 417, 418, 420, 423, 424, 425, 427, 428, 433, 434, 435
  (NPCl$_2$)$_3$·3SO$_3$, 431
  (NPCl$_2$)$_4$, 405, 406, 413, 416, 417, 418, 420, 422, 425, 428, 434
  (NPCl$_2$)$_4$·2HClO$_4$, 432
  (NPCl$_2$)$_5$, 405, 406, 410, 416, 417
  (NPCl$_2$)$_6$, 405, 406, 416
  (NPCl$_2$)$_7$, 405, 406
  (NPCl$_2$)$_8$, 406
  (NPCl$_2$)$_n$, 405, 406, 436, 437
2-Chloromethyl-2-methyl-1,3-dichloropropane, reaction with sodium diphenyl phosphide, 258
1-Chloro-2,2,3,4,4-pentamethylphosphetan (and oxide)
  synthesis from 2,4,4-trimethylpent-2-ene, 151
  stereochemistry of displacement reactions at P in, 152
Chlorophyll, 480–481

Chloroplast, 480

Chloroprene, reaction with P(III) compounds, 157–163

Chopard, P. A., O-acylation *vs* C-acylation of phosphorus ylids, 293

Chromium complexes
cis-Cr(CO)$_2$(PH$_3$)$_4$, 179, 180
Cr(CO)$_5$(PF$_3$), 183
Cr(CO)$_5$(PPh$_3$), 195
Cr(CO)$_5$\{P(OPh)$_3$\}, 195
trans-Cr(CO)$_4$\{P(OPh)$_3$\}$_2$, 195
Cr(CO)$_4$(Cl$_3$P.NMe)$_2$, 399

Chymotrypsin, inhibition by organophosphorus compounds, 505

Cis effect, 426

Cis-trans isomerization, 426–427

Cis and trans phosphetanium derivatives, 66

Citric acid, 483

Citrulline, 491

Clarke, R., concept of 'benign' war, 508

Clathrates of cyclopolyphosphazenes, 425

Cram, D. J., rule on asymmetric induction, 139

Coalescence temperatures, 388

Coates, G. E. and Livingstone, J. G., 447

Cobalt complexes
CoBr$_x$(PMePh$_2$)$_{4-x}$, 193
Co(CO)$_3$(PH$_3$)$_3$, 180
Co(CO)$_3$(NO)P, 191
Co(NO)(PF$_3$)$_2$(MeOPF$_2$), 183
Co\{P(OMe)$_3$\}$_5$ClO$_4$, 185
HCo(PF$_3$)$_4$, 183
[CoCl(NP(NMe$_2$)$_2$)$_6$]$^+$, 431
CoCl$_2$(Me$_3$PO)$_2$, 95

Coenzyme A, CoASH, 482, 487, 491, 493

Colas, 12

Copper complexes
Cu\{P(OMe)$_3$\}$_4$ClO$_4$, 185
CuCl(MeP)$_4$, 467
CuBr(MeP)$_5$, 467

Corbridge, D. E. C., 92–93, 388

Cotton, F. A. and Wilkinson, G., 195

Cotton-Kraihanzel force field technique, 191, 194

Coupled photosystems, 481

Coupling reactions, 456, 462

Craig, D. P., 412

Crandallite, 6

Crystal violet, indicator for HClO$_4$ titrations, 255

Cs$_6$P$_6$O$_{12}$, 455

β-Cyanoethylphosphonium salts, source of optically active phosphines, 254

Cyanomethylenetriphenylphosphorane
reaction with ethoxymethylene malonate, 295
reaction with tetracyanoethylene, 295

Cyanophosphines, R$_2$PCN, from phosphines and isonitriles, 356

Cyclic phosphinate esters
from 2-halo-1,3,2-dioxaphospholanes and dienes, 220
mechanistic scheme for hydrolysis of, 334

Cyclic phosphinyl halides, from ROPCl$_2$ and dienes, 220

Cyclic phosphites
synthesis by exchange reactions between acyclic phosphites and diols or triols, 147
synthesis from P(III) halides and glycols, 149

Cyclic phosphonates, mechanistic scheme for hydrolysis of, 333–334

Cyclic phosphonites
acid-catalysed disproportionation to phosphoranes, 220
base-catalyzed reaction with catechol, 220

Cyclic phosphoramidites
reaction with catechol, 221
reaction with pinacol, 221

Cyclic phosphordiamidates, rate of hydrolysis *vs* acyclic analogues, 330

Cyclization, of polyphosphazenes, 406

Cyclodiphosphazane, 386, 387, 395–401
i.r., 401
structure, 400
synthesis, 396

Cyclodiphosphazane derivatives (group on P)
anilido, 398, 400, 401
chloro, 387, 389, 395, 396, 397, 398, 399–401
diphenyl chloro, 395, 396
diphenyl fluoro, 401, 409
dimethylamino, 401
methylthio, 401
trifluoro, 396, 401

Cyclooctatetraene, by fragmentation of a phosphorane, 241

Cyclopentaphosphines
trifluoromethyl derivative, 461, 462, 464, 465
methyl, 462, 463, 464, 465, 466, 467
phenyl, *see* Phenylpolyphosphines
ethyl, 467

Cyclopentadienylidenetriphenylphosphorane
reaction with benzylidene malononitriles, 295
reaction with tetracyanoethylene, 296
reaction with tricyanovinylbenzene, 295
reaction with tricyanovinyl cyclohexyl ether, 296

Cyclophosphazenes, 408–433
see also under cyclopolyphosphazenes and name of groups attached to phosphorus, i.e., amido-, alkoxy-, aryloxy-, azido-, bromo-, chloro-, ethyl-, fluoro-, methyl-, perfluoro- and phenylcyclophosphazenes

Cyclopolyboraphosphanes, 449–454
bonding, 452
i.r. spectra, 453
physical properties and chemical behaviour, 451
synthesis, 449–451

Index

Cyclopolyphosphines, 461–471
　properties, 463–471
　synthesis, 461–463
Cyclopolyphosphazenes, 405–437
　bonding, 46, 410–412
　hydrolysis, 429–431
　i.r. spectra, 415–417
　nucleophilic substitution, 420–429
　physical properties, 412–415, 418–419
　reactions at nitrogen, 431–434
　synthesis, 405–409
Cyclopropanes, from benzylidene malononitriles and cyclic phosphoranes, 237
Cyclotriphosphazenes, see cyclophosphazenes
Cyclotriphosphine anion $(PhP)_3^{2-}$, 467
　adduct, 465
Cyclotetraphosphines, derivatives, 460
　cyclohexyl, 462, 463, 465, 466, 467
　ethyl, 464, 465, 466
　pentafluorophenyl, 465
　propyl, 464, 465
　trifluoromethyl, 460–461, 462, 464, 465, 466
Cytochromes, 482
Cytosine, 474

d-π Bonding, 44–46
　cyclopolyphosphazenes, 411–412, 419
　delocalized, 46
　donor, 383, 394, 399
　evidence for, 194–195
　to metals, 191–204
　monophosphazenes, 394
　pentacoordinate phosphorus, 55–57
　P=O bond, 42–44
　tetracoordinate phosphorus, 44–46
Darensbourg, D. J. and Brown, T. L., 197
Davies, A. G.
　rate of reaction of $(EtO)_3P$ with t-butoxy radicals, 364
　configuration of phosphoranyl radicals, 368
DDT, as an insecticide, 495
De Bruin, K. E., 66
Deflocculation, 12–13
Dehydrohalogenation reactions, 456
Delocalization energies, 412, 419
Demeton, as an insecticide, 497, 498
Denney, D. B.
　fragmentation of a phosphorance containing a 3-membered ring, 242
　Perkow reaction with acyclic and cyclic phosphites, 137
　reaction of optically active phosphine with t-butylhypochlorite, 123
　reaction of carbethoxytriphenylphosphorane with styrene oxide, 297
　reaction of 1,2-ethane diol with a spiro-pentaoxy-phosphorane, 245
　stereochemistry of reaction of epoxides with phosphines, 116

Deoxyribonucleic acid, DNA, 473, 488–489
Depolymerization
　curling-off and tailing-off mechanism, 435
　of polyphosphazenes, 435
　steric factors, 437
Desulphurization, of diphosphine disulphides, 457
Detergents, 12–13, 17
Deuterium-labelled olefins, by Wittig reaction, 280
Deuterium exchange
　in phosphorus ylids, 291
　in 'onium salts, 275
Dewar, M. J. S., Transition State Method for concerted reactions, 239
1,1-Diacetoxyphosphabenzoles, from phosphabenzoles and $Hg(OAc)_2$, 362
N,N-Dialkylaminofluorophosphoranes, reaction with HCl or HBr, 223
N,N-Dialkylaminotetrahalogenophosphoranes, disproportionation reactions, 216
Dialkyl disulphides, reaction with $(RO)_3P$, 367
Diamidophosphoric acid, 381–382
Diaroyl peroxides, reactions with P(III) compounds, 154
Diazo derivatives, 390
Diazoalkanes, in alkylation of phosphorus acids, 344
1,2-Dibenzoylethylene, reaction with P(III) compounds, 169–170
Dibenzoyl hydrogen tartrate, in separation of enantiomeric phosphonium salts, 254
Dibenzoyl peroxide, as initiator for radical reactions, 355
α,ω-Dibromoalkanes, in preparation of cyclic phosphonium salts, 258
Dicycloboraphosphane, 450
Dideuterioethylene (cis), by fragmentation of a phosphorane, 242
Dieldrin, as an insecticide, 495
Dienophilic reactivity
　isomerisation of 3-phospholenes to 2-phospholenes, 161
　preparation 3-phospholenes, scope of reaction, 158
　reaction of P(III) compounds with dienes, 157–163
　relative rate studies and mechanism, 162
　stereochemistry, 159
Diethyl alkylphosphonites, reaction with alkoxy and thioalkyl radicals, 365
Diethyl-N,N-dimethylphosphoramidate, 308
Diethyl-1-ethoxycarbonylethylphosphonate, use in the Horner-Emmons reaction, 282
Diethyl methyl phosphate, from $(EtO)_3P$ and t-BuOO·, 365
O,O-Diethyl methylphosphonate, 308

Diethyl peroxide
  reactions with P(III) compounds, 153, 216
  reactions with 3-phospholenes, 159
Diethyl phosphorochloridate, hydrolysis via hypothetical addition-elimination mechanism, 318
Difluorodiazirine, reactions with P(III) compounds, 215
Dihydroxyacetone phosphate, 485
α-Diketones
  reaction with P(III) compounds, 163–167
  in synthesis of phosphoranes, 163–165
2,2′-Dilithiobiphenyl
  reaction with PCl_5, 244
  reaction with spirophosphonium salts, 244
Dimefox
  as an insecticide, 496, 498
  in-vivo oxidative demethylation, 507
Dimethylaminocyclophosphazenes
  $(NP(NMe_2)_2)_3$, 433
  $(NP(NMe_2)_2)_4$, 414, 417, 418, 431, 432
  $(NP(NMe_2)_2)_6$, 431, 432
2-Dimethylamino-1,3,2-dioxaphospholan, reaction with t-butoxy radicals, 369
2,5-Dimethyl-2,5-dihydrothiophene-1,1-dioxide, pyrolysis to diene and sulphur dioxide, 159
N,N-Dimethyl-P,P-hexachlorocyclodiphosphazane $(Cl_3PNMe)_2$, 396, 401
  structure, 399
  complex, 399
1,1-Dimethylhexafluorocyclotetraphosphazene 1,1-$N_4P_4F_6Me_2$, 46, 415
2,6-Dimethylphenyl o-nitrophenyl ether, reaction with $(MeO)_3P$, 222
Dimethylphenylphosphine, electrolytic reduction, 373
Diphenyl phosphide anion Ph_2P^-, 459
Dimethyl phosphite, reaction with vinyl propionate, 358
2,6-Dimethylpyridine, as base catalyst in alcoholysis and hydrolysis of phosphate esters, 326
Dioxan, from bis-(2-hydroxyethyl) ether and $Ph_3P(OEt)_2$, 234
1,3,2-Dioxaphospholanes
  preparation, 167
  synthesis by exchange reaction, 147
1,4,2-Dioxaphospholane, preparation, 168
Diphenoxymethyltriphenylphosphonium chloride, reaction with phenyllithium, 260
2,5-Diphenyl-3,4-dicyanocyclopentadienone, addition of phosphines, 131
1,1-Diphenylethylene, by β-elimination from phosphonium salts, 260
1,1-Diphenylphosphabenzole, by reaction of phosphabenzole with $HgPh_2$, 373

Diphenylphosphinic acid, by electrolytic reduction of Ph_3P (via Ph_3P^-), 373
Diphenylphosphorochloridate, as a (protected) phosphorylating agent, 338
Diphenyltriethoxyphosphorane, ^1H nmr, 228
Diphos, $Ph_2PCH_2CH_2PPh_2$, 190
Diphosphate, $P_2O_7^{4-}$, 478
  $K_2P_2O_7$, 13
  $Na_2H_2P_2O_7$, 14
Diphosphazanes, see cyclodiphosphazanes
Diphosphazenes, linear derivatives (groups on P)
  amino, 402, 404, 408
  diphenyl, 402, 403, 404, 409
  trichloro, 401, 402, 404, 405, 406
  triphenoxy, 403
  triphenyl, 403, 404, 425
Diphosphide $K_2P_2Et_2$, 466
Diphosphine P_2H_4, 456, 462
  adducts, 460
Diphosphines, 456–461
  derivatives, 457, 458, 466
  dimethylamino $P_2(NMe_2)_4$, 384
  as ligands, 460
  miscellaneous, 456, 462
  phenyl, P_2Ph_4, 456, 458, 466
  properties, 458–461
  synthesis, 456–458
  trifluoromethyl, $P_2(CF_3)_4$, 459, 460
  $P_2Me_3(CF_3)$, 456
  $P_2H_2(CF_3)_2$, 459
  $P_2HMe(CF_3)_2$, 456
  $P_2F_2H_2$, 462
Diphosphine sulphides
  $Me_2P-P(S)Me_2$, 462
  MePhP(S)P(S)PhMe, 457
  $Ph_2P(S)-P(S)Ph_2$, 456
Diphosphorus tetrahalides
  P_2F_4, 462
  P_2I_4, 454, 458
Dipole moments (p), 38–40, 50
Dipyridyl, Cu^{2+} complexes as catalysts in hydrolysis of Sarin, 328
Dismutation of olefins, 190
Dithieten (bis-trifluoromethyl), 84
  reaction with P(III) compounds, 217
  reaction with cyclic P(III) compounds, 156
Doering, W. von E., rates of deuterium exchange in 'onium salts, 275
Dowco 199, as a fungicide, 500
Dynamic nmr, 212
  for determination of energy barriers to pseudorotation, 232

Ekatin, as an insecticide, 496, 499
Electric furnace process, 10
Electronegativity, 54, 57
  definition, 73

550 Index

Hinze and Jaffé, 73–74
of phosphorus, 73–74

Electronic effects of groups attached to phosphorus, 74–75

Electron spin resonance spectra (e.s.r.)
of Me$_3$ṖOBu^t, 352, 368
of Ph$_2$PNBu^t (O·), 357
of R$_2$Ṗ· @ 77° K, 357
of Ph$_2$Ṗ(O), 358
of (EtO)$_2$P(O)NBu^t (O·), 359
of radicals from Ph$_3$P and p-benzoquinones, 359
of R$_3$P·$^+$, 360
of phosphabenzoles + Hg(OAc)$_2$, 362
of phosphabenzoles + ArO· or Ar$_2$N·, 363
of phosphoranyl radicals from 1,3,2-dioxaphospholans and alkoxy radicals, 369
of bicyclic (spiro) phosphoranyl radicals, 370, 372
of acyclic phosphoranyl radicals, 371
of Me$_2$P̄P̄Ph, 373
of the phosphole radical anion, 373

Electron withdrawal from phosphorus, 39–40, 73–75

Electrophilic reactivity of P(III) compounds
hydrolysis of PCl$_3$, 145
P-C bond cleavage, 145
P-N bond cleavage, 146
P-O bond cleavage, hydrolysis and exchange reactions, 146
P-halogen bond cleavage, 148–152

Electropolishing, 12

Environmental phosphus, Chapter 1

Enzymes
aldolase, 485
creatine phosphokinase, 478–479
hexose phosphate isomerase, 484–485
pyrophosphatases, 493
triose phosphate isomerase, 485

Epoxides, reactions with P(III) halides, 152

Equatophilicity, 68

1,2-Ethane diol, reaction with a spiropentaoxyphosphorane, 245

Ethanolamine, reaction with Ph$_3$P(OEt)$_2$, 234

1-Ethoxy-1-phenyl-2,2,3,4,4-pentamethylphosphetanium hexachloroantimonate, stereochemistry of hydrolysis, 268

Ethyl benzenesulphenate, reactions with P(III) compounds, 157, 216

Ethyl cinnamate, from Wittig reaction, 285

Ethyl-m-chlorocinnamate, from Wittig reaction, 285

Ethylcyclophosphazenes, 417, 433

O-Ethyl-S-methyl ethylphosphonodithioate, 308

Ethylene oxide, tetramethyl, by thermal decomposition of a 1,3,2-dioxaphosphorane, 233

Ethylene, 1-phenyl-2-cyclopentyl, by the Wittig reaction, 297

Ethylene sulphide, from α-dicarbonyl compounds and ethylene phosphonothioites, 235

Ethylidenetriphenylphosphorane, preparation, 277

Eutrophication, 4, 18–19, 23–24

^{19}F nmr, 55, 58, 65, 67
of hexacoordinate phosphorus compounds, 245

α-Farnesene, by the Horner-Emmons reaction, 282

Fatty acids, 476, 490, 491

Feakins, D., 432

Fe complexes
FeCl$_3${P(O)(NMe$_2$)$_3$}, 385
Fe(CO)$_{12-x}${P(OMe)$_3$}$_x$, 185
Fe(PF$_3$)$_5$, 183
Fe(PF$_3$)$_n$(CO)$_{5-n}$, 182
cis-FeX$_2$(PF$_3$)$_4$, 183
Fe$_2$(CO)$_8$(C$_6$H$_{11}$P)$_4$, 467
Fe{P(O)(NMe$_2$)$_3$}$_4$2ClO$_4$, 96

Fermi resonance, 95

Ferredoxin, 493

Fertilizer, 9–10, 14–16

Five-coordinate phosphorus, *see* pentacoordinate phosphorus

Flameproofing, 12

Flavin adenine dinucleotide, FAD, 483, 484

Flavoprotein, 482

Florida phosphorite deposit, 6, 7

Fluorapatite, 5–6

9-Fluorenone, reaction with (EtO)$_3$P, 218

Fluorides in teeth, 7

Fluorocyclophosphazenes
(NPF$_2$)$_3$, 410, 417, 418, 422, 427, 428
(NPF$_2$)$_4$, 415, 417, 418, 419, 422, 427
(NPF$_2$)$_5$, 419, 422
(NPF$_2$)$_6$, 419, 422
(NPF$_2$)$_7$, 419, 422
(NPF$_2$)$_8$, 419, 422
(NPF$_2$)$_{3-17}$, 422
(NPF$_2$)$_n$, 436

Force constants (k), 191, 196–197, 204

Free energy of hydrolysis $\Delta G^{o'}$ of phosphates, 477, 478

Friedel-Crafts type arylation of cyclopolyphosphazenes, 428

Fructose-1,6-diphosphate, 485

Fructose-6-phosphate, 484

Fumaric acid, 483

Nucleophiles
  basicity, 328
  polarizability, 328
  α-effect, 328
Nucleophilic reactivity of P(III) compounds, 113
  at saturated carbon, 113
    alkyl halides, 113
    epoxides, 115
    Hammett ρ-values, 115
    reactivity vs N, 115
  at 'positive' halogen, 119
    halogen-carbon bonds, 123, 125
    halogen-halogen bonds, 121
    halogen-nitrogen bonds, 124
    halogen-oxygen bonds, 122
    halogen-sulphur bonds, 124
  at carbon-carbon multiple bonds, 126, 128
  at the carbonyl group, 130, 134
  the Perkow reaction, 133
  at nitro and nitroso groups, 139–142
  at $^+$N—O$^-$, As=O, S=O, P=O and P=S, 143
  at electrophilic oxygen, 144
  at unsaturated nitrogen
    the diazo ion and polyazines, 144
Nucleophilic substitution at P in phosphoryl and thiophosphoryl compounds
  acyclic phosphorus compounds, 309–329
    comparison with acyl compounds, 310
    the SN1(P) mechanism, 311–318
    the addition-elimination mechanism, 318
    the SN2(P) mechanism, 319–325
    catalysis, 325–328
      nucleophilic, 325
      general base, 326
      acid, 327
      electrophilic by metal ions, 327
      the nucleophile, 328
      the leaving group, 329
  cyclic phosphorus compounds, 329–334
    5,6, and 7-membered ring phosphates, 329–330
    phosphonates, 331
    phosphinates, 331
  hexacoordinate intermediates, 336–338
  phosphorylation, 338–342
Nucleophilic substitution at tetracoordinate phosphorus, 70–72
  apical vs equatorial, 71–72
  in cyclopolyphosphazenes, 420–431

Octahedral transition state, in hydrolysis of pentaaryloxyphosphoranes, 337
Ogata, Y. (and Yamashita, M.), reaction of benzil with (RO)$_3$P, kinetic data, 166
Oil additives, 12
Organophosphorus poisons
  general structure, 494
  administration
    by intravenous injection, 494
    by percutaneous absorption, 494
    by subcutaneous injection, 494
    oral, 494

Organolithium compounds, 427
Ornithine, 491
Ortho esters, in alkylation of phosphorus acids, 344
Orthophenylation, 185–186, 188
Ortho quinones, reaction with P(III) compounds, 163–167
Osmium complexes
  Os(CO)$_3$(PPh$_3$)$_2$, 190
  Os(CO)$_2$(PPh$_3$)$_3$ as oxygen carrier, 190
  OsH$_4$(PR$_3$)$_3$, 187
  OsX$_2${P(OPh)$_3$}$_4$, 185
Oulad-Abdoun phosphate deposit, 7
Oxaloacetic acid, 483
Oxalosuccinic acid, 483
1,3-Oxazolines, from isocyanates and 1,3,2-dioxaphosphoranes, 236
Oxetanes, reaction with P(III) halides, 152
Oxidative phosphonylation, 357
Oxidative phosphorylation, 477, 480–482
Oximes
  as nucleophilic catalysts in hydrolysis of Sarin, 326
  as reagents for the reactivation of inhibited AChE, 506
α-Oxoglutaric acid, 483, 487
Oxonium salts, from ethers and PF$_5$, 229
Ozone, reaction with phosphites, 144

P$_4$, 454, 455
^{31}P n.m.r. spectroscopy, see Phosphorus-31
Paddock, N. L., 412, 418, 419, 431
Paraoxon, as an insecticide, 496, 498
Parathion
  action of light, 352
  as an insecticide, 496, 497, 498
  as a nematicide, 501
  in vivo oxidation to Paraoxon, 497, 507
Paris Green, as an insecticide, 495
Partial rate constants, in hydrolysis of phosphonium salts, 262
Pentaaryloxyphosphoranes, hydrolysis of— kinetics and mechanism, 336
Pentacoordinate phosphorus, bonding in, 32, 48–72
  π-bonding, 54–57
  σ-bonding, 51–54
Pentacoordinate phosphorus derivatives
  amino, 55–56, 380, 381, 387, 388
  chloro, 50
  cyclo, 65, 66
  dimethylamino, 387
  fluoro, 50, 54–55, 55–56, 381, 387
  phenyl, 387
  pyridyl, 55

556 Index

Pentacoordinate phosphorus, reaction intermediates, 68–72
Pentafluorobenzaldehyde, reaction with (RO)$_3$P, 168
Pentahalogenophosphoranes
  reaction with AgF, 243
  reaction with dimethylurea, 243
  reaction with phosphine oxides, 243
  reaction with pyridine, 243
1,5-Pentane diol, reaction with Ph$_3$P(OEt)$_2$, 234
Pentaphenoxyphosphorane, from PCl$_5$ and PhOH, 222
Pentaphenylphosphorane
  reaction with n-butyllithium, 225
  reaction with tritium-labelled Ph$_5$P, 225
  from N-p-tosyliminoylid and PhLi, 228
Pentose cycle, 476
Perchloric acid, use in titration of phosphonium salts, 255
Perfluoro alkyl and aryl cyclophosphazenes, 407, 409, 417, 433, 437
Perkow reaction, 133–139
  kinetic studies, 136
  possible mechanistic schemes, 135
  product ratio (ketophosphonate: vinyl phosphate), 134
  stereochemical studies, 138
Peroxide
  di-t-butyl reaction with (EtO)$_3$P, 363, 365
  bis-trifluoromethyl, 90
Peroxy acids
  reactions with P(III) compounds, 154
  reactions with phosphorus ylids, 280
Pesticides, 495–502
  and the environment, 501
  discrimination between species (specificity), 497
  first-generation, 495
  organophosphorus, 496
    Table of, 498–499
  second-generation, 495
  systemic activity, 497
  third-generation, 496
Phenacylidenetriphenylphosphorane
  preparation by salt method, 277
  reaction with dimethylacetylene dicarboxylate, 296
Phenanthraquinone, reaction with ethylenephosphonothioite, 235
Phenoxyacetic acids, as defoliants, 502
Phenylcyclophosphines, 461, 465, 466, 467
  form A, 462, 463–464
  form B, 463–464
  form C, 463–464
  form D, 463
Phenylcyclophosphazenes
  (NPPh$_2$)$_3$, 407, 408, 409, 413, 417, 428, 433
  (NPPh$_2$)$_4$, 390, 407, 408, 409, 413, 417
p-Phenylene diamine, reaction with chloránil, 359

tris-o-Phenylenedioxy phosphate anion, hydrolysis, 246
2-Phenylethyldiphenylphosphine oxide, 260
Phenylphosphine (PhPH$_2$), addition to acrylonitrile, 127
Phenyl phosphinidene (PhP), 462–463
  reaction with benzil, 354
  reaction with diethyl disulphide, 354
Phenylphosphonous dibromide, reaction with dienes, 161
Phenylphosphonous dichloride
  reaction with dienes, 161
  reaction with zinc, 354
Phosdrin, 139
Phosphabenzoles
  reaction with ArO· and Ar$_2$N·, 363
  reaction with Hg(OAc)$_2$, 362
  reaction with Ph$_2$Hg, 373
Phosphate cycles, 2–4
  land based, 14–18
  primary inorganic, 5–14
  water-based, 18–26
L-Phosphatidic acid, 490
Phosph(III)azanes, 396
Phosphazenes, cyclic derivatives, see under name of groups attached to phosphorus or under cyclophosphazenes
Phosphepanium salts, stereochemistry of hydrolysis, 268
Phosphetans
  1-phenyl-2,2,4,4-tetramethyl, 90
  reaction with hexafluoroacetone, 212
Phosphetanium salts
  hydrolysis, 268–271
  intramolecular rearrangements during hydrolysis, 272
  preparation, 257
Phosphide anions, attack at saturated carbon, 113, 114
Phosphine imides or imines, see monophosphazenes
Phosphine, PH$_3$
  adducts, 448, 449
  ligand behaviour, 178, 179–180, 192
  steric size, 203
Phosphines, see trivalent phosphorus compounds
Phosphinium radical cations (R$_3$P$^{\ddot{+}}$)
  by electrolytic oxidation of phosphines, 360
  by γ-irradiation of phosphines, 360
  by reaction of phosphines with TCNE, 360
  by reaction of phosphines with TCNQ, 360
  configuration, 361
  definition, 353
  theoretical study of ground state configuration, 362

Index 557

Phosphino radicals (R$_2$Ṗ·)
　definition and general reactivity, 352
　half-life, 353
　preparative methods, 355
　reactions
　　addition to olefins, 355
　　H-abstraction, 355
　　with dienes, 356
　　with isonitriles, 356
　　with P(III) halides, 356
　trapping
　　with Bu^{t}N=O, 357
　　in benzene at low temperature, 357

Phosphinyl radicals (R$_2$Ṗ=O)
　definition and electronic structure, 353
　generation by photolysis or radical abstraction, 357
　reactions
　　with disulphides, 358
　　with olefins, 358
　　with vinyl propionate, 358

Phosphites, see under appropriate triorgano name

Phosphobenzene, 461, 463, 464
　see also phenylcyclophosphines

Phosphocreatine, 472, 478
　in conversion of ADP to ATP, 343

Phosphoenolpyruvate, 476, 478, 486
　in conversion of IDP to ITP, 343

1,3-Phosphoglyceric acid, in conversion of ADP to ATP, 343

Phospholan, 1-phenyl-3-methyl (cis and trans), 256

Phospholanium salts
　1-alkyl-1-phenyl-3-methyl, 256
　stereochemistry of hydrolysis, 265–267

$^3\Delta$-Phospholenes
　synthesis from dienes and P(III) compounds, 157–163
　isomerisation to $^2\Delta$ phospholenes, 161

Phospholenium salts, preparation from P(III) compounds and dienes, 257

Phosphonates, in the Horner-Emmons reaction, 343

Phosphonitrilic compounds, see cyclophosphazenes

Phosphonium cations
　PhPF(NMe$_2$)$_2$$^+$, 387
　substitution at, 46–47
　activation by, 47
　dimethylamino derivatives, 384

Phosphonium radical anions (R$_3$P$^-$)
　by electrolytic reduction of R$_3$P compounds, 373
　by reduction of phospholes with alkali metal, 373
　definition, 353

Phosphonium salts
　chiral,-separation of enantiomers, 254
　preparation, 255–259
　　by condensation of P(III) compounds with dienes, 257
　　by condensation of P(III) compounds with olefins, 257
　　by nucleophilic addition of P(III) compounds to polyenes, 257
　　by quaternization of P(III) compounds, 255
　　from α,ω-dihaloalkanes and RPH$_2$, 258
　　from α,ω-dihaloalkanes and red P, 259
　reactions, 259–274
　　general scheme for reactions with nucleophiles, 259
　　hydrolysis, 261–274
　　mechanistic scheme and kinetic order, 261
　　relative rates for various leaving groups, 262
　　kinetic study and partial rate factors, 262
　　stereochemistry for acyclic salts, 262
　　rate enhancements for 5-membered rings, 265
　　stereochemistry for 5-membered rings, 265
　　stereochemistry for 6-membered rings, 267
　　stereochemistry for 7-membered rings, 268
　　stereochemistry for 4-membered rings, 268
　　relative rates for sterically hindered acyclic salts, 271
　　intramolecular rearrangements during, 272
　　heteroaryl salts, 273
　reduction to phosphoranes
　　by LiAlH$_4$ or NaBH$_4$, 226
　　by PhLi or MeLi, 226

Phosphonous chlorides (RPCl$_2$), from P(III) halides and olefins, 356

Phosphonyl chlorides (RP(O)Cl$_2$), by oxidative phosphonylation of PCl$_3$, 357

Phosphoramidates
　hydrolysis via SN1(P) mechanism, 315
　preparation from amides and mixed anhydrides, 147

Phosphoranes
　chemistry of, 228–242
　　acyclic phosphoranes, 229–230
　　configuration, 210, 213
　　cyclic phosphoranes, 230, 463
　　pseudorotation and apicophilicity, 230
　　pseudorotation and diequatorial vs apical-equatorial disposition of 5 and 6 membered rings, 230–232
　　fragmentation of, 237
　　thermal decomposition of 1,3,2-dioxaphosphoranes, 233
　　fluorophosphoranes as Lewis Acids, 229
　preparation of
　　from P(III) compounds, 215–222
　　from P(V) compounds by exchange reactions, 222–225
　　from tetracoordinate P compounds, 225–227
　　from phosphorus ylids, 227

Phosphoranyl peroxy radicals, from phosphoranyl radicals and oxygen, 366

Phosphoranyl radicals (R$_4$Ṗ· and (RO)$_4$Ṗ· and thio analogues)

558   Index

chemistry
  configuration of, 368
  e.s.r. spectra, 368
  pseudoration of, 369–371
  relative stability of, 371
  α-scission of, 354, 365, 368, 371
  β-scission of, 354, 365, 368, 371
  preparative methods
    electrolytic reduction of phosphonium salts, 372
    from alkoxy radicals and phosphines, 365, 368
    from alkoxy radicals and phosphites, 363–365
    from alkoxy radicals and phosphoramidites, 369
    from alkyl radicals and P=O, 365
    from thioalkyl radicals and phosphites, 366
    photolysis of phosphoranes, 372
Phosphordiamidates
  hydrolysis of N,N-dialkylderivatives, 316
  relative rates of hydrolysis of tri- and tetra-alkyl derivatives, 316
Phosphoria phosphate deposit, 8
5-Phosphoribosylamine, 492
5-Phosphoribosyl-1-diphosphate, 492
Phosphoric acid, production, 11–12
Phosphoric acid mono esters, hydrolysis, 312–314
  activation parameters, 313
  displacement of optically active alkoxide groups, 312
  intramolecular proton transfer, 314
  k_{H_2O} vs k_{D_2O}, 313
  linear free energy relationship, 313
  ^{18}O studies, 312
  pH-rate profile, 312
  product distribution, 313
  via -dianions, 314
N-phosphorimidazoles, as phosphorylating agents, 341
Phosphorin, 36
Phosphorin complex, 178
Phosphorinanium salts, hydrolysis, 267
Phosphorites, 5
Phosphorus, the element
  phase diagram, 454
  production, 10
  polymorphs, 454
Phosphorus-31, nmr
  chemical shift (δ), 79
    of P(III) compounds, 80
    of tetracoordinate P compounds, 82
    of P(V) compounds, 83
    of hexacoordinate P compounds, 83, 246
    range, 84
    shielding of P nucleus, 81, 85
      group contributions to, 85
  coupling (spin-spin splitting) constant (J), 79, 85
    $J_{P/H}$ values, 86
    $J_{P/F}$ values, 87
    $J_{P/P}$ values in polyphosphines, 461, 462

$J_{P/}$ miscellaneous nuclei values, 87
    first-order spectra, 88
    non-equivalent P atoms, 88
    (n + 1) rule, 89
    stereospecific coupling, 89
    $^3\Delta$ phospholene, 89
    phosphetans, 89
    phosphorus ylids, 89, 90
  Fourier Transform, 78
  line shape and exchange processes, 79, 90
  magnetic moment, μ, 78
  magneto-gyric ratio, γ, 78
  spin-quantum number, I, 78, 88
$(NPCl_2)_{3,4}$ synthesis of, 406, 407
$P_3N_3Ph_4RH$, 408
  cyclopolyphosphazenes, 419
  aminoderivatives, 425
  cyclopolyphosphines, 465
Phosphorus amides, 380–383
Phosphorus amines, 383–388
Phosphorus (III) compounds, see trivalent phosphorus
Phosphorus (V) compounds, see pentacoordinate phosphorus
Phosphorus (III) ligands, Chapter 5, 178–207
Phosphorus-nitrogen compounds, Chapter 10, 380–443
  see also mono, cyclodiphosphazanes and cyclotriphosphazenes, etc.
Phosphorus oxide P_4O_6, 184
Phosphorus oxychloride, hydrolysis via $(HO)_2P(O)Cl$, 317
Phosphorus oxyfluoride, hydrolysis, 318
Phosphorus pentahalides
  reaction with catechol, 222
  reaction with R_2NH, 223
  reaction with SbF_5, 223
  reaction with SnR_4, 223
Phosphorus-phosphorus bond, 454–470
  see also polyphosphines
  $P_6O_{12}^{6-}$, 455, 459
  across cyclophosphazene ring, 419
Phosphorus sulphides, P_4S_3, 184, 455
Phosphorus trichloride, PCl_3
  ligand behaviour, 178, 192
  steric size, 203
Phosphorus trifluoride, 182
  bonding to metals, 191–192
  ligand behaviour, 178, 180–183
  preparation of complexes, 182
  reaction with F_2, 215
  reaction with $NaHF_2$, 243
  reaction with MoF_6, 215
  steric size, 203
Phosphorus ylids
  nomenclature, 274
  preparation of
    by alkylation, acylation, alkoxycarbonylation, silylation or halogenation of ylids, 279

Index

from activated olefins and P(III) compounds, 278
from benzyne and P(III) compounds, 278
from carbenes and P compounds, 278
from dihalophosphoranes and activated methylene compounds, 278
stabilized and non-stabilized ylids, 277
the 'salt' method, 276
reactions of
  acylation and alkylation, 291
  hydrolysis and alcoholysis, 290
  reduction, 293
  with carbon-carbon multiple bonds, 294
  with epoxides, 297
  the Wittig reaction, 279-290
    mechanism and stereochemistry, 282-290
    synthetic applications, 279-282
structure and bonding in, 274

Phosphoryl (and thiophosphoryl) acids (and esters)
  nomenclature, 306-308
  pK_a values, 307
  preparation, 308
  tautomeric equilibria within, 306

Phosphoryl amides and amines, 381-386

Phosphoryl (P=O) bond, 42-46
  bond dissociation energy, 112
  reactivity as nucleophiles, 344
  stretching vibration of, 92-95

Phosphoryl triamide, 381-382, 430

Phosphorylation
  classification of phosphorylating agents, 339
  design of phosphorylating agents, 339-342

Photochemical substitution of carbonyls, 179, 187

Photoelectron spectroscopy (ESCA), 194
  of PF_5, 211

Photosynthesis, 18-19, 23

Photosynthetic phosphorylation, 477, 480

π-acceptor ligands, 178-179, 191

π-acid complexes, 178

π-Bonding at phosphorus, 36
  boraphosphanes, 453
  cyclophosphazenes, 411-412, 415, 419
  donor N → P type, 383, 394, 399
  delocalized PNP, 404
  3d orbitals, 44-46
  pentacoordinate phosphorus, 54-57
  P=O, 42-46
  polyphosphines, 459, 464-465
  tetracoordinate phosphorus, 44-46
  and *trans* effect, 199

π-Constants, 94

Picolines, α, β and γ, as nucleophilic catalysts in hydrolysis of ATP, 326

Pidcock, A., 193, 201

Piperidine, from 5-aminopentan-1-ol and $Ph_3P(OEt)_2$, 234

pK_a' values for cyclopolyphosphazenes, 432-433

Plankton, 19-21
  phytoplankton, 19
  zooplankton, 21

Plants
  phosphorus uptake by, 16
  phosphorus content of, 16-17

Platinum complexes
  $PtCl_2(PPh_3)_2$, 189
  $PtCl_2(PR_3)_2$ isomers, 199
  $PtCl_2(PEt_3)_2$, 199
  $PtHCl(PEtPh_2)_2$, 199
  $PtHCl(PPh_3)_2$, 189
  $PtCl(PBu_3^n)_3$, 201
  $PtCl_2(PPhBu_2^i)_2$, 188
  $PtCl_2(PBu_3^n)_2$, 201
  $PtClMe(PEt_3)_2$, 201
  $PtCl_4(PBu_3^n)_2$, 201
  $Pt(PR_3)_2(NCS)_2$, 202
  $Pt(NCS)_2(Me_2N(CH_2)_3PPh_2)$, 202
  $Pt(PPh_3)_{2,3,4}$, 187
  $PtF_2(PF_3)_2$, 180
  $Pt(PF_3)_4$, 183

^{195}Pt n.m.r., 197, 201

Polymers of phosphazenes
  catalysts for, 435
  conductance measurements, 435
  cross-linking, 435
  end-of-chain groups, 436
  phosphazene, 405, 434-437
  steric factors, 437
  structure, 436

Polyphosphazenes, *see* linear polyphosphazenes or cyclopolyphosphazenes

Polyphosphines, 454-470
  see also cyclotetra-, cyclopenta-, di-, etc.

Polytopal rearrangements, *see* pseudorotation

Potentiation, of insecticides, 496

Pralidoxime, 522

Problems
  Chapter 1, 26
  Chapter 2, 75
  Chapter 3, 106
  Chapter 4, 171
  Chapter 5, 204
  Chapter 6, 248
  Chapter 7, 299
  Chapter 8, 345
  Chapter 9, 374
  Chapter 10, 437
  Chapter 11, 468

Protonation of cyclopolyphosphazenes, 432

Proton migration in cyclopolyphosphazenes, 429-430

Pseudorotation (ψ), 58, 60-68
  cycloboraphosphanes, 452
  cyclodiphosphazanes, 400-401, 430
  determination of apicophilicity, 66-67
  graphs, 61, 63, 64
  notation, 60-61

pentacoordinate reaction intermediates, 68-69
restricted, 62-68
ring compounds, 62-64
unrestricted, 60-62
3-H Pyrrolizine, by Wittig reaction, 280
Pyruvate, 476, 482-483

Quaternization, of phosphines by alkyl halides, 255

Raman spectroscopy, 97
Ramirez, F.
hexacoordinate intermediates in hydrolysis of phosphate esters, 337
reaction of aldehydes with $(MeO)_3P$, 167
reaction of chloranil with Ph_3P, 359
reaction of hexafluoroacetone with $(MeO)_3P$, 167
synthesis of phosphoranes from α-diketones, ortho-quinones and carbonyl compounds, 164-167, 217
synthesis of phosphoranes from PCl_5 and PhOH, 222
Rappoport, Z., reaction of benzylidene malononitriles with nBu_3P, 127
Red phosphorus, 454
Respiration, 18-19, 23
Restricted rotation about P-N bonds, 55-56, 386, 387-388
Retort process for phosphorus manufacture, 10
Rhenium complexes
$ReH(PF_3)_n(CO)_{5-n}$, 182
$ReH_5(PPh_3)_3$, 187
Rhodium complexes
$RhH(CO)(PPh_3)_3$, 189
$RhCl(PPh_3)_3$, 188-189
$RhMe(PPh_3)_3$, 188
$Rh_2Cl_2(C_8H_{16})\{P(OPh)_3\}_2$, 185
Ribonucleic acid (RNA), 474, 488
Ribulose-1,5-diphosphate, 479
Ring size and reactivity in cyclopolyphosphazenes, 421
Ring strain, in methyl ethylene phosphate, 331
Rivers, 17
Robinson, S. D., 179, 183, 185
Rodenticides, 495
Ruthenium complexes
$[RuCl(NO)_2(PPh_3)_2]PF_6$, 190
$RuHCl\{P(OPh)_3\}_4$, 185
$RuCl\{(C_6H_4O)P(OPh)_3\}_3$, 185
$RuHCl(PPh_3)_3$, 188

Salicoyl phosphate, hydrolysis (problem), 345
Salicyl phosphate, hydrolysis (problem), 345
Sarin
hydrolysis, catalyzed by oximes, 326
hydrolysis, catalyzed by Cu^{2+} ions, 327
reaction with catechol, 327
reaction with hydroxylamine, 329
reaction with HOO^-, 329
as a harassing agent, 502, 504
dose response, 503
Schenk, R. and Romer, G., 405
Schmitz-DuMont, O., 434
Schmutzler, R., 387
Schradan, as an insecticide, 496, 498
Schrader, G., Tabun, 496
Sea, 24
Atlantic, 25
Black, 24-25
Mediterranean, 24-25
Pacific, 25
Upwelling regions, 25
Sediment, 22-23
Sedimentary rocks, 5
Self-raising flour, 14
Semenov, N. N., theory of branching chain reactions, 352
Sesquimustard, as a lethal agent, 502
Sewage, 17-18
oxidation ponds, 18
treatment, 18
Shapes of phosphorus molecules, 32-33
Shaw, R. A., 432
σ framework about metal atom, 195
σ bonding in P-N compounds, 388, 394, 395, 399
σ framework in cyclopolyphosphazenes, 410-411
Silyl-substituted methylene phosphoranes, 279
Sodium phosphacyclopentadienide, 374
Soil phosphates, 16-17
$S_N1(P)$ mechanism, 70, 420-421
$S_N2(P)$ mechanism, 70-72, 420-421
Solubility product of $Ca_3(PO_4)_2$, 6
Speziale, A. J. (and Bissing, D. J.), kinetics of reactions of ylids with p-substituted benzaldehydes, 285
Spirocyclic-oxyphosphoranes, 48
phosphazenes, 424
spy, see square pyramidal
Squalene (all-trans), by the Wittig reaction, 281
Square pyramidal structure, 32, 48-49
Staudinger, H., 389
ylid chemistry, 274
reaction of diphenylmethylenetriphenylphosphorane with phenyl isocyanate, 279
Steric size of P(III) ligands, 202-203
Stilbene (cis and trans), from Wittig reaction, 288
Stokes, H. M., 398, 405
Stretching vibrations, see v under nu

Index

Styrene, by β-elimination from phosphonium salts, 260
Styrene oxide, by thermal decomposition of a 1,3,2-dioxaphosphorane, 233
Substituent constants (α and γ), 432–433
Substrate level phosphorylation, 484, 487
Succinic acid, 483
Sucrose, 478
Sulphonium salts, from R_2S and PF_5, 229
Sulphur hexafluoride, SF_6, 72
Superphosphate, 9–11, 16
Symmetry and vibrational spectroscopy, 97
Synapse, in nerve function, 503

Tabun,
    discovery, 496
    as a harassing agent, 502, 504
tbp, see trigonal bipyramid
Tebby, J. C., hydrolysis of phosphorus ylids, 291
Tepa, as a chemosterilant, 501
TEPP, as an insecticide, 496, 498
Tetraarylphosphonium salts, preparation, 256
Tetrachlorophosphonium hexafluorophosphide, vacuum sublimation, 225
Tetracoordinate phosphorus, bonding at
    π-bonding, 44–46
    σ-bonding, 43
Tetracyanoethylene (TCNE), reaction with phosphines, 360
Tetracyanoquinodimethane (TCNQ), reaction with phosphines, 360
Tetrafluorohydrazine, reaction with P(III) compounds, 215
Tetrahydrofuran, from reaction of 1,4-butane diol with $(EtO)_5P$, 234
Tetrahydropyran, from 1,5-pentane diol and $Ph_3P(OEt)_2$, 234
2,2,3,3-Tetramethylbutane, from reaction of Bu^tOOBu^t with phosphites, 363
Tetramethyldioxetane, reaction with Ph_3P, 156
Tetraphenylbiphosphine, in preparation of cyclic phosphonium salts, 258
Tetraphenyldiazocyclopentadienylide, reaction with Ph_3P, 144
Tetraphosphine, P_4H_6, 461
Thiols, reactions with $(RO)_3P$, 366
Thionazin, as a nematicide, 501
Thiophosphinites, from P(III) halides and thiols, 148
Thiophosphoryl amides and amines, 381–386
Thomas, L. C., 92–93, 94, 388, 400
Thymine, 474
Tin complex, $SnCl_4(NPMe_2)_3$, 431

Titanium complexes
    $TiCl_4(PH_3)$, 179
    $TiCl_4(NPMe_2)_3$, 431
Tolkmith, H., mechanism of fungicidal activity, 500
Tolman, C. A., 186, 191–192, 196
    χ values, 196
    ligand cone angles, 203
    steric factors, 202–203
Toothpaste, 7
Topological diagram (Cram), 61, 248
Topology of pentacoordinate phosphorus, 58–68
N-p-Tosylimino ylids, reactions with aryllithiums, 228
Toxicity
    LD_{50} (definition), 494
    LCt_{50} (definition), 494
    of Sarin, 502
    Table of
        for pesticides, 498–499
        for chemical warfare agents, 504
Trans effect and trans influence, 198–202
    bond length ratios in complexes, 200
    ^{31}P n.m.r. data, 200–201
    Gringberg's theory of, 198–199
Trialkyl phosphites, reaction with α-halocarbonyl compounds (Perkow reaction), 133
Triazo derivatives $X_3P=N-N=N-R^1$, 389
Tri-n-butylphosphine, reaction with di-t-butyl peroxide, 365
Tricarboxylic acid cycle, 476, 482–484
ω,ω,ω-Trichloroacetophenone, in the Perkow reaction, 137
Trichloromonophosphazene, $Cl_3P=NH$, 389, 390, 401, 406
Tricoordinate phosphorus, 35–42
    isomers and separation, 41
    inversion and energy barriers, 41
    compounds, see trivalent phosphorus compounds
    vibrational spectroscopy, 98–99
Triethyloxonium tetrafluoroborate
    in alkylation of phosphorus ylids, 292
    reaction with phosphetan oxides, 227
Triethyl phosphite, 384
    reaction with alkoxy radicals, 363
    complexes, 185
    ^{14}C labelling experiments, 364
    ligand behaviour, 178, 183–187, 192
    rate, 364
    reaction with t-butylperoxy radicals, 365
    steric size, 203
    u.v. irradiation of, 366
bis-Trifluoromethyl disulphide, reaction with P(III) compounds, 215
Trifluoromethyl hypofluorite, reaction with P(III) compounds, 215

bis-Trifluoromethyl peroxide, reaction with P(III) compounds, 215
tris-Trifluoromethylphosphine oxide, reaction with (Me₃Si)₂O, 227
Trigonal bipyramid structure, 32, 48–49
Triglycerides, 490
2,2,4-Trimethyl-2-pentene, reaction with PhPCl₂, 257
Trimethyl phosphate, reaction with aniline, 344
Trimethyl phosphite, P(OMe)₃
  ligand behaviour, 178, 183–187
  preparation of complexes, 185
  steric size, 203
Trimethyl silanol, in desilylation of ylids, 279
Trimethylsilyl derivatives, Me₃Si, 387, 389, 393, 395, 409, 450
Triphenylmonophosphazene Ph₃P=NH, 390, 391, 393
Triphenylphosphine, PPh₃
  ligand behaviour, 178, 187–190, 192
  steric size, 203
Triphenylphosphine oxide, Ph₃PO, 393
Triphenyl phosphite, P(OPh)₃, 184
  bonding to metals, 184
  ligand behaviour, 182, 183–187, 192
  preparation of complexes, 185
  steric size, 203
Triphosphate, P₃O₁₀⁵⁻, Na₅P₃O₁₀, 11
Triphosphazenes, linear derivatives, 402, 403, 404, 406
Triphosphine, P₃H₅, 461
Triphosphine derivatives, 456, 458, 461
Triple superphosphate, 9–10
Triphenylvinylphosphonium bromide, intramolecular rearrangement during hydrolysis, 273
Trippett, S.
  determination of apicophilicity by ¹⁹F nmr, 212
  hydrolysis of sterically hindered acyclic phosphonium salts, 271
  reaction of hexafluoroacetone with 1-substituted phosphetans, 169
  reversibility of betaine formation with non-stabilised ylids, 288
Tri-n-propyl phosphite, hydrolysis, 146
Trivalent phosphorus compounds
  amino, 380, 381
  bromo, 196, 203, 384
  butyl, 192, 203
  chloro, 178, 192, 203, 384, 388
  cyclohexyl, 196, 203
  dimethylamino, 383, 384, 388
  ethyl, 192, 203
  fluoro, 178, 180–183, 191–192, 203, 383–384
  iodo, 456
  methyl, 192, 203
  miscellaneous, 192, 196, 203

  phenyl, 388, see also triphenyl phosphine
  trialkyl, 178, 187–190, 203
  trifluoromethyl, 192, 203, 381, 388, 456
Tungsten complexes, see wolfram complexes
Turnstile rotation (τ), 59–60

Ultraviolet spectroscopy, 418
  of cyclopolyphosphazenes, 418
  and P–P compounds, 459, 464–465
α,β-Unsaturated carbonyl compounds, addition of P(III) compounds, 169
Uracil, 474
Uranium, 9
Urea, 476, 491
Uridine triphosphate, UTP, 474
US phosphate production, 7
USSR phosphate production, 7

Vibrational assignments, 97–102
Vibrational coupling, 94, 96
Vibrational spectra of phosphorus compounds, 91–108
Vinyl phosphates, formation in Perkow reaction, 133
Vinyl propionate, reaction with phosphinyl radicals, 358
Vinyltriphenylphosphonium bromide, reaction with 2-pyrrolealdehyde, 280
Violet phosphorus, 454
Vitamin A, synthesis via Wittig reaction, 281
VSEPR (Valence-shell electron-pair repulsion), 31–33, 57
  axioms, 31

Wadsworth, W. S. (and Emmons, W. D.), the Horner-Emmons reaction, 282
Walden cycle, for nucleophilic displacement at phosphorus in thiophosphoryl compounds, 322
Walling, C. (and Pearson, M. S.), reaction of (EtO)₃P with Buᵗ O·, 364
Walling, C. (and Rabinowitz, R.), reaction of (RO)₃P with thiols, 366
Warfare, chemical, 502–509
  classification of agents, 502
  ethics, 508
  historical aspects, 502
  symptoms of organophosphorus poisoning, 503
  Table of organophosphorus CW agents, 504
Wavellite, 6
Wepsyn, as a fungicide, 500
Westheimer, F. H., 58, 68
  rules governing hydrolysis of cyclic phosphorus esters, 331
  (and Archie, W. C. Jr), hydrolysis of pentaryloxyphosphorus, 336

Wet-acid process, 11
White phosphorus (α and β), 454
Whitlockite, 6
Wilkinson, G., 180, 181
Wilson, I. B., 522
Wittig, G.
  resolution of optically active phosphines, 260
  the Wittig reaction, 274, 279
    via phosphines and epoxides, 115
Wolfram complexes
  $W(CO)_5(PX_3)$, 183
  $WH_6(PMe_2Ph)_3$, 187
  $W(CO)_5(PR_3)$, 197
  $W(CO)_3(NPMe_2)_4$, 431
  $W(CO)_4(NP(NMe_2)_2)_4$, 431

Woodward-Hoffmann rules, 237
World phosphate-production and reserves, 8
Würster salts, from p-phenylene diamine and chloranil, 359

X-ray studies
  of hexacoordinate phosphorus including a diazaphosphetidine ring, 245
  of methyl ethylene phosphate, 331
  of phosphoranes (tbp and sqp), 210
  of phosphorus ylids, 275
  of tetrachlorophosphonium fluoride, 226
  of tri-catecholphosphate anion, 244

Zinc dialkylphosphorodithioates, 12
Zinc complex, $Zn\{P(O)(NMe_2)_3\}_4 2ClO_4$, 385